T0122044

CYBER-PHYSICAL SYSTEMS

A Computational Perspective

CYBER-PHYSICAL SYSTEMS

A Computational Perspective

Edited by

Gaddadevara Matt Siddesh

Ganesh Chandra Deka

Krishnarajanagar Gopalalyengar Srinivasa

Lalit Mohan Patnaik

CRC Press is an imprint of the
Taylor & Francis Group, an **informa** business
A CHAPMAN & HALL BOOK

CRC Press
Taylor & Francis Group
6000 Broken Sound Parkway NW, Suite 300
Boca Raton, FL 33487-2742

Printed on acid-free paper
Version Date: 20151103

International Standard Book Number-13: 978-1-4822-5975-9 (Hardback)

Visit the Taylor & Francis Web site at
http://www.taylorandfrancis.com

and the CRC Press Web site at
http://www.crcpress.com

Contents

Section III Robotics and Smart Systems in Cyber Context

Section IV Ubiquitous and Cloud Computing for Monitoring Cyber-Physical Systems

Section V Security Issues in Cyber-Physical Systems

Section VI Role of Cyber-Physical Systems in Big Data Analytics, Social Network Analysis, and Health Care

Preface

For so long the focus of the computer industry and research community has been on machines and processes that could do classic computation. Transforming the digital data into meaningful information using arithmetic and logical operations coupled with basic programming constructs like branching and iterations had been the dominant worldview. Though there were developments in networking the machines using wired/wireless methods to store data that grew exponentially, the focus was exclusively on computational devices. That those devices were changing shape and architecture did not matter as long as they adhered to the established notions of computation.

It is important to recognize that along with the advent of modern computers, there also came a plethora of devices whose primary intention was not computation. These devices, which earlier had an analog avatar, were acquiring digital behavior. In their applications in washing machines and medical equipment, these digital units were able to capture physical data and generate information in digital format. Added to this was the phenomenon of low-cost and portable sensors that could detect a variety of physical conditions, and the next stage of evolution began in the computing industry.

Though embedded systems had established themselves as a viable domain, they primarily focused on stand-alone devices. The breakthrough came when developments in diverse fields like sensors, networks, and embedded systems came together to create an ecosystem where physical processes could be controlled. This area is fittingly labeled as cyber-physical systems (CPS), because here sensors and embedded systems are networked together to monitor/manage a range of physical processes through a continuous feedback system. Today, CPS has become an umbrella term covering a host of specialized domains surrounding distributed computing using wireless devices, namely, heterogeneous, scalable, and modular system designs; quality service assurance; network security; and application of CPS into various domains like health care, manufacturing, governance, and the military.

This book attempts to cover the various developments of CPS in the chapters contributed by practitioners and researchers across the globe. The book is divided into six sections. Section I covers the physical infrastructure required for CPS and includes chapters that deal with sensor networks and embedded systems. Addressing energy issues in CPS with the use of supercapacitors and reliability assessment of CPS are covered in Section II. The modeling of CPS as a network of robots and issues regarding the design of CPS is the focus of Section III. The impact of ubiquitous computing and cloud computing in CPS is discussed in Section IV. Security issues in CPS are covered in Section V. Section VI discusses the role of CPS in big data analytics, social network analysis, and health care.

The editors are confident that the book will be a valuable addition to the growing knowledge about the importance and impact of CPS on our daily life. As digital devices become more intrusive and pervasive, there will be increasing interest in this domain.

It is hoped that this book will not only showcase the current art of practice but also set the agenda for future directions in this domain.

MATLAB® is a registered trademark of The MathWorks, Inc. For product information, please contact:

The MathWorks, Inc.
3 Apple Hill Drive
Natick, MA 01760-2098 USA
Tel: 508-647-7000
Fax: 508-647-7001
E-mail: info@mathworks.com
Web: www.mathworks.com

Editors

Dr. Gaddadevara Matt Siddesh earned his PhD in computer science and engineering from Jawaharlal Nehru Technological University, Hyderabad, India, in 2014. Currently, he is an associate professor in the Department of Information Science and Engineering, M. S. Ramaiah Institute of Technology, Bangalore. He earned his bachelor's and master's degrees in computer science and engineering from Visvesvaraya Technological University, Belguam, India, in 2003 and 2005, respectively. He is a member of the Institute of Electrical and Electronics Engineers, the Indian Society for Technical Education, and the Institution of Electronics and Telecommunication Engineers. He is the recipient of Seed Money to Young Scientist for Research (SMYSR) for the financial year 2014–2015 from the Government of Karnataka, Vision Group on Science and Technology (VGST). He has published a number of research papers in international conferences and journals. His research interests are distributed computing, grid/cloud computing, and the Internet of Things.

Ganesh Chandra Deka is the principal, Regional Vocational Training Institute for Women, Tura, Meghalaya, under DGT, Ministry of Skill Development and Entrepreneurship, Government of India. He is also in-charge of the Regional Vocational Training Institute for Women, Agartala Tripura, under DGT, Ministry of Skill Development and Entrepreneurship, Government of India. His previous assignments include assistant director of training in The Directorate General of Employment & Training, Ministry of Labor & Employment, Government of India, New Delhi (2006–2014); consultant (computer science), National Institute of Rural Development—North Eastern Regional Centre, Guwahati, Assam, under the Ministry of Rural Development (2003–2006), Government of India; and programmer (World Bank Project) at Nowgong Polytechnic, under the Directorate of Technical Education, Assam, India (1995–2003).

Dr. Deka's research interests include application of ICT in rural development, e-governance, cloud computing, data mining, NoSQL databases, and vocational education and training. He has published more than 57 research papers in various conferences, workshops, and reputed international journals, including the Institute of Electrical and Electronics Engineers (IEEE and Elsevier). He is editor-in-chief of the *International Journal of Computing, Communications, and Networking* (ISSN 2319-2720). So far, he has organized six IEEE international conferences as technical chair in India. He is a reviewer and member of the editorial boards of various journals and international conferences.

He is the coauthor for three textbooks on the fundamentals of computer science, and coeditor, with Pethuru Raj, of the *Handbook of Research on Cloud Infrastructures for Big Data Analytics* (IGI Global, Hershey, Pennsylvania, 2014) and chief editor of the *Handbook of Research on Securing Cloud-Based Databases with Biometric Applications* (IGI Global, Hershey, Pennsylvania, 2015).

Dr. Deka's academic qualifications include a three-year polytechnic diploma in computer engineering from the Assam Engineering Institute under the State Council for Technical Education, Assam, India; B Tech in computer engineering from North Eastern Hill University, Shillong, India; and PhD in computer science from Ballsbridge University, Roseau, Dominica. He is a member of the IEEE, the Institution of Electronics and Telecommunication Engineers, India, and an associate member, the Institution of Engineers, India. He can be reached at ganeshdeka2000@gmail.com.

Krishnarajanagar GopalaIyengar Srinivasa earned his PhD in computer science and engineering from Bangalore University, Karnataka, India, in 2007. He is a recipient of the All India Council for Technical Education's Career Award for Young Teachers, Indian Society of Technical Education's ISGITS National Award for Best Research Work Done by Young Teachers, Institution of Engineers (India)'s IEI Young Engineer Award in Computer Engineering, Rajarambapu Patil National Award for Promising Engineering Teacher Award from ISTE in 2012, and IMS Singapore's Visiting Scientist Fellowship Award. He has published more than 100 research papers in international conferences and journals. He has visited many universities abroad as a visiting researcher: the University of Oklahoma and Iowa State University (USA), Hong Kong University, Korean University, and the National University of Singapore are among his prominent visits. He has authored two books: *File Structures Using C++* by Tata McGraw-Hill and *Soft Computer for Data Mining Applications* in the LNAI Series by Springer. He has been awarded the BOYSCAST Fellowship by the DST for conducting collaborative research with CLOUDS laboratory at the University of Melbourne in the area of cloud computing. He is the principal investigator for many funded projects from UGC, DRDO, and DST. His research areas include data mining, machine learning, and cloud computing.

Prof. Lalit Mohan Patnaik earned his PhD in real-time systems in 1978 and DSc in computer systems and architectures in 1989, both from the Indian Institute of Science, Bangalore. During March 2008–August 2011, he was the vice chancellor, Defence Institute of Advanced Technology, Deemed University, Pune. Currently, he is an honorary professor with the Department of Electronic Systems Engineering, Indian Institute of Science, Bangalore, and INSA senior scientist and adjunct faculty with the National Institute of Advanced Studies, Bangalore. Prior to this, Dr. Patnaik was a professor with the Department of Computer Science and Automation at the Indian Institute of Science, Bangalore. During the last 40 years of his long service at the Indian Institute of Science, his teaching, research, and development interests have been in the areas of parallel and distributed computing, computer architecture, CAD of VLSI systems, theoretical computer science, real-time systems, soft computing, and computational neuroscience. In these areas, he has published over 1000 papers in refereed international journals and conference proceedings and has authored 29 technical reports. Dr. Patnaik is a coeditor/coauthor of 21 books and has authored 12 book chapters in the areas of VLSI system design and parallel computing. He has supervised more than 30 doctoral and more than 160 master's theses in these areas. He has been a principal investigator for a large number of government-sponsored research projects and a consultant to a few industries in the areas of parallel and distributed computing and soft computing. One of his research papers in the area of adaptive genetic algorithms has been cited over 2300 times by Google Scholar.

Dr. Patnaik has served as the president of the Advanced Computing and Communications Society and the Computational Intelligence Society of India. He is a distinguished lecturer of the Institute of Electrical and Electronics Engineers (IEEE) Region 10, and is a distinguished visitor of the IEEE Computer Society for the Asia-Pacific Region. As a recognition of his contributions in the areas of electronics, informatics, telematics, and automation, he received the following awards:

1989 Dr. Vikram Sarabhai Research Award
1992 Honorable Mention Award, Fourth CSI/IEEE International Symposium on VLSI Design, and the Dr. Ramlal Wadhwa Award from the Institution of

	Electronics and Telecommunication Engineers, for his contributions during the last ten years in the fields of electronics and telecommunication
1994	Platinum Jubilee Lecture Award (Computer Science), Indian Science Congress, and the Samanta Chandrasekhar Award from the Orissa Science Academy
1995	Certificate of Appreciation "For excellent service and dedication to the objectives of the IEEE Computer Society as a Distinguished Visitor"
1996	VASVIK Award in the field of electronic sciences and technology, Vividhlaxi Audyogik Samshodhan Vikas Kendra
1998	Hands for Indian IT Award for his contributions to the Indian IT (information technology) industry
1999	Distinguished Professional Engineer Award from the Computer Engineering Division of the Institution of Engineers (India); the IEEE Computer Society's "1999 Technical Achievement Award" for his contributions in the field of parallel, distributed, and soft computing and high-performance genetic algorithms; and the Pandit Jawaharlal Nehru National Award for Engineering and Technology
2000	Fourth Sir C. V. Raman Memorial Lecture Award and the First SVC Aiya Memorial Trust Award for Telecommunication Education, The Institution of Electronics and Telecommunication Engineers, Pune Centre
2001	Om Prakash Bhasin Award for contributions to the areas of electronic and information technology
2001–2002	FICCI Award for Innovation in Material Science, Applied Research and Space Science and the IEEE Computer Society's Meritorious Service Award
2003	Alumni Award for Excellence in Research for Engineering, the Indian Institute of Science
2004	Distinguished Engineer Award of The Institution of Engineers (India)
2005	Goyal Prize for Applied Science
2006	Honorary Fellow, the Indian Society for Technical Education
2007–2008	Indian Science Congress Association's Srinivasa Ramanujan Birth Centenary Award
2009	Biju Patnaik Award for Excellence in Science—2007
2010	Top Management Consortium, Pune, "Education—Defence Services Excellence Award—2009"
2012	Distinguished Alumnus Award, National Institute of Technology, Rourkela
2014	Felicitation of International Neural Network Society India Chapter for outstanding contributions to soft computing

Dr. Patnaik is fellow of the IEEE; The World Academy of Sciences (TWAS), Trieste, Italy; The Computer Society of India; the Indian National Science Academy; the Indian Academy of Sciences; the National Academy of Sciences; and the Indian National Academy of Engineering. He is a life member of the VLSI Society of India and the Instrument Society of India, a founder member of the Executive Committee of the Association for the Advancement of Fault-Tolerant and Autonomous Systems, and a fellow of the Institution of Electronics and Telecommunications Engineers and the Institution of Engineers. His name appears in Asia's *Who's Who of Men and Women of Achievement* and in the *Directory of Distinguished Computer Professionals of India*. He has served as general chair or program chair or steering committee chair for more than 150 international conferences; most of them sponsored by the IEEE. He has served on several significant committees of leading professional societies such as the IEEE, CSI, IETE, and the Institution of Engineers; and on the review and policy committees of government agencies, such as on the DST, DBT, MHRD, CSIR, DOS, UGC, AICTE, DRDO, DIT, and UPSC. His participation in the reviews of several engineering colleges in the states of Karnataka, Kerala, Andhra Pradesh, and Maharashtra, organized by the AICTE, UGC, NBA, and TEQIP has significantly improved the quality of technical education in the country.

As vice chancellor of the Defence Institute of Advanced Technology, deemed university, Dr. Patnaik was responsible for conducting numerous doctoral and master's short- and long-term courses in the field of defence technologies for service officers from the Army, Air Force, and Navy; DRDO scientists; officers from Indian Ordnance Factories; the director general of quality assurance; and service officers from several friendly foreign countries. He initiated three new postgraduate programs of relevance to defence services, including one on cybersecurity, a new center on nanotechnology for defence applications, and a collaborative program with the Naval Postgraduate School, California.

Dr. Patnaik has delivered several invited lectures in the United States, Canada, France, the United Kingdom, the former Yugoslavia, Hungary, Switzerland, Australia, New Zealand, Hong Kong, Singapore, Malaysia, Sri Lanka, and Japan.

It is worth mentioning that his training and research work are entirely indigenous. His contributions to higher education in the technical field in India are outstanding.

For publication and other details, please visit:

http://www.lmpatnaik.in/

http://www.diat.ac.in/index.php?option=com_content&view=article&id=156&%20Itemid=394

http://www.cedt.iisc.ernet.in/people/lmp/lmp.html

Contributors

Pothireddy Siva Abhilash
Department of ITG-BTO
3I-Infortech Consulting Services
Hyderabad, Telangana, India

Khaleel Ahmad
School of Computer Science and Information Technology
Maulana Azad National Urdu University
Andhra Pradesh, India

Tareq Alhmiedat
Department of Information Technology
Tabuk University
Tabuk, Saudi Arabia

B. Sathish Babu
Department of Computer Science and Engineering
Siddaganga Institute of Technology
Karnataka, India

Robin Singh Bhadoria
Discipline of Computer Science and Engineering
Indian Institute of Technology, Indore
Madhya Pradesh, India

Chitresh Bhargava
Department of Embedded System and Image Processing
Electrical and Electronics Research Center
Rajasthan, India

Genshe Chen
Intelligent Fusion Technology, Inc.
Germantown, Maryland

Yiyang Chen
Department of Electrical and Computer Engineering
University of Rochester
Rochester, New York

Punam Dutta Choudhury
Department of Electronics and Communication Technology
Gauhati University
Assam, India

Uma Datta
Central Mechanical Engineering Research Institute
Council of Scientific and Industrial Research
West Bengal, India

Sri Yogesh Dorbala
Discipline of Computer Science and Engineering
Indian Institute of Technology, Indore
Madhya Pradesh, India

Qinghe Du
Department of Information and Communications Engineering
Xi'an Jiaotong University
Xi'an, Shannxi, People's Republic of China

Fatih Erdem
Department of Electrical and Computer Engineering
University of Rochester
Rochester, New York

Afrah Fathima
School of Computer Science and Information Technology
Maulana Azad National Urdu University
Telangana, India

Nicholas Gekakis
Department of Electrical and Computer Engineering
University of Rochester
Rochester, New York

Moeen Hassanalieragh
Department of Electrical and Computer Engineering
University of Rochester
Rochester, New York

Srinidhi Hiriyannaiah
Department of Computer Science and Engineering
M.S. Ramaiah Institute of Technology
Karnataka, India

Grayson Honan
Department of Electrical and Computer Engineering
University of Rochester
Rochester, New York

G. Jagadamba
Department of Information Science and Engineering
Siddaganga Institute of Technology
Karnataka, India

Anita Kanavalli
Department of Computer Science and Engineering
M.S. Ramaiah Institute of Technology
Karnataka, India

Amit Konar
Artificial Intelligence Laboratory
Department of Electronics and Tele-Communication Engineering
Jadavpur University
West Bengal, India

Andrej Kos
Faculty of Electrical Engineering
University of Ljubljana
Ljubljana, Slovenia

Abhishek Kumar
Department of Computer Science and Engineering
M.S. Ramaiah Institute of Technology
Karnataka, India

Rajeev Kumar
College of Computing Sciences and Information Technology
Teerthankar Mahaveer University
Uttar Pradesh, India

Sumit Kundu
Department of Electronics and Communication Engineering
National Institute of Technology, Durgapur
West Bengal, India

Wenjia Li
Department of Computer Science
New York Institute of Technology
New York

Zhaojun Liu
Department of Electrical and Computer Engineering
University of Rochester
Rochester, New York

Amjad Mehmood
Institute of Information Technology
Kohat University of Science and Technology
Kohat, Pakistan

Kushagra Mishra
M.S. Ramaiah Institute of Technology
Karnataka, India

D.K. Mohanta
Department of Electrical and Electronics Engineering
Birla Institute of Technology
Jharkhand, India

Cherukuri Murthy
Department of Electrical and Electronics Engineering
National Institute of Science and Technology
Berhampur, India

Andrew Nadeau
Department of Electrical and Computer Engineering
University of Rochester
Rochester, New York

Charles C. Nguyen
Department of Electrical Engineering and Computer Science
The Catholic University of America
Washington, DC

Tien M. Nguyen
Department of Electrical Engineering and Computer Science
The Catholic University of America
Washington, DC

Monalisa Pal
Artificial Intelligence Laboratory
Department of Electronics and Tele-Communication Engineering
Jadavpur University
West Bengal, India

Khanh D. Pham
Spacecraft Component Technology Branch
Air Force Research Laboratory
Kirtland Air Force Base, New Mexico

Mitja Rakar
Faculty of Electrical Engineering
University of Ljubljana
Ljubljana, Slovenia

P.V.L. Narayana Rao
Department of Information System
College of Computing and Informatics
Wolkite University
Ethiopia, East Africa

K. Hemant Kumar Reddy
Department of Computer Science and Engineering
National Institute of Science and Technology
Odisha, India

and

Department of Electrical and Computer Engineering
University of Rochester
Rochester, New York

Pinyi Ren
Department of Information and Communications Engineering
Xi'an Jiaotong University
Xi'an, Shannxi, People's Republic of China

D. Sinha Roy
School of Computer Science
National Institute of Science and Technology
Berhampur, India

Anuradha Saha
Artificial Intelligence Laboratory
Department of Electronics and Tele-Communication Engineering
Jadavpur University
West Bengal, India

Sriparna Saha
Artificial Intelligence Laboratory
Department of Electronics and Tele-Communication Engineering
Jadavpur University
West Bengal, India

Kandarpa Kumar Sarma
Department of Electronics and Communication Technology
Gauhati University
Assam, India

Urban Sedlar
Department of Electrical Engineering
University of Ljubljana
Ljubljana, Slovenia

Gaurav Sharma
Department of Electrical and Computer Engineering
University of Rochester
Rochester, New York

M.K. Sharma
Department of Computer Science
Amrapali Institute
Uttarakhand, India

Gaddadevara Matt Siddesh
Department of Information Science and Engineering
M.S. Ramaiah Institute of Technology
Karnataka, India

Nabeel Siddiqui
Department of Computer Science and Engineering
M.S. Ramaiah Institute of Technology
Karnataka, India

Houbing Song
Department of Electrical and Computer Engineering
West Virginia University
Morgantown, West Virginia

B.J. Sowmya
Department of Computer Science and Engineering
M.S. Ramaiah Institute of Technology
Karnataka, India

Tolga Soyata
Department of Electrical and Computer Engineering
University of Rochester
Rochester, New York

Krishnarajanagar GopalaIyengar Srinivasa
Department of Computer Science and Engineering
M.S. Ramaiah Institute of Technology
Karnataka, India

Mayank Swarnkar
Discipline of Computer Science and Engineering
Indian Institute of Technology, Indore
Madhya Pradesh, India

Geetam Singh Tomar
Machine Intelligence Research Lab
Madhya Pradesh, India

Shikhar Verma
National Institute of Science and Technology
Berhampur, India

Abdul Wahid
School of Computer Science and Information Technology
Maulana Azad National Urdu University
Andhra Pradesh, India

Radhakishan Yadav
Discipline of Computer Science and Engineering
Indian Institute of Technology, Indore
Madhya Pradesh, India

Section I

Physical Infrastructures for Cyber-Physical Systems

1

Sensors and Sensor Networks with Applications on Cyber-Physical Systems

Tien M. Nguyen, Charles C. Nguyen, Genshe Chen, and Khanh D. Pham

CONTENTS

1.1 Introduction

1.1.1 Sensor Definition and the Use of Sensors

Sensors are devices that measure physical quantities of the environment around them and convert these quantities into electrical/optical/sound-wave/mechanical signals, which can be read or viewed by an observer or by an instrument. The physical quantity can be a movement of a human body or movement of an object or environmental temperature or wind velocity or gun shots. The signal can be in the form of electrical or mechanical or sound. In general, various sensor devices are typically used by wireless sensor networks (WSNs) and Mobile ad hoc networks (MANETs) that construct a cyber-physical system (CPS) for monitoring the physical quantities specified by a user. This section defines sensor networks and their uses in CPSs.

1.1.2 Sensor Network Definition and the Use of Sensor Networks

Sensor networks are wired or wireless networks of sensors, which can collect and disseminate environmental data. WSNs have applications on modern and emerging CPSs, such as in health care, environmental and structural monitoring in smart cities, smart battlefields, cyber space tracking, borderlines, platform location determination, platform self-navigation, and gathering sensing information remotely in both hostile and friendly locations. Sensor networks employ the types of sensor described in Section 1.2.1 in their general-purpose design approach that provides services to many aforementioned applications. The networks are designed and engineered according to specific plans with sensing devices and networks operating in a specified environment. Traditionally, the networks usually consist of a number of sensor nodes that are wired or wirelessly connected to a central processing station.

1.1.3 Traditional Sensor Networks vs. WSNs

Traditional sensor networks are generally designed to provide services for specific applications, for example, plant monitoring, home monitoring, and traffic monitoring. The networks are designed and engineered according to specific plans with sensing devices and networks operating in well-controlled environments. The networks usually consist of a small number of sensor nodes that are wired to a central processing station. The primary design concerns for traditional sensor networks are network performance and latencies; usually power and cost are not primary concerns. Unlike the traditional sensor networks, WSNs usually consist of a dense number of sensor nodes. Each sensor node is capable of only a limited amount of processing. But when coordinated with the information from a large number of other sensor nodes, they have the ability to measure a given physical environment in great detail. Thus, WSN employed by CPSs can be considered as a large collection of sensor nodes, which are working together in a coordinated manner to perform some specific action, such as movement monitoring in a remote area.

Presently, the researchers on sensor networking focus more on wireless, distributed, mobile sensing nodes for CPS applications when the exact location of a particular phenomenon is not known, thus distributed and mobile sensing allows for closer placement to the phenomenon than a single sensor would permit. By placing multiple sensor nodes around the phenomenon, the observer can overcome environmental obstacles like obstructions, line of sight constraints, etc. For wireless sensing applications on CPSs, the environment to be monitored usually does not have an existing infrastructure for either power or communications.

1.2 Sensors Employed by CPS

1.2.1 Types of Sensors

In general, sensors used by CPSs can be classified into 14 sensing types: (1) acoustic, sound, and vibration; (2) automotive and transportation; (3) chemical; (4) electric, magnetic, and radio; (5) environment, weather, moisture, and humidity; (6) flow and fluid velocity; (7) ionizing radiation and subatomic particles; (8) navigation instruments; (9) position, angle, displacement, distance, speed, and acceleration; (10) optical, light, and imaging; (11) pressure; (12) force, density, and level; (13) thermal, heat, and temperature; and (14) proximity and presence. Some examples describing these sensing types are given in the following for illustration purpose.

Acoustic, sound, and vibration sensors include microphone, geophone, seismometer, and accelerator. Automotive and transportation sensors are speedometer, map sensor, water sensor, parking sensor, and video sensor. Chemical sensors consist of sensing carbon, gas, hydrogen, oxygen, and smoke. Electric, magnetic, and radio sensors are magnetometer, metal detector, and telescope. Environment, weather, moisture, and humidity sensors are leaf sensor, rain/snow gauge, and pyranometer. Flow and fluid velocity sensors are air flow meter, flow sensor, and water meter. Ionizing radiation and subatomic particles sensors include cloud chamber, neutron detection, and particle detector. Navigation instruments sensors are air speed indicator, depth gauge, gyroscope, and turn coordinate. Position, angle, displacement, distance, speed, and acceleration sensors are accelerometer, position sensor, tilt sensor, and ultrasonic sensor. Optical, light, and imaging sensors consist of colorimeter, electro-optical sensor, infrared sensor, and photodiode. Pressure sensor includes barometer, boost gauge, pressure gauge, and tactile sensor. Force, density, and level sensors are force gauge, level sensor, load cell, and hydrometer. Thermal, heat, and temperature sensors are heat sensor, radiometer, thermometer, and thermistor. Proximity and presence sensors include motion detector, occupancy sensor, and touch switch.

1.2.2 Sensor Performance

Sensor performance can be characterized by the (1) range of values, from minimum to maximum value, that it can measure, (2) the measurement resolution, which is the smallest discernable change in the measured value, (3) sensor linearity or maximum deviation from a "straight-line" response, (4) sensor sensitivity or a measure of the change at the sensor output for a given change of sensor input, and (5) sensor accuracy or sensor precision or a measure of the difference between measured and actual values. A sensor network designer selects the sensors based on their performances to meet a specific requirement in designing a sensor network.

1.2.3 Smart Sensors

Smart sensors are currently employed by many contemporary CPSs. There are several definitions for smart sensors that basically describe a sensor that is capable of processing, manipulation, and computation of the sensor-derived data [1–4]. To perform "processing–manipulation–computation," the smart sensor requires interfacing circuit, logic functions, two-way communication device, and decision-making device. Ref. [4] describe three-key components of the smart sensor in action/reaction loop including Observe, Analyze-Make Decision, and Act. According to [4], the "Act" is connected to the "Observe" component by a "Process" that is provided as a Human-In-The-Loop (HITL) or an automated process. The smart sensor defined in this chapter is illustrated in Figure 1.1. It follows the OODA (Observe, Orient, Decide, and Act) loop developed by a military strategist and USAF Col. John Boyd [5,6]. As shown in Figure 1.1, the smart sensor is defined as the one that acts based on the "sensor goal" and "sensor control process." "Sensor goal" controls the "Orient" and "Decide" components through the "Analysis & Synthesis" process. The "Act" and Observe" components are controlled by the "Decide" component and "sensor control process, respectively."

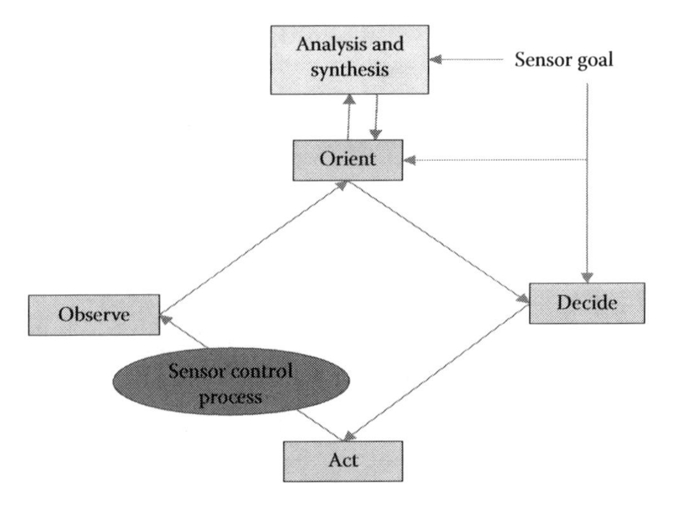

FIGURE 1.1
Definition of smart sensor using OODA loop for modern CPSs.

A smart sensor that employs the process shown in Figure 1.1 should include at least three baseline hardware/software components, including Sensor Networking Processor (SNP), Sensor Interface Module (SIM), and Sensor Control Processor (SCP). Basic SNP functions include two-way communication, message routing, message encoding and decoding, discover and control, and data correction interpretation. Basic SIM functions are storage, Analog Signal Conditioner (ASC), Analog-to-Digital Converter (ADC), trigger, command processor, data transfer, and two-way communication. SCP functions consist of sensor resource management and associated sensing control requirements to meet the sensor goal. Examples of smart sensors are accelerometers, optical angle encoders, optical arrays, infrared detector array, integrated multi-sensor, structural monitoring, and geological mapping. Applications of smart sensors include but are not limited to smart homes, smart cities, intelligent building, predictive maintenance, health care, industrial automation, energy saving, and defense.

1.2.4 Sensor Products

Presently, there are many types of sensor products available in the market. Examples of existing state-of-the-art sensor products are produced by Omoron Industrial Automation [7], Keyence Inductive Proximity Sensors [8], Pressure Profile Systems (PPS) [9], Harry G Security [10], Fujifilm Optical Devices for interferometer [11], Baumer Sensor Products for presence detection/distant measurement/angle measurement/process instrumentation/identification image processing [12], System Sensor for life and safety [13], SICK Sensor Intelligence for industrial sensors [14], Analog Devices for MEMS Accelerators/Gyroscopes/Internal Measurement Units (IMU)/Inertial sensors/Temperature sensors [15], Samsung for CMOS Image sensors [16], ST Life Augmented for MEMS Accelerometers/Gyroscopes/Digital Compasses/Pressure sensors/Humidity sensors and microphones/smart sensors/sensor hubs/temperature sensors/touch sensors [17], Linear Technology for wireless sensor networking [18], Honeywell Aerospace for magneto-resistive sensors [19], and GE Measurement & Control for MEMS Pressure sensors [20].

1.3 Sensor Networks and Associated Technologies for CPS Applications

Sensor networks can be classified into two categories, namely, wired sensor networks and WSNs. This section focuses on the WSNs that are used by modern and emerging CPSs described in Section 1.6. For the sake of completeness, this section also discusses the traditional centralized sensor networking approach using wired sensors. Typical sensor networks are used for monitoring and tracking objects, animals, humans, vehicles, structures, factory, etc.

1.3.1 WSNs—A Traditional Centralized Sensor Networking Approach

Traditional Centralized Sensor Network (TCSN) is a general-purpose design using centralized network management approach, which intends to serve many applications. The TCSNs are designed and built according to detailed plans with primary design concerns focused on sensor network performance and latencies that aligned with a required operational plan in a controlled environment. Control environment allows for easy access to sensor repair and maintenance with component failure and maintenance are addressed through sensor maintenance and repair plans. An example of TCSN is a sensor network for a plastic bottle manufacturing product line, where sensors are used to monitor temperature, pressure, and counting the bottles for making the bottles and packaging the bottles, respectively.

As opposed to TCSN, WSN is usually used for CPS applications, which is a single-purpose design using distributed network management to serve a specific application where power is the key design constraint for all sensor nodes and network components. WSN deployment for a CPS is often ad hoc without planning, and it usually operates in an environment with harsh conditions with very difficult physical access to sensor nodes. Unlike TCSN, WSN component failure is addressed through the design of the network. The focus of this chapter is on WSN with applications to CPSs, and the subsequent sections will describe the current state-of-the-art WSNs.

1.3.2 Distributed WSNs

Emerging CPSs employ MANET and distributed WSNs for collecting sensing data to gain knowledge on the activities of interest to users. This section discusses MANET and distributed WSNs that are currently used by emergent CPSs, including intelligent health-care cyber system, intelligent rescue cyber system, intelligent transportation cyber systems, and intelligent social networking cyber system.

1.3.2.1 Mobile Sensing Network

MEMS, NEMS, MANET technology, and advanced sensor products allow for the deployment of Mobile Sensing Network (MSN) with a collection of small, low cost, low power wireless sensor nodes that can move on their own and interact with the physical environment. These wireless sensor nodes are capable of sensing, local processing, wireless communication networking, and dynamic routing [21,22]. Ref. [22] has provided a good comparison among MSN, MANET, and CPS. A summary of the comparison is given in Table 1.1. A Wireless Network as Sensor Network described in Section 1.3.3 can be considered as a special case of MSN, since the wireless radio nodes are mobile.

1.3.2.2 Compressive Wireless Sensing Network

Recently, Compressive Wireless Sensing Network (CWSN) has been developed based on Compressive Sensing (CS) technology, which is a novel sampling technique to reduce the minimum samples required to reconstruct a signal by exploiting its compressibility property [23,24]. CS technology allows for less communication bandwidth, sensor processing, and power requirements imposed on the CWSN design and deployment. Hence, the key features of CWSN are: (1) processing and communications—they are combined into one distributed operation, (2) without or very little in-network processing and communications, and (3) consistent field estimation is possible even if little or no prior knowledge about the sensed data, while the total power required for the CWSN grows at most sub-linearly with the number of nodes in the network [25]. For a centralized CWSN, when there is no knowledge about the sensed data, Ref. [25] shows that there exists a power-distortion-latency, *D*, trade-off of the form:

$$D \approx \frac{1}{P_{tot}^{2\alpha/(2\alpha+1)}} \approx \frac{1}{L^{2\alpha/(2\alpha+1)}} \quad (1.1)$$

TABLE 1.1

Comparison of MANET, MSN, and CPS

Feature	Mobile Ad hoc Network (MANET)	Mobile Sensing Network (MSN)	Cyber-Physical System (CPS)
Network formation	Random and can support node mobility	Field-specific and allows less mobility	Crosses several fields. Connecting these fields usually relies on the Internet
Communication pattern	Supports arbitrary communication patterns, such as unicast, multicast, and broadcast	Support collective communications, for example, converge cast and query-and-response transactions. Requirements on routing capability are different from MANET	Very often required intra-WSN communications and cross-domain communications
Power management	Emphasizes on energy saving	Energy saving is critical, since sensors are usually deployed in unattended areas. Deeper sleeping modes and redundancy are required	Activation of sensors is usually mission oriented
Network coverage	Requires to meet some connectivity requirements	Requires to meet both connectivity and some coverage criteria	Same requirements for a WSN, but different levels of connectivity and coverage for different WSNs
Node mobility	Usually arbitrary	Requires both controllable and uncontrollable mobility	Requires both controllable and uncontrollable mobility
Knowledge mining	Emphasizes only on networking issues	Focuses more on collecting and managing sensing data	Emphasizes more on how to discover new knowledge across multiple sensing domains and to utilize intelligence properly
Quality of services	Quality of data transmissions is important	Quality of sensing data is important	Emphasizes on higher-level QoS, such as availability of networking and sensing data, security and confidentiality of sensing data, quality of knowledge/intelligence, etc.

Source: Wu, F.J. et al., *Pervasive Mobile Comput. J.*, 2011.

where

- α is a non-zero and positive coefficient, that is, $\alpha > 0$, and it is used to quantify the structural regularity of the centralized CWSN sensor network, which needs not to be known to the network
- P_{tot} is the total power consumption by CWSN
- L is the network latency. Note that when the approximation error exponent in Equation 1.1 for piecewise constant functions represented in a wavelet basis is: $2\alpha = 1$, then Equation 1.1 becomes

$$D \approx \frac{1}{P_{tot}^{1/2}} \approx \frac{1}{L^{1/2}} \tag{1.2}$$

When there is sufficient prior knowledge about the sensed data, the distortion D is given by:

$$D \approx \frac{1}{P_{tot}^{2\alpha}} \approx \frac{1}{L^{2\alpha}} \quad (1.3)$$

The distortion D can be used in the selection of optimum CS technique to achieve a desired network latency of a centralized CWSN.

1.3.2.3 IP-Based Sensor Network

IP-based sensor network (IP-BSN) allows the networks in a CPS to cross several sensor fields and connect these fields by Internet. The emergence of the IETF 6LoWPAN* (RFC 4944) standard for IP communication over low-power radio has made the design and implementation of IP-BSN a reality [26–31]. Figure 1.2 illustrates an example of IP-BSN. This network uses IETF 6LoWPAN adaptation layer that carries IPv6 addresses in a compact form using small IEEE 802.15.4 short addresses [30]. This network supports a variety of IP links while understanding the links, characteristics through the use of abstraction layer. A packet frame format using IEEE 802.15.4 standard is described in Figure 1.3 [31].

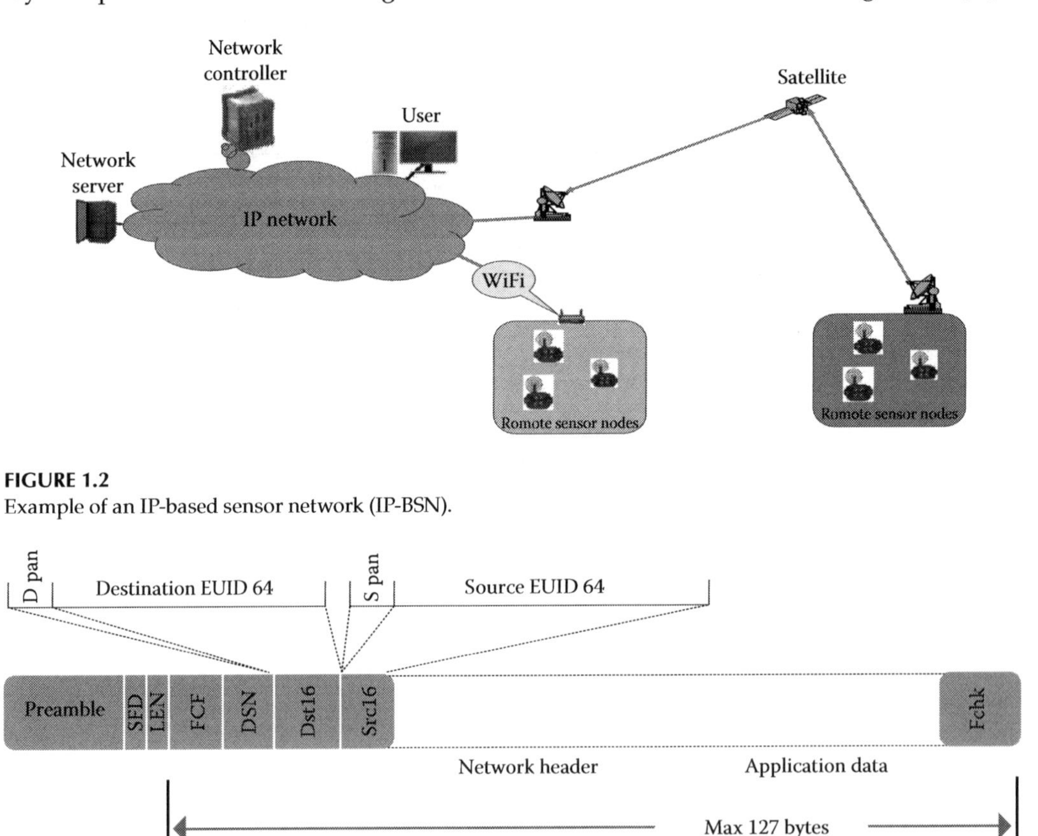

FIGURE 1.2
Example of an IP-based sensor network (IP-BSN).

FIGURE 1.3
IEEE 802.15.4 standard packet frame format. (From Arch Rock Corporation, IP-based wireless sensor networking: Secure, reliable, low-power IP connectivity for IEEE 802.15.4 networks, Website: http://www.cs.berkeley.edu/~jwhui/6lowpan/Arch_Rock_Whitepaper_IP_WSNs.pdf.)

* IPv6 over low power personal area networks.

1.3.2.4 Dynamic Spectrum Access (DSA) Sensing Networks

DSA Sensing Network (DSA-SN) consists of a set of wireless sensor nodes that are placed in a specified location for monitoring the RF (Radio Frequency) environment of interest. The purpose of DSA-SN is to perform spectrum sensing and detect unused spectrum for sharing the spectrum without harmful interference among users. A survey of the spectrum sensing techniques is provided in Table 1.2 [31]. The use of DSA-SN allows a CPS to achieve the same network coverage as typical WSNs but at different levels of connectivity and coverage for different WSNs.

The IP-BSN with a CS-based technique can be used in the design and implementation of a DSA-SN. DSA-SN is usually employed by a Cognitive Radio (CR) wireless network for frequency and bandwidth scheduling among coexisting network users [32–36].

1.3.2.5 Cognitive Radios (CR) Sensing Networks

CR Sensing Network (CRSN) adopts the CR capability in sensor networks [31,41]. Thus, a WSN comprises a number of sensor nodes equipped with CR that are likely to benefit from the potential advantages of the DSA features, such as (1) opportunistic channel usage for bursty traffic, (2) DSA, (3) using adaptability to reduce power consumption, (4) overlaid deployment of multiple concurrent WSN, and (5) access to multiple channels to conform to different spectrum regulations [31]. In general, CRSN can be defined as "a distributed network of wireless cognitive radio sensor nodes, which sense an event signal and collaboratively communicate their readings dynamically over available spectrum bands in a multi-hop manner ultimately to satisfy the application-specific requirements." CRSN applications can be classified into four categories [31]:

1. *Indoor Sensing*: Tele-medicine [34], home monitoring, emergency networks, and factory automation.
2. *Multimedia Sensing*: Video, still image, audio.
3. *Multi-class Heterogeneous Sensing*: Information is gathered through several WSNs and fused to feed a single decision support [42].
4. *Real-time Surveillance Sensing*: Military surveillance for target detection and tracking.

TABLE 1.2

Survey of Spectrum Sensing Techniques

Spectrum Sensing Method	Cons	Pros
Matched filter [37]	Requires a priori info on Primary User (PU) transmissions, and extra hardware on nodes for synchronization with PU	Best in Gaussian noise. Needs shorter sensing duration (less power consumption)
Energy detection [38]	Requires longer sensing duration (high power consumption). Accuracy highly depends on noise level variations	Requires the least amount of computational power on nodes
Feature detection [39]	Requires a priori knowledge about PU transmissions. Requires high computational capability on nodes	Most resilient to variation in noise levels
Interference temperature [40]	Requires knowledge of location PU and imposes polynomial calculations based on these locations	Recommended by FCC. Guarantees a predetermined interference to PU is not exceeded

Source: Akan, O.B. et al., Cognitive radio sensor networks, Website: http://nwcl.ku.edu.tr/paper/J20.pdf.

TABLE 1.3

A Survey of MAC Approach for CRSN

MAC Design Approach	Disadvantages in CRSN	Reasons to Adopt CRSN	Open Research Issues
On-demand negotiation [43]	Contention due to single channel for all negotiations	On-demand reservation is suitable for bursty traffic	Coordination of multiple control channels required for heavy traffic
Home channel [44]	Multiple transceiver requirement	Does not require negotiation for each packet (helps power conservation)	Mechanisms to make this scheme work with single transceiver needed
Time division-based negotiation [45]	Requires network-wide synchronization for negotiation intervals	Simple and very few rules imposed on nodes	Need for network-wide synchronization must be eliminated

Source: Azad, A.K.M. and Kamruzzaman, J., A framework for collaborative multi class heterogeneous wireless sensor networks, in *Proceedings of the IEEE ICC 2008*, May 2008, pp. 4396–4401.

CRSN requires complex Dynamic Spectrum Management (DSM) framework to regulate the spectrum access for the deployed wireless sensor nodes. The framework includes three key components, namely, Spectrum Sensing Component (SSC) (see Table 1.2), Spectrum Decision Component (SDC), and Spectrum Hand-off Component (SHC). A communication framework is required to support the DSM. The communication framework consists of Physical Layer (PL), Data Link Layer (DL2), Network Layer (NL), Transport Layer (TL), and Application Layer (AL). For CRSN, the design of Medium Access Control (MAC) within the DL2 to support wireless sensor nodes with access to medium in a fair and efficient manner is essential for minimum network latency. Table 1.3 summarizes MAC design approach for CRSN. A detailed description of these layers can be found in [42].

Emerging CPSs employ CRSN to allow for intra-WSN communications and cross-domain communications with emphasis on higher-level QoS, such as availability of networking and sensing data.

1.3.3 Wireless Networks as Sensor Networks

Extracting information from the variation in the strength of received signal from a wireless network can turn a wireless network into a sensor network. This type of sensor network is also referred to as Wireless Network as Sensor Network (WNaSN). The Received Signal Strength Indicator (RSSI) provided by a wireless network can be used for localization by position determination, motion detection, and velocity estimation of the wireless radio nodes in the network or position and motion of bodies external to the network [46–51]. The RSSI-based WNaSN can be divided into active and passive localization. Active localization is the practice of locating a person or asset that is carrying an RF device, such as Personal Communications Service (PCS) device or RF tag. Passive localization does not require the person or asset to carry any electronic device, sensor, or tag. The WNaSN employs passive localization, which is referred to as Device-Free Localization (DFL) or RF Tomography (RFT) [52]. For a narrowband receiver, the RSSI in dB (decibel) of a wireless radio node can be expressed mathematically as:

$$\text{RSSI (dB)} = P_T + 20\log_{10}\left|\sum_{i=1}^{N} s_i(t)\right| \tag{1.4}$$

TABLE 1.4

Typical Range of RSSI

RSSI Range (dB)	Signal Quality
Better than –40	Exceptional
–40 to –55	Very good
–55 to –70	Good
–70 to –80	Marginal
–80 and beyond	Intermittent to no operation

Source: Veris White Paper, Veris Aerospond Wireless Sensors: Received Signal Strength Indicator (RSSI), http://www.veris.com/docs/whitePaper/vwp18_RSSI_RevA.pdf.

where

$S_i(t)$ is the complex amplitude gain of the received component *i*th
N is the number of multipath components received at the wireless radio node
P_T is the transmitted power in dB

A typical range of RSSI expressed in dB is given in Table 1.4 [53].

1.3.4 Smart Sensor Networks

As discussed in Section 1.2.3, a smart sensor employs a process shown in Figure 1.1 with three basic components: SNP, SIM, and SCP. Thus, a Smart Sensor Network (SSN) consists of a network of smart sensors that are connected through a wireless network, and the network is designed to meet a specific application, such as smart home or smart city or intelligent building, etc. [54–57]. A typical system architecture for an SSN is presented in Figure 1.4. The wireless sensor networking nodes presented in this figure are assumed to use low power devices that meet IEEE 802.15.4 standard, which defines various layers for interconnecting these sensor nodes. The Operating System (OS) employed by SSN is the TinyOS operating system, which is a small core, multitasking. TinyOS was developed by the University of California [58] and used the NesC language [59]. The remote management and visualization system includes but is not limited to 4G Android or/and Personal Computer (PC) or/and Ipad with the smart sensor process described in Figure 1.1 incorporated into the system. SSN has been used by emerging CPS, such as intelligent health-care cyber system and intelligent transportation cyber system presented in Section 1.6.

1.3.5 Ubiquitous Sensor Networks for Internet of Things (IoT)

As defined in [60], the term "Ubiquitous Sensor Network" or USN is a network of intelligent sensors that are available "anywhere, anytime, by anyone and anything." USN is also referred to as "invisible," "pervasive" or "ubiquitous" computing or to describe "Internet of Things" or IoT. The network consists of three key elements including sensors, tags, and communication/processing capacity. By 2020, Gartner predicts that IoT would be made up of 26 billion "units" [61]. USN for IoT has been applied for CPS applications such as battle damage assessment described in the following and smart cities presented in Section 1.6.

Recently, Ref. [62] has defined IoT as the Networks of Sensors (NoSs) that are used by people through "Process," "Data," and "Things." Ref. [62] has also defined the "Process" as an approach to deliver the right information to the right person, and "Things" as the physical devices and object connected to the Internet and each other for intelligence decision making. In addition, Ref. [62] has envisioned an IoT "Connectivity" platform, including

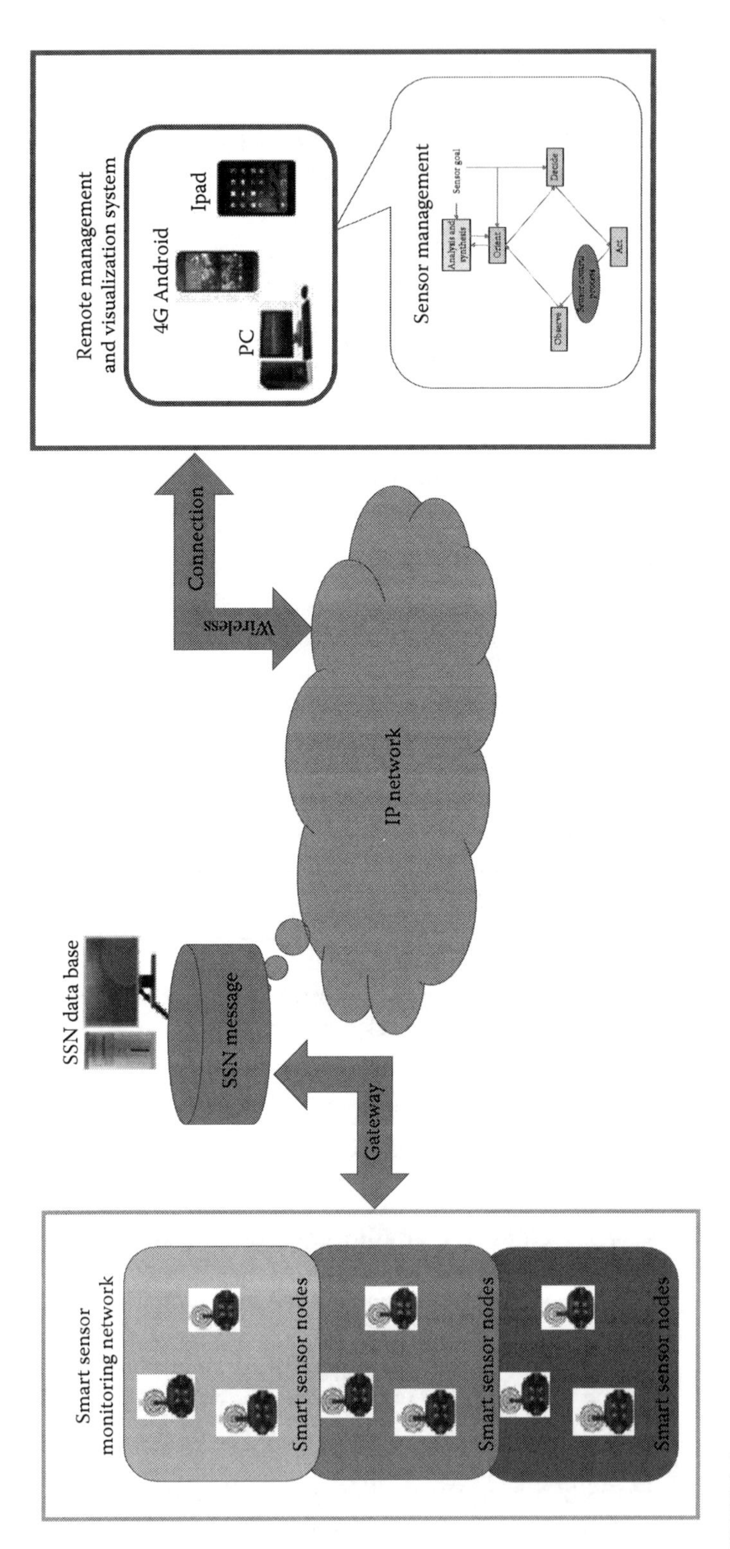

FIGURE 1.4
Typical SSN system architecture.

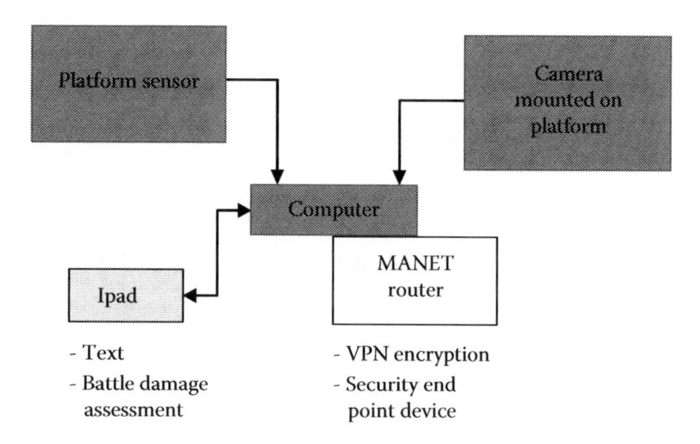

FIGURE 1.5
Example of cisco mobile ready net.

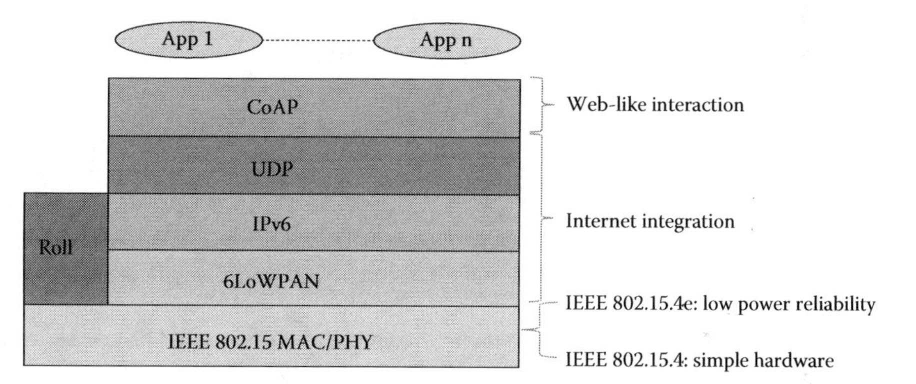

FIGURE 1.6
Recommended standard IoT protocol stack. (Revised from Liu, H., IoT, sensor networks, in *Workshop Proceedings on Future Research Needs and Advanced Technologies*, The Catholic University of America, Washington, DC, May 2, 2014.)

(1) operational technologies, (2) networks, "Fog" computing, storage, (3) data analysis, and (4) control system. The "Connectivity" includes data center and networks. "Fog" computing is defined as a computing layer to make simple determination of what information is needed for the mission. Figure 1.5 shows an IoT example of Cisco Mobile Ready Net developed for a mobile platform consisting of a platform sensor with an Ipad, a computer, and a MANET router for a battle damage assessment application. This Cisco Mobile Ready Net can also be used for a surveillance application.

As IoT technology evolves, standard organizations, such as IEEE, IETF, ETSI, etc., are working to make IoT protocols more robust, scalable, and efficient. The current enhancement task focuses on improving IETF Constrained Application Protocol (CoAP) and Routing Over Low Power and Lossy Networks protocol (ROLL) [63]. The CoAP and ROLL layers are currently incorporated in a standard protocol stack for IoT and it is illustrated in Figure 1.6.

1.3.6 Underwater Sensor Network

Underwater Sensor Network (UnSN) is also referred to as Underwater Acoustic Sensor Network (UW-ASN), which consists of a number of fixed sensor nodes and mobile sensor nodes deployed underwater. The data links use to connect these fixed and mobile sensor nodes

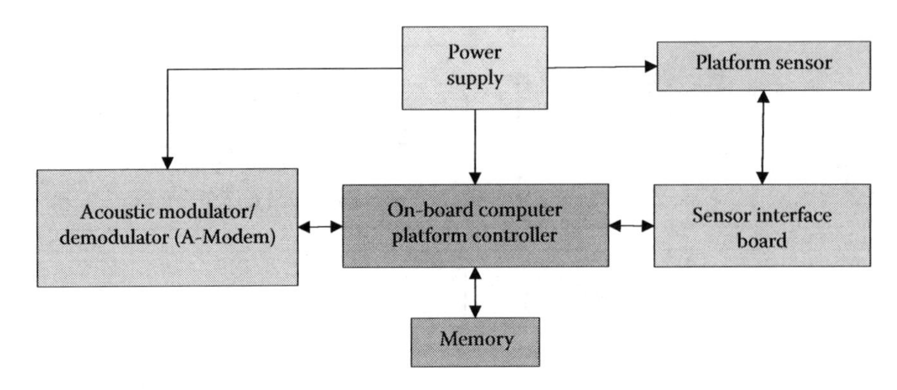

FIGURE 1.7
A typical underwater sensor system architecture.

employ acoustic wireless communications [64–67]. A typical underwater sensor system architecture using acoustic wireless communications is shown in Figure 1.7. The platform sensor can be placed in a fixed or mobile platform with an on-board computer that serves as the platform sensor controller providing sensor and communication resources management. The controller provides self-configuration of the network of underwater sensor nodes and adapts to harsh ocean environment. The Acoustic Modulator/Demodulator (A-Modem) connects the sensor nodes using underwater sensor network protocols. The four protocols employed by the existing A-Modem are flooding-based, multipath-based, cluster-based, and miscellaneous-based routing protocols [64]. The power supply is a limited battery, which cannot be replaced or recharged. The issue of energy conservation for UW-ASN is currently evolving to more power efficient underwater communication and networking techniques.

Applications of UW-ASN include scientific, Industrial, Military, and homeland security applications [65].

1.4 Architecture of WSNs for CPS Applications

A WSNs design for CPS applications can be very complex depending on their goal or mission. WSNs are broadly divided into infrastructured and infrastructureless WSNs [68]. The infrastructured WSN consists of a set of wireless nodes with a network backbone, and the infrastructureless WSN consists of distributed, independent, dynamic topology, low-power, and task-oriented wireless nodes. As shown in Figure 1.8, a basic network backbone includes a task manager for local management, a task manager for remote management, a router, a transit network gateway acting as a proxy for the sensor network on the Internet, a data storage, and an Internet satellite network. In either case, the architecture design should be scalable and flexible allowing for extending additional sensor nodes and preserving network stability.

The WSN architecture design goal for CPS applications is to maximize the network density, $D_N(d)$, within the WSN operating region R. If d is defined as the radio transmission range between the sensor nodes, and N is the number of wireless sensor nodes to be placed in the region R, the $D_N(d)$ is given by:

$$D_N(d) \approx \frac{(N \cdot d^2)}{R} \tag{1.5}$$

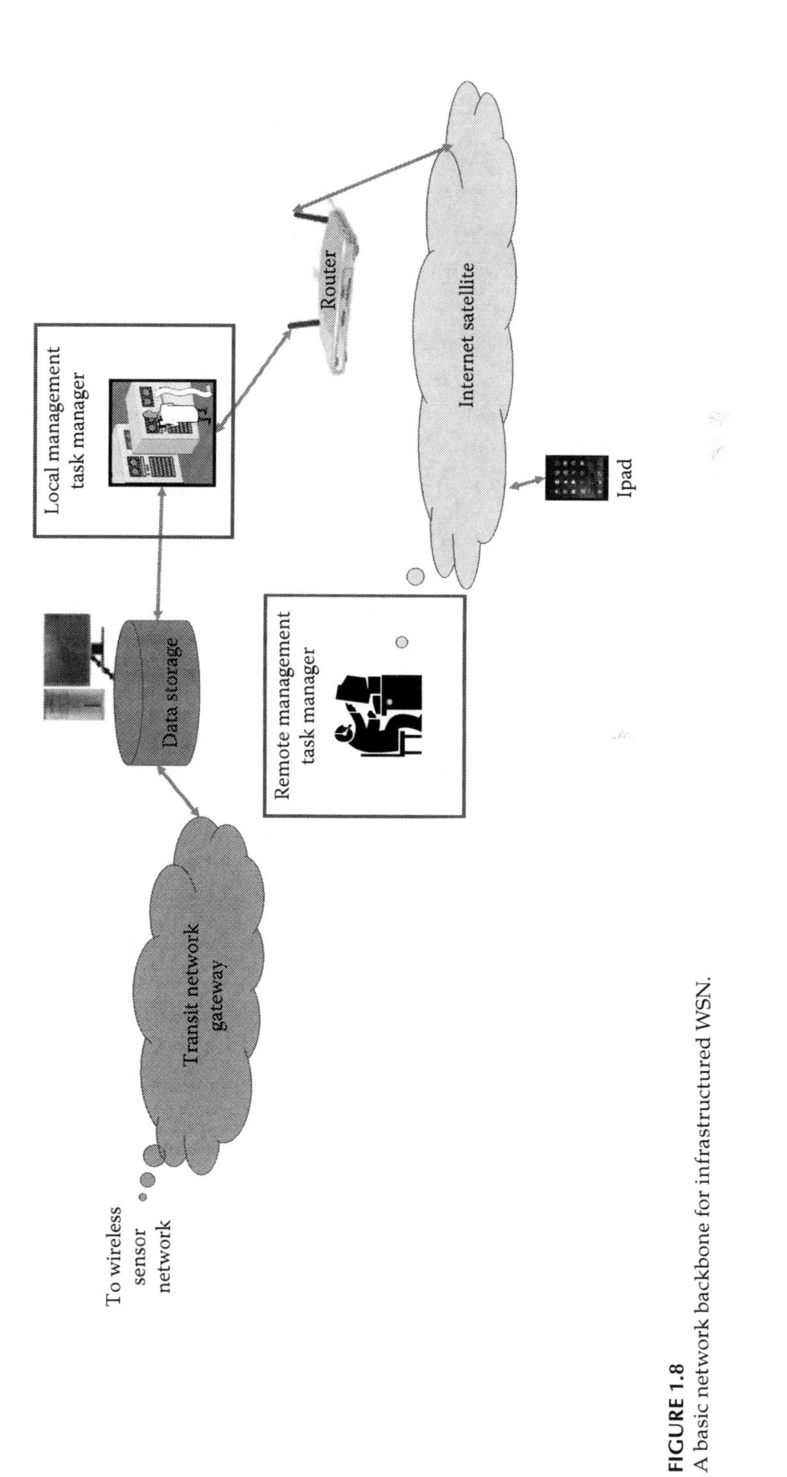

FIGURE 1.8
A basic network backbone for infrastructured WSN.

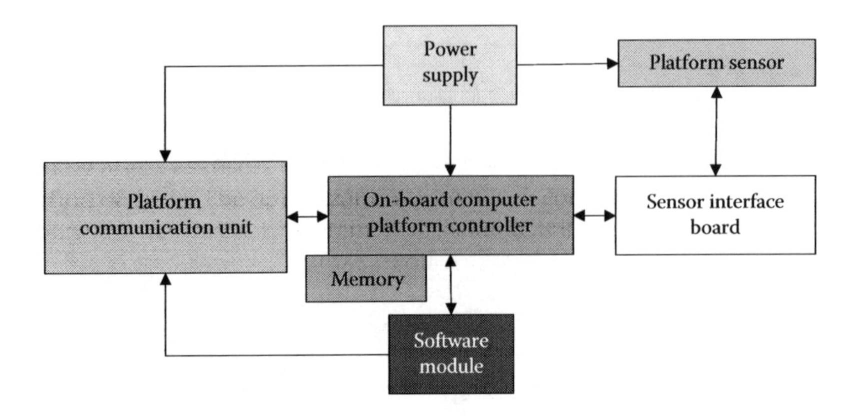

FIGURE 1.9
A typical SNoA.

The architecture design of the sensor network includes the sensor node architecture and overall network architecture designs. The sensor node architecture design focuses on (1) reducing the cost, (2) increasing flexibility, (3) conserving energy, and (4) providing fault-tolerance. Similar to underwater sensor system architecture shown in Figure 1.7, a typical Sensor Node Architecture (SNoA) is shown in Figure 1.9. The SNoA includes four key subsystems and a platform software module. The software module is used for the sensor and communication resources management. The four key subsystems consist of platform power, sensor, on-board computer, and communications. The detailed description and design issues for these subsystems can be found in Ref. [68].

Due to the wide range of CPS applications, the classical OSI (Open System Interconnect) paradigm may not be applicable for WSN; rather the cross-layer communications network design may be more appropriate. The communication network services supporting a sensor network can be classified into two categories, namely, data and control services. The data service provides sensor data messages that involve transfer of sensor data. The control service provides controlled messages used for configuration, topology detection, neighbor discovery, route discovery, time synchronization, reprogramming, reliability, and network management such as congestion and flow estimation and recovery. The control messages in general have higher priority than sensor messages. A standard protocol stack for WSNs is shown in Figure 1.10. Note that this protocol stack is mapped directly to the standard IoT protocol stack shown in Figure 1.6. The focus for the remaining of this section is on the network design including sensors as service-oriented architecture, sensor resource management and task scheduling, and sensor aware routing network.

1.4.1 Sensor Network as Service-Oriented Architecture (SOA)

Emerging CPSs employs sensor network that considers the Sensor Network and its contained services as a Service-Oriented Architecture (SN-SOA). This section discusses SN-SOA and its emerging trend for CPS applications.

The International Technology Alliance (ITA) Sensor Fabric, developed as part of the ITA in Network and Information Science, addresses challenges in the areas of sensor identification and discovery, sensor access and control, and sensor data consumability, by extending the message bus model commonly found in commercial IT infrastructures out to the edge of the network. Recently, Ref. [69] takes the message bus model at the edge and extends it to a service bus at the edge. The service bus at the edge is now extended into a semantically

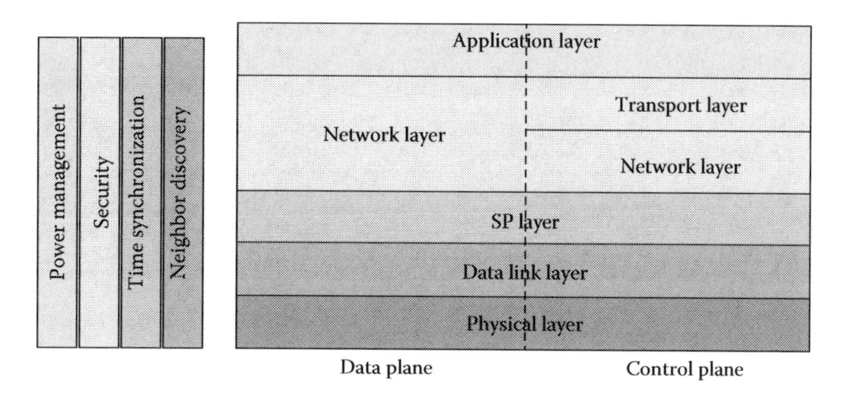

FIGURE 1.10
A standard protocol stack for WSNs. (From Liu, H., IoT, sensor networks, in *Workshop Proceedings on Future Research Needs and Advanced Technologies*, The Catholic University of America, Washington, DC, May 2, 2014; Kavar, J.M. and Wandra, K.H., *Int. J. Adv. Res. Electr. Electron. Instrum. Eng.*, 3, February 2014.)

rich, model-based design that considers the sensor network and its contained services as a SOA. Ref. [69] has addressed the medical use case for an SOA on a sensor network middleware* development. The approach is to modeling services on the sensor network using annotated UML activity diagrams and their associated semantics, which is expanded to include service composition, and ontology† for modeling the assets and services contained within the network. The service model may be transformed into alternative models, including formal models such as process algebras, for analysis with the results providing additional annotations to the core design model. The core model may also be transformed into a form that can be deployed onto a real sensor network using the ITA Sensor Fabric.

Another emerging trend for sensor network as SOA is the Service-Oriented Sensor Web (SOSW) being developed by the OpenGIS Consortium (OGC). OGC develops Sensor Web Enablement (SWE) standard, which is composed of a set of specifications, including SensorML, Observation & Measurement, Sensor Collection Service, Sensor Planning Service, and Web Notification Service. Ref. [70] also presents a reusable, scalable, extensible, and interoperable SOSW architecture that (1) conforms to the SWE standard; (2) integrates Sensor Web with Grid Computing, and (3) provides middleware support for Sensor Webs. Figure 1.11 provides a description of typical collaboration within SWE framework.

1.4.2 Semantic Modeling of Sensor Network and Sensor Attributes

Semantic Modeling of Sensor Network (SMoSN) uses declarative descriptions of sensors, networks and domain concepts to aid in searching, querying, and managing the network and data. As described in Section 1.4.1, SOSW is an OGC-style sensor web enriched with semantic annotation, querying, and inference [71]. SOSW relies on OGC standards and focuses on issues external to the network, although the use of semantics inside the network is not precluded. The Semantic Sensor Network (SeSN) employed SMoSN, which includes semantic sensor webs, semantic sensor networks that do not rely on OGC standards.

* Middleware is a computer software that provides services to software applications beyond those available from the operating system.

† Ontologies for sensors provide a framework for describing sensors, and allow classification and reasoning on the capabilities and measurements of sensors, provenance of measurements and may allow reasoning about individual sensors as well as reasoning about the connection of a number of sensors as a macro-instrument.

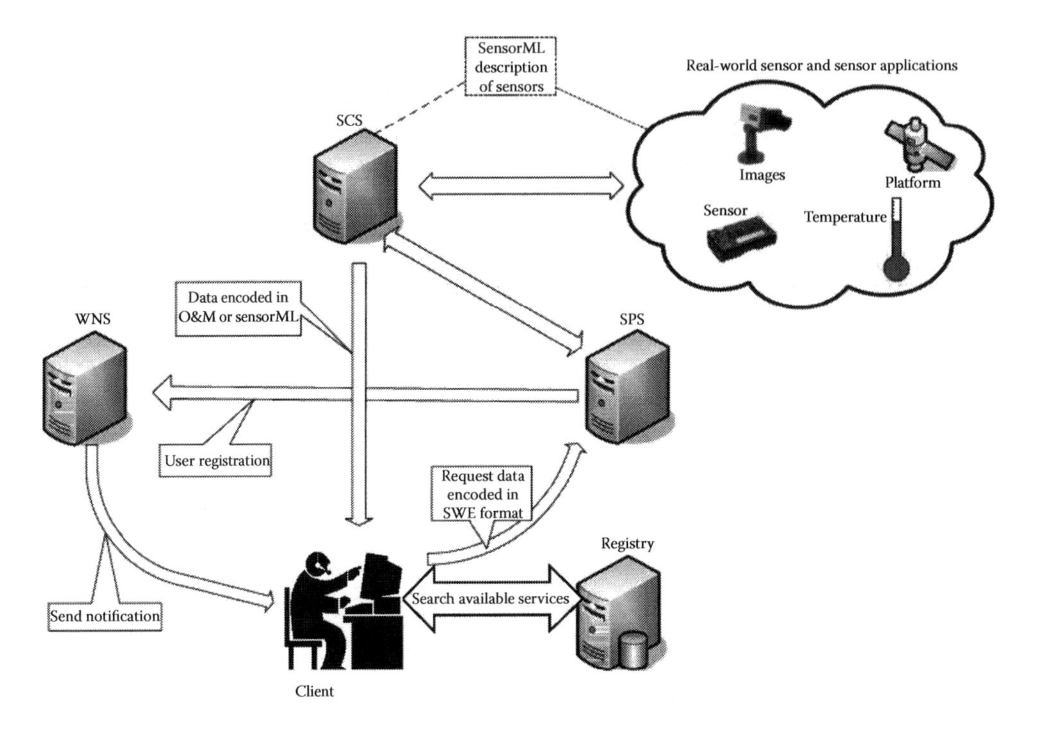

FIGURE 1.11
Typical collaboration within SWE framework. (From Chu, X. and Buyya, R., *Service Oriented Sensor Web*, The University of Melbourne, Melbourne, Victoria, Australia.)

SeSN explicitly allows the use of semantics to manage the network as well as its resulting data. Figure 1.12 illustrates a typical SeSN architecture [72]. The architecture includes three layers, namely, data, processing, and application layers, to support network-internal processing, inference and integration, and services, respectively.

Ref. [72] provides a good summary on sensor ontologies as a key enabling component of SeSN. The current state-of-the art SeSN can enable classification, and link of data and sensors, and the SeSN technology exists to construct virtual sensors as compositions of existing components. Ref. [73] addresses a semantic modeling scheme for large-scale sensor data streams emerging from IoT resources. The semantic modeling scheme deals with naming convention and data distribution mechanism for the sensor data streams. Table 1.5 compares sensor attributes between IoT data streams and conventional data streams.

1.4.3 Sensing Resource Management and Task Scheduling

Sensor Resource Manager (SRM) provides the control of available resources within a sensor network [74]. The sensor resources include sensor power, sensing time, sensing position, number of sensors, etc. SRM's objective is to provide the best situation awareness of the situation given the available sensor resources. As indicated in Ref. [74], the desired characteristics of a good sensor manager includes mission-objective oriented, adaptive to change of environment, anticipatory, requiring minimal interaction, accounting for sensor dissimilarities, and performing in near real time, etc.

Current SRM is designed based on the Markov Decision Process (MDP) assumption. In a SRM-MDP, the future states depend only on the current state, ignoring the total history of the state space. Thus, for a centralized sensor network, sensor nodes with complete

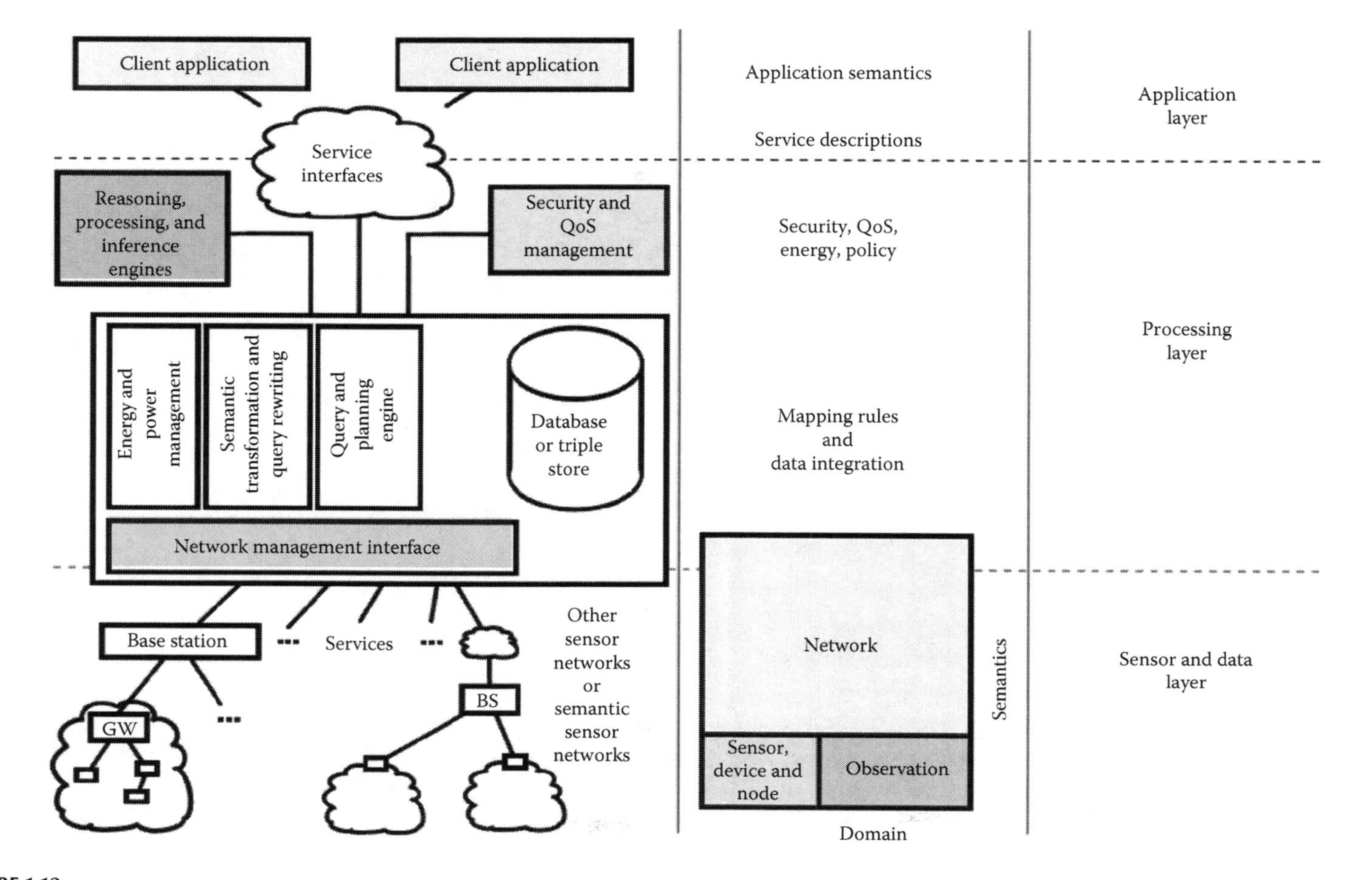

FIGURE 1.12
A typical SeSN architecture. (From Compton, M. et al., *A Survey of the Semantic Specification of Sensors*, ICT Centre, CSIRO, Canberra, Australian Capital Territory, Australia.)

TABLE 1.5

Sensor Attributes Comparison between IoT and Conventional Data Streams

Sensor Attributes	IoT Data	Conventional Data Streams
Size	Often very small; some IoT data can be a real number and unit of measurement; the meta-data is usually significantly larger than the data itself	Usually much larger than IoT data (video data)
Location dependency	Most of the time location dependent	Normally not location-dependent
Time dependency	Time-dependent; need to support various queries related to temporal attributes	Normally not time-dependent
Life span	Usually short-lived or transient	Long-lived
Number	Often very large	Usually smaller than IoT data items
Persistency	Some of the data needs to be archived	Usually persistent
Resolution	Names created from meta-data for resolution could be longer than conventional data (taking into account temporal and spatial dimensions)	Resolution is usually based on names

Source: Barnaghi, P. et al., *A Linked-Data Model for Semantic Sensor Streams*, Centre for Communication Systems Research, The University of Surrey, Guildford, U.K.

knowledge of the state space, that is, full awareness of sensor nodes to maintain the current state, and the network updates the state with incoming measurements or the passage of time. This new state can now be used to determine future states of the network. Figure 1.13 depicts a typical SRM for a centralized sensor network. This figure shows that the SRM is mission-objective oriented with a fusion manager to provide support to the mission by fusing the sensor tracks for identification purpose.

For a decentralized sensor network, each sensor node's state is only dependent on that node's prior state, input from the environment, and associated noise. Each node maintains its own individual status of the state space, but since inter-node communication is imperfect, these state representations are often dramatically different from node to node. A typical SRM for a decentralized WSN with centralized fusion manager is described in Figure 1.14.

A good SRM considers both (1) task-specific resource requirements and (2) available computational and communication resources. Ref. [75] presents a technique on how to (1) use a Task-Specific Quality of Service (TS-QoS) as a variable that is controlled by the sensor network, and (2) request for the resources that are generated based upon the feedback provided by the TS-QoS, where the request generator's parameters are adjusted using a controller. As described in Ref. [75], for sensor measurements to be meaningful, the QoS should be specific to the particular domain and task being performed. A QoS with this property is referred to as TS-QoS. The simulation results presented in Ref. [75] showed the SRM performance improvement using TS-QoS.

1.5 Design of WSNs for CPS Applications

One of the challenges in the SRM design for CPS applications is to determine a sensor field architecture that optimizes cost and provides high sensor coverage, and at the same time is resilience to sensor failures. The SRM designer must perform appropriate computation/communication trade-off for a desired CPS application. Optimum sensor deployment, which is also referred to as sensor placement, facilitates the unified WSN design

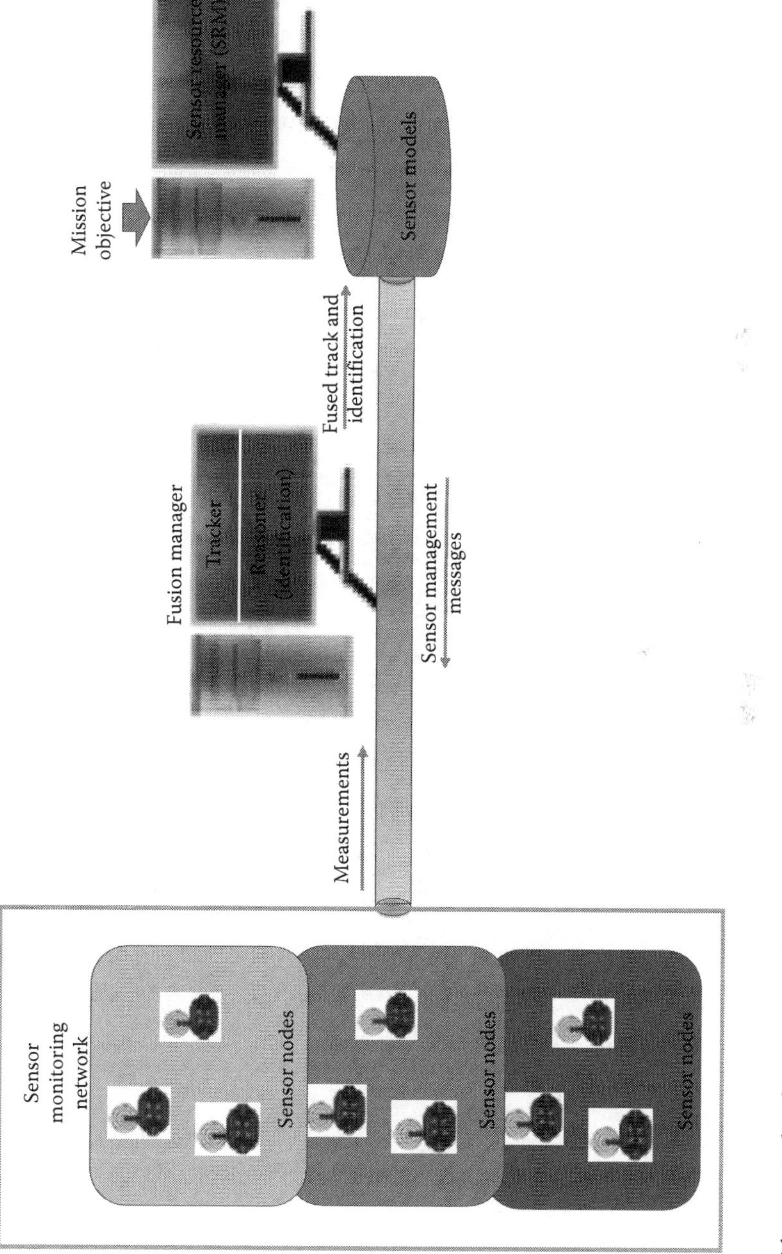

FIGURE 1.13
A typical SRM for a centralized sensor network.

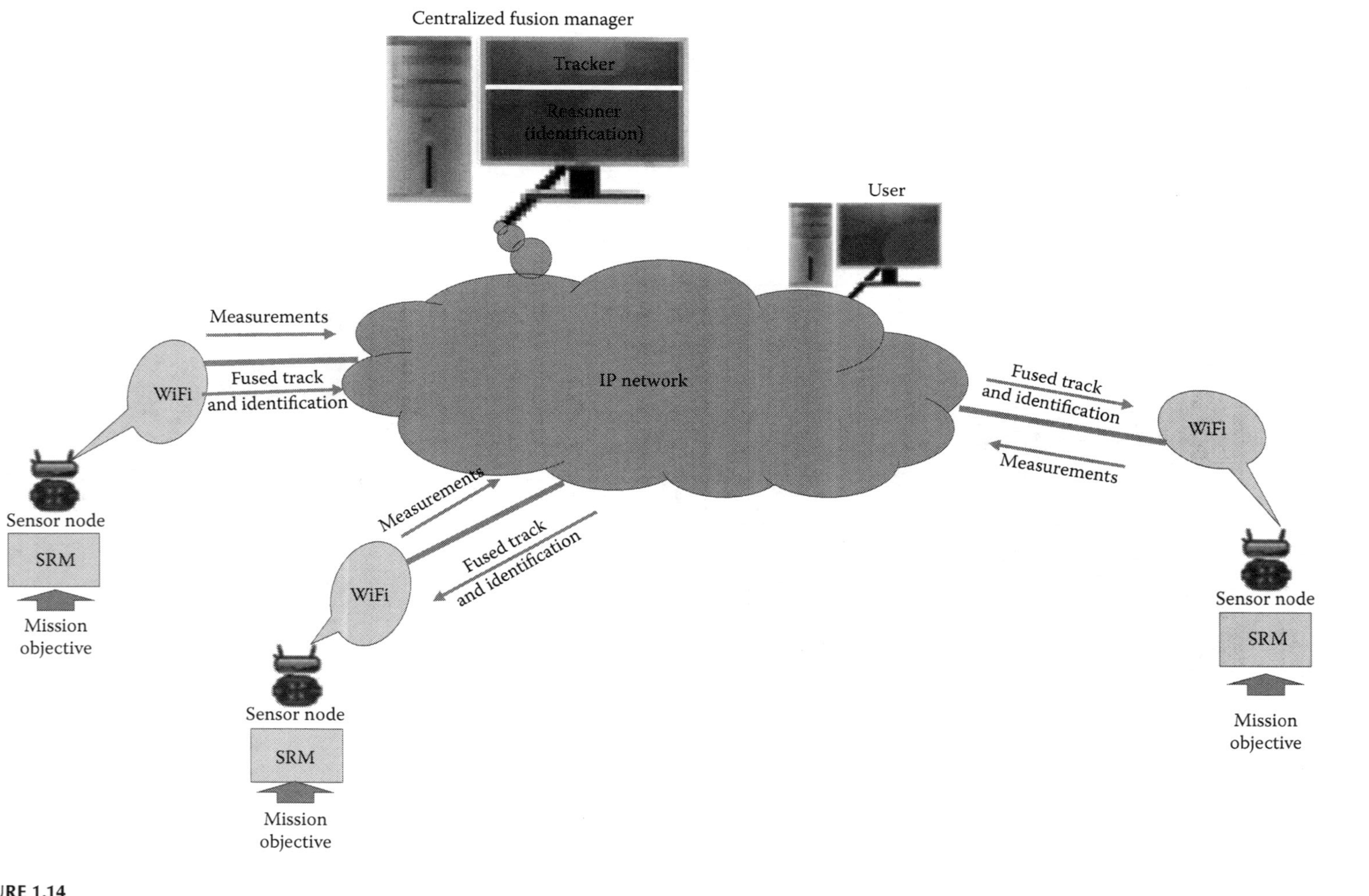

FIGURE 1.14
A typical SRM for a decentralized sensor network with a centralized fusion manager.

and operation and decreases the need for excessive network communication to support sensing tasks including surveillance, target location, and tracking. Thus, optimum sensor deployment can simplify the SRM processing that connects the front-end sensing processing and back-end exploitation. This section discusses sensing capacity, WSN design, and optimum deployment of sensor nodes of a WSN used by an emerging CPS.

1.5.1 Sensing Capacity of Sensor Networks

Sensing capacity of a sensor network employed by a CPS is defined as the minimum number of sensors necessary to monitor a given region to a degree of fidelity given on sensor noisy measurements. For a CSWN, sensing capacity provides a bound on the maximum number of signal components that can be identified per sensor under noisy measurements. The sensing capacity depends on Signal-to-Noise Ratio (SNR), signal sparsity, and sensing diversity. The signal sparsity is the inherent complexity/dimensionality of the underlying signal/information space and its frequency of occurrence, and the sensing diversity is the number of modalities of operation of each sensor. The sensor modality is the nature of the sensing signals and their associated SNRs. To visualize the sensor modality, let us consider a sensor network with N sensor nodes, each sensor node has a maximum power P. Each node can operate with a different modality. For this example, let us assume the sensor modalities include Narrow Beam High Resolution Modality (NB-HRM) and Wide Beam Low Resolution Modality (WB-LRM). For NB-HRM, each sensor node can form its beam to different region with a maximum permissible SNR. For WB-LRM, each sensor node can have a wider focus but with a corresponding decrease in SNR [76]. Ref. [76] shows that when the signal sparsity approaches zero, the sensing capacity also goes to zero irrespective of sensing diversity. In practice, this means that a large number of sensors can be required to reliably monitor a small number of targets in a given region.

1.5.2 Optimum Deployment of Wireless Sensor Nodes for CPS Applications

To deploy the sensors optimally within a CPS, the WSN designer must consider the following design factors:

- The nature of the terrain including obstacles between sensors (e.g., trees, hills, buildings, etc.), uneven surfaces, and elevations for hilly terrains, etc.
- Sensor redundancy due to the likelihood of sensor failures
- Required power to transmit information between deployed sensors
- Required power between a deployed sensor and the fusion center or cluster head

To simplify the design for optimum sensor deployment, the WSN designer can assume that the design is a resource-bounded optimization framework for sensor resource management under the constraints of sufficient grid coverage of the sensor field. For this type of bounded optimization problem, the distributed WSN has the following properties:

- A minimum number of sensors, N, to be deployed to support the required sensing objectives
- The sensors are assumed to transmit/report a minimum amount of sensed data
- Optimum sensor placement to ensure that the ensemble of these data contains sufficient information for the data fusion center to subsequently query a small number of sensors for detailed information (e.g., imagery and time series data, etc.)

For this optimization problem, the design goal is to aim at optimizing the number of sensors, N, and their placement for network provisioning and to support such minimalistic WSN. Ref. [77] tackles the problem by posing the sensor field as a grid of two- or three-dimensional of points with a target in the sensor field considered as a logical object, which is represented by a set of sensors that see it. Thus, an irregular sensor field is to be modeled as a collection of grids. The optimization framework considered by Ref. [77] is a probabilistic framework due to the uncertainty associated with sensor detections caused by the nature of terrain or sensor failures. Following is a summary of the framework presented in Ref. [77].

The probability of detection, P_{di}, of a target by the ith sensor is assumed to be varied exponentially with the distance, d_i, between the target and the sensor, that is,

$$P_{di} = e^{(-\alpha.d_i)} \tag{1.6}$$

where the parameter α is used to characterize the quality of the sensor and the rate at which its detection probability diminishes with distance d_i. For every two grid points i and j in the sensor field, the probability of detection will be modeled as:

- P_{ij} = Probability that a target at grid point j is detected by a sensor at grid point i, and
- P_{ji} = Probability that a target at grid point i is detected by a sensor at grid point j.

If some knowledge of the terrain is acquired prior to sensor placement through satellite imagery, then obstacles can be modeled by altering the detection probabilities for appropriate pairs of grid points, for example, if an object such as a building or foliage is present in the line of sight from grid point i to grid point j, then $P_{ij} = 0$. Partial occlusion can also be modeled by setting a non-zero, but small, value for the detection probability. Ref. [77] describes the algorithms for sensor placement for a given set of detection probabilities in a sensor field (both with and without obstacles). The goal of the sensor placement algorithms is to determine the minimum number of sensors and their locations such that every grid point is covered with a minimum confidence level. The coverage threshold T is used to refer to this confidence level, and it is provided as an input to the placement algorithm. The objective is to ensure that every grid point is covered with probability of at least T.

The algorithm starts by generating a sensor detection matrix $D = [P_{ij}]$ for all pairs of grid points in the sensor field. From the sensor detection matrix D, the miss probability matrix M is calculated:

$$M = m_{ij}, \quad \text{where}$$
$$m_{ij} = 1 - p_{ij} \tag{1.7}$$

Let the vector M^* be the set of miss probabilities for the $N = n^2$ grid points in the sensor field:

$$M^* = (M_1, M_2, \dots, M_N) \tag{1.8}$$

An entry M_i in this vector denotes the probability that grid point i is not collectively covered by the set of sensors placed in the sensor field. At the start of the placement algorithm, the vector M is initialized to the all-1 vector, that is, $M^* = (1, 1, \ldots, 1)$. Let $M_{min} = 1 - T$ be the maximum value of the miss probability that is permitted for any grid point, the pseudocode steps of the first sensor placement algorithm MAX_AVG_COV are outlined as follows [77].

```
Pseudocode Steps for MAX_AVG_COV (M, M*, Mmin)
begin
num_sensors:= 1;
repeat
for i:= 1 to N do
Σi = mi1 +mi2 + ... + miN;
Place sensor on grid point k such that Σk is minimum;
for i:= 1 to N do
Mi=Mimki; /* Update miss probabilities due to sensor on grid point k */
Delete kth row and column from the M matrix
num_sensors:= num_sensors + 1;
until Mi < Mmin for all i, 1 ≤ i ≤ N
or num_sensors > limit;
end
```

Note that the effectiveness of grid coverage due to an additional sensor is measured by the Σ_i parameter. The approach presented in Ref. [77] attempts to evaluate the global impact of an additional sensor by summing the changes in the miss probabilities for the individual grid points. The preferential coverage is modeled in Ref. [77] by assigning a different detection probability p_i to each grid point i. The miss probability threshold for a grid point i is then expressed as:

$$M^i_{min} = 1 - p_i \tag{1.9}$$

The Pseudo steps for MAX_AVG_COV are modified such that the termination criterion of the repeat/until loop is based on checking that the individual miss probability threshold of each grid point has been reached. The pseudocode steps of the second sensor placement algorithm MAX_MIN_COV are outlined as follows [77]:

```
Pseudocode Steps for MAX_MIN_COV (M, M*, Mmin)
begin
Place first sensor randomly
num_sensors:= 1;
repeat
for i:= 1 to N do
Mi=Mimki; /* Update miss probabilities due to sensor on grid point k */
Place sensor at grid point k such that Mk is max
Delete kth row and column from the M matrix
num_sensors:= num_sensors + 1;
until Mi < Mmin for all i, 1 ≤ i ≤ N
or num_sensors > limit;
end
```

Note that the Pseudocode steps described earlier assumed that the sensor detections are independent, that is, if a sensor detects a target at a grid point with probability p_1, and

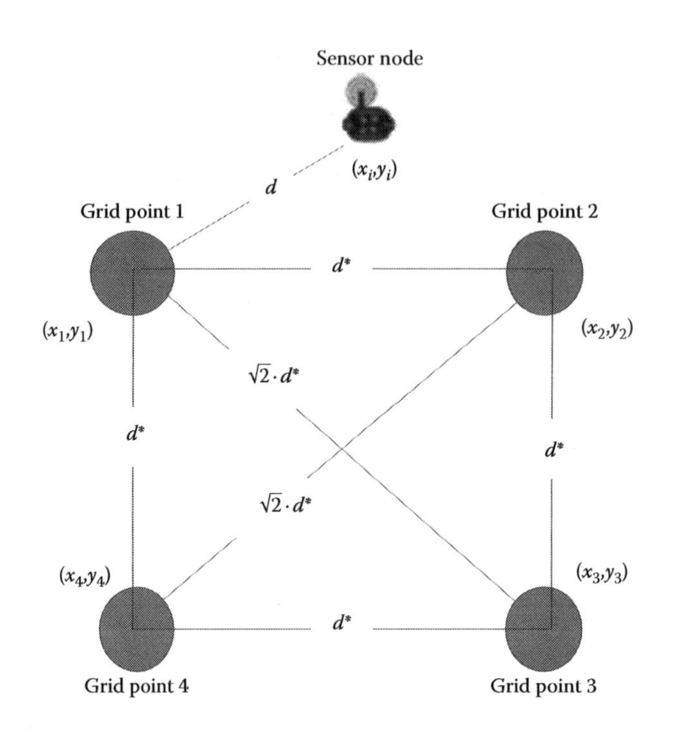

FIGURE 1.15
Sufficient condition to ensure sensor coverage.

another sensor detects the same target at that grid point with probability p_2, then the miss probability for the target is $(1 - p_1)(1 - p_2)$.

The WSN designer thus far has only considered the coverage of only the grid points in the sensor field. In order to provide robust coverage of the sensor field, the designer also needs to ensure that the region that lies between the grid points is adequately covered, that is, every non-grid point has a miss probability less than the threshold M_min. Consider the four grid points in Figure 1.15 that lie on the four corners of a square. Let the distance between these grid points be d^*. The point of intersection of the diagonals of the square is at distance d^*/SQRT(2) from the four grid points. To ensure that all non-grid points are adequately covered within the sensor network, Ref. [77] has derived sufficient condition under which the non-grid points are adequately covered by the MAX_AVG_COV and MAX_MIN_COV algorithms. For a two-dimensional space, the sufficient condition stated that if the distance between the grid point (x_1,y_1) and a potential sensor location (x_i,y_i) is d and the distance between adjacent grid points (x_1,y_1) and (x_2,y_2) is d^*, then a value of d + SQRT(2)·d^* can be used to calculate the coverage of grid point (x_1,y_1) due to a sensor at (x_i,y_i), and the number of available sensors is guaranteed to be adequate, the miss probability of all the non-grid points is less than the threshold $M_{\min}$ when the algorithms MAX_AVG_COV and MAX_MIN_COV terminate. Figure 1.15 illustrates this sufficient condition.

1.5.3 Routing Techniques

For CPS applications, routing protocols must be designed to allow intra-WSN communications and cross-domain communications. For IP-based WSNs, the routing techniques seem to be straight forward. But in practice, the number of sensor nodes in the network is usually very large and it is not possible to build a global addressing scheme for the deployment of a large number of sensor nodes as the overhead of ID maintenance is normally high. Thus, traditional IP-based protocols may not be applied to WSNs. In addition, the sensors can be deployed in an ad hoc manner and getting the sensing data is more important than knowing the ID of the sensor node that sent the data. Thus, the routing task in WSNs is not trivial, and this section addresses routing techniques employed by WSNs.

1.5.3.1 Negotiation-Based Protocols

The Negotiation-Based Protocols (NBP) employ high-level data descriptors in order to eliminate redundant data transmissions through negotiation [78,79]. The key of NBP is to suppress duplicate information and prevent redundant data from being sent to the next sensor or the base station by conducting a series of negotiation messages before the real data transmission begins.

1.5.3.2 Directed Diffusion

Directed Diffusion (DD) is a Data-Centric (DC) and application-aware paradigm in the sense that all data generated by sensor nodes is named by attribute-value pairs [80]. The main idea of the DD-DC is to combine the data coming from different sources enroute in-network aggregation by eliminating redundancy, minimizing the number of transmissions, thus saving network energy and prolonging its lifetime. DD-DC routing finds routes from multiple sources to a single destination that allows in-network consolidation of redundant data. DD-DC is different from NBP in two aspects. First, DD-DC issues on demand data queries as the base station sends queries to the sensor nodes by flooding some tasks. In NBP, sensors advertise the availability of data allowing interested nodes to query that data. Second, all communication in DD-DC is neighbor-to-neighbor with each sensor node having the capability of performing data aggregation and caching. Unlike NBP, there is no need to maintain global network topology in DD-DC.

1.5.3.3 Energy Aware Routing

The objective of Energy-Aware Routing protocol (EAR) is to increase WSNs lifetime [81]. EAR is similar to DD-DC, it differs in the sense that it maintains a set of paths instead of maintaining or enforcing one optimal path at higher rates. In EAR, the paths are maintained and chosen by means of a certain probability. The selected probability depends on how low the energy consumption of each path can be achieved. By having paths chosen at different times, the energy of any single path will not deplete quickly. The approach allows the energy to dissipate more equally among all nodes, and hence to achieve longer network lifetime. When compared to DD-DC, this protocol provides an overall improvement of 21.5% energy saving and a 44% increase in network lifetime [81].

1.5.3.4 Rumor Routing

Rumor Routing (RR) is also a version of DD-DC, and it is designed for applications where geographic routing is not feasible. In general, DD-DC uses flooding to inject the query to the entire WSN when there is no geographic criterion to diffuse tasks. In some cases, there is only a little amount of data requested from the sensor nodes and the use of flooding is unnecessary. The key idea of RR is to route the queries to the nodes that have observed a particular event rather than flooding the entire network to retrieve information about the occurring events. In order to flood events through the network, RR employs long-lived packets, called agents [82]. When a sensor node detects an event, it adds such event to its local table, called events table, and generates an agent. Agents travel the network to propagate information about local events to distant nodes. When a node generates a query for an event, the nodes that know the route may respond to the query by inspecting its event table. Hence, there is no need to flood the whole network, which reduces the communication cost. RR normally maintains only one path between source and destination as opposed to DD-DC where data can be routed through multiple paths at low rates. RR performs well only when the number of events is small [82].

1.5.3.5 Multipath Routing

Multipath Routing Protocols (MRPs) employ multiple-path rather than a single-path routing to enhance WSN performance. The fault tolerance or the resilience of a protocol is measured by the likelihood that an alternate path exists between a source and a destination when the primary path fails. MRP increases the resilience by maintaining multiple paths between the source and the destination at the expense of an increased energy consumption and traffic generation. These alternate paths are kept alive by sending periodic messages at the expense of increased energy consumption and overhead of maintaining the alternate paths. MRP concept is also described in EAR earlier, where the use of a set of suboptimal paths occasionally can increase the lifetime of the network. These paths are chosen by means of a probability, which depends on how low the energy consumption of each path is [81]. On the other hand, Ref. [83] proposes to route data through a path whose nodes have the largest residual energy. The path is updated whenever a better path is discovered. The primary path will be used until its energy falls below the energy of the backup path at which the backup path is used. Using this approach, the nodes in the primary path will not deplete their energy resources through continual use of the same route, hence achieving longer battery life.

MRP has also been proposed to enhance the reliability of WSNs [84]. The network reliability can be increased by providing several paths from source to destination and by sending the same packet on each path. This technique increases the WSN traffic significantly. Ref. [84] performs a tradeoff between the amount of traffic and the reliability of the network. The idea is to split the original data packet into sub-packets and then send each sub-packet through one of the available multipath. Ref. [84] shows that that even if some of these sub-packets are lost, the original message can still be reconstructed. Ref. [84] also finds that for a given maximum node failure probability, using higher multipath degree than a certain optimal value will increase the total probability of failure.

1.5.4 Sensor Network Security

For CPS applications, Sensor Network Security (SNS) is an essential element in the design of WSNs [85]. The WSN designer must consider the following security factors in the design of the WSNs:

- Confidentiality. This is the ability to conceal messages from a passive attacker. Any message communicated via WSN should remain confidential.
- Integrity. This is the ability to confirm a message has not been tampered with, altered or changed while in the network. This is the reliability of the message.

- Authentication. This is the reliability of the message to identify the origin of the message. Data authentication is the ability to verify the identity of senders.
- Availability. This is the ability to use the resources when they are available for messages to communicate.

There are various approaches for providing security service to WSNs [86,87]. These are encryption, steganography, securing access to the physical layer, frequency hopping, etc.

1.5.5 Power Management for WSNs

In practice, most of wireless sensor nodes are battery powered, hence have limited energy. Some sensor nodes are equipped with power harvesting units such as solar, vibrational, or wind energy. Energy in sensor nodes is always a critical design issue that should be addressed wisely in the design of WSNs. A WSN should be autonomous and self-sustainable, able to function for several years with a battery power supply. A node's lifetime is defined as the node's operating time without the need for any external intervention, like battery replacement. A WSN lifetime can be defined as the lifetime of the shortest living node in the network, but, depending on the application, density of the network and possibilities of reconfiguration, it can be defined as the lifetime of some other (main or critical) node [88].

The most commonly used Power Management (PM) technique is to minimize the duty cycling of a node's activity. In duty-cycled operation, a node follows a sleep–wakeup–sample–compute–communicate cycle, where the majority of the cycle is spent in the low power sleep state [89]. This process, which relies on hardware support for implementing sleep states, permits the average power consumption of a node (P_{avg}) to be reduced by many orders of magnitude. Duty cycle (D) of a node's activity is defined as the fraction of time when the node is active, that is,

$$D = \frac{t_{active}}{T} \tag{1.10}$$

where

t_{active} is the sensor active time

T is the sensor active time plus the sensor sleep time, t_{sleep}, that is,

$$T = t_{active} + t_{sleep} \tag{1.11}$$

Let P_{active} be the power consumption of the sensor node during active time, the average power consumption of a node, P_{avg}, is given by:

$$P_{avg} = D \cdot P_{active} \tag{1.12}$$

PM function is to minimize D by decreasing the active time t_{active} as much as possible. In practice, the duty cycle D is reduced to around 1%–2% and usually managed by the Medium Access Control (MAC) layer of the sensor communication protocol [89].

1.6 WSNs for CPS Applications

1.6.1 Transforming WSNs to Cyber-Physical Systems

Ad hoc communications and coverage of infrastructure networks enabled by MANETs when integrated with sensor-related data delivered by WSNs will allow for the interactions between the physical information collected by WSNs and cyber spaces. The interactions between physical and virtual worlds have allowed the WSNs to transition to Cyber-Physical Systems (CPSs). A typical architecture of CPS is described in Figure 1.16. The emergent CPS applications and platforms include, but are not limited to, health care, navigation and rescue, intelligent transportation systems, and social networking applications. The following sections describe these emerging CPSs and their applications.

1.6.2 Emerging Cyber-Physical Systems

1.6.2.1 Intelligent Health Care Cyber System: Heath Care Monitoring and Tracking

Intelligent Health Care Cyber System (IHCCS) employs cross-domain-sensing approach that provides future health-care applications [90–93]. Patients wear small sensors, such as ECG (Electrocardiogram), EMG (Electromyography), EEG (Electroencephalography), SpO2, accelerometer, and tilt sensors, for physical rehabilitation [91,92]. Security is essential in IHCCS where multiple Body Sensor Networks (BSNs) coexist. The telemedicine system employed by IHCCS uses ECG and PPG (Photoplethysmogram) sensor and an authentication mechanism to verify whether two nodes belong to the same BSN through the timing information of users' heartbeats that may uniquely represent an individual.

IHCCS is designed to allow for heterogeneous information flow such as the rehabilitation system described in Ref. [91]. This IHCCS system has a three-tier network architecture to monitor the physiological signal of remote patients. The lowest tier is a ZigBee BSN connecting to ECG, EMG, EEG, etc. The middle tier contains General Packet Radio Service (GPRS)/Bluetooth/Wide Local Area Network (WLAN) connecting to personal devices such as PDA, cell phone and PC, which serve as the sinks for local BSNs. The upper tier is a broadband network connecting servers to record and analyze collected data from individual patients and issue recommendations as necessary. As described in Ref. [91], the heterogeneous networks are used to connect video cameras, sound sensors, RFID (Radio-Frequency Identification) readers, and smart appliances such as smart door lock.

IHCCS is also designed to incorporate the decision/actuation system for providing recommendations to doctors and nurses [91]. For IHCCS that is designed to assist elders at home (which is also referred in literature to as ANGELAH [93]), there are six desired components that need to be incorporated into the decision/actuation system, namely, Sensing Entity (SEn), Actuator Entity (AEn), Home Manager (HMa), Surveillance Center (SCe), Local Responder (LRe), and Locality Manager (LMa). SEns are sensors deployed in elders' living spaces. AEns are actuators set up in elders' living spaces and controlled by HMa. For example, cameras in a living room are AEns, which can be triggered by RFID readers when elders' tags are detected. HMas are local servers located at elders' homes for gathering sensing data and controlling AEns. SCes are responsible for coordinating responses and selecting the most adequate volunteers to provide help promptly. When an emergency situation is detected, HMa informs SCe the emergency level, type, cause, and the kind of assistance that may be needed. LRes are usually the volunteers near elders' homes, who are willing to provide help immediately. ANGELAH defines the proximity of LRe by the coverage of network access points.

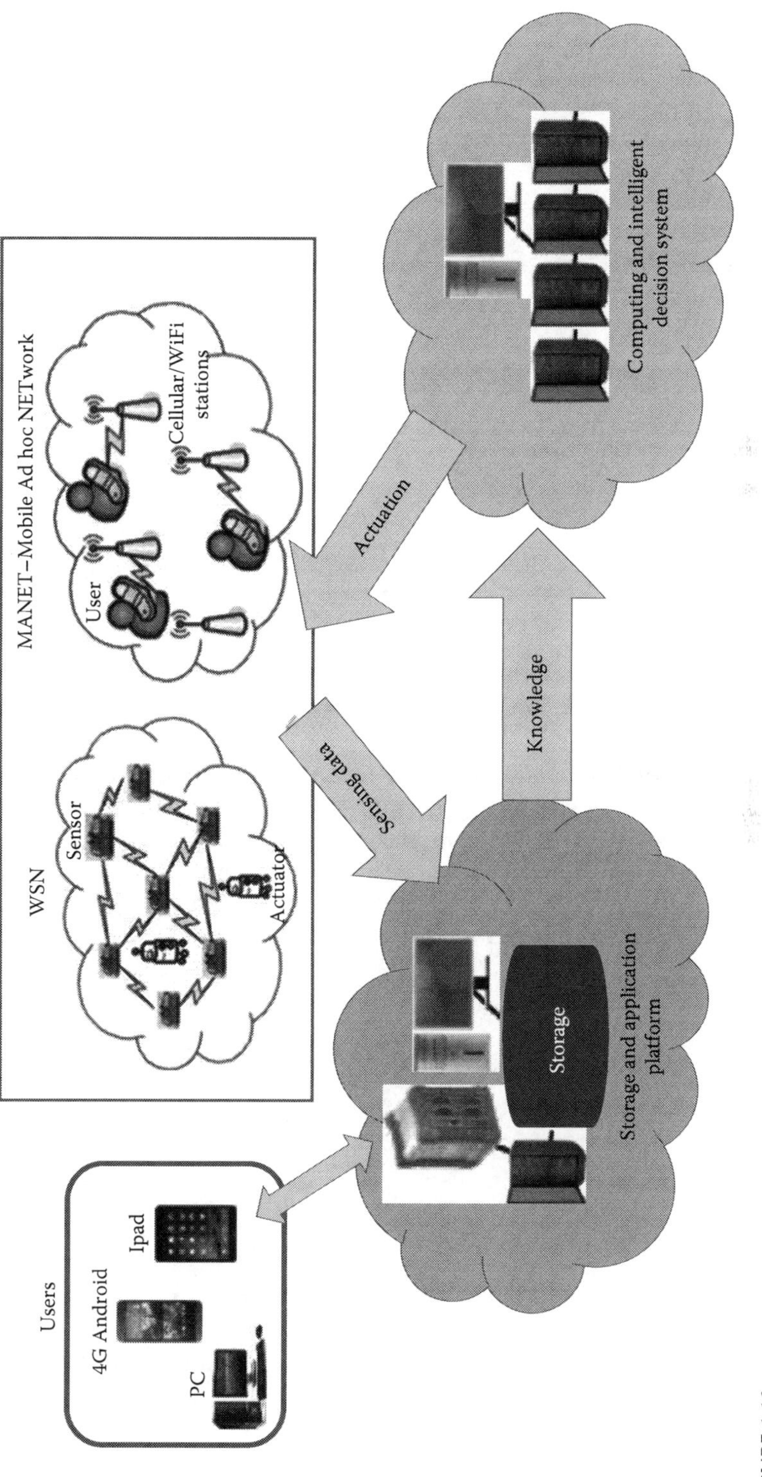

FIGURE 1.16
A typical architecture of CPS.

1.6.2.2 Intelligent Rescue Cyber System: Position–Navigation–Timing Monitoring and Tracking

Similar to IHCCS, the Intelligent Rescue Cyber System (IRCS) also employs the cross-domain sensing and heterogeneous information flow approach. The cross-domain sensing considered in the IRCS is a responsive navigation/rescue system that takes real-time sensing observations into consideration.

IRCS can be used to (1) navigate people in a dangerous region with multiple unsafe points and one safe exit, and (2) guide people in fire emergency to safe exits in 2D/3D environments, where a 3D environment involves multiple floors with stairs [94–96]. In practice, most IRCSs have smoke, temperature, and/or humidity sensors.

The heterogeneous information flows incorporated in IRCS are used to find safe navigation paths without passing any obstacle. IRCS described in Ref. [94] relies on exchanging attractive and repulsive potentials among sensors and exits. Similarly, exchanging "altitudes" is used in Refs. [95,96]. Ref. [97] addresses the navigation without location information. For this application, sensor nodes exchange information cooperatively to identify the medial axes of danger areas. These IRCSs require sensor-to-sensor, sensor-to-human, and sensor-to-infrastructure communications, and WiFi/ZigBee/IEEE 802.15.4 standard interfaces are commonly employed by the systems.

The design of next-generation large-scale IRCS has incorporated the Intelligent Response System (IRSY) with emergency response and evacuation supports, where both scalability and fault tolerance are the important design factors. Agent-based approach is used to address these factors [98]. This approach allows for computation tasks to be executed in distributed and parallel manners. For emergency support, IRCS can adopt high-performance computing grids such as the one proposed for FireGrid [99]. FireGrid has four major components: (1) data acquisition and storage component, (2) simulation component, (3) agent-based command-control component, and (4) grid middleware component. The simulation component has a set of computational models to interpret the current status and predict the future fire status. The computing grids provide access to processing units in parallel by emergency tasks. To support decisions for emergency responders, there are a query manager agent and a set of Command, Control, Communication, and Intelligence (C3I) user-interface agents to interact with users. FireGrid also has a Data Grading Unit (DGU) that is used to filter and validate accuracy and reliability of sensing data.

1.6.2.3 Intelligent Transportation Cyber System: Transportation Monitoring and Tracking

Intelligent Transportation Cyber System (ITCS) provides location tracking and road information sensing for transportation applications [100–102]. ITCS design is usually incorporated GPS receiver and other sensors, such as accelerometers and microphones in mobile phones for annotating traffic conditions in a map, braking, bumping, and honking. The information obtained from sensors allows user to search for driving directions with less traffic by avoiding chaotic roads and intersections. In GPS denied environment, ITCS uses WiFi interfaces on mobile phones to track nearby Access Points (APs) and their centroids for positioning and navigation [104]. ITCS can also be designed to track the locations of public vehicles, for example, buses, using a cooperative model between GPS and accelerometers [105]. Another type of ICTS is the BikeNet, where the system exploits inertial sensors and air sensors for bikers' experience sharing.

ICTS design also incorporated heterogeneous networking by collecting real-time information on objects such as car locations, which is proposed in CarWeb project [107].

CarWeb is used to mine the traffic conditions and waiting time at intersections. ICTS-CarWeb can also exploit inter-vehicle communication to prevent rear-end collision [108,109].

Currently, ICTS design has been incorporating the monitoring and provisioning systems for intelligent transportation applications. This is still a very challenging task to accurately monitor and measure traffic conditions. For this application, ICTS uses VTrack to exploit mobile phones to provide an estimate of traffic delay, detect hot spots during rush hours, and provide real-time route planning with sparely sampled GPS data [104]. VTrack applies an intelligent hidden Markov model to (1) estimate vehicle trajectories and eliminate outliers through map matching, and (2) associate each sample with the most likely road segment. Another type of the monitoring and provisioning system is ParkNet. ParkNet exploits sensing capability of drive-by vehicles to detect the occupancy of parking spots [110].

1.6.2.4 Intelligent Social Networking Cyber System: Social Networks Monitoring and Tracking

Traditional social networks, such as Facebook, Windows Live, and YouTube, are web-based. Intelligent Social Networking Cyber System (ISNCS) is designed to incorporate sensor-enhanced interaction allowing the social network to involve more physical inputs from WSNs that can enrich interactions among the network users. ISNCS uses CenceMe, allowing users to share their sensing information, such as audio, motion, acceleration, and location, via mobile phones [111]. ISNCS also allows users to share their moods, activities, mobility patterns, locations, etc. The users' information becomes exchangeable information in the cyber world. In addition, ISCNS can incorporate socialization tools to record and share water-drinking behaviors [112]. ISNCS using shared locations such as Micro-Blog [113] allows social participants to share and query information with each other via smart phones. Micro-Blog can be used to balance between localization accuracy and battery lifetime [112].

ISNCS is also designed to incorporate different networking scenarios for collecting and sharing sensing data. As an example, BSN collects the sensing data, and BSN's participants then use the Internet or 3G networks to interact with each other. Another example for the use of different networking scenarios is the smart mobile phone. It switches between two localization modes, WiFi and GPS. The WiFi mode is the default mode when the WiFi fingerprint does not change over time. Once it detects movement over a threshold, it switches to the GPS mode.

The design of ISNCS may also require a socialization and interaction platform. A social platform needs to connect many users around the world, thus demands huge data processing and storage capability. ISNCS such as CenceMe employs a split-level classification to process raw sensing data, and provides an audio classifier with an activity classifier to recognize human voices and activities such as sitting, standing, walking, and running [111]. A user's mobile phone first classifies data into high-level information, called primitives. When primitives arrive at the backend server, five complex classifiers are designed to retrieve second-level information, called "facts," to reflect users' social behaviors, such as on-going conversations, nearby buddies, user's mobility pattern (e.g., if he/she is in a vehicle), and location type (e.g., restaurant, library, etc., which is obtained via Wikimapia GIS databases [115]). The five complex classifiers are conversation classifier, social context, mobility mode detector, location classifier, and am-I-hot. The "am-I-hot" classifier can recognize users' social stereotypes (e.g., party animal, cultured, and healthy). "Facts" are stored in a database for retrieval and publishing. A user can also query a specific region. If there exists information in a database matching the query, the server will reply to this query. Otherwise, the server will select those phones located in the region and forward this query to them for possible responses. To encourage responses to socialized queries, a "give-and-take" scheme is adopted, where users earn/pay credits when they reply to/issue queries [111].

Interactive games in cyber space are very popular. The game adopts physical inputs, such as those from inertial sensors, and fuses the physical data with the game to enhance the cyber players' feeling [114].

1.6.3 Smart Cities

The urbanization trend is gradually growing. It is predicted that 60% of the population on earth will be living in cities by 2050 [116]. It arises a challenge as how to make smart cities to efficiently utilize and manage all the resources in the ecosystem to support a large amount of people and enhance the living condition. The Information and Communications Technology (ICT) is rapidly being developed, especially in which the WSN is an ideal candidate. There are several definitions of smart city, such as by Smart Cities Council [117], Frost & Sullivan [118], IEEE smart cities [119], etc. One definition of smart city is "A quasi-independent urban area with smart good environment, people with high level of education, smart governance, and smart transportation to support sustainable economic development and high quality of life." An example of a smart city is shown in Figure 1.17.

Smart cities include smart home and buildings with WSN to monitor people living conditions, such as temperature, sound, lightening, moisture, pollutants, and metering security and safety access, based on which actuators can improve human living conditions and greatly facilitate people's life. The intelligent transportation system consists of intelligent infrastructure management and intelligent vehicles. Smart cities can also include the

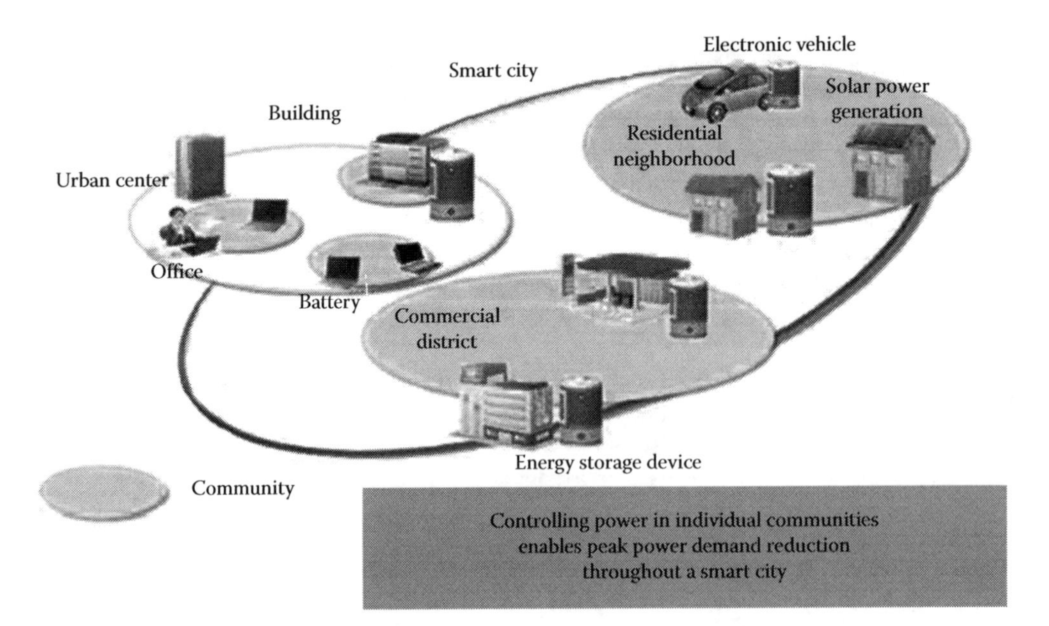

FIGURE 1.17
Example of a smart city. (From http://smartcitiescouncil.com/smart-cities-information-center/definitions-and-overviews; Singh, S., Smart cities—A $1.5 trillion market opportunity, Forbes, November 4, 2014; http://smartcities.ieee.org/about.html.)

intelligent infrastructure management consisting of emergency management, transit management, information management, crash prevention and safety, electronic payment and pricing, traveler information, commercial vehicle operations, traffic incident management, roadway operations, road weather information, and intermodal freight [120]. The intelligent vehicles include collision avoidance, driver assistance, and collision notification [120]. Smart Santander project [121] has been launched in the city of Santander, Spain, to provide information for drivers including traffic intensity, parking occupancy, and weather conditions, etc., with the help of WSN and big data analysis to make transportation system intelligent. The initiative smart Santander project involves 20,000 devices including sensors, actuators, cameras, and mobile terminals. To make smart city quasi-independent, WSN can be applied in agriculture. It can provide precision agriculture to maximize production efficiency with minimum environment impact [120]. With the help of sensed information, including the health of soil, water quality at various stages, plant/crop status, climate, and insect-disease-weed conditions, the warning to farmers and corresponding actions, including precision fertilization and insects killing can be provided [120].

1.6.4 Building Structural Health Monitoring

Structures are a very important part of human life, such as bridges, buildings, dams, pipelines, aircrafts, and ships, etc. These healthy structures provide the foundation of society's economic and industrial prosperity. However, long-term deterioration of these structures can be caused by harsh loading conditions and/or severe environmental conditions, such as earthquakes, winds, tornadoes, and hurricanes. Comparing with traditional coaxial wired Structural Health Monitoring (SHM) system, SHM using WSN (SHM-WSN) provides much cheaper solution. For instance, it is reported that the wired structural healthy monitoring system in tall buildings cost more than $5000 per sensing channel [122]. In Hong Kong, it costs more than $8 million to install over 350 sensing channels upon the Tsing Ma suspension bridge [123]. Therefore, for a single structure with traditional wired SHM system, only 10–20 sensors are installed in average to reduce the cost [124], which in contrast reduce the detection accuracy. With cheaper prices of different types of sensors for structure health monitoring, such as strain gages, accelerometers, linear voltage displacement transducers, inclinometers, etc., large amount of sensors can be deployed within a structure, providing accurate damage detection and structure status report, which makes SHM-WSN an ideal solution for SHM, especially considering the autonomous data processing capability of WSN, which further reduces the cost of human intervention. A typical SHM-WSN system architecture is depicted in Figure 1.18 [124].

Maser et al. [125] implemented the earliest field validation of wireless telemetry for highway bridges monitoring in 1996. Straser and Kiremidjian [126] monitored the status of Alamosa Canyon Bridge in 1998. Bennett et al. [127] made a series of field experiments in an actual asphalt highway surface in 1999. They all verified that SHM-WSN can provide both cheap and accurate civil infrastructure health monitoring [124]. Besides the SHM-WSN application for land-based structures, people also verified the application of SHM-WSN in sea-based structures. Schwartz has successfully validated the monitoring of sea-based environmental parameters, such as the ship spaces, the structural integrity of the hull, and the operational health of critical ship machinery in 2002 with the help of WSN [128].

SHM-WSN system has been developed to detect aircraft crack damage since 2002 [129]. Furthermore, SHM-WSN system can also be applied in railroad vehicles and railway structures. Ref. [124] reported that Nejikovsky and Keller had developed a comprehensive

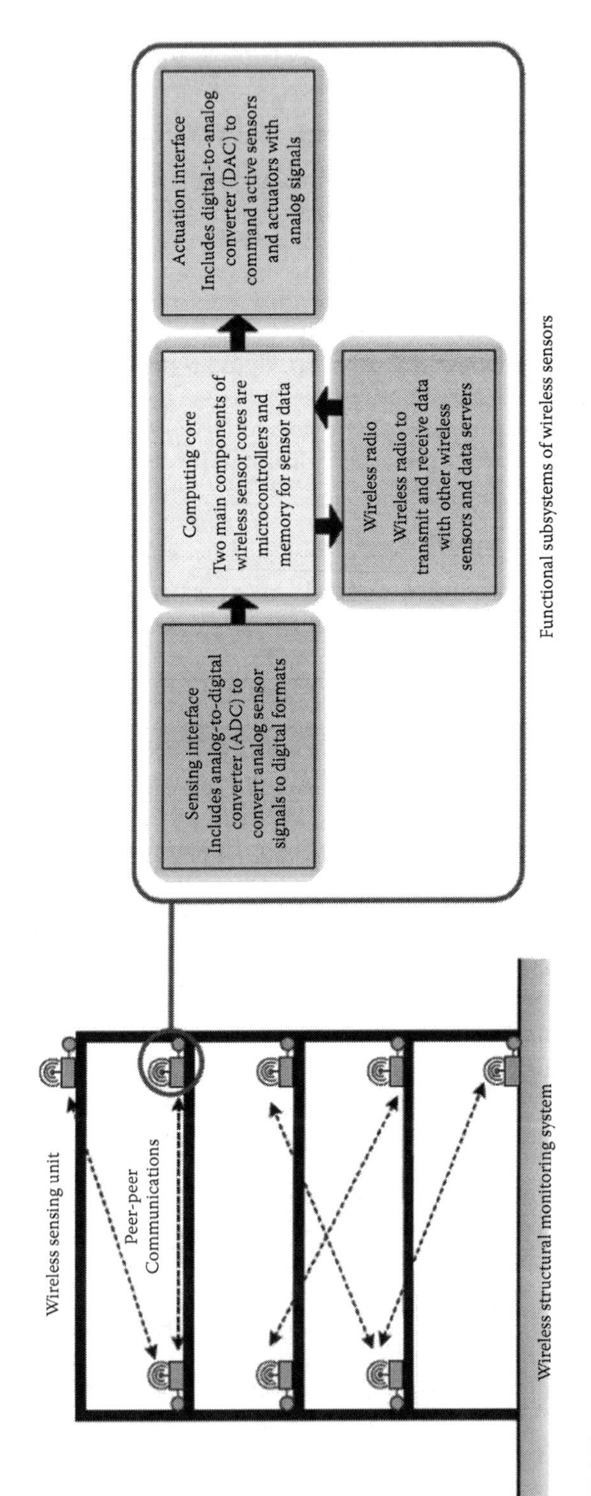

FIGURE 1.18
A typical SHM-WSN system architecture. (From Lynch, J.P. and Loh, K.J., *Shock Vib. Dig.*, 38, 91, March 2006.)

monitoring system for railway cars to monitor their structural conditions, and installed their onboard monitoring devices for railroad companies for remote rail monitoring system in 2000.

1.7 Summary and Conclusions

This chapter provides a solid foundation for understanding sensors, existing state-of-the-art sensor networks, and the design of WSNs for modern CPS applications. The advances of WSNs and MANET have allowed a transition to CPS with emergent applications and platforms. The transition is one of the most remarkable developments to have occurred during the last decade. And the end of this development is not yet in sight. The pattern of CPS growth may reflect the overall dynamics of intelligence conduits across multiple domains consisting of ad hoc communications and coverage of infrastructure networks enabled by MANETs as well as sensor-related data delivered by WSNs.

As described in this chapter, the rise of CPS has encouraged various additional sensors and networking technologies and closely related WSNs research and development accomplished to date. Among the most important of these accomplishments are sensor coverage, power management, and deployment of WSNs. The standard justification for coverage and connectivity is the decision problem where sensors are subject to fixed power with fixed sensing and communication ranges. On another front, yet not too far removed from energy-conserving and scheduling aspects, lies several thorny questions bearing on the scope of properly scheduling sensor on-duty time and maintaining the required coverage levels within a given power consumption level. Yet multiple mutually exclusive sets of sensor nodes by cover set scheme has been proved to be NP-complete.* Despite aforementioned positive technology developments, which would have exceeded all reasonable expectations for WSNs, this is not the time for celebration. Many technical challenges of CPS persist, and participate normally in modern CPS applications. Cross-domain sensor sources and data flows seem increasingly acceptable in heterogeneous networks. Beyond this, embedded and mobile sensing through carriers such as smart phones and vehicles will amount to uncertainty against sensing coverage and knowledge discovery. User contribution and cooperation through give-and-take-like models raise concerns over controversial and complex issues of privacy. Despite temptations to the contrary, however, it remains premature at this point to set forth "pay-as-you-go" concept in an emerging direction of load balancing for CPS over cloud-supported storage and computing platforms. And yet, it is too soon to offer some definite assessment of accumulated intelligence and knowledge via learning and data mining technologies in the presence of high dynamic and uncertain CPS data. Against such a background, pressures from sensor networking, computing, and security issues such as cross-domain interference avoidance, QoS, cloud computing, location-based services, monitoring services, security, and privacy challenges are pushing the applicability challenges of emerging CPS, as in relation to health care, navigation and rescue, intelligent transportation systems, social networking, and gaming applications.

* NP is "Nondeterministic Polynomial time." NP-complete problem is defined as the one that for a given solution to the problem, one can verify it quickly in polynomial time, and there is no known efficient way to locate a solution in the first place. The key characteristic of an NP-complete problem is that no fast solution to it is known. That is, the time required to solve the problem using any currently known algorithm increases very quickly as the size of the problem grows.

Acknowledgments

The authors appreciate the kind support from Taylor & Francis Group and Ganesh Chandra Deka of the Ministry of Skill Development and Entrepreneurship, Government of India, in the writing of this chapter.

The first author, Dr. Tien M. Nguyen, wants to thank his wife, Thu-Hang Nguyen, for her constant support during the writing of this chapter. He also thanks his colleagues at The Catholic University of America and Professor and Chair Ozlem Kilic for her constant support; Scott Shy of Raytheon Company for his unconditional support; Dr. Summner Matsunaga of The Aerospace Corporation for his guidance, valuable friendship, and continuous unconditional support and Mr. Andy Guilen for his support; and Dr. Zhonghai Wang of Intelligent Fusion Technology, Inc. for his careful review of this chapter.

List of Acronyms

A
ADC = Analog-to-Digital Converter
AEn = Actuator Entity
AL = Application Layer
A-Modem = Acoustic Modulator/Demodulator
APS = Access Points
ASC = Analog Signal Conditioner
B
BSNS = Body Sensor Networks
C
CCSDS = Consultative Committee for Data Space Systems
C2C = Command Control and Communications
C3I = Command, Control, Communication, and Intelligence
CoAP = Constrained Application Protocol
CPS = Cyber-Physical System
CR = Cognitive Radios
CRSN = CR Sensing Network
CS = Compressive Sensing
CWSN = Compressive Wireless Sensing Network
D
DD = Directed Diffusion
DD-DC = Directed Diffusion-Data-Centric
DC = Data-Centric
DFL = Device-Free Localization
DGU = Data Grading Unit
DL2 = Data Link Layer
DLST = Divisible Load Scheduling Theory
DSA = Dynamic Spectrum Access
DSA-SN = DSA Sensing Network
DSM = Dynamic Spectrum Management

E
EAR = Energy-Aware Routing protocol
ECG = Electrocardiogram
EMG = Electromyography
EEG = Electroencephalography
H
HITL = Human-In-The-Loop
HMa = Home Manager
I
ICT = Information and Communications Technology
ID = Identification
IHCCS = Intelligent Health Care Cyber System
IoT = Internet of Things
IP-BSN = IP-Based Sensor Network
IRCS = Intelligent Rescue Cyber System
IRSY = Intelligent Response System
ISNCS = Intelligent Social Networking Cyber System
ITA = International Technology Alliance
ITCS = Intelligent Transportation Cyber System
J
JPL = Jet Propulsion Laboratory
L
LMa = Locality Manager
LRe = Local Responder
M
MAC = Medium Access Control
MANET = Mobile Ad Hoc NETwork
MDP = Markov Decision Process
MEMS = Microelectromechanical System
MOSA = Modular Open System Architecture Approach
MRPs = Multipath Routing Protocols
MSN = Mobile Sensing Network
N
NBP = Negotiation-Based Protocols
NEMS = Nanoelectromechanical System
NoSs = Networks of Sensors
NL = Network Layer
O
OODA = Observe, Orient, Decide, and Act
OGC = OpenGIS Consortium
OSI = Open System Interconnect
P
PCS = Personal Communications Service
PL = Physical Layer
PM = Power Management
PNT = Position, Navigation, and Timing
PPG= Photoplethysmogram
PPS = Pressure Profile Systems
PU = Primary User

Q
QoAs = Quality of Attributes
QoS = Quality of Service
R
RFID = Radio-Frequency Identification
RFT = RF Tomography
ROLL = Routing Over Low Power and Lossy Networks protocol
RR = Rumor Routing
RRSI = Received Signal Strength Indicator
S
SEn = Sensing Entity
SeSN = Semantic Sensor Network
SCe = Surveillance Center
SCP = Sensor Control Processor
SHC = Spectrum Hand-off Component
SHM = Structural Health Monitoring
SHM-WSN = Structural Healthy Monitoring-Wireless Sensor Network
SIM = Sensor Interface Module
SMoSN =Semantic Modeling of Sensor Network
SNP = Sensor Networking Processor
SNoA = Sensor Node Architecture
SNR = Signal-to-Noise Ratio
SNS = Sensor Network Security
SOA = Service-Oriented Architecture
SOSW = Service-Oriented Sensor Web
SQRT = Square Root
SRM = Sensor Resource Manager
SSC = Spectrum Sensing Component
SWE = Sensor Web Enablement
T
TCSN = Traditional Centralized Sensor Network
TL = Transport Layer
TS-QoS = Task-Specific Quality of Service
U
UnSN = Underwater Sensor Network
USN = Ubiquitous Sensor Networks
UW-ASN = Underwater Acoustic Sensor Network
W
WNaSN = Wireless Network as Sensor Network
WSN = Wireless Sensor Network

References

1. PranayMondal, Introduction to smart sensor and its application, AEIE 6th SEM, Roll no,-47, http://www.slideshare.net/PranayMondal/introduction-to-smart-sensors-its-application.
2. Smart Sensor, http://en.wikipedia.org/wiki/Smart-Sensor.

3. W. Kester, B. Chestnut, and G. King, Smart sensors, Section 9, http://www.analog.com/static/imported-files/seminars_webcasts/706351365sscsect9.PDF.
4. A. Arshad, Smart sensors—Industrial instrumentation, Department of Chemical Engineering, University of Engineering & Technology Lahore, Punjab, India, http://www.urinepdf.org/t2ycl_ebooks-smart-sensors-for-industrial-applications.pdf.
5. OODA Loop, Wikipedia, http://en.wikipedia.org/wiki/OODA_loop.
6. J. Boyd, Destruction and creation, U.S. Army Command and General Staff College, September 3, 1976. Available at: http://www.goalsys.com/books/documents/DESTRUCTION_AND_CREATION.pdf.
7. OMRON Sensor. Available at: https://www.ia.omron.com/products/category/sensors/.
8. Inductive Proximity Sensors, Keyence website: http://www.keyence.com/products/sensor/proximity/index.jsp.
9. PPS Sensor. Available at: http://www.pressureprofile.com/?gclid=CL3KveD18sACFcZQ7AodLz4AAQ.
10. HarryG Security Sensor. Available at: http://store.harrygs.com/security-tags.html?_vsrefdom=adwords&gclid=CJKM0bX28s ACFeRj7AodvEYA3w.
11. Fujifilm. Available at: http://www.fujifilm.com/products/optical_devices/.
12. Baumer. Available at: http://www.baumer.com/us-en/products/.
13. System Sensor. Available at: http://www.systemsensor.com/en-us/Pages/welcome.aspx.
14. SICK Intelligence Sensor. Available at: http://www.sick.com/us/en-us/home/products/product_portfolio/Pages/product_portfolio.aspx.
15. Analog Devices. Available at: http://www.analog.com/en/mems-sensors/products/index.html.
16. Samsung Sensor. Available at: http://www.samsung.com/global/business/semiconductor/product/cmos-imaging/catalogue.
17. MEM and Sensors, ST Life. Augmented website: http://www.st.com/web/catalog/sense_power/FM89.
18. Linear Technology. Available at: http://www.linear.com/products/wireless_sensor_networks_-_dust_networks.
19. Honeywell Aerospace: Available at: http://aerospace.honeywell.com/en/products/sensors/non-inertial-sensors/magnetic-field-sensing-and-sensor-solutions/magnetic-sensor-components.
20. GE Measurement and Control. Available at: http://www.ge-mcs.com/en/pressure-mems.html.
21. I. Akyildiz, W. Su, Y. Sankarasubramaniam, and E. Cayirci, A survey on sensor networks, *IEEE Communication Magazine*, 40(8), 102–114, 2002.
22. F.-J. Wu, Y.-F. Kao, and Y.-C. Tseng, From wireless sensor networks towards cyber physical systems, *Pervasive and Mobile Computing Journal*, 2011, pp. 1–13, journal homepage: ww.elsevier.com/locate/pmc.
23. E. Candès and M. Wakin, An introduction to compressive sampling, *IEEE Signal Processing Magazine*, 25(2), 21–30, 2008.
24. D. Donoho, Compressed sensing, *IEEE Transactions on Information Theory*, 52(4), 1289–1306, 2006.
25. W. Bajwa, J. Haupt, A. Sayeed, and R. Nowak, Compressive wireless sensing, *IPSN'06*, Nashville, TN, April 19–21, 2006.
26. D. Yazar and A. Dunkles, Efficient application integration in IP-based sensor network, Swedish Institute of Computer Science, *BuildSys'09*, Berkerlys, CA, November 3, 2009.
27. A. Dunkles et al., The design and implementation of an IP-based sensor network for intrusion monitoring, Website: http://dunkels.com/adam/sncnw2004.pdf.
28. J.W. Hui and D.E. Culler, IP is dead, long live IP for wireless sensor networks, *SenSys'08*, Raleigh, NC, November 5–7, 2008.
29. J.P. and V.P. Thubert, IP in wireless sensor networks, Website: http://docbox.etsi.org/workshop/2008/2008_06_M2MWORKSHOP/CISCO_THUBERT_M2MWORKSHOP.pdf.
30. G. Montenegro, N. Kushalnagar, J. Hui, and D. Culler, Transmission of IPv6 Packets Over IEEE 802.15.4 Networks, RFC 4944 (Proposed Standard), September 2007.

31. O.B. Akan, O.B. Karli, and O. Ergul, Cognitive radio sensor networks, Website: http://nwcl.ku.edu.tr/paper/J20.pdf.
32. Arch Rock Corporation, IP-based wireless sensor networking: Secure, reliable, low-power IP connectivity for IEEE 802.15.4 Networks, Website: http://www.cs.berkeley.edu/~jwhui/6lowpan/Arch_Rock_Whitepaper_IP_WSNs.pdf.
33. Q. Zhao and B.M. Sadler, A survey of dynamic spectrum access-signal processing, networking, and regulatory policy, *IEEE ICASSP*, 2007, pp. 79–89.
34. S. Byun, I. Balasingham, and X. Liang, Dynamic spectrum allocation in wireless cognitive sensor networks: Improving fairness and energy efficiency, in *Proceedings of the IEEE VTC 2008*, pp. 1–5, September 2008.
35. I.F. Akyildiz, W.-Y. Lee, M.C. Vuran, and S. Mohanty, Next generation/dynamic spectrum access/cognitive radio wireless networks: A survey, Elsevier, *Computer Networks*, 50, 2127–2159, 2006.
36. L. Xiao, P. Wang, D. Niyato, and E. Hossain, Dynamic spectrum access in cognitive radio networks with RF energy harvesting. arXiv:1401.3502v2 [cs.NI] April 18, 2014. Available at: http://arxiv.org/pdf/1401.3502.pdf.
37. A. Sahai, N. Hoven, and R. Tandra, Some fundamental limits in cognitive radio, in *Proceedings of the 42nd Allerton Conference on Communication, Control, and Computing 2004*, October 2004.
38. A. Ghasemi and E.S. Sousa, Collaborative spectrum sensing for opportunistic access in fading environment, in *Proceedings of the IEEE DySPAN 2005*, pp. 131–136, November 2005.
39. A. Fehske, J.D. Gaeddert, and J.H. Reed, A new approach to signal classification using spectral correlation and neural networks, in *Proceedings of the IEEE DySPAN 2005*, pp. 144–150, November 2005.
40. FCC, ET Docket No 03–237 Notice of Inquiry and Notice of Proposed, ET Docket No. 03–237, November 2003.
41. B. Mercier et al., Sensor networks for cognitive radio: Theory and system design, Website: http://www.eurecom.fr/en/publication/2602/download/cm-publi-2602.pdf.
42. A.K.M. Azad and J. Kamruzzaman, A framework for collaborative multi class heterogeneous wireless sensor networks, in *Proceedings of the IEEE ICC 2008*, pp. 4396–4401, May 2008.
43. S. Wu, C. Lin, Y. Tseng, and J. Sheu, A new multi-channel MAC protocol with ON-demand channel assignment for multi-hop mobile Ad Hoc networks, in *Proceedings of the International Symposium Parallel Architectures, Algorithms and Networks* (*I-SPAN 2000*), pp. 232–237, 2000.
44. N. Choi, M. Patel, and S. Venkatesan, A full duplex multi-channel MAC protocol for multi-hop cognitive radio networks, in *Proceedings of the First International Conference on Cognitive Radio Oriented Wireless Networks and Communications*, pp. 1–5, June 2006.
45. J. So and N. Vaidya, Multi-channel MAC for AD-Hoc networks: Handling multi-channel hidden terminals using a single transceiver, in *Proceedings of the ACM MOBIHOC 2004*, pp. 222–233, 2004.
46. O. Kaltiokallio, M. Bocca, and N. Patwari, A multi-scale spatial model for RSS-based device-free localization. arXiv:1302.5914v1 [cs.NI] February 24, 2013. Available at: http://arxiv.org/pdf/1302.5914.pdf.
47. K. Subaashini, G. Dhivya, and R. Pitchiah, ZigBee RF signal strength for indoor location sensing—Experiments and results, *ICACT Transactions on Advanced Communications Technology (TACT)*, 1(2), September 2012.
48. M. Youssef and A. Agrawala, The horus WLAN location determination system, in *MobiSys'05: Proceedings of the Third International Conference on Mobile Systems, Applications and Services*, pp. 205–218, 2005.
49. H. Lemelson, T. King, and W. Eelsberg, Pre-processing of fingerprints to improve the positioning accuracy of 802.11-based positioning systems, in *MELT'08: Proceedings of the First ACM International Workshop on Mobile Entity Localization and Tracking in GPS-Less Environments*, San Francisco, CA, pp. 73–78, 2008.
50. Y. Chapre, P. Mohapatra, S. Jha, and A. Seneviratne, Received signal strength indicator and its analysis in a typical WLAN system (Short Paper).

51. K. Woyach, D. Puccinelli, and M. Haenggi, in *Sensorless Sensing in Wireless Networks: Implementation and Measurements*, University of Notredam, Notredam, IN.
52. M.A. Kanso and M.G. Rabbat, Compressed RF tomography for wireless sensor networks: Centralized and decentralized approaches, in *Fifth IEEE International Conference on Distributed Computing in Sensor Systems (DCOSS-09)*, Marina Del Rey, CA, June 2009.
53. Veris White Paper, Veris Aerospond Wireless Sensors: Received Signal Strength Indicator (RSSI), http://www.veris.com/docs/whitePaper/vwp18_RSSI_RevA.pdf.
54. N. Srivastava, Challenges of next-generation wireless sensor networks and its impact on society, *Journal of Telecommunications*, 1(1), 128–133, February 2010.
55. V. Weber and G. Vickery, OECD work on Green Growth, Smart Sensor Networks: Technologies and Applications for Green Growth, December 2009, www.oecd.org/sti/ict/green-ict. This report was also released under the OECD code DSTI/ICCP/IE(2009)4/FINAL.
56. H.Q. Luis Felipe, M.P. Francisco, R.M. Hector, and L.G. Carlos, Wireless Smart Sensors Networks, systems, trends and its impact in environmental monitoring, Website: https://www.dtic.ua.es/grupoM/recursos/articulos/LATINCOM-09.pdf.
57. A. Chu, F. Gen-Kuong, and B. Swanson, Smart Sensors and Network Sensor Systems, San Juan Capistrano, CA 92675, www.endevco.com, applications@endevco.com.
58. P. Levis et al., TinyOS: An operating system for sensor networks, in *Ambient Intelligence*, Springer, Berlin, Heidelberg, pp. 115–148, 2005.
59. D. Gay et al., The nesC language: A holistic approach to networked embedded systems, *ACM Sigplan Notices*, 38(5), 1–11, 2003.
60. ITU-T, Ubiquitous Sensor Networks (USN), ITU-T Technology Watch Briefing Report Series, No. 4, February 2008.
61. HP Report, Internet of Things Research Study, 4AA3-xxxxENW, July 2014, Website: http://fortifyprotect.com/HP_IoT_Research_Study.pdf.
62. L. Payne, The Internet of Every Thing (IOE), US Federal, CISCO Systems, Plenary Speech, 2013 MILCOM, San Diego, CA.
63. H. Liu, IoT, Sensor Networks, *Workshop Proceedings on Future Research Needs and Advanced Technologies*, The Catholic University of America, Washington, DC, May 2, 2014.
64. J.M. Kavar and K.H. Wandra, Survey paper on underwater wireless sensor network, *International Journal of Advanced Research in Electrical, Electronics and Instrumentation Engineering*, 3(2), 7294–7300, February 2014.
65. J. Heidemann, M. Stojanovic and M. Zorzi, Underwater sensor networks: Applications, advances and challenges, *Philosophical Transactions of the Royal Society A*, 370, 158–175, August 2, 2012.
66. J. Heidemann, Y. Li, A. Syed, J. Wills, and W. Ye, Underwater sensor networking: Research challenges and potential applications, in *Proceedings of the Technical Report ISI-TR-2005-603*, USC/Information Sciences Institute, 2005.
67. I.F. Akyildiz, D. Pompili, and T. Melodia, Challenges for efficient communication in underwater acoustic sensor networks, *ACM Sigbed Review* 1(2), 3–8, 2004.
68. A. Jangra and S.R. Priyanka, Wireless sensor network (WSN): Architectural design issues and challenges, *(IJCSE) International Journal on Computer Science and Engineering*, 2(9), 3089–3094, 2010.
69. J. Ibbotson et al., Sensors as a service oriented architecture: Middleware for sensor networks, in *The Sixth International Conference on Intelligent Environments—IE'10* KualaLumpur, Malaysia, pp. 209–214, July 19–21, 2010.
70. X. Chu and R. Buyya, *Service Oriented Sensor Web*, The University of Melbourne, Melbourne, Victoria, Australia.
71. A. Sheth, C. Henson, and S. Sahoo. Semantic sensor web, *IEEE Internet Computing*, 12(4), 78–83, 2008.
72. M. Compton, C. Henson, L. Lefort, and H. Neuhaus, *A Survey of the Semantic Specification of Sensors*, ICT Centre, CSIRO, Canberra, Australian Capital Territory, Australia.
73. P. Barnaghi, W. Wang, L. Dong, and C. Wang, *A Linked-Data Model for Semantic Sensor Streams*, Centre for Communication Systems Research, The University of Surrey, Guildford, U.K.

74. D. Khosla and J. Guillochon, Distributed sensor resource management and planning, signal processing, sensor fusion, and target recognition XVI, I. Kadar (ed.), *Proceedings of the SPIE*, 6567, 65670B, (2007).
75. Y. Xun, M.M. Kokar, and K. Baclawski, *Using a Task-Specific QoS for Controlling Sensing Requests and Scheduling*, Department of Electrical and Computer Engineering, Northeastern University, Boston, MA.
76. S. Aeron, M. Zhao, and V. Saligrama, *Sensing Capacity of Sensor Networks: Fundamental Tradeoffs of SNRs, Sparscity and Sensing Diversity*, Boston University, Boston, MA.
77. S. Singh Dhillon and K. Chakrabarty, *Sensor Placement for Effective Coverage and Surveillance in Distributed Sensor Networks*, Duke University, Durham, NC, 2003.
78. W. Heinzelman, J. Kulik, and H. Balakrishnan, Adaptive protocols for information dissemination in wireless sensor networks, in *Proceedings of the Fifth ACM/IEEE Mobicom Conference (MobiCom '99)*, Seattle, WA, pp. 174–185, August 1999.
79. J. Kulik, W.R. Heinzelman, and H. Balakrishnan, Negotiation-based protocols for disseminating information in wireless sensor networks, *Wireless Networks*, 8, 169–185, 2002.
80. C. Intanagonwiwat, R. Govindan, and D. Estrin, Directed diffusion: A scalable and robust communication paradigm for sensor networks, in *Proceedings of ACM MobiCom '00*, Boston, MA, pp. 56–67, 2000.
81. R.C. Shah and J. Rabaey, Energy aware routing for low energy Ad Hoc sensor networks, in *IEEE Wireless Communications and Networking Conference* (WCNC), Orlando, FL, March 17–21, 2002.
82. D. Braginsky and D. Estrin, Rumor routing algorithm for sensor networks, in *Proceedings of the First Workshop on Sensor Networks and Applications* (*WSNA*), Atlanta, GA, October 2002.
83. J.-H. Chang and L. Tassiulas, Maximum lifetime routing in wireless sensor networks, in *Proceedings of the Advanced Telecommunications and Information Distribution Research Program (ATIRP2000)*, College Park, MD, March 2000.
84. S. Dulman, T. Nieberg, J. Wu, and P. Havinga, Trade-off between traffic overhead and reliability in multipath routing for wireless sensor networks, in *WCNC Workshop*, New Orleans, LA, March 2003.
85. N. Sastry and D. Wagner, Security considerations for IEEE 802.15.4 networks, in *Proceedings of ACM Workshop on Wireless Security*, 2004.
86. T. Vanninen, H. Tuomivaara, and J. Huovinen, A Demonstration of Frequency Hopping Ad Hoc and Sensor Network Synchronization Method on WARP Boards, 2008 WinTech, ACM 978-1-60558-187-3/08/09.
87. J. Zyren, T. Godfrey, and D. Eaton, Does frequency hopping enhance security? http://www.packetnexus. com/docs/20010419_frequencyHopping.pdf.
88. C. Alippi, G. Anastasi, M. Di Francesco, and M. Roveri, Energy management in wireless sensor networks with energy-hungry sensors, in *IEEE Instrumentation and Measurement Magazine*, 12(2), 16–23, 2009.
89. P.K. Dutta and D.E. Culler, System software techniques for low-power operation in wireless sensor networks, in *Proceedings of the International Conference on Computer Aided Design* (*ICCAD*), San Jose, CA, pp. 925–932, November 2005.
90. V. Jeliˇci´, *Power Management in Wireless Sensor Networks with High-Consuming Sensors*, University of Zagreb, Faculty of Electrical Engineering and Computing, Zagreb, Croatia, April 2011.
91. J. Brutovsky and D. Novak, Low-cost motivated rehabilitation system for post-operation exercises, *Engineering in Medicine and Biology Society*, 6663–6666, 2006.
92. E. Jovanov, A. Milenkovic, C. Otto, and P.C. de Groen, A wireless body area network of intelligent motion sensors for computer assisted physical rehabilitation, *Journal of Neuroengineering Rehabilitation*, 6(2), 1–10, 2005.
93. C.C.Y. Poon, Y.-T. Zhang, and S.-D. Bao, A novel biometrics method to secure wireless body area sensor networks for telemedicine and M-health, *IEEE Communication Magazine*, 44(4), 73–81, 2006.
94. T. Taleb, D. Bottazzi, M. Guizani, and H. Nait-Charif, ANGELAH: A framework for assisting elders at home, *IEEE Journal of Selected Areas in Communication*, 27(4), 480–494, 2009.

95. Q. Li, M.D. Rosa, and D. Rus, Distributed algorithm for guiding navigation across a sensor network, *International Symposium on Mobile Ad Hoc Networking and Computing*, pp. 313–325, 2003.
96. Y.-C. Tseng, M.-S. Pan, and Y.-Y. Tsai, A distributed emergency navigation algorithm for wireless sensor networks, *IEEE Computing*, 39(7), 55–62, 2006.
97. M.-S. Pan, C.-H. Tsai, and Y.-C. Tseng, Emergency guiding and monitoring applications in indoor 3D environments by wireless sensor networks, *International Journal of Sensor Network*, 1 (1–2), 2–10, 2006.
98. M. Li, Y. Liu, J. Wang, and Z. Yang, Sensor network navigation without locations, *INFOCOM*, pp. 2419–2427, 2009.
99. N. Dimakis, A. Filippoupolitis, and E. Gelenbe, Distributed building evacuation simulator for smart emergency management, *Computer Journal*, 53(9), 1384–1400, 2010.
100. L. Han et al., FireGrid: An e-infrastructure for next-generation emergency response support, *Journal of Parallel and Distributed Computing*, 70(11), 1128–1141, 2010.
101. Intelligent Transportation Systems, http://www.its.dot.gov/.
102. California Center for Innovative Transportation, http://www.calccit.org/.
103. California Partners for Advanced Transit and Highways (PATH), http://www.path.berkeley.edu/.
104. P. Mohan, V.N. Padmanabhan, and R. Ramjee, Nericell: Rich monitoring of road and traffic conditions using mobile smartphones, *International Conference on Embedded Networked Sensor Systems*, pp. 323–336, 2008.
105. A. Thiagarajan et al., Vtrack: Accurate, energy-aware road traffic delay estimation using mobile phones, *International Conference on Embedded Networked Sensor Systems*, pp. 85–98, 2009.
106. A. Thiagarajan, J. Biagioni, T. Gerlich, and J. Eriksson, Cooperative transit tracking using smart-phones, *International Conference on Embedded Networked Sensor Systems*, pp. 85–98, 2010.
107. S.B. Eisenman et al., The BikeNet mobile sensing system for cyclist experience mapping, in *International Conference on Embedded Networked Sensor Systems*, pp. 87–101, 2007.
108. C.-H. Lo et al., CarWeb: A traffic data collection platform, in *International Conference on Mobile Data Management*, pp. 221–222, 2008.
109. L.-W. Chen, Y.-H. Peng, and Y.-C. Tseng, An infrastructure-less framework for preventing rear-end collisions by vehicular sensor networks, *IEEE Communication Letters*, 15(3), 358–360, 2011.
110. F. Ye, M. Adams, and S. Roy, V2V wireless communication protocol for rear-end collision avoidance on highways, in *International Conference on Communications*, pp. 375–379, 2008.
111. S. Mathur et al., ParkNet: Drive-by sensing of road-side parking statistics, in *International Conference on Mobile Systems, Applications, and Services*, pp. 123–136, 2010.
112. E. Miluzzo et al., Sensing meets mobile social networks: The design, implementation and evaluation of the CenceMe application, in *International Conference on Embedded Networked Sensor Systems*, pp. 337–350, 2008.
113. M.-C. Chiu et al., Playful bottle: A mobile social persuasion system to motivate healthy water intake, in *International Conference on Ubiquitous Computing*, pp. 185–194, 2009.
114. S. Gaonkar, J. Li, and R.R. Choudhury, Micro-Blog: Sharing and querying content through mobile phones and social participation, in *International Conference on Mobile Systems, Applications, and Services*, pp. 174–186, 2008.
115. C.-H. Wu, Y.-T. Chang, and Y.-C. Tseng, Multi-screen cyber–physical video game: An integration with body-area inertial sensor networks, in *International Conference on Pervasive Computing and Communications*, pp. 832–834, 2010.
116. United Nations, Department of Economic and Social Affairs, Population Division (2014). *World Urbanization Prospects: The 2014 Revision, Highlights (ST/ESA/SER.A/352).*
117. http://smartcitiescouncil.com/smart-cities-information-center/definitions-and-overviews.
118. http://ww2.frost.com/news/press-releases/frost-sullivan-global-smart-cities-market-reach-us156-trillion-2020.
119. http://smartcities.ieee.org/about.html.
120. Sarwant Singh, Smart cities - a $1.5 trillion market opportunity, Forbes, November 4, 2014.

121. http://www.smart-circle.org/smartcity/blog/smartsantander-city/
122. A. Zanella, N. Bui, A. Castellani, L. Vangelista, and M. Zorzi, Internet of things for smart cities, *IEEE Internet Things of Journal*, 1(1), 22–32, February 2014.
123. M. Celebi, Seismic instrumentation of buildings (with emphasis on federal buildings), Technical Report No. 0-7460-68170, United States Geological Survey, Menlo Park, CA, 2002.
124. C.R. Farrar, Historical overview of structural health monitoring, Lecture Notes on Structural Health Monitoring Using Statistical Pattern Recognition, Los Alamos Dynamics, Los Alamos, NM, 2001.
125. J.P. Lynch and K.J. Loh, A summary review of wireless sensors and sensor networks for structural health monitoring, *The Shock and Vibration Digest*, 38(2), 91–128, March 2006.
126. K. Maser, R. Egri, A. Lichtenstein, and S. Chase, Field evaluation of a wireless global bridge evaluation and monitoring system, in *Proceedings of the 11th Conference on Engineering Mechanics*, Fort Lauderdale, FL, Vol. 2, pp. 955–958, May 19–22, 1996.
127. E.G. Straser and A.S. Kiremidjian, A modular, wireless damage monitoring system for structures, Technical Report 128, John A. Blume Earthquake Engineering Center, Stanford University, Stanford, CA, 1998.
128. R. Bennett, B. Hayes-Gill, J.A. Crowe, R. Armitage, D. Rodgers, and A. Hendroff, Wireless monitoring of highways, Smart systems for bridges, structures, and highways, in *Proceedings of the SPIE*, Newport Beach, CA, Vol. 3671, pp. 173–182, March 1–2, 1999.
129. G. Schwartz, Reliability and survivability in the reduced ship's crew by virtual presence system, in *Proceedings of the IEEE International Conference on Dependable Systems and Networks (DSN'02)*, Washington, DC, pp. 199–204, June 23–26, 2002.
130. J.S. Kim, K.J. Vinoy, and V.K. Varadan, Wireless health monitoring of cracks in structures with MEMS–IDT sensors, Smart Structures and Materials, in *Proceedings of the SPIE*, San Diego, CA, Vol. 4700, pp. 342–353, March 18, 2002.

2

Architectural Analysis of Cyber-Physical Systems

Robin Singh Bhadoria, Radhakishan Yadav, and Geetam Singh Tomar

CONTENTS

2.1 Introduction

Cyber-physical system (CPS) is defined as the fusion of cyber world and the dynamic physical world. CPS perceives the physical world, processes the data by computers, and affects and changes the physical world. Tight coupling of cyber and physical resources and coordination between them is the prime differentiator of CPS, which yields novel capabilities. Moreover, CPS has significantly more intelligence in actuators and sensors with substantially stricter performance constraints.

The researches in the field of CPS spring up a rush after the report of the President's Council of Advisors on Science and Technology has placed CPS on the top of the priority list for federal research investment. CPS integrates computation and communication with monitoring and/or control of entities in the physical world.

Computation shall be embedded in all types of physical resources and applications, with powerful and robust computing devices and the cheaper and faster networking, which will make a big social impact and lead to economic benefit in time and across space. The tight conjoining of computational (or cyber) and physical resources, and the coordination between them, for achieving this kind of applications are the objectives of CPSs. Generally, a tight integration between computation, communication, and control (3Cs) is featured by CPSs in their operation and interactions with the task environment in which they are deployed.

Application of dedicated techniques for capability management and engineering, and reusability in very wide areas such as SOA, business process management (BPM), IoT (Internet of Things), and cloud computing will certainly boost up the research interest in this field.

Today, while the cyber-physical society describes the future environment of collaboration, service-oriented architecture (SOA) and business process management (BPM) are coined to specify how the computational resources can coordinate and collaborate, and (wireless) sensor networks (SNs) provide means to sensing the physical resources and interconnecting physical and cyber resources. Leveraging existing techniques like SOA, BPM, and SNs for achieving the vision of CPSs society is promising.

Let us start this chapter with the concept of 3C, which recently came into light. These 3Cs are: *Computation, Communication,* and *Control.* With "information" as the center, fusion the computation and communication and control, to achieve the real-time sensing, dynamic control, and information service in large-scale systems. CPSs have close relationships with embedded systems, sensors, and wireless network, but have their own characteristics, for example, the complexity and dynamics of the environment, the big problem space and solution space are closely related with the environment, the requirement for high reliability of the system. In the early stage, CPS had a two-tier structure inherently, the physical part and computing part. The physical part senses the physical environment, collects data, and executes the decision made by the computing part; the computing part analyzes and processes the data from the physical part, and then makes a decision. This is a kind of feedback control relation of the two parts. Recently, an architecture has been proposed as three-tier architecture of CPS:

1. *Environmental Tiers*: This consists of physical devices and a target environment, which includes end-users using the devices and their associated physical environment.
2. *Service Tiers*: A typical computing environment with services in SOA and CC (Cloud Computing).
3. *Control Tiers*: To receive monitored data that are gathered though sensors, to make controlling decisions, to find right services by consulting service framework, and to let the services invoked on the physical device.

Cyber world and physical world are essentially different, but they are connected and affect each other via information. One of the main features of the physical world is its dynamic nature; different properties are displayed by the same entity at different times. Therefore, when modeling the physical world entities, the dynamic features should be considered in particular. In the cyber world, changes are represented by state transitions, thus simulating the physical world may lead to state explosion. This is an important feature to be considered in the modeling and design process of CPS. As the base of CPS research, architecture is very critical, but currently, there is no unified framework or general architecture that can be used in most applications. In this chapter, we review the developments of CPS from the architecture perspective.

2.1.1 Characteristics of Cyber-Physical Systems

Different from traditional embedded systems, the CPS has its own characteristics and advantages, which makes it an efficient and promising solution for the integration of physical and cyber entities.

1. *Network Integration*: In CPS, the interoperability of Cloud Computing and WSNs provide the compliance with the networking standards. It involves several computational platforms, which interact over communication networks.
2. *Human–System Interaction*: For decision making, modeling and measuring contextual awareness, human perception of the system and its environmental changes in parameters are critical. This is an absolute requirement of dynamic and complex systems. With humans as a part of the system, some CPSs make the interaction easier because it is difficult to model humans using standalone systems.
3. *Dealing with Certainty*: Certainty provides the proof, which ensures that a design is valid and trustworthy. The evidence includes exhaustive tests in simulations or formal proofs. CPSs have been designed to function and evolve in a new and unreliable environment.
4. *Better System Performance*: As the sensors and cyber infrastructure in CPSs interact very closely, CPSs provide high system performance in terms of automatic redesign and feedback. The presence of multiple sensors, multiple data transfer mechanisms, end-user maintenance, and high-level programming language ensures better system performance in CPS.
5. *Scalability*: Utilizing the properties of Cloud Computing, these CPSs can scale the system on the basis of demand. End-users can acquire the necessary infrastructure without any investment in additional resources.
6. *Autonomy*: With sensor-cloud integration, CPS can facilitate autonomy. A typical CPS works in a closed loop system, in which sensors gather data of physical dynamics. These readings are transferred to cyber subsystems for further computation, which then drive actuators in the physical world.
7. *Flexibility*: Compared to WSN and Cloud Computing Systems individually, the CPS provides greater flexibility.
8. *Optimization*: Cloud infrastructure and other sensors like biomedical sensors have the possibilities of optimizations for various applications. This capability of physical devices can help CPS to optimize the system to a great extent.
9. *Faster Response Time*: The faster response time can help in early detection of remote failure and proper utilization of shared resources, for example, bandwidth. CPSs are able to provide faster response time due to faster processing of data and communication capability of sensors and cloud infrastructure.
10. *Heterogeneous*: For various types of terminal devices, the processing capability, communication mechanisms, and security requirements are quite different. So, this heterogeneity gives rise to huge challenges in system composition.

2.1.2 Challenges in CPS

A number of barriers hinder the progress of CPS design, development, and deployment. For instance, designing CPS for health care is a challenging task as it involves several issues such as reliability, interoperability, security, computational intelligence, privacy, and context awareness. Software is an essential part of health-care devices and hardware functions in close interaction with software. According to the characteristics of CPS mentioned in the previous section, some critical problems in CPS analysis and design are introduced as follows:

1. *Lack of CPS standard*: It is scientifically challenging to properly identify the knowledge-based definitions and measurement for the concepts like security, safety, and privacy. Even after the identification of the contextual definitions of these concepts, it remains a dilemma how utilization and reasoning will be performed. For example, doctors are bound to maintain medical data privacy of the patient. So the question arises what level of concern can be considered privacy violation.
2. *Lack of verification and validation tool*: The integration of WSN (Wireless Sensor Network) and Cloud Computing paradigm can be done in CPS applications. Although the tools for modeling and simulation of WSN and Cloud Computing are available, however, new tools are required to completely model the CPS as they collect, analyze, process, and actuate the heterogeneous sensors and other types of data that are a part of the system.
3. *Time management in architectural design*: The passage of time is unstoppable and concurrency is intrinsic in the physical world. Neither of these two properties is present in computing and networking abstractions of the modern era. In the CPS design, time will be a semantic property, not just a quality factor. Time synchronization is a critical issue in CPS for health care due to the application of heterogeneous sensors and cloud integration. The requirement of real-time CPS for many of its application, time management is of utmost importance for time synchronization, reference time definition, and time scaling, also considering the timing characteristics, the ways of communication with different sensors.
4. *CPS architecture*: The complexity of computing and physical dynamics is a huge challenge for the CPS, such as process integration, system structure, time management, and data correctness.
5. *Security assurance*: Three important aspects of system security that should to be considered are confidentiality, availability, and integrity. On the basis of this fact, the CPS is divided into two categories, one of which is the security critical system and the other, the non-security critical system.

For example, the confidentiality feature is important for military devices, but the real-time requirements are emphasized in the smart home system. Security of CPSs can be broadly categorized in three aspects: Perception security, which ensures the accuracy and security of the information gathered from the physical environment. Transport security, which ensures successful, complete, and accurate data transfer. Processing center security, which includes physical security and safety procedures in workstations and servers.

2.1.3 Requirement for Modeling, Simulation, and Verification of CPS

For modeling, simulation, and verification of the CPS, the key features required are mentioned as follows:

1. *Heterogeneous application support*: As CPS consists of different types of physical devices and hence any approach for modeling and design should be able to address heterogeneity logics.
2. *Physical modeling*: The physical modeling environment should incorporate specification logic, which facilitates formal verification, for example, model checking, and support mathematical expressions.

3. *Scalability support*: A unified modeling approach must provide support for both small-scale and large-scale CPS development.
4. *Mobility support*: To model the movement of mobile devices, mobility must be taken care of including the provision of suitable abstractions.
5. *Integration with simulation and verification tools in existence*: Modeling environment should be connected to existing simulation and verification tools with an easy-to-use support.

2.2 Background

Conventional Architecture of Embedded Systems: Current technologies that combine interactions and computations with the physical world typically originate from the fields of real-time and embedded systems.

Sensor and actuator units are tightly coupled with a higher level control unit in architecture shown in Figure 2.1. The timing properties at each system level are carefully measured so that control loops perform in a correct manner, both temporally and functionally.

A dramatic progress has been made in embedded systems design in the past decades. However, with the increase in system complexity and requirements, this centralized and tightly coupled architecture becomes problematic. In contrast, this gap of complexity has been successfully bridged by human society. In addition to interacting and manipulating the physical world, human society also interacts very well with the cyber world. In fact, human society possesses many identified attributes of a CPS. Thus, it is worth taking a closer look at human society to identify the factors that make the system successful. So, here a few basic observations:

1. A global reference time is the basis for all the system components to coordinate, communicate, and function correctly. Thus, components of system may interact with physical domain concurrently.
2. Event/information drives human society, which is abstraction of the physical world made by system components. Moreover, these abstractions have a lifespan property. Live events/information specifies current status of the physical world, whereas out-of-date events/information specifies past status of the physical world. By their very nature, events/information has a confidence property.
3. Different system components, for example, human individuals may assign different dependability and trustworthiness values to different input sources, which will be based on their personal likes and dislikes, experiences of the past, and knowledge. As a result, different system components might generate completely different outcomes/behaviors, even for the same input abstractions.
4. The two basic communication schemes used in human society are "Publish and Subscribe." Any Web service client can, either implicitly or explicitly, publish its abstraction of the physical entities by using a global time reference, which further can be subscribed and interpreted by intent individuals.

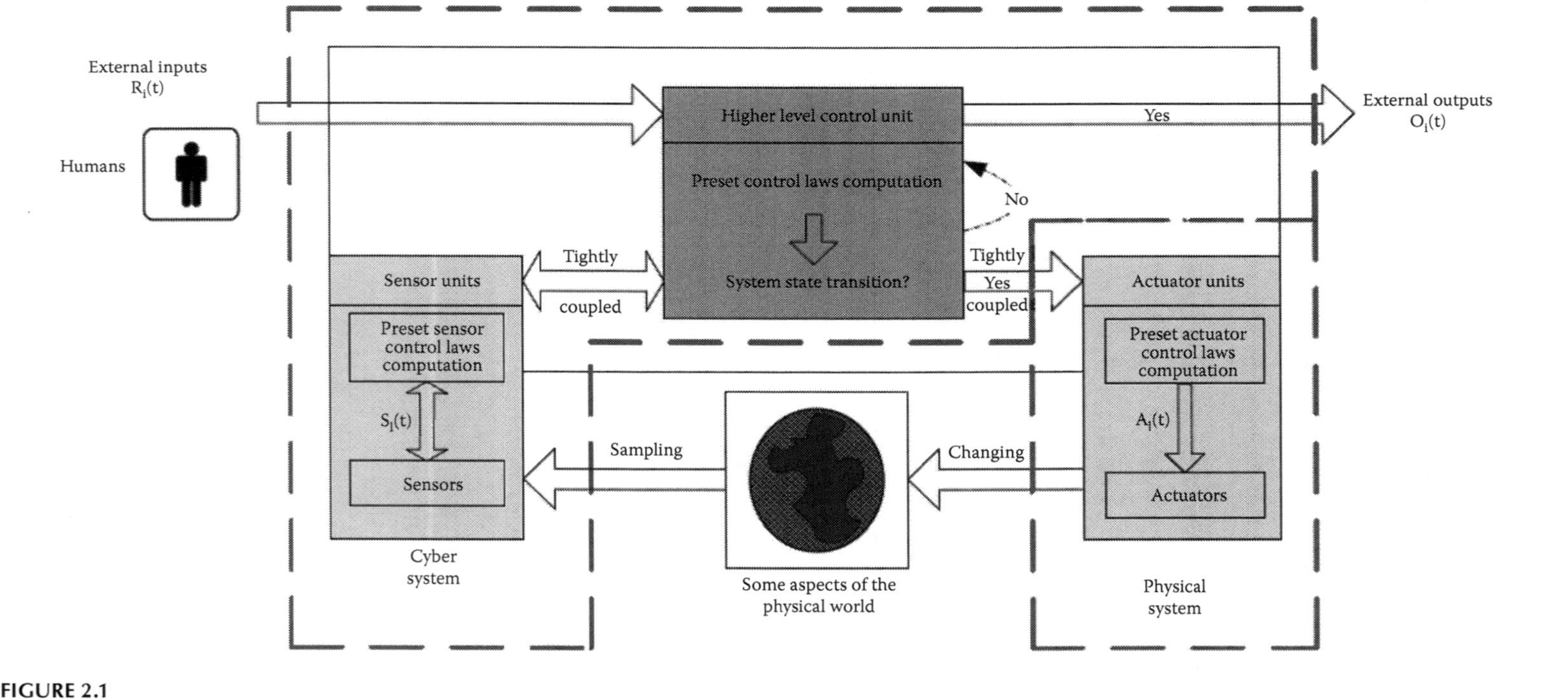

FIGURE 2.1
Typical architecture used in embedded systems. (Courtesy of Ying, T. and Steve, G., A prototype architecture for cyber-physical systems, in *Proceedings of ACM SIGBED*, New York, January 2008.)

One often asked question is: "Why are embedded system's technologies and real time concepts of present days insufficient for the CPSs of future?" The first step, for the time-sensitive distributed systems in system design requires a global reference time, which is either logical or physical that will help in achieving a global ordering of events. In this way, the information produced by any system component will be correctly interpreted by other system components.

However, embedded systems of present generation still follow isolated system-level time. Thus, whenever a complex system is built, the timing property has to be re-tested for all system components. Such methodology is a very expensive and time-consuming process. This process is also prone to errors for obvious reasons. Hence, this represents a major hurdle toward developing future CPSs.

The second question is: "If all system components are provided with a global reference time, is the problem solved?" Unfortunately, the answer is still "No." Lack of a unified mechanism to quantify the confidence of the output of system component at any time point is the reason to this.

A lack of public and private communication mechanisms for all system components to publish the event/information is the third problem, as a result of which the scalability of the system is limited. In summary, fundamental gaps between the technologies available today and the needs of future CPSs require reconsideration of system design available in a fundamental way. This includes quantified confidence and communication mechanisms at all system levels and a unified abstraction of time.

2.3 Architectural Analysis for Cyber-Physical System

Cyber-physical systems are a collection of loosely coupled distributed cyber systems connected through the next-generation network and physical systems monitored and controlled by user-defined semantic laws. Cyber systems mentioned here are collections of the control logic and a part of sensor units. On the other hand, physical systems are collections of actuator units.

2.3.1 Prototype Architecture for CPS

This architecture unifies the current machine-only computation model with the human-only computation model. It captures the essential attributes of a CPS. The highlights of this architecture include (Figure 2.2)

1. *Global Reference Time*: It is provided by the next-generation network. This must be accepted by all the system components including physical devices, humans, and the cyber logic in this architecture.
2. *Event or Information Driven*: Just as human society is driven by event or information, future CPSs should also use a similar mechanism for communication. In addition, Events and Information are differentiated. Events are either raw facts reported by the sensor units or the humans that are called Sensor Events or the actions performed by the actuator units or the humans that are called Actuator Events. On the other hand, Information is the abstraction of the physical world done either by CPS control units or humans through the event processing.

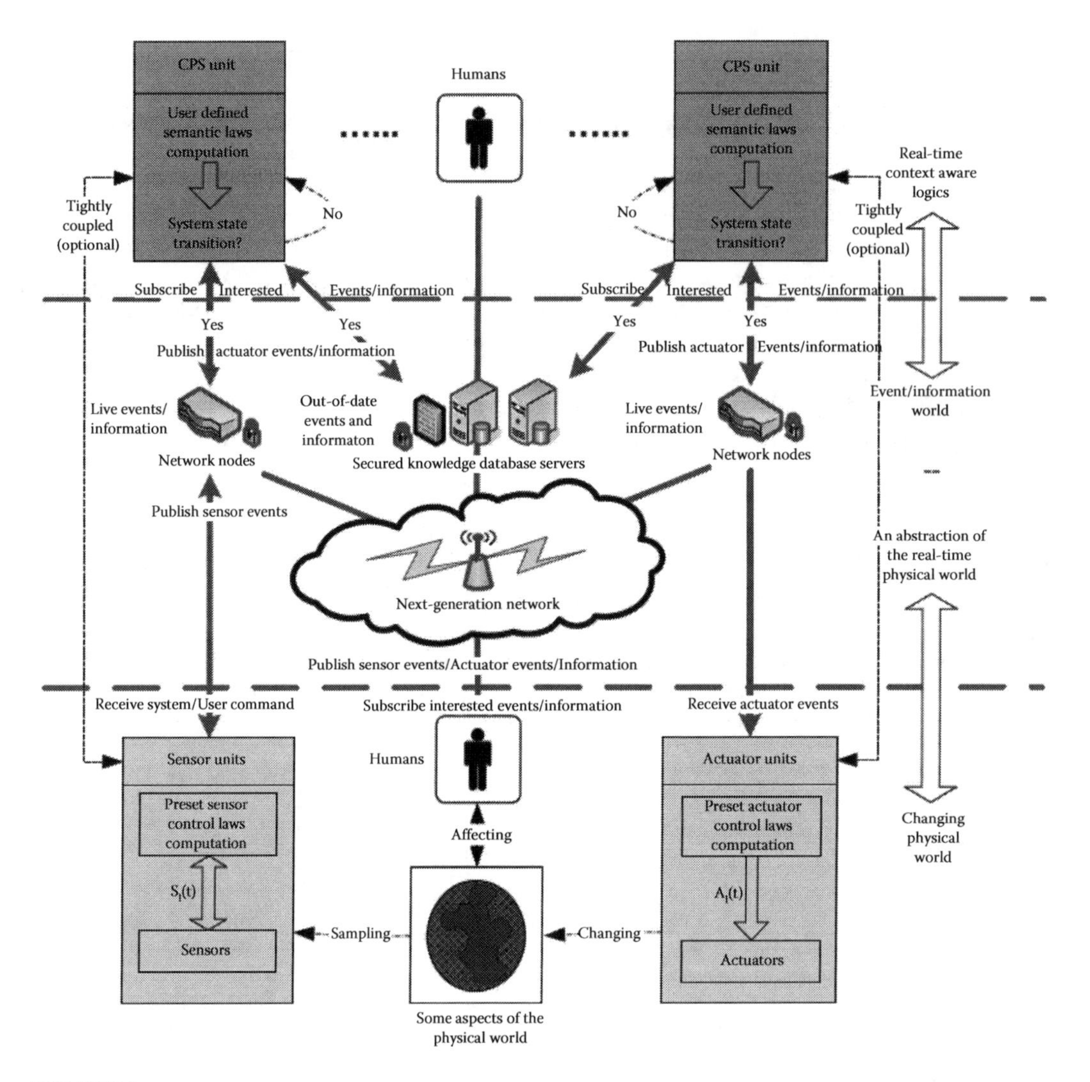

FIGURE 2.2
A prototype architecture for CPS. (Courtesy of Ying, T. and Steve, G., A prototype architecture for cyber-physical systems, in *Proceedings of ACM SIGBED*, New York, January 2008.)

3. *Quantified Confidence*: This design goal is archived using a Unified Event/Information model. In this architecture, any event/information should contain the following properties:
   a. *Global reference time*: This records the event/information occurrence/detected time.
   b. *Lifespan*: It specifies how long until the event/information's confidence level drops down to zero.
   c. *Confidence and its fading equation*: This specifies the event/information confidence level and how it fades over time. The confidence level and its equation are determined by a particular device and a control logic. This is to cater the subscriber with a standard method so that the confidence of the subscribed event or information can be calculated at any point of time.

d. *Digital signature and authentication code*: This represents who can publish and who can subscribe/access the event or information.
e. *Trustworthiness*: This represents how much subscriber trusts a particular publisher.
f. *Dependability*: This represents dependence of subscriber on the event/information provided by publisher to produce a particular outcome/information.
g. *Criticalness*: This represents the determination and urgency of every event/information to be subscribed as it published. With subscription, Web services client can access set of resources available through such system.

4. *Publish/subscribe scheme*: In this scheme, each Control Unit of CPS acts like a human in that it only subscribes to the interesting events or information based on the system goals. It also publishes event/information when necessary.
5. *Semantic control laws*: They form the core of each control unit of the CPS and usually defined in an Event-Condition-Action like form. System behaviors related to the environment context can be controlled according to user-defined conditions and scenarios very precisely with the abstraction of real-time physical world as shown in the Figure.
6. *New networking techniques*: In addition to a global reference time, next-generation networks should also provide new event/information routing and the data management schemes. Based on current confidence, to pass an event or information to its neighbor nodes, a "publish like" scheme should be used by each network node. Though confidence level for any such event or information falls down to certain threshold value or zero, accessibility for such event or information continues to "live." The Secured Network Knowledge Database Servers cater as a knowledge backup as they only accept data when it expires.

There have been many researches and experts, who are studying several aspects of the CPS with respect to multiple application domains. Here are mentioned of a few of them.

2.3.2 Module-Based Architecture for CPS

In CPSs, communication is required for providing sensor observations to controllers and actuators, which therefore makes the communication architecture design a critical requirement for the system functionality. Author Syed Hassan et al. proposed an architecture that contains several modules supporting the CPS. Every module in this architecture has its own significance and can be applied to various applications. These are the six modules mentioned here:

1. Sensing Module.
2. Data Management Module (DMM).
3. Next-Generation Internet.
4. Service Aware Modules (SAM).
5. Application Module (AM).
6. Sensors and Actuators.

As shown in Figure 2.3, five main modules of this architecture are described. The subsections afterward describe each module in more detail.

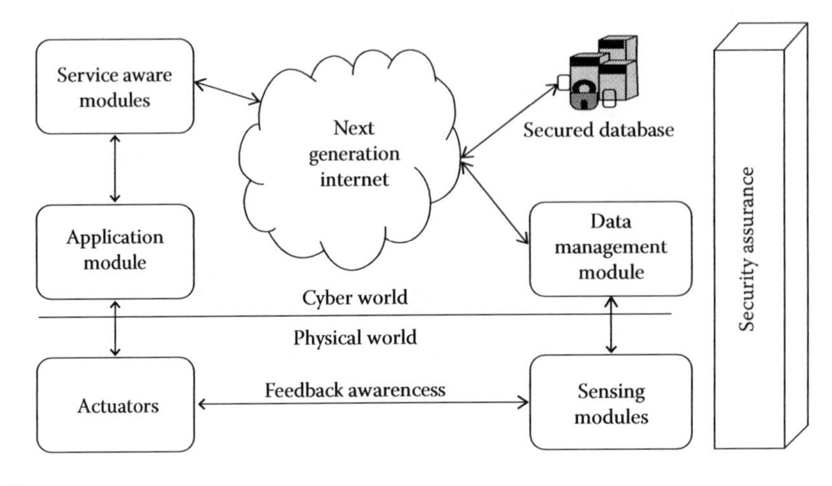

FIGURE 2.3
Standard CPS architecture. (Courtesy of Ahmed, S.H. et al., Cyber-physical system: Architecture, applications and research challenges, in *Proceedings of 6th IFIP/IEEE Wireless Days Conference* (*WDays*), Valencia, Spain, 2013.)

1. *Sensing Module*: The main function of this module is environment awareness, which is achieved by preliminary data preprocessing by data collection from physical world through sensors. The collected data are further transferred to Data Management Module (DMM). Depending on the nature of network deployed, multiple networks are supported by this sensing Module.

   For instance, each sensor node is equipped with a sensing module for real-time sensing in a WSN. Other network nodes can also operate with a part of this module in different scenarios. In case of vehicular cyber-physical system (VCPS), VANET (vehicular ad hoc network) nodes (i.e., cars) can be equipped with sensing module to sense data from physical world. In case of health CPSs, using BAN (Body Area network, which is a wireless network of wearable computing devices), sensors attached with the body or clothes of the patients are equipped with sensing module nodes to enable real-time monitoring and control.
2. *Data Management Module* (*DMM*): This module consists of the computational devices and storage media. Heterogeneous data processing such as normalization, data storage, noise reduction, and many other similar functions are provided by this module. Data management module acts as the bridge between dynamic environment and services as it is collecting the sensed data from sensors and forwards the data to service aware modules using Next-Generation Internet.
3. *Next-Generation Internet*: A common feature of emerging Next-Generation Internet is the ability for applications to select the path dynamically, which their packets follow between the source and destination. Designing CPSs require this dynamic nature of internet service. In the current Internet architecture, routing protocols find a single and the best path between a source and destination whereas the future Internet routing protocols will need to present applications with a choice of paths.

   Quality of Service routings does not scale to the size of the current Internet. It just provides applications with a path that meets the need of the application. Next-Generation Internet services trial is expected to include the ongoing projects such as IPv6, exploiting 802.16n and 802.16p.

4. *Service Aware Modules* (*SAM*): Typical functions of the whole system including the task analysis, decision making, task schedule, and so on are provided by this Service Aware Module (SAM). After receiving sensed data, this module recognizes and sends data to the services available.
5. *Application Module* (*AM*): A number of services are deployed and interact with NGI in this application Module (AM). Simultaneously, information is getting saved on secured database for QoS support. Local storage and cloud platforms are both used at the same time to maintain the database in order to keep data safe. The concept of NoSQL can be used for saving data. Despite a variety of different features in NoSQL systems, there are some common ones. First being the distributed data storage. The data that are distributed across multiple sites are managed by many NoSQL systems. These data that are saved over the Cloud system can be accessed from anywhere with authenticated access.
6. *Sensors and Actuators*: Sensing Modules and the Actuators are two different electronic devices that interact with the physical environment. The actuator may be some physical device or a lamp or watering pump or a car that receives and executes the commands from Application Module. The security assurance part is inherently important in a whole system, from access security to data and device security. CPS security is divided into different requirements in different scenarios. For example, in the Vehicle CPS or Health CPS, the real-time requirements are more emphasized but in military applications, the most important feature is the confidentiality of the data. Security of CPS can be divided into the following three phases:
   a. Awareness security to ensure the accuracy of the information collected from physical environment and its security.
   b. Transport security to prevent the data from being destroyed during the transmission processes.
   c. Physical security for safety procedures in servers or workstations.

Feedback Awareness is an advanced level service to minimize the data processing by communication between sensor and actuator for executing required actions directly.

2.3.2.1 Communication Topology of such Architecture

The interaction between modules is described in this subsection. First of all, the sensing module sends an association request to Data Management Module (DMM) and it replies with an acknowledgment packet (Figure 2.4).

Once association between DMM and Sensing module is completed, the nodes start transmitting the sensed data to DMM. Then, noise reduction and data normalization play their roles. Through QoS routing, data are transferred to Service Aware Modules using services of Next-Generation Internet. In application module, available services are assigned to different applications. During each network operation, data are stored in a local storage and also sent to a cloud platform to ensure the security and integrity of data.

2.3.2.2 Application Scenarios

Consider a deployment scenario of this architecture where various sensors are utilized to collect data while various actuators can make changes to the environment. Suppose that the sensed data S1 is to be sent by the sensing module to DMM as shown in Figure 2.2. The data

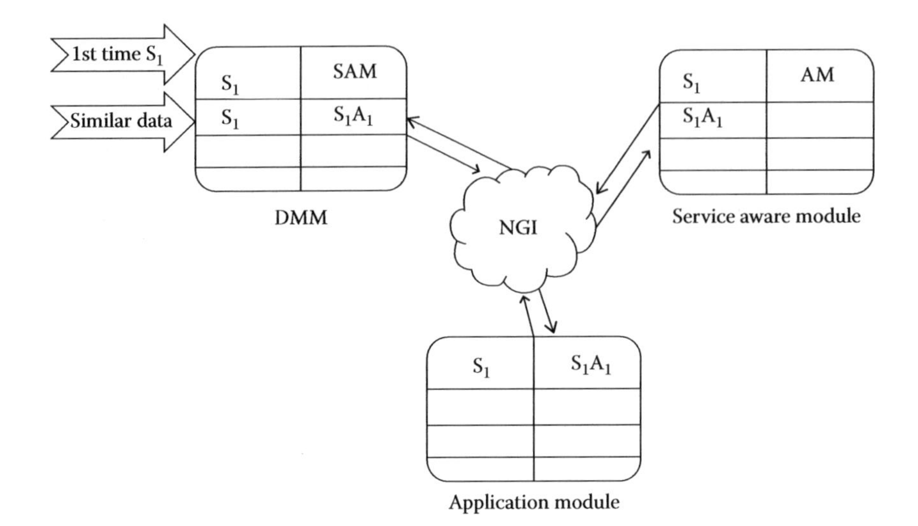

FIGURE 2.4
Dynamic communication. (Courtesy of Ahmed, S.H. et al., Cyber-physical system: Architecture, applications and research challenges, in *Proceedings of 6th IFIP/IEEE Wireless Days Conference* (*WDays*), Valencia, Spain, 2013.)

are then forwarded to service aware module by DMM, in addition to storing it in local storage. The service awareness module identifies and assigns an exact application A1 in response to the data S1. Now, A1 in the application module communicates with relevant actuators. Simultaneously, the application module sends association information between S1 and A1 (*S1A1*) to the service aware module and most importantly to DMM. DMM manages and look after the data and its associated application in local storage. In the future, when DMM receives a similar data, instead of forwarding that data to the service awareness module, it sends the data directly to the application module. Then, a required action request is sent to actuators by application module.

1. *Vehicular Scenario*: In vehicular CPS, suppose vehicles are connected to Road Side Unit, where Road Side Unit has access to Next-Generation Internet infrastructure providing different services through wired networks. Almost every car will be having different actuators like lights, brakes, speed controlling, etc., in the near future. This architecture is applicable for real-time control of those actuators. Communication and control are made efficient by Service aware module, which provides the best application in different services to a single car. For example, in case of folk, lights should be on automatically. In case of road damage ahead, breaks should be applied and in case of congestion, the best and shortest path is updated on GPS automatically.
2. *Agriculture Scenario*: A greenhouse scenario is considered for Agricultural CPS. In greenhouse technology, more parameters should be controlled due to a large variety of the crops. They are growing day by day because of the development in agriculture technology. In this situation, the wireless sensor network with additional hardware and software is an efficient solution for greenhouse control. Real-time control can be applied through this architecture in the future. After receiving a configuration of greenhouse by consumer, here this feedback awareness makes network efficient to control different services like watering, humidity, plant health monitoring, etc., with real-time control.

3. *Health Scenario*: The main concern in health scenario is patient safety and hospital liability. Several services are expected from Health CPS, for example, patient care administration, real-time tracking of Laboratory specimen, patients putting themselves at risk, for example, wandering patients or inability to alert staff of the patient's location for urgent needs and monitoring equipment exposure that may put patients at severe risk. It could be implemented with sensors and actuators are installed using BANs with patient and WSN within hospital building. Real-time monitoring and control can be achieved by the data management module and service awareness module. The service awareness module also plays an important role in providing QoS monitoring. While designing a system for hospital, sensitive applications are queued on the top priority.

2.3.3 Service-Based architecture for Cyber-Physical Systems

An application rebuild framework and lease protocol into CPSs is proposed by Author Chenguan Yu et al. to ensure adaptability and availability of applications. The atomicity property of lease protocol guarantees that an application either is granted the leases for all requested services or it gets no lease at all. Applications that are not granted the leases for all requested services will be rebuilt by application rebuild framework to ensure applications will be started in time.

Service-oriented architecture can be used to describe, manage, and compose physical devices in CPSs. Physical devices can be abstracted as service. So, SOA technologies such as service discovery, service composition can be used in the CPS. Compared with traditional application of internet, the application of the CPS has some new characteristics and requirements as mentioned in the following paragraph.

Service providers probably are physical devices with poor computing capability that have some states and characteristics. These physical devices may be used by just a service consumer at one time. The application of the CPS may use more than one physical device at the same time. Some application is critical system which should run in time. If any other application is using the device then the critical system which will use this device cannot run immediately. It would be dangerous, for instance, in the case of children monitoring system. Whenever the children are within danger, the children monitoring system will call their parents. If the children with danger and their parents are making a call at the same time, the children monitoring system cannot inform their parents. It would lead to serious consequences.

Taking all these characteristics and requirements into consideration, this architecture puts forward application rebuild framework and lease protocol into CPSs. The atomicity property of lease protocol guarantees that an application either is granted the leases for all requested services or it gets no lease at all (Figure 2.5).

The architecture shown in Figure 2.1 includes physical tier, service tier, and application tier. Physical tier connects a variety of physical devices by different communications protocol, such as Wi-Fi, Bluetooth, and Zig Bee. The control applications use messages which are collected by physical devices to produce control information. Drivers control physical devices according to the control information.

Device bundle: Device bundle can be produced by driver, which is provided by operating system and can directly control physical device. We can use the device bundle to control physical device or collect information from physical device.

Agent service: Consumers access services or service providers provide services through agents. This has the advantage that when a failure occurs in the service that is being used, the agent can redirect to other service with the same function. It would improve service's reliability.

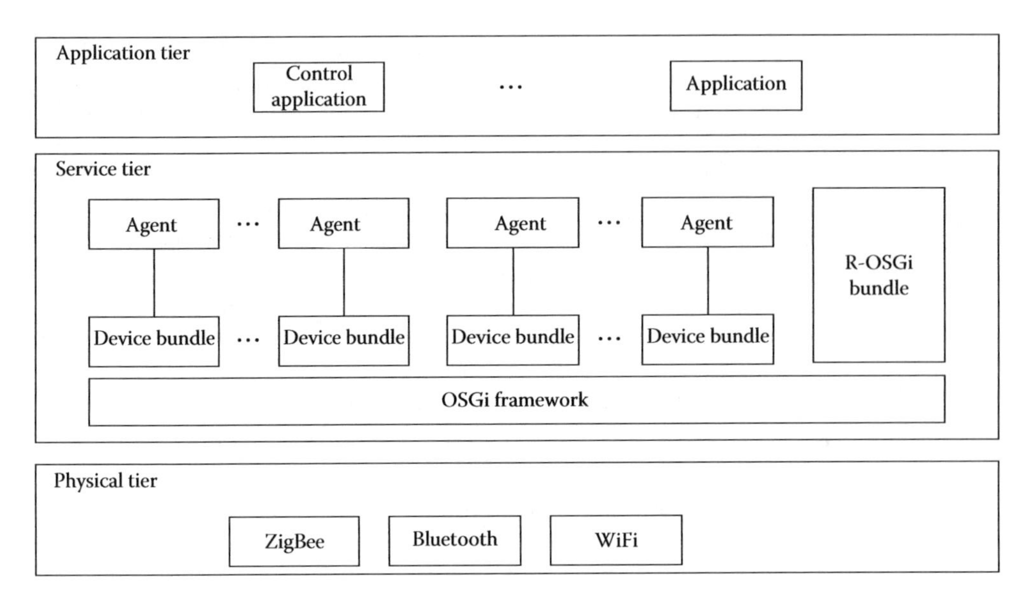

FIGURE 2.5
Service-based architecture for CPS. (Courtesy of Yu, C. et al., An architecture of cyber-physical system based on service, in *Proceedings of IEEE International Conference on Computer Science and Service System* (CSSS), Nanjing, China, August 2012.)

2.3.3.1 Rebuilding Application

All the resources must be acquired by the application before it runs. The application would run in time if the resources that will be used by this application are being used by an other application. It would be a danger if the application is a safety-critical application. For example, the application is a pressure detection system. If the pressure is higher than normal value, inlet valve should be turned down until the pressure becomes normal again. If the pressure is higher than the normal value while the inlet valve is being controlled by other application, pressure detection system cannot decompress pressure by turning down inlet value.

If pressure detection system does not run until the value is released, it is probable that the pipe will burst. In this case, it is necessary to rebuild the pressure detection system. For example, the pressure detection system can decompress pressure by turning up the outlet value. The application rebuilding framework is shown in Figure 2.6.

Rebuilding-Detecting Module is used to find which application should be rebuilt and then informs Rebuilding-Management Module. Service-Failure-Detecting Module is used to find the failure service which is being used. Rebuilding-Management Module is used to rebuild the application. Analysis Module uses Type-Check Tool and Semantics-Analysis Tool to check whether rebuilt application is the same as the old application.

Application rebuilding model before running: The process of rebuilding application is shown in Figure 2.7. Many service-based systems' service description includes domain knowledge and supports the automatic choice of services. Semantic services provide efficient support for discovering appropriate services.

Kalasapur proposed a dynamic service composition framework, which would automatically construct the application satisfying customer's requirement. So, based on the research results, application would be automatically constructed by semantic information. An obvious lease protocol has been introduced here into the CPS. It guarantees that an application either is granted the leases for all requested services or it gets no lease at all.

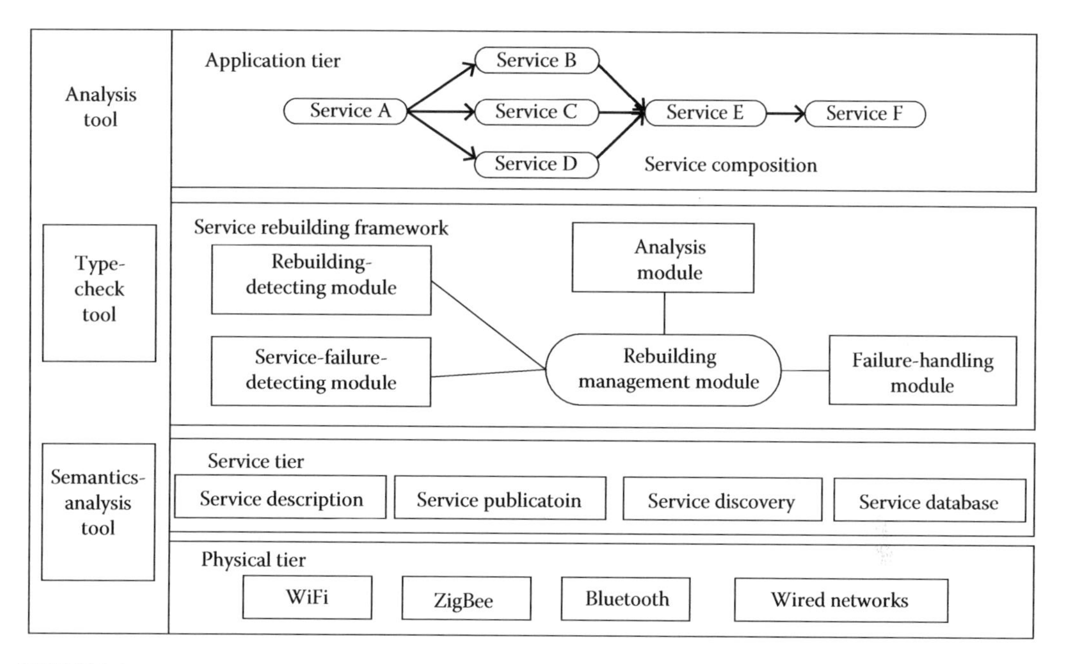

FIGURE 2.6
Application rebuilding framework. (Courtesy of Yu, C. et al., An architecture of cyber-physical system based on service, in *Proceedings of IEEE International Conference on Computer Science and Service System (CSSS)*, Nanjing, China, August 2012.)

Application rebuilding while running: The application would break if a failure occurs in the service, which is being used by the application. In the CPS, there are lots of resources that have similar function. Then we can find a new service to substitute the fail service in the CPS to guarantee that the application runs successfully. The model of rebuilding application while the application is running is shown in Figure 2.8.

Agents transmit service request to service provider and return results to service consumer. When service A becomes unavailable, the agent will send a message to Service-Failure-Detecting Module. Then Service-Failure-Detecting Module sends a message to Rebuilding-Management Module to find a new service, which has the same interface with the failed service. Then the agent redirects to the new service and sends the requesting message to the new service. If there is not a service that has the same interface with the failed service, the Failure-Handling Module will be activated, and then a warning is sent to the user.

Lease protocol: If a service consumer wants to access a service, it firstly sents a lease request. The service provider can accept or reject. When the lease is due, if the service consumer still wants to use the service, it must lease the service again.

The atomicity property of lease protocol guarantees that an application either is granted the leases for all requested services or it gets no lease at all. Application sends booking messages to all services that will be used. If all bookings are accepted, the application will send lease messages to all services. Otherwise, it will cancel all bookings. This architecture uses lease protocol to guarantee that the application acquires all resources before it runs. If there are some services that reject the application, the architecture will rebuild the application to guarantee that it will run in time.

Prototype: Water resource monitor CPS was implemented to test the availability of the architecture described here. The architecture is shown in Figure 2.5. It detects whether

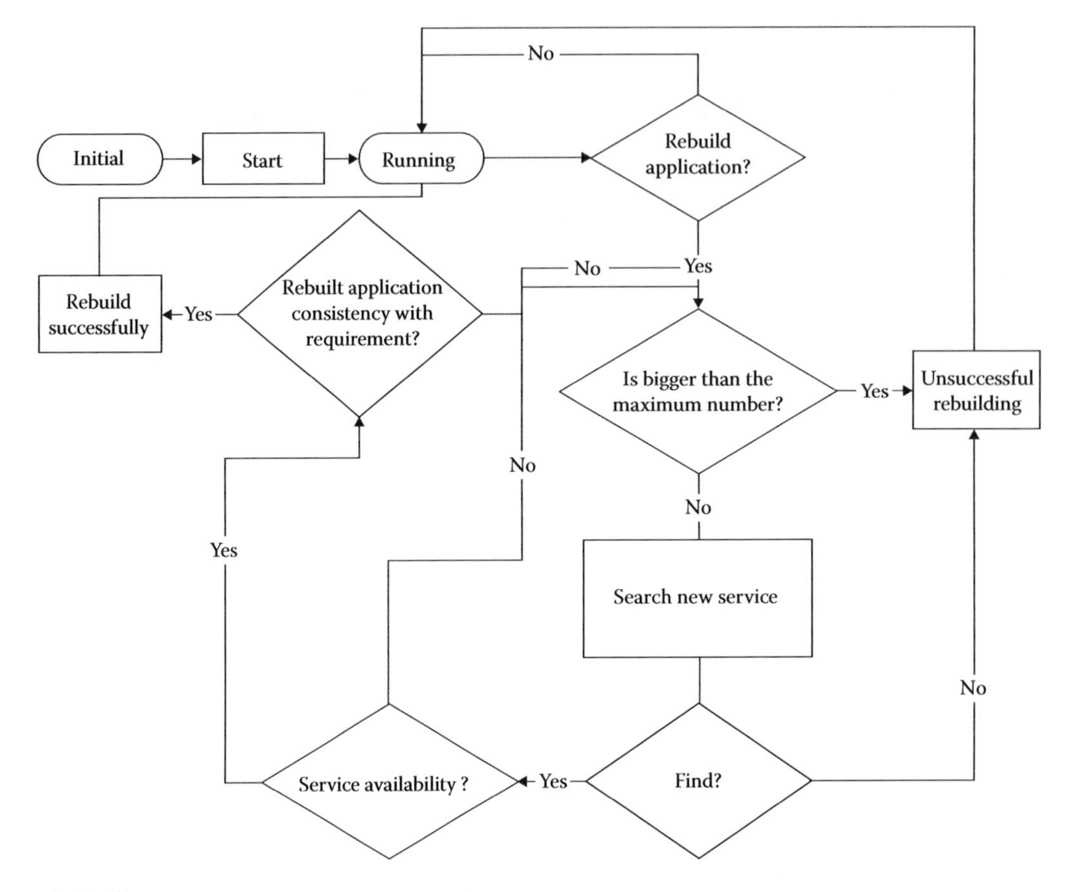

FIGURE 2.7
Process of rebuilding application. (Courtesy of Yu, C. et al., An architecture of cyber-physical system based on service, in *Proceedings of IEEE International Conference on Computer Science and Service System* (*CSSS*), Nanjing, China, August 2012.)

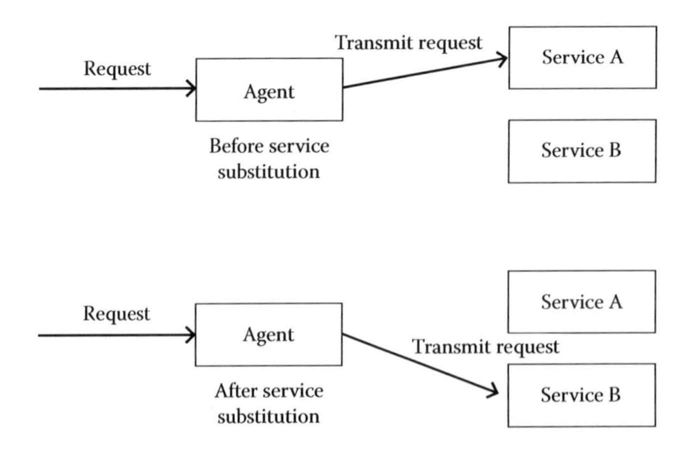

FIGURE 2.8
The model of rebuilding application while the application is running. (Courtesy of Yu, C. et al., An architecture of cyber-physical system based on service, in *Proceedings of IEEE International Conference on Computer Science and Service System* (*CSSS*), Nanjing, China, August 2012.)

there is water pollution, controls valve, and so on. In water resource monitor CPS, sensors are used to collect information and send it to the deliver point, which checks the received messages and then sends the handled messages to data server. Data server further handles received messages (Figure 2.9).

This system can avoid disaster by entire, quick, and precise monitoring water quality. If water pollution is detected, the system will turn off the inlet valve, so that polluted water will not be used by the citizens. If there is a failure occurring in the inlet valve, the system cannot turn off the inlet valve. It will result in serious consequence. If the architecture described is used, it will rebuild the application and can avoid the serious consequence by turning off the outlet valve.

2.3.4 A Service-Oriented Architecture of CPS

Author Peng Wang et al. proposed a SOA of the CPS as shown in Figure 2.10. In this framework, the CPS has been divided into four layers:

1. Service implementation layer
2. Service abstraction layer
3. Business process layer
4. Application layer

1. *Service implementation layer*: This layer serves as the foundation block for this architecture. The details about the process of its implementation are hidden for service users, and service providers can use various technology to implement the same Service Interface. The sense-actuate unit, the communications unit, and the computation control unit are contained in each of the CPS service implementation. Physical world is monitored by sense unit, which also transfers monitoring information to computation unit through communications unit. Strategies are determined by computation units and sent to control unit. Instructions are given by control unit to actuate unit for controlling physical processes, through communications unit. Each of these three units is described briefly in the following subsections:
    a. Actuators, sensors, and the terminal computation module fall inside sense-actuate unit. Physical entities and physical environment are monitored by sensors monitor whereas actuators control physical processes. The basic executive rules of actuator are contained inside terminal computation module which has small storage capacity of real-time data.
    b. Communications unit provides ubiquitous communication mechanism by the fusion of 2G, 3G, and 4G. It also involves heterogeneous networks integration, communication security, communication quality, and real-time interaction.
    c. Computation unit and control unit are contained in computation-control unit. Computation unit merges continuous domain and discrete domain into one. Control unit implements a strict spatial–temporal management. Cloud computing center and knowledge base can support the strategies determined by this unit.

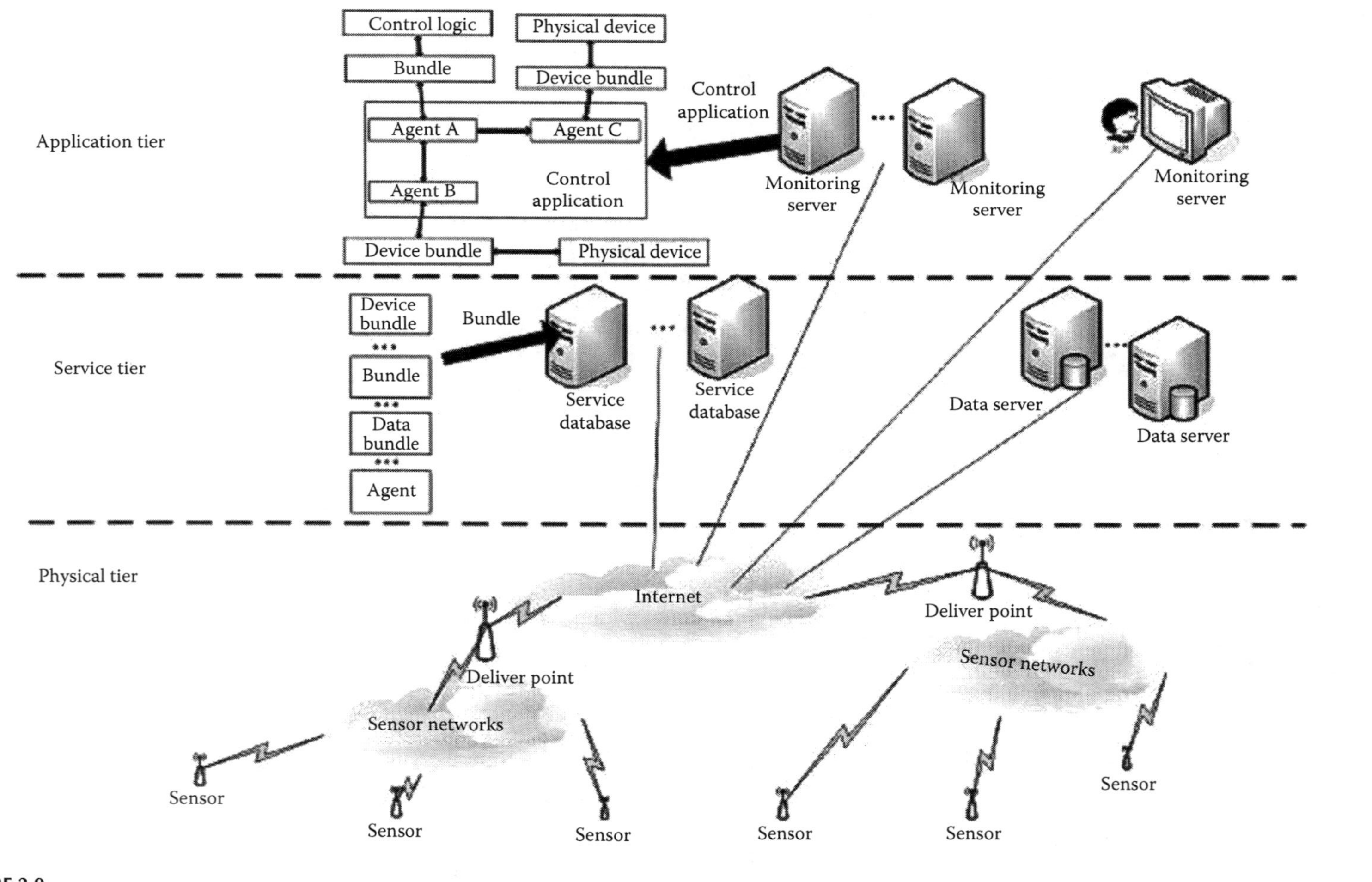

FIGURE 2.9
Water resources monitor CPS. (Courtesy of Yu, C. et al., An architecture of cyber-physical system based on service, in *Proceedings of IEEE International Conference on Computer Science and Service System* (*CSSS*), Nanjing, China, August 2012.)

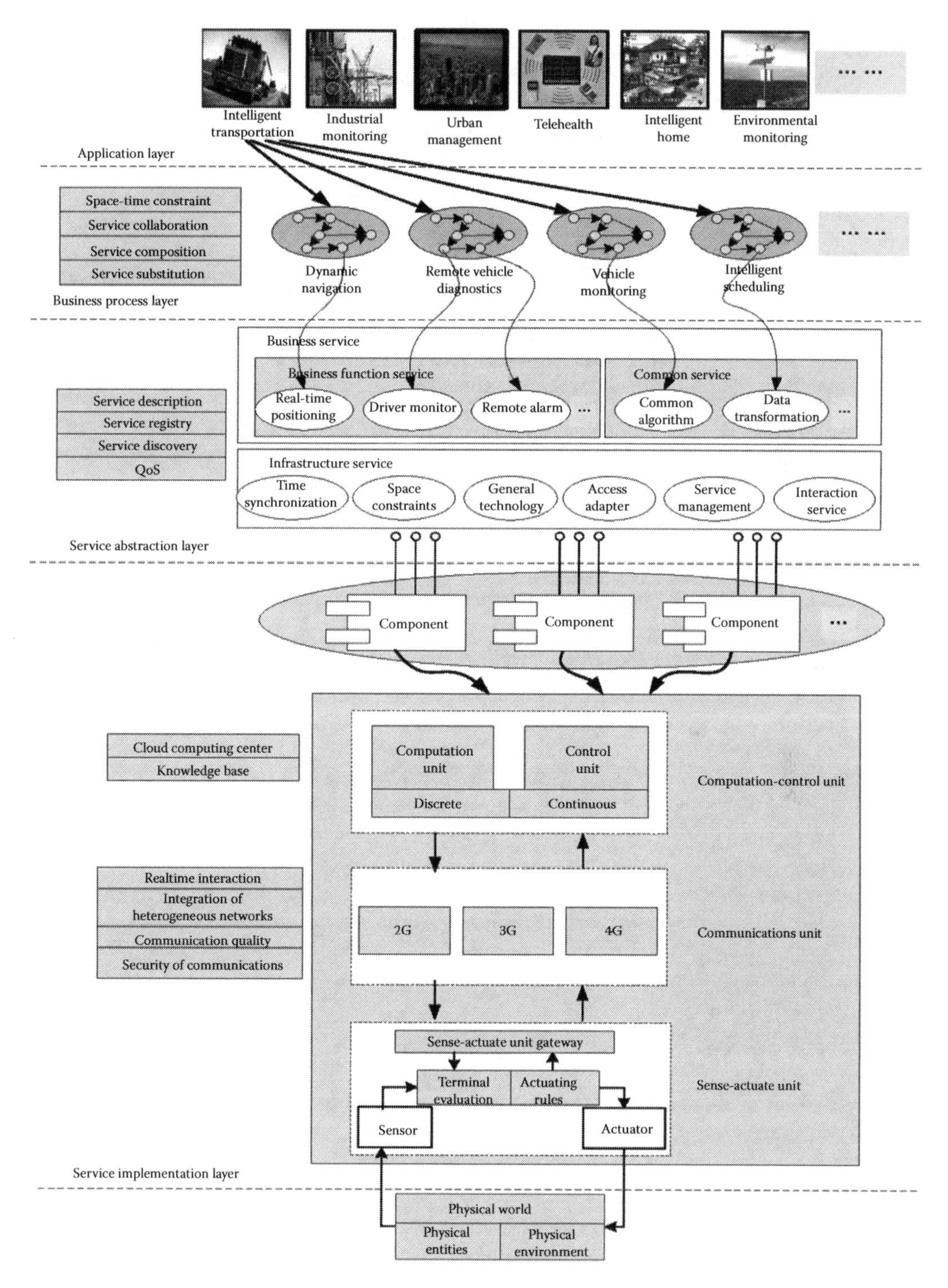

FIGURE 2.10
A service-oriented architecture (SOA) for CPS. (Courtesy of Wang, P. et al., Cyber-physical system components composition analysis and formal verification based on service-oriented architecture, in *Proceedings of Ninth IEEE International Conference on e-Business Engineering*, Hangzhou, China, September 2012.)

2. *Service abstraction layer*: Service functions accessed outside and how to access them are defined in this service abstraction layer. This layer chucks the details about how to implement service functions. Service description, service registry, service discovery, and QoS (quality of service) are also involved in this layer. Specifically, it describes characteristics of interface, usability of operation, data type, parameters, and access protocol. Modules or services outside know what can be done by the CPS service through this layer, also how to find it and how to exchange the message and how to invoke it and what returned results can be. Especially, physical properties must exist in the CPS service description (e.g., temporal and spatial attributes of physical entities), since service implementation layer contains the physical unit (computation unit and control unit) and monitored information without the temporal and spatial information is meaningless. Two types of services exist in this layer, that is, business service and infrastructure service. These two types are described as follows:
   a. *Business service*: Business service is part of business process and fine-grained subprocess of business requirement. It fulfills a specific business task automatically and is reused among different business processes. It includes two kinds: business function service and common service. Business function service is related to some business area, for example, real-time positioning, driver monitor, and remote alarm in an ITS (Intelligent Transportation System). The latter one can be used in different business areas, for example, common algorithm, data transformation, and many more.
   b. *Infrastructure service*: Infrastructure service is foundation of standardized integration of the CPS service. It involves space constraints, time synchronization, service management, general technology, and access adapter and interaction service. Time synchronization and space constraints are guaranteed to meet temporal and spatial condition when physical units and cyber units are mixed together in multiple scales. Technology infrastructure is provided by General technology for developing, delivering, and maintaining the CPS service, and the abilities of security, performance, availability, etc., as well. Access adapter changes the available resources of legacy systems into individual business service. Service management is to monitor the CPS service's state and provide support for abnormal condition, for example, SLA, capacity planning, cause analysis, etc. Interaction service is used for arranging the interfaces of the CPS service, not only for human–computer interaction but also into intelligent device.
3. *Business process layer*: A number of business processes are involved in this layer, where each business process consists of the CPS services following regular rules. Since a lot of fine-grained CPS services would lead to great cost and be ineffective, it is necessary to set up a properly complicated and reliable layer like this. Service composition, service collaboration, service substitution, and space–time constraints are also included in this layer.
4. *Application layer*: Many industry applications consisting of Business processes, which are cooperated with each other to fulfill higher level business goals, are included in this application layer. Compared with business process layer, this layer stresses more focus on integrating all kinds of application requirements through combining professional knowledge with business model in different industries.

2.3.5 Name-Centric Architecture

One of the main goals of SOAs is to enable easy coordination of a large number of computers and service orchestration connected via a network. However, SOA for the WSN and CPSs is still a challenging task. Consequently, for the development and design of very large CPS like WSNs connected to the clouds, SOAs have not yet evolved as an integral technology. One of the limiting issues is registration of service and their discovery.

In case of large CPS, discovery of services is tedious. This is mostly due to the fact that services are often semantically bound to an application function or a region while SOAs make service endpoints to be based on addresses of nodes. Moreover, SOA technologies are not used for service composition within and between the sensor nodes as of today. Also, there exist different methods for accessing the services in a WSN and in the backend. Service development differs, therefore, largely in WSN and cloud. To overcome this limitation, Hellbrueck et al. (2013) suggested a name-centric service architecture for the CPSs. The basis of this architecture is

1. Use of URNs instead of URLs in order to provide a service-centric architecture instead of service or location-centric networking.
2. Using a very common CCNx protocol as the basis for this architecture. It supports location and access transparency.
3. Employing CCN-WSN as the resource-efficient lightweight implementation for WSNs to build a name-based service bus for the CPSs.

While SOA is a powerful approach for building distributed applications based on a rather high-level programming paradigm, it is still difficult to implement it in wireless sensor networks (WSNs) or more generally in CPSs. The general idea of SOA is that whatever a node in the system offers will be made available through a service abstraction.

However, SOA solutions are often too resource-intensive for usual sensors and actuators. In addition to the complexity of the overall CPS environment, including mobility of nodes, heterogeneity of networks, or the sheer size of the system in terms of number of nodes often make it, as desirable it would be, still very difficult to use SOA.

One of the major issues when introducing SOA in CPSs is service registration and discovery. In traditional SOA approaches, service endpoints are bound to an address and/or a location. While this has proved useful in backend systems (for instance, based on clouds), it makes much less sense in the real-world part of the system where data and the access to it through a service are much more important than the node itself and its address. A service user is usually not at all interested in which sensor nodes offer a specific set of data; he rather likes to ask high-level questions such as "give me the average temperature in region x." This requires, however, an integrated approach for both accesses to service and composition of existing services to new, more powerful ones.

A transparent method for service access in both WSN and a backend system (clouds) does not yet exist, and it is to be doubted if such a method should be based on the traditional address-based SOA approach. Therefore, they believed that a content/data-centric approach for service registration and discovery in CPSs can remedy a lot of the problems one has come across so far.

Thus, in this architecture, they investigated how content-centric networking techniques can be used to realize the vision of an easy-to-use and SOA-based programming paradigm for CPSs. The architecture is based on a service bus, which allows transparent access to services in the CPS independent of whether services are located in the cloud or in a WSN.

2.4 Application of Cyber-Physical Systems from Architectural Perspective

Advantages of the CPSs, like efficiency and safety, opens up a wide scope of its application. CPSs allow individual systems to work together to form a heterogeneous systems with new capacities. Wide range of domains where CPSs can be applied includes next-generation transport, health, agriculture, social networking and gaming, manufacturing, alternative energy, environmental control, critical infrastructure control, telemedicine, assisted living, automotive, aviation, and defense. Critical infrastructure includes facilities for water supply (includes storage, treatment, waste water, transport, and distribution), gas production (includes distribution and transport), oil products (includes production, distribution, and transport), electricity generation (transmission and distribution) and telecommunication, which are necessary for the proper functioning of society and economy. Here are a few of these applications described with their critical requirement.

2.4.1 Vehicular CPS

With the increasing number of personal vehicles, lots of problems such as the air pollution, traffic congestion, and safety issues are drawing more attentions to be addressed. Advanced sensing and computing capabilities will be widely used in next-generation transportation systems, such as car control, railways, and air traffic, in order to improve safety and throughput.

Vehicular cyber-physical system (VCPS) is not a new concept. VCPS contributes to a wide range of integrated transportation management systems that should be efficient, accurate, and real-time as well. Based on modern technologies such as computers, sensors, networks, and electronics, the traditional modes of transportation are turning toward more intelligence. NAHSC (National Automated Highway Systems Consortium) is dedicated toward the study of VCPS and development of independent high-speed system "AHS" (Automated Highway systems), aiming to achieve an intelligent and more secure traffic. MIT students developed CarTel, which is one of the American NSF support projects. The CarTel project combines wireless networking, mobile sensing and computing, and the data-intensive algorithms running on cloud servers to address these challenges. CarTel helps system applications to easily collect, process, and then deliver the results. After this it analyzes and visualizes data from sensors located on mobile units. Traffic mitigation, road surface monitoring and hazard detection, vehicular networking, and so on are a few contributions of CarTel Project.

Perception, communication, computation, control, and service are the main characteristics of a Transport System. Considering these properties altogether, a SOA has been proposed by Yongfu et al. for Transportation CPSs. Tight fusion of transportation physical systems and transportation cyber systems is stressed in this architecture. This architecture lays the foundation for next generation of ITS (Intelligent Transportation Systems) (Figure 2.11).

Tradition transportation systems are ubiquitous spatial–temporal, strong coupling, and very large-scale nonlinear complex systems, which are expected to achieve full coordination and optimization, integrate humans with vehicles, traffic, and road conditions. These expectations are not met yet due to lack of widespread interconnection, intercommunication, and interoperability inside these systems.

This architecture follows 3C technology, that is, Computation, Communication, and Control to integrate cyber components and physical components, which fully embodies the depth fusion of the cyber world and physical world through the interaction and feedback between the cyber system and physical system. The core idea of this architecture is to transmit the information of transportation physical objects and the system state to the cyber

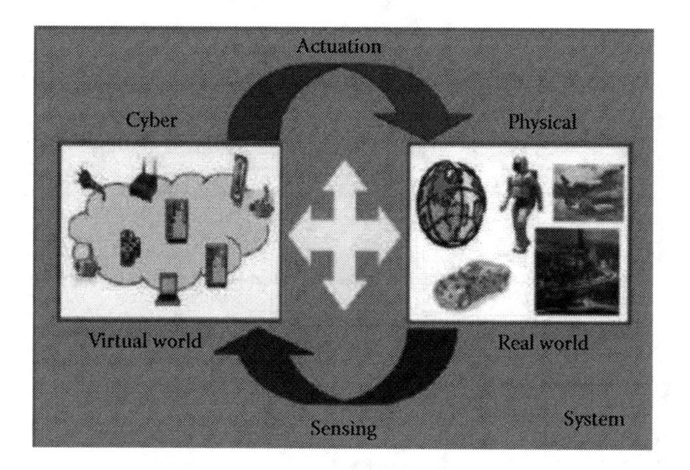

FIGURE 2.11
Interaction between cyber world and physical world. (Courtesy of Yongfu, L. et al., A service-oriented architecture for the transportation cyber-physical systems, in *Proceedings of 31st Chinese Control Conference* (CCC), Hefei, China, July 2012.)

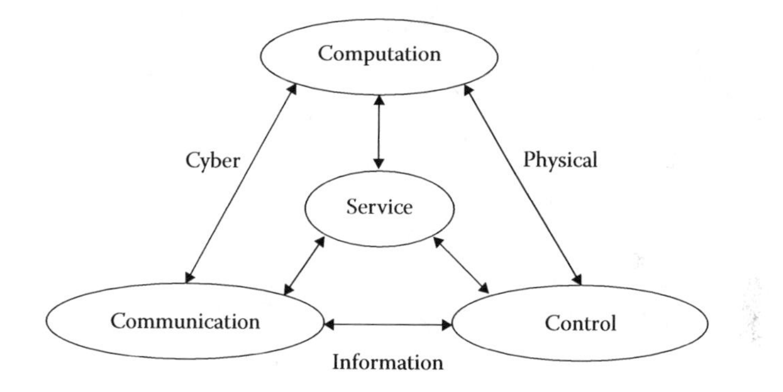

FIGURE 2.12
Relation among computation, communication, control, and service in T-CPS. (Courtesy of Yongfu, L. et al., A service-oriented architecture for the transportation cyber-physical systems, in *Proceedings of 31st Chinese Control Conference* (CCC), Hefei, China, July 2012.)

system, then to integrate the cyber components and physical components based on the fusion of computation, communications, and control technology in order to achieve information communication, system coordination, and optimal decision-making control of transportation system through the interaction and feedback between the physical system and the cyber system on the basis of accurate cognition for the transportation physical objects (Figure 2.12).

One purpose of T-CPS is to provide the efficient, safety, and high quality service for human life. Therefore, quality service is the requirement for the process of computation, communication, and control. Figure 2.12 shows the relation among 3Cs in T-CPS.

Considering the characteristics of transportation system, a SOA for T-CPS is shown in Figure 2.13, which consists of perception, communication, computation, control, and service. As shown in Figure 2.13, there are a large number of physical perception devices such as ultrasonic detectors, infrared detectors, radio frequency identification (RFID) tags, video sensors, engine coolant temperature (ECT) sensor, vehicle measurement system (VMS), light emitting diode (LED), microwave detectors, etc., and computing devices such as servers and computers in the T-CPS. Moreover, these devices were interconnected through the network (Figure 2.13).

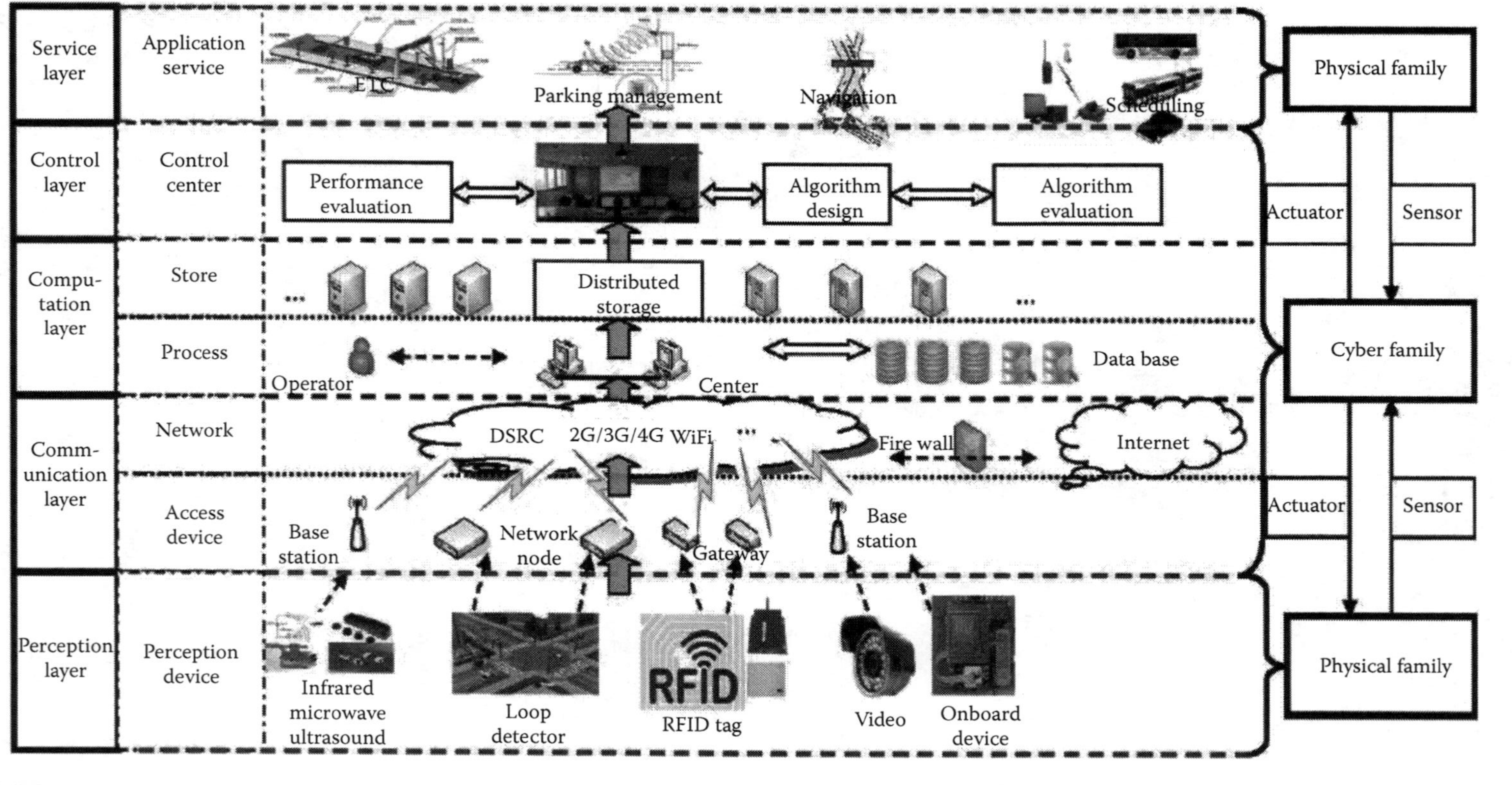

FIGURE 2.13
Conceptual architecture of T-CPS. (Courtesy of Yongfu, L. et al., A service-oriented architecture for the transportation cyber-physical systems, in *Proceedings of 31st Chinese Control Conference* (CCC), Hefei, China, July 2012.)

Functions of each layer:

1. *Perception layer*: This layer consists of some sensor nodes and collection nodes. It is responsible for the perception of some physical properties from the physical world, mainly including vehicles, infrastructures. The perception is the foundation of T-CPS.
2. *Communication layer*: It comprises some communication stations and network nodes. And its duty is to transmit the perceived original data to the information center, and meanwhile to ensure reliable communication between V2V, V2R, and R2S. The key is how to guarantee highly reliable and effective transmission.
3. *Computation layer*: Since T-CPS is with mass traffic data information, and there is interaction and feedback between the physical systems and cyber system, therefore there is a requirement of real-time performance for the analysis and simulation. So it is a challenge for the computing power and storage capacity of the T-CPS computation level.
4. *Control layer*: One goal of T-CPS is to strengthen the control ability of transportation physical systems. Based on the large amount of effective traffic information, T-CPS can provide scientific control algorithm and transmit the instructions to the control nodes or send a warning signal in an emergency via the actuators in order to carry out the precise control for the transportation physical system. However, how to ensure the robustness and stability of T-CPS is a key problem to achieve satisfactory control performance.
5. *Service layer*: T-CPS can obtain huge information of objects, tools, and infrastructures. Accordingly, it will provide the real-time traffic information service for the user terminals. This service holds very important practical significance. However, the main challenge remains in the process of perception and transmission. How to effectively deal with the incompleteness and the uncertainties while improving the accuracy and timeliness of the traffic information release.

2.4.2 Health CPS

In the CPS, the integration of active user input such as digital records of patient health analysis, smart feedback system, and passive user input such as biosensors and other smart devices can support data acquisition for an accurate and efficient decision making in health-care environment. However, this combination of decision-making system and data acquisition is yet to be deeply studied and explored in health-care applications. CPS utilization in health-care scenario includes the introduction of inter-operation of adaptive and autonomous devices with coordination and new concepts for managing medical physical systems with the help of computation and control, programmable materials, miniaturized implantable smart devices, BAN (body area networks), and new fabrication approaches.

Challenges of the CPS in health care are as follows:

1. *Software reliability*: Being an integral part of medical devices, software ensures device functionality, proper coordination between medical devices and the patients. Therefore, the safety and efficiency of the medical systems completely rely on the well-developed, well-designed, and well-managed software.
2. *Medical device interoperability*: Various medical devices may have different communication interfaces. Therefore, to integrate heterogeneous medical devices in a secured and certified manner, a well-maintained management system is required.

3. *Data extraction*: Medical devices collect multiple physiological parameters from the patients. Multiple physiological parameters widely can be collected from the patients, which are varied in nature. In addition to the general information of the patients, these parameters can provide an early prediction of future illness and help in nullifying an emergency situation. However, to design a system that can extract complex physiological parameters from patients is a challenging task.
4. *Security and privacy*: The collected patient data require very high privacy because illegal use of private data of the patient may damage reputation and thus cause mental unrest and abuse leading to further physical illness. So, it is the utmost requirement in health CPS.
5. *System feedback*: Development of feedback system is a very critical task for the CPS in health care because the design stability requires a proper feedback system. Healthcare CPS depicts a perfect feedback system with the smart alarm system that can notify the caretaker during any emergency situation or any possible illness.

   However, the next-generation alarm system has to essentially address the challenges such as types of physiological parameter, implementation ability, and system complexity. It is a challenging task to select right parameters with a wide range of possible illnesses, which can identify a specific illness from multiple possibilities. Any Complexity must not hinder the efficiency of feedback system.
6. *Complex query processing*: Battery-operated wireless sensors poses a few challenges such as low energy consumption and limited processing capabilities due to the presence of heterogeneous biosensors, thus complex query processing becomes difficult. Complex query processing can reduce transmission data amount and context-aware predictions. The access to multiple physiological parameters may enable the complex queries to predict possible illness of the patient. However, this approach requires complex computational skill sets and designs.

Health CPS (HCPS) will replace traditional health devices working individually, which we will face in the future. Through advancement of various health devices by using sensors and networking them to work together, patients' physical condition can be monitored in real time, especially for the patients in critical condition and require monitoring from moment to moment, such as heart patients. A portable terminal device carried by the patient will detect the patient's condition in real time and send timely prediction or alert in advance. In addition, the collaboration between health equipment and real-time data delivery would be much more convenient for patients.

Insup Lee and Oleg Sokolsky introduced the development and issues of the highly trustworthy health CPS system, including the dependence on software from the new function development, demand for network connections, and request for continuous monitoring of patients, and analyze the future development of Health CPS.

For health device interoperability issues, Cheolgi Kim et al. introduced a generic framework: the NASS (network-aware supervisory systems), which integrates health devices with such a clinical interoperability system that uses real networks. This framework provides a development environment in which health device supervisory logic can be developed based on the assumptions of a robust and ideal network. Concerning the particularity of the health applications, more features such as real-time, network delay, the higher requirement of security, will be considered in the design of Health CPS.

2.4.3 Agricultural CPS

Accuracy in agriculture was expected in the early 1980s from traditional agriculture, supported by information technology, to modern agriculture, implementing a full range of modern systems of agricultural management strategies and technologies. Design of precision agriculture includes fundamental geographic information of farmland, data management of production experiments, micro-climate information, and other data.

The project "underground wireless sensor network" was developed by the University of Nebraska-Lincoln Cyber-physical Networking Lab, where Agnelo R. Silva et al. developed a novel CPS, which was the integration of center pivot systems with wireless underground sensor networks and was called CPS^2 for precision agriculture. The WUSNs (wireless underground sensor networks) consist of underground sensor nodes connected wirelessly that communicate untethered through the soil. The results of these experiments demonstrated that the concept of CPS^2 is feasible. This concept can further be made highly reliable using commodity wireless sensor motes. This integration of the CPS and precision agriculture is one of the typical applications of the CPS.

Other application scenarios of the CPS are:

Automotive: Due to very complex traffic control algorithms, for example, the calculation of best route according to traffic situation, CPSs for the automotive industry require high computing power.

Environment: In a distributed wide and varied geographical area such as forests, rivers, and mountains, CPSs must operate for long time periods without human intervention and with minimal energy consumption. In such an environment, the real challenge is the accuracy and in-time data collection with low power consumption.

Aviation defense: High security, precise control, and high computing power are the main requirement of CPSs in aviation and defense. In this scope, the main challenge will be the development of the security protocols.

Critical infrastructure: CPSs for water resources management, energy control, etc., require precision and reliability in control, leading to application software methodologies, which ensure the quality of the software.

2.5 Conclusion

This chapter provides the readers with a basic introduction to architecture of the CPSs and analyzes them. It analyses and argues with the definition and showcases the current work in architecture for the CPSs. The architectures mentioned in this chapter serve the currently identified CPS requirements and characteristics in various fields. It also unifies the current human-only computation model and machine-only computation model.

There are many open research challenges for the CPSs. For instance, how should the publish/subscribe scheme be formulated? How can a global reference time be provided in large-scale heterogeneous systems? How should the formalized event or information model be specified? How is network routing achieved using event or information lifespan? Knowledge data management and dispatching form other challenges, while security assurance poses a challenge for any system.

References

Ahmed, S.H., Kim, G., and Kim, D. (November 2013). Cyber physical system: Architecture, applications and research challenges. In *Proceedings of Sixth IFIP/IEEE Wireless Days Conference (WDays)*, Valencia, Spain.

Hellbruck, H., Teubler, T., and Fischer, S. (December 2013). Name-centric service architecture for cyber-physical systems. In *Proceedings of IEEE Sixth International Conference on Service-Oriented Computing and Applications* (*SOCA*), Koloa, HI.

Petnga, L. and Austin, M. (December 2013). Cyber-physical architecture for modeling and enhanced operations of connected-vehicle systems. In *Proceedings of IEEE International Conference on Connected Vehicles and Expo* (*ICCVE*), Las Vegas, NV.

Wang, P., Xiang Y., and Zhang, S.H. (September 2012). Cyber-physical system components composition analysis and formal verification based on service-oriented architecture. In *Proceedings of Ninth IEEE International Conference on e-Business Engineering,* Hangzhou, China.

Ying, T. and Steve, G. (January 2008). A prototype architecture for cyber-physical systems. In *Proceedings of ACM SIGBED,* New York.

Yongfu, L., Dihua, S., Weining, L., and Xuebo, Z. (July 2012). A service-oriented architecture for the transportation cyber-physical systems. In *Proceedings of 31st Chinese Control Conference* (*CCC*), Hefei, China.

Yu, C., Jing, S., and Li, X. (August 2012). An architecture of cyber physical system based on service. In *Proceedings of IEEE International Conference on Computer Science and Service System* (*CSSS*), Nanjing, China.

Additional Readings

Abad, F.A.T., Caccamo, M., and Robbins, B. (August 2012). A fault resilient architecture for distributed cyber-physical systems. In *Proceedings of IEEE International Conference on Embedded and Real-Time Computing Systems and Applications* (*RTCSA*), Seoul, South Korea.

Benveniste, A. (March 2010). Loosely time-triggered architectures for cyber-physical systems. In *Proceedings of Design, Automation and Test in Europe* (*DATE*), Dresden, Germany, 2010.

Bhadoria, R.S., Dixit, M., Mishra, V., and Jadon, K.S. (December 2011). Enhancing web technology through wiki-shell architecture. In *Proceedings of IEEE World Congress on Information and Communication Technologies,* Mumbai, India.

Bhadoria, R.S. and Jaiswal, R. (November 2011). Competent search in blog ranking algorithm. In *Proceedings of Springer International Conference on Computation Intelligence & Information Technology,* Pune, India.

Bhadoria, R.S., Sahu, D., and Dixit, M. (2012). Proficient routing in wireless sensor networks through grid based protocol. *International Journal of Communication Systems and Networks,* 1(2), 104–109.

Bhave, A., Krogh, B.H., Garlan, D., and Schmerl, B. (April 2011). View consistency in architectures for cyber-physical systems. In *Proceedings of IEEE/ACM Second International Conference on Cyber-Physical Systems* (*ICCPS*), Chicago, IL.

Gang, L. and GuPing, Z. (March 2012). Self-reconfiguration architecture for distribution automation system based on cyber-physical. In *Proceedings of IEEE Asia-Pacific Power and Energy Engineering Conference* (*APPEEC*), Shanghai, China.

Haque, S.A., Aziz, S.M., and Rahman, M. (April 2014). Review of cyber-physical system in healthcare. In *Proceedings of International Journal of Distributed Sensor Networks,* Hindawi Publishing Corporation, Cairo, Egypt.

Hu, L., Xie, N., Kuang, Z., and Zhao, K. (April 2012). Review of cyber-physical system architecture. In *Proceedings of IEEE 15th International Symposium on Object/Component/Service-Oriented Real-Time Distributed Computing Workshops*, Guangdong, China.

Jung, H., Han, H., Yeom, H.Y., and Kang, S. (2010). CPS: Operating system architecture for efficient network resource management with control-theoretic packet scheduler. *Journal of Communications and Networks*, 12(3), 266–274.

Sanislav, T. and Miclea, L. (May 2012). An agent-oriented approach for cyber-physical system with dependability features. In *Proceedings of IEEE Automation Quality and Testing Robotics* (*AQTR*), Cluj-Napoca, Romania.

Seeber, S., Sehgal, A., Stelte, B., Rodosek, G.D., and Schönwälder, J. (October 2013). Towards a trust computing architecture for RPL in cyber physical systems. In *Proceedings of Ninth International Conference on Network and Service Management* (*CNSM*), Zurich, Switzerland.

Singh, A.K., Shrivastava, A., and Tomar, G.S. (June 2011). Design and implementation of high performance AHB reconfigurable arbiter for on-chip bus architecture. In *Proceedings of IEEE International Conference on Communication Systems and Network Technologies*, Jammu, India.

Zhu, Q., Rieger, C., and Basar, T. (August 2011). A hierarchical security architecture for cyber-physical systems. In *Proceedings of IEEE Fourth International Symposium on Resilient Control Systems* (*ISRCS*), Boise, ID, pp. 15–20.

3

Cyber-Physical Systems: Beyond Sensor Networks and Embedded Systems

Krishnarajanagar GopalaIyengar Srinivasa, Nabeel Siddiqui, Abhishek Kumar, and Anita Kanavalli

CONTENTS

3.1 Protocols and Architecture for Wireless Sensor Network

3.1.1 Architecture: Single Node Architecture—Objective

This chapter explains about the nodes that are a part of the Wireless Sensor Networks (WSNs). It describes the principal task of a node like computation, storage, communication, and sensing/actuation and the components required to perform these tasks. The nodes can store energy, they can gather energy from the environment (through ultraviolet rays) and saved intelligently to control the operation of the nodes as well as other components in the network. An operating system like software is required to control the operation, which is described in the last section of this chapter with examples of sensor nodes. At the end of this chapter, the reader will have a clear understanding of the capabilities and limitations of the nodes in a WSN. It serves as the foundation for the following chapter, which discusses the options available on how individual sensor nodes are connected into a WSN.

3.1.2 Hardware Components

While selecting the basic hardware components for a wireless sensor node, apparently the requirements laid by an application play an important role with regard to size, cost, and energy consumption by the nodes. Communication and computation done by the nodes is expected to be of acceptable quality, but the trade-offs between features and costs is a compromise. In some special cases, an entire sensor node should be smaller than 1 cc, weight (considerably) less than 100 g, and cheaper than US$1, and should dissipate less than 100 μW. In some extreme versions, the nodes are sometimes claimed to have reduced to the size of grain of dust. In some real-time applications the size of the node is not important rather, cost and power supply play a major role.

When we study different available hardware platforms, not a single uniform standard is present, and the available heterogeneous standard may not support all application types. In this section, different sensor node architectures are dealt in detail. The research projects focus on reducing the components in size, energy consumption, or costs, based on the fact that the available components do not live up to the requirement of the current requirements by the applications.

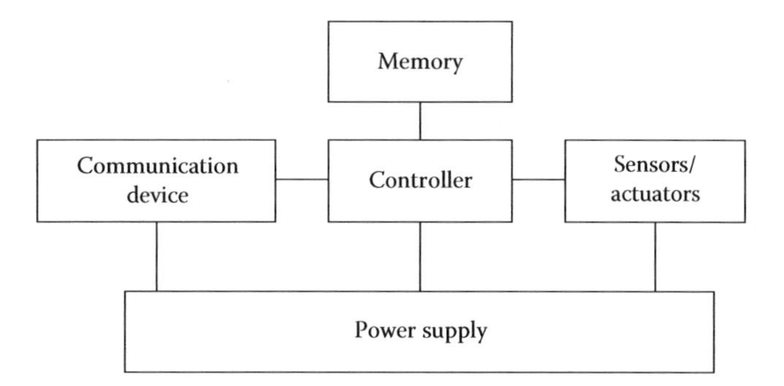

FIGURE 3.1
Basic sensor node components.

A basic sensor node comprises of five components as shown in Figure 3.1:

- *Controller*: Processes all the relevant data and capable of executing arbitrary code.
- *Memory*: Stores programs and intermediate data. Separate memory is used for programs and data (heterogeneous memory types used).
- *Sensors and actuators*: These are the actual interface to the physical world: devices that can gather information from the environment or control the physical parameters.
- *Communication*: Turning nodes send and receive information over a wireless channel.
- *Power supply*: No continuous power supply is available, batteries are necessary to provide energy. Sometimes, recharging by obtaining energy from the environment is available (Solar cells) [1].

Each of these components has to operate balancing the energy consumption on one hand and the need to fulfill their tasks on the other hand. The sensors are programmed to raise an interrupt if a given event occurs, for example, temperature value exceeds a given threshold or the communication device detects an incoming transmission. To support such an event there has to be an appropriate interconnection between individual components. Control and data information will be exchanged among these interconnections. This interconnection is very simple, for example, a sensor sends analog value to the controller or uses intelligence of its own to inform the controller. This is equivalent to pre-processing sensor data and only waking up the main controller if an actual event has been detected. Another example is detecting a threshold crossing in a simple temperature sensor. All preprocessing is highly customized to the specific sensor and ensure that the sensor operates continuously, resulting in improved energy efficiency.

3.1.2.1 Controller

The controller is the core component of a wireless sensor node. It basically collects data from all the sensors. These data are processed and sent to the relevant receivers and also decides on the actuator's behavior. It executes various programs like time-critical signal processing and communication protocols and application programs. Such a variety of processing tasks are performed by these controllers and different controller architectures are chosen keeping in mind the flexibility, performance, energy efficiency, and cost.

3.1.2.2 Memory

The random access memory (RAM) is used to store intermediate sensor data and packets. Program codes and system-related functions are stored in read-only memory (ROM) or, in electrically erasable programmable read-only memory (EEPROM) or flash memory. Sometimes flash memory serves as intermediate storage when RAM capacity is insufficient or when the power supply of RAM is shut down for some time. The access delays of read and write operation on flash memory and energy required for the operations are to be taken into account. Memory requirements are application-dependent and designing the RAM size is crucial with respect to power consumption and cost.

3.1.2.3 Communication Device

The communication device is used to exchange data between individual nodes in the network. In case of wireless communication, the first choice to make is that of selection of the transmission medium. The different options are radio frequency (RF), optical communication, and ultrasound. Other media like magnetic inductance are used in very special cases only. RF-based communication is most used and it best fits the requirements in most of the WSN applications. It provides:

- Long-range coverage
- High data rates
- Acceptable error rates at reasonable and less energy
- No line of sight between sender and receiver

3.1.2.4 Sensor and Actuators

The types of sensors and actuators available is very vast. It is only possible to provide an idea on the actual type that can be used in WSNs and are again application-dependent.

3.1.2.4.1 Sensors

Sensors are divided into three categories:

1. *Passive, Omnidirectional Sensors*: These sensors measure a physical quantity in their vicinity only and no direction involved and they do not probe the environment and hence are passive. These are self-powered and they obtain the energy required from the environment. The examples for such sensors are thermometer, light sensors, vibration sensors, microphones, humidity sensors, mechanical stress or tension in material sensing, chemical sensors, smoke detectors, and air pressure detection sensors.
2. *Passive, Narrow-Beam Sensors*: These sensors are passive and know the direction of measurement. An example is a camera, which can "take measurements" in a given direction, but also can be rotated as per the need.
3. *Active Sensors*: They probe the environment, for example, a sonar or radar sensor or some types of seismic sensors, which generate shock waves even for small explosions.

Sensors of various types are available in many different forms with different individual capabilities. They vary in terms of accuracy, dependability, energy consumption, cost, and size. Most of the theoretical work on WSNs considers passive, omnidirectional sensors. Narrow-beam-type sensors like cameras are used in some practical test beds. Active sensors are not discussed here in detail.

3.1.2.4.2 Actuators

Actuators are like sensors, all that a sensor node can do, for example, to open or close a switch or a relay or to set a value in some way can also be done by the actuators. The communication protocols design do not consider whether the operations are done by the sensors or the actuators. Hence, in this chapter, we shall treat actuators as sensors. But a good design practice for the embedded system applications is to pair an actuator with a correct controlling sensor.

3.1.2.5 Power Consumption of Sensor and Actuators

In case of passive light or temperature sensors, the power consumption is ignored in comparison to other devices on a wireless node. In case of active devices like sonar, power consumption is a concern and the size of the power sources on such sensor node is to be carefully chosen such that batteries are not overstressed. In addition, the active probing by the nodes and frequent sampling require more energy by the sensors. Finally, the energy is spent on data gathering, processing, and communication to the relevant intermediate components or the destination in the network.

3.1.2.6 Some Examples of Sensor Nodes

There are quite a number of nodes available commercially for use and to build a WSN and carry out research work and development. Depending on the requirement by the applications and considering their battery life, mechanical robustness of the node's housing, size, cost they are chosen. Table 3.1 lists the different sensors in use with their characteristics [1,2].

3.1.2.7 The "Mica Mote" Family

Starting in the late 1990s, an entire family of nodes have evolved out of research projects at the University of California at Berkeley, partially with the collaboration of Intel, over all these years. They are commonly known as the Mica motes11, with different versions (Mica, Mica2, Mica2Dot). Schematics for some of these designs are available in their data sheets. They are

TABLE 3.1

Example Characteristics of Sensor Nodes

Sensor	Accuracy	Sample Rate (Hz)	Startup (ms)	Current (mA)
Photoresistor	N/A	2000	10	1.235
Temperature	1 K	2	500	0.15
Barometric pressure	1.5 mbar	10	500	0.01
Humidity	2%	500	500–3000	0.775
Thermistor	5 K	2000	10	0.126

commercially available via the company Crossbow12 in different versions and grouped as kits. TinyOS is the operating system used in these nodes. These boards feature a microcontroller belonging to the Atmel family, a simple radio modem (usually a TR 1000 from RFM), and various connections to the outside interface. In addition, it is possible to connect additional "sensor boards," for example, barometric or humidity sensors to the node enabling a wider range of applications and experiments. Also, specialized enclosures have been built for use in rough environments, for example, monitoring bird and wild animal habitats. Sensors are connected to the controller via an I2C bus or via SPI, depending on the version. The MEDUSA-II nodes share the basic components and are quite similar in design [2,3].

3.1.2.8 EYES Nodes

The nodes developed by Infineon in the context of the European Union sponsored project "Energy-efficient Sensor Networks" (EYES) are another example of a typical sensor node. It is equipped with a Texas Instrument MSP 430 microcontroller, an Infineon radio modem TDA 5250, along with a SAW filter and transmission power control; the radio modem also reports the measured signal strength to the controller. The node has a USB interface to a PC and the possibility to add additional sensors/actuators.

3.1.2.9 BTnodes

The "BTnodes" are developed at the ETH Zurich through several research projects. They feature an Atmel ATmega 128L microcontroller, 64–180 kB RAM, and 128 kB FLASH memory. Unlike most other sensor nodes (but similar to some nodes developed by Intel), they use Bluetooth as their radio technology in combination with a Chipcon CC1000 operating between 433 and 915 MHz.

3.1.2.10 Commercial Solutions

Apart from these academic research prototypes, there are already a couple of sensor node-type devices commercially available, including appropriate housing and certification. Some of these companies include "ember" (www.ember.com), "Millenial" (www.millenial.net), and "Crossbow" (www.crossbow.com). The market here is more dynamic that can be reasonably reflected in a textbook and the reader is encouraged to watch for up-to-date developments.

This section has introduced the necessary hardware prerequisites for building WSNs, the nodes in particular. It has shown the principal ways of constructing the nodes and discussion on performance and energy consumption of its main components like the controller, the communication device, and the sensors. In summary a wireless sensor node consists of two separate parts, one part that is continuously vigilant, will detect and report events, and has small or even negligible power consumption. This is complemented by a second part that performs actual processing and communication, has higher, no negligible power consumption, and has therefore to be operated in a low duty cycle. This separation of functionalities is justified from the hardware properties as is it supported by operating systems like TinyOS [3].

3.1.3 Network Architecture: Sensor Network Scenarios

A sensor node is treated as a source and is any entity in the network that can provide information. It can also be an actuator node that provides information about an operation. A sink, on the other hand, is the entity where information is gathered. There are essentially three options for a sink, it could belong to the sensor network as such and be just another sensor/actuator node or it could be an entity outside this network. Secondly the sink could be an actual device, for example, a handheld or PDA used to interact with the sensor network. It can also act like a gateway to another larger network such as the Internet, where the actual request for the information comes from some node "far away" and only indirectly connected to such a sensor network.

3.1.3.1 Single-Hop versus Multi-Hop Networks

Due to inherent power limitation of radio communication, there is a limitation on the feasible distance between a sender and a receiver. Because of this limited distance, direct communication between source and the sink is not possible specifically in WSNs, which is intended to cover a lot of ground (e.g., in environmental or agriculture applications) or that operate in difficult radio environments with strong attenuation (e.g., in buildings). To overcome such limited distances, an obvious way out is to use relay stations, with the data packets taking multi-hops from the source to the sink. This concept of multi-hop networks is particularly attractive for WSNs as some of the sensor nodes can act like relay nodes, foregoing the need for additional equipment. Depending on the particular application, the likelihood of having an intermediate sensor node at the right place is actually been quite high. For example, when a given area has to be uniformly equipped with sensor nodes, but nevertheless, there is not always a guarantee that such multi-hop routes from source to sink exist, nor that such a route is particularly short. While multi-hop network is an evident and working solution to overcome problems with large distances or obstacles, it has also been claimed to improve the energy efficiency of communication.

3.1.3.2 Multiple Sinks and Sources

So far, only networks with a single source and a single sink have been illustrated. In many cases, there are multiple sources and/or multiple sinks present. In the most challenging case, multiple sources should send information to multiple sinks, where either all or some of the information has to reach all or some of the sinks. Figure 3.2 illustrates single-hop sensor network combinations.

3.1.3.3 Three Types of Mobility

In the scenarios discussed earlier, all entities/devices are stationary. But one of the main virtues of wireless communication is its ability to support mobile entities/devices. In WSNs, mobility is viewed in three main different forms.

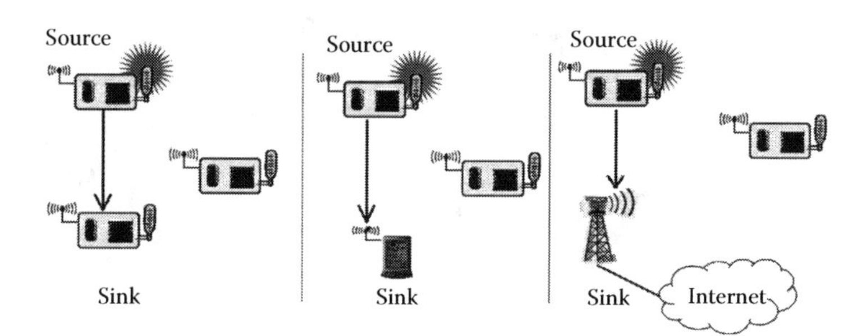

FIGURE 3.2
Three types of sinks in a very simple, single-hop sensor network.

3.1.3.3.1 Node Mobility

The wireless sensor nodes themselves can be mobile. The meaning of such mobility is highly application-dependent. In examples like environmental control, node mobility is strictly ruled out where as in livestock surveillance (sensor nodes attached to cattle) node mobility is evident.

In the case of node mobility, the network has to reorganize itself frequently to be able to function correctly. It is a choice between the frequency and speed of node movement and the energy required to maintain a desired level of functionality in the network.

3.1.3.3.2 Sink Mobility

The information sinks can be mobile as shown in Figure 3.3. While this can be a special case of node mobility, the important aspect is the mobility of an information sink that is not part of the sensor network, for example, a human user requested information via a PDA while walking in an intelligent building.

In a simple case, such a requester can interact with the WSN at one point and complete its interactions before moving on. In many cases, consecutive interactions are treated as separate, unrelated requests. Whether the requester is allowed interactions with any node or only with specific nodes is a design choice for the appropriate protocol layers. A mobile requester is very interesting, if the requested data are not locally available but to be retrieved from some remote part of the network, while the requester has communicated only with nodes in its vicinity, it might have moved to some other place. The network must make provisions that the requested data reach the node despite its movements with the help of the requester.

3.1.3.3.3 Event Mobility

In most applications like event detection and in particular the tracking applications, the occurrence of the events or the objects to be tracked can be mobile. In such scenarios, it is (usually) important that the observed event is covered by a sufficient number of sensors at all time. Hence, sensors will wake up around the object, engaged in higher activity to observe the present object, and then go back to sleep. As the event source moves through the network, it is nothing but the activity area [4].

3.1.3.4 Quality of Service

WSNs differ from other conventional communication networks mainly in the type of service they offer. These networks essentially only move bits from one place to another. Possibly, additional requirements about the offered Quality of Service (QoS) are made,

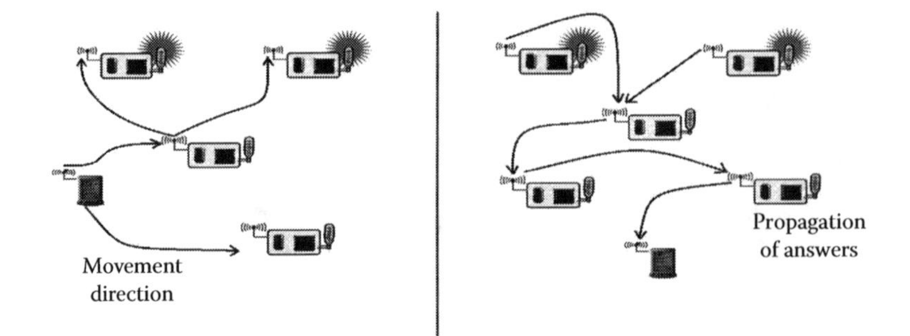

FIGURE 3.3
Movement of mobile node in a network.

especially in the context of multimedia applications. Such QoS are regarded as a low-level, networking device observable attributes like bandwidth, delay, jitter, and packet loss rate. At high-level, user-observable attributes like the perceived quality of a voice communication or a video transmission are considered for WSNs. While the first kind of attributes are applicable to a certain degree of WSNs as well (bandwidth, e.g., is quite unimportant), the second one clearly is not, but is really the most important one to consider. Hence, high-level QoS attributes corresponding to the subjective QoS attributes in conventional networks are required. But just like in traditional networks, the QoS attributes in WSN depend on the application in use [5,6]. Some generic possibilities are

Event detection/reporting probability: The probability that an event has occurred but is not detected or not reported to an information sink. For example, not reporting a fire alarm to a surveillance station would be a severe shortcoming. This probability depends on the overhead spent in setting up structures in the network that support the reporting of such an event (routing tables) or against the run-time overhead (sampling frequencies).

Event classification error: The events are to be detected and also to be classified, the error in classification must be small.

Event detection delay: It is the delay between detecting an event and reporting it to any/all interested sinks.

Missing reports: Some applications require periodic reporting, the probability that some reports are undelivered should be small.

Approximation accuracy: For function approximation applications like approximating the temperature as a function of location for a given area, the average/maximum absolute or relative error approximation with respect to the actual function is termed as accuracy. Similarly, for edge detection applications, it is the accuracy of edge descriptions.

Tracking accuracy: The reported position of the object being tracked should be as close to the real position as possible, and the error should be small. Other aspects of tracking accuracy are the sensitivity to sensing gaps.

3.1.3.5 Energy Efficiency

Much of the discussion has already shown that energy is the most important resource in WSNs and that energy efficiency will become an evident optimization goal. It is clear that with an arbitrary amount of energy, most of the QoS metrics defined earlier can be increased almost at will (approximation and tracking accuracy are notable exceptions as they also depend on the density of the network). Hence, putting the delivered QoS and the energy required to carry out operations into perspective will have to give a reasonable understanding of the term energy efficiency. The term "energy efficiency" is, in fact, rather an umbrella term for many different aspects of a system and is to be distinguished to form the actual, measurable figure of merit. The most commonly considered aspects are

Energy dissipated per correctly received bit: The energy, counting all sources of energy consumption at all possible intermediate hops, is spent on an average to transport one bit of information (payload) from the source to the destination. This is often a metric used by the periodic monitoring applications.

Energy dissipated per reported (unique) event: It is the average energy spent to report one event. The same event is reported from various sources in the network, therefore it is required to normalize this metric only for the unique events.

Delay/energy trade-offs: Some applications have a notion of "urgent" events, which can justify an increased energy investment for a speedy reporting of such events. In such cases there is a compromise between the delay and energy spent.

Network lifetime: The time for which the network is operational and is able to complete its tasks.

The possible timing durations are

Time taken for the first node death: The first node in the network run out of stored energy or fail and stop working.

Network half-life time: The 50% of the nodes run out of stored energy and stop operating. Any other fixed percentile is applicable as well.

Time to partition the network: The time taken for the first partition of the network in two (or more) disconnected parts. This can be as early as the death of the first node (if that was in a pivotal position) or may happen very late if the network topology is robust.

Time in loss of coverage: Usually, with redundant network deployment and sensors that covers a region instead of just a specific spot where the node is located, each point in the deployment region is covered by multiple sensor nodes. A possible figure of merit is thus the time when for the first time any specific small area in the deployment region is no longer covered by any node's observations. If k redundant observations are necessary (e.g., for tracking applications), the corresponding definition of loss of coverage would be the first time any small area in the deployment region is no longer covered by at least k different sensor nodes.

Time to failure of first event notification: An application-specific interpretation of partition is the inability to deliver an event. This can be due to an event not being noticed by the sensor node because it may be dead or failed or a partition between source and sink has occurred.

3.1.3.6 Scalability

The ability to keep the network alive and maintain the network performance characteristics irrespective of the size of the network is referred to as scalability. With WSN potentially consisting of thousands of nodes, scalability is an evidently indispensable requirement. Scalability is still served by any construct that requires globally consistent state, such as addresses or routing table entries that have to be maintained. Hence, the need to restrict such information is enforced by and goes hand in hand with the resource limitations of sensor nodes, especially with respect to memory. The protocol design is effected by the need for scalability and the network architectures and protocols designs should implement appropriate scalability support. Applications with a few dozen nodes might admit more efficient solutions than applications with thousands of nodes. These smaller applications might be more common in the first place. Nonetheless, a considerable amount of research has been invested into highly scalable architectures and protocols [7].

3.1.3.7 Design Principles for WSNs

Appropriate QoS support, energy efficiency, and scalability are important design and optimization goals for WSNs. But these goals themselves do not provide many hints on how to structure a network such that they are achieved. A few basic principles have emerged, which are useful when designing networking protocols. The general advice is to always consider the needs of a concrete application holds well for the basic principles [8].

3.1.3.8 Distributed Organization

Both the scalability and the robustness and to some degree also the other goals make it imperative to organize the network in a distributed fashion. This means that there should be no centralized entity in charge such an entity could, for example, control medium access or make routing decisions, similar to the tasks performed by a base station in cellular mobile networks. The disadvantages of such a centralized approach are obvious as it introduces exposed points of failure and is difficult to implement in a radio network, where participants only have a limited communication range. Rather, the WSNs nodes should cooperatively organize in the network using distributed algorithms and protocols. Self-organization is a commonly used term for this principle. When organizing a network in a distributed fashion, it is necessary to be aware of potential shortcomings of this approach. In many circumstances, a centralized approach can produce solutions that perform better or require fewer resources (in particular, energy). To combine the advantages, one possibility is to use centralized principles in a localized fashion by dynamically electing a specific node out of the set of nodes that assume the responsibilities of a centralized agent. Such elections result in a hierarchy, which is dynamic. The particular election rules and triggering conditions for re-election vary considerably, depending on the purpose for which these hierarchies are used.

3.1.3.9 In-Network Processing

When organizing a network in a distributed fashion, the nodes in the network are not only passing on packets or executing application programs, they are also actively involved in taking decisions about how to operate the network. This is a specific form of information processing that happens in the network, but is limited to information about the network itself. It is possible to extend this concept by also taking the concrete data that is to be transported by the network into account in this information processing, making in-network processing a first-rank design principle.

3.1.3.10 Interfaces of WSNs

3.1.3.10.1 Structure of Application/Protocol Stack Interfaces

A component-based operating system and protocol stack already enables one possibility to treat an application. It is just another component that can directly interact with other components using whatever interface specification exists between them (e.g., the command/event structure of TinyOS). The application could even consist of several components, integrated at various places into the protocol stack. This approach has several advantages: It is streamlined with the overall protocol structure, makes it easy to introduce application-specific code into the WSN at various levels, and does not require the definition of an abstract, specific service interface. Moreover, such a tight integration allows the application programmer a very fine-grained control over which protocols (which components) are chosen for a specific task;

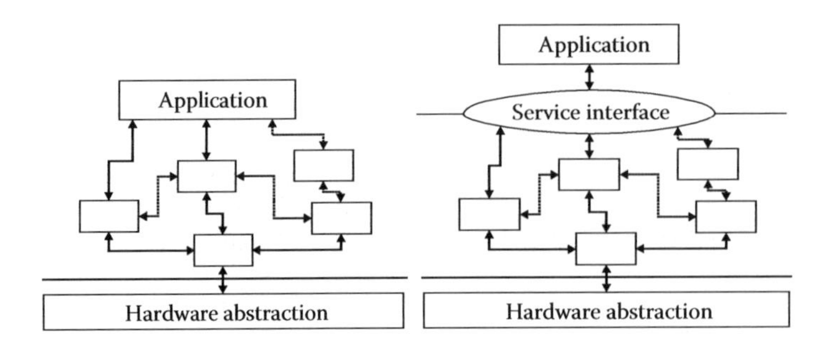

FIGURE 3.4
Two options for interfacing an application to a protocol stack.

for example, it is possible to select out of different routing protocols the one best suited for a given application by accessing this component's services. But this generality and flexibility is also the potential downside of this approach. In traditional networks such as the Internet, the application programmer can access the services of the network via a commonly accepted interface: sockets. This interface makes clear provisions on how to handle connections, how to send and receive packets, and how to inquire about state information of the network. This clarity is owing to the evident tasks that this interface serves—the exchange of packets with one (sometimes, several) communication peers. Therefore, there is the design choice between treating the application as just another component and designing a service interface that makes all components, in their entirety, accessible in a standardized fashion. These two options are outlined by Figure 3.4. A service interface would allow raising the level of abstraction with which an application can interact with the WSN, instead of having to specify which value to read from which particular sensor, it might be desirable to provide an application with the possibility to express sensing tasks in terms that are close to the semantics of the application. In this sense, such a service interface can hide considerable complexity and is actually conceivable as a "middleware" in its own right. Clearly, with a tighter integration of the application into the protocol stack, a broader optimization spectrum is open to the application programmer. On the downside, more experience will be necessary than when using a standardized service interface. The question is therefore on the one hand the price of standardization with respect to the potential loss of performance and on the other hand, the complexity of the service interfaces. In fact, the much bigger complexity and variety of communication patterns in WSNs compared to Internet networks makes a more expressive and potentially complex service interface necessary. To better understand this trade-off, a clearer understanding of expressibility requirements of such an interface is necessary.

3.1.3.11 Gateway Concepts

3.1.3.11.1 The Need for Gateways

For practical deployment, a sensor network only concerned with itself is insufficient. The network rather has to be able to interact with other information devices, for example, a user equipped with a PDA moving in the coverage area of the network or with a remote user, trying to interact with the sensor network via the Internet (the standard example is to read the temperature sensors in one's home while traveling and accessing the Internet via a wireless connection). Figure 3.5 shows this networking scenario. To this end, the WSN first of all has to be able to exchange data with such a mobile device or with some sort of gateway, which provides the physical connection to the Internet. This is relatively straightforward on the

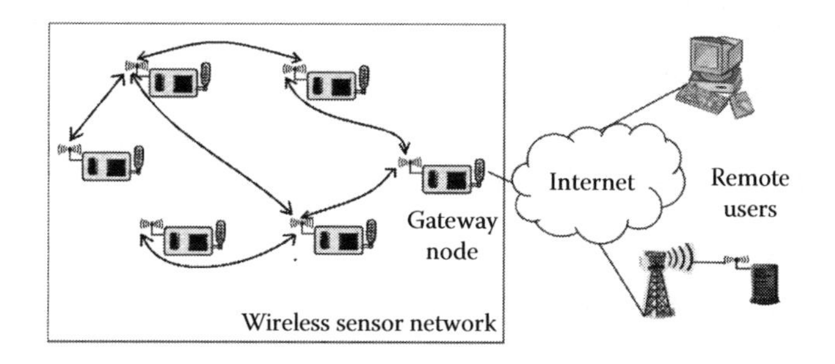

FIGURE 3.5
A wireless sensor network with gateway node, enabling access to remote clients via the Internet.

physical, MAC, and link layer; either the mobile device/the gateway is equipped with a radio transceiver as used in the WSN, or some (probably not all) nodes in the WSN support standard wireless communication technologies such as IEEE 802.11. Either option can be advantageous, depending on the application and the typical use case. Possible trade-offs include the percentage of multitechnology sensor nodes that would be required to serve mobile users in comparison with the overhead and inconvenience to fit WSN transceivers to mobile devices like PDAs. The design of gateways becomes much more challenging when considering their logical design. One option to ponder is to regard a gateway as a simple router between Internet and sensor network. While this option has been considered as well and should not be disregarded lightly, it is the prevalent consensus that WSNs will require specific, heavily optimized protocols. Thus, simple routers will not suffice as a gateway. The remaining possibility is therefore to design the gateway as an actual application level gateway. On the basis of the application-level information, the gateway will have to decide its action. A rough distinction of the open problems are made according to from where the communication is initiated.

3.1.3.11.2 Communication Protocols

Most common architecture for WSN follows the OSI Model. Basically, in sensor network we need five layers: application layer, transport layer, network layer, data link layer, and physical layer. Added to the five layers are the three cross layer planes as shown in Figure 3.6 [9].

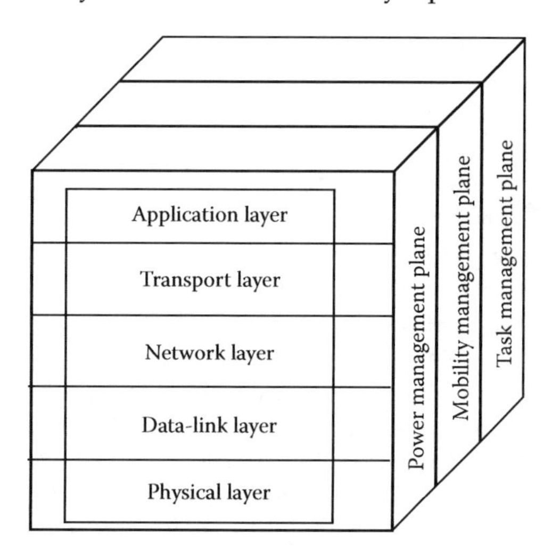

FIGURE 3.6
WSN architecture.

TABLE 3.2

Difference of Architectures between OSI, WLAN, and WSN

WSN	WLAN	OSI
Application	Application programs	Application layer
Middleware	Middleware, socket API	Presentation layer/session layer
Transport protocols	TCP/UDP	Transport layer
Routing protocols	IP	Network layer
Error control and MAC	WLAN adapter/device drivers	Data link layer
Transceiver	Transceiver	Physical layer

3.1.3.11.3 Cross Layers

The three cross planes or layers required in WSNs are: power management plane, mobility management plane, and task management plane. These layers are used to manage and maintain the network and make the sensor nodes to work collaboratively to increase the overall efficiency of the network [10]. The difference of architectures between OSI, WLAN, and WSN are shown in Table 3.2 [11].

Mobility management plane: detect sensor nodes movement. Node can keep track of neighbors and power levels (for power balancing). Task management plane: schedule the sensing tasks to a given area. Determine which nodes are off and which ones are on.

3.1.3.12 WSN OSI Layers

3.1.3.12.1 Transport Layer

The function of this layer is to provide node-to-node communication reliability and congestion avoidance schemes. The protocols designed for this layer provides this function and are applied on the upstream (user to sink, ex: ESRT, STCP, and DSTN), or downstream (sink to user, ex: PSFQ and GARUDA). These protocols use different mechanisms for packet loss detection (ACK, NACK, and Sequence number) and loss recovery (End to End or Hop by Hop). Providing a reliable hop-by-hop communication is more energy efficient than end-to-end delivery and that is one of the reason why TCP is not suitable for WSN. In general, Transport protocols are divided into

1. *Packet driven*: The packets sent by source must reach destination
2. *Event driven*: The event must be detected, but it is enough that one notification message reaches the sink. The following are some popular protocols in this layer with brief descriptions:
    a. *STCP* (*Sensor Transmission Control Protocol*): It is a upstream protocol; provides reliability, congestion detection and congestion avoidance. STCP function is applied at the base station. The node sends a session initiation packet to the sink that contains information about transmission rate, required reliability, and data flow. Then the sensor node waits for ACK before starting to send data. The base station estimates the arrival time of each packet; when there is a failure in packet delivery the base station checks whether the current reliability meets the required criteria. If current reliability is less than the required criteria, then sink sends NACK for retransmission, otherwise do nothing. The current reliability is computed by the packet fractions that are successfully received.

b. *PORT* (*Price-Oriented Reliable Transport Protocol*) [12,13]: This is a downstream protocol, assures that the sink receives enough information from the physical phenomena. PORT adapts a bias packet routing rate to increase sink information from specific region by two methods:
   i. *First method*: Node price is the total number of transmissions before the first packet arrives at the sink and this is used to define the cost of communication. Each packet is sent encapsulated with source price then the sink adjusts the reporting rate according to node price.
   ii. *Second method*: Use end-to-end communication cost to reduce congestion. When congestion occurs the communication cost is increased. The sink reduces the reporting rate for sources and increases the rate of other sources that have lower communication cost.

c. *PSFQ* (*pump slow fetch quick*): This is a downstream protocol, reliable, scalable, and robust. The three functions in this protocol are pump, fetch, and report.
   i. Pump uses two timers T_{min} and T_{max}, where the node waits T_{min} before transmission, to recover missing packets and remove redundant broadcast. Node waits for T_{max} if there are any packets or multiple packets lost.
   ii. Fetch operation requests a retransmission for the missing packets from neighbor.
   iii. Finally report the operation to provide a feedback to the user.

3.1.3.12.2 Network Layer

The major task of this layer is routing and the major challenges are in the power saving, limited memory and buffers, sensor does not have a global ID and should have self-organizing capabilities. There are different routing protocols designed for this layer, which can be divided as flat routing (e.g., direct diffusion) and hierarchal routing (e.g., LEACH) or can also be divided into time-driven, query-driven, and event-driven protocols. In time-driven protocol, the data are sent at specific times and in an event-driven and query-driven protocols, the sensor responds according to occurrence of event or request of user query [2,6].

Data aggregation and data fusion: In order to provide a full coverage for a certain area, even when there is a failure, redundant sensors are deployed. These redundant sensors provide a repeated data, in addition to sensors that are sending data on multi-hop style (from sensor to another till it reaches the sink) and sometimes as in case of sensors using flooding protocols. One node may receive a huge amount of repeated data from different neighbors and these data could be generated from the same origin node or even generated by redundant nodes. Since the data processing consumes less power than data transmission, we can solve the power problem by data aggregation and data fusion to remove the redundant data [6]. Data aggregation is described as "a set of automated methods combining the data that comes from many sensor nodes into a set of meaningful information and eliminate the duplication." This is basically used in flat routing [11]. Data fusion is described as "when the nodes do some more processing on the aggregated data to produce more accurate output for example reducing the noise in the signals" [11].

Data centric routing protocols: The first type of data centric protocols are SPIN and directed diffusion, but before that we have flooding and gossiping protocols [11], where each node receives data, re-broadcasts it again to all nodes. This re-broadcast causes two problems; implosion, that is, duplicate messages sent to the same node, and overlap, when two nodes sensing the same region, will send the same message to the same neighbor. Spin: broadcast ADV message to advertise for data availability, where the only interested node sends a REQ to receive the data, then the transmission starts. Direct Diffusion [11]: where the sink broadcasts a query, then certain node replies with the data by broadcasting it to the neighbors, the sink then chooses the best path and forces others to turn off, but if the current path is no longer efficient then the sink sends a negative reinforcement to reduce the rate or implement time out.

Hierarchy protocol: Under this type of routing protocols there are large number of suggested protocols for routing and considering the power consumption problem at the same time. For example, PEAS (Probing Environment and Adaptive Sleeping), GAF, SPAN, ASCENT, AFECA, CLD (Controlled Layer Protocol), MTE (Minimum Transmission Energy), LEACH (The Low-Energy Adaptive Clustering Hierarchy). All of these protocols solve routing and energy problems by using clustering and distributing methods. The most popular hierarchy routing protocol is LEACH [4,11]. This divides the network into clusters and randomly selects the cluster head for it to do the routing job from cluster to the sink after carrying out data aggregation.

3.1.3.12.3 Data Link Layer

The protocols working in this layer are responsible for multiplexing data streams, data frame detection, MAC, and error control, ensure reliability of point–point or point–multipoint transmission. Errors or unreliability comes from [12,13]:

- Co-channel interference at the MAC layer and this problem is solved by MAC protocols
- Multipath fading and shadowing at the physical layer and this problem is solved by forward error correction (FEC) and automatic repeat request (ARQ)

  ARQ: not popular in WSN because of additional re-transmission cost and overhead. ARQ is not efficient to frame error detection so all the frame has to retransmit if there is a single bit error [14].

  FEC: decreases the number of retransmission by adding redundant data on each message so the receiver can detect and correct errors. By that we can avoid re-transmission and wait for ACK [15].

  MAC layer: This layer is responsible for channel access, scheduling of events, buffer management, and error control [16]. In WSN we need a MAC protocol design to address energy efficiency, reliability, low access delay, and high throughput as major factors.

3.1.3.12.4 Physical Layer

The physical layer is mostly concerned with modulation and demodulation of digital data. This task is carried out by the so-called transceivers. In sensor networks, the challenge is to find modulation schemes and transceiver architectures that are simple, low cost, but still robust enough to provide the desired service.

3.1.3.12.5 Application Layer

This layer is responsible to transform the data into a standard format understandable by all the platforms for different applications and also send queries to obtain certain information. Sensor networks are used in various applications like military, medical, environment, agriculture fields [9,13].

3.2 Integration of Embedded System and CPS

3.2.1 Introduction

Cyber-physical system (CPS) is an integration of computation programs with the physical processes. Embedded computers and networks monitor and control the physical processes and these effect the computations. In today's computing and networking abstractions, concurrency and time are not addressed and hence there is a scope for research and investment in CPS. There are technical approaches that bridge the real-time operating systems, middleware technologies, embedded processor architectures, and networks partially and hence there is considerable room for improvement of these technologies.

Many of the examples are using a structure as shown in Figure 3.7. There are three main parts in this figure. First, the physical plant is the "physical" part, comprising the mechanical parts or biological or chemical processes, or human operators. Second, the computation part includes sensors, actuators, and different OS. Third, the network part dictates how to compute and communicate. In the figure, Platform 2 controls the physical part via Actuator 1. It measures the processes in the physical plant using Sensor 2. The box labeled Computation 2 implements a control unit, which determines the commands to issue to the actuator based on the data sent by the sensors. Platform 1 makes additional measurements using Sensor 1, and sends messages to Platform 2 via the network fabric. Computation 3 realizes an additional control law, which is merged with that of Computation 2, possibly pre-empting it [17,18].

3.2.2 Applications of CPS

The CPSs have made way in medical equipment, traffic control systems, automotive systems, process control in mechanical industry, energy preservation, environmental monitoring, avionics, instrumentation, critical infrastructure control (e.g., electric power, water resources, and communications systems), distributed robotics (telepresence,

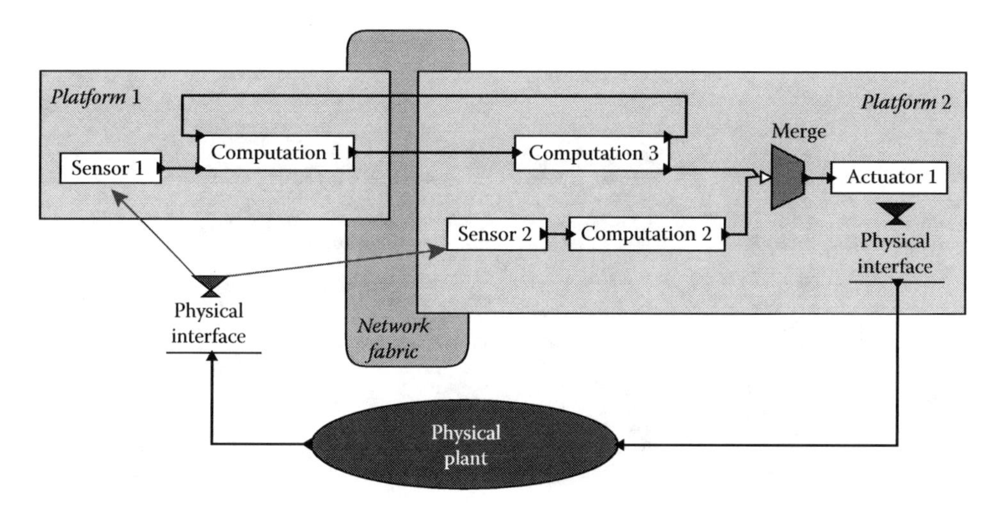

FIGURE 3.7
Structure of cyber-physical system.

telemedicine), defense systems, manufacturing, and smart structures. Embedded intelligence in automobiles improves safety and efficiency in modern transportation systems. Networked autonomous vehicles help disaster recovery techniques and in military applications. In communications, cognitive radio could gain enormously from distributed unanimity about currently available bandwidth and from distributed control technologies. Distributed real-time games that integrate sensors and actuators could change the (relatively passive) nature of online social interactions. Many of these applications may not be achievable without remarkable changes in the core abstractions [19].

3.2.3 Integration of Cyber-Physical System with Embedded Systems

Most of the embedded applications like communication systems, aircraft control systems, automotive electronics, home appliances, games, and toys have integration of physical processes with computing. In all these applications the computing details are not known. The embedded software is subjected to benchmark testing before integrating in the network. If the functionality of the networked system fails the entire software is rewritten. These kind of challenges exist still in embedded systems. The most widely used networking techniques today introduce a great deal of timing variability and stochastic behavior. Embedded computing also exploits concurrency models other than threads. For the next generation of CPSs, it is required to build concurrent models of computation that are far more deterministic, predictable, and understandable [20].

3.2.4 Motivating Example

In this section, we describe a motivating example of a CPS. Our goal is to use this example to illustrate the importance of the breadth of topics covered in this text. The specific application is the Stanford testbed of autonomous rotorcraft for multi-agent control (STARMAC), developed by Claire Tomlin and colleagues as a cooperative effort at Stanford and Berkeley (Hoffmann et al., 2004). The STARMAC is a small quadrotor aircraft; it is shown in flight in Figure 3.8. Its primary purpose is to serve as a testbed for experimenting with multivehicle autonomous control techniques [19].

FIGURE 3.8
The STARMAC quadrotor aircraft in flight.

3.3 Smart Structural Monitoring and Control Using Sensors, Actuators, and ICT Technologies

3.3.1 Introduction

Recent advances in wireless communication, and in embedded computing, have opened new opportunities for WSNs. Among them is smart structural technology, an active research area that promises enhancing infrastructure management and safety. A smart structure means specially equipped structure (e.g., buildings, bridges, dams, etc.) that can monitor and react to event occurring in the surrounding environment and the structure's own conditions, and provide information of interest.

Smart structural technology has two fields: structural health monitoring (SHM) and structural control. A SHM system measures structural responses and predicts, identifies, and locates the onset of structural damage, examples are

- Damage due to deterioration
- Damage due to hazardous events
- Structural sensors collect important information buildings

On the other hand, structural control technology aims to mitigate adverse effects in real time due to excessive dynamic loads. The traditional wired systems have been replaced by wireless sensors, which have their own advantages and disadvantages. The purpose of this chapter is to bring out the important issues and metrics for adopting WSNs in smart structural systems. In a SHM system, passive sensors are used for measuring structural responses. Structural control systems, on the other hand, need to respond in real time to mitigate excess dynamic response of structures. However, the wireless systems have limited bandwidth, latency, communication delay, and energy. These constraints and reliability need to be considered carefully using an integrated system approach and testing the prototype. The hardware and software components of such a CPS need to be carefully designed [21].

3.3.2 Design and Implementation of a Prototype Wireless Sensing and Control Unit

The prototype wireless unit is designed in such a way that the unit can serve as either a sensing unit (that collects data from sensors and wirelessly transmits the data), a control unit (that calculates optimal control decisions and commands control devices), or a unit for both sensing and control. Figure 3.9 shows the functional diagram of the prototype wireless sensing and control unit. The wireless sensing unit shown in the top part of Figure 3.9 serves as the core component, with which off-board modules for signal conditioning and signal generation can be easily incorporated [22].

3.3.3 Wireless Structural Health Monitoring

The prototype wireless unit is first investigated for applications in wireless structural health. This section provides an overview to the wireless SHM system, and then introduces the communication protocol design for reliable data management in the prototype system. A large-scale field deployment of the wireless SHM system is summarized at the end of the section.

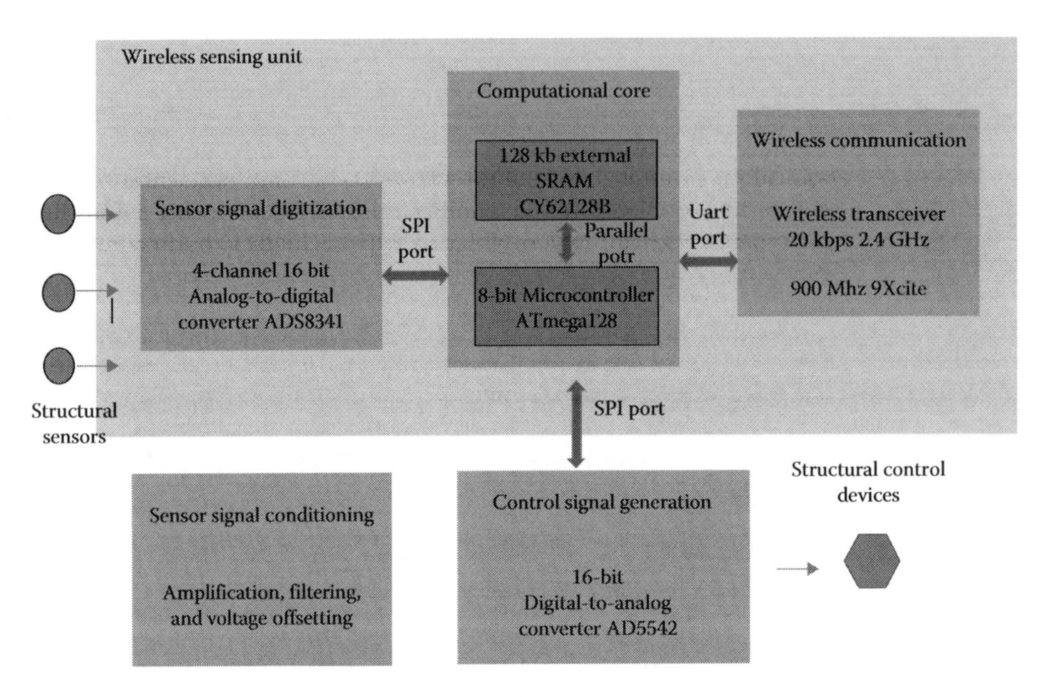

FIGURE 3.9
Functional diagram detailing the hardware design of the wireless sensing unit.

3.3.4 Overview of the Wireless Structural Health Monitoring System

A simple star-topology network is adopted for the prototype wireless sensing system. The system includes a server and multiple structural sensors, signal conditioning modules, and wireless sensing units as shown in Figure 3.10. The server is used to organize and collect data from multiple wireless sensing units in the sensor network. The server is responsible for: (1) commanding all the corresponding wireless sensing units to perform data collection or interrogation tasks, (2) synchronizing the internal clocks of the wireless sensing units, (3) receiving data or analysis results from the wireless network, and (4) storing the data or results. Any desktop or laptop computer connected with a compatible wireless transceiver can be used as the server. The server can also provide Internet connectivity so that sensor data or analysis results can be viewed remotely from other computers over the Internet. Since the server and the wireless sensing units must communicate frequently with each other, portions of their software are designed in tandem to allow seamless integration and coordination. At the beginning of each wireless structural sensing operation, the server issues commands to all the units, informing the units to restart and synchronize. After the server confirms that all the wireless sensing units have restarted successfully, the server queries the units one by one for the data they have thus far collected. Before the wireless sensing unit is queried for its data, the data are temporarily stored in the unit's onboard SRAM memory buffer.

A unique feature of the embedded wireless sensing unit software is that it can continue collecting data from interfaced sensors in real time as the wireless sensing unit is transmitting data to the server. In its current implementation, at each instant in time, the server can only communicate with one wireless sensing unit. In order to achieve real-time continuous data collection from multiple wireless sensing units with each unit having up to four analog sensors attached, a dual stack approach has been implemented to manage

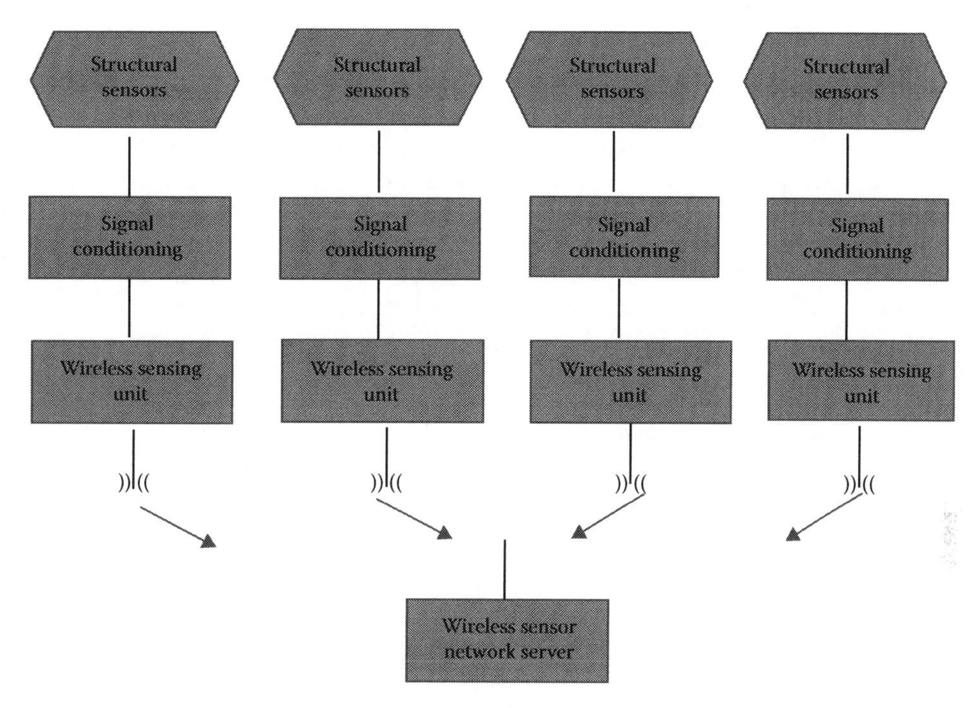

FIGURE 3.10
An overview of the prototype wireless structural sensing system.

the SRAM memory. When a wireless sensing unit starts collecting data, the embedded software establishes two memory stacks dedicated to each sensing channel for storing the sensor data. For each sensing channel, at any point in time, only one of the stacks is used to store the incoming data stream. While incoming data are being stored into the dedicated memory stack, the system transfers the data in the other stack out to the server. For each sensing channel, the role of the two memory stacks alternate as soon as one stack is filled with newly collected data.

3.3.5 Communication Design of the Wireless Structural Health Monitoring System

To ensure reliable wireless communication between the server and the wireless units, the communication protocol needs to be carefully designed and implemented. TCP is not suitable for this type of communication. Latency of packet transmission is another drawback in the wireless system, therefore a protocol need to be designed that addresses the problem of data loss. The protocol designed here addresses data packetizing, sequence numbering, timeout checking, and retransmission just like in TCP. Finite state machine concepts are used in designing the communication protocol for the wireless sensing units and the server. During each round of data collection, the server collects sensor data from all of the wireless units for the computation and communication.

3.3.6 Motivating Example

Another example in the field of wireless sensing is ICU monitoring system, which is based on KGSAN architecture invented by Dr. K.G. Srinivasa, Abhishek, and Nabeel patented under Indian Patent Act 1970 bearing patent number 2271/CHE/2014.

3.3.7 Medical (ICU Monitoring)

Medicine field today is a combination of traditional science and technology. Intensive Care Units are the most modernized and technologically advanced centers in a hospital. These centers require continuous monitoring in terms of humidity, temperature, luminosity, pressure, and air quality. These parameters effect the functioning of the technologically advanced machines like Electrocardiogram, which monitors heart rate. In a hospital facility, usually a number of people are involved in ICU monitoring and they need live feed of data to keep track of events. With this prototypic deployment, we centralize the process of monitoring by sending regular updates of these parameters on the personal mobiles of hospital employees concerned with ICU. We propose to add graphs, histograms to demark changes in the data collected by the sensors and thus adding a visual attribute to the process. The graphs can be plotted as shown in Figure 3.11 keeping Y-axis as the parameter and X-axis to keep track of time interval [23].

The architecture followed by MED SENSE Application is KGSAN, which itself is an integration of embedded systems and sensing control units. This five part architecture is depicted in Figure 3.12.

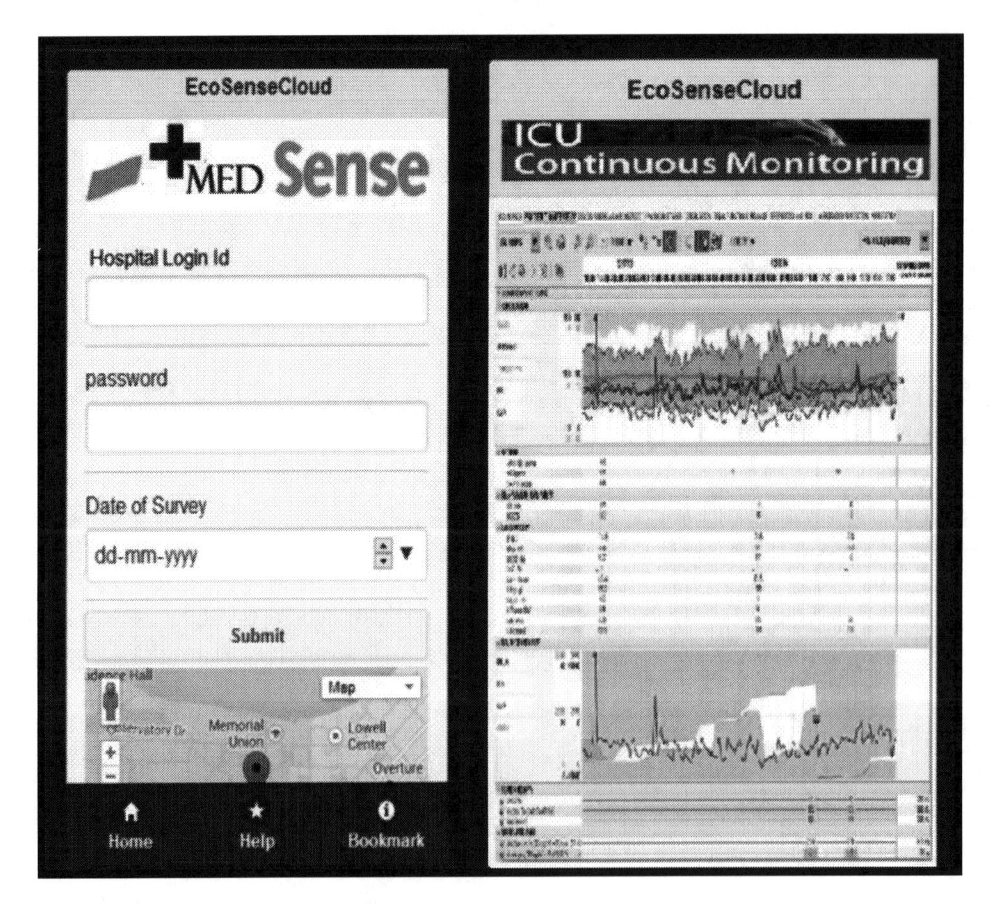

FIGURE 3.11
ICU monitoring handheld application.

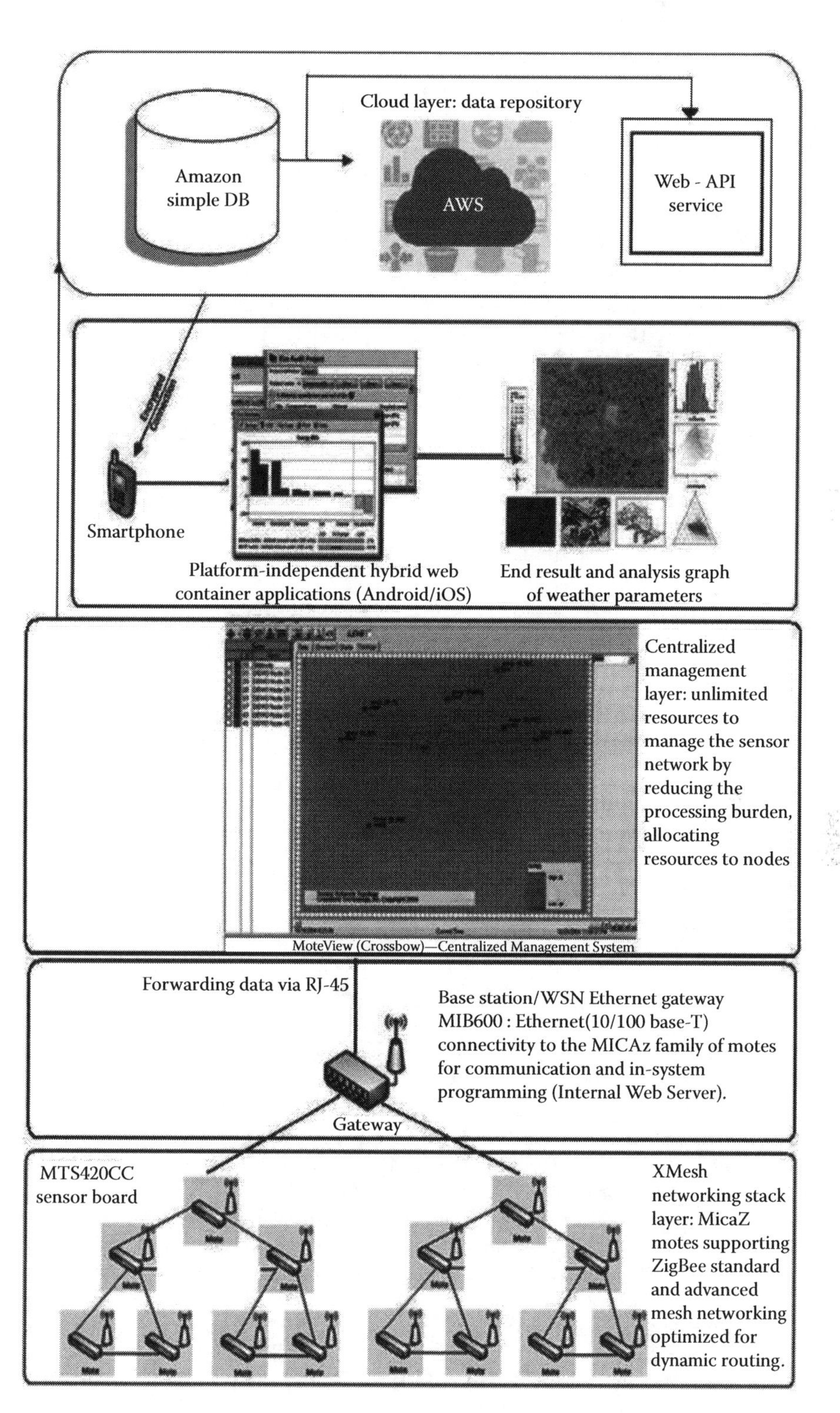

FIGURE 3.12
KGSAN architecture.

The functions of these five layers are as follows:

Application Layer: Accessing weather data from the cloud layer through an encrypted connection between the new age portable devices and the generic API to extract the information and use it to generate the end result and analysis graph of weather parameter.

Centralized Management System Layer: Displaying graphical analysis of real-time prediction of weather data in the centralized management layer to the user.

Cloud Layer: Developing a web-based API (application program interface), which provides the user with an abstraction of the ontology to make cross platform interoperable applications and centrally storing information of the collected data.

WSN Ethernet Gateway Layer: Transmitting the real-time prediction of weather data from the X-mesh networking stack layer to a centralized management layer through a WSN Ethernet gateway layer.

Xmesh Networking Stack Layer: Gathering the real-time prediction of weather data in a X-mesh networking stack layer.

3.4 Summary

In this chapter, various issues of applying WSNs to modern smart structural technologies, including SHM and structural control, are discussed. A smart structural system can be built with wireless sensor units and embedded computing. The design concepts proposed here address the information constraints in a WSN, such as bandwidth, latency, range, and reliability. Communication protocol design for centralized and decentralized information architectures using state machine concepts is proposed for managing the information flow in the wireless network. The laboratory and field validation tests have been conducted to validate the efficacy and robustness of the information management schemes implemented in the wireless structural monitoring and control system.

References

1. M. Martino and J. Varley, A Wireless Sensor Node Powered by a PV/Supercapacitor/Battery Trio. University of Toronto, Toronto, 2012. Available at: www.ti.com/corp/docs/university/docs/University_of_Toronto_Wireless_Sensor_Node_MatthewMartino_JordanVarle_y.pdf
2. JiZhong Zhao, Wei Xi, Yuan He, YunHao Liu, Xiang-Yang Li, LuFeng Mo, and Zheng Yang, Localization of Wireless Sensor Networks in the Wild: Pursuit of Ranging Quality. IEEE/ACM Transactions on Networking, 2012. Available at: https://iit.edu/arc/workshops/pdfs/SENSORS.pdf.
3. K. Holger and W. Andreas, *Protocols and Architectures for Wireless Sensor Networks*, John Wiley & Sons, New York, 2005.
4. P. Ciciriello, L. Mottola, and G. P. Picco, Efficient routing from multiple sources to multiple sinks in wireless sensor networks, in *Wireless Sensor Networks*, Lecture Notes in Computer Science, vol. 4373, 2007, pp. 34–50.

5. B. Bhuyan, H. Sarma, N. Sarma, A. Kar, and R. Mall, Quality of Service (QoS) provisions in wireless sensor networks and related challenges, *Wireless Sensor Network*, 2, 2010, 861–868.
6. W.K.G. Seah and H. X. Tan, Quality of Service in Mobile Ad Hoc Networks, IGI Global National University of Singapore, Singapore, 2008.
7. C. Fischione et al., Design principles of wireless sensor networks protocols for control applications, in S. K. Mazumder, ed., *Wireless Networking Based Control*, Springer, New York, 2011, pp. 203–238.
8. I. F. Akyildiz, W. Su, Y. Sankarasubramaniam, and E. Cyirci, Wireless sensor networks: A survey, *Computer Networks*, 38(4), 2002, 393–422.
9. R. Kazi, A survey on sensor network, *Journal of Convergence Information Technology* 1.1, 2010.
10. K. Mauri, M. Hännikäinen, and T. Hämäläinen, A survey of application distribution in wireless sensor networks, *EURASIP Journal on Wireless Communications and Networking*, 2005, 774–788.
11. P. R. Pereira, A. Grilo, F. Rocha, M. S. Nunes, A. Casaca, C. Chaudet, P. Almström, and M. Johansson, End-to-end reliability in wireless sensor networks: Survey and research challenges, in P. Pereira, ed., *EuroFGI Workshop on IP QoS and Traffic Control*, December 6–7, 2007, Lisbon, Portugal, vol. 54, pp. 67–74.
12. J. Yick, M. Biswanath, and D. Ghosal, Wireless sensor network survey, *Computer Networks*, 52(12), 2008, 2292–2330.
13. I. F. Akyildiz, T. Melodia, and K. R. Chowdhury, A survey on wireless multimedia sensor networks, *Computer Networks*, 51(4), 2007, 921–960.
14. A. Koubaa, M. Alves, and E. Tovar, Lower protocol layers for wireless sensor networks: A survey, IPP-HURRAY Technical Report, HURRAY-TR-051101, November 2005.
15. A. A. A. Alkhatib and G. S. Baicher, MAC layer overview for wireless sensor networks, in *2012 International Conference on Computer Networks and Communication Systems (CNCS) 2012*, April 7–8, 2012, Kuala Lumpur, Malaysia, pp. 16–19.
16. Foundations for Innovation in Cyber-Physical Systems, Prepared by Energetics Incorporated Columbia, Maryland, National Institute of Standards and Technology, January 2013. Available at: Web. www.nist.gov/el/upload/CPS-WorkshopReport-1-30-13-Final.pdf.
17. E. A. Lee, Cyber-physical systems-are computing foundations adequate, in *Position Paper for NSF Workshop on Cyber-Physical Systems: Research Motivation, Techniques and Roadmap*, October 16–17, 2006, Austin, TX, vol. 2.
18. E. A. Lee and S. A. Seshia, *Introduction to Embedded Systems: A Cyber-Physical Systems Approach*, 2011.
19. P. Marwedel, *Embedded System Foundations of Cyber-Physical Systems*, Springer, New York, 2011.
20. E. Oruklu et al., Smart and sustainable wireless sensor networks for structural health monitoring, Illinois Institute of Technology, Illinois, 2009.
21. B. F. Spencer Jr., M. Ruiz-Sandoval, and N. Kurata, Smart sensing technology for structural health monitoring, in *Proceedings of the 13th World Conference on Earthquake Engineering*, August 1–6, 2004, Vancover, British Columbia, Canada, paper no. 1791.
22. Y. Wang and K. H. Law, Wireless sensor networks in smart structural technologies, in J.-C. Chin, ed., *Recent Advances in Wireless Communications and Networks*, INTECH Open Access Publisher, Rijeka, Croatia, 2011, pp. 405–434.
23. ParaSense - A Sensor Integrated Internet of things Prototype for Real Time Monitoring Applications by Srinivasa K G, Nabeel Siddiqui and Abhishek Kumar, IEEE Tensymp, June 2016. Available at: Web. projectwsn.weebly.com/uploads/2/6/9/0/26900759/final_report.pdf.

4

Wireless Sensor Networks toward Cyber-Physical Systems

Tareq Alhmiedat

CONTENTS

4.1 Introduction

Cyber-physical systems (CPSs) can be found in diverse areas, including aerospace, automobiles, energy, health care, transportation, and consumer appliances. Common applications of CPS normally fall under sensor-based communications. Recent development in wireless technology and the invention of small, inexpensive, low-power microprocessors have led to the emergence of networked, embedded systems named wireless sensor networks (WSNs). A sensor network is an infrastructure consisting of sensing, computing, power-source, and communication elements, which offer the ability to observe, instrument, and react to events and phenomena in a specified environment. WSNs have emerged as an essential technology for monitoring and exploring remote, hostile, and hazardous environments. WSNs may be used in a variety of everyday activities or services. For instance, monitoring is a common application of WSNs, in which sensor nodes are distributed over an area of interest in order to monitor some phenomena. A practical use of WSNs could be found in the military in detecting intrusion (of soldiers, tanks, or vehicles). In this chapter, we explain the fundamental aspects of WSNs and cover WSN technology, applications, communication protocols, routing, clustering, and security.

4.2 Overview of Wireless Sensor Networks

Recently, a large number of research activities have been dedicated to the field of WSNs and the CPS has emerged as a promising track to enrich the interaction between physical and virtual worlds. WSNs have emerged as an essential technology for monitoring and exploring remote, hostile, and hazardous environments. Normally, a sensor network consists of sensor nodes capable of collecting information from the surrounding environment and communicating with each other via wireless transceivers. The collected sensed data will be delivered to one or more sink nodes through multi-hop communication. These sensor nodes are usually expected to operate with batteries and are often installed to not-easily-accessible or hostile environment. Therefore, it can be difficult or impossible to replace the batteries of sensor nodes.

As presented in Figure 4.1, with sensor networks technology, sensor devices can be deployed close to the phenomenon that must be observed. Since the sensor nodes are physically small, battery operated, and contain a wireless radio, deploying such a WSN disturbs the environment minimally and reduces the installation and maintenance costs. Furthermore, the inexpensive nature of these devices attracts scientists to place a high-resolution node grid in the field and obtain frequent measurements, providing an extremely rich data set.

A WSN is a collection of sensor nodes organized into a cooperative network. Sensor nodes are small in size, low in cost, and have short communication range. Usually, a sensor node consists of the following four subsystems (Figure 4.2):

1. *A computing subsystem*: Responsible for main functions such as execution of the communication protocols and control of onboard sensors.
2. *A sensing subsystem*: Environment characteristics are sensed through a wide range of sensors (temperature, humidity, light, etc.).

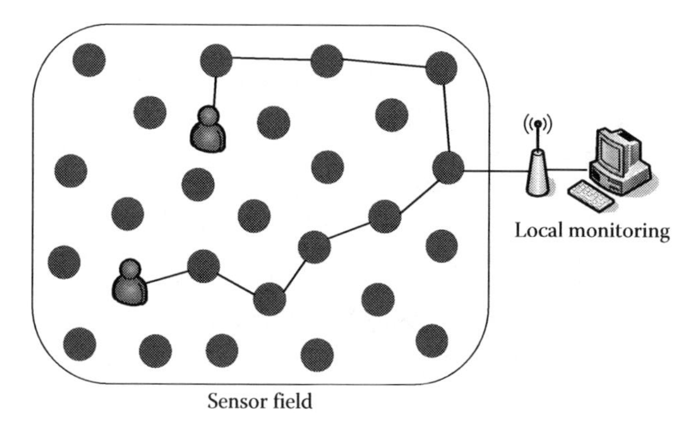

FIGURE 4.1
An example of WSN.

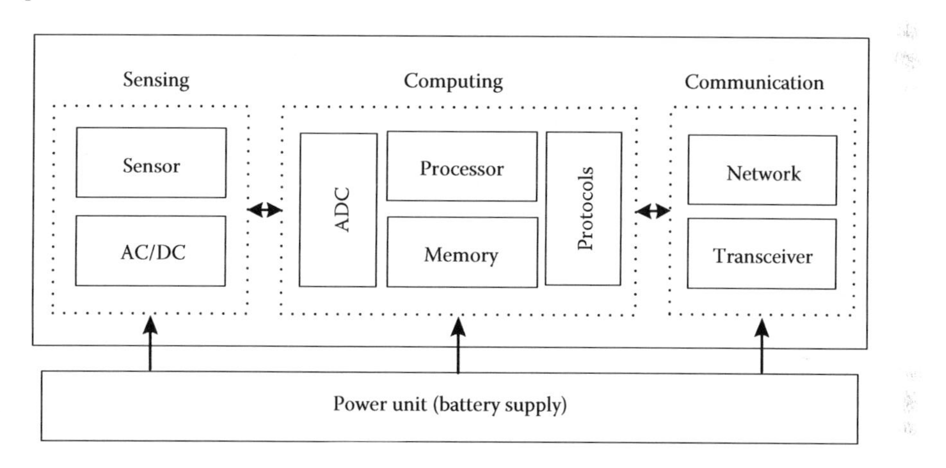

FIGURE 4.2
Sensor node architecture.

3. *A communication subsystem*: This includes a short radio range used to communicate with neighboring nodes.
4. *A power supply subsystem*: This includes a battery source that feeds computing, onboard sensors, and communication subsystems.

4.3 Applications of Wireless Sensor Networks

4.3.1 Range of Applications

WSNs are a collection of inexpensive and compact computational nodes that measure local environment conditions (such as humidity, temperature, gas level, etc.) and forward such information to a central point for processing. WSNs can sense the environment through onboard sensors, communicate with neighboring nodes through transceiver, and perform basic computations on the data being collected. WSNs support a wide range of valuable applications.

4.3.1.1 Military Application

WSNs have become an efficient tool for military applications, including intrusion detection, perimeter monitoring, and smart logistic support in an unknown area. The work presented in [1] paid attention to the secure communication in WSNs. A survey on threat detection and tracking for military applications is presented in [2]. The work presented in [3] includes a sensor network architecture based on cluster-tree multi-hop model with an optimized cluster head election and the corresponding node design to meet the tactical requirements. A soldier sensor system was proposed by DARPA [4], which aimed to provide a report about any events the soldiers may encounter during their missions, in addition to using the reported information for future operations and missions. The U.S. Army's Future Combat System use the unattended ground sensor nodes to detect, locate, and identify enemies. These sensor nodes can be deployed by hand or randomly using airplane or vehicles [5–8].

4.3.1.2 Industrial Application

Due to the distributed nature of WSN, it has a great potential in industrial monitoring and control applications, which demand real-time data collection or detection of exceptional events. For instance, sensor nodes can be used to measure a number of physical phenomena such as heat, vibration, noise, light, level of gas, etc. A number of existing WSN-based systems have been proposed and implemented, which target industrial applications, as in [9], which include developing and implementing a group of algorithms for wireless collaborative sensing and actuation. The proposed algorithms are validation for testing control of lighting or temperature in industrial applications. The work in [10] presents development of a distributed collaborative estimation and control approach for WSNs. This approach has been designed based on a locally collaborative control algorithm, which utilizes the collaboration between sensors and actuators. In many industrial applications, tracking and localization products, people, and items are significant tasks; the work presented in [11–15] targeted localization and tracking in WSNs.

4.3.1.3 Road Safety Application

WSNs present a breakthrough technology for traffic monitoring, which combines inexpensive tiny detectors with scalable self-configuring wireless sensors. In WSNs, the wireless connectivity reduces installation cost and maintenance with respect to the existing monitoring technologies and at the same time presents a much greater flexibility of deploying a large number of sensor nodes throughout the road network. Traffic sensors were integrated with new standards for wireless connectivity for vehicle-to-vehicle and vehicle-to-infrastructure communication [16–18]. The high spatial resolution of wireless sensor nodes for traffic monitoring systems can improve safety by making the road infrastructure capable to quickly recognize, warn, and react to dangerous situations such as accidents or traffic jams [19–22].

4.3.1.4 Home Automation Application

Recently, the scope of WSNs has been expanded to places such as home automation and remote control applications, in order to provide residents with diverse services. Wireless lightening can be easily accomplished with WSN in general and ZigBee technology in particular (e.g., controllable light switches and energy saving on bright days).

A feasible and practical home automation system based on ZigBee protocol was proposed in [23], in which hardware and software were presented. A ZigBee-based intelligent self-adjusting system was proposed in [24] to address the trade-off between performance and cost.

The use of WSNs in home automation presents attractive benefits; however, it introduces a number of technological challenges. The authors of [23] discussed the technological challenges in home automation systems. The work presented in [25] includes developing heterogeneous network architecture through designing and implementing a home gateway among IEEE 802.11 and ZigBee networks, in which the gateway facilitates the interoperability between a home automation and video surveillance network. A ZigBee-based home automation system is integrated with a Wi-Fi network through a common home gateway [26], in which the home gateway provides network interoperability, a simple and flexible user interface, and remote access to the system.

4.3.1.5 Medical Application

One of the most important applications of WSN technology is the medical applications. One of the critical WSN medical projects is the wireless body area sensor network, which consists of several physiological sensors (ECG, temperature, pressure, oximetry, etc.) scattered on the body and transmitting the collected measurements for sensors to personal terminal, which is used to collect, analyze, and visualize data [27]. A number of existing applications include support and rehabilitation [28], collecting and processing information from attached sensors and transmitting them to a remote location [29], and prevention of ulcers [30]. Another example of medical applications includes systems that cover a wide area, for instance, distributing sensor nodes through hospital in order to track medical stuff and equipment [31].

4.4 Medium Access Control Protocols for Wireless Sensor Networks

WSNs are essentially composed of a large number of sensor nodes deployed over a geographical area. In WSN applications, usually sensor nodes require to transmit the collected data from onboard sensors to the base station through multi-hop communication. The establishment of multi-hop wireless network infrastructure for data transfer necessitates establishing communication links between neighboring sensor nodes. In order to achieve this goal, a medium access control (MAC) protocol must be utilized [32].

4.4.1 Background

The choice of MAC protocol is the main determining factor in WSN performance. In WSN, communication among sensor nodes is usually achieved by means of a single channel. This is valid when a single sensor node requires transmitting a message at any given time. Therefore, shared access to the channel necessitates the establishment of MAC protocol in order to access the shared wireless medium regularly. Therefore, the main objective of MAC protocol is the regular access to the shared wireless medium. In open systems interconnection (OSI) model, the MAC protocol functionalities are offered by the lower sublayer of the data link layer (DLL), whereas the higher sublayer of the DLL is referred as the logical link control layer. Figure 4.3 presents the OSI reference model and the DLL logical

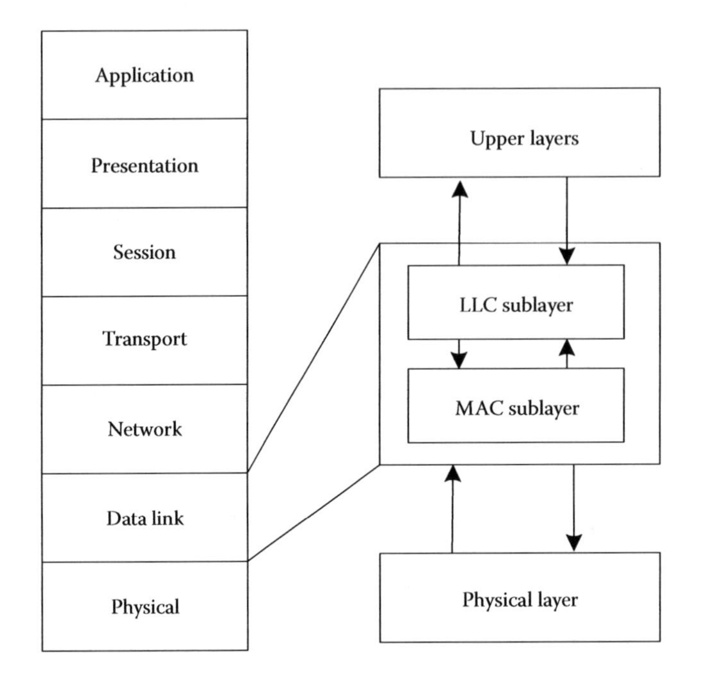

FIGURE 4.3
The OSI reference model and DLL architecture.

architecture for shared medium access in WSNs. The MAC sublayer is located above the physical layer, and it offers the following functionalities [32]:

- Assembly of data into a frame for transmission by adding a header field, which contains address information and a trailer field that supports error detection
- Disassembly of a received frame to extract addressing information and error control information
- And finally management of accessing the shared transmission medium

4.4.2 Fundamentals of MAC Protocols

The major difficulty in designing successful MAC protocol for shared access media arises from the spatial distribution of the communicating nodes. In order to reach an agreement as to which sensor node can access the communication channel at any certain time, the sensor nodes must exchange some coordinating information. There are two main factors affecting the aggregation behavior of a distributed multiple access protocols: the intelligence of the decision made by the access protocol and the overhead involved.

The performance of MAC protocols can be determined through a number of issues, including delay, throughput, robustness, scalability, and fairness. *Delay* refers to the total time required by a data packet in the MAC layer before it is transmitted successfully. *Throughput* is defined as the rate at which messages are serviced by a communication system. It is generally measured either in messages per second or in bits per second. *Robustness* includes reliability, availability, and dependability requirements and reflects the degree of the protocol insensitivity to errors. *Scalability* refers to the capability of a communication system to meet its performance characteristics apart from the number of nodes and the size of the network. *Stability* involves the ability of a communication system to handle variations of the traffic load over a continuous period of time. *Fairness* refers to distributing the allocated channel evenly among competing communication nodes without reducing the network throughput. And finally, *energy efficiency* is one of the most important issues in the design of MAC protocol, since sensor nodes are powered using batteries with small capacities [32].

There are a number of power consumption factors in the MAC protocol. First is collision, which occurs when any two sensor nodes try to transmit simultaneously. Second is the energy waste in idle listening, which occurs when the sensor node enters a mode when it is listening for traffic. Third is overhearing, which happens when a node receives a packet destined to another node. Fourth is control packet overhead, in which a number of control packets are required regularly to access the transmission channel. The final one is frequent switching between different operation modes, which results in high power consumption.

4.4.3 MAC Protocols for WSNs

Conservation of energy is the most significant issue in designing and implementing a MAC protocol for WSNs. As discussed earlier, a number of factors contribute to energy waste, and so the main goal of most MAC layer protocols in WSN is to reduce the energy waste caused by idle listening, collision, excessive overhead, and overhearing. MAC protocols can be categorized into two main groups: schedule-based MAC protocols and contention-based MAC protocols. Schedule-based MAC protocols are a class of deterministic protocols in which the access to the communication channel is based on a schedule. The channel access is limited to a single sensor node at a time. On the other hand, contention-based MAC protocols allocate a single radio channel, which is shared by all nodes and assigned on demand. More than one node attempting to access the communication channel will result in collision. The main goal of contention-based MAC protocols is to minimize the occurrence of collisions.

4.4.3.1 Schedule-Based Protocols

In schedule-based protocols, a schedule is required to regulate access to network resource in order to avoid contention between sensor nodes. The main goal of schedule-based protocols is to offer high level of energy efficiency in order to extend the WSN lifetime. Most of the existing schedule-based WSN protocols use a variant of a time division multiple access (TDMA) scheme in which the communication channel is divided into time slots, as presented in Figure 4.4. A set of N continuous slots—N is a system parameter—form a logical frame. The logical frame repeats regularly over time, in which each logical frame is assigned a set of specific time slots. The schedule can be either fixed, constructed on demand, or hybrid.

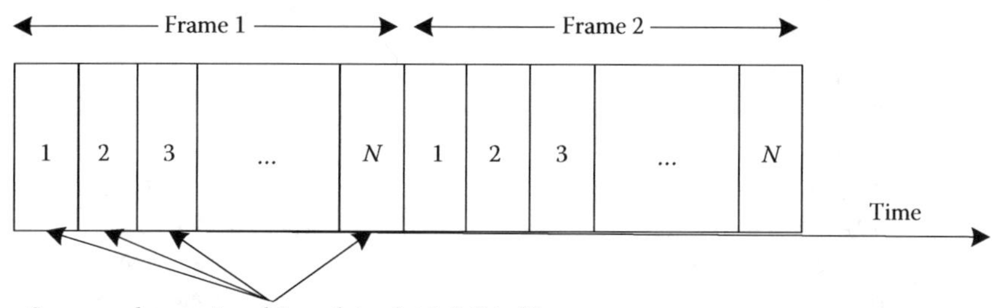

FIGURE 4.4
TDMA-based MAC protocol for WSNs.

The sensor node alternates between modes of operation (active mode and sleep mode) based on its assigned schedule. Through the active mode, the sensor node uses its assigned slots within a logical frame in order to transmit and receive data frames, whereas in the sleep mode, the sensor node switches their radio transceiver off to save energy. A diverse number of TDMA protocols have been proposed for MAC in WSNs, including [33,34]. A brief description of some of these protocols is as follows:

- *Self-organizing medium access control for sensor nets* (*SMACS*): The main goal of SMACS is to enable the formation of random network topologies with no need to establish global synchronization among the networks. A significant feature of SMACS is its use of a hybrid TDMA/FH method referred to nonsynchronous schedule communication to permit links to be established and scheduled concurrently throughout the network with no need for costly exchange of global connectivity information or time synchronization [35].
- *Low-energy adaptive clustering hierarchy* (*LEACH*): It intends to organize sensor nodes in the WSN into clusters (groups), in which a cluster head is selected for each cluster. The cluster head forward the messages received from the base station to its cluster nodes and vice versa. LEACH deploys TDMA to accomplish communication between sensor nodes and their cluster heads. In LEACH protocol, in order to prevent collisions among data messages, the cluster head establishes a TDMA schedule and broadcasts this schedule to all sensor nodes in its cluster. Each node determines the time slots in which it must be active using the received schedule from its cluster head. Therefore, this allows each cluster node, except the cluster head, to turn off their radio transceivers until its allocated time slots. In order to reduce the inter-cluster interference, LEACH adopts a transmitted-based code assignment scheme. Communications between sensor node and its cluster head are accomplished using direct-sequence spread spectrum, where every cluster is assigned a unique spreading code, which is employed by all sensor nodes to transmit their data to the cluster head [36].

4.4.3.2 Contention-Based Protocols

Contention-based protocols are also known as random access protocols, with no need for coordination among the sensor nodes accessing the channel. A collision occurs, and then the colliding sensor nodes back off for a random duration of a time before trying again to access the channel. These protocols are not well appropriate for WSN environments. The development of contention-based protocols includes collision avoidance as well as request-to-send (RTS) and clear-to-send (CTS) methods, which improve their performance and make them robust to the hidden terminal problem. However, the energy efficiency remains low due to collisions, overhearing, idle listening, and excessive control overhead. Large efforts in the design of random access MAC layer protocols focused on minimizing energy consumption to extend the WSN lifetime.

The power aware multicast protocol with signaling (PAMAS) avoids overhearing between neighboring sensor nodes through using a separate signaling channel [37]. The sparse topology and energy management protocol deals with latency for energy efficiency. This is accomplished by using two radio channels: a data radio channel and a wake up radio channel [38]. The Berkeley media access (B-MAC) is a low-power carrier-sense media access protocol for WSNs. The B-MAC protocol includes a small core of media access functionality, in which B-MAC employs clear channel assessment and packet back off for channel arbitration, listening for low power consumption, and link layer acknowledgments for reliability [39]. The timeout-MAC (T-MAC) is a connection-based MAC layer protocol developed for applications characterized by low message rate and low sensitivity to latency. Through T-MAC, sensor nodes use RTS, CTS, and acknowledgment packets to communicate with each other in order to avoid collision and guarantee reliable transmission [40].

4.4.4 Sensor MAC Protocol

The sensor MAC (S-MAC) protocol intends to minimize the energy waste caused by collision, control overhead, overhearing, and idle listening. Its main goal is to enhance energy efficiency while achieving a high level of stability and scalability. S-MAC protocol was designed to deal with a high number of sensor nodes with limited storage capacity, communication, and processing capabilities. In S-MAC protocol, sensor nodes are configured in an ad hoc manner, self-organized, and self-managed wireless network. Data produced by sensor nodes are processed and communicated in a store-and-forward way [32].

4.5 Routing in Wireless Sensor Networks

In WSN systems, data from individual sensor nodes must be sent to a central base station via multi-hop communication, often located from the sensor network, through which the end user can access the data. In order to deliver data efficiently to their destination in WSN systems with the minimum cost possible, an energy-efficient routing function is required in such applications. The research and development of WSN-based routing protocols was initially driven by the defense applications. According to [41], routing protocols are divided into four main categories (Figure 4.5): network structure, communication model, topology-based protocols, and reliable routing schemes. Each category is discussed as follows.

4.5.1 Network Structure

The network structure can be classified based on node uniformity. This category concerns with the way that the nodes are connected and information is routed based on the network's architecture. This addresses two types of node deployments: flat protocols and hierarchal protocols. The former includes that all nodes in the network play the same role, and this presents a number of advantages, including minimal overhead to maintain the infrastructure between communicating nodes [42–45], whereas the routing protocol on the hierarchal protocols imposes a structure on the network to accomplish energy efficiency, stability, and scalability. In this type of protocol, nodes are organized into clusters [46–49].

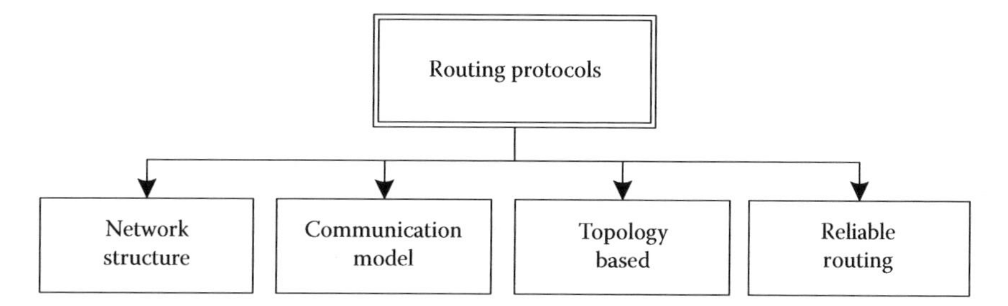

FIGURE 4.5
Categorization of routing protocols.

4.5.2 Communication Model

In the communication model, the main operation of routing protocol is followed in order to route packets in the network. The routing protocols falling under this category can deliver more data for a certain amount of energy. The problem with such protocols is that they do not have high delivery rate for the data that they send to a destination. The communication model protocols can be classified into three main categories: query-based protocols, coherent and noncoherent protocols, and negotiation-based protocols. In query-based protocols, the destination nodes propagate a query for data from a node through the network and the node that owns this data will send the data that match the query back to the node in which the query is initiated [50,51]. In coherent and noncoherent protocols, the data are forwarded to aggregators after a minimum processing in coherent routing, whereas nodes locally process the raw data before they are sent to other nodes in the WSN in noncoherent routing [38–40]. In negotiation-based protocols, a metadata negotiation is used to minimize the number of redundant transmissions in the WSN [52–54].

4.5.3 Topology-Based Protocols

Topology-based protocols adopt the principle in which every node in a WSN maintains topology information, and the main process of the protocol operation on the network topology. Topology-based protocols can be further classified into location-based and mobile-agent-based protocols. The former is based on the position information for each node in the WSN in order to forward the received data to only certain regions and not to the whole WSN. The routing protocols in this class find the path from the source to the destination and then reduce the energy consumption of the sensor nodes. The problems in such protocols are limited stability in case those sensor nodes are mobile, and each sensor node must identify the locations of other nodes in the WSN [55–58]. On the other hand, mobile-agent-based protocols are used to send data from the sensed area to the destination area. Mobile agent protocols provide the WSN with additional flexibility [59–61].

4.5.4 Reliable Routing

The protocols in this category are more flexible to route failures either by achieving load-balancing routes or by satisfying convinced quality of service (QoS) metrics. Reliable routing protocols are classified into multipath-based protocols and QoS-based protocols. Reliable routing protocols achieve load balancing and are more flexible to route failures. On the other hand, in QoS-based protocols, the network has to achieve balancing between energy consumption and data quality. Before transmitting a number of packets, the transmission process has to meet a specific level of quality [62–66].

In conclusion, routing protocols are categorized into four main categories. The flat protocols might be an efficient routing solution for small networks with static nodes; however, flat protocols are inefficient in large networks due to the link and processing overhead. The hierarchal protocols try to solve this issue and offer scalable and efficient solutions, in which the network is divided into clusters in order to maintain the energy consumption for sensor nodes in addition to perform data aggregation in order to minimize the number of transmitted messages through the WSN. Therefore, hierarchal protocols are more suitable for WSNs with heavy load and wide coverage area.

On the other hand, location-based protocols employ the location information in order to calculate the distance between sensor nodes and, therefore, reduce the total energy consumption for WSNs. The negotiation-based protocols offer efficient solutions for both point-to-point and broadcast networks; however, they do not guarantee the successful delivery of data. The multipath protocols offer several paths from nodes to sink, and this offers fault tolerance and easy recovery; however, they suffer from the overhead of maintaining the tables and states at each sensor node.

In the query-based routing protocols, the destination nodes send a query asking for the data through the network, and then the node having the data sends the requested data to the destination nodes. Query-based routing protocols are useful in dynamic network topologies, and they support multiple route replies. QoS routing protocols ensure optimized QoS metrics such as energy efficiency, delay bound, and low bandwidth consumption. The coherent-based routing protocols are an energy-efficient mechanism with the minimum processing at the sensor node, whereas in noncoherent routing protocols, the sensor nodes process the data locally at each sensor node and then send the processed data to other nodes for further processing.

4.6 Clustering Techniques in Wireless Sensor Networks

In WSNs, it is essential to collect the sensed data from static sensor nodes distributed over an area of interest. The collection process may be accomplished in an energy-efficient manner through a number of different approaches, including adopting mobile sinks to gather data from static nodes [67–69], using data aggregation method, deploying clustering algorithms, or deploying a system that integrates the previous approaches. In this section, we focus on clustering algorithms. The process of grouping sensor nodes into small groups is called clustering. Dividing sensor nodes into clusters has been widely pursued by the research community in order to reach the network scalability objective. Each cluster would have a leader, usually referred to as the cluster head, which may collect data from sensor nodes in its group, aggregate the gathered information, and transmit it to a central base station, as presented in Figure 4.6.

An extensive research effort has been made on the clustering techniques. The existing clustering algorithms can be divided into four main categories depending on the cluster formation criteria and parameters used for the selection of the cluster head. The four categories are identity-based clustering, neighborhood information-based clustering, probabilistic clustering, and biologically inspired clustering algorithms, as presented in Figure 4.7.

Identity-based clustering algorithms include the *linked cluster algorithm* (LCA) [70], in which each sensor node is assigned a unique ID number and has the possibility to become a cluster head in two ways: first, when a node has the highest ID number in the neighboring set, and second, when none of the neighboring nodes is a cluster head. LCA2 [71], an extension of LCA, was implemented to eliminate the election of redundant number of cluster heads. In LCA2 the number of cluster heads is less than the number in LCA. Both LCA and LCA2 are limited clustering algorithms for WSN, because both of them do not address the issue of limited energy of sensor nodes. In addition, both algorithms require clock synchronization with time complexity.

The *neighborhood information-based clustering algorithms* include the following algorithms: highest connectivity cluster algorithm, max-min D-cluster algorithm, weighted clustering algorithm (WCA), and clustering algorithm via waiting timer (CWAT). In the highest connectivity cluster algorithm [72], the sensor node broadcasts the number of neighboring

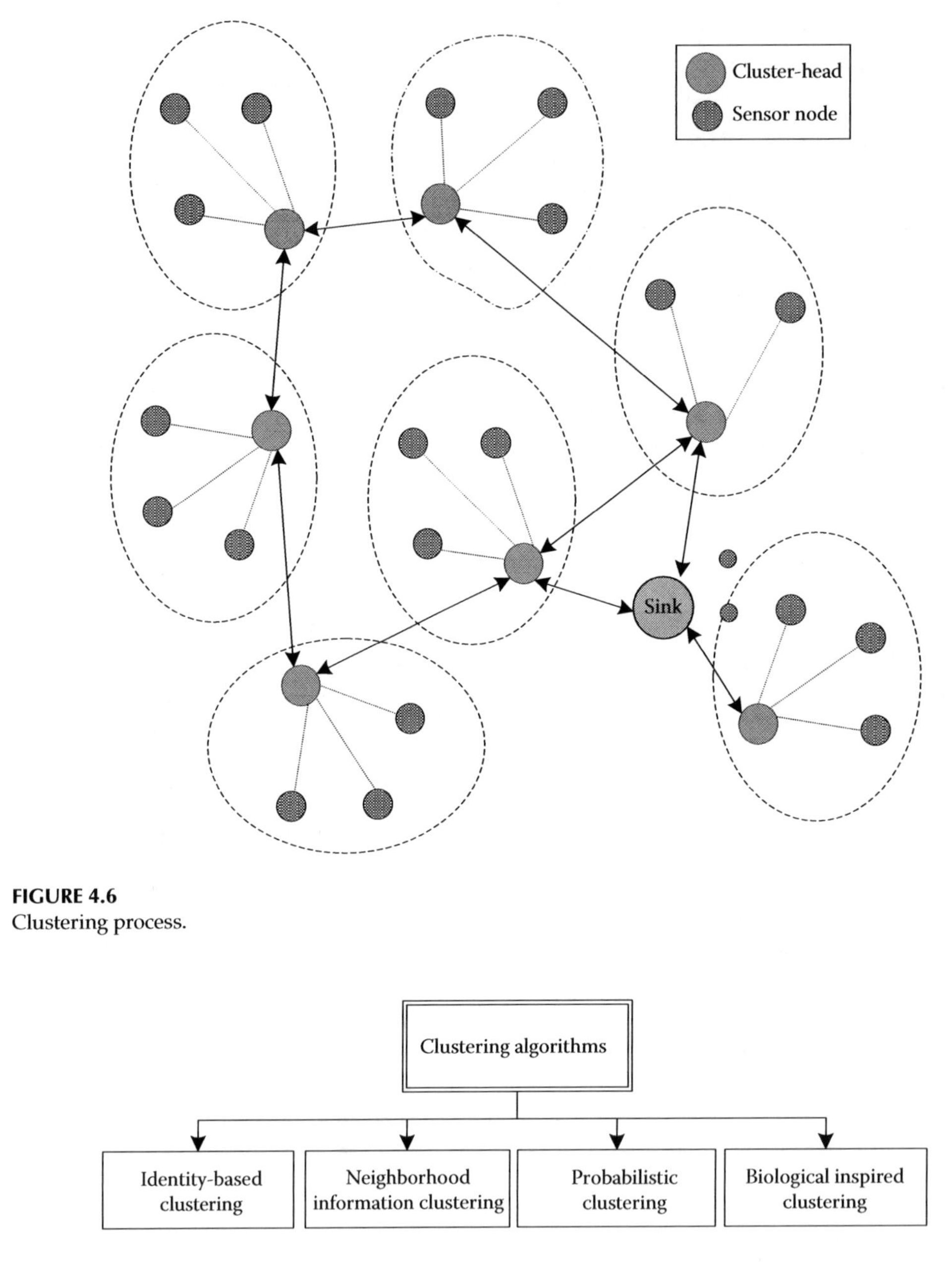

FIGURE 4.6
Clustering process.

FIGURE 4.7
Classification of clustering algorithms.

nodes to the surrounding nodes. The sensor node with the maximum connectivity is elected to be a cluster head. The sensor nodes that have been selected as a cluster head before will not be considered again. The sensor node with the maximum number of one-hop neighbors will be selected as a cluster head. The requirement of one-hop cluster and clock synchronization limits the usage of the highest connectivity cluster algorithm.

Max-min D-cluster algorithm [73] is a distributed clustering algorithm; cluster heads are required to be in one hop far from their sensor nodes. In max-min D-cluster algorithm, cluster heads are selected based on their node IDs. There is no requirement for clock synchronization in max-min D-cluster algorithm, and it offers better load balancing compared to LCA and LCA2 algorithms.

WCA [74] is a nonperiodic procedure for cluster head selection. WCA is called on demand every time a reconfiguration of the network's topology is inescapable. Anytime when the sensor node loses its connection with the cluster head, the WCA is called, therefore saving energy. WCA is based on a combination of metrics in order to select the best cluster head; these parameters include ideal node degree, mobility, transmission power, and the remaining energy of sensor nodes. The election process is based on a global parameter, which is called a combined weight.

CWAT [75] is a decentralized algorithm for homogeneous networks using idea transmission range. Through simulation experiments, it was observed that the generalization of the proposed algorithm is required to monitor its performance with respect to load balancing, cluster head reelection, and energy consumption for WSN.

Probabilistic clustering algorithm includes the following algorithms: LEACH, two-level LEACH (TL-LEACH), energy-efficient clustering scheme (EECS), hybrid energy-efficient distributed clustering (HEED), time-controlled clustering algorithm (TCCA).

LEACH algorithm is a major improvement on clustering [76]. It involves random rotation of selecting the cluster head. LEACH algorithm employs data fusion in order to reduce the amount of data transmission. In LEACH algorithm, a scheduling scheme is deployed to reduce energy consumption because the sensor nodes can turn off their radio during the scheduled time slot. TL-LEACH [77] is an extension of LEACH algorithm and utilizes two levels of cluster heads (primary and secondary). The primary cluster head in each cluster communicates with the secondary cluster heads, and in turn secondary cluster heads communicate with the sensor nodes in their sub-cluster. As in LEACH, the communication process is still scheduled using TDMA time slots. TL-LEACH reduces the total number of nodes that need to transmit to the base station and hence effectively minimizing the total energy usage.

EECS algorithm [78] is similar to LEACH with additional improvements in cluster formation and cluster head selection process. A constant number of cluster heads is selected based on the residual energy of sensor nodes using localized competition process with no need for iteration process. The cluster size is determined based on the distance between cluster head and the base station. The EECS protocol offers much lower message overheads and uniform distribution of cluster heads compared to LEACH protocol.

HEED is multi-hop clustering algorithm for WSNs [79]. In the HEED protocol, cluster heads are selected based on two factors: the residual energy and intra-cluster communication cost. The residual energy factor is used to choose the initial set of cluster heads, whereas the intra-cluster communication cost reflects the node degree to the neighbor, which is used by each node to decide to which cluster may join. In the HEED protocol, the knowledge of the entire network structure is required to determine intra-cluster communication cost; however, the configuration of those parameters might be difficult in real experiments.

The TCCA protocol [80] is similar to LEACH. In TCCA the process is divided into rounds, enabling better load distribution among sensor nodes. Through each round, a cluster setup phase is processed, which targets building the cluster and selection of cluster head. And then a steady-state phase is processed, which focuses on cyclic collection, aggregation and transfer of data at cluster head to base station. As soon as the cluster head is selected, it advertises its selection as the cluster head to its neighboring nodes through transmitting advertisement message, which includes the cluster head node ID, initial time to live, its residual energy, and a timestamp.

Biologically inspired clustering algorithms include ANTCLUST-based clustering [81], which consists of swarm-intelligence-based clustering algorithm. Swarm intelligence is a technique used to model the collective behavior of social insects and shows the properties of robustness and distributed problem-solving capabilities. ANTCLUST is a clustering method in which cluster heads with high residual energy become cluster head independently. The sensor nodes that meet each other will be clustered in a random way. Each sensor node with low residual energy will choose a cluster head based on the residual energy of the cluster head and an estimation of the cluster size.

4.7 Security Protocols for Wireless Sensor Networks

In this section, we describe the security issues in WSNs. The energy-constrained nature of sensor nodes makes the problem of integrating security very challenging. Each sensor node has limited processing power, storage, energy, and bandwidth; therefore, the design of the security protocols for WSNs should be general toward conservation of the sensor node resources. The designed security solution for WSN must satisfy the following requirements [82]:

- *Confidentiality*: Sensor networks must not share sensor readings to neighboring networks. Therefore, the security protocol must allow only authorized users to have access to the secret keys and other confidential information.
- *Authenticity*: Through data authentication, the receiver verifies that the data were really sent from the claimed sender. Furthermore, the access to the shared key should be restricted to only authenticated parties.
- *Integrity*: It ensures that the received data by the receiver were not altered or tampered in transition by an adversary.
- *Freshness*: The received data are recent, and it ensures that no adversary replayed old messages.
- *Availability*: The services offered by security solutions in WSNs must ensure that confidentiality and authentication are available to authorized parties. To guarantee the availability of message protection, the sensor network must protect its resources from unnecessary processing of key management messages in order to minimize energy consumption.
- *Self-organization*: The security protocol must be able to adapt itself to the corresponding environment, since security solutions cannot make any assumptions on sensor nodes with regard to network deployment.
- *Flexibility*: WSNs will be used in dynamic situations where environmental conditions, threats, and tasks may change rapidly. Security protocols must be flexible enough to offer solutions to all possible situations a sensor network may encounter.

4.7.1 Attacks on Wireless Sensor Networks

Attacks on WSNs can be considered from two different levels: attacks on security mechanisms and attacks on basic mechanisms. The major attacks on WSNs include denial of service, attacks on information transition, Sybil attack, blackhole/sinkhole attack, hello flood attack, and wormhole attack [83].

4.7.2 Attacks and Defense Suggestions in OSI Model

Through the layer network architecture, security issues can be easily analyzed, and the robustness can be improved by circumscribing layer interactions and interfaces. Figure 4.8 presents the typical layered networking model of sensor network. Each layer is subject to different attacks. In this section, we briefly describe attacks and defenses for the transport and below layers.

4.7.2.1 Physical Layer

The physical layer is responsible for signal detection and modulation, frequency selection, and carrier frequency generation. Jamming and tampering are the major attacks on the physical layer. Jamming can be defended via various forms of spread-spectrum or frequency-hopping communication. However, this requires greater design complexity and more power. Other defense methods against jamming include switching to low duty cycle and saving power. On the other hand, tamper protection falls into two categories: passive and active [84].

4.7.2.2 Data Link Layer

The DLL or MAC is responsible for data frame detection, multiplexing of data streams, medium access, and error control. It supports point-to-point and point-to-multipoint connections and channel assignment for neighbor-to-neighbor communication. Collision, unfairness, and exhaustion are the main attacks on DLL. Error-correcting code can handle the collision attacks; however, the result is still limited because malicious node may corrupt more data than the network can correct. TDMA is an efficient method to prevent collision, but it requires more control resources.

Random back-offs can decrease the probability of an inadvertent collision, but it may be ineffective for preventing this attack. Time division multiplexing provides each

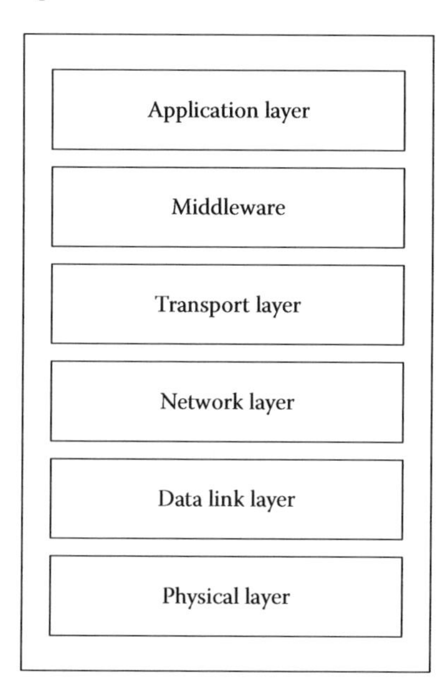

FIGURE 4.8
Layered networking model for WSN.

sensor node with a slot for transmission with no need for arbitration for each frame. This can help in solving the indefinite postponement problem in back off algorithm; however, it remains susceptible to collision. An efficient solution is rate limiting in MAC admission control; however, still additional work is required [85]. Adversaries can maximize their own transmission time in nonpriority MAC mechanism so that other safe nodes do not have any time to transmit packets, and this will cause unfairness, a weaker form of DoS.

4.7.2.3 Network Layer

Special multi-hop wireless routing protocols between the sink node and the router nodes are required to deliver data throughout a WSN. Authors of [86] listed the attacks on the network layer as follows: altering, spoofing, or replacing routing information, selective forwarding, Sybil attacks, sinkhole attacks, and acknowledgment spoofing.

Encryption and authentication, identity verification, multipath routing, authentication, and bidirectional link verification can guard sensor network routing protocols against Sybil attacks, external attacks, bogus routing information, and acknowledgment spoofing. However, sinkhole attacks and wormholes cause considerable challenges to secure routing protocol design.

4.7.2.4 Transport Layer

The transport layer protocols present reliability and session control for sensor network application. The transport layer is strongly needed when the network system is required to be accessed through the Internet or other external networks. Attacks on the transport layer include flooding and de-synchronization, which can threaten security. Although limiting the number of connections may prevent flooding, it can also prevent legitimate clients from accessing the victim. Connectionless protocols can easily resist this type of attacks, but they might not provide satisfactory transport-level services of the WSNs [82].

4.7.3 Security Schemes for WSNs

A number of studies have targeted security methods in WSNs. Authors of [86] proposed an analysis of secure routing in WSNs, and the work presented in [87] studies how to design distributed sensor networks with multiple supply voltages in order to reduce energy consumption. The system presented in [86] aims at increasing energy efficiency for key management in WSNs.

4.8 Conclusions

WSNs have been deployed widely in several applications, including military, industrial, environmental, and medical applications. In this chapter, we provided an overview on WSNs, including applications, routing, clustering, and security issues.

The essential properties of singular sensor nodes pose additional challenges to the communication protocol in terms of power consumption. Most of the existing WSN applications and communication protocols are mainly designed to deliver high energy efficiency.

Each sensor node has limited power sources. Thus, while the traditional networks are designed to enhance the performance metrics such as throughput and delay, WSN protocols focus on power conservation. Another factor considered in developing WSN protocols is deployment of sensor nodes. Moreover, the density of sensor nodes is also exploited in WSN protocols because of the short transmission range of sensor nodes. This chapter reviews the term WSN and discusses the foremost WSN applications. Moreover, it discusses the MAC protocols for WSN, reviews the routing protocols used in WSNs, and argues the clustering algorithms deployed for WSNs systems. Finally, it discusses the security protocols deployed for WSNs and the attacks on WSNs.

References

1. Shi, E. and A. Perrig, Designing secure sensor networks, *IEEE Wireless Communications* 11(6) (2004): 38–43.
2. Alhmiedat, T., A. A. Taleb, and M. Bsoul, A study on threats detection and tracking systems for military applications using WSNs, *International Journal of Computer Applications* 40(15) (2012): 12–18.
3. Lee, S. H., S. Lee, H. Song, and H. S. Lee, Wireless sensor network design for tactical military applications: Remote large-scale environments, in *IEEE Military Communications Conference, 2009, MILCOM 2009*, Boston, MA, October 18–21, 2009, pp. 1–7.
4. Steves, M. P., Utility assessments of soldier-worn sensor systems for assist, in *Proceedings of Performance Metrics for Intelligent Systems Workshop* (*PerMIS'06*), Gaithersburg, MD, August 2006, pp. 165–171.
5. Marquet, L. C., J. A. Ratches, and J. Niemela, Smart sensor networks and information management for the Future Combat System (FCS), *Proceedings of the SPIE*, 4396 (2001): 8–14.
6. Nemeroff, J., L. Garcia, D. Hampel, and S. DiPerro, Application of sensor network communications, in *Proceedings of IEEE Military Communications Conference* (*MILCOM'OI*), Vienna, VA, October 2001, pp. 336–341.
7. Nemeroff, J., P. Angelini, and L. Garcia, Communication for unattended sensor networks, *Proceedings of the SPIE*, 5441 (2004): 161–167.
8. Nemeroff, J. and L. Garcia, Networked sensors for the future force (NSFF) advanced technology demonstration (ATD) communications system, *Proceedings of the SPIE*, 5820 (2005): 83–91.
9. Nakamura, M., A. Sakurai, S. Furubo, and H. Ban, Collaborative processing in Mote-based sensor/actuator networks for environment control application, *Signal Processing*, 88(7) (2008): 1827–1838.
10. Chen, J., X. Cao, P. Cheng, Y. Xiao, and Y. Sun, Distributed collaborative control for industrial automation with wireless sensor and actuator networks, *IEEE Transactions on Industrial Electronics*, 57(12) (2010): 4219–4230.
11. Alhmiedat, T., F. Omar, and A. Abu Taleb, A hybrid tracking system using ZigBee WSNs, in *International Conference on Computer Science and Information Technology*, Amman, Jordan, March 26–27, 2014, pp. 71–74.
12. Alhmiedat, T., A. O. Abusalem, and A. Abu Taleb, An improved decentralized approach for tracking multiple mobile targets through ZigBee WSNs, *International Journal of Wireless and Mobile Networks*, 5(3) (2013): 61–76.
13. Alhmiedat, T., G. Samara, and A. Abu Salem, An indoor fingerprinting localization approach for ZigBee wireless sensor networks, *European Journal of Scientific Research*, 105(2) (2013): 190–202.
14. Alhmiedat, T. A. and S. Yang, Tracking multiple mobile targets based on ZigBee standard, in *Proceedings of the 35th Annual Conference of the IEEE Industrial Electronics Society*, Porto, Portugal, November 2009, pp. 2726–2731.

15. Alhmiedat, T. A. and S. Yang, A ZigBee-based mobile tracking system through wireless sensor networks, *International Journal of Advanced Mechatronic Systems*, 1(1) (2008): 63–70.
16. Sukuvaara, T. and P. Nurmi, Wireless traffic service platform for combined vehicle-to-vehicle and vehicle-to-infrastructure communications, *IEEE Wireless Communications*, 16(6) (2009): 54–61.
17. Hartenstein, H. and K. P. Laberteaux, A tutorial survey on vehicular ad hoc networks, *IEEE Communications Magazine*, 46(6) (2008): 164–171.
18. IEEE draft standard for wireless access in vehicular environments (WAVE)—Multi-channel operation, IEEE P1609.4/D9, August 2010, August 17, 2010, pp. 1–61.
19. SAFESPOT-Cooperative Networks for Intelligent Road Safety. Integrated Project of the European Union's Sixth Framework Programme of Research, available online at http://www.safespot-eu.org/, accessed February 2010.
20. P. R. Coopers, *Co-Operative Networks for Intelligent Road Safety*, Tra-transport research arena, Europe, 2006, Goeteborg, Sweden, June 12th–15th, 2006. Greener, Safer and Smarter Road Transport for Europe. Proceedings, 2006.
21. CVIS—Cooperative vehicle-infrastructure system: Integrated Project of the European Union's Sixth Framework Programme of Research, available online at http://www.cvisproject.org/, accessed January 2010.
22. Williams, B. M. and A. Guin, Traffic management center use of incident detection algorithms: Findings of a nationwide survey, *IEEE Transactions on Intelligent Transportation Systems*, 8(2) (2007): 351–358.
23. Li, Y., J. Maorong, G. Zhenru, Z. Weiping, and G. Tao, Design of home automation system based on ZigBee wireless sensor network, in *First International Conference on Information Science and Engineering* (*ICISE*), Nanjing, China, December 26–28, 2009, pp. 2610–2613.
24. Byun, J., B. Jeon, J. Noh, Y. Kim, and S. Park, An intelligent self-adjusting sensor for smart home services based on ZigBee communications, *IEEE Transactions on Consumer Electronics*, 58(3) (2012): 794–802.
25. Gill, K., F. Yao, and S.-H. Yang, Transparent heterogeneous networks for remote control of home environments, in *IEEE International Conference on Networking, Sensing and Control, 2008, ICNSC 2008*, Sanya, China, April 6–8, 2008, pp. 1419–1424.
26. Gill, K., S.-H. Yang, F. Yao, and X. Lu, A ZigBee-based home automation system, *IEEE Transactions on Consumer Electronics*, 55(2) (2009): 422–430.
27. Furtado, H. and R. Trobec, Applications of wireless sensors in medicine, in *MIPRO, 2011 Proceedings of the 34th International Convention*, Opatija, Croatia, May 23–27, 2011, pp. 257–261.
28. Jovanov, E., A. Milenkovic, C. Otto, and P. C. De Groen, A wireless body area network of intelligent motion sensors for computer assisted physical rehabilitation, *Journal of NeuroEngineering and Rehabilitation*, 2(1) (2005): 6.
29. Van Laerhoven, K., B. Lo, J. Ng, S. Thiemjarus, R. King, S. Kwan, H.-W. Gellersen et al., Medical healthcare monitoring with wearable and implantable sensors, in *Proceedings of the Third International Workshop on Ubiquitous Computing for Healthcare Applications*, Tokyo, Japan, September, 2004, pp. 11–16.
30. Dabiri, F., A. Vahdatpour, H. Noshadi, H. Hagopian, and M. Sarrafzadeh, Ubiquitous personal assistive system for neuropathy, in *Proceedings of the Second International Workshop on Systems and Networking Support for Health Care and Assisted Living Environments*, Breckenridge, CO, June 17, 2008, p. 17.
31. Hamalainen, M., P. Pirinen, and Z. Shelby, Advanced wireless ICT healthcare research, in *Mobile and Wireless Communications Summit, 2007, 16th IST*, Budapest, Hungary, July 1, 2007, pp. 1–5.
32. Sohraby, K., D. Minoli, and T. Znati, *Wireless Sensor Networks: Technology, Protocols, and Applications*. John Wiley & Sons, New York, 2007.
33. Sohrabi, K. and G. J. Pottie, Performance of a novel self-organization protocol for wireless ad hoc sensor networks, in *Proceedings of the IEEE 50th Vehicular Technology Conference* (*VTC'99*), Amsterdam, the Netherlands, September 1999, pp. 1222–1226.
34. Bharghavan, V., A. Demers, S. Shenker, and L. Zhang, MACAW: A media access protocol for wireless LANs, in *Proceedings of the ACM SIGCOMM*, Portland, Oregon, October, 1994, pp. 212–225.

35. Sohrabi, K., J. Gao, V. Ailawadhi, and G. J. Pottie, Protocols for self-organization of a wireless sensor network, *IEEE Personal Communications*, 7(5) (2000): 16–27.
36. Heinzelman, W. R., A. Chandrakasan, and H. Balakrishnan, Energy efficient communication protocols for wireless microsensor networks, in *Proceedings of the 33rd Hawaii International Conference Systems Sciences (HICSS'00)*, Maui, HI, January 2000, pp. 3005–3014.
37. Singh, S. and C. S. Raghavendra, PAMAS: Power aware multi-access protocol with signalling for ad hoc networks, *ACM Computers in Communications Review*, 28(3) (1998): 5–26.
38. Schurgers, C., V. Tsiatsis, S. Ganeriwal, and M. Srivastaval, Optimizing sensor networks in the energy-latency density design space, *IEEE Transactions on Mobile Computing*, 1(1) (2002): 70–80.
39. Polastre, J., J. Hill, and D. Culler, Versatile low power media access for wireless sensor networks, in *Proceedings of the Second ACM Conference on Embedded Networked Sensor Systems (SenSys'04)*, Baltimore, MD, November 3–5, 2004, 13 pp.
40. Dam, T. V. and K. Langendoen, An adaptive energy-efficient MAC protocol for wireless sensor networks, in *Proceedings of the First ACM Conference on Embedded Networked Sensor Systems (SenSys'03)*, Los Angeles, CA, November 2003, pp. 171–180.
41. Pantazis, N. A., S. A. Nikolidakis, and D. D. Vergados. Energy-efficient routing protocols in wireless sensor networks: A survey, *IEEE Communications Surveys & Tutorials*, 15(2) (2013): 551–591.
42. Arce, P., J. C. Guerri, A. Pajares, and O. Láro, Performance evaluation of video streaming over ad hoc networks using flat and hierarchical routing protocols, *Mobile Networks and Applications*, 13(3–4)(2008): 324–336.
43. Ogier, R., F. Templin, and M. Lewis, Topology dissemination based on reverse-path forwarding (TBRPF), RFC 3684, Internet Engineering Task Force, February 2004.
44. Pucha, H., S. Das, and Y. Hu, The performance impact of traffic patterns on routing protocols in mobile ad hoc networks, *Computer Networks*, 51(12) (2007): 3595–3616.
45. Ma, M., Y. Yang, and C. Ma, Single-path flooding chain routing in mobile wireless networks, *International Journal of Sensor Networks*, 1(2) (2006): 11–19.
46. Vidhate, D. A., A. K. Patil, and S. S. Pophale, Performance evaluation of low energy adaptive clustering hierarchy protocol for wireless sensor networks, in *Proceedings of the International Conference and Workshop on Emerging Trends in Technology (ICWET 2010) TCET*, Mumbai, India, February 26–27, 2010, pp. 59–63.
47. Almazaydeh, L., E. Abdelfattah, M. Al-Bzoor, and A. Al-Rahayfeh, Performance evaluation of routing protocols in wireless sensor networks, *Computer Science and Information Technology*, 2(2) (2010): 64–73.
48. Lotf, J., M. Bonab, and S. Khorsandi, A novel cluster-based routing protocol with extending lifetime for wireless sensor networks, in *Proceedings of the Fifth IFIP International Conference on Wireless and Optical Communications Networks (WOCN08)*, Surabaya, Java, May 5–7, 2008, pp. 1–5.
49. Kandris, D., P. Tsioumas, A. Tzes, G. Nikolakopoulos, and D. D. Vergados, Power conservation through energy efficient routing in wireless sensor networks, *Sensors*, 9(9) (2009): 7320–7342.
50. Pantazis, N. A., D. J. Vergados, and D. D. Vergados, Increasing intelligent wireless sensor networks survivability by applying energy-efficient schemes, in *Proceedings of the Third IFIP International Conference on Artificial Intelligence Applications and Innovations*, Athens, Greece, June 7–9, 2006, Vol. 204, pp. 657–664.
51. Sadagopan, N., B. Krishnamachari, and A. Helmy, Active query forwarding in sensor networks (ACQUIRE), *Ad Hoc Networks*, 3(1) (2005): 91–113.
52. Kulik, J., W. Rabiner, and H. Balakrishnan, Adaptive protocols for information dissemination in wireless sensor networks, in *Proceedings of the Fifth Annual ACM/IEEE International Conference on Mobile Computing and Networking*, Seattle, WA, August 1999, pp. 174–185.
53. Bokhari, F., Energy-efficient QoS-based routing protocol for wireless sensor networks, *Journal of Parallel and Distributed Computing*, 70(8) (2010): 849–857.
54. Cordeiro, C. M. and D. P. Agrawal, *Ad hoc and Sensor Networks: Theory and Applications*. World Scientific, Singapore, 2006.

55. Leong, B., B. Liskov, and R. Morris, Geographic routing without planarization, in *Proceedings of the Third Conference on Networked Systems Design and Implementation*, San Jose, CA, May 8–10, 2006, Vol. 3, pp. 25–39.
56. Zimmerling, M., W. Dargie, and J. M. Reason, Energy-efficient routing in linear wireless sensor networks, in *Proceedings of the Fourth IEEE International Conference on Mobile Ad hoc and Sensor Systems (MASS 2007)*, Italy, Pisa, October 8–11, 2007, pp. 1–3.
57. Chen, D. and P. Varshney, On-demand geographic forwarding for data delivery in wireless sensor networks, *Computer Communications*, 30(14–15) (2007): 2954–2967.
58. Zou, L., M. Lu, and Z. Xiong, PAGER-M: A novel location-based routing protocol for mobile sensor networks, in *Proceedings of the 2007 ACM SIGMOD International Conference on Management of Data*, Beijing, China, June 11–14, 2007, pp. 1182–1185.
59. Chen, M., S. Gonzalez, and V. Leung, Applications and design issues of mobile agents in wireless sensor networks, *IEEE Wireless Communications*, 14(6) (2007): 20–26.
60. Chen, M., S. Gonzalez, Y. Zhang, and V. Leung, Multi-agent itinerary planning in wireless sensor networks, *Computer Science*, 22(10) (2009): 584–597.
61. Chen, M., L. Yang, T. Kwon, L. Zhou, and M. Jo, Itinerary planning for energy-efficient agent communication in wireless sensor networks, *IEEE Transactions on Vehicular Technology*, 60(7) (2011): 1–8.
62. Chen, M., M. Guizani, and M. Jo, Mobile multimedia sensor networks: Architecture and routing, in *Proceedings of the Mobility Management in the Networks of the Future World*, Shanghai, China, April 2011, pp. 409–412.
63. Chen, M., T. Kwon, S. Mao, Y. Yuan, and V. Leung, Reliable and energy-efficient routing protocol in dense wireless sensor networks, *Sensor Networks*, 4(1) (2008): 104–117.
64. Chen, Y., N. Nasser, T. El Salti, and H. Zhang, A multipath QoS routing protocol in wireless sensor networks, *Sensor Networks*, 7(4) (2010): 207–216.
65. Yang, Y. and M. Cardei, Delay-constrained energy-efficient routing in heterogeneous wireless sensor networks, *Sensor Networks*, 7(4) (2010): 236–247.
66. Haider, M. B., S. Imahori, and K. Sugihara, Success guaranteed routing in almost delaunay planar nets for wireless sensor communication, *Sensor Networks*, 9(2) (2011): 69–75.
67. Taleb, A. A., T. Alhmiedat, O. Al-haj Hassan, and N. M. Turab. A survey of sink mobility models for wireless sensor networks, *Journal of Emerging Trends in Computing and Information Sciences*, 4(9) (2013): 679–687.
68. Taleb, A. A., T. A. Alhmiedat, R. A. Taleb, and O. Al-Haj Hassan, Sink mobility model for wireless sensor networks, *Arabian Journal for Science and Engineering*, 39(3) (2014): 1775–1784.
69. Taleb, A. A. and T. Alhmiedat. Depth first based sink mobility model for wireless sensor networks, *International Journal of Electrical, Electronics, and Computer Systems*, 19(2) (2014): 892–897.
70. Baker D. J. and A. Epheremides, The architectural organization of a mobile radio network via a distributed algorithm, *IEEE Transactions on Communications*, 29(11) (1981): 1694–1701.
71. Epheremides, A., J. E. Wieselthier, and D. J. Baker, A design concept for reliable mobile radio networks with frequency hopping signalling, *Proceedings of IEEE*, 75(1) (1987): 56–73.
72. Tsigas, P., Project on mobile ad hoc networking and clustering for the course EDA390 Computer Communication and Distributed Systems, Manual for University Course.
73. Amis, A., R. Prakash, T. Vuong, and D. Huynh, Max-min d-cluster formation in wireless ad hoc networks, in *IEEE INFOCOM*, Tel Aviv, Israel, March 26–30, 2000, pp. 32–41.
74. Chatterjee, M., S. K. Das, and D. Turgut, WCA: A weighted clustering algorithm for mobile ad hoc networks, *Cluster Computing*, 5 (2002): 193–204.
75. Selvakennedy, S. and S. Sinnappan, An energy efficient clustering algorithm for multihop data gathering in wireless sensor networks, *Journal of Computers*, 1(1) (2006): 40–47.
76. Ye, Wei, John Heidemann, and Deborah Estrin. An energy-efficient MAC protocol for wireless sensor networks. In *Proceedings of the Twenty-First Annual Joint Conference of the IEEE Computer and Communications Societies (INFOCOM)*, June 2002, vol. 3, pp. 1567–1576, New York.
77. Loscri, V., G. Morabito, and S. Marano, A two-level hierarchy for low-energy adaptive clustering hierarchy, in *Proceedings of Vehicular Technology Conference*, Stockholm, Sweden, May 30–June 1, 2005, Vol. 3, pp. 1809–1813.

78. Ye, M., C. Li, G. Chen, and J. Wu, EECS: An energy efficient clustering scheme in wireless sensor networks, in *24th IEEE International Conference on Performance, Computing, and Communications*, April 7–9, 2005, pp. 535–540.
79. Younis, O. and S. Fahmy, HEED: A hybrid energy-efficient distributed clustering approach for ad hoc sensor networks, *IEEE Transactions on Mobile Computing*, 3(4) (2004): 366–379.
80. Selvakennedy, S. and S. Sinnappan, An adaptive data dissemination strategy for wireless sensor networks, *International Journal of Distributed Sensor Networks*, 3(1) (2007): 1–13.
81. Ye, M., C. Li, G. Chen, and J. Wu, An energy efficient clustering scheme in wireless sensor networks, *Ad Hoc & Sensor Wireless Networks*, 1 (2006): 1–21.
82. Chen, X., K. Makki, K. Yen, and N. Pissinou, Sensor network security: A survey, *IEEE Communications Surveys & Tutorials*, 11(2) (2009): 52–73.
83. Pathan, A. S. K., H.-W. Lee, and C. S. Hong, Security in wireless sensor networks: Issues and challenges, in *The Eighth International Conference on Advanced Communication Technology, 2006, ICACT 2006,* Dublin, Ireland, February 20–22, 2006, Vol. 2, 6pp.
84. Carman, D. W., P. S. Kruus, and B. J. Matt, Constraints and approaches for distributed sensor network security, NAI Labs Technical Report 00-010, 2000.
85. Wood, A. D. and J. A. Stankovic, Denial of service in sensor networks, *IEEE Computer*, 35 (2002): 54–62.
86. Karlof, C. and D. Wagner, Secure routing in wireless sensor networks: Attacks and countermeasures, *Elsevier's Ad Hoc Networks Journal*, 1 (2003): 293–315, special issue on *Sensor Network Applications and Protocols*.
87. Yuan, L. and G. Qu, Design space exploration for energy-efficient secure sensor network, in *Proceedings of the IEEE International Conference on Application-Specific Systems, Architectures and Processors*, San Jose, CA, July 17–19, 2002, pp. 88–97.

5

Role of CDMA-Based WSN for Effective Implementation of CPS Infrastructure

Uma Datta and Sumit Kundu

CONTENTS

5.1 Introduction

In the past two decades, one of the fastest-growing research areas has been wireless communication and network. Significant progress has been made in the fields of mobile ad hoc network (MANET) and wireless sensor network (WSN) [Lin, 2012; Wu, 2011]. More recently, the cyber-physical system (CPS) has emerged as a promising direction to enrich human-to-human, human-to-object, and object-to-object interactions in the physical world as well as in the virtual world. Apparently, CPS would adopt and even nurture the areas of MANET and WSN because more sensor inputs and richer network connectivity may be needed.

5.1.1 Cyber-Physical Systems and Wireless Sensor Network Technology

WSN is designed particularly for delivering sensor-related data. CPS typically involves multiple dimensions of sensing data, crosses multiple sensor networks and the Internet, and aims at constructing intelligence across these domains. The increasing pervasiveness of WSN technologies in many applications makes them an important component of emerging CPS designs. The most important design requirements of CPS architectures are reliability and predictability. This is because the quality of service (QoS) of CPS applications, such as emergency real-time system and health care, highly depends on these two factors. When sensor technologies are integrated with CPSs, it is a challenge for the decision-making system to ensure reliability and predictability. Deployment involves how to place sensor nodes over the given monitoring region in an efficient way, while localization approach aims at providing location information for sensor nodes. Any coverage method requires that the region of interest, where the interesting events might happen, has to be covered by sensors. The data gathering scheme ensures that the collected information can be successfully delivered from sensors to the sink node (which can be treated as the real-time decision-making system). Moreover, to ensure the negotiation of any two neighboring sensors and to conserve energy consumption, the communication (medium access control) support should also be considered in CPS designs. As most of the nodes of CPS are battery-operated devices, energy-efficient operation may be important criteria in CPS design apart from latency, fault tolerance, and scalability. A few CPS application domains that depend on WSNs in their design architectures and implementations need a design with efficient mesh networking protocol.

5.1.2 Role of CDMA-Based WSN in CPS

Several medium access control (MAC) protocols, defining rules for orderly access to the shared medium, are investigated to minimize the energy consumption of nodes. Theoretically, low latency, fault tolerance, scalability, and energy efficiency should be achieved using single perfect WSN MAC. Code division multiple access (CDMA) achieves many of them such as latency, fault tolerance, connectivity, and scalability at the expense of power consumption. So, CDMA may be useful for some cases of WSN application. Moreover, to handle a large number of nodes, CDMA is a good choice as a MAC protocol. Though CDMA has been widely used in cellular networks with base stations to coordinate the access and distribute codes to the mobile terminals, these methods are not directly applicable to sensor networks that are distributed and autonomous in nature without any central infrastructure. The reason for the unpopularity of CDMA technology in WSN is the nonavailability of low-power CDMA transceiver. Developments in the field of very large-scale integration technology aim to reduce power consumption in digital transceiver, while energy-harvesting technology augments the lifetime of battery further. Research is going on for the implementation of CDMA in next-generation WSN in air and under-water applications. CDMA-based networks are inherently limited by "multiple access interference" (MAI) and require "power control." However, when applied to WSN with battery-operated and simple transceivers, perfect power control is difficult to achieve. The performance of the network is affected by the generation of different types of interferers such as MAI and "node interference" (NI), which increase with node density, though increase in node density improves connectivity. Moreover, interferences due to shadow fading may be correlated at the receiver. Therefore, analysis of different parameters in WSN using CDMA technology is essential in designing efficient CDMA-based WSN. This chapter investigates the possibility of designing efficient single- and multi-hop CDMA WSNs in terms of energy and latency where channel impairments such as "shadow-faded correlated interferers" and "power control error" (pce) are considered. Several performance metrics of the networks such as bit error rate (BER), energy, latency, and lifetime of the network are evaluated. Impacts of node density, correlation among interferers, and pce on these metrics are estimated. The optimum packet size for data transfer is evaluated using the maximization of energy efficiency as the objective function. These results will be useful in designing efficient CDMA-based WSN associated with CPS.

5.1.3 Topology and Protocols

Maximizing the network lifetime is a common objective of sensor network research, since sensor nodes are assumed to be disposed when they are out of battery. Under these circumstances, the "topology," that is, the pattern of placement of nodes, and the MAC protocol play an important role to reduce the potential energy wastes [Demirkol, 2006]. Several topologies are used in WSN. Typically, in some cases, the network has a regular topology with sensor nodes deployed following a specific geometry, while in some others the network is assumed to have random topology, that is, deployed randomly. In this chapter, regular topology is considered for analysis, where the nodes are placed in a deterministic fashion, and the distances among the nodes participating in multi-hop communication are the same. Nodes are selectively activated in such a way that the set of active nodes at any time lies on the vertices of a regular polygon, as shown in Figure 5.1 [De, 2004].

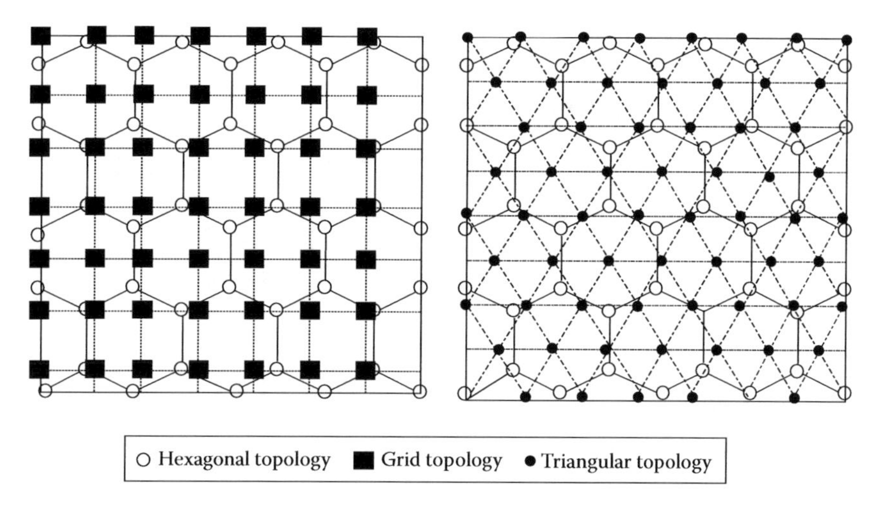

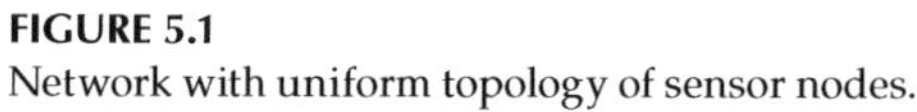

FIGURE 5.1
Network with uniform topology of sensor nodes.

The protocol stack used by the sink and sensor nodes is given in Figure 5.2 [Akyildiz, 2002]. This protocol stack combines power and routing awareness, integrates data with networking protocols, and communicates in a power-efficient manner through the wireless medium, while promoting cooperative efforts of sensor nodes. The protocol stack consists of physical layer, data link layer, network layer, transport layer, application layer, power management plane, mobility management plane, and task management plane.

The physical layer is responsible for frequency selection, carrier frequency generation, signal detection, modulation, and data encryption. The data link layer is responsible for the multiplexing of data streams, data frame detection, medium access, and error control. It ensures reliable point-to-point and point-to-multipoint connections in a communication network. The network layer takes care of routing the data and directing the process of selecting paths along which data will be sent in the network. The transport layer helps to maintain the flow of data if the sensor networks application requires it. This layer is especially needed when the system is planned to be accessed through the Internet or other external networks.

Depending on the sensing tasks, different types of application software can be built and used on the application layer. It defines a standard set of services and interface primitives available to a programmer independently on their implementation on every kind of platform.

In addition, the power, mobility, and task management planes monitor the power, movement, and task distribution among the sensor nodes, respectively. These planes help the sensor nodes coordinate the sensing task and lower the overall power consumption.

Since thousands of sensor nodes are densely scattered in a sensor field, the MAC scheme must establish communication links for data transfer and efficiently share communication

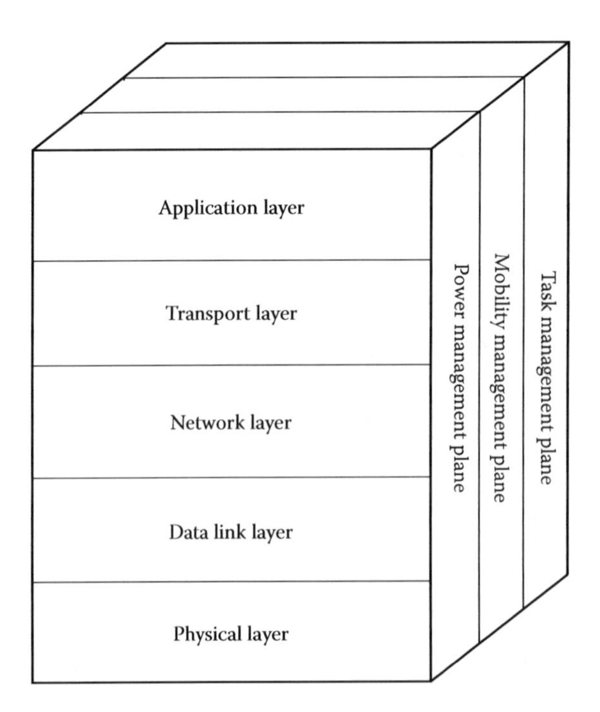

FIGURE 5.2
The sensor networks protocol stack. (From Akyildiz, I.F. et al., *IEEE Commun. Mag.*, 40(8), 102, 2002.)

resources between sensor nodes. Existing MAC protocols used in cellular systems and other wireless network systems cannot be adopted in sensor network scenario since there is no central controlling agent such as the base station in cellular systems. This makes network-wide synchronization a difficult task. Thus, a distributed MAC is more suitable for most of the applications in WSN.

In contrast to Bluetooth and the MANET systems, the sensor network may have a much larger number of nodes. The transmission power (~0 dBm) and radio range of a sensor node are much less than those of Bluetooth or MANET. Topology changes are more frequent in a sensor network, CPS, and can be attributed to both node mobility and node/link failure.

To design a good MAC protocol for WSN, the following attributes must be considered, as described by Ye [2004] and Demirkol [2006]. The first attribute is energy efficiency. We have to define energy-efficient protocols in order to prolong the network lifetime. Other important attributes are scalability and adaptability to changes. Changes in network size, node density, and topology should be handled rapidly and effectively for successful adaptation. Other important attributes such as latency, throughput, and bandwidth utilization may be secondary in sensor networks. Contrary to other wireless networks, fairness among sensor nodes is not usually a design goal, since all sensor nodes share a common task.

Due to the lack of central authority and constantly changing topology, interesting tradeoffs and interactions between physical, MAC, and network layers of the protocol stack are required. In particular, the impairments introduced by the wireless channel make the choice of the MAC protocol critical. The use of scheduling mechanisms—such as time division multiple access (TDMA), frequency division multiple access, or CDMA—in large ad hoc wireless networks is quite complex, due to synchronization reasons. Random access (RA) then seems a more appealing choice for large ad hoc wireless networks. A wide range of MAC protocols as used in sensor networks are reviewed by Demirkol [2006], stating the essential features of the protocols. Most of the protocols are based on TDMA or RA or a hybrid of RA and TDMA, and CDMA.

5.1.4 Major Issues in Design of CDMA WSN

The major issues involved in designing a CDMA WSN are power consumption, coverage, connectivity, lifetime of the network, delay, cost, and ease of deployment. However, many of these evaluation metrics are interrelated. Often it may be necessary to decrease performance in one metric, such as sample rate, in order to increase another, such as lifetime. Taken together, the set of metrics, described as follows, forms a major multidimensional space that can be used to describe the capabilities of a CDMA WSN.

- *Power consumption*: To meet the long lifetime requirements, individual sensor nodes must use incredibly low power. This ultra-low-power operation can be achieved only by utilizing low-power hardware components. Also transmission protocol must be designed to achieve low power consumption. The power consumed by a sensor node has several components such as transmit and receive power, as well as power consumed in sensing and processing.
- *Scalability and robustness*: Depending on the application, the number of sensor nodes deployed in studying a phenomenon may be of the order of hundreds or thousands. In a typical deployment, hundreds of nodes would have to work in harmony for years. To achieve this, the system must be constructed so that it can

tolerate and adapt to individual node failure. Additionally, each node must be designed to be as robust as possible. In addition to increasing the system's robustness to node failure, a WSN must also be robust to external interference, which can further be greatly increased through the use of multi-channel and spread spectrum radios.

- *Lifetime*: Critical to any WSN deployment is the expected lifetime. In general, network lifetime is the time span from the deployment to the instant when the network is considered nonfunctional.
- *Delay*: Like energy consumption, delay (D) for the successful delivery of a message is also an important factor to determine the quality of the WSN, specially delay critical services, for example, security-monitoring region.
- *Energy efficiency*: A suitable metric that captures the energy and reliability constraints is the energy efficiency.

5.2 CDMA-Based WSN

It is well known that MAI is a key factor in determining the performance (e.g., BER, throughput, etc.) of a CDMA network [Rappaport, 2002; Viterbi, 1995]. Even if each node transmits at the lowest possible power to its intended receiver, the lack of coordination and/or decentralized control mechanisms will result in a significant amount of interfering power from neighboring nodes. The interference problem becomes more severe as the node density increases, although apparently a higher node density might otherwise help improve connectivity and network performance.

CDMA has been advocated for WSN by Muqattash [2003], De [2004], Pompili [2007], and Kang [2008], where distribution of interference power in randomly distributed nodes is discussed. Interferences shadowed by the same obstacles near a receiver tend to be correlated, and such correlation has a significant impact on the signal-to-interference ratio (SIR) [Abu-Dayya, 1994; Pratesi, 2000]. The presence of such correlation affects the SIR and thereby might affect the performance of a CDMA-based WSN. Transmitter power control minimizes the near–far problem. However, in the case of WSN, without any central control station, pce cannot be minimized beyond a certain level. In this chapter, the analysis of interference, as presented by De [2004], is extended considering the presence of such statistically correlated interferers due to shadow fading and pce.

5.2.1 Architecture of WSN

In the present study, a layered architecture is considered, where the sink is at the center of concentric layers. The layers of sensor nodes surrounding it correspond to the sensor nodes that have the same hop count at the sink, as shown in Figure 5.3 [Kang, 2008]. The sensor nodes that are not located within the one-hop layer of the sink cannot directly communicate with the sink. Their information should be relayed through the sensor nodes located in the lower hop layers in order to be delivered to the sink. In other words, multi-hop transmission is needed to deliver data from the sensor node to the sink.

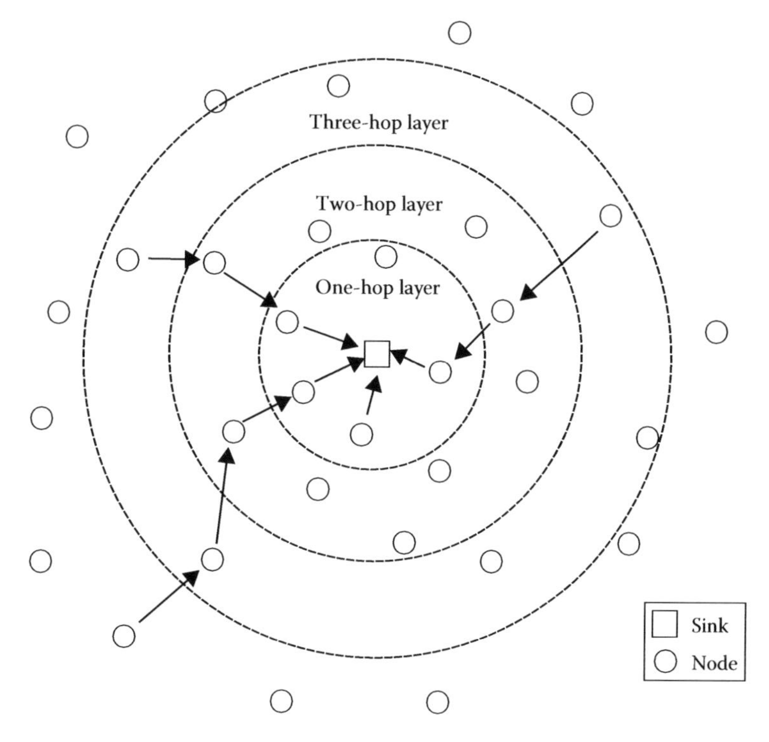

FIGURE 5.3
Layered architecture.

The advantage of layered architecture is that each sensor node is involved only in short distances. A short transmission range can reduce energy consumption and interference in highly dense sensor networks [Yang, 2003].

Placement of nodes may be regular or random in nature. This chapter considers regular placement of nodes, where inter-node distances are considered to be same.

5.2.2 Single-Hop Communication with Correlated Interferers

A wireless CDMA sensor network with layered architecture, as shown in Figure 5.3, is considered. Different layers are formed by the sensor/relay nodes that have the same hop count at the sink. Information from a source node reaches the final destination node (i.e., sink) via a suitable path in multi-hop through several intermediate nodes in shadowed environments. It is assumed that the information can be routed only from a higher hop layer node to a lower hop layer node. Each node is assigned a unique binary signature code. In this section, only the last hop is considered, that is, the single-hop communication between any node lying in the first layer and the sink, as shown in Figure 5.4. The sink s receives information from its immediate predecessor source/relay node d. Since only the nodes located within one-hop layer can directly communicate with the sink, the concurrent nodes that are sending their information to the sink within the one-hop layer of area πr_R^2 would contribute to MAI, where r_R is the receiving range of each node. On the other hand, the source/relay node k, while sending information to its own destination node k' (it can be any sensor, except the sink), causes

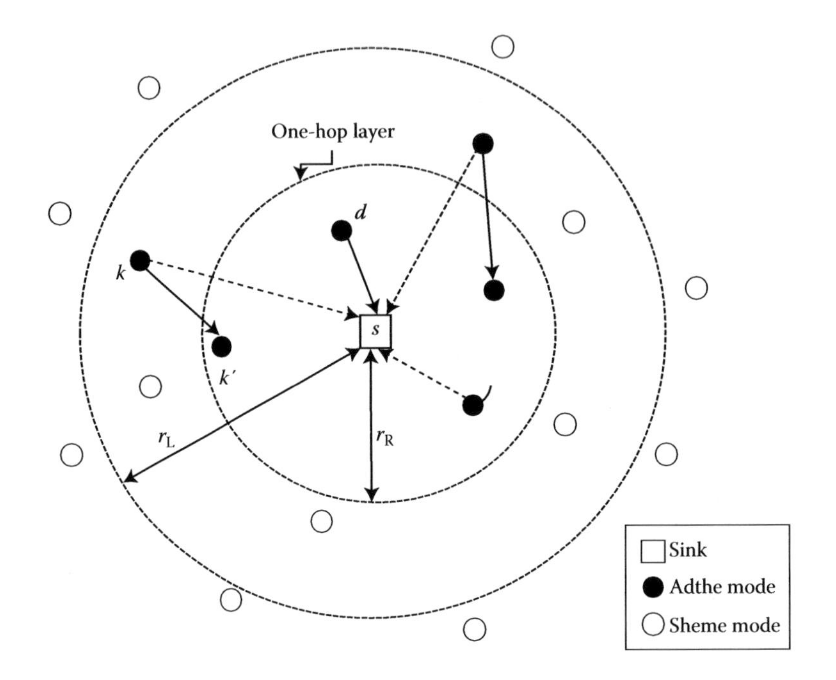

FIGURE 5.4
Interference model at the sink.

interference to the sink, as the sink lies within the interference range. In Figure 5.4, node j can be regarded as causing MAI while node k causes NI.

To obtain the interference power distribution, consider the following assumptions [De, 2004]:

- Sensor nodes are static and uniformly distributed over a sensor field with omnidirectional transmit and receive antenna of the same gain.
- The receiving distance and interference distance are r_R and r_I, respectively. Typically, $r_I \approx 2r_R$.
- The minimum distance between two nodes is r_0. When $r_0 \ll r_R$, we can safely assume that the spatial distribution of active nodes remains uniform.
- Imperfect power control.

5.2.2.1 BER, Latency, Throughput, and Energy in Single-Hop Scenario

It is assumed that nodes are power controlled, and the target received power level is P_r. To achieve the desired received power P_r at a local neighbor k', the required transmit power from a node k is

$$P_{t,kk'} = P_r \cdot d_{kk'}^{\alpha} \cdot e^{-S_{11}} \cdot e^{R_{11}} \tag{5.1}$$

where

$d_{kk'}$ is the distance separating k and k'

$m_{S_{11}}, \sigma_{S_{11}}$ and $m_{R_{11}}, \sigma_{R_{11}}$ are the mean and standard deviation of shadowing and pce, respectively, for the path kk'

Depending on receiver sensitivity, P_r is a constant

$P_{t,kk'}$ is a random variable (RV)

As shown in Figure 5.4, the distance (d_{ks}) of the interfering transmitter k from the node of interest s (i.e., sink) is an RV in $(r_0, r_I]$ and the interference power received at sink due to transmitter k is given by

$$P_I = P_{t,kk'} \cdot d_{ks}^{-\alpha} \cdot e^{S_{21}} = P_r \cdot \left(\frac{d_{kk'}}{d_{ks}}\right)^{\alpha} e^{S_{21}} \cdot e^{-S_{11}} \cdot e^{R_{11}} \tag{5.2}$$

where

$r_0 < d_{kk'} \leq r_R$

$r_0 < d_{ks} \leq r_I$

mS_{21}, σS_{21} are the mean and standard deviation of shadowing of the path between the nodes k and s

$E[p_i] = \eta$ in tune with Equation 5.3

The mean value of the collected interference power η' from an interfering node is derived as described by Datta [2009]:

$$\eta' = E[P_I] = E\left[P_r\left(\frac{d_{kk'}}{d_{ks}}\right)^{\alpha}\right] \cdot E[e^{S_{21}}] \cdot E[e^{-S_{11}}] \cdot E[e^{R_{11}}]$$

$$= \eta \cdot e^{(m_{S21} + (\sigma_{S21}^2/2))} \cdot e^{-(m_{S11} + (\sigma_{S11}^2/2))} \cdot e^{(m_{R11} + (\sigma_{R11}^2/2))} \tag{5.3}$$

In this case, η is given as [De, 2004]:

$$\eta = \frac{4 \cdot P_r \cdot \left(r_R^{\alpha+2} - r_0^{\alpha+2}\right) \cdot \left(r_I^{\alpha-2} - r_0^{\alpha-2}\right)}{(\alpha^2 - 4) \cdot r_0^{\alpha-2} \cdot r_I^{\alpha-2} \cdot \left(r_I^2 - r_0^2\right) \cdot \left(r_R^2 - r_0^2\right)} = \gamma \cdot P_r \tag{5.4}$$

where $2 < \alpha < 6$.

For large number of nodes (N) in the area (A) and a very small area of interference region of a receiver as compared to the total area (A), following De [2004], the expected numbers of nodes between the receiving and interference ranges of the receiver (n_1) and within the receiving range (n_2) are expressed as follows:

$$n_1 = \rho\pi\left(r_I^2 - r_R^2\right) \tag{5.5}$$

$$n_2 = \rho\pi r_R^2 \tag{5.6}$$

where ρ is the node density. Nodes present in between the receiving and interference ranges of the receiver contribute to NI, while the nodes present within the receiving range of the sink contribute only to MAI. A node will contribute interference only when it is active. Activity factors determine the number of active nodes at any instant in the two layers contributing MAI and NI. Thus, t_1, a fraction of n_1 as given in Equation 5.5, will be effective in contributing NI, and t_2, a fraction of n_2 as given in Equation 5.6, will be effective in contributing to interference due to MAI. Activity of the nodes contributing the MAI is assumed to be higher than the activity of the nodes contributing the NI, since sensor nodes near the sink are more active in sending data toward the sink because of multi-hop relaying.

Interference at sink is composed of t number of correlated interferers. Out of such t, it is considered that t_1 number interferers (a fraction of n_1) cause NI while t_2 number interferers (a fraction of n_2) cause MAI. As the sum of lognormal is well approximated by an equivalent lognormal, the total interference power due to NI is expressed as follows:

$$P_{NI} = \sum_{i=1}^{t_1} p_i \cdot e^{Z_i} \cong \eta \cdot (e^{Z_1} + e^{Z_2} + \cdots + e^{Z_{t1}}) \cong \eta \cdot e^{Z} \tag{5.7}$$

where

$Z_i = S_{2i} - S_{1i} + R_{1i}$, is a Gaussian RV
$p_i = P_r \cdot (d_{1i}/d_{2i})^{\alpha}$

The variable subscript "1" indicates the signal path, whereas the variable with subscript "2" indicates the interference path. For example, in Figure 5.4, $d_{kk'} = d_{11}$ represents the distance corresponding to the signal path and $d_{ks} = d_{21}$ represents the distance corresponding to the interference path between the node k and the sink S.

The total interference power due to NI and MAI is expressed as follows [Datta, 2009, 2010c]:

$$I = P_{NI} + P_{MAI} = \eta \cdot \sum_{i=1}^{t_1} e^{Z_i} + P_r \cdot \sum_{i=1}^{t_2} e^{R_{ki}} \tag{5.8}$$

where R_{ki} denotes the RV associated with pce in the path between ith MAI node such as j and sink s. Approximating the summation (I) by an equivalent lognormal RV ($P_r \cdot e^{\phi}$),

$$I = P_r \cdot e^{\phi} = \eta \cdot \sum_{i=1}^{t_1} e^{Z_i} + P_r \cdot \sum_{i=1}^{t_2} e^{R_{ki}} \tag{5.9}$$

Applying Wilkinson's approach [Abu-Dayya, 1994; Cardieri, 2000], the mean m_{ϕ} and the standard deviation σ_{ϕ} of ϕ in Equation 5.9 are estimated [Datta, 2009]. Two variables u_1 and u_2, the first and second moment of I, are used as intermediate variables in the analysis:

$$\begin{aligned} u_1 = E[I] = E[P_r \cdot e^{\phi}] &= P_r \cdot e^{m_{\phi} + (\sigma_{\phi}^2/2)} \\ &= \eta \cdot \sum_{i=1}^{t_1} e^{m_{Z_i} + (\sigma_{Z_i}^2/2)} + P_r \cdot \sum_{i=1}^{t_2} e^{m_{Rki} + (\sigma_{Rki}^2/2)} \end{aligned} \tag{5.10}$$

$$\begin{aligned} u_2 = E[I^2] = E\left(P_r^2 \cdot e^{2\phi}\right) &= P_r^2 \cdot e^{2m_{\phi} + 2\sigma_{\phi}^2} \\ &= \eta^2 \cdot \left[\sum_{i=1}^{t_1} e^{2m_{Z_i} + 2\sigma_{Z_i}^2} + 2 \cdot \sum_{i=1}^{t1-1} \sum_{j=i+1}^{t_1} e^{m_{Z_i} + m_{Z_j} + \frac{1}{2}\left(\sigma_{Z_i}^2 + \sigma_{Z_j}^2 + 2r_{Z_iZ_j}\sigma_{Z_i}\sigma_{Z_j}\right)} \right] \\ &+ P_r^2 \cdot \left[\sum_{i=1}^{t_2} e^{2m_{Rki} + 2\sigma_{Rki}^2} + 2 \sum_{i=1}^{t2-1} \sum_{j=i+1}^{t_2} e^{m_{Rki} + m_{Rkj} + \frac{1}{2}\left(\sigma_{Rki}^2 + \sigma_{Rkj}^2 + 2r_{RkiRkj}\sigma_{Rki}\sigma_{Rkj}\right)} \right] \\ &+ 2 \cdot \eta \cdot P_r \cdot \left[\sum_{i=1}^{t_1} \sum_{j=1}^{t_2} e^{m_{Z_i} + m_{Rkj} + \frac{1}{2}\left(\sigma_{Z_i}^2 + \sigma_{Rkj}^2 + 2r_{Z_iRkj}\sigma_{Z_i}\sigma_{Rkj}\right)} \right] \end{aligned} \tag{5.11}$$

where

$r_{Z_iZ_j}$, $r_{R_{ki}R_{kj}}$ are the pair-wise correlation coefficients among the interferers contributing NI and MAI, respectively

$r_{Z_iR_{kj}}$ is the correlation coefficients among NI and MAI interferers

Solving Equations 5.10 and 5.11 for m_ϕ and σ_ϕ yields

$$m_\phi = 2\ln\left(\frac{u_1}{P_r}\right) - \frac{1}{2}\ln\left(\frac{u_2}{P_r^2}\right) \tag{5.12}$$

$$\sigma_\phi^2 = \ln\left(\frac{u_2}{P_r^2}\right) - 2\ln\left(\frac{u_1}{P_r}\right) \tag{5.13}$$

For simplicity all interferers are assumed to have identical statistics, that is, $m_{R_{ki}} = m_{R_{kj}} = m_{Z_i} = m_{Z_j} = 0$ and $\sigma_{R_{kj}} = \sigma_{R_{ki}}$.

With the direct sequence binary phase shift keying transmission of "spreading bandwidth" W and for constant received signal power levels, the probability of error under the Gaussian approximation is [Jerez, 2000]

$$P_e = Q\left(\sqrt{2\frac{E_b}{\chi}}\right) = Q\left(\sqrt{\frac{2P_r \cdot e^R / R_b}{I/W}}\right) = Q\left(\sqrt{\frac{2P_r \cdot e^R / R_b}{P_r \cdot e^\phi / W}}\right) = Q(e^\psi) \tag{5.14}$$

Since $e^\psi = \sqrt{2(E_b/\chi)}$,

so

$$2\psi = \ln\left(\frac{(2 \cdot W)}{R_b}\right) + (R - \phi) = \ln(2 \cdot pg) + (R - \phi) \tag{5.15}$$

$$\psi = \frac{1}{2}[\ln(2 \cdot pg) + (R - \phi)] \tag{5.16}$$

The mean and variance of ψ are

$$m_\psi = E(\psi) = \frac{1}{2}[\ln(2 \cdot pg) - m_\phi] \tag{5.17}$$

$$\sigma_\psi^2 = \text{var}(\psi) = \frac{1}{4}\left(\sigma_R^2 + \sigma_\phi^2\right) \tag{5.18}$$

Following Jerez [2000], the mean probability of error (BER) can be approximated by

$$\overline{P_e} = \frac{2}{3}Q(e^{m_\psi}) + \frac{1}{6}Q(e^{m_\psi + \sqrt{3}\sigma_\psi}) + \frac{1}{6}Q(e^{m_\psi - \sqrt{3}\sigma_\psi}) \tag{5.19}$$

A received packet is not accepted until all bits of that packet are received without error. The packet is retransmitted on getting a negative acknowledgment (NACK) from the sink, and it continues until the packet is received correctly. Assuming n_f bits/packet for forward transmission as information and n_b bits/packet for reverse acknowledgment (ACK)/NACK, following Kleinschmidt [2007], the average probabilities of error at packet level for forward transmission of packet and reverse transmission of ACK/NACK are, respectively

$$\overline{P_{pf}} = 1-(1-\overline{P_e})^{n_f \cdot r_c} \tag{5.20}$$

$$\overline{P_{pr}} = 1-(1-\overline{P_e})^{n_b \cdot r_c} \tag{5.21}$$

where r_c is the "code rate." Thus, the average packet error probability for automatic repeat request (ARQ) packets is

$$P_f = 1-[(1-\overline{P_{pf}})\cdot(1-\overline{P_{pr}})] \tag{5.22}$$

Considering the probability of error in ACK/NACK to be very low, as it is a very small packet, that is, $\overline{P_{pr}} \ll \overline{P_{pf}}$, and the ACK/NACK to be instantaneous, one has $P_f = \overline{P_{pf}}$. A continuously active data service model following Kim [2000] is considered, without queuing. A packet is transmitted only when the preceding packet is transmitted successfully. Assuming perfect error detection with infinite ARQ, let n be the number of retransmissions required for successful delivery of a packet. The probability of success at nth transmission is given by [Kleinschmidt, 2007]

$$p_N[n] = (1-P_f)\cdot(P_f)^{n-1} \tag{5.23}$$

As an "infinite ARQ" scheme is considered, the average number of retransmissions for successful transmission of a packet [Kleinschmidt, 2007] is

$$\overline{N} = \sum_{n=1}^{\infty} p_N[n]\cdot n = \frac{1}{(1-P_f)} \tag{5.24}$$

The time required for successful delivery of a packet from a node to the sink constitutes the delay. Considering only the transmission delay without any queuing and assuming $n_f \gg n_b$, the average packet delay for successful transmission of packet is obtained as follows [Kim, 2000]:

$$D_{av} = \frac{T_i}{(1-P_f)} = \frac{n_f/R_b}{(1-P_f)} \tag{5.25}$$

where R_b is the "bit rate." In the present case, throughput is defined as the number of bits successfully transferred from the source node to the sink node per unit time. Considering R_c and r_c as the "chip rate" and code rate, respectively, and following Kim [2000], the average throughput is expressed as follows [Datta, 2010c]:

$$G_{av} = \frac{n_f \cdot r_c}{D_{av}} = \frac{r_c \cdot R_c \cdot (1-P_f)}{pg} \tag{5.26}$$

The energy E_b required to communicate one bit of information in one hop is [Sakhir, 2008]

$$E_b = \frac{(P_t + P_r)}{R_b} \tag{5.27}$$

where
R_b is the bit rate
P_t is the mean of transmitted power in Equation 5.1 for the hop length r_R and is represented by [Datta, 2009]:

$$P_t = P_r \cdot r_R^{\alpha} \cdot e^{-(m_{S_{de}} + (\sigma_{S_{de}}^2/2))} \cdot e^{(m_{R_{de}} + (\sigma_{R_{de}}^2/2))} \tag{5.28}$$

where *de* represents the distance of length r_R.

The energy consumption of "startup transients" can be significant sometimes in the context of energy-constrained sensor nodes. This is taken into account while estimating the energy consumption for data communication in WSN. It is observed that startup energy consumed in the transmitter/receiver varies from 10 to 45 μJ [*ASH Transceiver Designer's Guide*; *ATMEL Transceiver Designer's Guide*]. In the present analysis, while estimating the total energy consumption for successful delivery of information from the source to the sink, two conditions are considered separately, that is, with and without considering startup energy of source/transmitter node. The energy consumed per packet from the source to the sink, that is, single-loop transmission of information from the source to the sink via single hop, with ACK/NACK from sink to source and considering startup energy, is

$$(E_{pkt})_{strtup} = E_b \cdot n_f + E_b \cdot n_b + E_{strtup} \tag{5.29}$$

while the same parameter without startup energy is

$$(E_{pkt}) = E_b \cdot n_f + E_b \cdot n_b \tag{5.30}$$

where E_{strtup} is the startup energy of the source/transmitter node. We consider that the sink is not an energy-constrained node and it is always in active state. Further, we assume that other nodes will wake up from sleep state to active state and will remain in that state until the information is received correctly at the sink. Since on the average, each packet requires $\overline{N}$ number of retransmissions from source to destination for successful delivery, the average energy consumed for successful delivery of a packet is

$$(E_{av})_{strtup} = \overline{N} \cdot E_{pkt} + E_{strtup} \tag{5.31}$$

Without considering startup energy at each node, the average energy consumed for successful delivery of a packet is

$$E_{av} = \overline{N} \cdot E_{pkt} \tag{5.32}$$

5.2.2.2 Energy-Efficiency-Based Packet Size Optimization

A suitable metric that captures the energy and reliability constraints (i.e., the energy efficiency ξ) is discussed by Sankarasubramaniam [2003]. It is an important objective function and is considered for optimization of packet length to minimize energy consumption. Following the system model shown in Figure 5.4 and the energy model considered in Equation 5.27, the energy efficiency of the system is evaluated and the variation of energy efficiency with packet length is observed under both the conditions, that is, with and without considering startup energy of sensor nodes [Datta, 2010b]. We consider stop-and-wait ARQ with cyclic redundancy check (CRC), where the received packet is checked by CRC decoding, followed by retransmission in the case of erroneous data. Let $\ell_{payload}$ be the payload length and n_f be the length of the packet to be transmitted after encoding with CRC, that is, $n_f = (\ell_{payload} + \beta_1)$ is the size of the packet. Here $\ell_{payload}$ includes the actual message ($\ell_{message}$) and header (∂), and β_1 is the length of the frame check sequence, as shown in Figure 5.5.

Energy consumed only by the payload in single transmission, that is, the effective energy, including the startup energy, is

$$(E_{eff})_{startup} = \frac{(P_t + P_r)\cdot(n_f - \beta_1)}{R_b} + E_{strtup} \tag{5.33}$$

The effective energy without startup energy at each node is

$$E_{eff} = \frac{(P_t + P_r)\cdot(n_f - \beta_1)}{R_b} \tag{5.34}$$

Considering successful reception of a packet from the source to the sink, the energy efficiency with startup energy is expressed by

$$\xi_{strtup} = \frac{(E_{eff})_{strtup}}{(E_{av})_{strtup}} \tag{5.35}$$

where $(E_{av})_{strtup}$ is given in Equation 5.31.

Without considering startup energy,

$$\xi = \frac{E_{eff}}{E_{av}} \tag{5.36}$$

where E_{av} is given in Equation 5.32.

Header (∂)	Message ($\ell_{message}$)	Frame check sequence (β_1)

FIGURE 5.5
MAC frame format of DATA in 802.15.4. (From IEEE Standard for Information Technology, Telecommunications and information exchange between systems—Local and metropolitan area networks—specific requirement Part 15.4: Wireless Medium Access Control (MAC) and Physical Layer (PHY) Specifications for Low-Rate Wireless Personal Area Networks (WPANs), 2006.)

The optimal packet size to maximize energy efficiency is derived from these expressions. Without considering startup energy, after simplification, we derive the optimized packet length as follows:

$$L_{opt}^{eff} = \frac{\sqrt{c_0^2 + (4c_0 / (\ln(1-(\overline{P_e})_H)))} - c_0}{2} \tag{5.37}$$

where $c_0 = \beta_1 + n_b$. However, considering startup energy, which is significantly high as compared to the combined energy consumption due to transmission and reception, a closed form solution to the optimized packet length by maximizing energy efficiency ξ_{strtup} is not straightforward to obtain. We evaluate it with the help of simulation under such startup energy included case.

5.2.2.3 Performance of Single-Hop Scheme

The simulation model is developed in MATLAB® for the validation of analytical results on BER, throughput, and delay considering pce, shadowing, and presence of correlated MAI and NI for single hop. Parameters used in the analysis and simulation are shown in Table 5.1. In this section, Figures 5.6 through 5.8 show the results using analytical and simulation models, and the other figures show the results using the analytical model.

Pair-wise correlation coefficients between different interferers are assumed to be same, that is, $r_{z_i z_j} = r_{R_{ki} R_{kj}} = r_{Z_i R_{kj}} = r$, for $i \neq j \neq k$. In a typical case, 25% of total nodes within receiving distance (r_R) are assumed to be active for MAI, that is, constituting t_2 number interferers, while 12.5% of total nodes within the receiving and interfering distances (r_R,r_I) are active for NI, constituting t_1 number interferers, which are used in Equation 5.9. E_{pkt} is calculated assuming a receiving distance of 15 m between transmitter and sink.

TABLE 5.1

Parameters Used in Analysis of Single-Hop CDMA WSN with Correlation among Interferers, Applying Infinite ARQ

Parameter	Value
Receiving distance (r_R)	15 m
Node density (nodes/m^2)	0.008, 0.016, 0.024, 0.032
Minimum distance between two nodes (r_0)	1 m
Processing gain (pg)	128
Constant receive power (P_r)	1.0e^{-07} mW
Path loss parameter (α)	3
Transmission rate (R_b)	20.0 kbps
Forward transmission (n_f)	128 bits/packet
NACK/ACK (n_b)	2 bits/packet
Correlation coefficient (r)	0.0, 0.3, 0.6
Standard deviation of pce (σ_R)	0.5, 1.0 dB
Standard deviation of shadowing (σ_S)	3 dB
Code rate (r_c)	1
Startup energy of each node	10×10^{-3} mJ
Spread bandwidth (W)	5 MHz

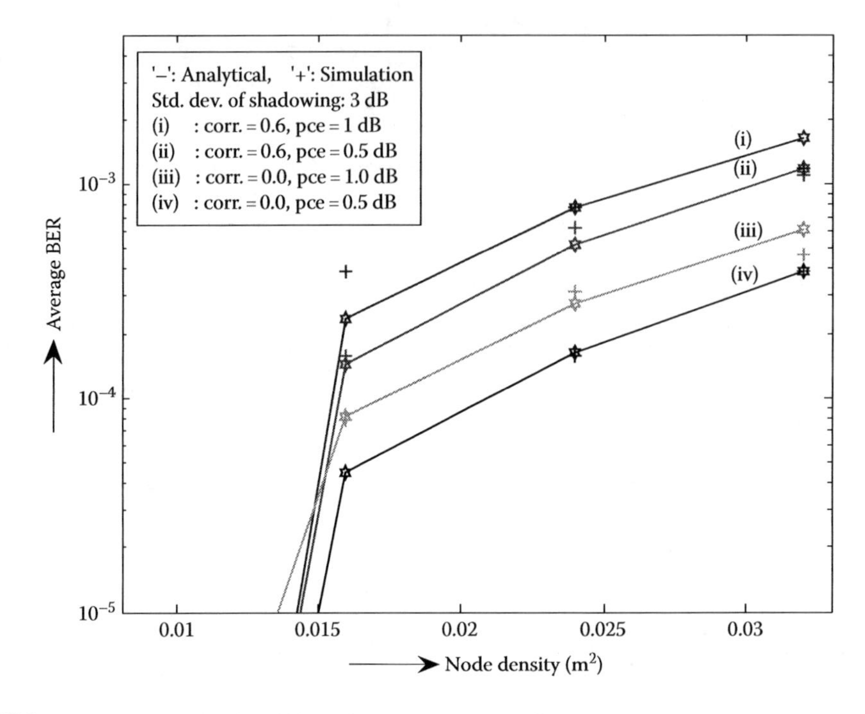

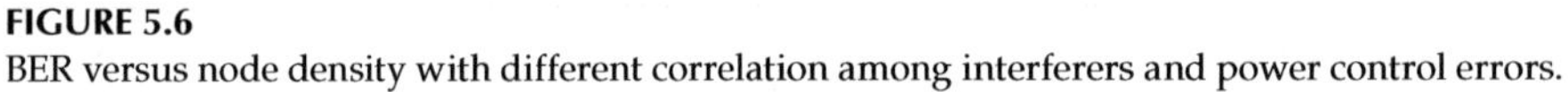

FIGURE 5.6
BER versus node density with different correlation among interferers and power control errors.

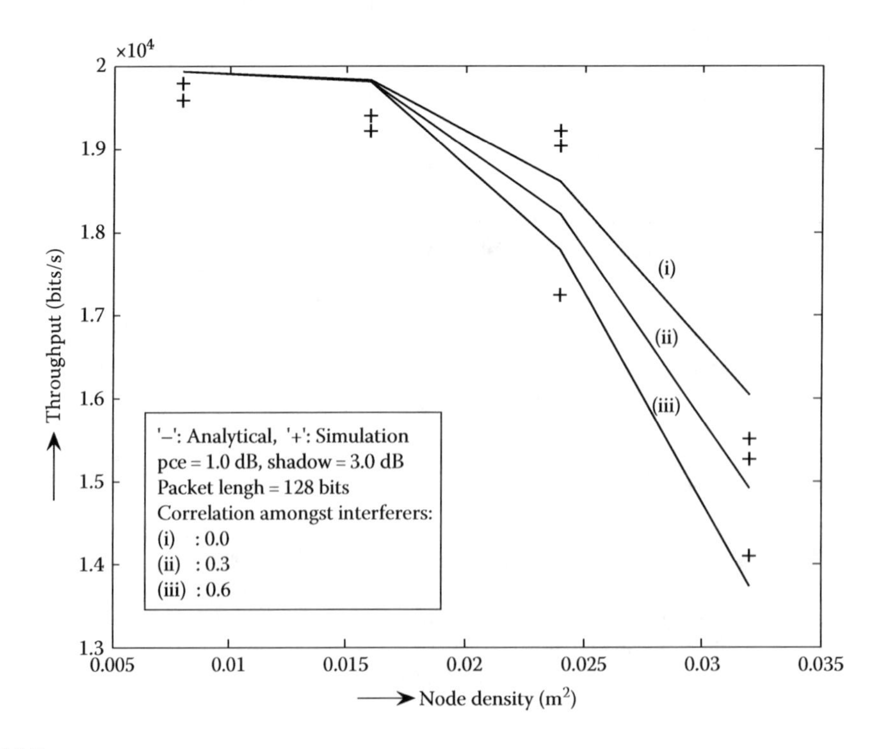

FIGURE 5.7
Average throughput versus node density with different correlation among interferers.

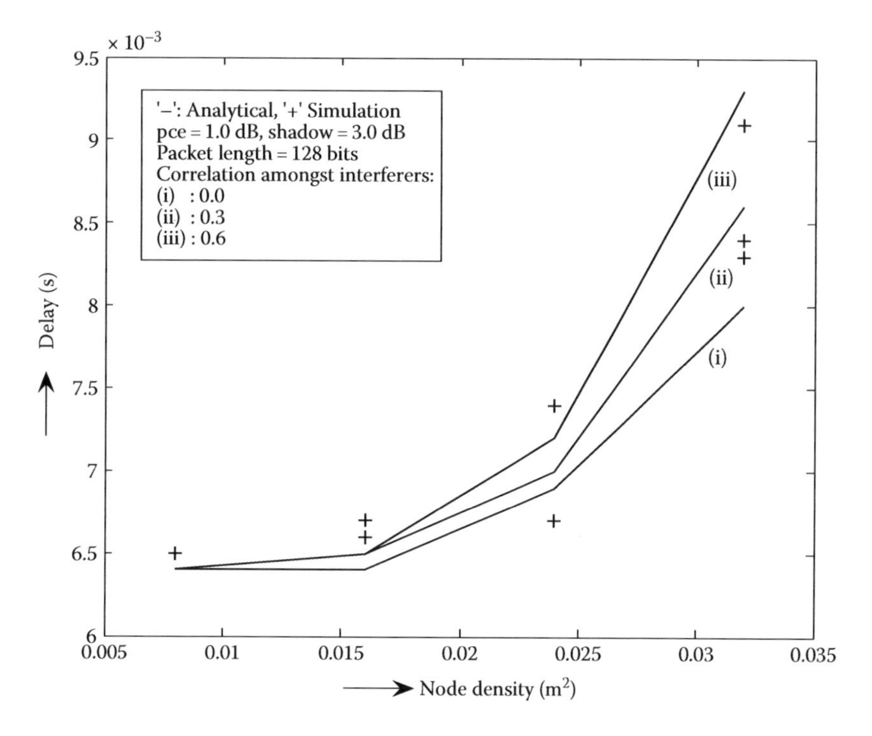

FIGURE 5.8
Average delay versus node density with different correlation among interferers.

Figure 5.6 depicts the variation of BER with node density. The impact of correlation and pce on BER is also shown. With the increase in node density, the number of interferers within a certain region increases, which results in the degradation of BER. It is seen that BER increases with increase in correlation as SIR degrades due to increase in correlation among interferers [Abu-Dayya, 1994]. Further, as pce increases, total interference at sink increases, which leads to increase in BER. For a node density of $0.024/m^2$ and pce = 1.0 dB, BER increases by 37% as correlation r increases from 0.3 to 0.6. For a node density of $0.016/m^2$ and $r = 0.3$, BER increases by 85% as pce increases from 0.5 to 1.0 dB.

Figure 5.7 depicts the variation of average throughput G_{av} with node density for several values of correlation. The packet size is 128 bits and pce is fixed at 1.0 dB. For a particular node density and pce, as BER degrades with increase in correlation, the average number of retransmissions required for successful reception of a packet increases. This in turn reduces the average number of information bits successfully transferred per second, that is, the average throughput G_{av} also reduces. For a node density of $0.032/m^2$, G_{av} decreases by 30% as correlation increases from 0.3 to 0.6.

Figure 5.8 shows the variation of average delay (D_{av}) in receiving a packet of 128 bits successfully with node density for different correlations among the interferers. Higher correlation increases BER followed by an increase in packet error rate (PER), which requires more number of retransmissions. This, in turn, increases D_{av} in receiving a packet successfully for a particular node density. For a node density of $0.016/m^2$, D_{av} increases by 16% as the correlation increases from 0.3 to 0.6.

Figures 5.9 and 5.10 depict the variation of average energy consumption for the successful delivery of a packet of length 128 bits to the sink with node density for different correlations among interferers. Both the cases, that is, with and without considering the

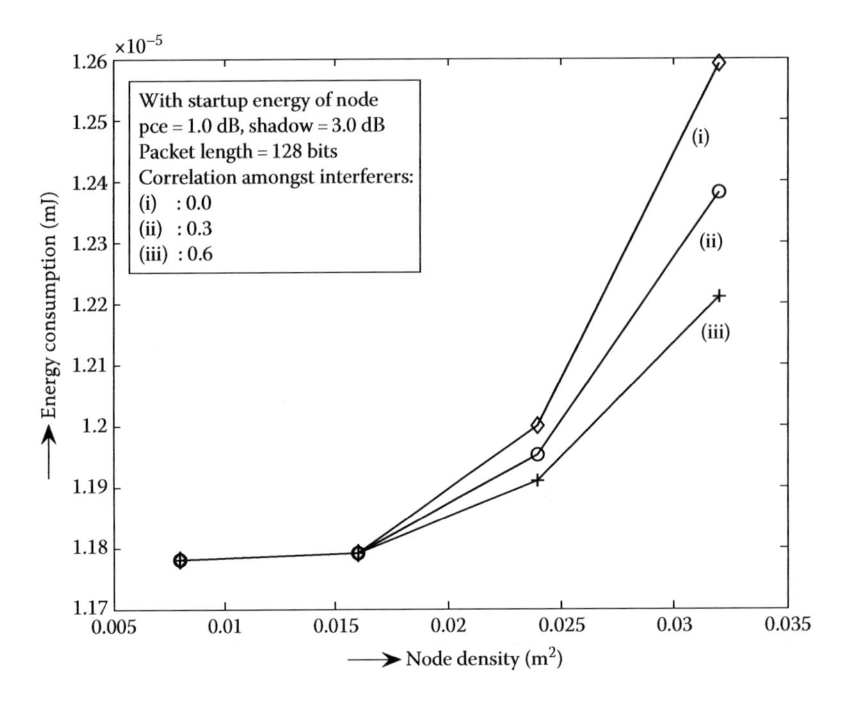

FIGURE 5.9
Average energy consumption versus node density with different correlation among interferers, considering startup energy of sensor nodes.

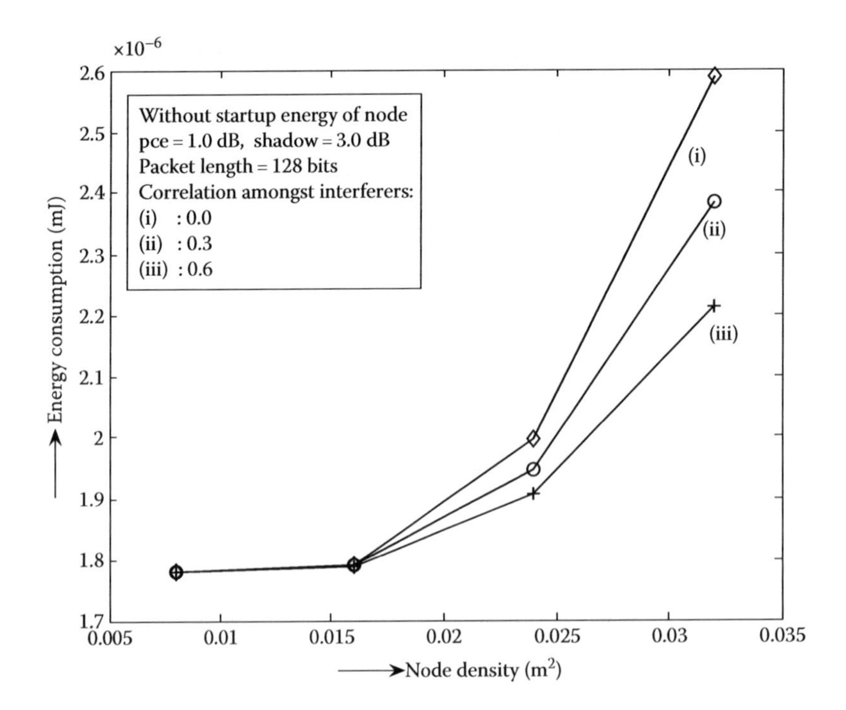

FIGURE 5.10
Average energy consumption versus node density with different correlation among interferers, without startup energy of sensor nodes.

startup energy of nodes in the network, are shown. It is seen that the startup energy consumes a significant amount of energy. Though the nature of variation of energy consumption with node density remains the same in both the schemes, that is, with and without startup energy, the percentage change in energy consumption for the same variation of node density is very low with startup energy as compared to without startup energy. For example, on changing the node density from $0.024/m^2$ to $0.016/m^2$, energy consumption is decreased by 1.3% in the case with startup energy, while energy consumption is decreased by 7.2% without startup energy. Increase in correlation increases BER followed by an increase in PER, and consequently the number of retransmissions required for successful delivery of a packet increases. Thus, more energy has to be spent for the successful delivery of a packet. For a node density of $0.032/m^2$, in the case without startup energy, the average energy consumption increases by 8.6% when correlation increases from 0.3 to 0.6. However, considering startup energy, the average energy consumption increases by only 1.6% for the same change in correlation, that is, the impact of correlation on energy consumption is more significant for the model without startup energy.

Figures 5.11 and 5.12 depict the variation of energy efficiency with packet length for several values of node densities, with and without considering startup energy at each node, while other parameters remain fixed as shown in the figures. In the present case also, startup energy plays a significant role in controlling the nature of energy efficiency curve of the system. As the startup energy at each node consumes a significant amount of energy in comparison to the transmission and reception energy associated with each node, no optimal packet length is available to maximize energy efficiency, as depicted in Figure 5.8. With increase in packet length, PER increases, which reduces energy efficiency.

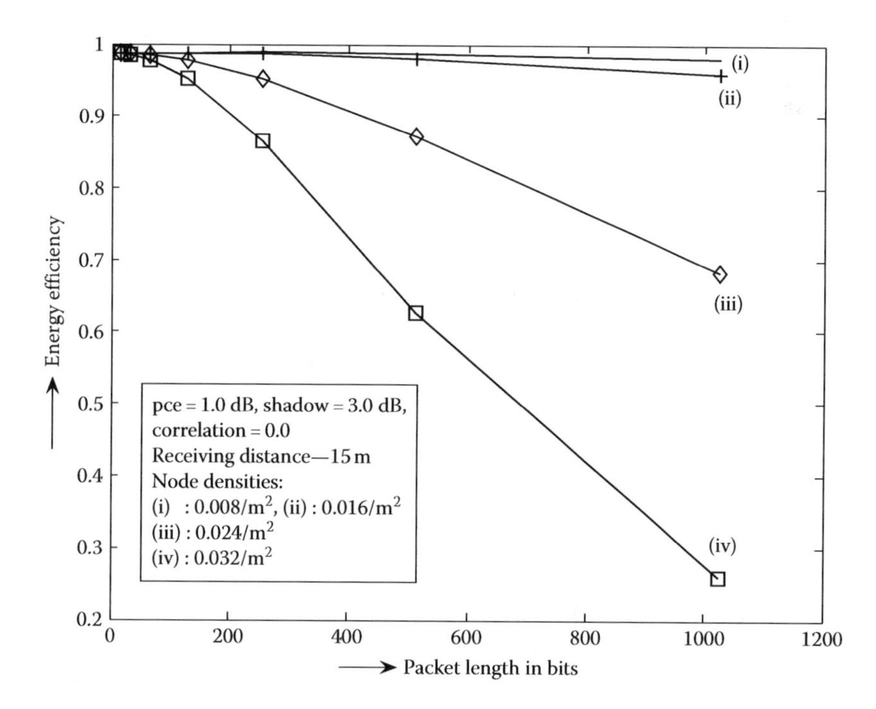

FIGURE 5.11
Energy efficiency versus packet lengths under several node densities, considering startup energy of sensor nodes.

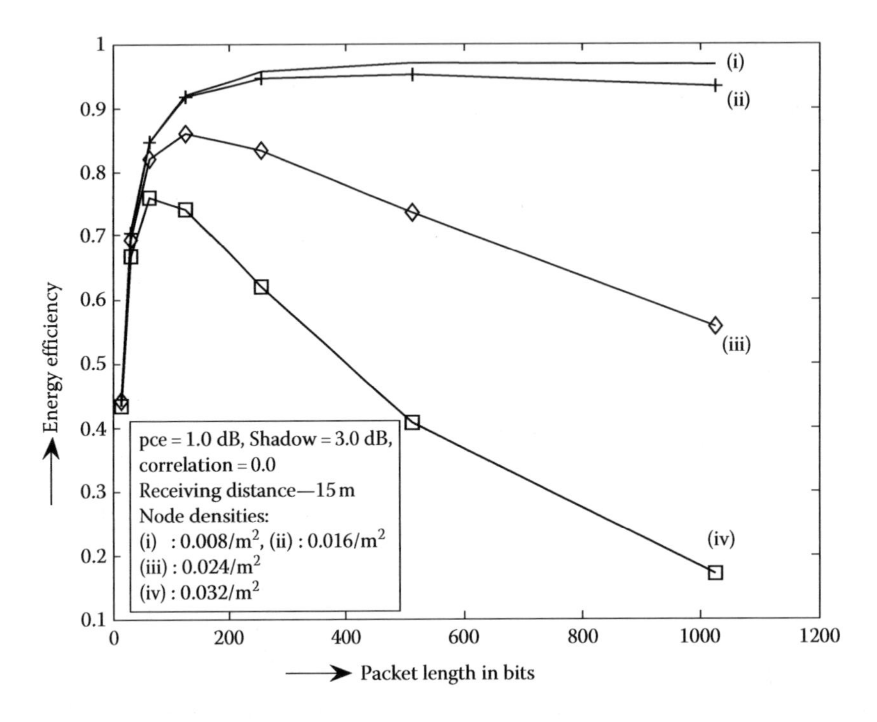

FIGURE 5.12
Energy efficiency versus packet lengths under several node densities, without startup energy of sensor nodes.

With increase in node density, degradation of BER causes increase in energy consumption for successful delivery of packet and decrease in energy efficiency. Without considering startup energy, both the attainable energy efficiency and the optimized packet length, as stated in Equations 5.36 and 5.37, increase with decreasing node density since BER decreases with decrease in node density. Steep drop of energy efficiency for smaller payload length is due to considerable overhead for smaller packets. For a fixed node density, as packet length increases larger than the optimized one, increase in PER causes decrease in energy efficiency as indicated in Equation 5.36. Thus, we have an optimized packet length in the case of the model without startup energy but no such optimal packet length in the other case. For a 128-bit packet length, considering startup energy, energy efficiency decreases by only 2.4% (curves iii, iv; Figure 5.11) as node density increases from $0.024/m^2$ to $0.032/m^2$; whereas without startup energy, energy efficiency decreases by 13.8% (curves iii, iv; Figure 5.12) under the same change of node density.

Figure 5.13 shows the comparison of total energy consumed for successful transmission of a file of size 4096 bits based on optimized packet length and using several other arbitrary packet sizes. It is observed that energy consumption using optimum packet sizes at different node densities is significantly lower as compared to that of other packet lengths. It may be noted that in the case of a transmission based on optimized packet length, we use optimum packet size corresponding to that particular node density. Thus, the transmission packet size needs to be changed adaptively with change in node densities.

Figure 5.14 depicts the delay to transmit a file of size 4096 bits successfully based on optimized packet length and using several other packet lengths. It is observed that at low node densities, delay is low at different packet lengths, including the optimized one; whereas at higher node densities, increase in delay using optimized packet lengths is not significant as compared to other fixed packet lengths, for example, 256, 128 bits.

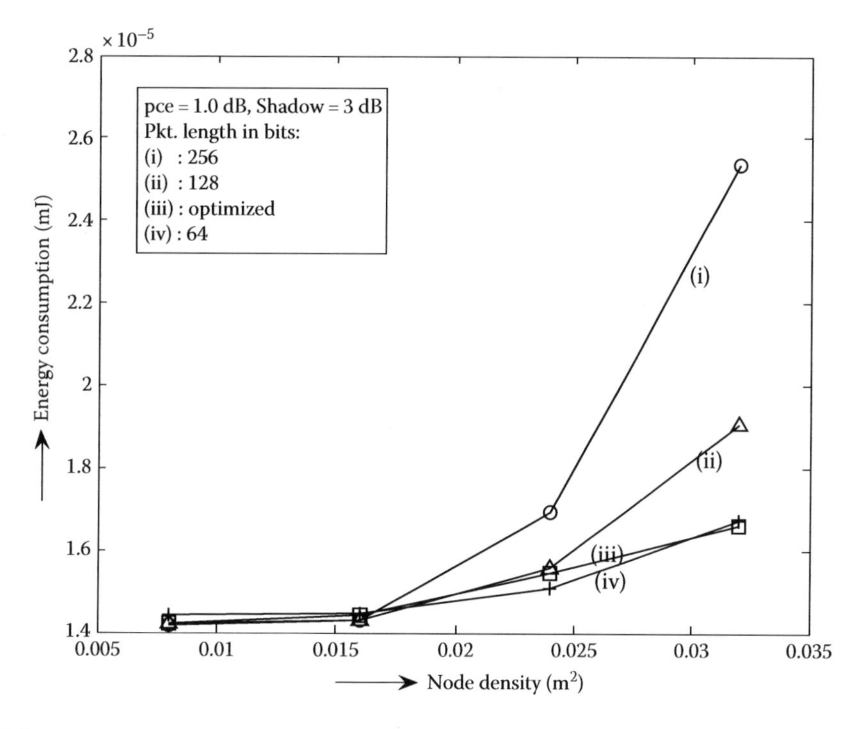

FIGURE 5.13
Energy consumption for successful transmission of a file, size 4096 bits with node densities, using different packet lengths.

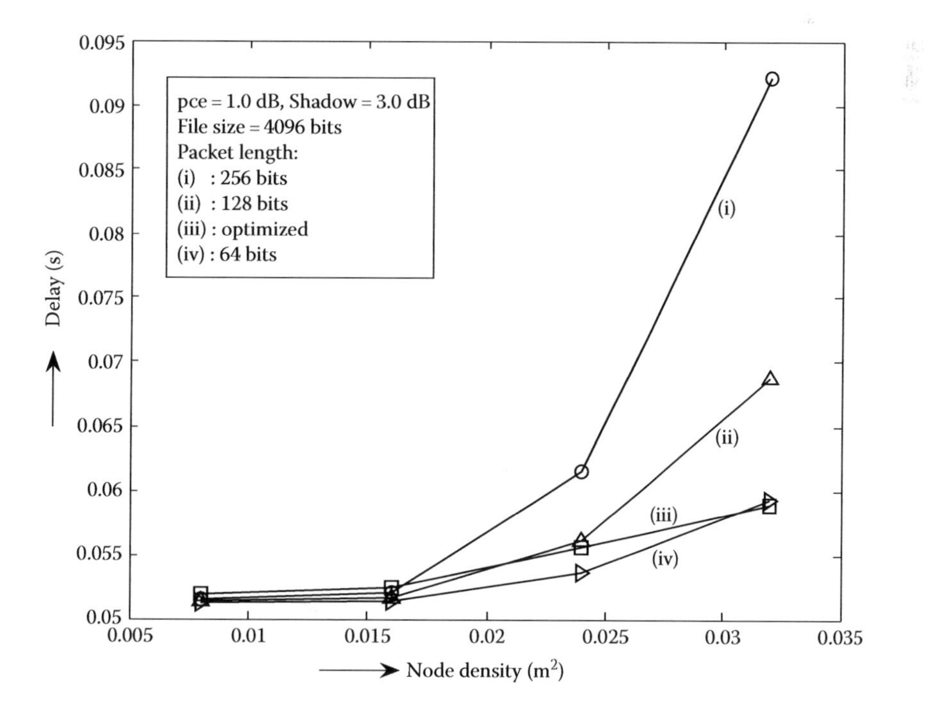

FIGURE 5.14
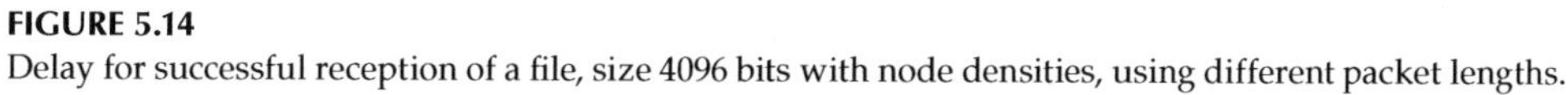
Delay for successful reception of a file, size 4096 bits with node densities, using different packet lengths.

5.2.3 Multi-Hop Communication

In wireless distributed sensor networks, relayed transmission, as shown in Figure 5.15, is a promising technique that helps in attaining broader coverage while combating the impairment of the wireless channel. Further relaying information on several hops reduces the need to use a large power at the transmitter and distributes the use of power throughout the hops. This results in extended battery life and lowered level of interference [Laneman, 2000], maintaining QoS issues such as outage and BER, which are important metrics for WSN.

In this section, transmission of information via multi-hop communication in a CDMA-based sensor network is considered, and the end-to-end performance in terms of BER over independent, identically distributed lognormal shadowed channels is evaluated using appropriate modeling of MAI and NI for each hop considering correlated interferers and imperfect power control. Multi-hop communication is realized through regenerative relays where fresh bits are regenerated from the received signal and transmitted with power control. Energy consumption using three information-delivery mechanisms, as described in Section 5.2.3.2, is investigated and compared. Simulation is carried out to support analytical results.

5.2.3.1 System Model with Fixed Hop Lengths

A CDMA WSN with uniformly distributed nodes in layered architecture has the sink at the center of concentric layers. Different layers are formed by the sensor/relay nodes that have the same hop count at the sink, as shown in Figure 5.15. A network is considered where information from a source node reaches the final destination node (i.e., the sink)

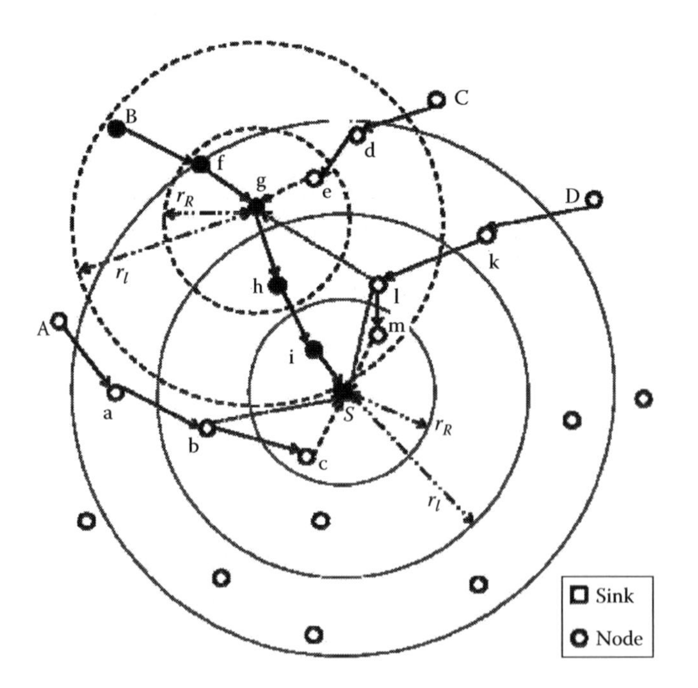

FIGURE 5.15
Interference model at a node at an instant in multi-hop layered architecture.

via a suitable path in multi-hop. Information can be routed only inward from a higher hop layer node to a lower hop layer node with respect to the sink. A node can identify its intended neighbor and a proper routing and node-scheduling algorithm ensures the data flow from the source to the sink in multi-hop via several intermediate relay nodes. For example, in Figure 5.12, the destination node (i.e., sink S) receives information from the source nodes A, B, C, D, etc., which are situated at higher layers with respect to sink, via multi-hop communication using digital relaying as employed in any regenerative systems. Relays regenerate the received signal and transmit the same with power control to the next hop. As we are considering a CDMA sensor network, any node can transmit to any other node, so that the source information finally reaches the destination, that is, the sink. At a particular instant, nodes f and e use the same intermediate relay node g to route their information to the sink. Since only the nodes located within the receiving distance r_R (i.e., one-hop layer of g) can directly communicate with the intended receiver node g, the concurrent nodes that send their information to g within the area πr_R^2 would cause MAI. The concurrent transmitted signal power generated by the source/relay nodes situated within the interference range r_I of g, which send information to their respective destination nodes, might be sensed by g, which causes NI. In Figure 5.15, at any instant, node e can be regarded as causing MAI, and nodes h, i, d, l, and source node B can be regarded as causing NI to g. During the propagation of signal from the source to the sink via different nodes at different hop layers, the desired signal at each receiver node is accompanied with MAI and NI. To obtain the interference power distribution at each receiving node, the assumptions as described in Section 5.2.2, including the following, are used:

- As all information is received by the sink, sensor nodes within the one-hop layer of the sink will have higher level of activity than those in other layers. Thus, the numbers of nodes creating MAI to the sink are more in comparison to those causing MAI to other intermediate relay nodes.
- Homogeneous network is considered, that is, on an average, circular areas of the same radius contain the same number of nodes. This assumption justifies taking the number of NI nodes as the same for sink as that for other intermediate relay nodes.

For analysis of the average end-to-end BER, first modeling of BER at each hop is done following Equation 5.19. The number of hops between a source node and the destination node (i.e., the sink) is assumed to be H with the same channel conditions throughout all hops. The average end-to-end BER for H hops without any error correction mechanism, applied at the intermediate relay nodes, is expressed by the relation [Tonguz, 2004]

$$\overline{(P_e)}_H = 1 - \prod_{i=1}^{H}\left(1 - \overline{P_e}^{(i)}\right) \tag{5.38}$$

where $\overline{P_e}^{(i)}$ is the mean probability of error at the ith hop. The average BER for each hop will remain same at all hops, except at the final hop, where higher activities of nodes surrounding the sink, that is, more number of active nodes in this hop. BER is calculated separately for this hop using Equation 5.19, with necessary changes in parameters (m_ψ, σ_ψ) due to changes in the number of active interferers.

5.2.3.2 Different Transmission Strategies

A multi-hop path with H hops between the source and the destination (i.e., the sink) is considered, where n_f bits per packet in forward transmission of information while n_b bits per packet for NACK/ACK with an assumption of error-free reception of NACK/ACK are assumed. Assuming perfect error detection of a CRC code, an infinite ARQ mechanism is used for error correction. Three alternative mechanisms are considered, as shown in Figure 5.16:

Scheme I is based on hop-by-hop (HbH) retransmission, as shown in Figure 5.16a following She [2009], where at every hop the receiver checks the correctness of the packet and requests for a retransmission with a NACK packet to previous node until a correct packet is received. An ACK packet is sent to the transmitter, indicating a successful transmission.

Scheme II is based on end-to-end (E2E) delivery with intermediate nodes, performing as digital repeaters [Taddia, 2004] as shown in Figure 5.16b. The packet is checked only at D for correctness; retransmissions are requested to the source, with a NACK coming back from the destination (sink) to the source through a multi-hop path.

Scheme III is based on E2E delivery with intermediate nodes, performing as digital repeaters [She, 2009], as shown in Figure 5.16c. The packet is checked at the destination for correctness, and retransmissions are requested to the source, with a NACK coming back to the source directly from the destination (sink) instead of a multi-hop path in scheme II. Energy consumptions for different schemes are calculated.

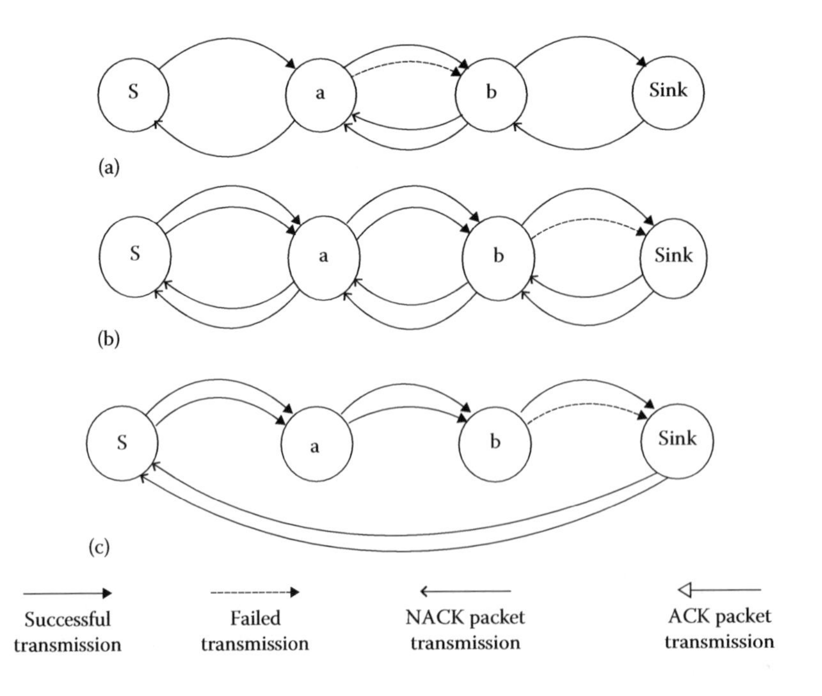

FIGURE 5.16
Information-delivery schemes. (a) Scheme I, (b) Scheme II, and (c) Scheme III.

5.2.3.3 Energy Consumption Using Different Transmission Strategies

Scheme I: The average probability of error at packet level in each hop is expressed by Equation 5.22.

The average number of retransmissions $\overline{(N)}_I$ for successful delivery of packet through each link using scheme I is expressed by Equation 5.24. The energy consumed per packet at each hop, $(E_{pkt})_I$, is considered the energy spent only in the communication of data packet in both forward direction and transmission of ACK/NACK in reverse direction, as expressed by Equation 5.30. Thus, the average energy consumed for the successful delivery of packet through H hop communication is expressed as follows [Datta, 2013]:

$$(E_{av})_I = (H-1)\cdot\overline{(N)}_I\cdot(E_{pkt})_I + (\overline{(N)}_I)_{lasthop}\cdot(E_{pkt})_I \tag{5.39}$$

The first term represents the average energy consumption for all hops, excluding the last hop, where sink is the receiver. The second term represents the average energy consumption for the successful delivery of information at the last hop. $(\overline{(N)}_I)_{lasthop} > \overline{(N)}_I$ due to the higher node activity at the last hop.

Scheme II: The average probability of error at packet level for H hops is

$$\overline{(P_f)}_H = 1-(1-(\overline{P_e})_H)^{n_f} \tag{5.40}$$

where $(\overline{P_e})_H$ is the average route BER, as expressed in Equation 5.38. Following Kleinschmidt [2007], the average number of retransmissions for the successful delivery of a packet from the source to the destination using scheme II is

$$\overline{(N)}_{II} = \frac{1}{\left(1-\overline{(P_f)}_H\right)} \tag{5.41}$$

The energy consumed per packet for a single-loop communication from the source to the destination through H hops is considered the energy spent in the communication for both forward and reverse transmission [Datta, 2010a; Sakhir, 2008]:

$$(E_{pkt})_{II} = H\cdot(P_t+P_r)\cdot\frac{n_f}{R_b} + H\cdot(P_t+P_r)\cdot\frac{n_b}{R_b} \tag{5.42}$$

where

R_b is the data rate
P_t is the transmit power for one bit, as given in Equation 5.28
P_r is the receive threshold for one bit

The first and second terms represent the energy spent only in the communication of data packet through multi-hop in the forward direction and the multi-hop transmission of ACK/NACK in the reverse direction. The average energy consumed for the successful delivery of information through a multi-hop communication of H hops using the second mechanism is

$$(E_{av})_{II} = \overline{(N)}_{II}\cdot(E_{pkt})_{II} \tag{5.43}$$

Scheme III: In this scheme, the average PER over H hops is same as that of scheme II. The average number of retransmissions for the successful delivery of packet from the source to the sink, $(\overline{N})_{III}$, is given by Equation 5.41. The energy consumed per packet, $(E_{pkt})_{III}$, for a single-loop communication from the source to the destination is considered the energy spent in the forward communication from the source to the sink through H hops and reverse communication for NACK/ACK from the sink to the source through a single hop [Datta, 2010a; Sakhir, 2008]:

$$(E_{pkt})_{III} = H \cdot (P_t + P_r) \cdot \frac{n_f}{R_b} + (P_{tr} + P_r) \cdot \frac{n_b}{R_b} \tag{5.44}$$

where

$$P_{tr} = P_r \cdot (H \cdot r_R)^{\alpha} e^{-(m_S + (\sigma_S^2/2))} \cdot e^{(m_R + (\sigma_R^2/2))} \tag{5.45}$$

is the mean transmitted power for transmitting ACK/NACK from the sink to the source directly over a length of $H \cdot r_R$

r_R is the length of each hop

α is the path loss component

The average energy consumed for the successful delivery of information through a multi-hop communication of H hops using scheme III is

$$(E_{av})_{III} = (\overline{N})_{III} \cdot (E_{pkt})_{III} \tag{5.46}$$

5.2.3.4 Network Lifetime

Critical to any WSN deployment is the expected lifetime. In general, network lifetime is the time span from the deployment to the instant when the network is considered nonfunctional. A network should be considered nonfunctional, for example, the instant when the first sensor dies, a percentage of sensors dies, or the loss of coverage occurs [Dietrich, 2009; Kang, 2008]. Panichpapiboon [2006] presented optimal solutions for maximizing the time to the first node failure for a static broadcast tree. Following Panichpapiboon [2006], a simple analysis is presented for computing the average lifetime of a node. Every node has an identical initial finite battery energy denoted by E_{batt}, and packets are transmitted with the average rate λ_t packets per second without queuing. The average energy depleted per second due to transmission and reception is simply $\lambda_t \cdot E_{packet}$, where E_{packet} is the average energy consumed per node for the successful transmission of a packet. Further, $E_{packet} = E_{av}/H$, where E_{av} is the average energy consumed for the successful transmission of a fixed-length packet from the source to the sink through H hops in each scheme. Finally, the total time it takes to completely exhaust the initial battery energy is

$$\tau = \frac{E_{batt}}{(\lambda_t \cdot E_{packet})} \tag{5.47}$$

Due to the assumption of uniform traffic, on an average, all nodes exhaust their battery at the same time. This simple analysis does not take into account the energy consumed when a node is processing packets. Thus, the actual lifetime of a node will be shorter than what is predicted by our analysis.

5.2.3.5 *Performance of Multi-Hop Schemes with Fixed Hop Lengths*

The parameters shown in Table 5.2 are used in analysis. In this section, Figures 5.17 and 5.18 show the results using analytical and simulation models, and the other figures show the results using the analytical model.

The means of all shadowing and pce components are considered to be zero. The pair-wise correlation coefficients between different interferers are assumed to be equal to r. We assume that in the last hop involving sink, 10% of total nodes are active for MAI,

TABLE 5.2

Parameters Used in Analysis of Multi-Hop CDMA WSN with Fixed Hop Length, Applying Infinite ARQ

Parameter	Value
Node density (nodes/m^2)	0.008, 0.016, 0.024, 0.032
Receiving distance (r_R)	15 m
Minimum distance between two nodes (r_0)	1 m
Processing gain (pg)	128
Constant receive power (P_r)	1.0e^{-07} mW
Path loss parameter (α)	3
Transmission rate (R_b)	9.6 kbps
Forward transmission (n_f)	128, 512 bits/packet
NACK/ACK (n_b)	2 bits/packet
Correlation coefficient (r)	0.0, 0.3, 0.6
Standard deviation of pce (σ_R)	0.5, 1.0 dB
Standard deviation of shadowing components (σ_S)	2, 2.5 dB
Number of hops	5, 8

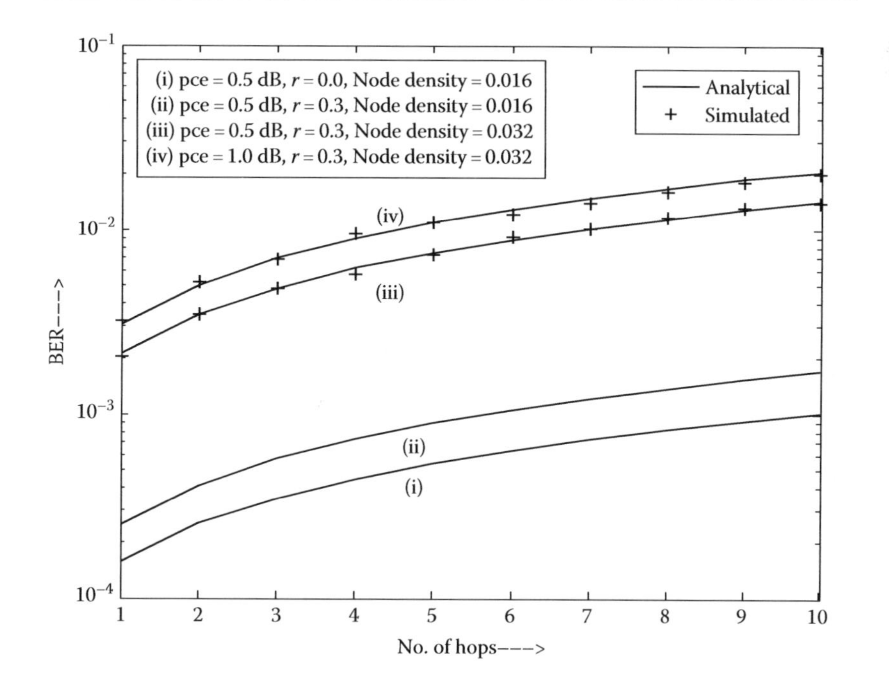

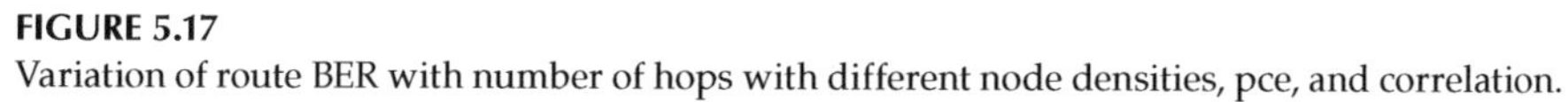

FIGURE 5.17
Variation of route BER with number of hops with different node densities, pce, and correlation.

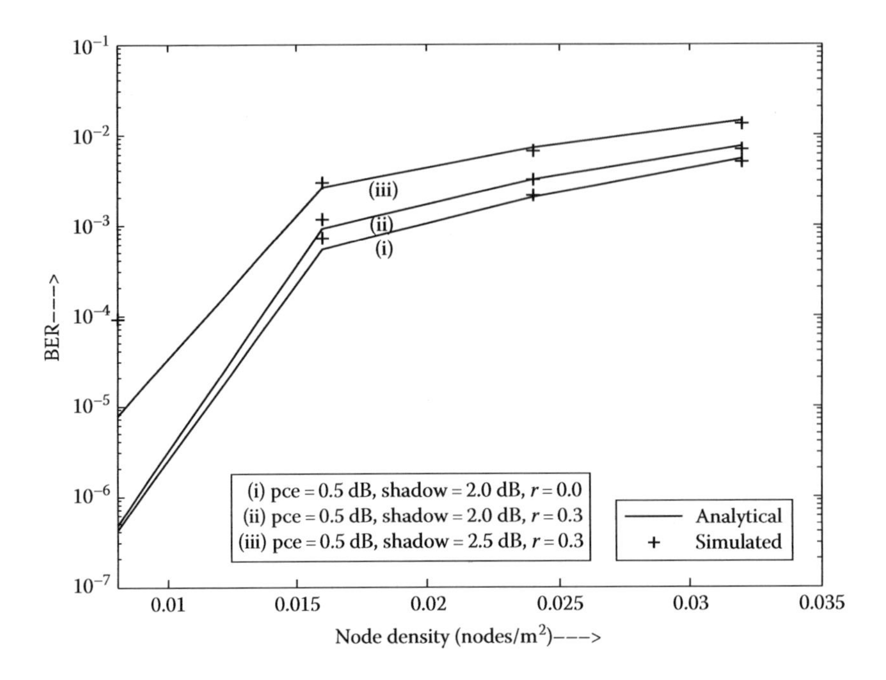

FIGURE 5.18
Variation of route BER in scheme II or scheme III with node density for different correlation among interferers and standard deviation of shadowing.

while 20% of MAI nodes are active for NI. As we consider uniform distribution, 10% of total nodes are considered to be active for both MAI and NI at other nodes in the network at any instant. The energy consumed per packet (E_{pkt}) is estimated considering the distance between any two consecutive nodes, $r_R = 15$ m, that is, the hop length is 15 m.

Figure 5.17 shows the variation of route BER, pertaining to scheme II and scheme III, with the number of hops for different values of pce and correlation among interferers having a fixed hop length $r_R = 15$ m. In a multi-hop communication, E2E BER is determined by the cumulative effects of BER of individual hops. Thus, with an increasing number of hops, route BER increases. For a fixed number of hops and same channel condition, route BER increases with an increase in pce. At individual hops, SIR decreases as interference increases with an increase in pce, which increases link BER as well as the route BER. For a number of hops equal to 5, node density 0.016/m^2, shadowing standard deviation of 2 dB, and pce of 0.5 dB, BER increases by 66% for an increase in correlation (r) from 0.0 to 0.3. For the same number of hops, node density 0.032/m^2, shadowing of 2 dB, and correlation (r) of 0.3, BER increases by 45% for an increase in pce from 0.5 to 1.0 dB.

Figure 5.18 shows the variation of route BER with node density in scheme II and scheme III for different values of shadowing and correlation for five-hop communication, that is, H is equal to 5, and pce of 0.5 dB. It may be noted that route BER is same for scheme II and scheme III. The severity of shadowing increases the required transmit power for correct reception at the nodes, which increases the interference at the receiver. This results in an increase in BER at each hop and hence the route BER increases.

Figure 5.19 depicts the variation of average energy consumption with node density for several correlations among interferers under different information-delivery schemes. With an increase in correlation among the interferers only, BER increases. Thus, for a particular packet length, the average number of retransmissions for the successful delivery of packets

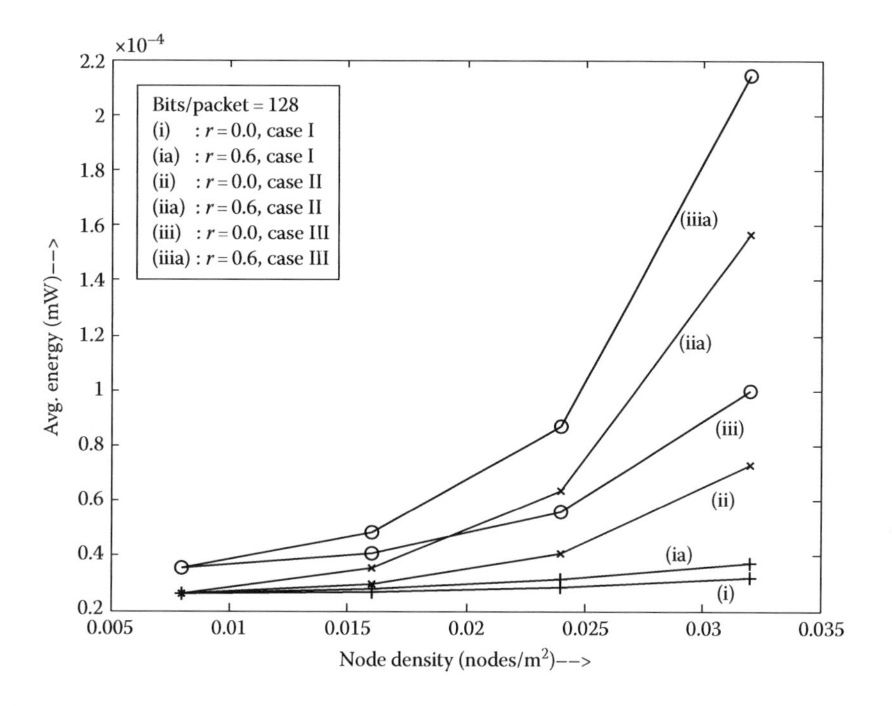

FIGURE 5.19
Average energy consumed versus node density with different correlation among interferers; shadowing standard deviation σ = 2 dB, pce = 1 dB, number of hops = 5, and number of bits/packet = 128.

increases. Scheme I/case I, that is, HbH retransmission shows best performance. Due to the high requirement of energy to transmit ACK/NACK in scheme III/case III, the maximum amount of energy is required for the successful delivery of a packet in scheme III.

Figure 5.20 depicts the variation of average energy consumption with node density for different number of hops (H) with a fixed packet length of 128 bits, pce of 1 dB, shadowing of 2 dB, and correlation (r) of 0.3. It is observed that with an increase in the number of hops with a fixed hop length, energy consumption increases. Change in the consumption of energy with increasing hop count is less in scheme I as compared to other two mechanisms. This is due to the higher energy consumption involved in the degradation of route BER associated with the higher number of retransmissions in scheme II and scheme III. Scheme III shows a significant increase in energy with an increase in the number of hops. It is due to a sharp increase in energy consumption for ACK/NACK bits from the sink to the source in a single hop. For a node density of 0.032/m^2, increasing H from 5 to 8 increases the average energy consumption by 58% in scheme I, whereas for scheme II and scheme III, they are 240% and 600%, respectively.

5.2.3.6 Multi-Hop Transmission between a Fixed Source and Sink

In the analysis of multi-hop communication as investigated in Sections 5.2.3.1 to 5.2.3.5, increase in the number of hops increases the distance between the source and the sink. In practical cases, for a static WSN, the distance between the source and the sink is kept fixed. Thus, increase in the number of hops results in a reduction in hop length. Moreover, the startup energy of the transmitter and receiver from the sleep mode to the active mode, which consumes a significant amount of energy at each hop, needs to be considered.

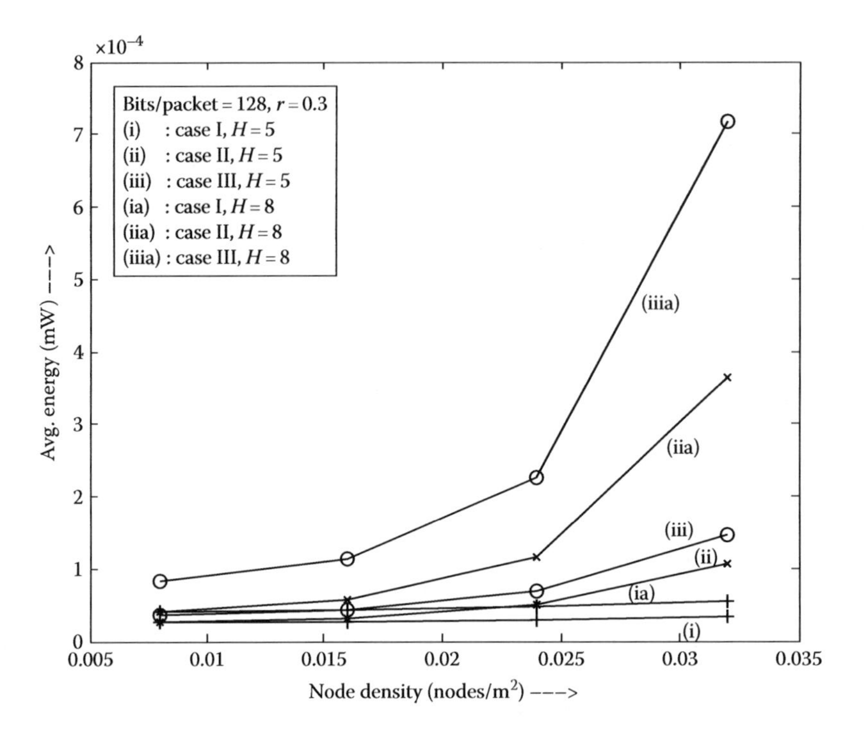

FIGURE 5.20
Average energy consumed versus node density with different number of hops with equal hop length under different schemes.

Due to the significant amount of startup energy of transceiver while wakening from the sleep mode at each hop, an increasing number of hops (i.e., decreasing hop length) beyond a limit will not be an efficient mechanism for multi-hop communication in energy-constrained WSN. In this context and considering the fact that power is a critical resource, optimizing the usage of this resource is essential. Estimation of the optimized number of hops for a fixed source-to-sink distance under different conditions of channel, physical layer, and link layer is an important task to design an energy-efficient WSN system.

To analyze the system under such a scenario of fixed distance between the source and the sink, system modeling is done as described in Section 5.2.3.1, followed by evaluation of the average link BER and route BER.

We consider a multi-hop path between a source and destination (i.e., the sink), as shown in Figure 5.15. Further, n_f bits per packet is considered in forward transmission of information, while n_b bits per packet is considered for NACK/ACK with an assumption of error-free reception of NACK/ACK. Three different schemes of information-delivery mechanism are discussed in Section 5.2.3.2. However, it is observed in Figures 5.19 and 5.20 that the scheme, which uses direct transmission of NACK/ACK from the destination to the source, consumes maximum power. This section concentrates only on the other two energy-efficient schemes:

- Scheme I based on HbH ARQ (Figure 5.16a).
- Scheme II based on E2E ARQ (Figure 5.16b). There is no error checking in intermediate hops. The packet is checked only at the sink for correctness; retransmissions are requested to the source, with a NACK coming back from the sink to the source through a multi-hop path.

5.2.3.7 Energy Consumption and Delay in Variable Hop Length

5.2.3.7.1 Energy and Delay in Scheme I

The average probability of error at the packet level at each link is expressed by Equation 5.22. The average number of retransmissions $(\overline{N})_I$ for the successful delivery of packet through each link, using scheme I, is expressed by Equation 5.24. Assuming the average startup energy from sleep mode to either transmit/receive mode is equal to 30 μJ, the energy consumption for forward transmission and reception of a packet per hop involving two adjacent nodes is expressed by [Datta, 2013]

$$(E_{pkt})_I = E_{bI} \cdot n_f + E_{bI} \cdot n_b + 30 \times 10^{-6} \tag{5.48}$$

Here E_{bI} is the energy required to communicate one bit of information in one hop, expressed by Equation 5.27. The number of MAI is assumed to be higher at the sink than in other intermediate nodes. Thus, the average number of retransmissions is expected to be more at the final hop, where the sink is the receiver, as compared to other hops. Further, it is assumed that each node will wake up from the sleep state to the active state and will remain in that state until the information is successfully transmitted to the next hop. Considering this, the average energy consumption for successful delivery of packet through H number of hops is expressed by [Datta, 2013]

$$E_{av,I} = (H-1)\cdot(\overline{N})_I \cdot (E_{bI} \cdot n_f + E_{bI} \cdot n_b) + ((\overline{N})_I)_{lasthop} \cdot (E_{bI} \cdot n_f + E_{bI} \cdot n_b) + 30 \times 10^{-6} \cdot H \tag{5.49}$$

The first term represents the average energy consumption for all hops, excluding the last hop, where the sink is the receiver. The second term represents the average energy consumption for the successful delivery of information at the last hop. Further, $((\overline{N})_I)_{lasthop} > (\overline{N})_I$ due to the higher node activity at the last hop. The third term represents the total energy consumed by all nodes due to startup energy.

The time required for the successful delivery of a packet from a source node to the sink constitutes the delay. We consider only the transmission delay and no queuing. Assuming $n_f \gg n_b$ and delay associated with transmission of ACK/NACK is negligible, the average packet delay for the successful transmission of a packet of length n_f in scheme I is obtained as follows [Datta, 2013]:

$$D_{av,I} = (H-1)\cdot(\overline{N})_I \cdot \frac{n_f}{R_b} + ((\overline{N})_I)_{lasthop} \cdot \frac{n_f}{R_b} \tag{5.50}$$

5.2.3.7.2 Energy and Delay in Scheme II

Following Equations 5.40 and 5.41 in Section 5.2.3.3, and considering startup energy of each node, the energy consumed per packet from the source to the sink, that is, single-loop transmission of information from the source to the sink via H hops, with ACK/NACK via multi-hop is

$$(E_{pkt})_{II} = E_{bII} \cdot n_f + E_{bII} \cdot n_b + 30 \times 10^{-6} \cdot H \tag{5.51}$$

where E_{bII} is the energy required to communicate one bit of information from the source to the sink through H number of hops, and it is expressed as

$$E_{bII} = \frac{(P_t + P_r)\cdot H}{R_b} \tag{5.52}$$

It is assumed that the nodes will change from the sleep to the active state only once during the transmission of a packet and will remain in the active state until the information is transmitted successfully at the destination (i.e., the sink). On an average, each packet requires $(\overline{N})_{II}$ number of retransmissions from the source to the destination for successful delivery, as expressed by Equation 5.41. Hence, the average energy consumed by a packet through multi-hop communication of H hops is [Datta, 2013]:

$$E_{av,II} = (\overline{N})_{II} \cdot (E_{bII} \cdot n_f + E_{bII} \cdot n_b) + 30 \times 10^{-3} \cdot H \tag{5.53}$$

The average packet delay for the successful transmission of a packet is obtained as follows [Datta, 2013]:

$$D_{av,II} = (\overline{N})_{II} \cdot \frac{n_f}{R_b} \tag{5.54}$$

5.2.3.8 Results with Variable Hop Length

Parameters used in analysis are shown in Table 5.3. Figure 5.21 shows analytical and simulated results. Figures 5.22 through 5.25 are based on analytical results.

The values of means of all shadowing and pce components are considered to be the same as in the previous cases. We assume that 25% of the total nodes within the receiving distance (r_R) of the sink are active for MAI, while 12.5% of nodes between r_R and r_I of the sink are active for NI. As we consider uniform distribution of nodes and $r_I \approx 2r_R$, 12.5% of the total nodes within r_R of other intermediate relay nodes are active for MAI, while the same percentage of nodes between r_R and r_I of other intermediate relay nodes are active for NI at any instant. All parameters at each hop are calculated considering the distance between two consecutive nodes as $r_R = L/H$ meter, where L and H are, respectively, the distance and number of hops between the source and the sink.

Figure 5.21 shows the variation of the average link BER with node densities for different number of hops using scheme I for a fixed distance between the source and the sink,

TABLE 5.3

Parameters Used in Analysis of Multi-Hop CDMA WSN with Fixed Source and Sink, Applying Infinite ARQ

Parameter	Value
Distance between source and sink (L)	100 m
Minimum distance between two nodes (r_0)	1 m
Number of hops (H)	2–7
Processing gain (pg)	128
Constant receive power (P_r)	1.0×10^{-10} mW
Path loss parameter (α)	3
Transmission rate (R_b)	20.0 kbps
Bits/packet in forward transmission (n_f)	128, 256 bits/packet
NACK/ACK bits (n_b)	2 bits
Correlation coefficient (r)	0.0, 0.3, 0.6
Standard deviation of pce (σ_R)	1 dB
Standard deviation of shadowing (σ_S)	3 dB

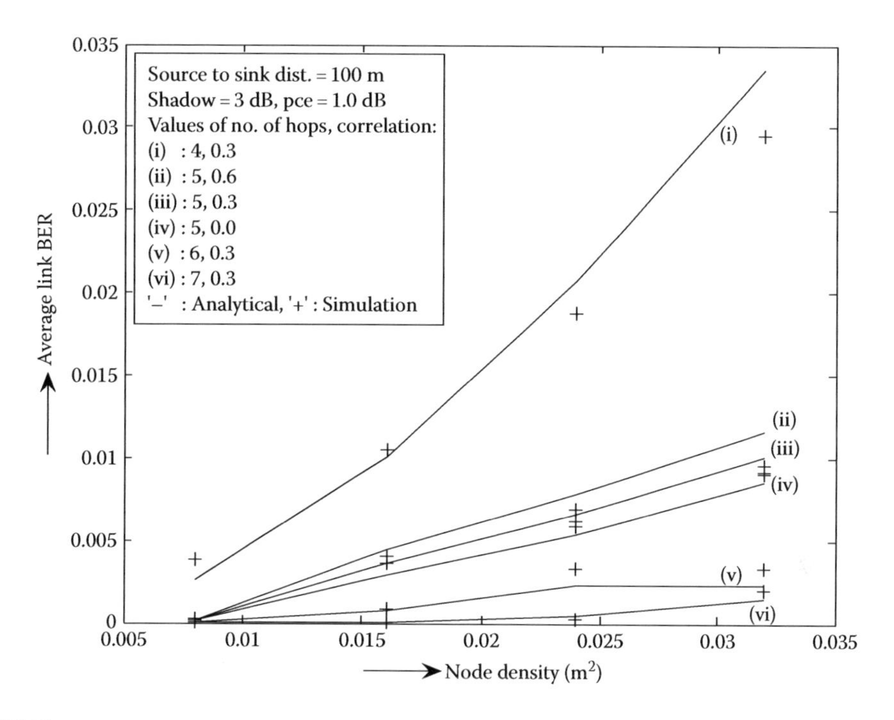

FIGURE 5.21
Variation of average link BER with node densities for different number of hops between a fixed source and sink using HbH ARQ (scheme I).

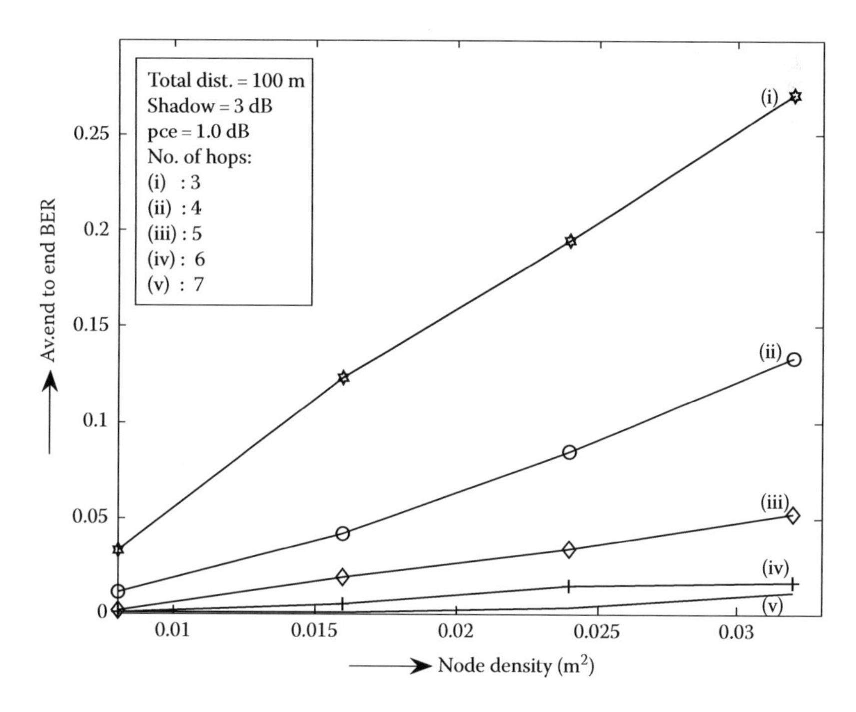

FIGURE 5.22
Variation of average E2E BER with node densities for different number of hops between a fixed source and sink using E2E ARQ (scheme II).

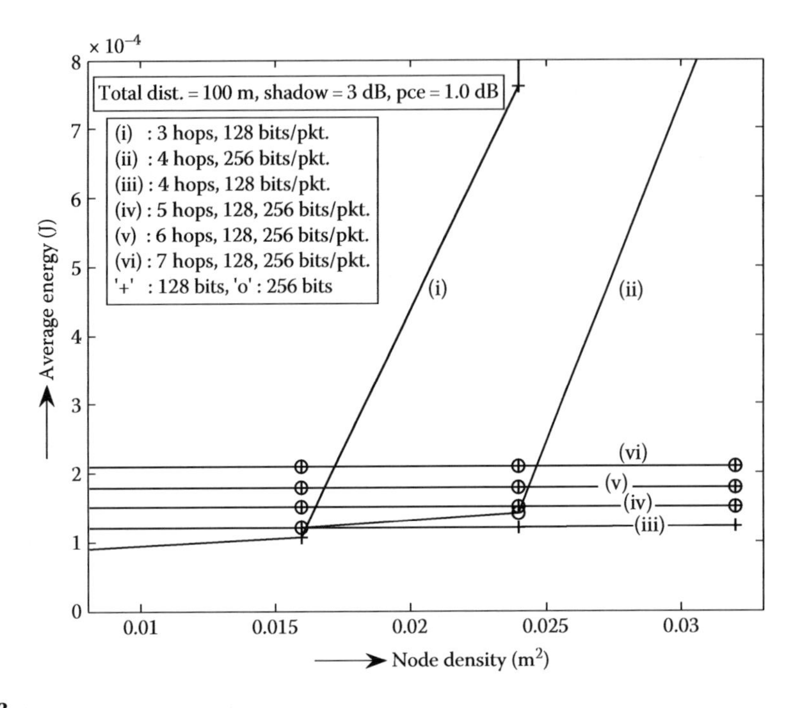

FIGURE 5.23
Variation of average energy consumption for successful transmission of information from a fixed source to sink with node densities under different number of hops, packet lengths using HbH ARQ (scheme I).

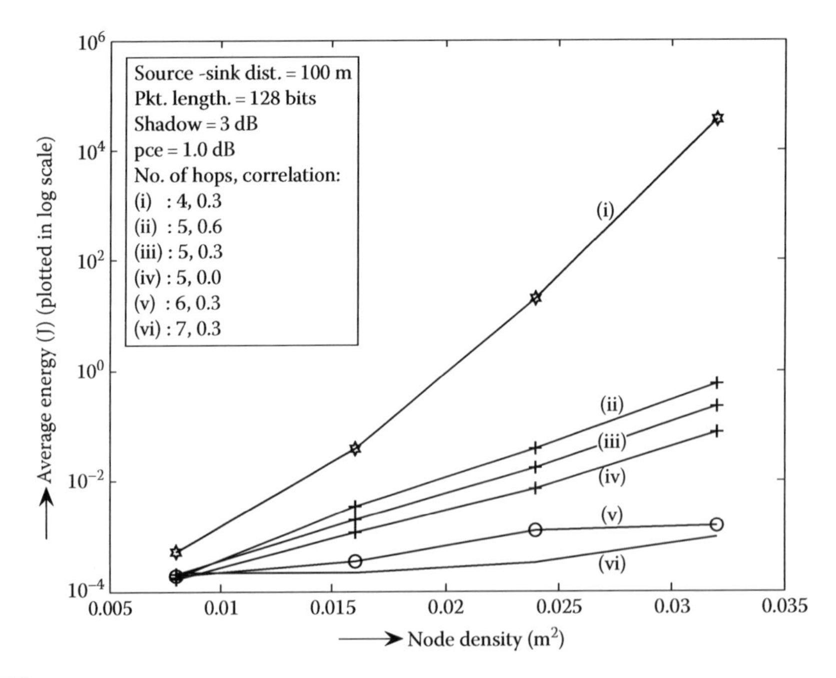

FIGURE 5.24
Variation of average energy consumption for successful delivery of information from a fixed source to sink with node densities for different number of hops using E2E ARQ (scheme II).

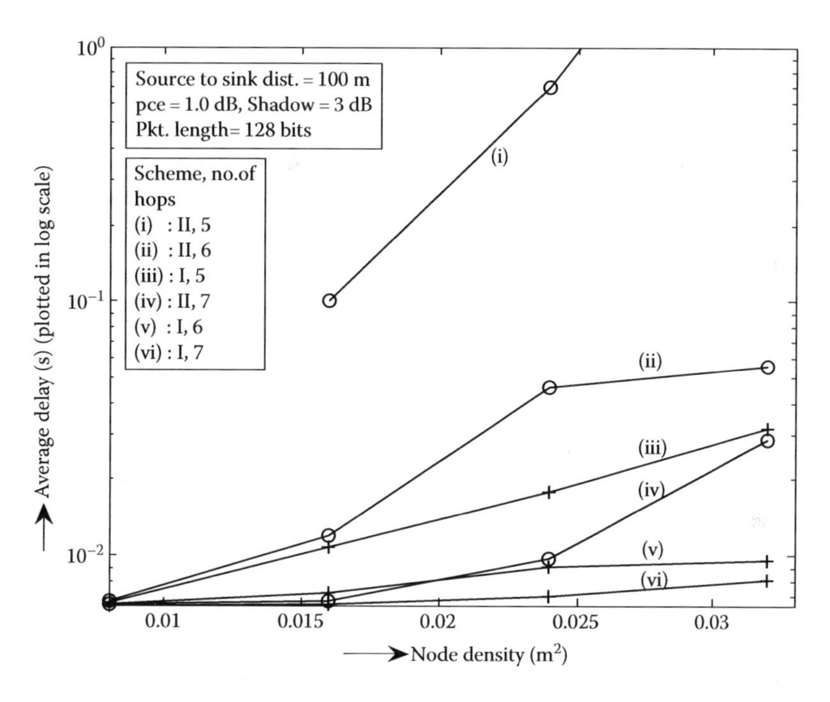

FIGURE 5.25
Variation of average delay for successful delivery of information from a fixed source to sink with node densities for different number of hops using HbH ARQ and E2E ARQ.

shadowing, and pce. BER improves with increase in the number of hops, while other parameters are fixed. As the number of hops increases between the fixed source node and the sink, the distance between two consecutive nodes, that is, the hop length decreases. Increase in hop count (i.e., decrease in hop length) is associated with a decrease in the number of interferers, which improves BER at each hop. It is seen that for a node density of 0.024 nodes/m^2, 68% improvement of link BER is obtained when the number of hops is increased from 4 to 5. For a particular number of hops, as the number of interferers decreases with a decrease in node density, BER at each hop improves. It is seen that 44% improvement of link BER is observed for a change in node density from 0.024 to 0.016 nodes/m^2, while the number of hops is fixed at 5. The effects of correlation among interferers (*r*) on performance is presented for five numbers of hops. In this condition, it is seen that with a decrease in (*r*) from 0.6 to 0.0, BER improves by 32% due to the overall improvement in SIR on the receiver. It is also observed that incremental improvement of BER at each hop reduces with increasing number of hops (comparison of curves i, iii with curves iii, v or curves iii, v with curves v, vi).

Figure 5.22 shows the variation of average E2E BER, that is, route BER with node densities for different number of hops using scheme II. In a multi-hop communication, E2E BER is determined by the cumulative effects of BER at individual hops. In our case, increasing the number of hops leads to a decrease in hop length. Under this situation, BER at each hop improves considerably due to a decrease in interferers, as shown in Figure 5.18, and consequently E2E BER improves. However, incremental improvement in E2E BER diminishes with a higher number of hops, as seen in Figure 5.18. At a node density of 0.024 nodes/m^2, increasing the number of hops from 4 to 5 leads to 79% improvement in route BER (curves ii and iii), while 59% improvement in BER is seen by increasing the number of hops from 6 to 7 (curves iv and v).

Figure 5.23 shows the variation of average energy consumption for different number of hops using scheme I for packet sizes of 128 and 256 bits, respectively. It is observed from the curves that at very low node densities, that is, at low interference condition, the average energy required for the successful delivery of a packet from the source to the sink increases with an increase in the number of hops. In this region, the energy consumption is mainly attributed to the energy consumed by the transceivers in changing their states from sleep mode to transmit/receive mode at each hop, which increases with an increase in the number of hops. As the node density increases, the number of interferers increases, which leads to an increased number of retransmissions. Increase in the number of retransmissions is significantly high when the distance between two successive nodes (r_R) is high, that is, at low hop count, due to an increase in the number of interferers within the receiving distance r_R. Thus, energy consumption increases significantly with higher node density at lower number of hops, for example, curve i with 128 bits per packet and the number of hops equals to 3. The number of hops, which gives optimum performance in terms of energy consumption, can be predicted from the figure for a fixed source-to-sink distance and a particular channel condition. It is seen that at a typical condition, four hops give the optimum performance in terms of energy consumption with a packet length of 128 bits (curve iii). It is further observed that packet size has a significant impact on energy consumption when the number of hops is low. For example, in this case, a higher packet size of 256 bits consumes significantly higher energy as compared to transmission based on 128 bits per packet (curves ii, iii) at four hops. However, as the number of hops is increased beyond 4, no significant difference in energy consumption is seen for two different packet sizes under consideration (curves iv, v, and vi). It is due to the significant increase in the total startup energy as compared to the total communication energy by all nodes.

Figure 5.24 shows the variation of average energy consumption with node density for different number of hops using scheme II with 128 bits packet length. Comparing Figures 5.23 and 5.24, it is observed that scheme I is more energy efficient than scheme II. For example, 92% decrease in energy consumption is seen in scheme I than the other at 0.016 nodes/m^2 for five hops. This difference significantly depends on the node density and number of hops, keeping other parameters such as shadowing, pce, and source-to-sink distance fixed. The impact of correlation among interferers on energy consumption using scheme II for a particular channel condition and pce with five hops is also observed. In this condition, it is seen that with an increase in correlation among interferers from 0.0 (i.e., uncorrelated) to 0.6, energy consumption increases due to an increase in E2E BER. A significant higher rate of increase in energy consumption is observed beyond a node density (e.g., 0.016/m^2 in this figure) for hop count such as four. However, the rate of increase in energy consumption reduces with further increase in hop count beyond four. It is because increase in the number of interferers with increase in node densities will be more at lower number of hops due to higher receiving distance. As the number of hops increases, the increase in the number of interferers with node density will be less as compared to the case of lower number of hops.

Next, the latency behavior of the network is shown with two schemes. Figure 5.25 shows the variation of average delay for successful E2E delivery of information with node densities under different number of hops for two schemes under consideration, schemes I and II. It is observed that with an increase in the number of hops, the average delay in the successful delivery of a packet from the source to the sink decreases. Increase in the number of hops for a fixed source-to-sink distance improves the average BER at individual hop, resulting in a significant decrease in the average number of retransmission. Thus, the average delay at each hop decreases and finally the E2E average delay reduces. Next, with an increase in node density, the delay at each hop increases significantly as BER increases due to an increase in interference. It is also observed that a decrease in delay is not significant for further increase in the number of hops beyond 6, while standard deviation of shadowing is 3 dB, pce is 1 dB, and packet length is 128 bits. This is due to an insignificant change in BER beyond six hops, as explained in Figure 5.22. Further, scheme I shows better performance in terms of average delay (curves iii, v, and vi), as compared to scheme II.

5.3 Conclusion

CPS designs show great promise in enabling human-to-human, human-to-object, and object-to-object interactions between the physical and virtual worlds. The increasing pervasiveness of WSN technologies in many applications makes them an important component of emerging CPS designs. Like mobile networks, in the future, the impact of efficient CDMA WSN may be high on CPS design.

The impact of several network parameters, such as node density, hop count, pce, correlation among shadow-faded interferers only as well as between signal and interferers, on the performance of a CDMA-based WSN with a layered architecture is investigated. The performance has been evaluated in terms of link BER, energy consumption in successful packet transmission, energy efficiency, delay, etc. under several packet transmission schemes based on single and multiple hops. Important findings are summarized.

It is observed that BER performance both at link and route level degrades with the increase in node density, pce, and correlation among shadow-faded interferers only. Similar observation is noted for energy consumption and delay for the successful delivery of a packet from the source to the sink. Energy efficiency degrades with an increase in node spatial density. Optimum packet size corresponding to maximum energy efficiency increases with a decrease in node density. Transmission based on optimum packet size saves energy significantly as compared to fixed packet-based transmission. It is observed that the impact of startup energy at each node on the overall energy consumption for the successful delivery of a data packet, energy efficiency and lifetime of the network is significant. In the case of a multi-hop communication with a fixed distance between the source node and the sink, the optimum hop count to design an energy-efficient system is assessed, while considering startup energy at each node. Performance of different information-delivery schemes in multi-hop communication is assessed. HbH ARQ scheme shows the highest energy efficiency as compared to other E2E ARQ schemes. Numerical results demonstrated that system parameters should be carefully designed to obtain the desired energy efficiency and delay, crucial for different applications.

Glossary

Bit error rate: Probability of error in transmitting 1 bit of information associated with electronic communication system. Bit error rate reveals the channel condition.

Correlation: There may be correlation among received signals if these are obstructed near the receiver end. Correlation among received signals depends on the angle between the signal paths from the obstacle and the path length of each signal from the obstacle.

Multiple access interference: When multiple nodes simultaneously want to send information to a particular node, which is the intended receiver to each sending node, the actual received signal at the receiver suffers from multiple access interference. In this condition, the multiple senders are located within the receiving distance of the receiver.

Node interference: In this situation, multiple nodes simultaneously want to send information to their respective receivers, and the receiver of interest is simultaneously receiving a signal of interest. If the multiple nodes, as described earlier, are within the interference range of the receiver of interest, the signal of interest will suffer from node interference due to transmitted signals from those multiple nodes.

Optimization of packet length: Packet length can be optimized in wireless communication, especially in WSN system, to maximize energy efficiency for longer lifetime. Packet length may be optimized for other parameters also, for example, packet throughput associated with delay and resource utilization associated with both energy consumption and delay.

Power control: In wireless communication, transmitting power control is done to minimize near–far problem.

Wireless CDMA Sensor Network: A WSN system where the underlying MAC protocol is based on code division multiple access, where a large number of nodes can be accessed simultaneously and efficiently.

List of Index Terms

Bandwidth utilization: Proper utilization of network bandwidth improves network capacity.

BER: Probability of error in transmitting 1 bit of information associated with electronic communication system.

Bit rate: Number of bits per second.

CDMA: Code division multiple access.

Chip rate: Number of pulses per second (chips per second) at which the code is transmitted in spread spectrum communication. The chip rate is larger than the symbol rate, meaning that one symbol is represented by multiple chips. The ratio is known as the "spreading factor" or processing gain.

Code rate: Data rate.

Coverage: Coverage in wireless sensor networks is usually defined as a measure of how well and for how long the sensors are able to observe the physical space. Coverage is important for a sensor network to maintain connectivity.

Fault tolerance: The ability of a system to deliver a desired level of functionality in the presence of faults.

Infinite ARQ: Automatic request from receiver to sender to send the information repeatedly until successful delivery of information.

Latency: Delay in reliable receipt of a data packet after transmission.

MAC protocols: Communication protocols used in the medium control access layer.

Mesh networking protocols: Robust mesh networking with power-optimized protocol, which supports sleeping routers for power-sensitive or battery-dependent applications.

Multiple access interference: Performance of a deployed wireless sensor network (WSN) is greatly influenced when multiple nodes want access to a particular node during operation.

Node interference: Due to the interference range of receivers, the transmitting signal power generated by any source designated for a particular receiver may cause interference to other nodes, which act as receivers for corresponding designated sources.

Power control: Communication is usually the most energy-consuming event in WSNs. One way to significantly reduce energy consumption is applying transmission power control techniques to dynamically adjust the transmission power.

Power control error: Efficient transmission power control algorithms cannot be applied on WSN nodes, as these are battery operated. Hence, received signals are associated with some error due to transmission power control.

Scalability: A WSN must perform well as the network grows larger or as the workload increases.

Shadow-faded correlated interferers: Correlation among interferers due to shadow fading.

Spreading bandwidth: In spread spectrum service, the transmitted signal is spread over a wide frequency band, much wider in fact than the minimum bandwidth required to transmit the information being sent. This is done by taking a carrier of a certain occupied bandwidth and power and "spreading" the same power over a wider occupied bandwidth. The effect is a reduction in the carrier's spectral density (dB W/Hz).

Startup transients: Startup current transients are common in many modern devices. The startup current of transceiver chip can be a function of time between power cycles, temperature, power supply sequencing, etc. These currents can be large, often several times higher than normal running condition.

Throughput: Data delivery rate at the destination.

Topology: Which node is able/allowed to communicate with other nodes. Topology control needs to maintain connectivity.

References

[IEEE Standard for Information Technology, 2006] IEEE Standard for Information Technology, Telecommunications and information exchange between systems—Local and metropolitan area networks—specific requirement Part 15.4: Wireless Medium Access Control (MAC) and Physical Layer (PHY) Specifications for Low-Rate Wireless Personal Area Networks (WPANs), 2006.

[Rappaport, 2002] Rappaport T. S., *Wireless Communications Principles and Practice*, Prentice Hall, Inc., Upper Saddle River, NJ, 2002.

[Tonguz, 2004] Tonguz O. K. and Ferrari G., *Adhoc Wireless Networks: A Communication—Theoretic Perspective*, John Wiley & Sons, West Sussex, England, 2004.

[Viterbi, 1995] Viterbi A. J., *CDMA: Principles of Spread Spectrum Communications*, Addison-Wesley, Boston, MA, 1995.

Journal Articles

[Abu-Dayya, 1994] Abu-Dayya A. A. and Beaulieu N. C., Outage probabilities in the presence of correlated log-normal interferers, *IEEE Transactions on Vehicular Technology*, 43(1), February 1994, 164–173.

[Akyildiz, 2002] Akyildiz I. F., Su W., Sankarasubramaniam Y., and Cayirci E., A survey on sensor networks, *IEEE Communications Magazine*, 40(8), 2002, 102–114.

[Datta, 2010c] Datta U., Kundu C., and Kundu S., Performance analysis of CDMA wireless sensor networks in shadowed environment, *Sensors and Transducers Journal*, 118(7), July 2010, 87–100.

[Datta, 2013] Datta U., Sahu P. K., and Kundu S., On performance of multi hop CDMA wireless sensor networks with correlated interferers, *The International Journal of Communication Networks and Distributed Systems*, 10(3), 2013, 280–302.

[De, 2004] De S., Qiao C., Pados D. A., Chatterjee M., and Philip S. J., An integrated cross-layer study of wireless CDMA sensor networks, *IEEE Journal on Selected Areas in Communications*, 22(7), September 2004, 1271–1285.

[Demirkol, 2006] Demirkol I., Ersoy C., and Alagoz F., MAC protocols for wireless sensor networks: A survey, *IEEE Communications Magazine*, 44(4), 2006, 115–121.

[Dietrich, 2009] Dietrich I. and Dressler F., On the lifetime of wireless sensor networks, *ACM Transactions on Sensor Networks*, 5(1), February 2009, 5–39.

[Jerez, 2000] Jerez R., Ruiz-Garcia M., and Diaz-Estrella A., Effects of multipath fading on BER statistics in cellular CDMA networks with fast power control, *IEEE Communications Letters*, 4(11), November 2000, 349–351.

[Kang, 2008] Kang H., Hong H., Sung S., and Kim K., Interference and sink capacity of wireless CDMA sensor networks with layered architecture, *ETRI Journal*, 30(1), February 2008, 13–20.

[Kim, 2000] Kim J. B. and Honig M. L., Resource allocation for multiple classes of DS-CDMA traffic, *IEEE Transactions on Vehicular Technology*, 49(2), March 2000, 506–519.

[Lin, 2012] Lin C., Zeadally S., Chen T., and Chang C., Enabling cyber physical systems with wireless sensor networking technologies, *International Journal of Distributed Sensor Networks*, 2012, 1–21, Article ID 489794.

[Panichpapiboon, 2006] Panichpapiboon S., Ferrari G., and Tonguz O. K., Optimal transmit power in wireless sensor networks, *IEEE Transactions on Mobile Computing*, 5(10), October 2006, 1432–1447.

[Pratesi, 2000] Pratesi M., Santucci F., Graziosi F., and Ruggieri M., Outage analysis in mobile radio systems with generically correlated log-normal interferers, *IEEE Transactions on Communications*, 48(3), March 2000, 381–385.

[Wu, 2011] Wu F., Kao Y., Tseng Y., From wireless sensor networks towards cyber physical systems, *Pervasive and Mobile Computing*, 7, 2011, 397–413.

[Ye, 2004] Ye W., Heidemann J., and Estrin D., Medium access control with coordinated adaptive sleeping for wireless sensor networks, *IEEE/ACM Transactions on Networking*, 12(3), June 2004, 493–506.

Conference Proceedings

[Cardieri, 2000] Cardieri P. and Rappaport T., Statistics of the sum of lognormal variables in wireless communications, *Proceedings of the 51st IEEE Vehicular Technology Conference 2000*, May 15–18, 2000, Tokyo, Japan, Vol. 3, pp. 1823–1827.

[Datta, 2009] Datta U., Kundu C., and Kundu S., BER and energy level performance of layered CDMA wireless sensor network in presence of correlated interferers, *Proceedings of the Fifth IEEE International Conference on Wireless Communications and Sensor Networks (WCSN 2009)*, December 15–19, 2009, Allahabad, India, pp. 180–185.

[Datta, 2010a] Datta U., Sahu P. K., Kundu C., and Kundu S., Energy level performance of multihop wireless sensor networks with correlated interferers, *Proceedings of the IEEE International Conference on Industrial and Information Systems (ICIIS 2010)*, July 29–August 1, 2010, Mangalore, India, pp. 35–40.

[Datta, 2010b] Datta U., Kundu C., and Kundu S., Performance of an optimum packet based CDMA wireless sensor networks in presence of correlated interferers, *Proceedings of the IEEE International Conference on Computer & Communication Technology (ICCCT 2010)*, September 17–19, 2010, Allahabad, India, pp. 22–27.

[Kleinschmidt, 2007] Kleinschmidt J. H., Borelli W. C., and Pellenz M. E., An analytical model for energy efficiency of error control schemes in sensor networks, *Proceedings of IEEE International Conference on Communications*, June 24–28, 2007, Glasgow, Scotland, pp. 3895–3900.

[Laneman, 2000] Laneman J. N. and Wornell G. W., Energy-efficient antenna sharing and relaying for wireless networks, *Proceedings of IEEE Wireless Communications and Networking Conference 2000*, September 23–28, 2000, Chicago, IL, Vol. 1, pp. 7–12.

[Muqattash, 2003] Muqattash A. and Krunz M., CDMA-based MAC protocol for wireless ad hoc networks, *Proceedings of the ACM MobiHoc*, June 1–3, 2003, Annapolis, Maryland, pp. 153–164.

[Pompili, 2007] Pompili D., Melodia T., and Akyildiz I. F., A distributed CDMA medium access control for underwater acoustic sensor networks, *Proceedings of the Sixth Annual Mediterranean Ad Hoc Networking Workshop*, June 13–15, 2007, Corfu, Greece, pp. 63–70.

[Sakhir, 2008] Sakhir M., Ahmed I., Peng M., and Wang W., Power optimal connectivity and capacity in wireless sensor network, *Proceedings of the 2008 International Conference on Computer Science and Software Engineering*, December 12–14, 2008, Wuhan, China, Vol. 4, pp. 967–970.

[Sankarasubramaniam, 2003] Sankarasubramaniam Y., Akyildiz I. F., and McLaughlin S. W., Energy efficiency based packet size optimization in wireless sensor networks, *Proceedings of the First IEEE International Workshop on Sensor Network Protocols and Applications*, May 11, 2003, Anchorage, AK, pp. 1–8.

[She, 2009] She H., Lu Z., Jantsch A., Zhou D., and Zheng L.-R., Analytical evaluation of retransmission schemes in wireless sensor networks, *Proceedings of the IEEE 69th Vehicular Technology Conference*, April 26–29, 2009, Barcelona, Spain, pp. 1–5.

[Taddia, 2004] Taddia C. and Mazzini G., On the retransmission methods in wireless sensor networks, *Proceedings of the IEEE VTC*, September 26–29, 2004, Los Angeles, CA, pp. 4573–4577.

[Yang, 2003] Yang Y. and Prasanna V. K., Energy-balanced multi-hop packet transmission in wireless sensor networks, *Proceedings of the IEEE GLOBECOM*, December 1–5, 2003, San Francisco, CA, Vol. 1, pp. 480–486.

Product Manuals

[*ASH Transceiver Designer's Guide*] RFM, Tr-1000 product technical information sheet, available online on http://www.rfm.com/products/data/

[*ATMEL Transceiver Designer's Guide*] AT86RF212 Transceiver, available online on http://www.atmel.com/images/doc8168.pdf.

Section II

Energy and Reliability Issues in Cyber-Physical Systems

6

Modeling of Supercapacitors as an Energy Buffer for Cyber-Physical Systems

Nicholas Gekakis, Andrew Nadeau, Moeen Hassanalieragh, Yiyang Chen, Zhaojun Liu, Grayson Honan, Fatih Erdem, Gaurav Sharma, and Tolga Soyata

CONTENTS

6.1 Introduction

Supercapacitors have been established as a compelling solution to high-power buffering applications due to their ability to bank and supply power at levels an order of magnitude beyond the capabilities of electrochemical battery technologies per unit weight. This superior power density has been utilized for regenerative breaking (Rotenberg et al., 2011), elevator (Rufer and Barrade, 2002), and automated starting systems for combustion engines (Catherino et al., 2006). Additionally, recent developments have also begun using supercapacitors for energy storage applications in order to take advantage of their excellent charge discharge efficiency as well as their power density capabilities. Energy efficiency is especially critical for self-sustaining environmentally powered systems, where efficient storage/use of a limited energy supply can prolong the time of operation and improve the quality of service. Another useful characteristic of supercapacitors is their relationship between terminal voltage and remaining stored energy. This relationship provides more accurate energy awareness for systems with dynamic supply and usage of power.

However, reliance on these efficiency, power density, and energy-awareness benefits for the design of supercapacitor-based systems must be tempered by the fact that supercapacitors do not operate as ideal devices. Classical concepts of capacitance apply much more closely to parallel plate or electrolytic devices. Observed supercapacitor behavior differs significantly from theoretical ideal capacitor performance. These operational differences between supercapacitors and their much weaker conventional cousins are a direct result of the physical phenomena governing supercapacitor behavior. Dynamic system and equivalent circuit models have been developed to characterize supercapacitor performance. However, focus has been on accurately predicting supercapacitor frequency response for power-buffering applications using techniques such as electrochemical impedance spectroscopy (Bertrand et al., 2010) or characterizing long-term storage efficiency for low-power applications (Zhang and Yang, 2011) (Table 6.1).

An emerging category of applications, cyber-physical systems (CPSs) hold potential to benefit from supercapacitor energy buffering. CPSs have the ability to deploy significant computational resources into the field at the location of data collection. For example, these computational resources can enable face recognition without the need to transmit high bandwidth video streams back to a base station. While communication overheard and the need for infrastructure such as base stations are reduced for CPSs, these systems can require much higher power to sustain their computational capabilities as opposed to wireless sensor nodes. Additionally, these systems are many times remotely deployed, making maintenance costly. Supercapacitors can be an ideal fit for CPSs due to their long operational lifetimes and peak power capabilities.

This chapter begins by introducing an accepted model for supercapacitor behavior and then presents the analysis of this model relevant to supercapacitors used in energy-buffering equations. Specifically, the analysis describes how the three-branch model is implemented to provide energy awareness and track a system's remaining available energy. The model is also simulated to characterize energy storage efficiency trade-offs for supercapacitors. By implementing a simulated supercapacitor model, which describes nonideal behaviors, this chapter offers insight into the most significant factors affecting efficiency and the utility of supercapacitors for energy storage applications. The primary factors influencing storage efficiency are found to be supercapacitor size, stored energy level, and power at which energy must be delivered. The three-branch equivalent circuit model (Zubieta and Bonert, 2000), based on the physical phenomena governing supercapacitor behavior, is used to simulate these performance trade-offs. These simulations demonstrate how the exceptional charge–discharge efficiency benefits of supercapacitor-based storage can be severely degraded depending on the power level, supercapacitor voltage, and supercapacitor size. These implications for energy efficiency must be considered in system design to realize the benefits of supercapacitor-based energy storage over alternatives such as rechargeable batteries.

TABLE 6.1

Possible Significant Benefits of Supercapacitor-Based Systems

	Conventional Capacitors	Supercapacitors	Electrochemical Batteries
Energy density (W·h/kg)	10^{-2} to 10^{-1}	10^{0} to 10^{1}	10^{1} to 10^{2}
Power density (W/kg)	10^{3} to 10^{4}	10^{3} to 10^{4}	10^{1} to 10^{2}
Efficiency (%)	≥95	≥95[a]	70–99

[a] Efficiency performance can be jeopardized by naive disregard for nonideal supercapacitor behavior.

Our simulations implementing the three-branch equivalent circuit model for supercapacitor behavior with parameters both measured from physical devices and taken from literature demonstrate:

- Superior charge–discharge efficiency for supercapacitor-based energy storage
- The importance of power awareness to maintain this superior charge–discharge supercapacitor performance
- The dependence between supercapacitor size and reasonable power level limits on the application within which near-ideal performance can be expected
- The dependence of the near-ideal operation power region limits on supercapacitor terminal voltage, as well as capacitor size

The organization of this chapter begins by justifying the use of the three-branch model as an equivalent circuit for the physical phenomena that govern supercapacitor operation (Sections 6.2 through 6.5). Section 6.6 then details the procedure for measuring the model's parameters from a physical supercapacitor device. Then Section 6.7 explains how this model can provide an accurate estimate of the remaining energy to the system. Section 6.8 lays out the model parameters and the techniques used to simulate a varied collection of supercapacitors and explore efficiency relationships for each. Results showing the significance of design considerations for supercapacitor-based systems are given in Section 6.9. To conclude, findings are summarized in Section 6.10.

6.2 Background

To motivate the significance of efficiency trade-offs inherent in supercapacitor energy storage, we first outline the principles and construction of supercapacitors. Relevant supercapacitor properties include (1) orders of magnitude greater energy storage capacity than conventional or electrolytic capacitors, (2) superior peak power performance in relation to electrochemical batteries, (3) long operational lifetime and low environmental impact, (4) internal charge redistribution among an array of internal time constants, (5) equivalent series resistance (ESR), and (6) self-discharge leakage current.

In contrast to conventional capacitors in which opposing electric charge collects on electrode plates separated by a dielectric layer, the physical design of supercapacitors is shown in Figure 6.1. Instead of storing energy through the electric field created within a dielectric, supercapacitor electrodes are made of porous materials such as activated carbon with extremely high surface area. Both the mechanisms by which supercapacitors store charge, electric double layers (EDL) and pseudocapacitance (Conway et al., 1997), contribute to energy storage proportional to the surface area of electrodes. These phenomena, linking electrode surface area to capacitance, explain how such high capacitance is possible by using activated carbon electrodes with surface areas in the range of thousands of square meters per gram. The EDL and pseudocapacitance mechanisms also account for the relationships between power consumption, supercapacitor voltage level, and efficiency, developed later in this chapter.

The first charge storage mechanism, EDL, has been used to refer to supercapacitor devices as electric double-layer capacitors (DLCs) and results from the accumulation of opposing charge on the surfaces of the two electrodes. This surface charge accumulates because the ion-permeable membrane prevents the terminal voltage from driving direct

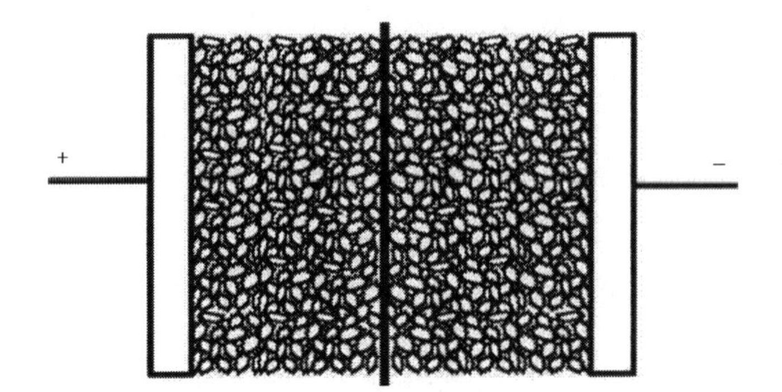

FIGURE 6.1
Typical supercapacitor construction includes two electrodes consisting of high surface area activated carbon immersed in an electrolyte. Positive and negative terminals of the supercapacitor are connected to distinct regions of the electrode material separated by a membrane. This membrane is only permeable to the electrolyte and allows charge-carrying ions to pass through but keeps the regions of electrode material isolated, preventing current from directly flowing between electrodes.

current flow between the two regions of activated carbon electrode material connected to the opposite terminals. The EDL phenomena occur because opposing electrode surface charges on either side of the membrane selectively attract charged ions of the opposite polarity from the electrolyte. As the electrolyte ions flow through the electrode material pores toward the opposing surface charges, concentration gradients build up resulting in Helmholtz layers. Helmholtz layers refer to the high concentrations of oppositely charged ions from the electrolyte that counteract the electrode surface charges by collecting along the porous electrode material surface. Supercapacitor charge storage by means of this EDL, relying on diffusion of ions, contrasts electrochemical batteries where redox (reduction oxidation) reactions store charge but limit the charge and discharge power according to the speed of reaction kinetics. This EDL capacitance along the porous electrode material surface and flow of ions through electrode pores have motivated the use of RC transmission line elements to model the electrical behavior of the porous electrode supercapacitor design (Bertrand et al., 2010; Buller et al., 2001). These models predict supercapacitor charge redistribution over a network of RC time constants. Another important effect of the torturous path that the ions must flow through to reach the electrode surfaces is significant supercapacitor ESR, resulting in wasted energy expended to support the ion currents.

The second mechanism by which charge is stored in supercapacitors, pseudocapacitance, relies on charge transfer in redox reactions that occur at the electrolyte–electrode surface interface (Conway et al., 1997). However, parasitic side reactions can also spontaneously dissipate stored energy (Niu et al., 2006). This energy loss can be modeled as a leakage current and depends on many factors such as initial charge, storage duration, and temperature (Yang and Zhang, 2011; Zhang and Yang, 2011). Another source of supercapacitor charge leakage is direct ohmic pathways by which current flows between terminals through the membrane.

Both the significant sources of energy waste that degrade supercapacitor efficiency, ESR and leakage, are accounted for in the three-branch model equivalent circuit (Zubieta and Bonert, 2000), as shown in Figure 6.2. Supercapacitor charge redistribution is modeled by the three resistive–capacitive branches, each with progressively greater time constant. Supercapacitor ESR is modeled by the series resistors R_1, R_2, and R_3, and leakage is modeled by the parallel resistor R_{leak}. Using this three-branch model, results show how *instantaneous* supercapacitor efficiency depends on both application power and supercapacitor voltage levels. This chapter builds on our previous study of how *net* supercapacitor efficiency and

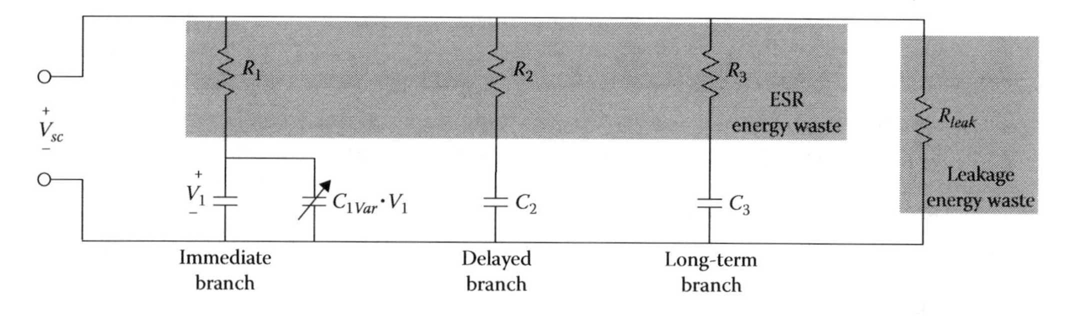

FIGURE 6.2
The three-branch equivalent circuit supercapacitor model accounts for charge redistribution between the immediate, delayed, and long-term capacitive branches of increasing time constant. Resistances in the three capacitive branches model internal supercapacitor ESR, while the fourth purely resistive branch models supercapacitor charge leakage. The initial branch also models voltage-dependent capacitance observed for supercapacitors. (From Zubieta, L. and Bonert, R., *IEEE Trans. Ind. Appl.*, 36(1), 199, 2000.)

effective capacity vary for charge–discharge profiles of varying speed. Supercapacitor efficiency and capacity have been shown to depend on the different distributions of charge that result from different charging and discharging speeds. By using the same measured parameters for the supercapacitor model previously validated against experimental results, this work uses the three-branch model to simulate how instantaneous supercapacitor efficiency changes with power and voltage level, controlling for charge redistribution by fixing equal voltage levels across all capacitive branches of the three-branch model to remove the effects of charge redistribution.

6.3 Equal Series Resistance

For electrical current to charge or discharge a supercapacitor, ions must diffuse through the electrolyte into the torturous pathways in the porous electrode material. Any movement of these ions results in waste heat just as electrical current flowing through a resistance. In a double-layer supercapacitor, the resistivity of carbon particles within the electrode materials also contributes to the supercapacitor's ESR (Zubieta and Bonert, 2000). The ESR plays an important role in applications where the power density is a major concern (Hassanalieragh et al., 2014). Examples of such applications are elevators and electric vehicles. In such cases, a large amount of energy must be supplied in a very short amount of time, which in turn leads to a high current. High current across the series resistance leads to a voltage drop, which decreases the energy delivery efficiency. This problem can be mitigated by having several supercapacitors in parallel.

6.4 Leakage and Charge Redistribution

High leakage, also known as self-discharge, is the loss of energy stored in a device and is frequently cited as a major drawback of supercapacitors. It is widely reported that leakage increases exponentially with a device's terminal voltage; however, research has shown that real leakage is not as significant a problem as was previously believed and that most of the observed leakage effects are actually due to charge redistribution within the supercapacitor. This process is illuminated by models of supercapacitors such as the three-branch model. Charge redistribution occurs when a supercapacitor has stopped charging and one of the capacitors in the model had significantly more charge put into it than the others; it can be thought of as similar to battery relaxation. This occurs frequently because of the large difference in time constants of the three branches, leading each branch to take in energy at widely different timeframes, with the first branch getting charged up in seconds, while the third branch can take hours. In other words, once charging stops, the branches with larger time constants continue to charge, taking their energy from the branches with lower time constants.

Consequently, it has been shown that the longer a supercapacitor charges for, the slower its exhibited leakage. The redistribution of charge leads to a reduction in the terminal voltage of the supercapacitor and has, therefore, been perceived as leakage instead of redistribution. The redistribution is not without energy loss, however; as the charge redistributes itself, energy is lost in each branch's resistance. Energy loss in supercapacitors, therefore, has two sources: leakage and redistribution loss. Research has shown that supercapacitors charged for a very short time, for example, 0.1 h, lose the majority of their energy through redistribution, while supercapacitors charged for a long time, for example, 10 h, lose the majority of their energy through leakage. Increased energy loss due to leakage in supercapacitors that have been charged for a long time can be attributed to the fact that they maintain high voltage longer. Meanwhile, the quickly charged supercapacitors redistribute their charge very quickly, thus lowering their voltage and minimizing leakage. Thus, in situations that require a minimum voltage, it should be kept in mind that the more time a supercapacitor is given to charge, the longer it is able to maintain a high voltage (Merrett and Weddell, 2012).

Leakage is relevant in only those applications that use supercapacitors as long-term storage, such as field systems that need supercapacitors as a long-term energy buffer (Fahad et al., 2012; Hassanalieragh et al., 2014). Alternatively, other usage scenarios are nearly agnostic to leakage. Each computer on the data center racks could have a peak power demand, which is supplied from a supercapacitor block. In this case, the energy is stored and buffered within less than a minute, leaving no time for the leakage to take effect. Typically, the leakage time constant is in the order of hours. Therefore, any application that uses the stored energy in a supercapacitor (or a block of supercapacitors) will not be affected from leakage. Leakage can be thought of as being an equivalent parallel resistor connected in parallel to the three branches.

6.5 Three-Branch Model

Researchers at the University of Toronto set out to develop a simple model that accurately illustrates a supercapacitor's behavior in the first 30 min of a charge or discharge cycle. Additionally, they desired for their model to require only parameters that could be measured at the terminals of the supercapacitor. An RC circuit is used to best model the behavior of supercapacitors; however, more than one, each with a unique time constant, is necessary due to the different timeframes over which supercapacitors respond (Zubieta and Bonert, 2000). This range in response time is due to the same feature of supercapacitors that allows them to have such high capacitance: the incredible porousness of the materials they are made form. This is because the charge has to navigate through the caverns of the material, which results in many different paths of varying length to be taken. The result is an uneven charging for rapid charge cycles, which in turn brings about charge redistribution effects. Charge redistribution, it should be noted, is also responsible for drop in terminal voltage and is part of the reason energy estimation in supercapacitors should not be based on their terminal voltage (Nadeau et al., 2014).

An arbitrary number of branches can be used for the model; however, three has been suggested as it is the least required to obtain a reasonably accurate model of supercapacitor behavior over a period of 30 min. Each branch consists of a resistor to model the energy lost in the materials that form the double-layer charge distribution, and a capacitor to model the capacitance between the electrode and electrolyte. Each branch's time constants should vary widely, ideally being at least one order of magnitude apart from each other, so that only the behavior of a single branch is dominating the overall behavior of the model at any given time. This, in turn, makes it significantly easier to determine the parameters of each branch that allow the model to be simpler to use as well as more accurate. The first branch, the immediate branch, has the smallest time constant and models the immediate response, in the first seconds, to charging. The second branch, which the authors call the delayed branch, models the bulk of what occurs in the first minutes of charging. The final branch, the long-term branch, models what occurs past 10 min. The first branch contains an additional voltage-dependent capacitor in order to simulate the voltage dependency behavior exhibited by supercapacitors. Out of interest for simplicity, the other two branches do not contain this additional capacitor. Finally, in addition to the three RC branches in the model, a parallel resistor is included to simulate leakage and self-discharge effects in the supercapacitor (Zubieta and Bonert, 2000).

The three-branch model is just one of the many models proposed to model the behavior of supercapacitors; however, it has been shown to be the one that most adequately addresses long-term behavior. That being said, the three-branch model is not without its drawbacks: it was developed to model larger supercapacitors and loses some accuracy when applied to smaller ones that might be used in CPSs or wireless sensor networks. Additionally, it takes each branch to be completely independent, which, as research has shown, is not an accurate reflection of the supercapacitor behavior. Despite this, the three-branch model remains the most accurate, even for small supercapacitors, due mainly to the fact that the majority of the research and literature concerns larger supercapacitors and there is not as thorough an understanding of smaller supercapacitors (Weddell et al., 2011).

6.6 Measuring Three-Branch Model Parameters

In this section, we elaborate on measuring the three-branch model parameters based on the approach originally introduced by Zubieta and Bonert (2000). A precisely timed and controllable current source is needed for conducting the experiment in order to be able to keep track of the injected charge. This method presumes distinct time constant of three branches. Initially, the supercapacitor is charged to the rated voltage in a very short amount of time. The charging period must be short enough to make sure the initial charge is placed only on the first branch. Observing the terminal voltage in this period leads to identifying the first-branch parameters. After that, the current source is turned off, and analyzing the terminal voltage over a longer period of time will lead to measuring the parameters of the other two branches. The supercapacitor must be completely discharged prior to conducting the experiment, which needs the supercapacitor to be in short circuit state for a long time.

Throughout this section, we refer to "instantaneous capacitance" many times. The usual definition of capacitance is actually the relation between supplied charge and the voltage of the capacitor, which is = (Q/V). When working with variable capacitance, this definition must be revisited: if we inject a small amount of charge dQ into the capacitor, there will be dV voltage change. The instantaneous capacitance at this specific voltage determines the relation between the injected charge and the voltage difference. Specifically we have

$$C_{inst}(V') = \frac{dQ}{dV}|_{V'}.$$

6.6.1 First-Branch Parameter Measurement

If we denote the supercapacitor terminal voltage shortly after turning on the current source by V_1 (taking into account the rise time of the current source) and the charging current by I_{ch}, we have

$$R_1 = \frac{V_1}{I_{ch}}.$$

We can use the instantaneous capacitance of the supercapacitor during charging to the rated voltage to determine C_1 and C_{1Var}. Instantaneous capacitance (C_{inst}) is defined by

$$C_{inst} = I_{ch}\frac{\Delta t}{\Delta V}.$$

where ΔV is the supercapacitor voltage change during a small time period Δt. After finding the instantaneous capacitance in rated operating voltage range using a high current, we can determine C_1 and C_{1Var} by using a simple least square line fitting method.

6.6.2 Second-Branch Parameter Measurement

After turning off the current source, there is an immediate voltage decrease due to R_1. We denote the terminal voltage by V_1 at this time stamp. As in the case of turning on the current source, the fall time of the current source must also be taken into account. We continue monitoring the terminal voltage afterward. The voltage starts decreasing because of charge redistribution. If there is ΔV voltage change in Δt amount of time, we can assume there is a virtually constant current from the first to the second branch (I_{tr}), given by

$$I_{tr} = \frac{V_1 - (\Delta V/2)}{R_2}.$$

ΔV must be small enough so that this holds true. ΔV is commonly chosen to be 50 mV. Choosing a smaller voltage difference will lead to better approximation of the constant current, but the precision of the measurement device (both sampling frequency and voltage measurement precision) prevents us from choosing a very small voltage difference.

We can relate I_{tr} to the first-branch instantaneous capacitance (C_{inst}), which is measured at $V_1 - \Delta V/2$ by $I_{tr} = C_{inst} * \Delta V/\Delta t$. Using the two derived equations for I_{tr}, we have

$$R_2 = \frac{(V_1 - (\Delta V/2)) * \Delta t}{C_{inst} * \Delta V}.$$

For measuring the second-branch capacitance C_2, we need to wait long enough so that charge redistribution from the first to the second branch has already taken place. Typically, the time constant of the second in a double-layer supercapacitor is of the order 100 s. Waiting for three times the time constant will be adequate. C_2 can be easily calculated by taking into account the charge conservation principle; specifically, we have $Q_{total} = Q_1 + Q_2$, where Q_{total} is the total charge supplied to the supercapacitor and Q_1 and Q_2 are the amount of charge present in the first and second branches, respectively. Q_{total} is easily determined by considering the charging period and the charging current.

If the terminal voltage is V_2 at the end of this period, we can write

$$Q_1 = \int_0^{V_2} (C_1 + C_{1Var} * V)dV = V_2\left(C_1 + \frac{C_{1Var}}{2} * V_2\right),$$

$$Q_2 = V_2 * C_2.$$

So we can easily derive

$$C_2 = \frac{Q_{total}}{V_2} - \left(C_1 + \frac{C_{1Var}}{2} * V_2\right).$$

6.6.3 Third-Branch Parameter Measurement

When charge redistribution between the first and second branches has taken place, there is still no charge in the third branch due to its very long time constant.

For determining R_3, we will wait Δt amount of time until the terminal voltage reaches $V_3 = V_2 - \Delta V$, where ΔV is a small value, for example, 50 mV, as it was chosen for calculating R_2. Since ΔV is pretty small, there is a virtually constant transfer current to the third branch given by $I_{tr} = (V_2 - (\Delta V/2))/R_3$, while holding the assumption $R_1 \ll R_2 \ll R_3$. Since the time constant of the first branch is quite shorter compared to the second branch, I_{tr} is mostly supplied by the first branch for this time period. If C_{inst} is the first-branch capacitance at $V_2 - (\Delta V/2)$, we can write

$$R_3 = \frac{(V_2 - (\Delta V/2)) * \Delta t}{C_{inst} * \Delta V}.$$

For calculating C_3, we have to wait long enough so that the charge redistribution between the three branches has finished. If we assume the terminal voltage to be V_4 at the end of this period, we can calculate C_3 by taking advantage of the charge conservation principle:

$$C_3 = \frac{Q_{total}}{V_4} - \left(C_1 + \frac{C_{1Var}}{2} * V_8\right) - C_2.$$

The measurement of the parameters must be conducted on several test units in order to mitigate the possibility of measurement errors and variation in supercapacitor parameters.

6.7 Kalman Filtering

The three-branch model is used by CPSs and energy-buffering applications to predict a supercapacitor's behavior. One of the most important aims is to provide a measure of the amount of energy buffered within a supercapacitor that is available to the system. However, predictions and measurements that use the three-branch model require the knowledge of the internal voltages across the capacitors in the equivalent circuit's three branches. These three branch voltages are referred to as the supercapacitor's internal state and cannot be directly observed from a single measurement of the supercapacitor's terminal voltage. For example, if a supercapacitor is left at rest after it has been charged up, charge redistribution will cause the current to flow to any branch with a lower voltage than the others. All three branches will settle to an equilibrium at the terminal voltage. In this equilibrium case, the state is directly observable. However, the same observed terminal voltage could also be produced by rapidly charging the supercapacitor such that most of the charge is stored in the first branch. The knowledge of the internal state distinguishes this rapid charging case from the supercapacitor at equilibrium even when the terminal voltages observed in both cases are identical. Tracking the internal state is important because while both cases are identical to an observer, the energy buffered in the supercapacitor is significantly greater for the first case because the long-term branches store more energy at equilibrium.

It has been shown that tracking a supercapacitor's internal state provides much greater accuracy than treating a supercapacitor as an ideal device of capacitance C, for which the buffered energy E is directly observable from the terminal voltage V_{sc}, as

$$E = \frac{1}{2} C V_{sc}^2.$$

Simulations show that this simple observation-only energy-awareness scheme underestimates the buffered energy in a supercapacitor by a root mean square error of 31% over a test profile, including charging and discharging at various current levels (Nadeau et al., 2014). The stored energy is underestimated because approximating the supercapacitor as an ideal device neglects the long-term branches that store additional energy when the supercapacitor is charged slowly. A supercapacitor's rated capacitance normally represents the device's average instantaneous capacitance measured while it is quickly charged. The ideal model can be adjusted for applications that operate within a narrow power range by increasing C in proportion to the extra energy stored on the long-term branches at that specific power level. This strategy of fitting a supercapacitor with a single constant capacitance value dependent on the operating power fails for applications with variable power supply and demand, such as CPSs that rely on solar power. Solar power can vary day to day, hour to hour, and even minute to minute in the case of variable cloud cover, causing varying portions of charge to be stored in the three branches, depending on how the supercapacitor is charged and discharged.

An alternative to modeling the supercapacitor as a simple ideal capacitor is to treat the supercapacitor as a black box and integrate the net power into and out of the supercapacitor without the need for any modeling. Buffered energy is found as the difference between the energy inputted, determined by the current into the supercapacitor over time, $I_{in}(t)$, and the energy outputted to the application, determined by the current out of the supercapacitor over time, $I_{out}(t)$:

$$E = \int I_{in}(t)V_{sc}(t)dt - \int I_{out}(t)V_{sc}(t)dt.$$

Because the aforementioned equation does not rely on any model for the supercapacitor, it is easily implemented without the need to set any parameters such as the capacitance C. However, neglecting internal loss treats the supercapacitor as perfect energy storage and overestimates E in the long term. In the short term, supercapacitors can operate at an efficiency close to 100% because of their small ESR (Maxwell Technologies, Inc., 2012). However, over the long term, leakage and series resistance losses accumulate and degrade the accuracy of this technique. Additionally, the measured quantities $I(t)$ and $V(t)$ are always subject to measurement noise. This noise accumulates over time in this model resulting in significant inaccuracy. For the simulation profile mentioned, this energy-awareness scheme is found to produce a root mean square error of 79.3%, due to the long duration of the simulation, which allows error due to internal losses to accumulate (Nadeau et al., 2014).

As opposed to the two energy-awareness schemes described prior, best results are produced by using the three-branch model to determine the energy buffered in a supercapacitor. Assuming that the supercapacitor's internal state $x = \begin{bmatrix} V_1 & V_2 & V_3 \end{bmatrix}^T$ is known, the contribution of each branch to the total buffered energy is calculated as

$$E = \frac{1}{2}C_1V_1^2 + \frac{1}{3}C_{1Var}V_1^3 + \frac{1}{2}C_2V_2^2 + \frac{1}{2}C_3V_3^2.$$

A simple technique to track the state x is to recursively predict x each time step according to the dynamics of the equivalent circuit. This method provides acceptable root mean square error of 4.8% in the simulation because it is only subject to the accumulation of measurement error, which is set to be small and zero mean (Nadeau et al., 2014). However, outside of simulation, modeling error is also present and the estimates of circuit parameters are imperfect. These inaccuracies introduce systematic error that accumulates in E over time and would cause greater error in the energy estimate. This prediction-only method also fails to utilize information in the observed supercapacitor voltage.

To use both the observed input current into the supercapacitor and the observed terminal voltage to estimate x, the Kalman filter is used. The discrete Kalman filter provides an optimal estimate of the supercapacitor's state, $\hat{x}$, taking into account all observations, including the present and previous values of $V_{sc}(t)$. Given that the three-branch equivalent circuit is a linear system of the form

$$\dot{x}(t) = F \cdot x(t) + B \cdot I_{sc}(t),$$

$$V_{sc}(t) = H \cdot x(t) + D \cdot I_{sc}(t),$$

the prediction for the supercapacitor's internal state $\tilde{x}$ over any discrete time step can be found by matrix exponentiation of F:

$$\tilde{x}(t+\Delta t) = \frac{e^{F \cdot \Delta t} \cdot x(t) + F}{(e^{F \cdot \Delta t} - I)B \cdot I_{sc}(t)},$$

$$\tilde{V}_{sc}(t+\Delta t) = H \cdot \tilde{x}(t+\Delta t) + D \cdot I_{sc}(t+\Delta t).$$

The Kalman filter is an efficient iterative solution that uses an update step to incorporate the information of each new voltage observation into the next estimate of the internal state $\hat{x}(t+\Delta t)$. Each new observation is incorporated by distributing the error residual between the predicted terminal voltage and the actual observation into the predicted state according to the Kalman gain K:

$$\tilde{x}(t+\Delta t) = \tilde{x}(t+\Delta t) + K \cdot \{V_{sc}(t+\Delta t) - \tilde{V}_{sc}(t+\Delta t)\}.$$

The Kalman gain can be calculated analytically for the discrete linear system of the three-branch model, but often it is approximated using a sigma-point Kalman filter. Alternatively, more complex models can be simplified by using a linear approximation as in the extended Kalman filter. Of all the energy-awareness techniques mentioned, Kalman filtering provides the lowest root mean square error at less than 1% (Nadeau et al., 2014) and is best suited for energy awareness in CPSs.

6.8 Energy Efficiency Modeling

One reason why supercapacitors are a good choice for energy buffering, especially in CPSs, is that supercapacitors provide high charge–discharge efficiency over a wide range of power levels without the need for more complex battery management techniques required for electrochemical batteries. However, the three-branch model includes series resistors and leakage that consume power and result in less than 100% efficiency. This section applies the three-branch model to determine a power range that a supercapacitor can comfortably operate within with near 100% efficiency. For example, CPSs commonly rely on solar energy harvesting and experience large fluctuations in the input power. A supercapacitor must efficiently buffer the energy that is harvested regardless of whether the weather is sunny and provides high power, or the weather is cloudy and energy harvesting is much slower. The operating efficiency of a supercapacitor is determined by the two sources of internal energy loss: series resistance in each capacitive branch and current loss through the parallel resistor in the leakage branch. At high power, high currents flow in and out of the supercapacitor and make power lost in the series resistances a more significant source of loss. At low power, losses in the series resistances are less significant, but leakage consumes a more significant portion of the power transferred to or from the supercapacitor.

To test a supercapacitor's efficiency limits, the three-branch model is simulated using various equivalent circuit parameters. Circuit parameters are measured from an Illinois Capacitor 10 F–2.7 V-DCNQ (Illinois Capacitor, Inc., 2012) and Maxwell BCAP0050 (Maxwell Technologies, Inc., 2012) and also simulated from literature (Zubieta and Bonert, 2000) for the 470 and 1500 F DLCs. These parameters are listed in Table 6.2.

TABLE 6.2

Three-Branch Model Parameters Found for 10 F Illinois Capacitor and 50 F Maxwell Supercapacitor

	Illinois Capacitor 10 F–2.7 V-DCNQ	Maxwell BCAP0050	DLC (Zubieta and Bonert, 2000) 470 F	DLC (Zubieta and Bonert, 2000) 1500 F
C_{Rated} (F)	10	50	470	1500
C_1 (F)	2.05	42.5	270	900
C_{1Var} (F/V)	6.03	5.1	190	600
R_1 (mΩ)	56	205	2.5	1.5
C_2 (F)	9.43	10.5	100	200
R_2 (Ω)	4.0	112	0.9	0.4
C_3 (F)	6.76	4	220	330
R_3 (Ω)	77.5	628	5.2	3.2
R_{leak} (kΩ)	90	36	9	4

Source: Zubieta, L. and Bonert, R., *IEEE Trans. Ind. Appl.*, 36(1), 199, 2000.

Parameters for 470 F DLC and 1500 F DLC supercapacitor from Zubieta and Bonert (2000) were also used to test efficiency.

Supercapacitor efficiency is tested over a wide range of power by simulating the three-branch model charged to voltage V. Efficiency is then measured by discharging the model through a range of different load resistances R. For each different R, the supercapacitor's efficiency η is the ratio between the useful power delivered to the load, P, and the total power, including the internally wasted power, P_W, within the three-branch model:

$$\eta = \frac{P}{P + P_W} \times 100\% .$$

Depending on how a supercapacitor is charged, the distribution of charge storage represented in the three-branch model by the voltages across the three capacitances can vary significantly from the terminal voltage V. Simulations remove any ambiguity in the supercapacitor's internal state by assuming an equilibrium where the voltages across all capacitors in the three-branch model are the same as V. This assumption gives intermediate results for efficiency: if the supercapacitor's internal state is distributed with more charge stored in the long-term branches, leakage losses are reduced, but series resistance losses become worse. If the supercapacitor's internal state is distributed toward short-term storage, series and leakage losses are skewed in the opposite directions. As the supercapacitor discharges, its terminal voltage V_{sc} is lower than the internal branch voltages V. The total wasted power P_W across all branches is

$$P_W = \frac{(V - V_{sc})^2}{R_1} + \frac{(V - V_{sc})^2}{R_2} + \frac{(V - V_{sc})^2}{R_3} + \frac{V_{sc}^2}{R_{leak}},$$

and the power delivered to the load is

$$P = \frac{V_{sc}^2}{R} .$$

The terminal voltage V_{sc} is determined by solving the node equation for the currents in all of the model's branches in addition to the current to the load:

$$\frac{V - V_{sc}}{R_1} + \frac{V - V_{sc}}{R_2} + \frac{V - V_{sc}}{R_3} - \frac{V_{sc}}{R_{leak}} = \frac{V_{sc}}{R}.$$

6.9 Evaluation

The energy-buffering efficiency η for the 10, 50, 470, and 1500 F supercapacitors is evaluated using the equations given in the previous section and shown in Figures 6.3 through 6.6. Each figure plots percent efficiency in relation to the power delivered to the load as resistance value R for the load varies. For example, a large load resistance is used to test supercapacitor efficiency at low power. The large load resistance prevents the flow of high current and limits the power delivered to the load. As the load resistance is decreased, more current is drawn from the supercapacitor and the power increases as $P = I_{sc}^2 R$. For each sequence of load resistances, there is an inflection point of maximum power that can be drawn from the supercapacitor. Beyond this maximum power, further decreasing the load resistance no longer delivers greater power. This inflection point happens because such small R draws very high current from the supercapacitors, and a large amount of wasted power is consumed by the supercapacitors' ESR. The efficiency of this maximum power operating point is below the range of efficiencies shown in Figures 6.3 through 6.6. Due to the low efficiency of drawing maximum power and because the simulated current significantly exceeds the maximum rating of physical devices, these cases do not occur in typical energy aware operation.

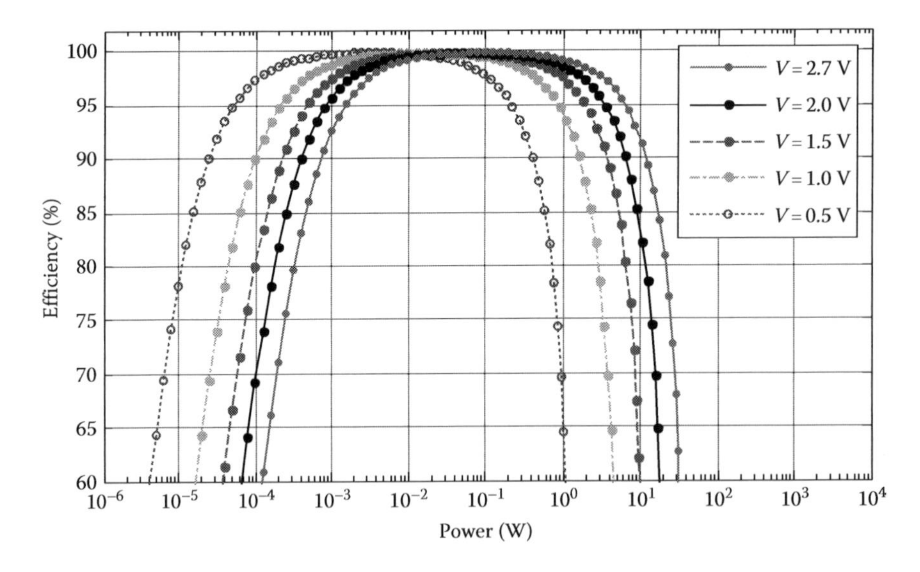

FIGURE 6.3
Predicted efficiency for discharging a 10 F supercapacitor is modeled by the three-branch model.

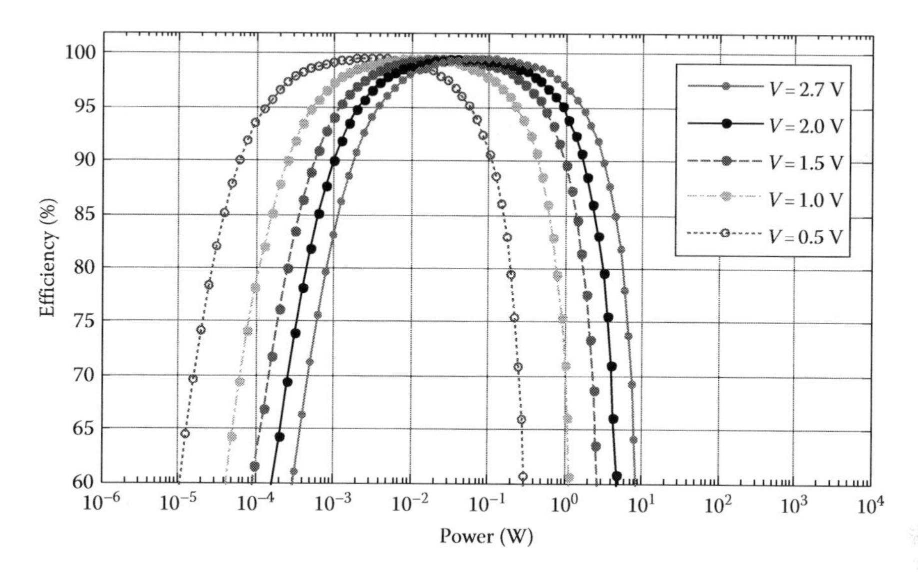

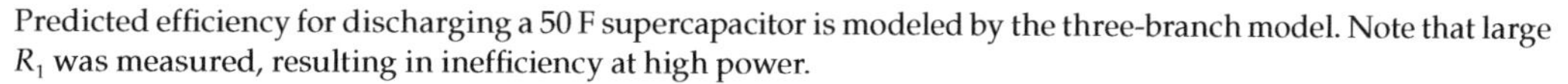

FIGURE 6.4
Predicted efficiency for discharging a 50 F supercapacitor is modeled by the three-branch model. Note that large R_1 was measured, resulting in inefficiency at high power.

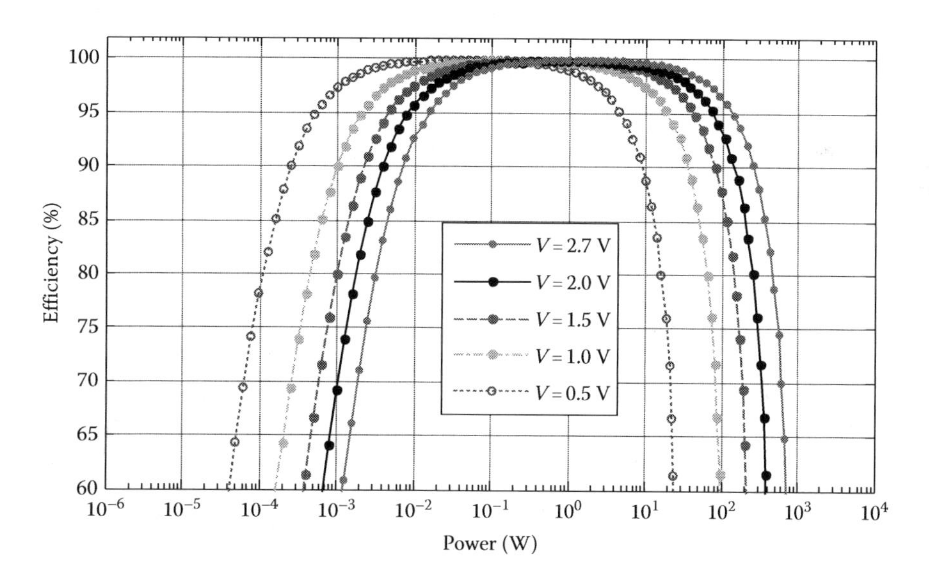

FIGURE 6.5
Predicted efficiency for discharging a 470 F supercapacitor is modeled by the three-branch model.

As power delivered to the load increases, efficiency initially increases due to diminishing leakage: a greater portion of the power flows to the load rather than to the parallel leakage branch in the three-branch model. As in Table 6.2, leakage resistance is shown to be inversely proportional to supercapacitor size resulting in more severe inefficiency for larger supercapacitors at low power. Inefficiency due to leakage is also influenced by the voltage of the supercapacitor. Because leakage is modeled as current through a resistor, the waste power is V_{sc}^2/R_{leak}. Consequently, leakage waste increases with the square of voltage.

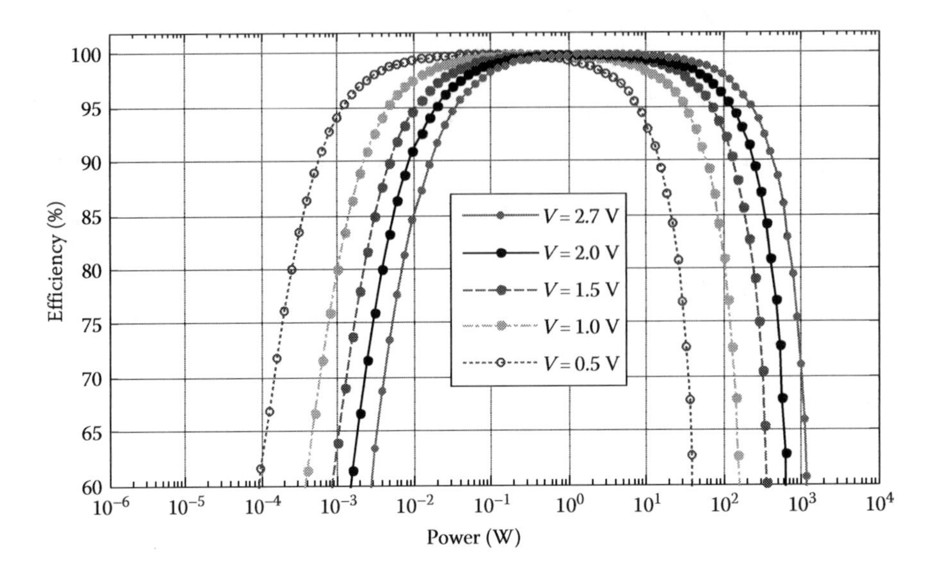

FIGURE 6.6
Predicted efficiency for discharging a 1500 F supercapacitor is modeled by the three-branch model.

The second important relationship between efficiency and power is the inefficiency of supercapacitors while delivering high power to load resistances. Inefficiency at high power results from high current flowing through the ESR of the supercapacitor. Consequently, as the current increases to deliver more power at a certain voltage or to maintain power at a lower V, efficiency suffers. It can be seen that larger supercapacitors perform better at high power due to their smaller internal ESR, represented by the R_1 parameter in the three-branch model.

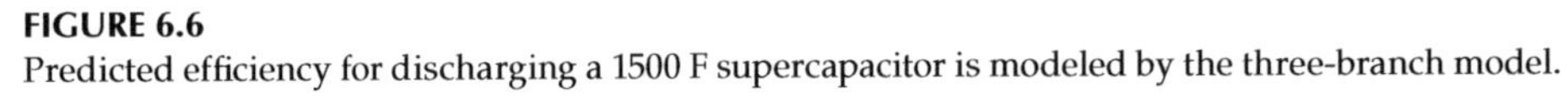

6.10 Conclusions and Future Work

This chapter demonstrated the feasibility of using supercapacitors as the energy-buffering mechanism for CPSs in which operational longevity and self-sustainability are of high priority. Modeling, parameter measurement, and energy-tracking methods have been introduced for the precise assessment of buffered energy in supercapacitors. Also, design considerations that impact the relative advantages of supercapacitors over electrochemical batteries have been discussed. Demonstrations use a three-branch model equivalent circuit for supercapacitors with capacitances spanning a wide range. Results show that reasonable efficiency can be maintained by operating the supercapacitor devices within suitable power limits that depend on the supercapacitor size and voltage level. Operation outside these limits can severely degrade supercapacitor efficiency.

The three-branch model can characterize supercapacitor behavior over a wide range of operating power levels. Combined with a Kalman tracker, this model provides a method for accurately tracking buffered energy in supercapacitor systems.

Precise energy estimation of supercapacitors by the proposed Kalman filtering approach can allow future CPSs to incorporate better energy management techniques, which, in turn, can increase the operational lifetime of the system. A high density of energy storage, with nearly unlimited charge–discharge cycles, makes supercapacitors an attractive choice for CPSs where self-sustainability and maintenance-free operation are key requirements.

Glossary

Charge redistribution: The evening out of charge across the supercapacitor that occurs most dramatically when a supercapacitor has been charged over a short period of time. Charge redistribution leads to a reduction in terminal voltage, which is sometimes mistaken for voltage.

DC–DC converter: A switching converter that takes energy from a power source and transfers it to another power source. Specifically, since both of these sources might have different DC voltage levels, the configuration of the DC–DC converter might have to be chosen specifically depending on the input/output voltage levels (i.e., *Vin* vs. *Vout*). A boost converter transfers energy in the *Vout* > *Vin* case, whereas a buck converter transfers energy in the opposite scenario (i.e., *Vout* < *Vin*). A SEPIC converter can transfer energy in both cases.

Double-layer capacitance: Capacitance at the interface of the electrode and electrolyte inside a supercapacitor. Due to the extremely high surface area of the materials used for electrodes in supercapacitors, double-layer capacitance can be very large. It is one of two contributing factors to the overall capacitance of a supercapacitor.

Energy density: The amount of energy stored per unit volume for a given energy-buffering device (e.g., supercapacitor or rechargeable battery). Supercapacitors are known to have around 10 times lower energy density than rechargeable batteries. For example, while supercapacitors have an average 10 W·h/kg energy density, a rechargeable battery typically has 100 W·h/kg density.

Energy efficiency: Quantifies what percentage of the energy that is input into a DC–DC converter is actually transferred to the output. For example, if 1 W is being applied to the input and a constant 0.85 W is being transferred to the output (e.g., to the supercapacitor storage), this system has an 85% efficiency. In other words, 15% of the incoming energy is wasted.

Equivalent parallel resistance: The three-branch model approximates the supercapacitor behavior as three parallel-connected RC pairs. The leakage aspect of supercapacitors is modeled as a resistor that is connected to these three branches in parallel, thereby continuously wasting energy. For this reason, the leakage resistor can be thought of as being an equivalent parallel resistor.

Equivalent series resistance: The first branch of the three-branch model has a very small resistor connecting to the main capacitor, which is the dominant storage element in the entire supercapacitor. The resistance of this first branch is termed equivalent series resistor (ESR) and limits the amount of current that can flow from the main terminals into this first-branch capacitor. ESR is an important parameter in determining the power density of the supercapacitor.

Kalman filter formulation: A technique used to continuously track the state of charge of the capacitors in the three-branch model. It utilizes recursive operations to continuously estimate the state of charge. This technique yields accuracy within 1%.

Latent variables: In Kalman filtering, the latent variables are the ones that the model tries to estimate, since they cannot be directly measured. In the case of supercapacitor modeling using Kalman filtering, the latent variables are the individual voltages and currents of the three branches. When the voltages of three individual branches deviate, they have to eventually equalize in the long term. So the primary advantage of using Kalman filtering is to estimate the latent variables during the iterative Kalman filtering process.

Leakage: Energy lost internally in the supercapacitor; increases exponentially with terminal voltage. Leakage is considered to be high in supercapacitors and is sometimes referred to as self-discharge.

Observable variables: These are the variables that can be measured and fed back into the Kalman filter. In the case of supercapacitor modeling, the only two observable variables are the terminal voltage of the supercapacitor and the terminal current (i.e., the current that is being fed into the supercapacitor).

Power density: The amount of power that can be supplied to the load. Supercapacitors are known to have around 10–100 times better power density than rechargeable batteries. Primarily determined by their ESR, supercapacitors can supply a real high peak power to the load. This allows supercapacitors to be used in applications with very high power demand, such as hybrid electric cars and elevators. It is measured in watts per kilogram.

Pseudocapacitance: Electrochemical storage of energy in supercapacitors by means of redox reactions. Like, double-layer capacitance, it increases with the surface area of the electrode. It is one of two contributing factors to the overall capacitance of a supercapacitor.

Rechargeable battery: A type of battery whose charge can be replenished by applying a charge in the reverse (charging) direction. This is the opposite direction in which the stored energy is being consumed.

State of charge: The level of charge of each branch in the three-branch model. There is no way to directly observe the state of charge for each of the three branches.

Supercapacitor: A device used to store electrical energy through two processes: double-layer capacitance and pseudocapacitance. They are made of highly porous materials such as activated carbon, which allows them to achieve capacitance, energy density, and power density levels orders of magnitude higher than electrolytic capacitors. Compared to rechargeable batteries, supercapacitors have much lower energy density but much higher power density. They are also known as ultracapacitors and electric double-layer capacitors.

Three-branch model: The most accurate model for supercapacitors that consists of three branches, each containing a resistor and a capacitor. The time constant in each branch is significantly different from that of the other branches so that only one branch dominates the supercapacitor behavior at any one time. The three-branch model also consists of a parallel resistor to simulate leakage and a voltage-dependent capacitor in the first branch to replicate supercapacitors' voltage-dependent behavior.

Acknowledgment

This work was funded in part by the National Science Foundation grant CNS-1239423 and a gift from Nvidia Corporation.

References

Bertrand, N., Sabatier, J., Briat, O., and Vinassa, J. M. (2010). Embedded fractional nonlinear supercapcitor model and its parametric estimation method. *IEEE Transactions on Industrial Electronics*, 57(12): 3991–4000.

Buller, S., Karden, E., Kok, D., and De Doncker, R. (2001). Modeling the dynamic behavior of supercapacitors using impedance spectroscopy. *Industry Applications Conference 2001. Thirty-Sixth IAS Annual Meeting. Conference Record of the 2001 IEEE* (pp. 2500–2504). September 30–October 4, Chicago, IL: Hyatt Regency Hotel.

Catherino, H. A., Burgel, J. F., Shi, P. L., Rusek, A., and Zou, X. (2006). Hybrid power supplies: A capacitor-assisted battery. *Journal of Power Sources*, 162(2): 965–970.

Conway, B., Birss, V., and Wojtowicz, J. (1997). The role and utilization of peudocapacitance for energy storage by supercapacitors. *Journal of Power Sources*, 66: 1–14.

Fahad, A., Soyata, T., Wang, T., Sharma, G., Henzelman, W., and Shen, K. (2012). SOLARCAP: Super capacitor buffering of solar energy for self-sustainable field systems. *Proceedings of the 25th IEEE International System-On-Chip Conference* (pp. 236–241). Niagra Falls, NY: IEEE.

Hassanalieragh, M., Soyata, T., Nadeau, A., and Sharma, G. (2014). Solar-supercapacitor harvesting system design for energy-aware applications. *Proceedings of the 27th IEEE International System-On-Chip Conference*. September 2–5, Las Vegas, NV: IEEE.

Illinois Capacitor, Inc. (2012). Super capacitor products. Retrieved 2012, from Illinois Capacitor Inc. Web site: http://www.illinoiscapacitor.com/ic_search/_super_products.aspx.

Maxwell Technologies, Inc. (2012). Maxwell technologies K2 series ultracapacitors high capacity cells. Retrieved 2012, from Maxwell Technologies, Inc. Web site: http://www.maxwell.com/products/ultracapacitors/products/k2-series.

Merrett, G. V. and Weddell, A. S. (2012). Supercapacitor leakage in energy-harvesting sensor nodes: Fact or fiction? *2012 Ninth International Conference on Networked Sensing Systems* (pp. 1–5). June 11–14, Antwerp, Belgium: IEEE.

Nadeau, A., Sharma, G., and Soyata, T. (2014). State-of-charge estimation for supercapacitors: A Kalman filtering formulation. *Proceedings of the 2014 IEEE International Conference on Acoustics, Speech, and Signal Processing* (pp. 2213–2217). Florence, Italy: IEEE.

Niu, J., Pell, W. G., and Conway, B. E. (2006). Requirements for performance characterization of c double-layer supercapacitors: Applications to a high specific-area c-cloth material. *Journal of Power Sources*, 156(2): 725–740.

Rotenberg, D., Vahidi, A., and Kolmanovsky, I. (2011). Ultracapacitor assisted powertrains: Modeling, control, sizing, and the impact on fuel economy. *IEEE Transactions on Control Systems Technology*, 19(3): 576–589.

Rufer, A. and Barrade, P. (2002). A supercapacitor-based energy storage system for elevators with soft commutated interface. *IEEE Transactions on Industry Applications*, 38(5): 1151–1159.

Weddell, A. S., Merrett, G. V., Kazmierski, T. J., and Al-Hashimi, B. M. (2011). Accurate supercapacitor modeling for energy harvesting wireless sensor nodes. *IEEE Transactions on Circuits and Systems II: Express Briefs*, 58(12): 911–915.

Yang, H. and Zhang, Y. (2011). Self-discharge analysis and characterization of supercapcitors for enviornmentally powered wireless sensor network applications. *Journal of Power Sources*, 196(20): 8866–8873.

Zhang, Y. and Yang, H. (2011). Modeling and characterization of supercapacitors for wireless sensor network applications. *Journal of Power Sources*, 196(8): 4128–4135.

Zubieta, L. and Bonert, R. (2000). Characterization of double-layer capacitors for power electronics. *IEEE Transactions on Industry Applications*, 36(1): 199–205.

7

Energy Harvesting and Buffering for Cyber-Physical Systems: A Review

Grayson Honan, Nicholas Gekakis, Moeen Hassanalieragh, Andrew Nadeau, Gaurav Sharma, and Tolga Soyata

CONTENTS

7.1 Introduction

Cyber-physical systems (CPSs) rely on generating their own energy to operate autonomously in the field without requiring maintenance. Solar, radio frequency (RF), wind, and vibration energy are among the sources that have been used or proposed for powering CPS. Each of these sources comes with its own advantages and disadvantages. For example, while ambient RF energy can be harvested in the dark, or where no solar energy is available, it requires a large area for the antenna, which could make CPS impractical. Alternatively, solar energy can provide the largest power output; however, it is only available in areas where there is constant sunshine (Chalasani and Conrad, 2008). Wind energy is widely available, but it requires a mechanical transmission that can be expensive to maintain; wind also tends to be extremely variable in terms of strength. Therefore, the optimal choice for energy harvesting is not obvious and requires detailed analysis of the location and operation modes of the CPS. After harvesting, the energy must be buffered using a storage device to provide power for the CPS operations during time periods where the energy source is unavailable. For example, in equatorial regions where sunlight cycles on and off on a 12 h cycle, the energy buffer can provide the operational power during dark hours. Combining different energy sources (e.g., solar and RF) can reduce the demands on the buffering, but cannot completely eliminate the need for buffering. Therefore, different mechanisms for buffering must be carefully considered before deploying a CPS. Options for buffering include rechargeable batteries, such as Li ion, lead acid, NiCd, and Ni-Mh or supercapacitors. While supercapacitors are virtually maintenance-free, for CPS with moderate to large power requirements, they take up a large amount of space posing a challenge for CPS deployment (Buchmann, 2011). Batteries can solve the energy density problem, at the expense of 3- to 4-year maintenance cycles (as compared to 10-year cycles for supercapacitors). This chapter provides a review of existing energy sources and energy buffering mechanisms and compares each option.

7.2 Energy Sources

Today, solar energy harvesting using photovoltaic (PV) cells is the mainstream energy harvesting technology for CPS. However, other sources of energy exist that may be better suited for harvesting in particular environments. In situations where solar needs to be supplemented or cannot be used at all, RF, wind, or vibration energy harvesting systems can be utilized. In this section, we provide an overview of each harvesting technology and conclude with an analysis that identifies which technology is best suited to a given environment (Table 7.1).

TABLE 7.1

Footprint Area Densities of Common Energy Harvesting Technologies

Harvesting Technology	Power Density
Solar	
Solar cells (outdoors at noon)	15 mW/cm^2
RF	
Dedicated source at short range	50 μW/cm^2
Ambient RF	2 μW/cm^2
Wind	
Small-scale turbine	83.3 μW/cm^3
Vibration	
Piezoelectric (shoe inserts)	330 μW/cm^3
Vibration (magnetostrictive Metglas material)	606 μW/cm^3

Source: Wang, L. and Yuan, F.G., Energy harvesting by magnetostrictive material (MsM) for powering wireless sensors in SHM, *Sensors and Smart Structures Technologies for Civil, Mechanical, and Aerospace Systems 2007*, San Diego, CA, SPIE, 2007.

7.2.1 Solar

PV cells are devices that convert light energy to electrical energy; they are used in applications ranging from low-power devices to large-scale harvesting for homes and commercial buildings.

The most common choice for PV materials is a silicon-based cell. These devices offer a reasonable price-to-performance ratio and represent a relatively mature technology, with the installation price per watt dropping 5%–7% annually for the past decade (Feldman et al., 2012). There are three main categories for silicon-based PV cells: monocrystalline (mono-Si or single-crystal-Si), amorphous (a-Si), and polycrystalline (mc-Si or p-Si) (Figure 7.1) (Green et al., 2014). Monocrystalline silicon is the most expensive of the three, but typically reaches efficiencies of around 25% (Green et al., 2014). Polycrystalline types are cheaper than monocrystalline, but only reach around 20% efficiency (Green et al., 2014). Amorphous silicon is the cheapest of these three categories and is used regularly in lower-power devices such as simple four-function calculators. In a typical outdoor environment, amorphous cells can reach efficiencies of around 10%, but indoors can maintain an efficiency between 3% and 7% where monocrystalline variants can only reach 1%–2% efficiency while indoors (Hande et al., 2007). Amorphous cells are well suited to absorb indoor light spectra, but other cell types are less efficient at converting the low intensity of artificial light into electricity (Randall and Jacot, 2002). The vast majority of PV systems are installed outdoors, where a square meter of harvesting space can yield between 100 and 1000 W, but for indoor systems, on the other hand, about 10 W/m^2 can be expected (Hande et al., 2007).

Wireless sensor networks (WSNs) are the major focus of research in low-power solar energy harvesting, with applications in medicine, environmental monitoring, and military operations. The smart dust program at the University of California, Berkeley, was one of the first major investigations into the use of PV cells in WSNs. The smart dust program used the energy harvested from PV panels to communicate optically with a laser-based transceiver system (Atwood et al., 2000; Kahn et al., 2000).

The power needs of a given CPS installation dictate the size of PV panel needed. The power density for solar cells harvesting outdoors at noon is approximately 15 mW/cm^2 (Chalasani and Conrad, 2008). Based on this information, an appropriately sized panel can be selected. Because the thickness of PV panels is negligible in most applications, PV

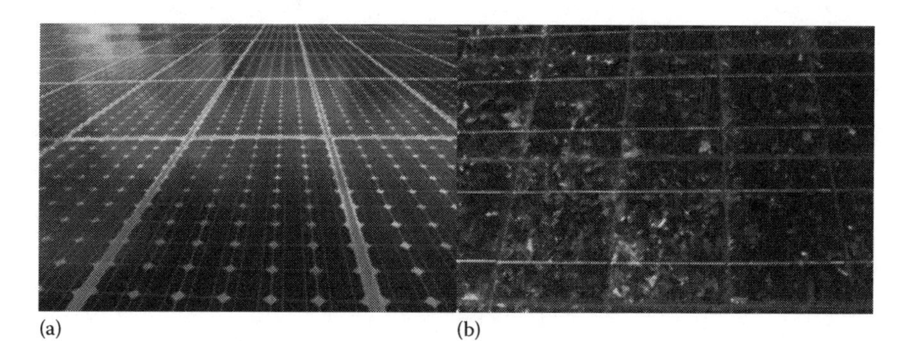

FIGURE 7.1
Monocrystalline cells (a) have a distinctive rounded geometric shape and appear smooth, while polycrystalline cells (b) are rectangular with a jagged internal structure. (From Andrewatla, Solar panel. Retrieved August 27, 2014, from freeimages: http://www.freeimages.com/photo/873825, September 19, 2007; DebbieMous, Solar panels reflecting the sky, Retrieved August 27, 2014, from freeimages: http://www.freeimages.com/photo/1364625, September 16, 2011.)

panel power density is typically considered in terms of power per area rather than per volume. For low-power PV panels (around 1 W), the thickness is around 4 mm or less. Larger panels (around 250 W) have a thickness of approximately 4.5 cm. At the time of writing, the cost of consumer PV panels in the 1–3 W range is around $10 and 5–10 W PV panels around $30.

Overall, solar energy harvesting is a very strong option for CPS and is extremely effective in comparison with other harvesting methods provided that the environment has sunlight available. The efficacy of PV panels is obviously very dependent on the location's light availability. The quality of a light source can be thought of in terms of intensity (typically, measured in lumens) and interruptions. For outdoor environments, intensity is determined by the angle of incidence for the sunlight on the panel and the thickness of the atmosphere traversed. This makes areas that are further from the equator less likely candidates for solar energy harvesting. Interruptions are the second factor we consider in the quality of a light source. When the sun is the light source, intermittent cloud coverage and nightfall are the primary causes of interruptions. For artificial light, the lifetime of the bulb and the efficiency of the solar panel within the emitted spectral bands from the bulb need to be considered. Regardless of the light source being used, one also needs to consider the possibility of dust or other debris blocking the light from efficiently entering the PV cells.

In addition to light source quality, one also needs to consider the solar panel's ability to harvest solar energy. As explained earlier, indoor situations call for amorphous PV panels since they maintain a higher efficiency (Hande et al., 2007). And in outdoor situations, monocrystalline has ~25% efficiency, polycrystalline has ~20% efficiency, and amorphous has around 10% efficiency. Installations of PV panels may also include the ability to automatically adjust their orientation to track the light source and harvest more energy than static installations.

7.2.2 Radio Frequency

RF energy harvesting is an attractive method for powering wireless devices that range from consumer electronics, such as mobile phones, to sensor nodes used by researchers to collect data in remote environments. The utility of RF harvesting comes from the fact that energy in this form is readily available, consistent, and reliable when compared to energy harvesting from other environmental sources such as solar radiation, wind, or vibrations (Figure 7.2).

The most important apparatus in harvesting RF signals is the rectenna, a combination of rectifier circuitry, most often in the form of either one Schottky diode or several cascaded Schottky diodes, and an antenna, most often a dipole antenna (Visser and Vullers, 2013).

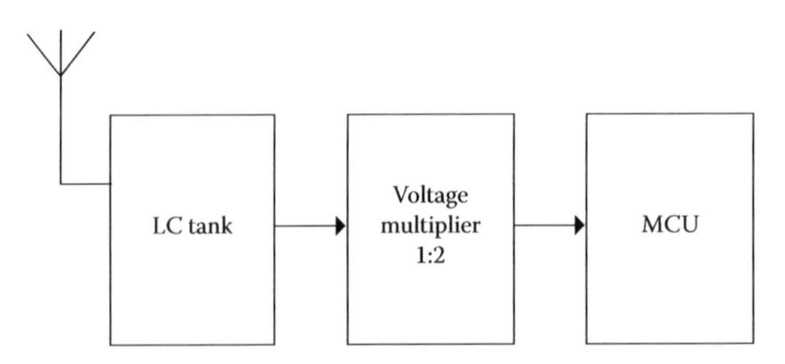

FIGURE 7.2
A block diagram for a standard radio-frequency energy harvester.

When in the presence of a suitable RF signal, that is, electromagnetic waves within an appropriate frequency band, a voltage is induced in the antenna as an alternating current (ac) signal. The rectifier circuitry then converts the signal to direct current (dc), which can then be used to drive a target device. In addition to the rectifier and antenna, the rectenna will frequently also include some form of dc to dc converter because, in the majority of cases, the output voltage of the rectifying circuitry will be too low to be utilized directly in an application (Jabbar et al., 2010).

The downside to RF harvesting is that the area density of recoverable energy is very low, reaching at most 50 $\mu W/cm^2$ (Roundy et al., 2004). This low-power yield is a result of the relative low density of ambient RF energy in most locations and inefficiencies in the antennas used during energy harvesting (Hudak and Amatucci, 2008). In ideal conditions, the antennas used during energy harvesting only reach a maximum of 45% efficiency (Hudak and Amatucci, 2008). One method to increase the energy harvested from RF signals is to use a dedicated RF source, rather than harvesting energy from ambient RF signals. A dedicated source can offer higher energy density because of its close proximity to the device (Visser and Vullers, 2013). In the context of CPS, this solution loses some key advantages over ambient RF energy harvesting because the dedicated source needs its own energy supply. Such a system would also require the harvester to stay within close proximity of the dedicated RF source. Due to the low-power yield of ambient RF harvesting, dedicated RF harvesting is the more widely used method, with its main applications in radio-frequency identification (RFID) (Hudak and Amatucci, 2008).

In RFID, a small device stores identifying data and remains powered off until it needs to be read. During a read, RF is emitted to power on the RFID device. The RFID device can then send its data for reading through one of several backscattering methods. When the reading device ceases transmitting its RF signal, the RFID device powers off once again (Curty et al., 2005). This is not strictly considered energy harvesting because the energy gathered is used to directly power the RFID devices rather than to charge a battery (Hudak and Amatucci, 2008). It is also important to note that RFID devices are not a form of CPS because, on their own, they do not do any computational tasks.

Another challenge in RF energy harvesting is impedance matching. Because RF signals exhibit variable frequency (ambient RF frequencies typically range from 1 to 3.5 GHz), the impedance of the circuit needs to be adjusted to continuously ensure maximum power (Shameli et al., 2007). Currently, most research on RF energy harvesting efficiency is focused on improving the antennas that gather the RF signals (Thomas et al., 2006). One breakthrough in improving the space efficiency of RF antennas showed that four antennas can be placed in the area that would normally be occupied by just two (Mi et al., 2005).

One of the great advantages of RF energy harvesting is that the system can be made very small; typically, the greatest amount of area is being used by the antenna itself. Usually, an RFID device will only be slightly larger than a grain of rice, and this includes circuitry for the data storage and transmission (Foster and Jaeger, 2007). If more power is required, the antenna must be made larger, but usually RF energy harvesting systems occupy at most few square inches (Hudak and Amatucci, 2008).

The major disadvantage of RF energy harvesting is its low-power yield; in order to get a power yield on the order of solar panels, a very large antenna would be needed. When using a dedicated RF supply at close range, 50 $\mu W/m^2$ can be harvested. If ambient RF signals are being used, only about 2 $\mu W/m^2$ can be harvested (Visser and Vullers, 2013).

RF harvesting devices require very little maintenance because no moving parts are involved and because they do not need to be exposed to the environment, unlike a solar or wind harvester.

FIGURE 7.3
A wind energy harvester at the University of Rochester.

If a dedicated RF supply is being used, RF energy harvesting can be done virtually anywhere; however, due to its low-power yield, RF harvesting should not be considered if access to direct sunlight is available. If the device relies on ambient RF waves, then an urban environment would yield far greater power than other locations (Figure 7.3).

7.2.3 Wind

Wind energy harvesting exploits the mechanical energy of blades or cups of a windmill spun by ambient wind, which is converted into electrical energy via a generator (Goudarzi and Zhu, 2013). Modern wind turbines are capable of providing power on the order of 2.5 MW or more in favorable wind conditions (Goudarzi and Zhu, 2013). When scaled down to very small sizes, generators cease to be practical in converting the mechanical energy harvested from the wind into electrical energy. This is because when the cross-sectional areas of the wind turbine's blades are very small (2–5 in.^{2}), the force created is on the order of 0.1 N or less under typical wind speeds that range between 3 and 10 mph. This is not enough force to efficiently operate a generator (Myers et al., 2007). Power per area can be increased by equipping the harvester with a diffuser (Ohya et al., 2008); however, to achieve the greatest efficiency at the smallest possible scale, a different method of energy conversion is required.

By replacing the usual generator found within a wind harvesting device with bimorph piezoelectric transducers, researchers from the University of Texas at Arlington created a successful small-scale wind energy harvesting device (Myers et al., 2007). When subjected to mechanical stress, piezoelectric transducers generate a voltage difference between two electrodes, thereby converting mechanical energy into electrical energy (Myers et al., 2007). Instead of using the rotational energy provided by a wind turbine to spin a conductor in a magnetic field, the rotational energy is used to spin a shaft with stoppers fixed on one end. As the shaft turns, the stoppers strike the piezoelectric transducers that are evenly spaced around the shaft, just within reach of the stoppers. The mechanical stress introduced when a stopper hits one of the transducers creates a voltage difference in the transducer, which can then be used to power a load. After optimizing the design for metrics such as space efficiency, minimizing damage done to the transducers, and synchronizing the stress put on the individual transducers, this new design yielded 5 mW of continuous power when operating in average wind speeds of 10 mph. This design used 18 piezoelectric bimorph transducers, but more transducers can be used to generate more power. If more transducers are added to the design, stronger winds are required (Myers et al., 2007).

One drawback for wind harvesting, especially for CPS, is that the space required for wind energy harvesting is quite large relative to other forms of energy harvesting. This is because a large cross-sectional area is needed to capture the wind's energy and because the mechanism used to convert the energy, either an electric generator or piezoelectric transducers, requires its own casing. Additionally, because wind harvesting systems use many moving parts, maintenance is typically necessary over the lifetime of the harvester to deal with the mechanical wear and tear.

7.2.4 Vibration

Vibration energy can be harvested from a variety of sources including raindrop impacts, industrial machinery, transport such as subways and cars, low-frequency seismic activity, and human motion (Guigon et al., 2008b). Research into ambient vibration energy harvesting can be classified into three main categories: smart material, electromagnetic, and electrostatic generation (Beeby et al., 2006; Wang and Yuan, 2007). Smart materials can be further classified into the categories of piezoelectric and magnetostrictive materials (Wang and Yuan, 2007). Piezoelectric materials are much more common, but magnetostrictive materials are an emerging candidate for energy harvesting applications. Mechanical deformation of a piezoelectric material typically produces power on the order of microwatts, as does the mechanical deformation of a magnetostrictive material while possibly yielding higher average power and power density (Beeby et al., 2006; Wang and Yuan, 2007). The efficacy of piezoelectric generators relies heavily on the type of piezoelectric material being used, the shape and size of the selected material, and the material's orientation in relation to the mechanical forces that are deforming it (anisotropy) (Beeby et al., 2006). As with all transduction techniques, the efficiency with which mechanical energy is converted into electrical energy relies strongly on the apparatus being used to capture the mechanical energy, and its potential strengths or weaknesses in effectively translating mechanical energy in a given axis of motion (Chalasani and Conrad, 2008).

Piezoelectric generators are simple to implement in microelectromechanical systems (MEMS), but are fragile and can allow leakage of charge. Magnetostrictive materials are difficult to implement in MEMS, but do not suffer from the depolarization or charge leakage problems that piezoelectric generators face (Wang and Yuan, 2007).

Electromagnetic generators work on the principle of electromagnetic induction. A spring–mass system, where the magnetic mass provides changing magnetic flux in the presence of a conductor coil, produces microwatts of power on average (Beeby et al., 2006). Relative to other vibration-driven generators, electromagnetic generators are large and difficult to implement in MEMS (Chalasani and Conrad, 2008). The main disadvantages of electromagnetic generators include their size and low peak voltage of only around 0.1 V; smart materials and electrostatic generators have voltages of 2–10 V (Wang and Yuan, 2007).

Electrostatic generators harvest energy using charged capacitor plates moving back and forth relative to each other, oscillating between minimum capacitance and maximum capacitance (Beeby et al., 2006). As the capacitance oscillates with constant charge on the plates, the voltage on the plates varies inversely with the distance between the plates (Shad et al., 2002). According to Beeby et al. (2006), the work done by vibrations "against the electrostatic force between the plates provides the harvested energy" (p. R187). There are three types of electrostatic generators—each one characterized by its unique structure and orientation of capacitor plates (Shad et al., 2002). As illustrated in Figure 7.4, out-of-plane gap closing electrostatic generators work with two relatively large capacitive plates separated by a gap that closes due to vibrations perpendicular to the plates. In order for this generator to produce large capacitance changes and thus be effective, the gap between the plates must become very small (Shad et al., 2002). As Shad et al. explain, this type is also prone to shorting due to the plates becoming permanently stuck together via surface interaction forces, and thus this is a less desirable design. The second type of electrostatic generator is the in-plane gap closing method. The capacitive plates in this design are fingerlike projections interwoven with capacitive plates extending from the moving part of the design (Shad et al., 2002). As the apparatus vibrates in plane with the movable section of the devices, the fingers get closer together and further apart, but the overlap remains the same. In-plane gap closing designs can easily incorporate mechanical stops to prevent shorts, so this design does not suffer from the problems of out-of-plane gap closing generators (Shad et al., 2002). Finally, in-plane overlap varying electrostatic generators work with a similar, interwoven finger design, but rather than varying the gap between the plates, vibrations in this scheme change the overlap area of the fingers (Shad et al., 2002).

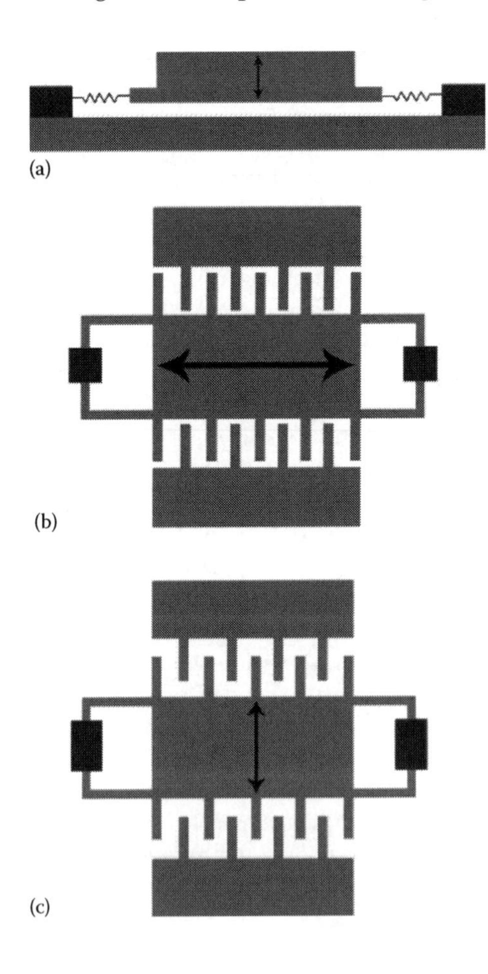

FIGURE 7.4
Out-of-plane gap closing type (a). In-plane gap closing type (b). In-plane overlap type (c). (Based on Shad, R. et al., Micro-electrostatic vibration-to-electricity converters, *Proceedings of IMECE 2002*, New Orleans, LA, ASME, November 17–22, 2002.)

Electrostatic devices are easily implemented in MEMS and do not need a smart material, although they do require an additional voltage source to provide the initial charge to the capacitive plates (Wang and Yuan, 2007).

As shown in Table 7.2, Beeby et al.'s survey on vibration energy harvesting indicates the average electromagnetic device is approximately 4220 mm³, the average piezoelectric device is approximately 3960 mm³, and the average electrostatic device is approximately 740 mm³ (Beeby et al., 2006).

Vibration energy typically produces power on the order of microwatts, although power on the order of milliwatts can be harvested in extreme conditions. Raindrop impacts, for example, may be capable of producing 12 mW in downpour conditions with large-sized drops (Guigon et al., 2008a).

Electrostatic, electromagnetic, and piezoelectric vibration energy harvesters were the subject of Beeby et al.'s survey. Table 7.3 shows the power generation capability of the

TABLE 7.2

Volume of Vibration Energy Harvesting Devices

	Electromagnetic (mm³)	Electrostatic (mm³)	Piezoelectric (mm³)
	5.4	75	16,000
	240	750	214
	2,100	1000	25,000
	840	800	125
	30,000	1800	1,000
	4	32.4	1,000
	2,000		1,947
	250		60
	1,000		2,185
	1,000		2
	9,000		0.9
			0.0027
Average	4,222	743	3,961

Source: Beeby, S.P. et al., *Meas. Sci. Technol.*, 17(12), R175, 2006.

TABLE 7.3

Power Generated per mm³ for Vibration Energy Harvesting Devices

Electromagnetic (μW/mm³)	Electrostatic (μW/mm³)	Piezoelectric (μW/mm³)
0.056	0.8533	0.08125
2.21	0.0049	0.08879
0.00000019	0.11	0.336
0.214	0.0075	0.0168
0.1333	0.5844	0.21
0.625	2.16	0.375
0.000002		0.006112
0.00576		0.0000167
0.01		0.03661
0.83		0.05
0.7778		0.67
		37

Source: Beeby, S.P. et al., *Meas. Sci. Technol.*, 17(12), R175, 2006.

various devices surveyed. Note that magnetostrictive materials were not part of Beeby's survey, most likely due to the few number of papers on the subject in the context of energy harvesting.

Vibration energy harvesting devices are generally reliable and easy to maintain. Electromagnetic devices, in particular, are quite reliable as there is very little chance for physical contact between the moving and stationary parts of the device, which could cause a short or physical damage to the device (Chalasani and Conrad, 2008).

Piezoelectricity is a suitable transduction method in areas where there is a consistent, significant source of vibration. Other energy harvesting techniques can produce more power, such as solar harvesting, but if sunlight is not present or is unreliable, vibration converters can provide the energy for low-power embedded CPS. Some suitable sources for vibration energy harvesting include train tracks, automobile engines, cardiac tissue, geological fault lines, raindrop impacts, areas of high foot traffic, and factory machinery.

7.3 Energy Buffering

In order to investigate energy buffering, it is important to be familiar with how batteries and supercapacitors are rated in terms of capacity and energy. The capacity of energy buffers is expressed in ampere-hours (Ah), or frequently milliampere-hours (mAh). This is equal to the amount of charge transferred at a steady current over 1 h. In order to accurately calculate energy, one needs to integrate the power delivered over the discharge interval. In slowly discharging batteries, voltages tend to linearly decrease as they drain, so the energy of a battery can be easily estimated by assuming a constant average voltage:

$$\text{Energy}(\text{J}) \approx \text{Average voltage}(\text{V}) * \text{Capacity}(\text{Ah}) * 3600\frac{\text{s}}{\text{h}}$$

Battery energy can also be thought of in terms of kilowatt-hours. Assuming a constant power, a kilowatt-hour measures the total energy consumed in 1 h:

$$\text{kW} * \text{h} = \text{kW} * 3600\ \text{s} = \frac{\text{kJ}}{\text{s}} * 3600\ \text{s} = 3600$$

7.4 Batteries

7.4.1 Lead Acid

Lead-acid batteries are the oldest form of rechargeable batteries and have maintained their position as the most widely used rechargeable battery for over 100 years. This is true despite their relatively small storage capacity thanks to their dominance in the automotive market, where they provide cars with the energy required for lighting and ignition functions associated with starting (Kularatna, 2011).

Lead-acid batteries are manufactured with capabilities ranging from less than 1 Ah to greater than 3000 Ah (Bullock, 1994). At 25°C, the standard cell potential is 2.048 V and common lead-acid batteries are capable of supplying between 30 and 55 Wh/kg (Bullock, 1994). Normally, the battery's capacity will remain very near its expected voltage for most of its life and then will eventually lose some capacity due to age and usage (Table 7.4) (Kularatna, 2011). Lead-acid batteries are also highly versatile: different lead-acid battery designs have achieved up to a 70-year float life, more than 1000 deep-discharge cycles, power of up to 200 W/kg, and have been made with as little 6 mm thickness (Bullock, 1994) (Figure 7.5).

Lead-acid batteries are categorized both by the type of electrode they use and their electrolyte configuration. There are three electrode types: Planté, Fauré or pasted, and tubular (Bullock, 1994; Kularatna, 2011). The Planté electrodes consist of high-surface-area plates that have been corroded in acid to form lead dioxide on the surface. They are used exclusively for industrial applications in which long life is prioritized over high specific energy (Bullock, 1994).

Fauré plates are made by applying a lead paste to a lead grid. Due to the softness of pure lead, a lead alloy is often used to prevent deformation of the plates; antimony has historically been added to prevent grid growth, but antimony accelerates water loss. Lead–calcium alloys are a suitable replacement for antimony, as they can be used to prevent grid

TABLE 7.4

Comparison of Common Rechargeable Battery Types in Terms of Cell Voltage, Life Cycles, and Self-Discharge Rate

Rechargeable Battery Type	Typical Cell Voltage (V)	Average Life Cycles (Cycles)	Average Self-Discharge Rate (Month)
Sealed lead acid	2.0	1250	3
NiCd	1.2	750	20
NiMH	1.2	650	23
Li-ion	3.6	1100	8

Source: Kularatna, N., *IEEE Instrum. Meas. Mag.*, 14(2), 20, 2011.

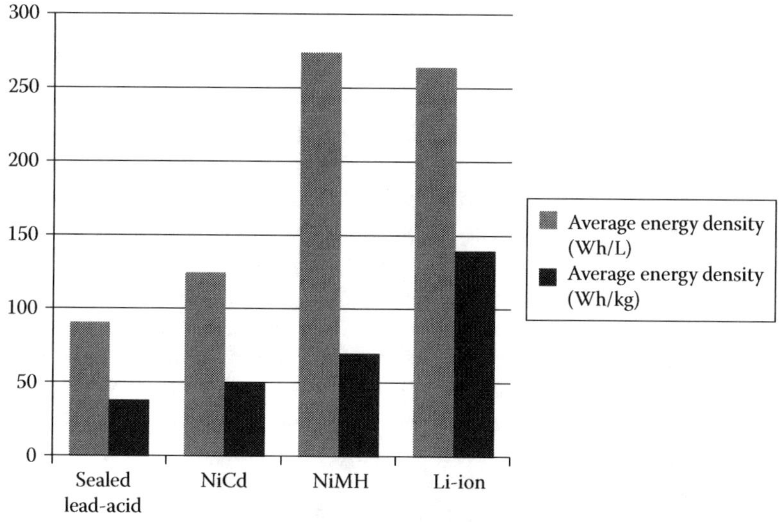

FIGURE 7.5
A comparison of common rechargeable battery types in terms of both volumetric and gravimetric energy density. (From Kularatna, N., *IEEE Instrum. Meas. Mag.*, 14(2), 20, 2011.)

corrosion and do not cause water loss. High-quality lead–calcium Fauré electrodes have very low maintenance and do not require water replacement throughout the lifetime of the battery (Bullock, 1994). The paste applied to Fauré electrodes consists of powdered lead oxide, sulfuric acid, and water. It gets applied to the grid at a high temperature and humidity to form a hardened porous structure. Fauré electrodes are excellent when low-cost or high specific energy or power is required.

Tubular electrodes are made by packing lead oxide powder into porous tubes made of chemically inert fibers; lead rods are then inserted down the center of each tube and connected together in parallel. In general, tubular electrodes can only be used as the positive electrode and are used in combination with a negative Fauré electrode; they make up for their lack of versatility by having very high deep cycle life.

Two types of electrolytes are used in lead-acid batteries: flooded and valve-regulated lead-acid (VRLA) (Bullock, 1994). VRLAs can be further broken down into gelled and separated implementations. In a flooded lead-acid battery, the entire system is flooded with 30%–40% sulfuric acid; however, this design requires regular maintenance. When a flooded lead-acid battery reaches 85%–90% of its charge capacity, the recharge reaction becomes less efficient, and water starts to break down into hydrogen and oxygen gas. Because of this, the batteries cannot be sealed, or pressure will start to accumulate. As this water is lost as gas over time, water needs to be readded to the system.

VRLA batteries were designed to minimize the problem of water loss by focusing on a design that promotes the chemical recombination of oxygen at the negative electrode (Bullock, 1994). Some VRLA designs achieve up to 99% oxygen recombination efficiencies. The remaining gas that does not get reconverted to water is vented out via a one-way pressure relief vent, which leads to the names for this design, VRLA or sealed lead-acid (SLA).

There are two kinds of VRLA batteries: the gelled electrolyte and the retained (or absorbed) system. In the gelled electrolyte system, the electrolyte is suspended in a silica gel; in the retained system, a glass fiber separator absorbs and retains the liquid electrolyte (Kularatna, 2011). While the automotive industry still primarily employs the basic flooded type of lead-acid battery, newer VRLA batteries are becoming popular in other fields, such as uninterruptible power supplies for the telecommunications industry (Bullock, 1994).

7.4.2 Nickel–Cadmium

Nickel–cadmium (NiCd) batteries are a very mature and well-understood technology. For many years, NiCd batteries were the preferred battery type for many applications ranging from emergency medical equipment to power tools (Buchmann, 2011). However, in recent years, NiCd batteries have been largely replaced in nearly every application by emerging battery technologies such as lithium ion (Li ion) and nickel–metal hydride (NiMH). This is largely due to three drawbacks of NiCd batteries: low specific energy, a high self-discharge rate, and a strong presence of the memory effect. NiCd batteries are capable of providing just 40–60 W/kg, one of the lowest gravimetric energy densities of any battery, only beating out lead-acid batteries. Additionally, because cadmium is a toxic, heavy metal, disposal of NiCd batteries is difficult and environmentally unfriendly. This negative feature of cadmium has spurred research into cleaner technologies and has led to the widespread adoption of NiMH batteries (Kularatna, 2011). NiCd battery's high self-discharge rate of 15%–25% per month means that it must be recharged after storage (Kularatna, 2011). Finally, one of the most detrimental aspects of NiCd batteries is their strong susceptibility to the "memory effect" or, more correctly, the voltage depression effect (Kularatna, 2011). After repeated partial discharges, a NiCd battery will lose its cell voltage prematurely during normal use (discharge). This often causes systems being buffered by the cell to become inoperable until a sufficient cell voltage is returned. Batteries with the voltage depression effect must be fully discharged and recharged before they will return to normal operation.

Nevertheless, NiCd batteries are not completely out of use; many features of NiCd batteries make them highly useful in niche applications (Buchmann, 2011; Kularatna, 2011). Specifically, NiCd batteries can withstand a wide range of temperatures, handling low temperatures particularly well. NiCd batteries also benefit from a very fast charge, and they also have a very pronounced peak in their charging profile at 100% charge. This makes charging relatively simple when compared to NiMH batteries. Finally, NiCd batteries boast a large amount of cycle lives—up to 1000 cycles—and thus they remain the cheapest battery in terms of cost per cycle.

7.4.3 Nickel–Metal Hydride

Nickel–metal hydride (NiMH) batteries have been in use since 1989 and are still widely used thanks to an array of highly desirable features that make them ideal for many different applications (Kularatna, 2011). NiMH batteries are largely based on nickel–cadmium (NiCd); however, they have two important characteristics that have caused them to largely eliminate the use of their predecessor. First, they have about a 40% increase in specific energy over NiCd, and they are made of much more environmentally friendly materials and are easier to recycle (Buchmann, 2011; Fatcenko et al., 2007). Second, the cadmium of NiCd batteries is replaced with a metal alloy improving the environmental friendliness of the batteries.

In addition to these characteristics, other notable features of NiMH batteries include wide range of cell sizes, safety in operation at high voltages and during charge and discharge, comparatively high energy and power densities, very low maintenance, and inexpensive charging and electronic control circuits (Fatcenko et al., 2007).

7.4.4 Lithium Ion

Li-ion batteries are commonly found in consumer electronics and are popular due to their high energy density, slow loss of charge, and minimal voltage depression effect (Buchmann, 2011). The self-discharge rate of Li-ion batteries is 5% in the first 24 h after being fully charged and then 1%–2% per month. An additional 3% loss per month can be attributed to the protection circuitry that is present in nearly all Li-ion batteries. Li-ion batteries typically have specific energies between 90 and 190 Wh/kg, energy density ranging from 200 to 330 Wh/L, and a capacity ranging from 1 to 4 Ah.

Li-ion batteries are considered to be dangerous in some situations due to their inclusion of a flammable electrolyte and the fact that they are kept pressurized. When considering energy buffering solutions, Li-ion batteries can only be safely charged at temperatures ranging from 0°C to 45°C. Li-ion batteries can only be safely discharged at temperatures ranging from −20°C to 60°C. Charging or discharging a Li-ion battery outside of its rated temperature limits can cause irreparable damage to the battery's capacity or can cause ruptures and leakage of electrolyte.

Li-ion batteries are of six main types: lithium cobalt oxide (LCO), lithium manganese oxide (LMO), lithium iron phosphate (LFP), lithium nickel manganese cobalt oxide (NMC), lithium nickel cobalt aluminum oxide (NCA), and lithium titantate (LTO). According to Buchmann (2011), LCO has high capacity and is primarily used for cell phones, cameras, and laptops.

LMO, LFP, and NMC are safer and have high specific power and long life spans, but they have a lower capacity than LCO. NCA and LTO are less frequently used, but are becoming increasingly important in electric power trains and grid storage. NCA has high specific energy and power with a long life span, but poor relative safety and cost. LTO batteries are extremely safe, but are rather costly and have low energy density.

7.5 Supercapacitors

Supercapacitors, sometimes called ultracapacitors or electric double-layer capacitors, are electrochemical devices that represent a transition between two distinct device classes: conventional capacitors and rechargeable batteries (Figure 7.6). Supercapacitors have the highest energy density among capacitors, but are only ~10% as energy dense as conventional batteries. What supercapacitors lack in energy density they make up for in terms of power density. Compared to conventional batteries, supercapacitors are generally 10–100 times more power dense.

Supercapacitors can derive their capacitance from two principles—the first of which is the electric double layer (EDL). The EDL is a storage structure in which charged ions from an electrolyte surface are attracted to a conductive electrode surface by electrostatic forces or selective absorption of ions into the surface (Conway, 1991). The second principle leading to energy storage in supercapacitors is pseudocapacitance, allowing chemical storage of energy, rather than electrostatic storage found in conventional batteries (Conway, 1991).

FIGURE 7.6
Pictured are four supercapacitors. From left to right, they are 5, 50, 350, and 3000 F.

The amount of energy stored in a supercapacitor can be naïvely calculated based on its capacitance and terminal voltage. For example, taking a 2.7 V, 3000 F supercapacitor,

$$E = \frac{1}{2}CV^2 = \frac{1}{2} * 3000\,F * 2.7\ V^2 = 10{,}935\ J$$

This calculation, while acceptable as a rough estimation, does not take into account the physics of charge storage in supercapacitors, often modeled via the commonly accepted three-branch model of supercapacitors (Zubieta and Bonert, 2000). To more accurately predict the state of charge in a supercapacitor, a Kalman filtering approach has been used (Nadeau et al., 2014). Assuming ideal capacitance or using a recursive calculation of stored energy yields up to 85% error in estimating state of charge, but by using Kalman filtering to estimate the true parameters of the three-branch model, this error can be reduced to just 1%.

When considering supercapacitors as a possible energy buffering solution, the energy harvesting method is also a key part of the decision. If solar energy is being harvested, for example, cloud coverage can lead to short-duration power spikes that can easily be handled by supercapacitors, but would overwhelm conventional batteries (Nadeau et al., 2014). Supercapacitors can also be charged and discharged a greater number of times than conventional batteries, so long-term CPS installations would benefit from supercapacitors' life span (Buchmann, 2011).

Compared to lithium-ion batteries, supercapacitors offer some great advantages. Given an adequate voltage, supercapacitors are capable of charging on the order of seconds (Buchmann, 2011). This superior *power* density characteristic allows supercapacitors to be used in data centers that serve very-compute-intensive applications (Kocabas et al., 2013; Kocabas and Soyata, 2014; Kwon et al., 2014; Page et al., 2014; Soyata et al., 2012a,b, 2014; Wang et al., 2014). Since these applications cause major spikes in power demand, supercapacitors are an excellent tool to locally store the energy and provide it to the rack computers that demand it.

Lithium-ion batteries, on the other hand, commonly take up to an hour or more to reach a full charge. Supercapacitors allow for a much greater range in charge and discharge temperature as well, between −40°C and 65°C, compared to lithium-ion batteries' range of 0°C–45°C for charging and −20°C to 60°C for discharging (Buchmann, 2011). Supercapacitors' durability can also be seen in their life cycle that can reach one million cycles, while lithium-ion batteries are on the order of several hundred. Supercapacitors' greatest advantage over batteries is their specific power, which can be as high as 10,000 W/kg, compared to just 3,000 W/kg for lithium-ion batteries.

The disadvantages of supercapacitors are their low specific energy, which is usually around 5 Wh/kg, whereas lithium-ion batteries can achieve between 100 and 200 Wh/kg (Buchmann, 2011). Additionally, they are more expensive per watt. The price of a typical supercapacitor is around $20/Wh, while a large lithium-ion battery can be as little as $0.50/Wh.

7.6 Cyber-Physical Energy Harvesting/Buffering Circuit Design

In this section, we describe the circuitry that is employed by energy sources and buffering devices. Common circuit structures to harvest energy from commonly used power sources as well as charge circuitry for energy buffering devices are provided.

7.6.1 Solar Energy Harvesting

When harvesting solar energy, it is important to note that a solar panel's I–V relationship is nonlinear (Fahad et al., 2012). This implies that solar panels connected in series will not yield the sum of their individual advertised power ratings. In order to use solar panels efficiently and to yield the maximum power from solar panels, one must use a maximum power point tracking (MPPT) circuit or a software implementation to sample the output of the cells and apply the proper load resistance to ensure the solar panel is working at its maximum power point (MPP) (Femia et al., 2005). The MPP of a solar panel depends on the intended load, the current irradiance levels hitting the cells, and the panel's temperature (Hassanalieragh et al., 2014). An example of MPP curve is shown in Figure 7.7. An MPPT circuit or software solution simply realizes an algorithm that operates close to the MPP of solar panels. Without MPPT circuitry/control software, solar panels can see significant degradation in efficiency.

Two common gradient descent algorithms are used for low-cost MPPT implementation: these are perturb and observe (P&O) and incremental conductance (INC) (Femia et al., 2005). P&O is particularly easy to implement, but also includes some inherent oscillation around the MPP at steady state. Essentially, P&O varies the operating voltage of the PV panel in a given direction, and the power from the PV panel is observed. If the power

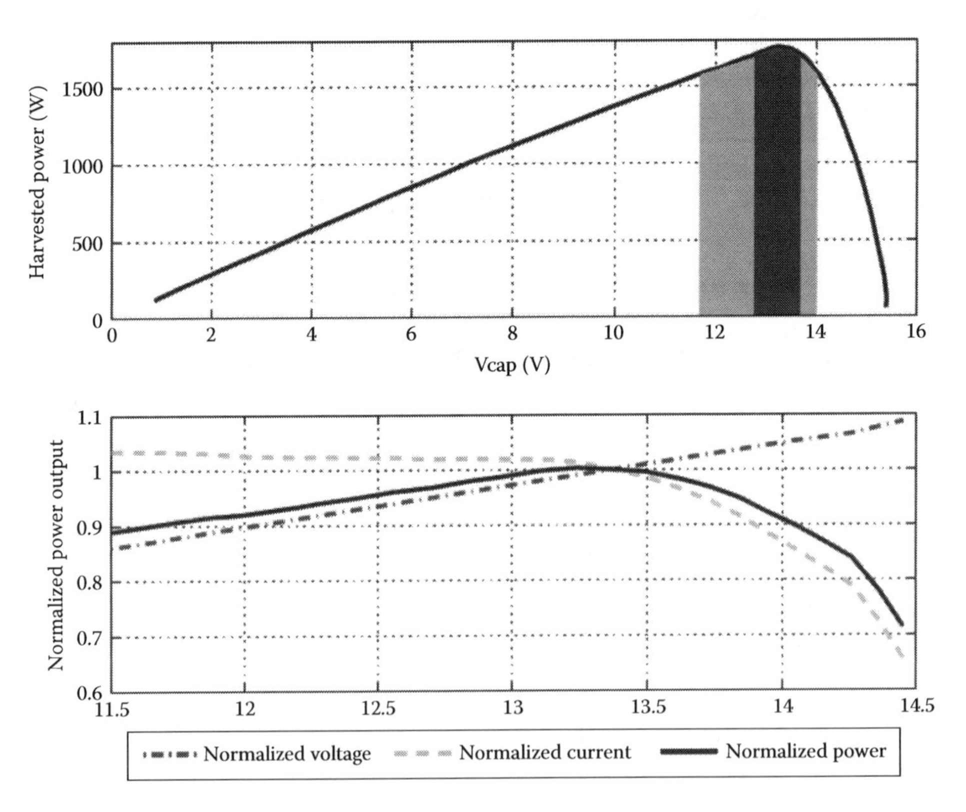

FIGURE 7.7
Nonlinear I–V curve of three solar panels in series. The maximum power point is reached at 13.3 V. (Reprinted with permission of the authors Hassanalieragh, M., Soyata, T., Nadeau, A., and Sharma, G., Solar-supercapacitor harvesting system design for energy-aware applications, *Proceedings of the 27th IEEE International System-On-Chip Conference (IEEE SOCC 2014)*, Las Vegas, NV, IEEE, September 2–5, 2014.)

drawn increases, this means that the operating point has become closer to the MPP, and thus the operating voltage is further perturbed in that direction until the power drawn decreases. At the point where the power drawn decreases, the operating voltage is perturbed in the opposite direction. As the perturbation approaches a very small value, and the cycle time for this algorithm decreases, the problem of oscillating around the MPP is reduced but never fully eliminated.

INC attempts to solve the oscillation problem, but is algorithmically more complex, which translates to slower sampling rates and increased cost in terms of power and hardware. INC is based on the observation that at MPP, $di_{PV}/dv_{PV} + i_{PV}/v_{PV} = 0$, where i_{PV} is the PV panel current and v_{PV} is the PV panel's voltage (Femia et al., 2005).

If $di_{PV}/dv_{PV} + i_{PV}/v_{PV} < 0$, this means the operating point is to the right of the MPP on the V–P plane, and if $di_{PV}/dv_{PV} + i_{PV}/v_{PV} > 0$, this means the operating point is to the left of the MPP. INC causes a perturbation in the correct direction based on the sign of the quantity $di_{PV}/dv_{PV} + i_{PV}/v_{PV}$, and when the quantity = 0 within a reasonable tolerance, the algorithm stops all perturbation to prevent the oscillation problem.

Figure 7.8 shows one way to design a simple P&O or INC circuit. A major issue with both P&O and INC is that they may are not well suited to determine local versus global maximums and minimums and thus may be stuck at a local optimum operating point. A third and final algorithm that is relatively simple to implement and solves this local optimum problem is the fractional open-circuit voltage method. This method approximates the MPP as a fraction of the open-circuit voltage, usually around 75%, but this differs based on environmental conditions (Ahmad, 2010). In this method, a measurement is made of the open-circuit voltage, and then the circuit attempts to keep the input solar power at 75% (or whatever fraction is correct for the environmental conditions) of that measurement. There are still drawbacks with the fractional open-circuit method: the method is approximate rather than exact; also, to measure the open-circuit voltage, harvesting must be stopped for a small duration of time.

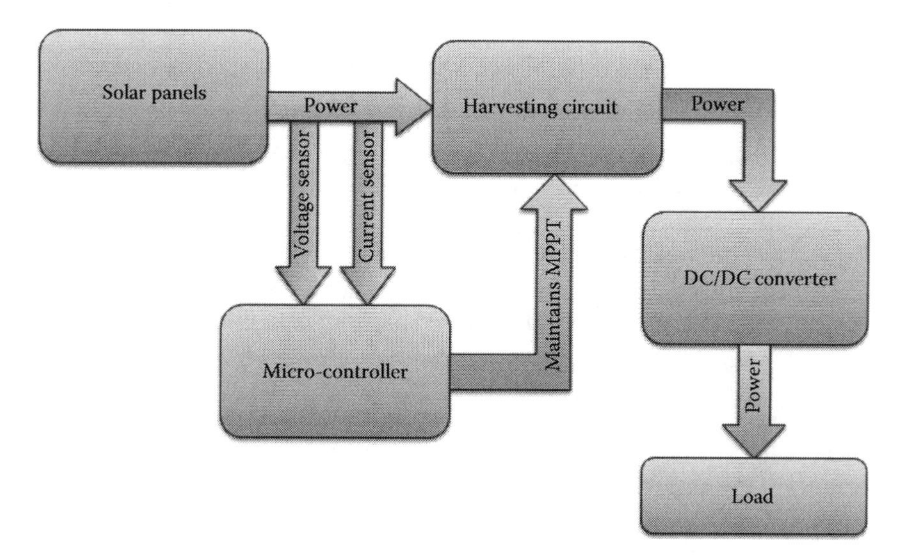

FIGURE 7.8
Block diagram of solar energy harvesting circuit. (From Buckley, B., n.d., MPPT—Maximum power point tracking, Retrieved from Bryan Buckley: http://bryanwbuckley.com/projects/mppt.html.)

7.6.2 Energy Storage with Supercapacitors

Supercapacitors are very easy to charge as they do not require any specific circuitry that ensures a specific charging pattern; simply applying a terminal voltage that causes current to flow into the supercapacitor allows it to store energy. Depending on the source of the input energy, circuitry may be required to ensure energy is being transferred efficiently. As an example, Figure 7.9 depicts circuitry used to buffer energy from solar panels.

The most important function of the circuitry shown in Figure 7.9 is to ensure that the solar panel input is kept at the MPP. The circuit in Figure 7.9 achieves this goal through the use of a SEPIC dc–dc converter (Hassanalieragh et al., 2014). After measuring the voltage of the attached solar panels, a switch adjusts the duty cycle for the circuit, thus increasing or decreasing the average current demanded from the solar panels. As the duty cycle increases, there is increased average current demand from the solar panel due to the leftmost inductor in the SEPIC configuration. As this inductor is turned on for a greater percentage of time, more power is demanded, and the less time it is on, the lower the average current demand. This allows the I–V curve of the solar panel to be adjusted, thus maintaining the optimum point for harvesting maximum power.

The SEPIC dc–dc converter is used for the following reasons (Hassanalieragh et al., 2014): First, SEPIC is capable of both upconverting and downconverting, whereas boost and buck designs can only do one or the other. Second, the SEPIC has a continuous-input and discontinuous-output current draw. This implies that at high frequencies, capacity losses at the supercapacitor block can be eliminated with the proper electrolytic capacitors on the output side of the system. And third, since this system is meant to operate in the field in a variety of harsh environmental conditions, having a graceful short-circuit response of SEPIC is invaluable.

Current sensing is a vital part of MPPT and is generally done through a small resistor called a sense resistor, typically in the range of 5–100 mΩ. Because the voltage drop across the sense resistor is so small, the circuit shown in Figure 7.9 includes a current sense amplifier to amplify the voltage before it is sent to the ADC of a microcontroller. The correct current sense amplifier for a circuit depends on the amplifier's placement within the circuit. High-side sensing, in which the sense resistor is placed directly after the input source, requires a different current sense amplifier than low-side sensing because the sense resistor will be subjected to a high common mode voltage. The circuit in Figure 7.9 uses a MAX4372H, which is a high-side current sense amplifier. A final consideration when choosing current sense amplifiers is their intended use for measuring dc. Because current sense amplifiers are designed for dc, their effective bandwidth is not high (Hassanalieragh et al., 2014).

When working with supercapacitors, one should also be aware of the variable manufacturing tolerances. Due to these manufacturing tolerances, supercapacitors can easily be overcharged when there are multiple supercapacitors arranged in series. For example, among eight supercapacitors rated to be 3000 F, researchers at the University of Rochester measured actual capacitances ranging from 2855 to 3139 F. In this situation, charging the eight supercapacitors as a block to the theoretical maximum of 21.6 V would overcharge the smaller supercapacitors beyond their maximum of 2.7 V. This is because the same current flowing through each would charge the smallest supercapacitor (2855 F) to 2.87 and the largest supercapacitor (3139 F) to only 2.61 V. Although complicated circuitry could be implemented to control the charging of each supercapacitor individually, the simplest solution is to charge the supercapacitor block to a voltage below its rated maximum. This leads

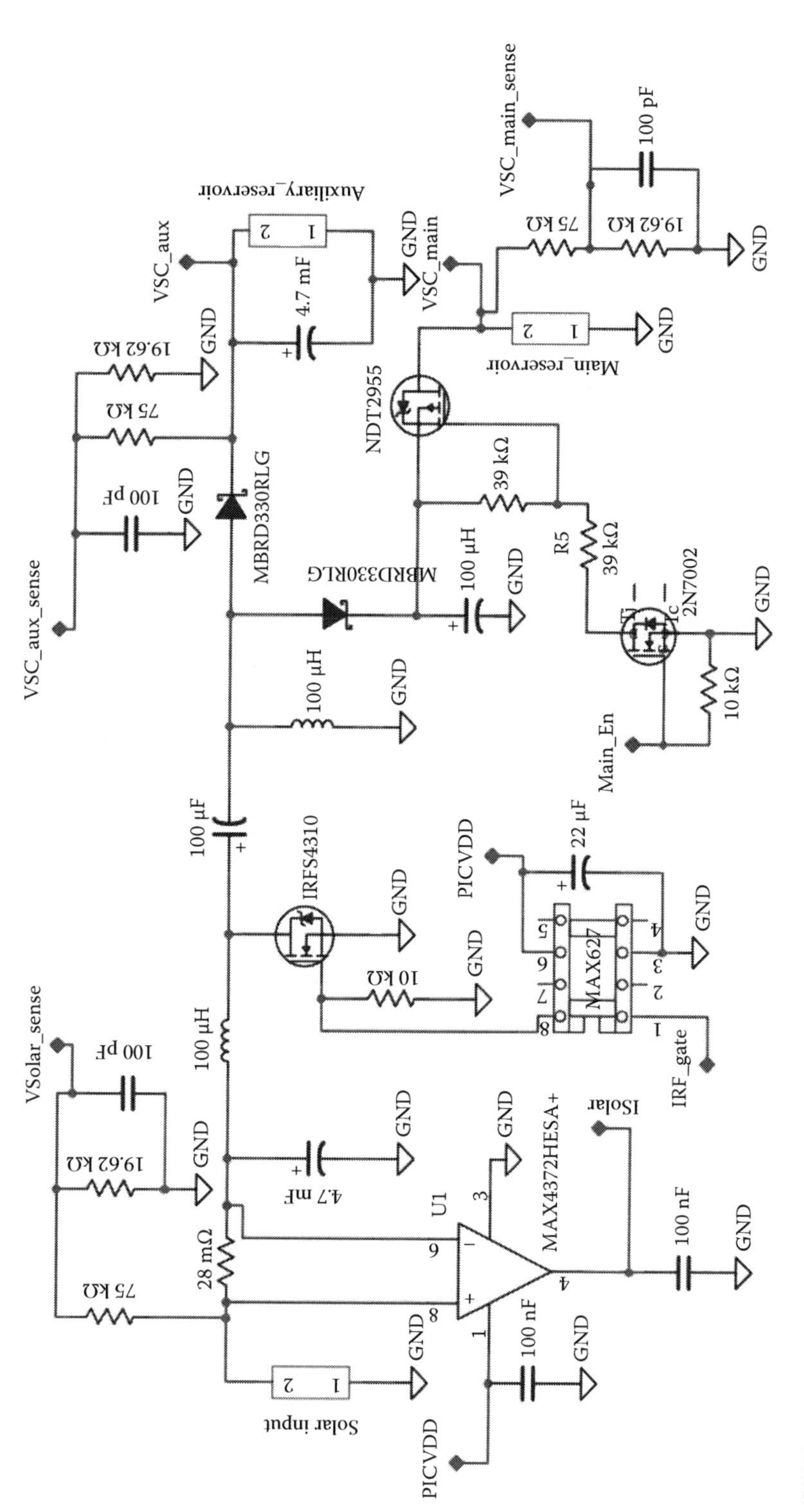

FIGURE 7.9
Solar energy harvesting circuit built around a SEPIC dc–dc converter. (Reproduced with permission of the authors Hassanalieragh, M., Soyata, T., Nadeau, A., and Sharma, G., Solar-supercapacitor harvesting system design for energy-aware applications, *Proceedings of the 27th IEEE International System-On-Chip Conference (IEEE SOCC 2014)*, Las Vegas, NV, IEEE, September 2–5, 2014.)

to inefficient use of the larger supercapacitors; for Hassanalieragh et al., this technique yielded a total voltage of 20.3 V, or 94% efficiency.

7.6.3 RF Energy Harvesting

In a system designed to utilize RF energy, such as an RFID device, the power harvester has two important functions: rectifying the receiving signal and generating the supply voltage (Shameli et al., 2007). The power harvesting circuitry consists of a series of rectifier cells, similar to the one shown in Figure 7.10 with each cell building on the voltage of the one before it, in order to accumulate the supply voltage.

This functionality is commonly achieved through a charge pump. The circuit in Figure 7.10 shows a series of rectifier cells, each consisting of a dc-level shifter and a peak detector. In the first half of the signal's period when voltage is low, the sampling capacitor C_1 will be charged up to a certain level depending on the amplitude of the input signal, the voltage drop across the diode, and the voltage of the cell preceding it in the chain.

In order to maximize the efficiency of the harvesting circuitry, the voltage drop across the diode must be minimized. A simple and frequently suggested solution is to use high-efficiency Schottky diodes, but such diodes are not available in standard CMOS processes and therefore would need to be specially manufactured, greatly increasing the cost of the circuit (Jabbar et al., 2010; Shameli et al., 2007). Therefore, a method of reducing voltage losses that is available in standard CMOS processes is highly desirable.

Currently, no way exists to increase the efficiency of the diodes using standard CMOS processes that yields efficiency comparable to Schottky diodes at very low voltage levels; however, a passive network used to increase the amplitude of the input signal at the input of the charge-pump circuit can mitigate some of the challenges of operating at very low voltage levels (Shameli et al., 2007).

Because the circuit is so sensitive to low voltages, voltage losses in the circuit must be minimal in order to achieve relatively high efficiency with the power harvester. One way to minimize losses is through an impedance matching circuit as shown in Figure 7.11. This circuit matches the load impedance to the input impedance, which results in the ideal

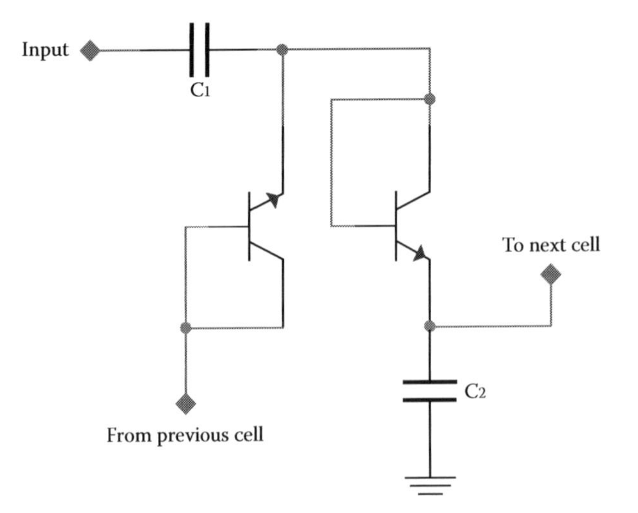

FIGURE 7.10
Charge-pump circuit that might be found in a radio-frequency harvesting system. (Based on Shameli, A. et al., *IEEE Trans. Microw. Theory Tech.*, 55(6), 1089, 2007.)

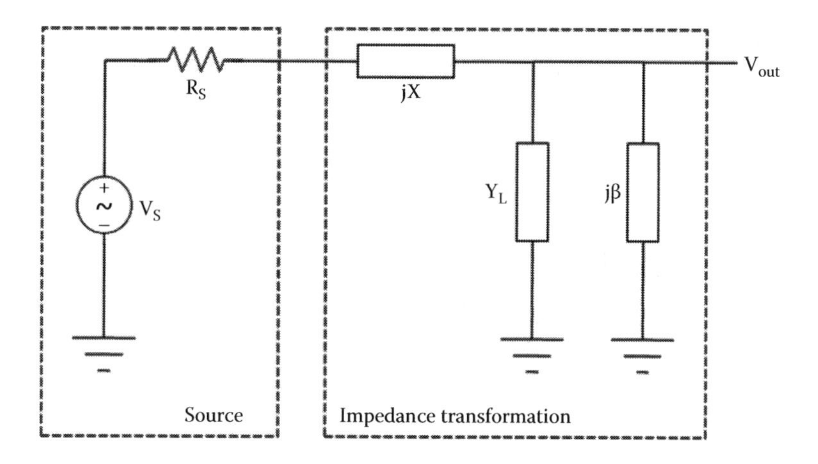

FIGURE 7.11
Impedance transformation circuit. (Based on Shameli, A. et al., *IEEE Trans. Microw. Theory Tech.*, 55(6), 1089, 2007.)

conditions for power transfer from the source to the load. In order to achieve the greatest possible output voltage, the input impedance should be equal to the source impedance.

7.6.4 Charge Circuitry for Rechargeable Batteries

Charge circuitry for NiCd batteries relies on full-charge detection based on either temperature or voltage signature (Buchmann, 2011). Low-cost charging circuitry simply measures the external surface temperature of the battery and compares the measurement to a known cutoff value (Buchmann, 2011). Most low-cost NiCd chargers use 50°C as cutoff temperature, at which point the charger turns on its "ready" light to indicate full charge. Temperatures exceeding 45°C can damage NiCd cells, but brief exposure to 50°C to detect a full charge is standard. Higher-cost temperature-based chargers involve using a microcontroller to measure the rate dT/dt of temperature change. Charging is stopped upon detecting a high rate of temperature increase characteristic of the end of a charge cycle. Charger manufacturers commonly use 1°C per minute as the cutoff rate, but if the battery's dT/dt never reaches this rate, maximum temperature-based shutoff is also in place to prevent charging after the battery reaches 60°C. Temperature-based full-charge detection is not ideal because when a fully charged battery is inserted into a charger, it becomes overcharged, thus damaging the battery.

A more advanced method of full-charge detection is via voltage signature. A microcontroller can be used to poll for voltages and compare recent values to detect a specified voltage signature. Once the microcontroller detects the end of a charge cycle via a characteristic voltage drop, or negative delta V (NDV), charging stops. NDV detection typically looks for a drop of 10 mV per cell to indicate full charge, but this method requires a charge rate of 0.5C or higher. The C-rate here is a measure of the rate at which a battery is charged or discharged relative to its maximum capacity. For example, a 1C rate means that a 1.0 Ah battery would be discharged in 1 h at a discharge rate of 1.0 A, or at 0.5C, the same battery would provide 500 mA for 2 h. In addition to NDV detection, charging microcontrollers also include plateau detection that stops charging after the voltage has been stuck in a steady state.

Charging of NiMH batteries relies on the same principles, but NDV detection is much more subtle for NiMH batteries. The voltage drop at the end of a charge cycle is only 5 mV, so NiMH chargers use electronic filtering to provide a clean signal to the microcontroller. A combination of many detection methods is common in modern chargers, including NDV, voltage plateau, dT/dt, absolute temperature, and timers—whichever method triggers first is used to shut off the charging.

Charging lead-acid batteries is very slow compared to other rechargeable batteries. An SLA battery takes 12–16 h to charge, but for large stationary batteries, it can take up to 48 h. This time can be reduced to 10 h or less with special charging methods, but these are not precisely characterized and the battery may not be fully charged as a result. The method used in charging lead-acid batteries involves three distinct phases, the first of which is a constant-current phase that takes about half of the charging time and provides most of the final charge. The second stage is the topping phase, which continues at a lower constant current and saturates the cell. Finally, the float charge stage recharges the battery, as needed, due to self-discharge.

Stage one brings the cell charge to 70% and takes 5–8 h, and in the next 7–10 h, stage two brings the battery up to 100%. The transition between stage one and two occurs when the battery has reached a set voltage limit. Full charge is considered reached when the current drops to 3% of the rated current for the battery. If the battery does not reach this low saturation current, a plateau timer is used to stop charging. The float charging phase is merely present to keep the battery fully charged as it slowly self-discharges.

Li-ion batteries are charged in a manner similar to lead-acid batteries, and the charging pattern is illustrated in Figure 7.12. The major differences between Li-ion charging and lead-acid charging are that (1) Li-ion cells have higher voltage than their lead-acid counterparts, (2) Li-ion charging must be more precise because Li-ion batteries have much less

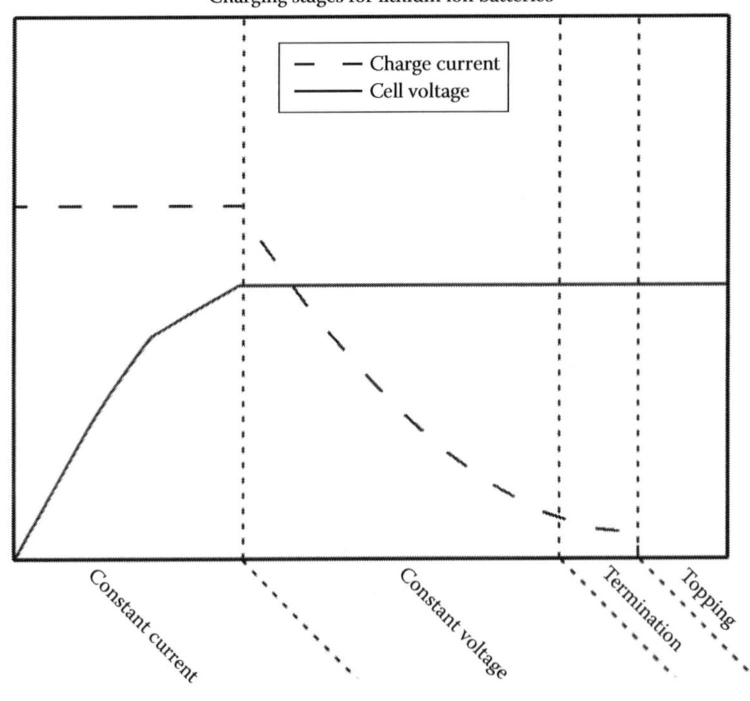

FIGURE 7.12
Charging stages for lithium-ion batteries.

tolerance for overcharging, and (3) Li-ion charging does not include a float charging phase. Most Li-ion cells charge to 4.2 V with a tolerance of around 50 mV; overshooting this tolerance is unsafe and will most likely damage the battery.

The constant-current phase of charging for Li-ion cells has a rate of around 1C, and this takes around 3 h. The charging then transitions into phase two where the cell is kept at a constant voltage. Charging stops when the current reaches 3% of the rated maximum for the cell, and then finally topping charging occurs as needed. High current can be used to rapidly reach 70% charge, but the topping phase will take longer. Using only 70% of the battery's capacity is recommended unless the full capacity is truly required. Fully charging a Li-ion battery reduces the battery's ability to fully charge again.

7.6.5 Sample Use Cases for CPS

In this section, we survey a few examples of energy sources and buffering mechanisms commonly used in CPS. ZebraNet (Juang et al., 2002) is a mobile sensor network primarily indented to collect data and study the biological behavior of zebras in central Kenya. ZebraNet is an example of a CPS used for wildlife monitoring and analysis. Each ZebraNet node contains a GPS system with user-programmable CPU, nonvolatile memory storage for data logging, and radio transceivers for data communication. Based on the nature and location constraints of the application, solar cells are used as the energy source of the system, and rechargeable lithium-ion batteries are employed for energy buffering.

Everlast (Simjee and Chou, 2006) is another example of a CPS that is powered by solar energy. Each Everlast node contains a light sensor and an accelerometer bundled with a low-power microprocessor and a wireless transceiver. Since extended unsupervised operation (20 years without maintenance) is a primary goal in Everlast, the system uses supercapacitors as the only energy buffering mechanism. On the sensing part, Everlast contains a light sensor with a pin-selectable light aperture for measuring a wide dynamic range of light intensity and an ultralow-power dual accelerometer consuming 600 μA.

A wireless sensor node for distributed active vibration control in automotive applications has been proposed by Zielinski et al. (2014). The piezoelectric element in the system turns vibrations into electrical energy to power up the storage, conversion, and processing units. The work proposed in Parks et al. (2013) is an example of a wireless sensor node utilizing ambient RF energy to power up an entire sensing, processing, and communication system. The energy harvesting part consists of an antenna, impedance matching, and a charge pump for converting the RF energy into electrical energy. The application platform includes board sensors for gathering information, a microcontroller, and a 2.4 GHz radio. While the implementation used a ceramic chip capacitor, the authors do point out that the use of rechargeable batteries and high-density supercapacitors can increase the system lifetime to 10–20 years, albeit with increased system complexity and cost.

Exploiting multiple energy sources and a hybrid energy buffering mechanism has been proposed as a solution to increase the reliability and longevity of operation in CPS. For example, AmbiMax (Park and Chou, 2006) is an autonomous energy harvesting platform using solar and wind energy and a hybrid architecture of supercapacitors and lithium polymer batteries to sustain the operation of an Eco wireless sensor node (Park, 2005). Eco is an ultracompact wireless sensor node for real-time motion monitoring that combines a two-axis accelerometer and 2.4 GHz GFSK radio for data transmission.

CPSs that require long-term maintenance-free operation are increasingly moving toward supercapacitor-based energy buffering because of the advantages supercapacitors offer over batteries. Also, because of the relative simplicity of supercapacitors energy modeling, these devices are also an attractive choice for intelligent CPSs, where precise energy management and control is essential.

Glossary

Amorphous silicon cell: This form of photovoltaic cells consists of silicon atoms randomly bonded with no crystal lattice.

Cyber-physical system (CPS): A networked system integrating computational algorithms and physical entities. Current areas of research in CPS include autonomy, capability, and safety.

Kalman filter: An algorithm used to estimate unknown parameters by using a series of measurements over time. Commonly used for guidance and navigation of vehicles.

Magnetostrictive effect: An effect specific to ferromagnetic materials leading to a change in shape during magnetization. The inverse effect is also true: during mechanical stress, ferromagnetic materials will undergo a change to their magnetic susceptibility.

Maximum power point tracking (MPPT): A method used to extract the maximum power from a photovoltaic (PV) device—without MPPT, PV cells rarely operate optimally.

Microelectromechanical systems (MEMS): The technology of miniaturized mechanical systems made using microfabrication techniques.

Monocrystalline silicon cell: Photovoltaic cells of this form are created from a single, continuous crystal lattice.

Piezoelectric effect: The ability of certain materials to generate a voltage when mechanically deformed. The inverse effect is also observed: when an electric field is applied to a piezoelectric material, it mechanically deforms.

Polycrystalline silicon cell: Photovoltaic cells of this type consist of many small silicon crystals creating the larger cell, as opposed to the single large crystal of monocrystalline types.

Rectifier: An electrical device that converts alternating current to direct current.

Radio-frequency identification (RFID): A tracking and identification system consisting of a "tag" device and a "reader" device. The reader emits a specific radio frequency used to communicate and/or power the RFID tag, which may in turn send data to the reading device.

Supercapacitor: A class of electrochemical devices that combines the properties of both rechargeable batteries and capacitors. Supercapacitors derive their energy storage properties from an electric double layer and pseudocapacitance.

Acknowledgment

This work was supported in part by the National Science Foundation grant CNS-1239423 and a gift from Nvidia Corp.

References

Ahmad, J. (2010). A fractional open circuit voltage based maximum power point tracker for photovoltaic arrays. In *2010 Second International Conference on Software Technology and Engineering (ICSTE)*, October 3–5 (Vol. 1) (pp. V1-2247–V1-2250). San Juan, Puerto Rico: IEEE.

Andrewatla. (September 19, 2007). Solar panel. Retrieved August 27, 2014, from freeimages: http://www.freeimages.com/photo/873825.

Atwood, B., Warneke, B., and Pister, K. S. (2000). Preliminary circuits for smart dust. In *2000 Southwest Symposium on Mixed-Signal Design, 2000. SSMSD*, February 27–29 (pp. 87–92). San Diego, CA: IEEE.

Beeby, S. P., Tudor, M. J., and White, N. M. (2006). Energy harvesting vibration sources for microsystems applications. *Measurement Science and Technology*, 17(12): R175–R195.

Buchmann, I. (2011). *Batteries in a Portable World.* Vancouver, British Columbia, Canada: Cadex Electronics Inc.

Buckley, B. (n.d.). MPPT—Maximum power point tracking. Retrieved from Bryan Buckley: http://bryanwbuckley.com/projects/mppt.html.

Bullock, K. R. (1994). Lead/acid batteries. *Journal of Power Sources*, 51(1): 1–17.

Chalasani, S. and Conrad, J. M. (2008). A survey of energy harvesting sources for embedded systems. In *IEEE Southeastcon, 2008*, April 3–6 (pp. 442–447). Huntsville, AL: IEEE.

Conway, B. E. (1991). Transition from "supercapacitor" to "battery" behavior in electrochemical energy storage. *Journal of the Electrochemical Society*, 138(6): 1539–1548.

Curty, J.-P., Joehl, N., Dehollain, C., and Declercq, M. J. (2005). Remotely powered addressable UHF RFID integrated system. *IEEE Journal of Solid-State Circuits*, 40(11): 2193–2202.

DebbieMous. (September 16, 2011). Solar panels reflecting the sky. Retrieved August 27, 2014, from freeimages: http://www.freeimages.com/photo/1364625.

Fahad, A., Soyata, T., Wang, T., Sharma, G., Heinzelman, W., and Shen, K. (2012). SOLARCAP: Super capacitor buffering of solar energy for self-sustainable field systems. In *Proceedings of the 25th IEEE International System-on-Chip Conference* (*IEEE SOCC*), September 12–14 (pp. 236–241). Niagara Falls, NY: IEEE.

Fatcenko, M., Ovshinsky, S., Reichman, B., Young, K., Fierro, C., Koch, J., Zallen, A., Mays, W., and Ouchi, T. (2007). Recent advances in NiMH battery technology. *Journal of Power Sources*, 165(2): 544–551.

Feldman, D., Barbose, G., Margolis, R., Darghouth, N., and Goodrich, A. (2012). *Photovoltaic (PV) Pricing Trends: Historical, Recent, and Near-Term Projections.* Golden, CO: National Renewable Energy Laboratory.

Femia, N., Petrone, G., Spagnuolo, G., and Vitelli, M. (2005). Optimization of perturb and observe maximum power point tracking method. *IEEE Transactions on Power Electronics*, 20(4): 963–973.

Foster, K. and Jaeger, J. (2007). RFID inside. *IEEE Spectrum*, 44(3): 24–29.

Goudarzi, N. and Zhu, W. (2013). A review on the development of wind turbine generators across the world. *International Journal of Dynamics and Control*, 1(2): 192–202.

Green, M. A., Emergy, K., Hishikawa, Y., Warta, W., and Dunlop, E. D. (2014). Solar cell efficiency tables (version 44). *Progress in Photovoltaics: Research and Applications*, 22(7): 701–710.

Guigon, R., Chaillout, J.-J., Jager, T., and Despesee, G. (2008a). Harvesting raindrop energy: Theory. *Smart Materials and Structures*, 17: 25–39.

Guigon, R., Chaillout, J.-J., Jager, T., and Despesse, G. (2008b). Harvesting raindrop energy: Experimental study. *Smart Materials and Structures*, 17: 15–39.

Hande, A., Polk, T., Walker, W., and Bhatia, D. (2007). Indoor solar energy harvesting for sensor network router nodes. *Microprocessors and Microsystems*, 31(6): 420–432.

Hassanalieragh, M., Soyata, T., Nadeau, A., and Sharma, G. (2014). Solar-supercapacitor harvesting system design for energy-aware applications. In *Proceedings of the 27th IEEE International System-On-Chip Conference* (*IEEE SOCC 2014*), September 2–5. Las Vegas, NV: IEEE.

Hudak, N. S. and Amatucci, G. G. (2008). Small-scale energy harvesting through thermoelectric, vibration, and radiofrequency power conversion. *Journal of Applied Physics*, 103: 101301.

Jabbar, H., Song, Y. S., and Jeong, T. T. (2010). RF energy harvesting system and circuits for charging mobile devices. *IEEE Transactions on Consumer Electronics*, 56(1): 247–253.

Juang, P., Oki, H., Wang, Y., Martonosi, M., Peh, L.-S., and Rubenstein, D. (2002). Energy-efficient computing for wildlife tracking: Design trade-offs and early experiences with ZebraNet. In *Proceedings of the 10th International Conference on Architectural Support for Programming Languages and Operating Systems* (pp. 96–107). San Jose, CA.

Kahn, J. M., Katz, R. H., and Pister, K. S. (2000). Emerging challenges: Mobile networking for "smart dust". *Journal of Communications and Networks*, 2: 188–196.

Kocabas, O. and Soyata, T. (2014). Medical data analytics in the cloud using homomorphic encryption. In P. Chelliah and G. Deka (eds.), *Handbook of Research on Cloud Infrastructures for Big Data Analytics* (pp. 471–488). Hershey, PA: IGI Global.

Kocabas, O., Soyata, T., Couderc, J., Aktas, M., Xia, J., and Huang, M. (2013). Assessment of cloud-based health monitoring using homomorphic encryption. In *Proceedings of the 31st IEEE International Conference on Computer Design* (pp. 443–446). Asheville, NC.

Kularatna, N. (2011). Rechargeable batteries and their management. *IEEE Instrumentation and Measurement Magazine*, 14(2): 20–33.

Kwon, M., Dou, Z., Heinzelman, W., Soyata, T., Ba, H., and Shi, J. (2014). Use of network latency profiling and redundancy for cloud server selection. In *Proceedings of the Seventh IEEE International Conference on Cloud Computing*, June 27–July 2 (pp. 826–832). Anchorage, AK.

Mi, M., Mickle, M. H., Capelli, C., and Swift, H. (2005). RF energy harvesting with multiple antennas in the same space. *IEEE Antennas and Propagation Magazine*, 47(5): 100–106.

Myers, R., Vickers, M., Kim, H., and Priya, S. (2007). Small scale windmill. *Journal of Applied Physics*, 90: 054106.

Nadeau, A., Sharma, G., and Soyata, T. (2014). State-of-charge estimation for supercapacitors: A Kalman filtering approach. In *Proceedings of the 2014 IEEE International Conference on Acoustics, Speech and Signal Processing (ICASSP 2014)*, May 4–9 (pp. 2213–2217). Florence, Italy: IEEE.

Ohya, Y., Karasudani, T., Sakurai, A., Abe, K.-i., and Inoue, M. (2008). Development of a shrouded wind turbine with a flanged diffuser. *Journal of Wind Engineering and Industrial Aerodynamics*, 96(5): 524–539.

Page, A., Kocabas, O., Soyata, T., Aktas, M., and Couderc, J. (2014). Cloud-based privacy-preserving remote ECG monitoring and surveillance. *Annals of Noninvasive Electrocardiology, 20.*

Park, C. (2005). Eco: An ultra-compact low-power wireless sensor node for real-time motion monitoring. In *Proceedings of the Fourth International Symposium on Information Processing in Sensor Networks*. Los Angeles, CA: IEEE Press.

Park, C. and Chou, P. (2006). AmbiMax: Autonomous energy harvesting platform for multi-supply wireless sensor nodes. In *3rd Annual IEEE Communications Society on Sensor and Ad Hoc Communications and Networks, 2006. SECON '06* (pp. 168–177). Reston, VA: IEEE.

Parks, A. N., Sample, A. P., Zhao, Y., and Smith, J. R. (2013). A wireless sensing platform utilizing ambient RF. In *Topical Conference on Wireless Sensors and Sensor,* January 20–23 (pp. 127–129). Austin, TX: IEEE.

Randall, J. F. and Jacot, J. (2002). The performance and modelling of 8 photovoltaic materials under variable light intensity and spectra. In *World Renewable Energy Congress VII & Expo*. Cologne, Germany.

Roundy, S., Steingart, D., Frechette, L., Wright, P., and Rabaey, J. (2004). Power sources for wireless sensor networks. In H. Karl, A. Wolisz, and A. Wilig (eds.), *Wireless Sensor Networks* (pp. 1–17). Berlin, Germany: Springer.

Shad, R., Wright, P. K., and Pister, K. S. (2002). Micro-electrostatic vibration-to-electricity converters. In *Proceedings of IMECE 2002*, November 17–22. New Orleans, LA: ASME.

Shameli, A., Safarian, A., Rofougaran, A., Rofougaran, M., and De Flaviis, F. (2007). Power harvester design for passive UHF RFID tag using a voltage boosting technique. *IEEE Transactions on Microwave Theory and Techniques*, 55(6): 1089–1097.

Simjee, F. and Chou, P. (2006). Everlast: Long-life, supercapacitor-operated wireless sensor node. In *Low Power Electronics and Design, ISLPED'06*, October 4–6 (pp. 197–202). Tegernsee, Germany: IEEE.

Soyata, T., Ba, H., Heinzelman, W., Kwon, M., and Shi, J. (2014). Accelerating mobile-cloud computing: A survey. In H. T. Mouftah and B. Kantarci (eds.), *Communication Infrastructures for Cloud Computing* (pp. 175–197). Hershey, PA: IGI Global.

Soyata, T., Muraleedharan, R., Ames, S., Langdon, J., Funai, C., Kwon, M., and Heinzelman, W. (2012a). COMBAT: Mobile-Cloud-based cOmpute/coMmunications infrastructure for BATtlefield applications. In *Proceedings of SPIE* (pp. 84030K-13). Baltimore, MD.

Soyata, T., Muraleedharan, R., Funai, C., Kwon, M., and Heinzelman, W. (2012b). Cloud-vision: Real-time face recognition using a mobile-cloudlet-cloud acceleration architecture. In *Symposium on Computers and Communications (ISCC)* (pp. 59–66). Capadoccia, Turkey: IEEE.

Thomas, J. P., Qidawi, M. A., and Kellogg, J. C. (2006). Energy scavenging for small-scale unmanned systems. *Journal of Power Sources*, 159(2): 1494–1509.

Visser, H. J. and Vullers, R. J. (2013). RF energy harvesting and transport for wireless sensor network applications: Principles and requirements. *Proceedings of the IEEE*, 101(6): 1410–1423.

Wang, H., Liu, W., and Soyata, T. (2014). Accessing big data in the cloud using mobile devices. In P. Chelliah and G. Deka (eds.), *Handbook of Research on Cloud Infrastructures for Big Data Analytics* (pp. 444–470). Hershey, PA: IGI Global.

Wang, L. and Yuan, F. G. (2007). Energy harvesting by magnetostrictive material (MsM) for powering wireless sensors in SHM. In *Sensors and Smart Structures Technologies for Civil, Mechanical, and Aerospace Systems 2007.* San Diego, CA: SPIE.

Zielinski, M., Mieyeville, F., Navarro, D., and Bareille, O. (2014). *A Low Power Wireless Sensor Node with Vibration Sensing and Energy Harvesting Capability* (pp. 1065–1071). Warsaw, Poland: IEEE.

Zubieta, L. and Bonert, R. (2000). Characterization of double-layer capacitors for power electronics. *IEEE Transactions on Industry Applications*, 36(1): 199–205.

8

Hybrid DTDNN-MMSE-Based Channel Tracking and Equalization in Multiantenna OFDM System for Application in Mobile Cyber-Physical Framework

Punam Dutta Choudhury and Kandarpa Kumar Sarma

CONTENTS

8.1 Introduction

The exponential growth of technology has driven innovations in communication systems and has ensured that mobile and handheld devices become reliable means of conveying information. The recent growth of cyber-physical system (CPS)-based applications have made them common objects in areas like transpiration and process control, and infrastructure domains like civil construction and energy, health care, manufacturing, entertainment, consumer appliances, and so on. The growth of mobile communications has facilitated the deployment of wirelessly connected CPS frameworks. The wireless links of mobile CPS setups are prone to fluctuating link reliability due to random variations in the propagation medium (Rajkumar et al., 2010). This is due to the stochastic behavior of the wireless medium. Solutions are available to mitigate these effects and there is still scope to further raise the levels of link reliability. Over the years, the expansion of mobile communication in all parts of the globe has necessitated the design of systems that provide continuously improving performance. This is noticed distinctively in wireless communication–based applications like mobile CPS, which continues to be a challenging arena. The challenges in wireless channel are related to its stochastic nature. Transmissions from source to destination resort to multipath propagation. As a result, a range of effects are observed. Mostly, these are related to the variations in the waveform magnitudes, phase changes, and related distortions. Further, problems like intersymbol interference (ISI), intercarrier interference (ICI), and cochannel interference (CCI) are related to multipath fading, time-varying nature of the wireless channel and its frequency-related aspects. So, there is a range of issues related to the design of systems that needs to be considered to achieve better quality of service (QoS) (Suthatharan, 2003).

The continuous growth in wireless communication has further placed demands on the capacity of systems. As a result, the shift has been toward the use of multicarrier systems. One such method is the use of orthogonal frequency division multiplexing (OFDM). Multiple antenna structures at transmitter and receiver ends generate multiple input multiple output (MIMO) setups that not only combat fading but are also found to enhance the system capacity. MIMO in combination with OFDM has been accepted to be a reliable framework for high data rate communication. Certain fundamental aspects are discussed in the following text.

Single-carrier modulation is based on one sinusoidal carrier, while multicarrier techniques use several carriers simultaneously. In wireless systems, frequency-selective fading is a common occurrence where different frequency components of a signal suffer different amounts of fading during transmissions. In single-carrier systems, adaptive equalizers are used to mitigate the effect, which is not the case with multicarrier systems. As data rates increase (which is a common occurrence in present-day wireless systems), ICI become severe. One solution to ICI was found to be incorporation of guard bands. But it contributes toward waste of bandwidth. Hence, the necessity of nonoverlapping, adjacent subchannel without the possibility of ICI despite not using guard bands came into the limelight. This has been made possible by OFDM. OFDM provides orthogonality among multiple carriers forming the subbands. It also enhances throughput of the system by ensuring multicarrier modulation with better QoS than conventional techniques. But OFDM suffers from ISI for which cyclic prefixes (CP) are used in the OFDM block. To ensure proper QoS, CP-OFDM further uses unused and additional subcarriers called guard bands. The selection of spacing of the subcarriers in OFDM is a critical aspect. The constraint is that in a wireless channel, each subcarrier should only suffer flat fading. Normally, subcarrier spacing is determined by the channel coherence time. OFDM uses a parallel mode of transmission. In such a method of communication using various subcarriers, the single-carrier symbol durations are much shorter compared to the resultant OFDM symbol periods. Further, the performance of the longer symbol duration OFDM transmissions under time-selective fading is constrained by the duration of channel stationarity. The symbol period is determined by taking the inverse of the subcarrier spacing. Therefore, the rate of change of the channel state determines the minimum subcarrier spacing. Hence, for OFDM-based systems, the channel state information (CSI) is critical (Langton, 2004).

As MIMO-OFDM systems are accepted options of high data rate communication, application of innovative solutions to achieve better QoS has been one of the critical issues of present-day research. In case of MIMO, computational complexity during channel estimation is a crucial factor in determining the design complexity due to the fact that there is a requirement of finding several unknowns at the same time. There is a constant demand for use of methods that can reduce this computational complexity. One of the reliable means is to use CSI better in both MIMO and OFDM systems. This can be done by using learning-based techniques like artificial neural networks (ANN). These can learn from the surrounding, retain the learning, and use it subsequently. The use of ANN in OFDM has several advantages. These are like use of the complete CSI in symbol recovery, reduction of the cyclic prefix, and other guard band sizes that shall contribute toward better QoS and saving of the bandwidth.

In the past, various techniques have been developed for dealing with these distortive effects of the channel so that the baseband signal appears at the receiver in the same way as it is produced at the transmitter and without any significant variation. Different transmit and receive diversity techniques have been developed in order to exploit the diversity in space, time, and frequency so that the transmission reliability can be improved. The deployment of space-time or space-frequency codes in such cases has also yielded efficient results. Statistical tools like feedback equalizers and Kalman filters have also made effective contributions in this genre and so has the domain of neural networks. However, there is further scope of improvement that can ensure authentic increase in the transmission reliability of the wireless communication links irrespective of its physical or environmental conditions without compromising with the required high data rates.

During the past few years, the ANN and related techniques have been widely adopted for the purpose of CSI estimation and equalization. In a reported work (Bhuyan and Sarma, 2012), the authors have presented an arrangement based on certain dynamic ANN topologies, which not only contributed in reducing the overhead imposed by the pilot symbols but also found to efficiently track the CSI for a 2 × 2 MIMO channel and the system developed had been found to outperform a widely popular statistical method, namely, least square estimation technique. The same authors, in 2014, have proposed a method for real-time identification and prediction of time-varying mobile radio channel where a Feedforward ANN structure was combined with decision feedback equalization (DFE) and the proposed system was found to effectively identify and equalize an autoregressive system underlying the fading mechanism (Bhuyan and Sarma, 2014).

Authors Sarma and Mitra (2011a,b) proposed a MIMO channel modeling and estimation technique based on complex time delay fully recurrent neural network in combination with self-organizing map–based optimization that not only exhibited improvement in the overall system but was also found to offer considerable processing time saving. In 2012, system models based on two ANN structures, namely, multilayer perceptron (MLP) and recurrent neural network (RNN) were proposed that had been employed for estimating space-time block coded (STBC) MIMO channel and its associated experimental results showed that the system developed excelled in terms of its training and bit error rate (BER) performances due to its nonlinear dynamic adaptive behavior (Gogoi and Sarma, 2012). In the same year, another significant work based on ANN had been reported that described the use of ANN and coding to improve the performance of an equalizer in Nakagami-m fading channel (Baruah and Sarma, 2012).

In this chapter, we discuss the use of a hybrid system based on distributed time delay neural network (DTDNN) and minimum mean square error (MMSE) algorithm for tracking and equalizing wireless channels in mobile CPS framework realized as OFDM channels in single input single output (SISO) and MIMO configurations impaired by both noise and fading components. Here, the channel variations are tracked using the dynamic ANN structure called DTDNN and the distortions incorporated by the channel are compensated using MMSE algorithm. DTDNN has the capability of tracking temporal variations, while MMSE is an effective tool to deal with several channel-related distortions. Therefore, their combination is a learning-based efficient approach to deal with the variations observed in the wireless channels and to ensure effective data recovery.

The rest of the chapter is organized into the following sections. Section 8.2 describes the background theory and the related details of the topics that are needed to understand the basic concepts of wireless communication, mobile CPS as well as channel tracking and equalization methodologies. Section 8.3 presents an elaborate description of the proposed hybrid system and its associated experimental details. Section 8.4 provides a discussion on the simulation results, and Section 8.5 highlights the conclusions drawn from the proposed work and the possible improvements that can be made into the system in the upcoming time. The final section concludes the chapter.

8.2 Theoretical Background

In order to explain the working of the proposed channel tracking and equalization method, it is important to get acquainted with the basics of wireless communication and other related topics. This section provides a brief overview on such background topics. Within this broad section of theoretical background, the following describe the basics of mobile CPS, wireless communication, OFDM, MIMO, DTDNN, MMSE, and channel tracking and equalization. Next, there is an elaborate review on three significant works that have been done on mobile CPS as well as wireless channel tracking and equalization.

8.2.1 Mobile CPS

A CPS is a combination of process control devices, actuators, and switches designed as part of certain applications regulated and monitored by a computer system, which most of the time is connected to a network. A generic CPS is a collaborative framework with computation, networking, and physical processes that can be shared, regulated, and monitored remotely. Figure 8.1 shows a generic CPS. It can be viewed as a network of interacting elements with physical input and output instead of as standalone devices.

At times, the CPS maybe in a mobile form and connected wirelessly. Such systems are popularly termed as mobile CPS. The rise in popularity of smartphones has increased the interest in the area of mobile CPS. Moreover, such systems are also commonly observed in intelligent traffic management and transportation systems. Figure 8.2 shows such an arrangement. In the mobile CPS case, the ill effects observed due to stochastic wireless channels play a detrimental role. Therefore, innovative solutions are required to improve the link reliability of such systems (Rajkumar et al., 2010).

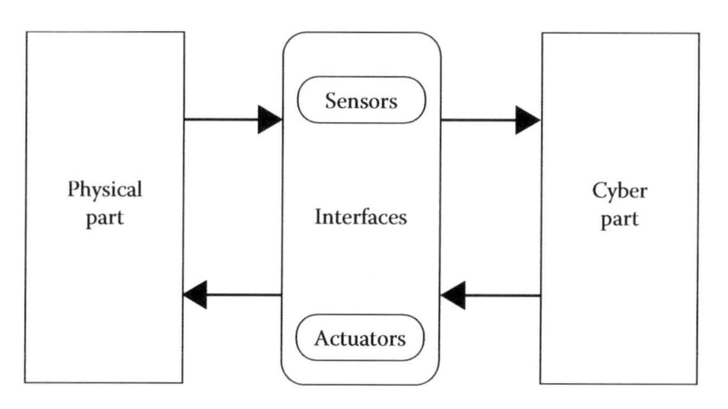

FIGURE 8.1
A generic cyber-physical system architecture.

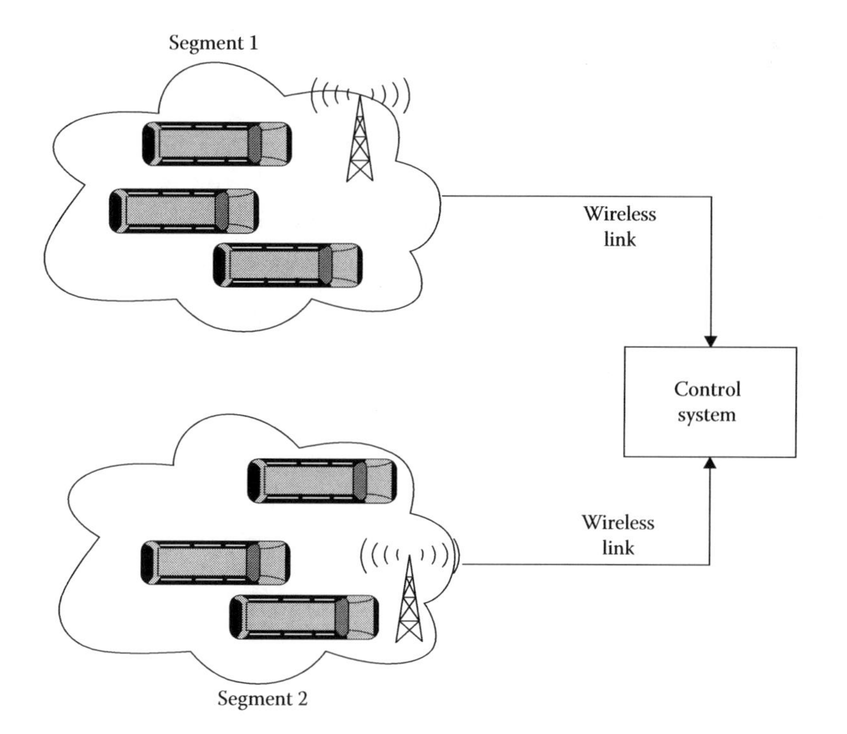

FIGURE 8.2
A traffic management and transportation system.

8.2.2 Wireless Communication

The principles of wireless communication are employed for transmitting various forms of data such as text, numbers, audio, images, or video over a distance by making use of electromagnetic (EM) waves. For instance, a printer connected wirelessly to a computer receives the data to be printed through EM waves and does not require bulk of wires that consumes lot of space and requires large installation time and cost. There is an unending list of applications such as cell phones, wireless gears, and high-speed internet access designed and developed using principles of wireless communication, and they have formed an integral part of our daily life (Stallings, 2005).

In communication scenario, the received signal is found to differ from that of its transmitted counterpart due to various impairments incorporated by transmitter and channel segments. In case of analog signals, these impairments introduce random modifications in the transmitted signal that distort its waveform and degrade the signal quality. In case of digital signals, they introduce bit errors. The different categories of such transmission impairments are attenuation, free space loss, noise, and multipath.

In case of attenuation, the signal strength decreases along the distance over any transmission medium. The attenuation–distance relationship is exponential for guided media, while for unguided media, it is a more complex function. A transmission engineer needs to take care of three factors while dealing with attenuation: first, the strength of the received signal must be sufficiently high so that the receiver can detect and interpret the signal; second, the signal level must be sufficiently higher than the noise signal in order to be received without error; and last, attenuation is greater at higher frequencies that causes distortion and is termed as attenuation distortion (Rappaport, 1997).

It should be noted that for a point-to-point transmission, the signal must be strong enough to be detected intelligibly but these process of signal strengthening must not overload the circuitry of the transmitter or the receiver. Beyond a certain distance, the problem of attenuation becomes unacceptably great and such situations are tackled with the deployment of repeaters or amplifiers that boost the signal at regular intervals. However, the use of amplifiers needs one important attention regarding the amplifier noise that hampers the signal-to-noise ratio (SNR). Moreover, in order to reduce attenuation distortion, such amplifiers are used that amplify higher frequencies more than lower frequencies.

Free space loss is a wireless propagation problem associated with signal dispersion while traveling through the channel. Due to signal dispersion, an antenna with a fixed area receives lesser signal power depending on its distance from the transmitting antenna. A transmitted signal gets attenuated over distance because it is being spread over a larger area. This form of attenuation is known as free space loss. Free space loss is defined as the ratio of the radiated power P_t to the received power P_r and for an ideal isotropic antenna, it is given by

$$\frac{P_t}{P_r} = \frac{(4\pi d)^2}{\lambda^2} = \frac{(4\pi f d)^2}{c^2} \tag{8.1}$$

where

P_t is the signal power at the transmitting antenna
P_r is the signal power at the receiving antenna
λ is the carrier wavelength
f is the carrier frequency
d is the propagation distance between antennas
c is the speed of light, which is equal to 3×10^8 m/s (Stallings, 2005)

Message signals get contaminated on their way to receiver by undesirable signals that are termed as noise and such signals are random and unpredictable in nature. They can be broadly classified under four categories, namely, thermal noise, intermodulation noise, crosstalk, and impulse noise. Thermal noise is due to thermal agitation of electrons and is found to occur in all electronic devices. Since it is uniformly distributed across the frequency spectrum it is often called white noise. The amount of thermal noise found in a bandwidth of B Hz in any device or conductor is given by

$$N_0 = kTB \tag{8.2}$$

where
- N_0 is the noise power density in watts per 1 Hz of bandwidth
- k is the Boltzmann constant, which is equal to 1.38×10^{-23} J/K
- T is the absolute temperature (in Kelvin)

Intermodulation noise is produced due to the presence of some nonlinearity in the transmitter, receiver, or intervening transmission system, and it produces signals at a frequency that is the sum or difference of the two original frequencies or multiples of those frequencies. For instance, the mixing of signals at frequencies f_1 and f_2 might produce energy at the frequency $f_1 + f_2$. This derived signal interferes with the originally transmitted signals at the frequency $f_1 + f_2$. Crosstalk is an unwanted coupling between signal paths that is usually encountered in case of ISM bands that are unlicensed frequency bands. All the three types of noises discussed earlier are predictable to a certain extent and have relatively constant amplitudes; however, impulse noise is one such type of noise that consists of irregular spikes of short duration and of relatively high amplitudes. It is generated from a variety of causes including external EM disturbance such as lightening.

Apart from the issues discussed earlier, the wireless channels are found to consist of the distortions incorporated by multipath fading environment. Perhaps, it is the most challenging problem that the communication engineers are facing in a mobile environment. The multipath fading environment is a result of three propagation mechanisms, namely, reflection, diffraction, and scattering and is illustrated in Figure 8.3. Reflection is found to occur when an EM signal encounters a surface that is large relative to the signal

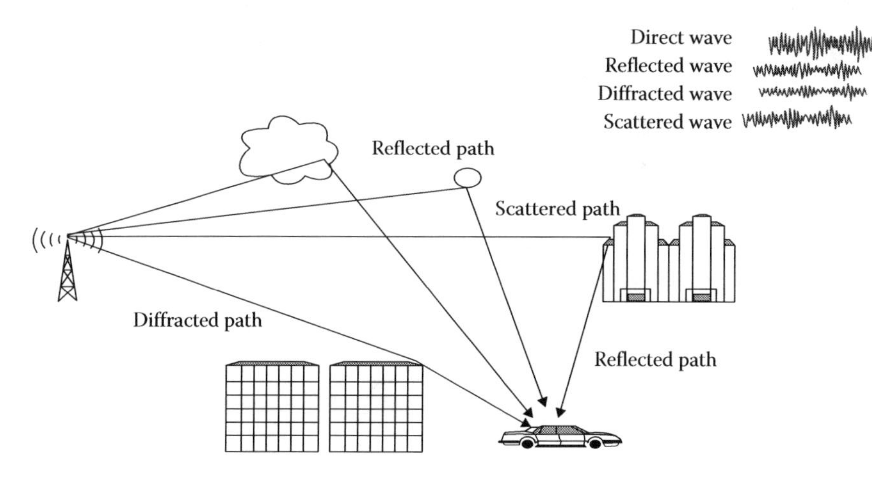

FIGURE 8.3
Multipath fading environment.

wavelength, whereas diffraction occurs at the edge of an object that cannot be penetrated and is large as compared to the signal wavelength. If the size of the obstacle is less than or of the order of signal wavelength, scattering takes place. These propagation mechanisms make an impact on the system performance in various ways depending on local conditions and as the mobile unit moves within a cell. Since in wireless channels, the line-of-sight (LOS) path between the transmitter and the receiver is found to be rare due to the presence of various obstacles like buildings, hills, and cars; therefore, due to the propagation mechanisms discussed before, multiple copies of the transmitted signal are generated and are obtained at the receiver. These replicas of the transmitted signal vary from each other in terms of attenuation and time delay. The time delays cause phase shifts. These received signal components with different phase shifts interfere constructively or destructively to form the main received signal component, and such interaction causes fluctuations in the signal strength. Figure 8.4 shows the difference between the magnitudes of the transmitted and the received pulses. In the lower line of the figure, the first white and black pulses are LOS-received pulses and the remaining short pulses are the ones received from non-LOS paths. Altogether, they give rise to a main received signal that consists of either deep gains or deep fades. This phenomenon leads to the problem of fading. Therefore, the wireless channels are mainly found to be multipath fading channels. The multipath fading channels can be modeled as either Rayleigh fading or Rician fading channels based on their respective distribution functions. Rayleigh fading is mainly encountered in urban areas where the presence of a large number of obstacles makes LOS path rare, whereas Rician fading is found in rural areas or in satellite channels where LOS paths are more probable. Fading effects can be classified as fast or slow depending on the nature of variation of the channel characteristics with time. They can also be classified as frequency flat or frequency-selective fading. In the former, all frequency components of the received signal fluctuate in the same proportion simultaneously, whereas in the latter, different spectral components of the radio signal undergo different amount of fluctuations at a time.

Phase shifts also result in delay spread, which can be viewed as the splash of previously transmitted symbol on the upcoming symbol; thus, the received signal is found to have the problem of ISI. Moreover, the communication system suffers from ICI where the subcarriers carrying the transmitted information interfere with one another, and their interference affects the information carried by them. Consequently, the received signal appears to be a

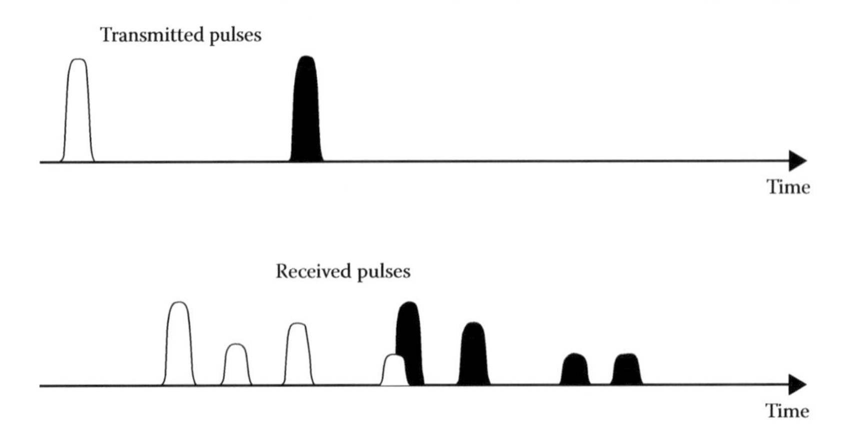

FIGURE 8.4
Comparison between transmitted and received pulses.

degraded form of the originally transmitted signal. The problem of CCI occurs when two channels having same frequencies but serving different cells interfere due to improper distribution of transmission powers in mobile stations or base stations over cellular networks (Langton, 2004).

8.2.3 Orthogonal Frequency Division Multiplexing

Multiplexing is a method of sharing the bandwidth by independent analog message signals or digital data streams generated by different sources. This technique helps in proper utilization of a very important resource, that is, bandwidth and effectively reduces its wastage. Frequency division multiplexing (FDM) is a type of multiplexing technique in which the available bandwidth is split into a number of nonoverlapping frequency subbands, and each of them is used to carry an analog message signal or a digital data stream.

OFDM is a special case of FDM. In FDM, the independent signals from different sources are modulated using separate carrier frequencies and are multiplexed to form a single FDM signal. However, in OFDM, an independent signal is split into its constituent components. These independent components are modulated using different carrier frequencies and are multiplexed to form a single OFDM signal. Therefore, although, both of them carry same amount of data, they respond differently to interference. This concept can be understood as such. Let us suppose there are three signals A, B, and C obtained from three different sources. In FDM, these three signals are modulated using three different carrier frequencies and are multiplexed to form a single FDM signal. If the channel restricts or distorts any of the given carrier frequency, the signal corresponding to that frequency will be lost. Now, let us check the scenario of OFDM. Here, each of the signals A, B, and C will be broken up into its constituent blocks. These blocks will be modulated with different carrier frequencies and will be multiplexed again to form OFDM signals $\hat{A}$, $\hat{B}$, and $\hat{C}$. In case of any channel disturbance affecting a particular carrier frequency, the block corresponding to that frequency will be lost instead of the total signal as in the case of FDM.

The main concept of OFDM is the orthogonality of its subcarriers. It is known that carrier signals are either sine or cosine wave, and the area under one period of a sine or cosine wave is zero. In order to understand the concept of orthogonality of the subcarriers in OFDM, let us multiply a sine wave of frequency C_1 by a sine wave of frequency C_2, where, both C_1 and C_2 are integers (Langton, 2004). Then, we have

$$f(t) = \sin(C_1\omega t) * \sin(C_2\omega t) \tag{8.3}$$

or

$$f(t) = \frac{1}{2}\cos(C_1 - C_2)\omega t - \frac{1}{2}\cos(C_1 + C_2)\omega t \tag{8.4}$$

When Equation 8.4 is integrated over one period, then both the terms become zero. Thus, when a sinusoid of frequency C_1 is multiplied by an another sinusoid of frequency C_1 or C_2, then the area under their product is zero. This is the principle of orthogonality that allows simultaneous transmission of a number of subcarriers through a single frequency band without interference; thus, it solves the problem of ICI.

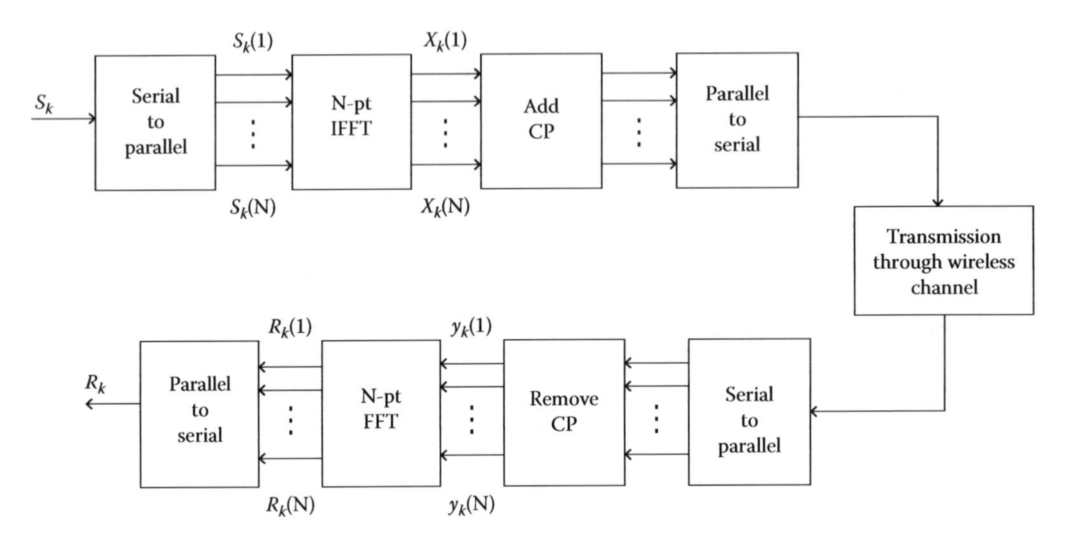

FIGURE 8.5
Typical OFDM transmitter and receiver structure.

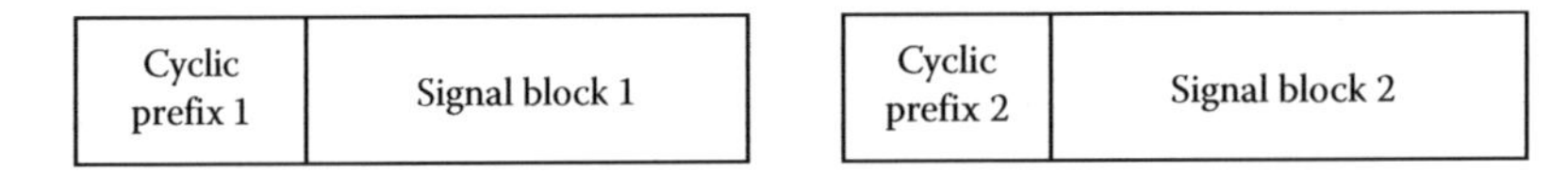

FIGURE 8.6
Cyclic prefixed signal blocks.

The details of a generic OFDM transmitter and receiver are presented in the block diagram shown in Figure 8.5. The OFDM system, basically, involves the transmission of a CP signal over a multipath fading channel. Here, the input signal S_k is split into N independent signals, and N-point inverse fast Fourier transform (IFFT) is performed on each of the N signal components. A CP of length ν, say, is added to each block of IFFT signal and, prior to transmission, they are multiplexed to form a single OFDM signal. CP is a copy of the tail of the signal block to be transmitted, that is, to form a CP of length ν, a copy of last ν bits of the signal block is replicated. It is appended over the head of that signal block as shown in Figure 8.6. Figure 8.6 shows two consecutive CP signal blocks. When ISI occurs between these two consecutive blocks, the CP guarding the signal block 2 gathers the splash of the previous block (signal block 1) on itself and saves its parent signal block. This is how OFDM solves the issue of delay spread that causes ISI.

8.2.4 Multiple Input Multiple Output System

When multiantenna systems were first described in the 1990s by Gerard Foschini and others, the surprising bandwidth efficiency provided by them seemed to be in violation of the Shannon limit. But later, it was proved that they are in complete accordance with the Shannon limit. They became popular because of their ability to provide higher data rates as well as spectral efficiency. Multiantenna setups can be viewed in single input multiple output, multiple input single output, and MIMO configurations. Among them, MIMO is the most popular form that makes use of multiple antennas both at the transmitter and the receiver end as shown in Figure 8.7. International Telecommunications

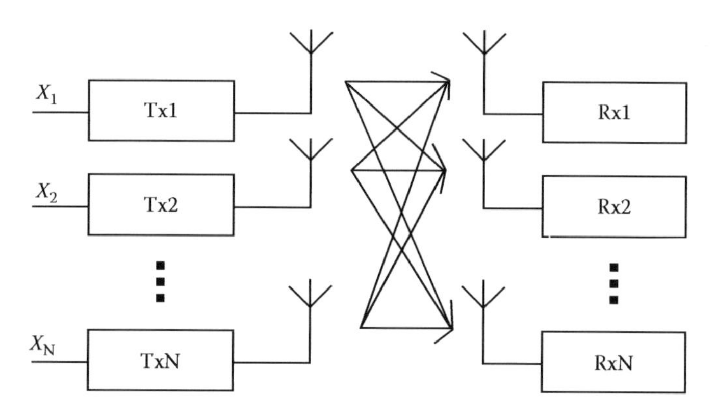

FIGURE 8.7
A generic $N \times N$ MIMO setup.

Union uses MIMO in the high-speed downlink packet access part of the Universal Mobile Telecommunication System standard. The long-term evolution standard also makes use of MIMO (Suthatharan, 2003).

MIMO offers certain benefits over a traditional SISO channel. In order to explore this fact, let us look at the capacity of a SISO link. The capacity of a SISO link is called information-theoretic capacity and is given by

$$C = \log_2(1 + SNR) \tag{8.5}$$

A substantial increase in the channel capacity is observed when MIMO setup is used in the communication system. This increase in the channel capacity translates to higher data throughputs. It also ensures improved link reliability by lowering the BER. The most significant factor is that MIMO provides these benefits without any additional bandwidth or increased transmit power (Kim et al., 2011). The capacity of a $N_t \times N_r$ Rayleigh fading MIMO channel is given by

$$C_{nn}(SNR) = E\left[\log\det\left(\mathrm{I}_{N_t} + \frac{SNR}{N_t} HH^*\right)\right] \tag{8.6}$$

where N_t and N_r are the number of antennas in the transmitter and the receiver ends respectively. In an $N \times N$ MIMO channel as shown in Figure 8.7, the channel capacity increases linearly with N over the entire SNR range (Langton, 2011).

The input–output relationship in a SISO or a MIMO channel is written as

$$R = HS + N \tag{8.7}$$

where

R is the received signal
H is the channel impulse response
S is the transmitted signal
N is the noise component

In case of a SISO channel, the parameters in Equation 8.7 take single values, while in case of MIMO channel, they hold values in the form of matrices (Oestges and Clerckx, 2007). The channel matrix H is called channel information in MIMO literature.

8.2.5 Distributed Time Delay Neural Network

An ANN is an information processing system that is inspired by biological nervous system and is developed with an aim to mimic human brain. It is defined as a massively parallel distributed processor made up of artificial neurons that provide it with the ability of learning. These neurons are highly interconnected and work in unison to solve specific problems. During operation, neural networks are trained to perform any function by adjusting the synaptic weights between the neurons. Generally, they are provided with a pattern set and a target set and are trained until the network output matches the target as shown in Figure 8.8. After training, they can produce reasonable output for such inputs that are not introduced into the network at the time of training. This is possible due to the ANN's ability of learning and generalization (Haykin, 2005).

Figure 8.9 shows a simple neuron with R-element input vector. Its output a is evaluated as

$$a = f(n) = f(WP + b) \tag{8.8}$$

where

P is the input vector
b is the bias value
W is the weight matrix

Here, f is an activation function that limits the value of a within the desired range. This is how, the network output is evaluated. A network may have a number of layers of such neurons.

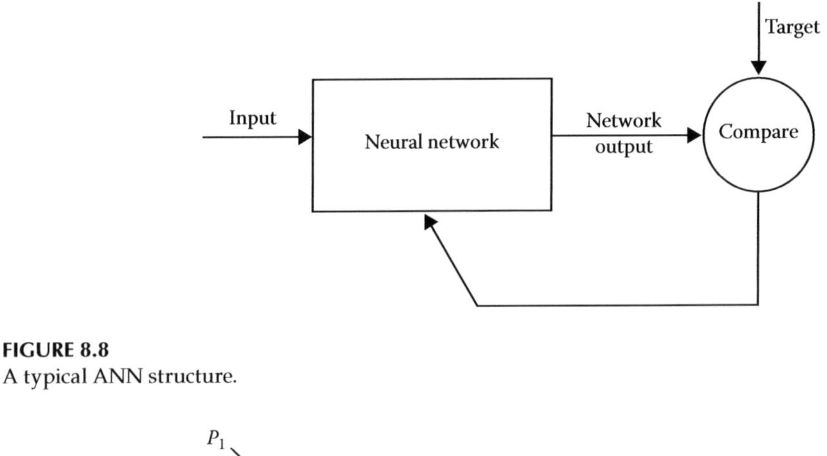

FIGURE 8.8
A typical ANN structure.

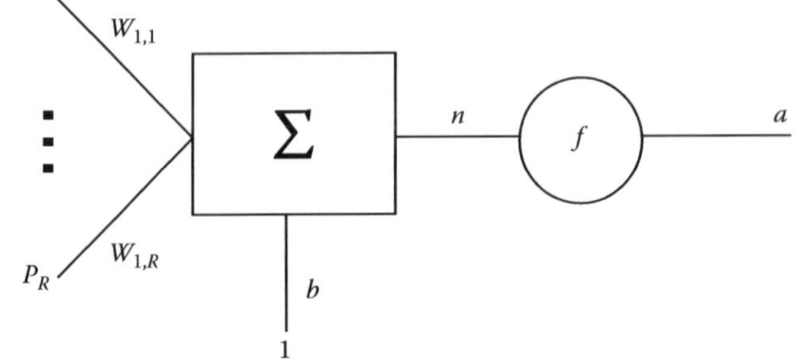

FIGURE 8.9
A neuron with R-element input vector.

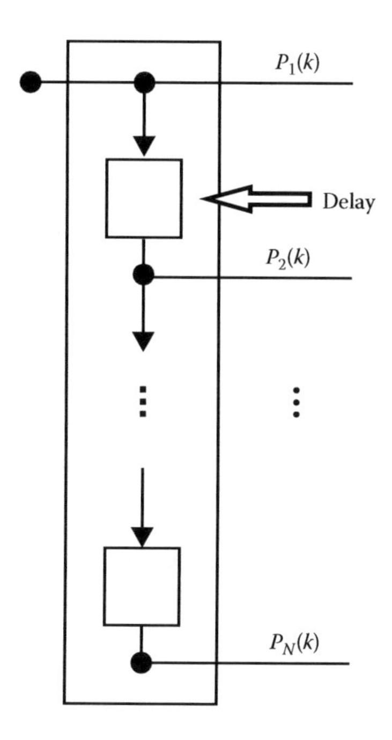

FIGURE 8.10
A tapped delay line.

Neural networks can be classified into static and dynamic categories. Static networks are also known as feedforward networks, and they have neither feedback elements nor delays. Here, the output is calculated directly from the input through feedforward connections. On the other hand, dynamic networks are the networks that can be further classified into two categories, namely, feedforward-dynamic and recurrent-dynamic networks, and in such networks, the output depends not only on the current input to the network but also on the current or previous inputs, outputs or states of the network. Time delay neural network (TDNN) is a type of dynamic ANN whose structure is optimized for dealing with real-time data. In this structure, the delayed version of the continuous input is also fed as input to the network and the use of such delayed inputs allows temporal processing. It is, basically, a multilayer feedforward network where hidden and output neurons are replicated across time. The two most commonly used TDNN structures are focused TDNN and distributed TDNN.

DTDNN is a type of TDNN in which the tapped delay lines are distributed throughout the network. A tapped delay line is shown in Figure 8.10. Each input and layer weights of DTDNN have a tapped delay line associated with it. This allows the network to have a finite dynamic response to time series input data.

8.2.6 Minimum Mean Square Error Algorithm

MMSE algorithm is defined as an estimation method, which is based on the minimization of the mean square error (MSE) of the fitted values of a variable. Bayes theorem is the main concept behind this algorithm. Here, there is some prior information available about the parameter to be estimated. In this approach, such prior information is captured by the prior probability density function of the available parameters, and it enables to evaluate better posterior estimates by employing Bayes theorem.

Let x be a $n \times 1$ unknown random variable and y be a $m \times 1$ known random variable. The estimation $\hat{x}$ of x is any function of y. The estimation error vector is given by

$$e = \hat{x} - x \tag{8.9}$$

and its MSE is given by the error covariance matrix

$$\text{MSE} = E\{(\hat{x} - x)(\hat{x} - x)^T\} \tag{8.10}$$

where the expectation E is taken over both x and y. The MMSE estimator is then defined as the estimator that works to achieve minimum MSE.

8.2.7 Channel Tracking and Equalization

The phenomenon of tracking a channel is needed to acquire CSI that provides the statistics of variation of the magnitude and the phase of the channel. Taking the channel variations to be constant for a particular interval of time, the knowledge of CSI can help in inverting the effect of the channel from the received signal. Here, the knowledge of channel variations is implemented as inverse filter on the received signal. This phenomenon of correcting the distortive impacts of channel is known as channel equalization.

In the past, various techniques have been employed for tracking and equalizing MIMO-OFDM channels. In one of the literatures, a channel estimation technique based on ANN structures, namely, RNN and MLP, is used for STBC-MIMO-based wireless system with binary phase shift keying (BPSK) modulation scheme over Rayleigh fading channel where, RNN structures are found to outperform MLP, both in terms of training performance and BER results (Gogoi and Sarma, 2012). In an another work, an effective channel estimation technique that can tackle statistical nature of wireless channels has been discussed, where two MLP architectures with temporal characteristics are used, which are found to be better suited for time-varying channel conditions especially for slow fading conditions for applications in indoor networks with MIMO-OFDM system (Ciflikli et al., 2009). In an another potential work, a channel estimation and equalization algorithm using three-layered ANN with feedback for MIMO wireless communication systems is discussed (Ling and Xianda, 2007). Here, a turbo iteration process between the different algorithms is formed, which is found to improve the estimation performance of the channel equalizer. Simulation results showed that this channel equalization algorithm has better computational efficiency and faster convergence than higher-order-statistics-based algorithms. ANN structures such as MLP and RNN have been widely used for accomplishing the task of channel estimation (Naveed et al., 2004; Nawaz et al., 2009; Potter et al., 2009; Saikia and Sarma, 2012; Sarma and Mitra, 2012). Statistical methods have also been employed for achieving the purpose. A blind channel estimation algorithm for STBC systems is discussed in one of the papers that exploit the statistical independence of sources before space-time encoding (Choqueuse et al., 2011). The channel matrix is estimated by minimizing a kurtosis-based cost function after zero-forcing equalization. Moreover, statistical methods like maximum likelihood receiver, Kalman filter, and DFE are widely popular for the task of MIMO-OFDM channel estimation (Al-Naffouri and Quadeer, 2008). The proposed hybrid system that has been discussed in this chapter is a combination of an ANN structure, that is, DTDNN and a statistical algorithm called MMSE technique;

therefore, it exploits the benefits of both the genres and forms a robust system for tracking and equalizing SISO-OFDM and MIMO-OFDM channels.

8.2.8 Traditional Channel Tracking and Equalization Approaches: A Literature Review

In the past few years, there has been a lot of work done in the genre of channel estimation and equalization, which has been recognized as one of the most significant issue in wireless communication. A number of works reported in this regard are cited before. Among them, two such works have been chosen for case study and are elaborately discussed in the following text.

Ciflikli et al. (2009) had proposed an ANN-based channel estimation method that had been implemented for tracking an OFDM system over Rayleigh fading channel. Here, the ANN structures were trained using Levenberg–Marquardt (LM) training algorithm. This method was implemented as an alternative to pilot-based techniques and thus was instrumental in achieving bandwidth efficiency. In this work, at the transmitter end, the required binary information was mapped using baseband modulation scheme quadrature amplitude modulation, and then serial to parallel conversion was done on the modulated data followed by IFFT and addition of cyclic prefix. The resultant OFDM signal was made to pass through a transmission medium realized in the form of frequency-selective fading channel in combination with additive white Gaussian noise (AWGN). At the receiver end, the first set of processing involved removal of cyclic prefix and fast Fourier transform (FFT). For channel estimation, MLP structures were employed consisting one input layer, two hidden layers, and one output layer. These MLP structures were split into real and imaginary segments. The real and imaginary parts of the complex signal are separately fed into the real and imaginary segments of the MLP, respectively. The estimates obtained from the implemented structure of ANN were combined to form the required received signal, and this complex signal was demodulated to recover the transmitted bits. In order to minimize the error in the aforementioned neural network structure, LM algorithm was implemented. The performance of the proposed system was found to outperform a popularly known pilot-based statistical method based on DFE.

Nawaz et al. (2009) had also proposed a neural network–based channel equalizer that had been employed for MIMO-OFDM channel. In this case, comb-type pilot arrangement was used to feed the channel information to the neural network during its training phase. In the testing session, the real and imaginary parts of the signal received from the channel are separately fed to two neural network structures that are trained separately for obtaining real and imaginary estimates of the transmitted signal. These estimates are used at the receiver for data recovery.

Besides MIMO-OFDM channel tracking and equalization, there has been a lot of work done in the genre of managing network QoS in mobile cyber-physical frameworks. One work proposed by Xia et al. (2008) identified various challenges related to the network QoS management in wireless sensor/actuator networks (WSANs) (WSANs play an essential role in CPS). The authors have proposed an effective feedback scheduling framework that tackles some of these identified challenges. The issues identified are service-oriented architecture, QoS aware communication protocols, resource self-management, and QoS aware power management. In order to deal with these issues, the authors have proposed a feedback scheduling framework that dynamically adjusts specific scheduling parameters of relevant traffic so as to maintain a desired QoS level.

8.3 Channel Tracking and Equalization Approaches Based on DTDNN and DTDNN-MMSE System

The proposed channel tracking and equalization work has been divided into a number of sections. In the first part, the design of a DTDNN system for application in a SISO-OFDM framework is discussed. This part appears as Section 8.3.1 and the associated results are described in Section 8.4.1. The working of the proposed system model based on DTDNN structure has been improved using MMSE algorithm with it, and the resultant hybrid system model based on DTDNN-MMSE is discussed in Section 8.3.2. Sections 8.3.2 and 8.3.3 present the implementation of the composite DTDNN-MMSE system on SISO-OFDM and MIMO-OFDM channels, respectively. Their respective simulation results and associated inferences are described in Sections 8.4.2 and 8.4.3. Experiments are performed in channel conditions following Rayleigh distribution with Doppler shifts of 10–100 Hz with 64-bit blocks of transmission for a cycle of 6400 information bits. Specific details are provided in relevant sections.

8.3.1 Implementation of DTDNN System in SISO-OFDM Channel

This section discusses the implementation of a tracking and equalization system designed and developed using DTDNN structure solely. Here, the design of the transmitter and receiver structure employed for realizing SISO-OFDM channel, its specifications, the proposed system model based on DTDNN structure, and its working principle are described.

8.3.1.1 SISO-OFDM Transmitter and Receiver

Figure 8.11 shows the transmitter and receiver structure employed for realizing SISO-OFDM channel. Here, at the transmitter end, the data to be transmitted have been modulated using BPSK modulation scheme. The modulated signal is split into a number of

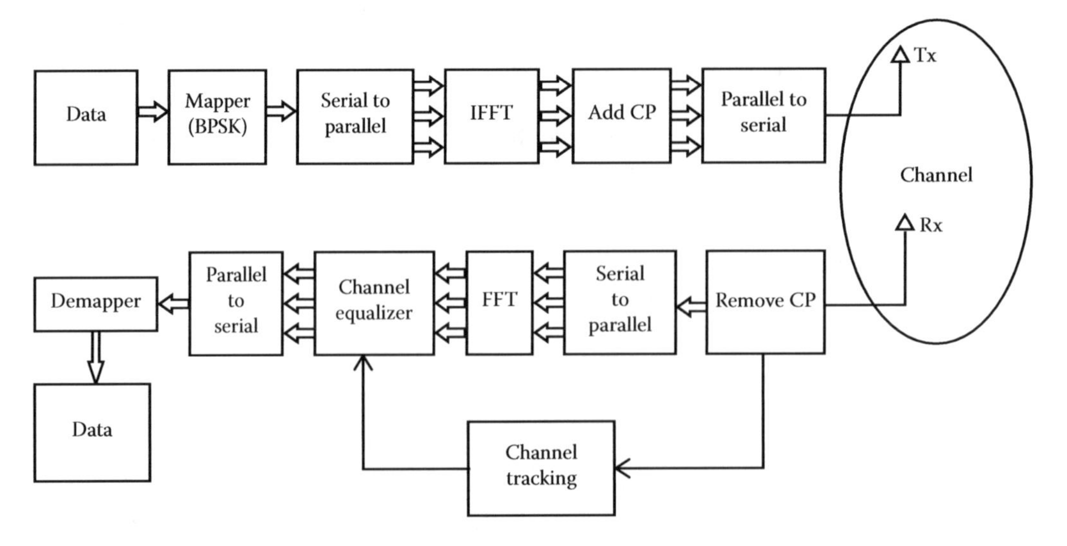

FIGURE 8.11
Block diagram of SISO-OFDM transmitter and receiver.

signal blocks. IFFT is performed on each of the signal blocks followed by the addition of CP to each of the blocks. The length of the CP is calculated on the basis of block size. These CP signal blocks are added to form a single OFDM signal. These OFDM signal is made to pass through a transmission medium realized as Rayleigh fading channel in combination with AWGN.

At the receiver end, the CP portion of the received signal blocks is eliminated. Then, the remaining received signal is made to pass through channel tracking block that consist of DTDNN structure. In the channel equalizer block, the trained DTDNN structure is used to obtain the estimate of the transmitted bits. These estimates are demodulated and are used to recover the actual data bits. The overall working of the channel tracking and equalization blocks is discussed in the subsequent sections.

8.3.1.2 Specifications of the SISO-OFDM Tx-Rx Structure

Table 8.1 shows the specifications of the SISO-OFDM system. Here, cycles of 6400 bits are used as the information that is modulated using BPSK modulation scheme. Since the deployed system is in SISO configuration, it consists of one transmitter and one receiver. The block size of each OFDM block is 64. In each block, there are 5 pilot bits embedded along with the information bits. CP of length 7 bits is used. The multipath fading channel that has been used as the transmission medium is assumed to follow Rayleigh distribution. The SNR over which the system performance is evaluated ranges from −10 to 20 dB.

8.3.1.3 System Model Based on DTDNN and Its Specifications for SISO-OFDM

Figure 8.12 shows the system model based on DTDNN structure proposed for tracking and equalizing SISO-OFDM channels. Table 8.2 represents the specifications of the DTDNN structure. Channel tracking and equalizer block shown in the receiver block diagram comprises DTDNN structure as shown in Figure 8.12. Here, the real and imaginary parts of the received pilot signal are fed to DTDNN1 and DTDNN2, respectively. The system initially is in a training mode. After the training, it enters into a validation (testing) mode where fresh data blocks not used previously are applied to ascertain its ability to make discrimination. These two phases have certain latency, but the testing time is comparatively smaller than other ANN structures employed for the similar tasks. Training and validation are carried out using several configurations for DTDNN structures till the MSE is reached. MSE is used as a cost function to determine the system performance.

TABLE 8.1

System Specifications of SISO-OFDM Tx and Rx

Total number of bits in a cycle	6400
Block size	64
Modulation	BPSK
Number of transmitters	1
Number of receivers	1
Pilot type	Block type
Pilot interval	5
Channel type	Rayleigh fading
SNR range	−10 to 20 dB

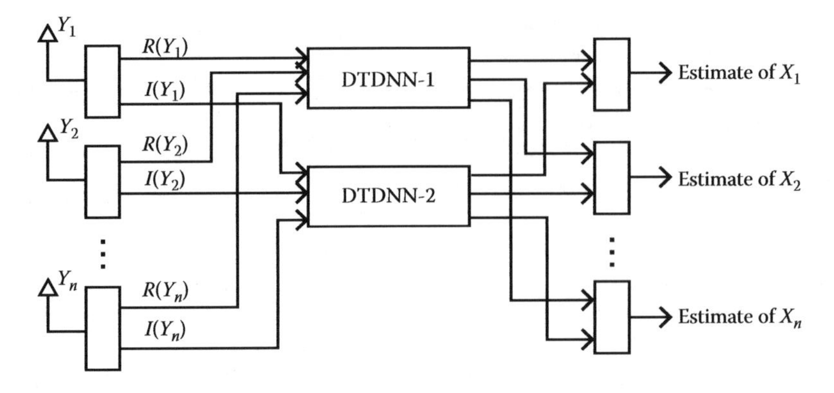

FIGURE 8.12
Proposed system model for SISO-OFDM channel tracking and equalization.

TABLE 8.2
Specifications of DTDNN Structure

ANN type	DTDNN
Training algorithm	traingda
Maximum number of epochs	60
Input layer size	Depends on block size
MSE goal	0.001
Training time	40.7884 s
Validation time	26.5462 s

The length of the hidden layers is fixed by trial and error method. Although the LM training algorithm is the default algorithm for DTDNN structure, gradient descent with adaptive learning rate back propagation algorithm has been used because the former, though, known for its fast training speed but suffers from lack of memory when the length of the data sequence is increased. Till MSE is reached, the network is trained such that for the pattern set given by the received pilot bits, the network output reaches the target set, which is nothing but the corresponding transmitted pilot bits. In the testing phase, the network is simulated using the real and imaginary parts of the received information data, and here, it provides the real and imaginary estimates of the transmitted information data. The real and imaginary parts are combined as $r = a + jb$. Now, we know that the received signal can be modeled as

$$y = x * h + n \tag{8.11}$$

where
- x is the transmitted signal
- h is the channel coefficient
- n is the noise component

Thus, replacing x by $\hat{x}$ in Equation 8.11, the channel coefficients can be evaluated. At this instant of time, CSI is known at the receiver. Taking the channel variations to be constant for a finite amount of time, the information bits are recovered. The training and validation time are tabulated in Table 8.2.

8.3.1.4 Working Principle of DTDNN Structure in SISO-OFDM

In the proposed system, the steps involved in conducting the whole procedure have been enumerated as follows:

- At first, information bits are generated and are modulated using BPSK modulation scheme.
- Pilot signal is generated and added to the original signal in block-type format with pilot interval of five.
- Taking the block size to be 64, the original signal is split into a number of signal blocks. These blocks are made to undergo IFFT and a CP of length 7 is added to each of them. As mentioned earlier, the length of CP is calculated on the basis of the block size. The signal blocks are added to form a single OFDM signal.
- The OFDM signal is made to pass through the channel consisting of Rayleigh fading (Doppler shifts of 10–100 Hz are considered) and AWGN noise.
- At the receiver end, CP is removed from the received signal. The received bits corresponding to the transmitted pilot bits are extracted and are fed to the channel tracking block. This block consist of DTDNN structure and its working has been explained in Section 8.3.1.3. It provides the estimates of the transmitted pilot bits that are used to evaluate the channel coefficients; thus, CSI is known at the receiver.
- Considering the channel to be constant for a finite amount of time, this channel information is used to invert the effect of the channel from the received information bits in the channel equalizer block; thus, estimates of the data bits are recovered.
- Finally, these estimated bits are demodulated and original data bits are recovered. Equalized data bits are compared with the originally transmitted bits in order to calculate BER of the system.

Details of the associated results are discussed in Section 8.4.1. Though the results provided by DTDNN are found to be better compared to another ANN structure widely used for the task of channel estimation, namely, MLP, the limitation encountered in slow learning convergence in multiantenna environment leads to the incorporation of MMSE algorithm with DTDNN structure. This improvisation in the system model showed improvement in its performance as a composite estimator and equalizer, and the relevant composite system is discussed in the next section.

8.3.2 Implementation of Improvised System Based on Hybrid DTDNN-MMSE Structure on SISO-OFDM Channel

This section discusses the changes that have been made in the proposed system model for SISO-OFDM channel in order to improve the system performance. Here, the MMSE algorithm has been combined with DTDNN structure.

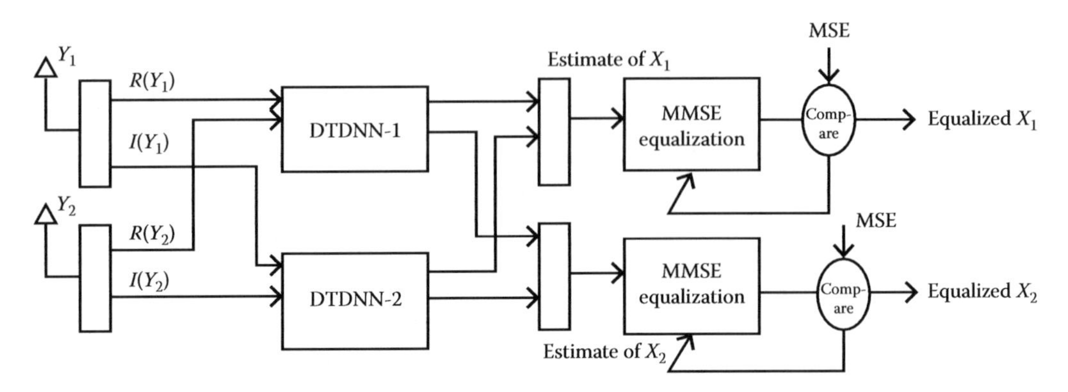

FIGURE 8.13
Proposed system model (improvised) for SISO-OFDM channel tracking and equalization.

8.3.2.1 System Model Based on DTDNN-MMSE System for SISO-OFDM

Figure 8.13 shows the improvised form of the proposed system model used for tracking and equalizing SISO-OFDM channels in which MMSE algorithm has been combined with DTDNN structure. Here, in the channel equalizer block, MMSE algorithm is incorporated along with DTDNN. The extracted pilot sequences are not only used to train the neural network but also employed in MMSE algorithm. MMSE exploits the pilot information to recover the information bits more efficiently. The relevant simulation results are described in Section 8.4.2.

8.3.3 Implementation of DTDNN-MMSE System on MIMO-OFDM Channel

In this section, DTDNN-MMSE-based MIMO-OFDM channel tracking and equalization has been discussed. Here, the DTDNN structure has been trained extensively to track the variations of a MIMO-OFDM channel, and the estimates of the transmitted bits obtained from the network have been employed to recover the originally transmitted data bits. MMSE algorithm exploits the pilot information and improves the BER scenario for recovering information bits more efficiently.

8.3.3.1 Transmitter and Receiver Structure for MIMO-OFDM Channel

Figure 8.14 shows the transmitter and receiver structures employed for realizing MIMO-OFDM channel. Here, at the transmitter end, the data to be transmitted are broken into two streams and are modulated using BPSK modulation scheme. Each of the modulated signals is split into a number of signal blocks. IFFT is performed on each of the signal blocks; thereby, CP is added to each of the blocks. The length of the CP is calculated on the basis of block size. These CP signal blocks are added to form a single OFDM signal. Thus, we have two such OFDM signals that are transmitted through the transmitters Tx1 and Tx2.

8.3.3.2 Specifications of the MIMO-OFDM Tx-Rx Structure

Table 8.3 shows the specifications of the MIMO-OFDM system. Here, cycles of 6400 bits are used as the information that is modulated using BPSK modulation scheme. The deployed MIMO system consists of two transmitters and two receivers. The block size of each OFDM

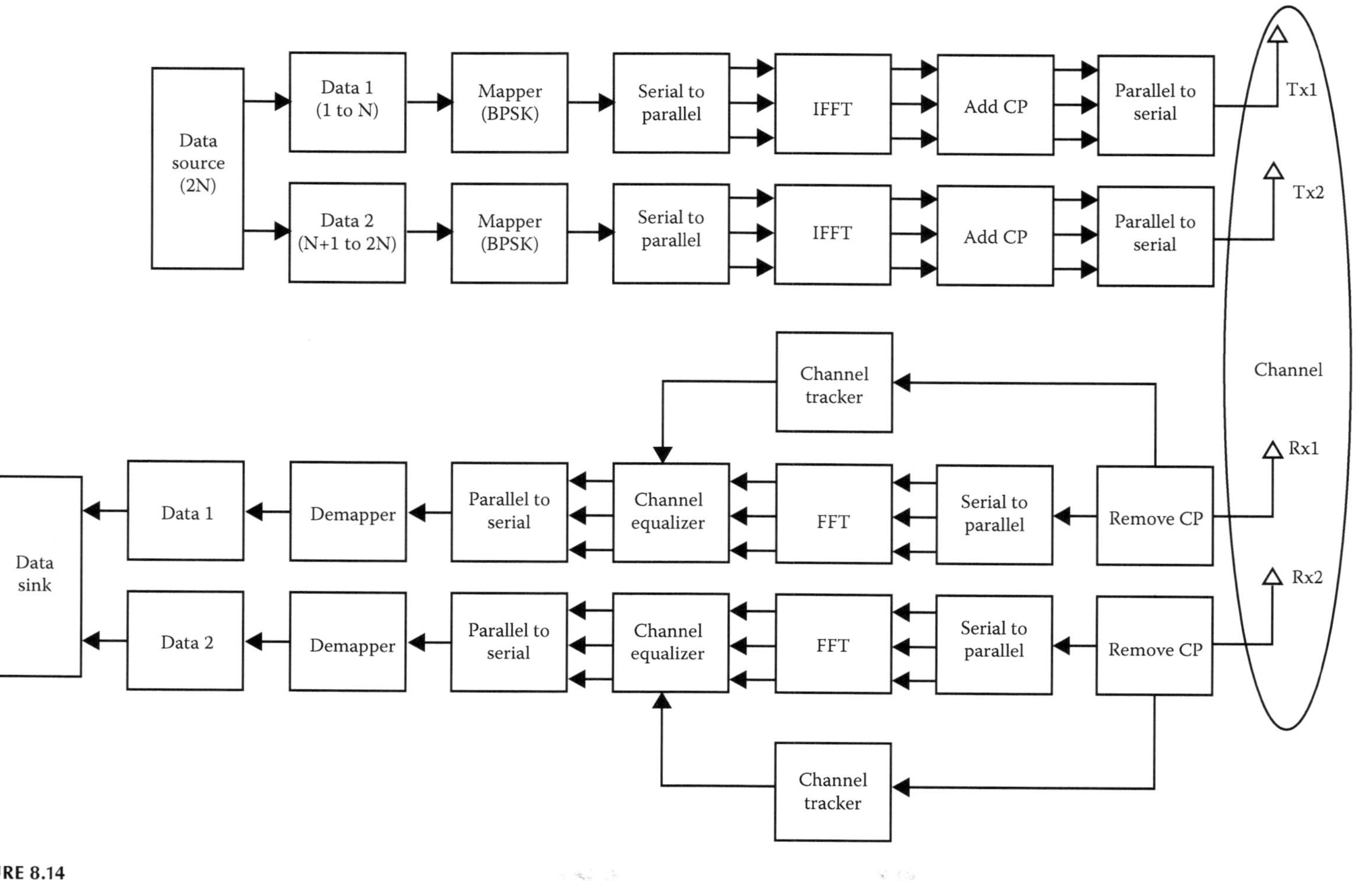

FIGURE 8.14
Block diagram of MIMO-OFDM transmitter and receiver.

TABLE 8.3

System Specifications

Total number of bits in a cycle	6400
Block size	128
Modulation	BPSK
Number of transmitters	2
Number of receivers	2
Pilot type	Block type
Pilot interval	5
Channel type	Rayleigh fading
SNR range	–10 to 20 dB

TABLE 8.4

Specifications of DTDNN Structure

ANN type	DTDNN
Training algorithm	traingda
Maximum number of epochs	60
Input layer size	Depends on block size
MSE goal	0.001
Training time	146.0511 s
Validation time	31.8607 s

block is 128. In each block, there are 5 pilot bits embedded along with the information bits. CP of length 13 bits is used. The multipath fading channel that has been used as the transmission medium is assumed to follow Rayleigh distribution. The SNR over which the system performance is evaluated ranges from –10 to 20 dB.

8.3.3.3 System Model Based on DTDNN-MMSE System and Its Specifications for MIMO-OFDM

This system model based on DTDNN and MMSE is already explained in Figure 8.13. Its specifications are shown in Table 8.4. Training and validation time of the proposed system employed for MIMO-OFDM channel are tabulated in Table 8.4.

8.3.3.4 Working Principle of DTDNN-MMSE in MIMO-OFDM

In the proposed system, the steps involved in conducting the whole procedure have been enumerated as follows:

- At first, information bits are generated, are broken into two streams, and each of the data streams are modulated using BPSK modulation scheme.
- Pilot signal is generated and added to the modulated signals in block-type format with pilot interval of five.
- Taking the block size to be 128, both the signals are split into a number of signal blocks. These blocks are made to undergo IFFT, and a CP of length 13 is added to each of them. As mentioned earlier, the length of CP is calculated on the basis of the block size. The signal blocks are added to form a single OFDM signal. Thus, we have two such OFDM signals that are transmitted using 2 × 2 MIMO transmitters.

- The OFDM signals are made to pass through the channel consisting of Rayleigh fading (Doppler shifts of 10–100 Hz are considered) and AWGN noise.
- At the receiver end, CP is removed from the received signal. The received bits corresponding to the transmitted pilot bits are extracted and are fed to the channel tracking block. This block consist of DTDNN structure and its working has been already explained earlier. It provides the estimates of the transmitted pilot bits that are used to evaluate the channel coefficients; thus, CSI is known at the receiver.
- Considering the channel to be constant for a finite amount of time, this channel information is used to invert the effect of the channel from the received information bits in the channel equalizer block consisting MMSE system; thus, estimates of the data bits are recovered.
- Finally, these estimated bits are demodulated and original data bits are recovered. Equalized data bits are compared with the originally transmitted bits in order to calculate BER of the system.

The simulation results corresponding to the implementation of DTDNN-MMSE system in MIMO-OFDM channel are discussed in Section 8.4.3.

8.4 Simulation Results and Discussion

This section describes the simulation results and provides the necessary discussion associated with the proposed system models discussed in the previous section.

8.4.1 Simulation Results and Discussion: DTDNN-MMSE System in SISO-OFDM Channel

Figure 8.15 shows the comparison between the unfaded and faded signal powers for 250 (out of 6400) information bits. Unfaded signal power is evaluated for originally transmitted bits, and faded signal power is evaluated for the same data obtained after passing through the fading channel. It is clear from the figure that there is a significant decrease in the signal power after passing through the channel. Thus, the signal strength gets faded due to the impact of channel. Sometimes, such fades become so severe that there is a sudden drop in the channel SNR that leads to the temporary failure in the communication link.

Error performances are evaluated as BER against the SNR range. The BER versus SNR curve is obtained for the proposed system as well as for MLP structure, which is another potential ANN structure used widely for the task of channel tracking and equalization. The systems are simulated over the range of SNR −10 to +20 dB. Figure 8.16 shows the BER versus SNR scenario for theoretical case, DTDNN and MLP. It clearly shows that the DTDNN structure outperforms the MLP network. This is due to the temporal processing generated by the time delay blocks present in this specific ANN. Thus, a reliable framework for tracking time-induced variations in the SISO-OFDM channel has been developed using DTDNN.

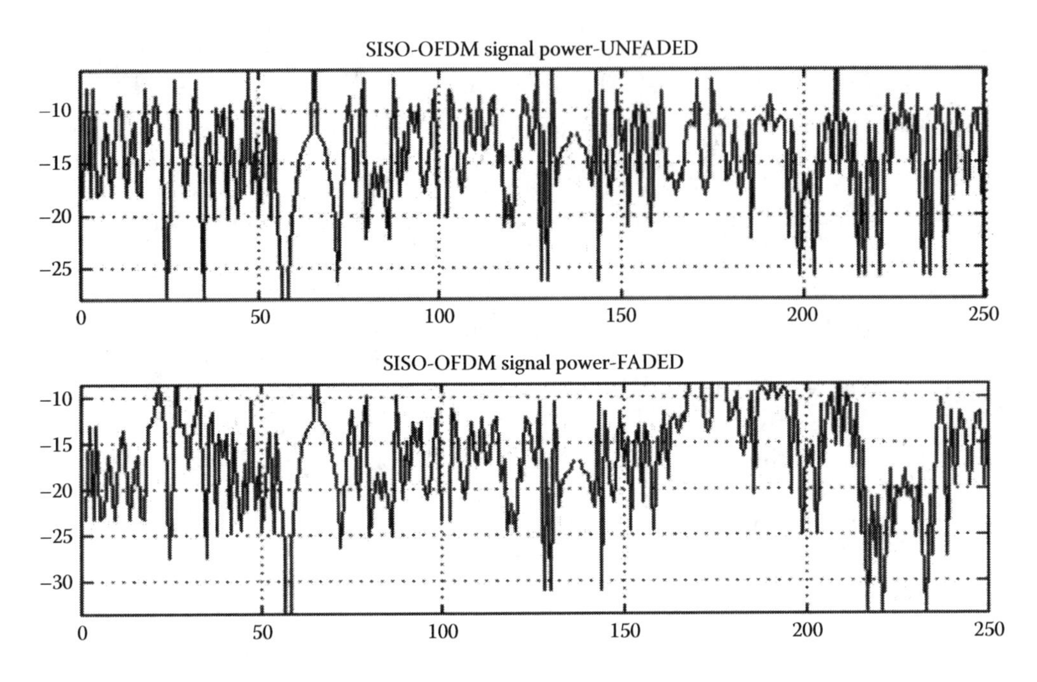

FIGURE 8.15
Comparison between unfaded and faded signal powers (SISO-OFDM).

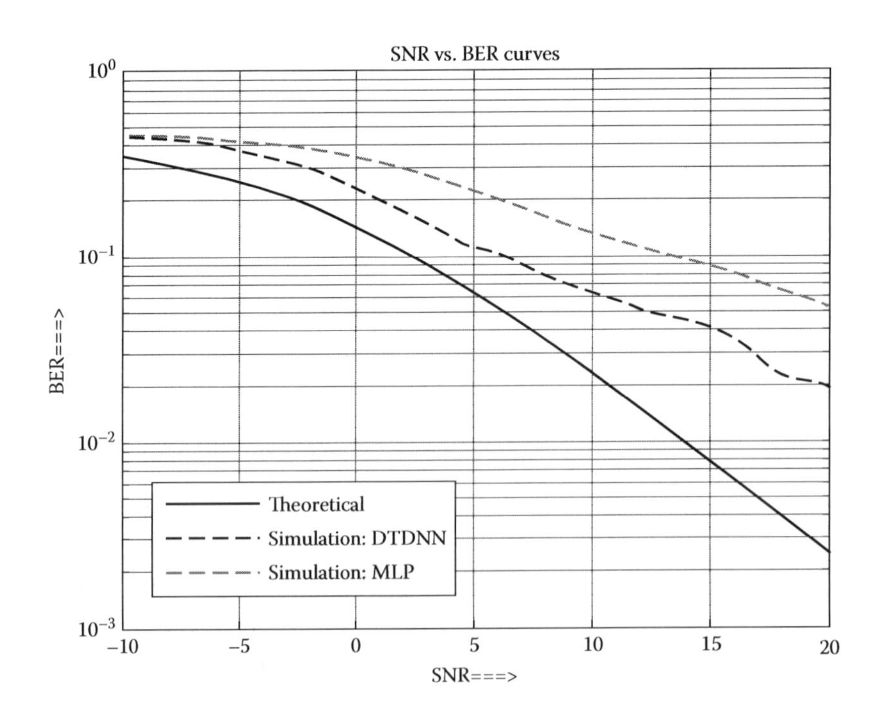

FIGURE 8.16
BER versus SNR curve for DTDNN system employed for SISO-OFDM channel.

8.4.2 Simulation Results and Discussion: DTDNN versus DTDNN-MMSE in SISO-OFDM Channel

The incorporation of MMSE with DTDNN has improved the error performance of the system. Figure 8.17 shows the comparison of the BER versus SNR performance of the system based on DTDNN with that of the system based on DTDNN-MMSE. It clearly shows that the performance of the hybrid DTDNN-MMSE system has outperformed the DTDNN system by a huge margin.

8.4.3 Simulation Results and Discussion: DTDNN-MMSE System in MIMO-OFDM Channel

Figures 8.18 and 8.19 show the comparison between the unfaded and faded signal powers for 1800 (out of 6400) information bits transmitted through Tx1 and Tx2, respectively. It is clear from the figure that there is a significant decrease in the signal power after passing through the channel. Thus, the signal strength gets faded due to the impact of channel.

Error performances are evaluated as BER against the SNR range. The BER versus SNR curve is obtained for the proposed system as well as for DFE-LMS system, which is another potential statistical method used widely for the task of channel tracking and equalization. The systems are simulated over the range of SNR –10 to +20 dB. Figure 8.20 shows the BER versus SNR scenario for theoretical case, DTDNN-MMSE and DFE-LMS. It is clear from the figure that the proposed system outperforms the DFE-LMS system by a huge margin. Thus, a reliable framework for tracking time-induced variations in the MIMO-OFDM channel has been developed using hybrid DTDNN-MMSE setup.

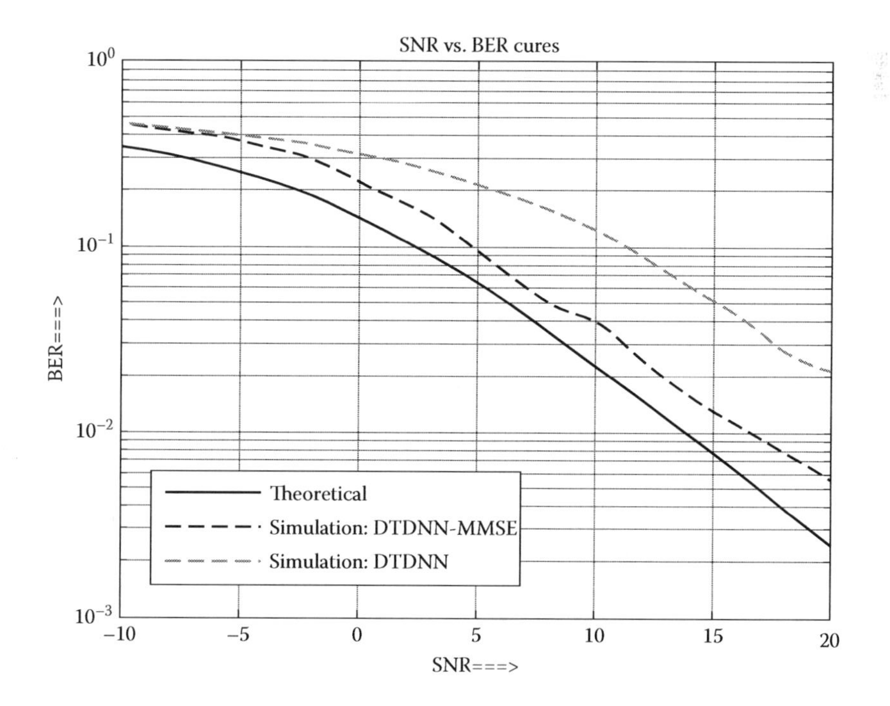

FIGURE 8.17
BER versus SNR curve for DTDNN-MMSE system employed for SISO-OFDM channel.

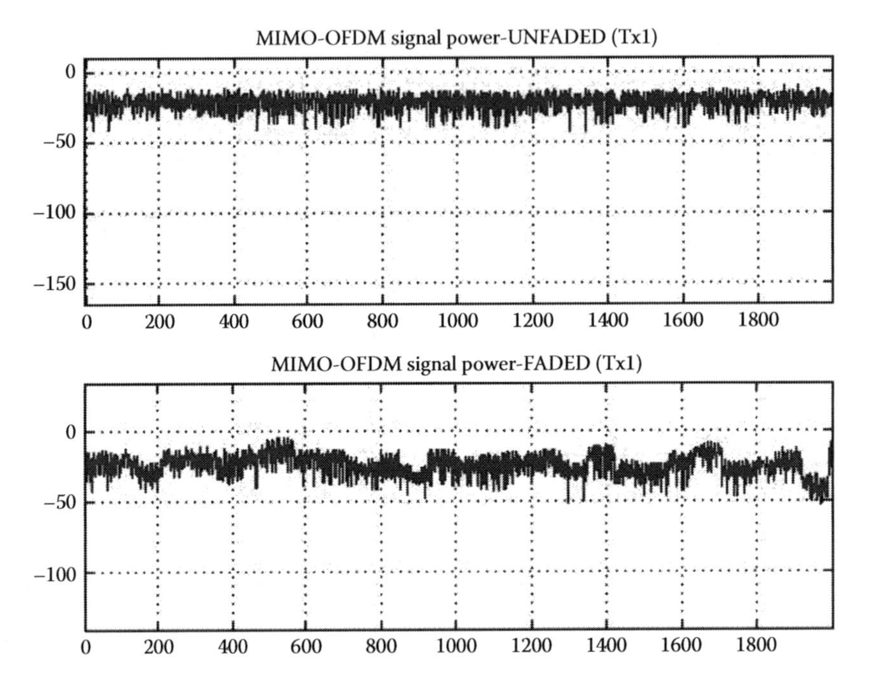

FIGURE 8.18
Comparison of unfaded and faded signal powers of MIMO-OFDM system (Tx1).

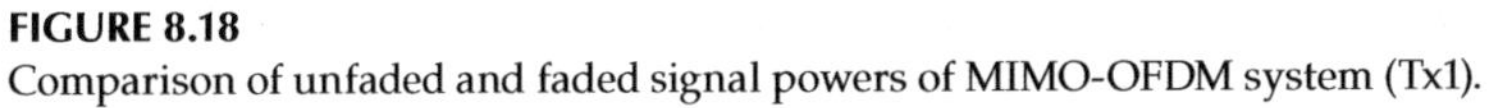

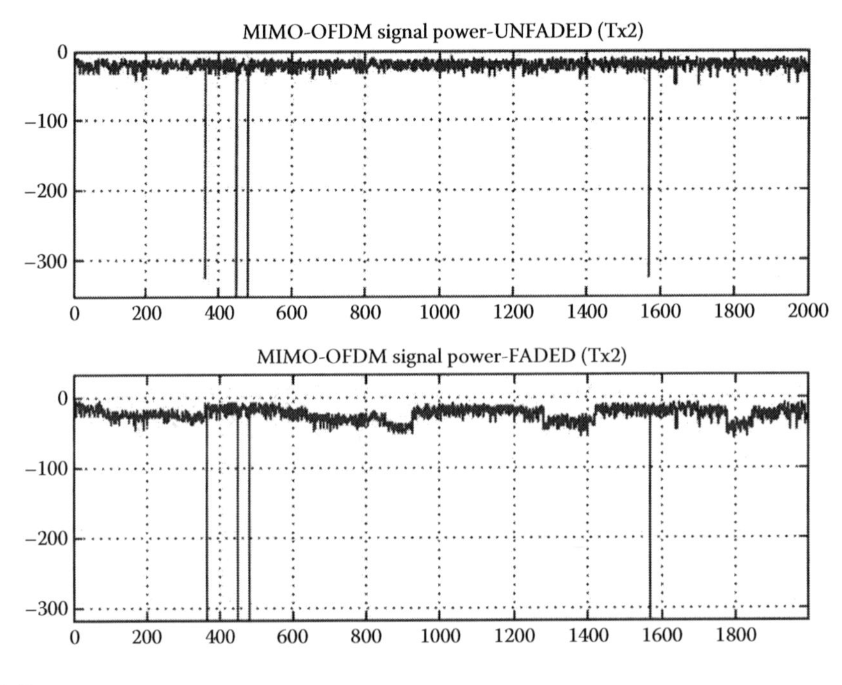

FIGURE 8.19
Comparison of unfaded and faded signal powers of MIMO-OFDM system (Tx2).

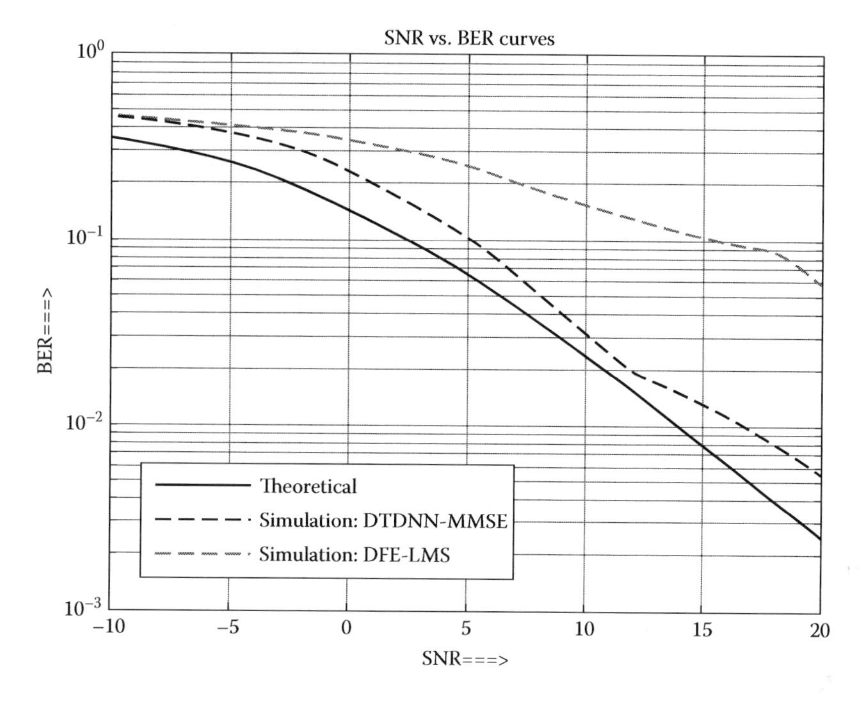

FIGURE 8.20
BER versus SNR curve for DTDNN-MMSE system employed for MIMO-OFDM channel.

TABLE 8.5
Comparison of Normalized Processing Time

Techniques	Training Time (s)	Testing Time (s)	Total Processing Time (s)	Normalized Processing Time (%)
DTDNN system in SISO-OFDM	40.7884	26.5462	67.3346	37.84
DTDNN-MMSE system in SISO-OFDM	40.7884	23.6401	64.4285	36.21
DTDNN-MMSE system in MIMO-OFDM	146.0511	31.8607	177.9118	100

8.4.4 Computational Complexity

Table 8.5 shows a comparison in the processing time required by the proposed system models. It is clear from the table that the incorporation of MMSE with DTDNN in SISO-OFDM channel has reduced the processing time; however, the implementation of the composite system in MIMO-OFDM channel requires a bit larger computational time. It is because of the fact that incorporation of multiple antennas increases the system complexity. This increase in the processing time can be optimized by making use of specialized hardware techniques.

8.5 Conclusion

In this chapter, a new and efficient system has been presented for tracking and equalizing OFDM-based wireless communication system in SISO and MIMO configurations that can be effectively implemented in mobile cyber-physical frameworks. The experimental results so obtained by employing DTDNN-MMSE system have clearly depicted that the proposed system has tracked both SISO-OFDM and MIMO-OFDM channels efficiently, and the tracked channel information has been utilized effectively for correcting the channel distortions. Thus, the proposed system models are an effective way to deal with the channel distortions and to improve link reliability in mobile cyber-physical frameworks. The training and testing time of the proposed system during the course of the work are found to be marginally large in comparison to other ANN structures due to the presence of tapped delay lines, but since it shows much improved ability to track temporal variation of the channel, this issue needs to be solved. The slowing down in training and testing can be removed considerably by using specialized hardware and DSP processor framework and this part of the system improvement has been taken up as its future work.

References

Al-Naffouri, T. Y. and Quadeer, A. A. (2008). A forward-backward Kalman filter-based STBC MIMO-OFDM receiver. *EURASIP Journal on Advances in Signal Processing*, Article ID 158037, 1–14, doi: 10.1155/2008/158037.

Baruah, B. B. and Sarma, K. K. (2012). ANN based equalization using coded inputs in Nakagami-m faded channel. *International Journal of Smart Sensors and Ad Hoc Networks*, 1(3), 83–86.

Bhuyan, M. and Sarma, K. K. (2012). MIMO-OFDM channel tracking using a dynamic ANN topology. *Journal of World Academy of Science, Engineering and Technology*, 71, 1321–1327.

Bhuyan, M. and Sarma, K. K. (2014). Nonlinear model based prediction of time varying SISO-MIMO channels using FANN-DFE combination. *Proceedings of IEEE First International Conference on Emerging Trends and Applications in Computer Science*, Shillong, India.

Choqueuse, V., Mansour, A., Burel, G., Collin, L., and Yao, K. (2011). Blind channel estimation for STBC systems using higher-order statistics. *IEEE Transactions on Wireless Communications*, 10(2), 495–505.

Ciflikli, C., Ozsahin, A. T., and Yapici, A. C. (2009). Artificial neural network channel estimation based on Levenberg-Marquardt for OFDM systems. *Wireless Personal Communication*, 51, 221–229, doi: 10.1007/s11277-008-9639-2.

Gogoi, P. and Sarma, K. K. (2012). Channel estimation technique for STBC coded MIMO system with multiple ANN block. *International Journal of Computer Applications*, (0975 8887), 50(13), 10–14.

Haykin, S. (2005). *Neural Networks—A Comprehensive Foundation*, New Delhi, India: Pearson Education.

Kim, Y. H., Cho, B. Y., and Kim, J. Y. (2011). *Wireless Communication: Trend and Technical Issues for MIMO-OFDM System*, Seoul, Korea: Kwangwoon University and Nokia-Siemens Networks.

Langton, C. (2004). Orthogonal Frequency Division Multiplex (OFDM) Tutorial. Retrieved from www.complextoreal.com.

Langton, C. (2011). Finding MIMO. Retrieved from www.completoreal.com.

Ling, Z. and Xianda, Z. (2007). MIMO channel estimation and equalization using three-layer neural networks with feedback. *Tsinghua Science and Technology*, ISSN-1007-021405/20, 12(6), 658–662.

Naveed, A., Qureshi, I. M., Cheema, T. A., and Jalil, A. (2004). Blind equalization and estimation of channel using artificial neural networks. *IEEE International Multitopic Conference*, Islamabad, Pakistan: IEEE, pp. 184–190.

Nawaz, S. J., Mohsin, S., and Ikram, A. A. (2009). Neural network based MIMO-OFDM channel equalizer using comb-type pilot arrangement. *International Conference on Future Computer and Communication*, Kuala Lampur, Malaysia, pp. 36–41.

Oestges, C. and Clerckx, B. (2007). *MIMO Wireless Communications*, Oxford, U.K.: Elsevier.

Potter, C., Venayagamoorthy, G. K., and Kosbar, K. (2009). *RNN Based MIMO Channel Prediction*, Rolla, MO: Real-Time Power and Intelligent Systems Laboratory, Missouri University of Science and Technology and Department of Electrical and Computer Engineering, Missouri University of Science and Technology.

Rajkumar, R., Lee, I., Sha, L., and Stankovic, J. (2010). Cyber physical systems—The next computing revolution. *Design Automation Conference*, Anaheim, CA.

Rappaport, T. S. (1997). *Wireless Communications—Principles and Practice*, Upper Saddle River, NJ: Pearson Education.

Saikia, S. J. and Sarma, K. K. (2012). ANN based STBC-MIMO set-up for wireless communication. *International Journal of Smart Sensors and Ad Hoc Networks (IJSSAN)*, ISSN No. 2248-9738, 1(3), 87–90.

Sarma, K. K. and Mitra, A. (2009). ANN based Rayleigh Multipath fading channel estimation of a MIMO-OFDM system. *Proceedings of First IEEE Asian Himalayas International Conference on Internet AH-ICI 2009—The Next Generation of Mobile, Wireless and Optical Communications Networks*, Kathmandu, Nepal, 3–5 November, pp. 1–5.

Sarma, K. K. and Mitra, A. (2011a). Estimation of MIMO channels using complex time delay fully recurrent neural network. *IEEE Second National Conference on Emerging Trends and Applications in Computer Science*, Shillong, India, pp. 6–10.

Sarma, K. K. and Mitra, A. (2011b). Estimation of MIMO wireless channels using artificial neural networks. *Cross-Disciplinary Applications of Artificial Intelligence and Pattern Recognition: Advancing Technologies*, Hershey, PA: IGI Global, pp. 509–545.

Sarma, K. K. and Mitra, A. (2012). Modeling MIMO channels using a class of complex recurrent neural network architectures. *Elsevier International Journal of Electronics and Communications*, 66(4), 322–331.

Stallings, W. (2005). *Wireless Communications and Networks*. Upper Saddle River, NJ: Pearson Education.

Suthatharan, S. (2003). *Space-Time Coded MIMO-OFDM Systems for Wireless Communications: Signal Detection and Channel Estimation*, Singapore: National University of Singapore.

Xia, F., Ma, L., Dong, J., and Sun, Y. (2008). Cyber-physical systems: The next computing revolution. *IEEE International Conference on Embedded Software and Systems Symposia*, Sichuan, China, pp. 302–307.

9

Reliability Assessment of Cyber-Physical Systems: A Hardware–Software Interaction Perspective

D. Sinha Roy, Shikhar Verma, Cherukuri Murthy, and D.K. Mohanta

CONTENTS

9.1 Introduction

Cyber-physical systems (CPSs) refer to a category of engineered systems that envisions usage of advanced computation, communication, information, and control technologies for controlling and monitoring of physical world devices for next-generation industrial systems. Processing, analyzing, and storing of the generated voluminous, heterogonous, real-time data from CPSs necessitate a powerful, reliable, secure, and cost-effective information-handling paradigm. Cloud computing with its huge computation and storage capabilities has been hinted as a key component that needs to be integrated with CPSs.

The primary contribution of this chapter is to provide a unified hardware–software reliability model of the CPS taking into account the effect of hardware–software interaction failures in addition to conventional hardware and software failures. The same has been delivered in this chapter by means of an illustrative CPS, namely, the smart grid. For the sake of simplicity, the acronym SCPS has been adopted to refer to smart grid as a CPS.

Power networks have pervaded across all terrains of habitation, and their mere geographical spread is awe inspiring. Thus, to satisfactorily provide services to such a huge and divergent user base, appropriate monitoring and control schemes become a prime necessity. The concept of smart grid has been conceived in recent times, and it has received a lot of hype [1–4]. The envisioned smart electric grid comprises of physical infrastructures such as power generation, transmission, and distribution infrastructures, as well as an expansive information and communication technology (ICT) infrastructure [5]. Risks pertaining to the physical power system components have been well studied in the literature over the past decades. On the contrary, cyber security-related risks pertaining to the smart grid have been investigated only recently [6–9]. Owing to the constant and perpetual demand of electric power, ensuring reliable operation of the smart grid is of paramount importance and has evolved as one of the major research challenges of late.

Reliability modeling of the electric power grid has received ample research interest owing to its critical role in delivering smart grid services [11,12]. Unfortunately, these models are restricted in terms of possible failure modes. In fact, none of them accommodate hardware–software interaction failures in modeling reliability of the power grid, which has been an area of keen research interest recently, partly because of its recent identification and partly because of the consequences of such failures in some of the contemporary safety critical systems.

In recent years, however, there has been a lot of emphasis on identifying, modeling, and quantifying the effect of hardware–software interaction failures on system reliability in myriad cases, including integrated circuit fabrication technologies and jet propulsion systems [13]. On the contrary, traditional methods of analyzing physical infrastructures assume the software subsystem to be in series with the hardware subsystem from a reliability perspective. This assumption imposes a serious limitation for systems, particularly for embedded systems, where hardware and software work in close proximity with each other. In smart grid, numerous embedded devices work in tandem, and thus its hardware and software subsystems bear close correlation for proper functioning. In fact, this interaction between hardware and software subsystems is present throughout the smart grid ecosystem.

The primary contribution of this chapter is to provide a unified hardware–software reliability model of the smart grid physical infrastructure that takes into account the effect of hardware–software interaction failures in addition to conventional hardware and software failures. To this end, a Markov model has been presented that identifies partial degradations and paves the path for identifying hardware–software interaction failures based on historical data. Smart grid is a safety critical system, having very small failure rates. Besides, it is yet to attain widespread deployment status. These two facts combined together lead to the reality that available failure data from the field are scanty. Thus, a Monte Carlo simulation technique has been employed to identify the extent and effect of hardware–software interaction failures.

The remainder of this chapter is organized as follows. Section 9.2 presents a brief overview of the smart grid as a cyber-physical infrastructure, briefing its different modules and failure modes. In Section 9.3, an overview of smart grid's failure modes has been discussed with a special focus on hardware–software interaction failures. A unified Markov model for smart grid failures has also been presented and a detailed methodology for arriving at hardware–software interaction failure has also been corroborated in Section 9.4. Section 9.5 presents the results and analysis with a detailed account for SCPS reliability analysis based on the proposed unified reliability model. Section 9.6 concludes the chapter.

9.2 Smart Grid as a Cyber-Physical System

CPS, a name coined recently, is a boost for next-generation engineered systems that envisions the use of advanced computation, communication, information, and control technologies for controlling and monitoring physical world devices, including systems ranging in variety between industrial automation to aerospace exploration [14]. The evolution of CPSs depends on proper usage and integration of latest ICT. Physical systems are becoming smarter like smart grid development of new applications and services, which are enabled by advanced information technologies. The powerful, reliable, secure, and cost-effective information paradigm is required for processing, analyzing, and storing generated heterogonous real-time data in CPS.

Cloud computing is a futuristic computing paradigm based on massive scale data centers (DC) having huge computation and storage capabilities provided by cloud providers to deliver computing as a service [15]. Therefore, cloud computing should be integrated with CPS.

Recently, there has been a lot of research on integration of cloud computing with CPS applications, like smart grid. Fang et al. [15] have presented a novel smart grid information management paradigm called cloud service-based smart grid information management, which analyzed the benefits and opportunities from the perspective of the smart grid realm. A more refined model of smart grid data management in cloud has been presented in [16] that focus on real-time aspects of processing and accessing smart grid information with a focus on advantages and benefits of cloud computing in smart grid deployment. Therefore, smart grid infrastructure is heavily dependent on its cyberinfrastructures, that is, cloud computing infrastructures. In order to provide reliable, secure, and fault-tolerance controlling and monitoring of smart grid, cloud computing should be reliable and secure.

Therefore, adequate measures should be studied for reliability analysis of such complex systems, which includes both complex power grid with advanced information technology, that is, cloud. Cloud is the key element of such a system because the entire physical system is dependent on cyberinfrastructure for controlling and monitoring.

For functioning of SCPS, that has been referred to as SCPS, both the cloud computing realm and smart grid realm depends on DC as shown in Figure 9.1. Modern day DC host hundreds of thousands of servers networked via hundreds of switches/routers that communicate with each other to coordinate tasks in order to deliver highly available cloud computing services [17]. Numerous software need to be deployed across servers, and many application programs need to run across both the cloud and smart grid domains. Those application programs from both realms send requests to servers to get cloud services. The primary contribution of this chapter is to provide a unified hardware–software reliability model of the smart grid physical infrastructure that takes into account the effect of hardware–software interaction failures in addition to conventional hardware and software failures.

9.3 Brief Account of Failures in the SCPS

According to the National Institute of Standards and Technology, cloud computing is a model for enabling ubiquitous, convenient, on-demand network access to a shared pool of configurable computing resources (e.g., network, servers, storage, application, and services) that can be rapidly provisioned and released with minimal management effort or service provider interactions [18]. So, numerous computing resources are needed to deliver highly available cloud computing services. The computing resources may be hardware such as servers, networks, and storage and also software parts such as application programs or programs across the servers. The computing resources can be divided into two parts, that is, hardware and software.

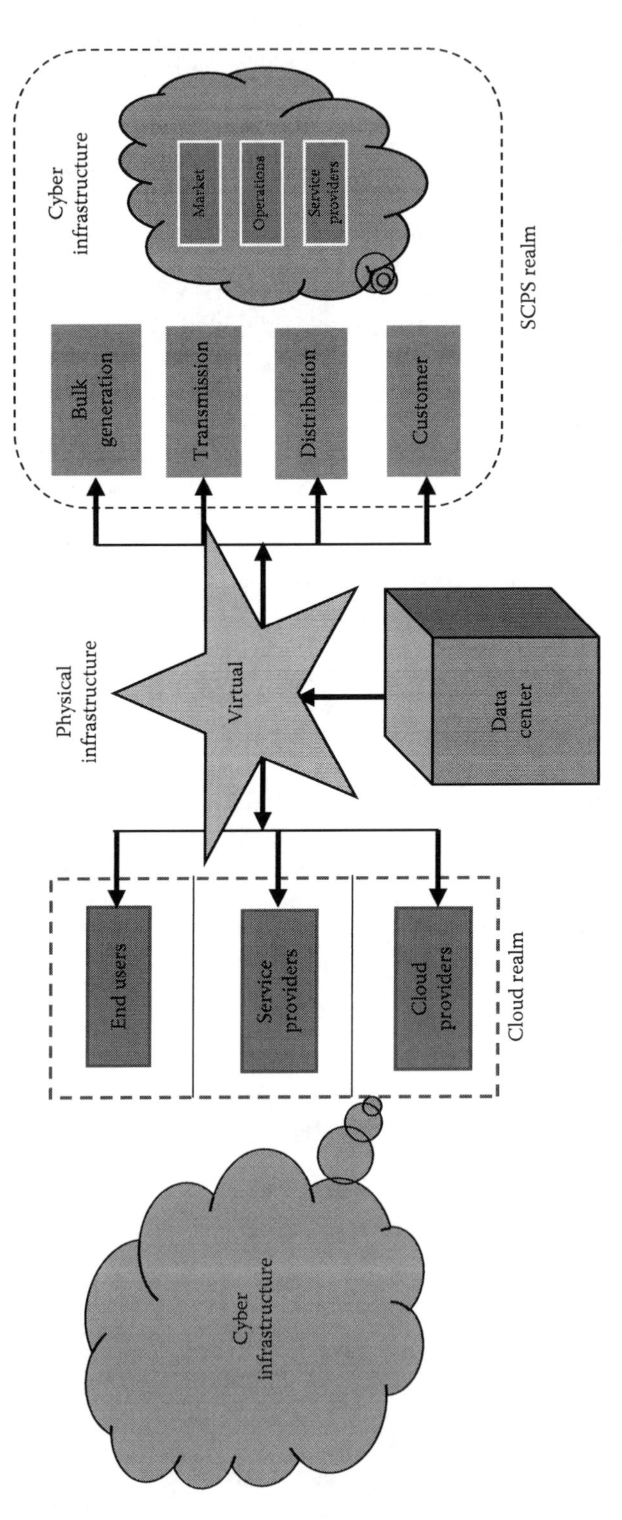

FIGURE 9.1
Architecture of smart grid as a cyber-physical system.

9.3.1 Brief Overview of Hardware Failures in Cloud-Based SCPS (C-SCPS)

Both the cloud domain and smart grid domains interact with DC [19]. The cloud providers in the cloud realm of the SCPS have massive DCs capable of mass scale computation and storage capacities. The modality of service interactions among the smart grid and cloud computing realms occur in terms of exchange of abstracted service requirements. In the cloud computing realm, service providers in fact take the role of an end user [15]. Therefore, each actor in a smart grid domain will act as end user in cloud computing domain and interact with service providers, and service providers interact with cloud providers. All the actors in SCPS are dependent on cloud infrastructure and DCs as key components.

This chapter has focused on DCs as hardware for reliability analysis of SCPS. Today, DCs host hundreds of thousands of servers networked via hundreds of switches/routers that communicate with each other to coordinate tasks in order to deliver highly available cloud computing services [17]. These servers consist of hard disk, memory modules, network cards, and processors, which are capable of failing though carefully designed. The probability of such failures in its lifetime in an industry (typically 3–5 years) can be small. However, the probability of component failures across servers hosted in a DC can be very high [20].

Hardware failure can lead to degradation in the performance of the system because the unavailability of the system to provide the services to the end users can result in economic or any disastrous losses to society. The hardware will fail due to a number of factors like age of servers and number of hard disk. Hard disks are not only the most replaced component, but they are also the most dominant reason behind server failure [20]. Vishwanath and Nagappan [20] has analyzed the server faults and given the data sets of server failures in normalized manner.

9.3.2 Brief Overview on Hardware–Software Interaction Failures

As the demand for complex hardware–software systems is increasing, it is becoming evident that care should be taken to study the interaction among hardware and software failures. A number of studies have revealed ample evidence of such interactions. For instance, Butner and Krishnan Iyer [10] demonstrated that almost 33% of software failures were hardware. This study was conducted on a particular system, namely, Multiple Virtual Storage. Endres [21] established that around one-third of all e-mail outages were caused by failures in server hardware. Huang et al. [22] studied the effect of software operations profile leading to a system encountering the likelihood of hardware–software interactions.

The impact of hardware failure is that it puts an increased burden on software stack by adding complexity for dealing with frequent hardware failures [23]. So, there are different types of software and application programs that are running in cloud and accessing DCs at a fraction of seconds. The industry often has to deal with systems that include sizes, which are in the order of 500K lines of noncommented source code [12].

Therefore, an improved reliability is constructed, which mainly focuses on hardware, software, and hardware–software interaction failures of such complex system of systems.

9.4 Unified Reliability Model For SCPS

To account for the hardware–software interaction failures, a unified hardware–software reliability model is presented, which captures a system-level abstraction of the SCPS to address hardware and software interaction failures. Figure 9.2 depicts the Markov model

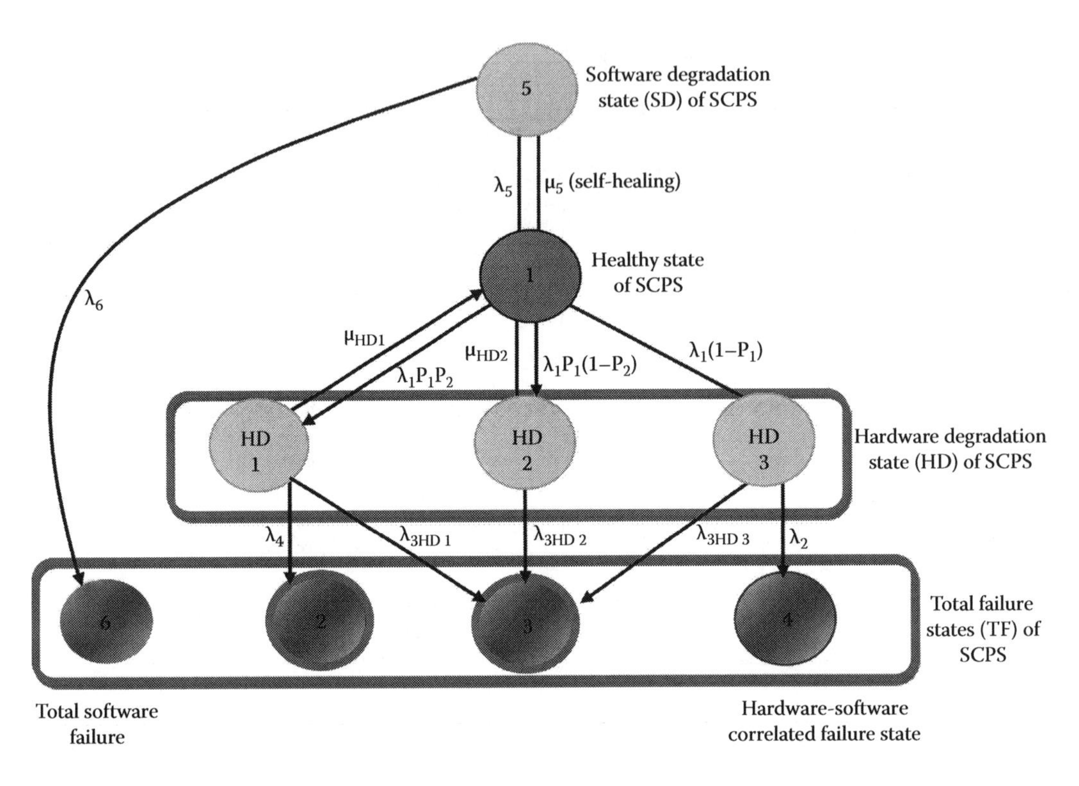

FIGURE 9.2
Markov state-space diagram for phasor measurement unit's unified hardware–software reliability model.

and describes how hardware failures impact software and system reliability. It reveals seven different states in the earlier state transition diagram for estimating hardware, software, and hardware–software interaction failures in SCPS devices. State 1 describes the normal working or the healthy state of SCPS. From this state, on partial hardware degradation (HD), SCPS may transit to a partial degradation state with a transition probability λ_1, which has been depicted by the shaded rectangle in Figure 9.2 as HD. Of course, such partial degradations may be detected but not recovered, or detected then recovered by a software (with repair rate μHD2), or go undetected. These three possibilities have illustrated by states HD1, HD2, and HD3. It has to be noted that a transition from state HD1 to state 1 with μHD1 implies replacement of degraded hardware. Upon any further hardware failures from any of the three HD states, the SCPS system transits to a total SCPS failure state, as depicted in Figure 9.2 as total failure (TF). It has to be noted in Figure 9.2 that transition from state HD3 to state 4 represents the fraction of hardware–software interaction failures. State HD3, undetected HD, does not give any signal to SCPS system regarding the SCPS's health status. Thus, the operation of the software uses the normal usage profile. However, since that time the hardware is operating in degraded mode, hence it might not be able to provide requisite hardware resources for SCPS software. Therefore, SCPS software might not give desired output. This path in the state transition diagram from state 1 to HD3 to state 4 represents a hardware–software interaction failure. State 5 describes the partial degradation state of software as software degradation (SD). Bugs in the software might increase as it is in partial degradation state, so it may lead to TF, that is, TF state.

9.4.1 Analysis of the Proposed Unified Reliability Model for C-SCPS

The transition states of the model presented in Figure 9.2 are assumed to be a Markov process. The fundamental principle underlying a Markov process is independent of the future state, irrespective of the past state, and the present state is known. Reliability of hardware, software, and hardware–software interaction failures has also been assumed to be independent of each other. In other words, transition paths 1 to HD1/HD2–2,3 of TF; 1 to HD3 to 4 of TF; and 1 to SD to TF have been assumed to be independent of each other. Based on the assumptions, the reliability of C-SCPS can be derived using Equation 9.1.

$$R_{CSG}(t) = R_{hw}(t)R_{sw}(t)R_{hw/sw}(t) \tag{9.1}$$

Reliability of hardware can be defined by the following equation, which is given by the Weibull model:

$$R_{hw}(t) = e^{-\lambda t^{\beta}} \tag{9.2}$$

where

λ is the failure rate of hardware components
β is the shape parameter or slope of Weibull distribution function

Reliability of the software can be estimated using the nonhomogeneous Poisson process model, which is defined by the equation

$$\lambda_{sw} = ab\exp(-bt) \tag{9.3}$$

where

t is the testing time
parameters a and b are number of faults to be detected and fault detection rate, respectively

Assuming that the probability of SCPS being in state i at time t is Pi(t), differential equations can be defined from the state transition diagram as shown in Figure 9.2 for finding out the state probabilities based on the property of the Markov process as follows:

$$\begin{aligned}
Q_0'(t) &= -\lambda_1 Q_0(t) - \lambda_5 Q_0(t) + \mu_{HD1}(t)Q_{HD1}(t) + \mu_{hd2}(t)Q_{HD2}(t) + \mu_5 Q_5(t) \\
Q_{HD1}'(t) &= \lambda_1 p_1 p_2 Q_0(t) - (\mu_{HD1} + \lambda_{3HD1} + \lambda_4)Q_{HD1}(t) \\
Q_{HD2}'(t) &= \lambda_1 p_1 (1 - p_2) Q_0(t) - (\mu_{HD2} + \lambda_{3HD2})Q_{HD2}(t) \\
Q_{HD3}'(t) &= \lambda_1 (1 - P_1) Q_0(t) - (\lambda_4 + \lambda_{3HD3})Q_{HD3}(t) \\
Q_2'(t) &= \lambda_4 Q_{HD3}(t) \\
Q_4'(t) &= \lambda_2 Q_{HD3}(t) \\
Q_3'(t) &= \lambda_{3HD1} Q_{HD1}(t) + \lambda_{3HD2} Q_{3HD2}(t) + \lambda_{3HD3} Q_{3HD3}(t) \\
Q_5'(t) &= \lambda_5 p_1 Q_0(t) - \mu_5 Q_5(t) \\
Q_6(t) &= \lambda_6 Q_5(t)
\end{aligned} \tag{9.4}$$

In this set of equations, states 2, 3, 4, and 6 are failure states, and states HD1, HD2, and HD3 are SCPS's partial degradation states, whereas state 1 is the normal SCPS working state. Assuming that SCPS is initially in working state, that is, $Q_0(0) = 1$, solutions of the equations mentioned earlier can be obtained [12]. The probability of the number of permanent hardware–software interaction failures at time t can be represented by

$$R_{HW/SW}(t) = Q_0(t) + Q_{1a}(t) + Q_{1b}(t) + Q_{1c}(t) \tag{9.5}$$

Therefore, combined reliability in Equation 9.1 can be calculated as

$$\begin{aligned} R_{system}(t) &= R_{hw}(t) R_{sw}(t) R_{hw/sw}(t) \\ &= e^{-\lambda t^{\beta}} e^{-(m(t+T)-m(T))} * \left(Q_0(t) + Q_{1a}(t) + Q_{1b}(t) + Q_{1c}(t)\right) \end{aligned} \tag{9.6}$$

Equation 9.5 reveals the extent of hardware–software interaction failures. Thus, from the SCPS failure data, it is possible to account for the hardware–software interaction failures. From [20], this chapter has concluded the server failure data in cloud computing.

9.5 Results and Analysis

This section provides a detailed account of the results obtained for SCPS reliability analysis based on the proposed unified reliability model. In the subsequent sections, the methodologies for fitting the SCPS failure data to the proposed reliability model and further analysis are presented.

9.5.1 Fitting the Cloud Failure Data to the Unified Reliability Model

The estimated number of SCPS hardware failures each year obtained from [20] during the mission time of 9 years is shown in Table 9.1. Column 1 shows the number of years. For each year, the exposure time of a system running an SCPS software (column 2) is noted. Column 4 shows the total hardware failures across all the systems in each year as obtained from [20]. Table 9.1 does not include hardware–software failures because the hardware failure is more observable than related software failures.

TABLE 9.1

Failure Data of Cloud Servers

Years	Number of Systems	Exposure Time of Systems	Hardware Failure Over All Systems
9	36,000	36,000	4,554
8	32,000	32,000	10,224
7	19,000	19,000	8,094
6	5,500	5,500	3,124
5	5,500	5,500	3,905
4	1,000	1,000	852
3	500	500	497
2	300	300	341
1	200	200	256
Total			**31,837**

Source: Obtained from Vishwanath, K.V. and Nagappan, N., Characterizing cloud computing hardware reliability, *Proceedings of the First ACM symposium on Cloud computing*, ACM, New York, pp. 193–204, 2010.

Therefore, only the hardware failure data are used. To calculate hardware–software failures, the data of hardware failures are analyzed, and thereafter, a value of F is assumed, where F denotes the fraction of observed hardware failures that signify hardware–software interaction failures. To fit the model, the SCPS systems were divided into six groups. In this case, hardware failures are assumed to be undetected, which leads to total hardware failures or hardware-related software failures from the model shown in Figure 9.2. The undetected hardware failures are considered as they might lead to hardware-related software failures. Therefore, except λ_1 and λ_2, all failure rates are considered to be zero. Therefore, $R_{HW/SW}$ (t) in Equation 9.5 reduces to

$$R_{HW/SW}(t) = \frac{\lambda_2 e^{-\lambda_1 t} - \lambda_1 e^{-\lambda_2 t}}{\lambda_2 - \lambda_1} \tag{9.7}$$

Equation 9.7 gives the hardware–software interaction failures from the proposed model. The hardware–software interaction failures are modeled for a SCPS system as a renewal process with an interarrival distribution [12]. Renewal process means that the model of an item in a continuous operation is replaced at each failure in a negligible amount of time, by a new, statistically identical item. Assuming T_i and X_i to be the aggregate exposure time and number of hardware–software interaction failures of each system group that begins to run software at the ith year, respectively, the expected value of X_i can be found out as follows:

$$X_i = \frac{T_i}{\mu} \tag{9.8}$$

where $\mu = 1/\lambda_1 + 1/\lambda_2$. The estimated value of (λ_1, λ_2) can be calculated by a nonlinear sum of square estimates (SSE)

$$SSE(\lambda_1, \lambda_2) = \sum_{i=1}^{8} \left[\frac{X_i - T_i}{((1/\lambda_1) + (1/\lambda_2))} \right]^2 \tag{9.9}$$

Since (λ_1, λ_2) cannot be obtained uniquely in the estimation process, reparameterization is done for S (λ_1, λ_2) to (G, λ_2), where $G = \lambda_2/\lambda_1$. The varying value of G is assumed to find λ_2 as λ_1 can be obtained by minimizing SSE and thereafter putting the value of G. Therefore, the least-square estimates of λ_1 is

$$\lambda_1 = \left(1 + \frac{1}{G}\right) \frac{\left(\sum_{i=1}^{8} X_i T_i\right)}{\sum_{i=1}^{8} T^2 i} \tag{9.10}$$

Thus, $\lambda_2 = G\,\lambda_1$ can be obtained.

The hardware part of failure is more apparent than related software failures. Therefore, only hardware failure data are used. The flow chart depicting the steps in fitting the phasor measurement unit failure data is shown in Figure 9.3, where N, Nfailure, and Nsystem denote the number of observations in years, number of failures per year,

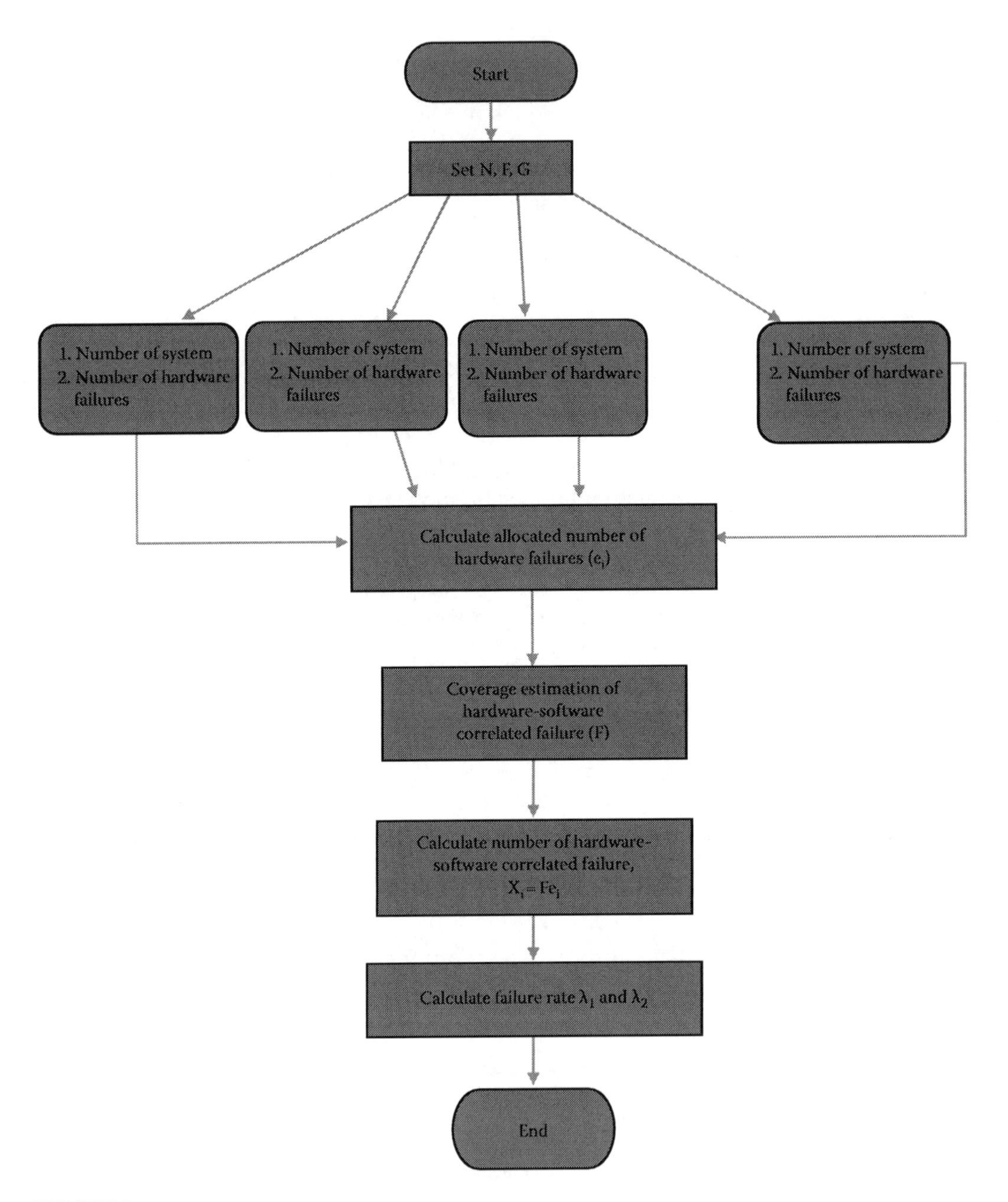

FIGURE 9.3
Flowchart depicting the steps in fitting the phasor measurement unit failure data.

and number of systems deployed at each year, respectively. To calculate the extent of hardware–software interaction failures, the hardware failure data are analyzed, and a fraction F of hardware failures is assumed with respect to hardware–software interaction failures. This may vary. To fit the model, the system is divided into different groups. In this case, those hardware failures were assumed to be undetected, which may lead to total hardware failures or hardware-related software failures. This has been depicted in Figure 9.2 by the state transitions: HD3 to 4 of TF. The undetected hardware failures have been focused on as they might lead to hardware-related software failures. Therefore, except λ_1 and λ_2, all failure rates are considered zero. Thereafter, the obtained number of hardware failures of each system set has been calculated by proportionally allocating total number of hardware failures using the ratio obtained by dividing aggregate exposure time by total exposure time. Out of these hardware failures, there may be a fraction F of hardware–software failures. Therefore, $X_i = Fe_i$ where e_i are the values of aggregate hardware failures. Since the values of F and G are unknown, the various values of F and G have been assumed in this chapter. Thereafter, the value of λ_1 for different values of F and G and $\lambda_2 = \lambda_1$ G can be calculated. The reliability of hardware and software can be obtained because of the interactions with the system. To calculate λ_1, the values of T_i and X_i are obtained first. From column 9, value of T_i beginning with the ith year is found out. Columns 1 and 2 of Table 9.2 show the system groups and number of systems in each group. System group is decided according to the number of new systems added each year. Therefore, from column 2 of Table 9.1, column 2 of Table 9.2 can be obtained by subtracting the number of systems at the ith year from that of (i − 1)th year, where i ≥ 0. Columns 3 through 8 of Table 9.2 show the partition of exposure time associated within each group. Column 9 in Table 9.2 is the sum of columns 3 through 8 and represents the aggregate exposure time of each system set. Column 10 of Table 9.2 is the derived number of hardware failures of each system set and can be calculated by proportionally allocating total number of hardware failures from Table 9.1 using the ratio obtained by dividing aggregate exposure time (column 9 of Table 9.2) by total exposure time. Therefore, 10 number of hardware failures were found. Out of these 10 hardware failures, there may be F fraction of hardware–software failures. Therefore, $X_i = Fe_i$, where ei are the values in column 10 of Table 9.2 of each system group. As the values of F and G are unknown, so in this chapter, a number of possible values for F and G have been assumed. Thereafter, the value of λ_1 is calculated from Equation 9.10 for different values of F and G and from Equation 9.9 by the relation $\lambda_2 = \lambda_1 G$. Table 9.3 shows the values of λ_1 and λ_2 along with the assumed values of F and G. The reliability of SCPS because of hardware–software interaction failure can be calculated using Equation 9.7 by putting different values of λ_1 and λ_2 obtained in Table 9.3.

TABLE 9.2

Hardware Failures Pertaining to Each System Groups

System Groups	No. of Systems	Number of Years									Aggregate Exposure Time T_i	Allocated Number of Hardware Failures
		1	2	3	4	5	6	7	8	9		
1	200	200	200	200	200	200	200	200	200	200	1,800	600
2	100		100	100	100	100	100	100	100	100	800	267
3	200			200	200	200	200	200	200	200	1,400	467
4	500				500	500	500	500	500	500	3,000	1,000
5	4,500					4,500	4,500	4,500	4,500	4,500	18,000	6,001
6	0						0	0	0	0	0	0
7	13,500							13,500	13,500	13,500	40,500	13,502
8	13,000								13,000	13,000	26,000	8,668
9	4,000									4,000	4,000	1,333
Total											**95,500**	**31,837**

TABLE 9.3

Least Square Estimate Value of (λ_1,λ_2)

F	G	λ_1	λ_2
0.10	2	3.333×10^{-2}	6.667×10^{-2}
0.15	2	5.0007×10^{-2}	1.00015×10^{-1}

9.6 Conclusion

This chapter presents a novel reliability model for CPSs that takes into account the effect of hardware–software interaction failures in addition to pure hardware and pure software failures in a massive scale, distributed, complex CPS. To demonstrate the intricacies of the proposed model, this chapter delves into smart grid as a case study of CPS, referred to as the SCPS. The model presented herein works well for any cyber-physical system, though the results presented are restricted to a smart grid system.

References

1. Li, F., Qiao, W., Sun, H., Wan, H., Wang, J., Xia, Y., and Zhang, P. (2010). Smart transmission grid: Vision and framework. *IEEE Transactions on Smart Grid*, *1*(2), 168–177.
2. Xie, K., Liu, Y. Q., Zhu, Z. Z., and YU, E. K. (2008). The vision of future smart grid. *Electric Power*, *41*(6), 19–22.
3. Farhangi, H. (2010). The path of the smart grid. *IEEE Power and Energy Magazine*, *8*(1), 18–28.
4. Amin, S. M. and Wollenberg, B. F. (2005). Toward a smart grid: Power delivery for the 21st century. *IEEE Power and Energy Magazine*, *3*(5), 34–41.
5. Karnouskos, S. (2011). Cyber-physical systems in the smart grid. In *2011 Ninth IEEE International Conference on Industrial Informatics (INDIN)*, July 26–29 (pp. 20–23). Caparica, Lisbon: IEEE.
6. Ericsson, G. N. (2010). Cyber security and power system communication—Essential parts of a smart grid infrastructure. *IEEE Transactions on Power Delivery*, *25*(3), 1501–1507.
7. Hadley, M., Lu, N., and Deborah, A. (2010). Smart-grid security issues. *IEEE Security and Privacy*, *8*(1), 81–85.
8. Sridhar, S., Hahn, A., and Govindarasu, M. (2012). Cyber–physical system security for the electric power grid. *Proceedings of the IEEE*, *100*(1), 210–224.
9. Dagle, J. E. (2012). Cyber-physical system security of smart grids. In *Innovative Smart Grid Technologies (ISGT)*, January 16–20 (pp. 1–2). Washington, DC: IEEE.
10. Butner, S. E. and Krishnan Iyer, R. (1980). *A Statistical Study of Reliability and System Load at SLAC*. Stanford, CA: Center for Reliable Computing, Computer Systems Laboratory, Stanford University.
11. Wang, Y., Li, W., and Lu, J. (2010). Reliability analysis of wide-area measurement system. *IEEE Transactions on Power Delivery*, *25*(3), 1483–1491.
12. Teng, X., Pham, H., and Jeske, D. R. (2006). Reliability modeling of hardware and software interactions, and its applications. *IEEE Transactions on Reliability*, *55*(4), 571–577.
13. Iyer, R. K. and Velardi, P. (1985). Hardware-related software errors: Measurement and analysis. *IEEE Transactions on Software Engineering*, *2*, 223–231.

14. Al-Hammouri, A. T., Branicky, M. S., and Liberatore, V. (2008). Co-simulation tools for networked control systems. *Hybrid Systems: Computation and Control.* Berlin, Germany: Springer, pp. 16–29.
15. Fang, X., Misra, S., Xue, G., and Yang, D. (2012). Managing smart grid information in the cloud: Opportunities, model, and applications. *Network, IEEE, 26*(4), 32–38.
16. Rusitschka, S., Eger, K., and Gerdes, C. (2010). Smart grid data cloud: A model for utilizing cloud computing in the smart grid domain. *2010 First IEEE International Conference on Smart Grid Communications (SmartGridComm)*, October 4–6 (pp. 483–488). Gaithersburg, MD: IEEE.
17. Barroso, L. A., Dean, J., and Holzle, U. (2003). Web search for a planet: The Google cluster architecture. *IEEE Micro, 23*(2), 22–28.
18. Mell, P. and Grance, T. (2009). The NIST definition of cloud computing. *National Institute of Standards and Technology, 53*(6), 50.
19. Mohsenian-Rad, A.-H. and Leon-Garcia, A. (2010). Coordination of cloud computing and smart power grids. *2010 First IEEE International Conference on Smart Grid Communications (SmartGridComm)*. Gaithersburg, MD: IEEE.
20. Vishwanath, K. V. and Nagappan, N. (2010). Characterizing cloud computing hardware reliability. *Proceedings of the First ACM Symposium on Cloud computing* (pp. 193–204). New York: ACM.
21. Endres, A. (April 1975). An analysis of errors and their causes in system programs. *ACM Sigplan Notices, 10*(6), 327–336.
22. Huang, B. et al. (2011). Hardware error likelihood induced by the operation of software. *IEEE Transactions on Reliability, 60*(3), 622–639.
23. Patterson, D. et al. (2002). Recovery-oriented computing (ROC): Motivation, definition, techniques, and case studies. Technical Report UCB//CSD-02-1175, UC Berkeley Computer Science.

Section III

Robotics and Smart Systems in Cyber Context

10

Applications of Robotics in Cyber-Physical Systems

Chitresh Bhargava

CONTENTS

10.1 Introduction

In the past few years, humanity experienced an incredible growth in the robotics area: one in exploration and the other by taking the vision of the common person to its surely limitless and myriad outlook. Various fields of robotics have presently grown into *standard* as regards the workforce, and while the manuscript of this section of the book is being processed, the department is close to the edge of a fresh access to the sectors of development. In this chapter, we discuss the basic concept of robotics and the rules that we have to keep in mind while working on robotic systems in cyber-physical system (CPS), with a lot of data that guide everyone about human and robot interaction. To represent the state of wisdom of a branch and a huge range of graduates who analyze this text, we sense the requirement of a chapter that introduces a kind of extra conceptual information on robotics architecture and designing knowledge and robotics fields and their applications in the CPS. This chapter is our positive statement that sets a positive thinking about the trends and researching concepts of robotic system, and set some free time in the distant future. The topics of this chapter have been selected from the author's perspective in order that they will stay elementary over a period of several years in the light of a various advanced high-tech modernization and improved visions in robotics. All the sections that are included in the chapter are discussed in brief as follows:

- The first section of this chapter is the basic concepts in robotics. Most of the books, magazines, and articles on robotics begin with a dictionary definition of a robot. Fundamentally, it is a good option to define used terms at the start and be specific as to what we are discussing about. Unfortunately, there is no standard definition of a robot, and many popular dictionaries contain no definition at all. The most common definition says, "a robot is a machine that is capable of being reprogrammed." The fact that a robot can be reprogrammed is important: it is a characteristic of robots. But, of course, this given definition is crude. Fortunately, we present a polished description of a true robot definition from this chapter section. Similarly, if you are curious about what the early robot and robotic design ever created was like, when it was created, who created it, and what it performs, the answer to all the queries is: history of robotics.
- The second section introduces the fundamental guidelines of robotics, which are proposed as an elementary structure to locate the action of robot layout having a strength of freedom. The most acknowledged series of law is that scriptural law by Isaac in 1942, but the Engineering and Physical Sciences Research Council (EPSRC) and Arts and Humanities Research Council (AHRC) have planned more series of law and few outstanding information notes in 2013.
- The third section introduces the miscellaneous concepts to robotics architecture and designing knowledge. It presents a brief historical advancement in the field of robot construction and the skills required to design a robot. It also introduces the basic organization of a cyber-physical robotic system that are utilized to construct, study, and regulate different field architectures. All the elementary building blocks are enclosed in this section: power supply, programming intelligence, control system (kinematics and dynamics), mechanisms and actuation, and sensing and estimation techniques for assignment preparation and knowledge.

- The final section defines the various kinds of the latest technological trends in robotics section in terms of CPS, like factory automation, aerial science, space science, agriculture, medical science, infrastructure for smart grid and utilities, and smart building and infrastructure.

10.2 Basic Concepts in Robotics

10.2.1 Robots: Everywhere around Us

Robotics is a course that fascinates many young talents, generally due to the encyclopedic representation of robots in many sci-fi movies like the Transformers, Start Wars, Avatar, Robocop, and Wall-E, which have publicized these concepts faster than anything else. These films are magnificent works of the imagination and present us with an appealing outlook to wonder about. Nevertheless, the reality is a lot different but is equally as appealing as the ones imagined. If you glimpse around, you will identify robots are all around; they appear in a category of various configurations and diameters, created to perform a remarkable spectrum of tasks. Robots are found almost everywhere such as in space and on the moon, in oceans and hospitals, and in factories making goods, devices, and merchandises, saving time and lives. Presently robots have a huge impact on many conditions of modern life from industrial production, packaging, and assembly to health management, travel in space and deep sea, and weaponry to transplantation. The vision to design gadgets that are proficient and very smart is part of humankind's inquisitive thinking from the beginning of time. This thought is now becoming part of one world's striking reality.

10.2.2 True Robot Definition

One of the fundamental questions most of the people want to know about robotics is, "What is a robot?" This is followed instantly by the another question "What can a robot do?" The well-known picture of a robot is usually related to a critter having structures or actions that look or are like those of Homo sapiens. The Homo sapiens-like machines in most conditions have arms, legs, and a brain that can understand, and occasionally they even exhibit some feelings. The researcher and inventor who operate with designs in robotics, the reality a functioning robot is different. The fundamental concept of a realistic and efficient robot is to adopt the character of living souls in monotonous, troubling and hazardous, or exhausting assignments. Because the most monotonous and exhausting task found on the real planet today is the mass production of gadgets in a manufacturing line, the first concept of realistic robotics was related to the manufacturing industry. The impressive situation to acknowledge is that a realistic robot on manufacturing and production line has a presence that, in nearly all conditions, doesn't duplicate a human being in any manner. These equipments and gadgets have many arms that are outlined in suitable structures and diameters for particular assignments. A robot or an automation machine in reality is not as quick as an individual in the majority of the operations; it has, nevertheless, retained its pace over a very long duration. Accordingly, the work rate and capacity rise if the total amount of items to be created is huge.

Additionally, the understanding of the agility of today's leading robot is moderately close to human brilliance. Thus, the understanding of robot without an actual understanding of its advantages will be destructive and is not recommended. So, over the same inquiry, what is a robot? According the International Standard of Organization (ISO), a robot is characterized as a programmable device of multioption structure to carry components or particular devices through changeable programs for the accomplishment of a range of tasks. But a true robot is "an independent arrangement that stands in the real world, can be aware of its surroundings, and can behave in a certain way to carry out some objectives." This may sound like a pretty vast characterization, but piece by piece, each sector is valuable and essential. Let's undergo step by step to identify why:

"*A robot is an independent arrangement ...*": As an independent arrangement, it means that a robot operates on the actions of its judgments and is not regulated by a living person. There are obviously a lot of illustrations of gadgets that are not independent arrangements but are alternately regulated by a living person. They are said to be teleoperation means to easily manage a gadget or an arrangement over a range. These gadgets, on the other hand, are not real robots. Real robots operate independently. They could be ready to take in data and guidance from a human being but are not entirely supervised by them.

"A robot is an independent arrangement *that stands in the real world ...*": Being able to stand in the real world means that the identical environment that human beings, creatures, gadgets, plants, and a lot of other stuffs exist in is a basic characteristic of robots. Possession of understanding with that natural world and its die-hard environmental rules and threats is what creates robotics a real dare. Robotics that continues to use desktops are simulation software; they do not have to deal with the real characteristics of the natural environment because simulation software is not as complicated as the natural world. As a result, there is a set of such gadgets in the computer world; a real robot stands in the natural world.

"A robot is an independent arrangement that stands in the real world, *can be aware of its surroundings ...*": Becoming aware of the surroundings means the robot has a tool or some ways of observing (tools like the ability to perceive sound, to see objects, to feel physical contact, and to sense breath) and discovering facts or changes in circumstances and providing a parallel output. An artificial robot, in comparison, can simply be given the data or intelligence, as if it is a supernatural power. A genuine robot can understand its universe only through its sensing devices, just as human beings and animals do. Thus, if an arrangement does not understand the marvelously given data, we do not believe it is a genuine robot, because it cannot acknowledge what goes in the vicinity.

"A robot is an independent arrangement that stands in the real world, can be aware of its surroundings, and *can behave in a certain way ...*": The step in recognizing logical intake and carrying out what is requested is an essential component of being a robot. A gadget that does not perform is not a true robot. As we will identify, operation in nature happens in a very miscellaneous fashion, and that is the specific logic why the territory of robotics is so vast.

"A robot is an independent arrangement that stands in the real world, can be aware of its surroundings, and can behave in a certain way to *carry out some objectives*": Without delay, we ultimately take part of the brilliance or somewhat the efficiency of a robot. An arrangement or gadget, which although stands in the natural world and is identified, that performs aimlessly or meaninglessly is not at all a true robot, because it does not utilize the identified data and is not capable of performing in order to carry out something valuable for others. Thus, we want a real robot to contain one or more objectives and to perform so as to carry out the objectives. Objectives can be much easier and understandable, for example, "execute whatever it leads to keep your partner secure."

A closer examination of this definition indicates that robotics is the study of learning robots (machines and gadgets), which means it is the analysis and learning of the independent, purposeful, and calculated observations and their performances in the natural world.

Note that robotics is a quickly developing area whose description has been expanded over time, along with the area itself.

10.2.3 History of Robotics

Robots have their backgrounds as much as old fictitious stories and traditions are concerned. New approaches were established and developed when mechanical transformation permitted the usage of more additional machines like tools. The past events of robotics are briefly explained as follows, from the ancient moment to the existing time, century after century.

1495 AD (*mechanical knight*): The earliest well-known humanoid robot was created and outlined by Leonardo Da Vinci in 1495 (Figure 10.1). The robot machine design is a knight in a representation of an Italian–German armor. The robot machine design consisted of two functioning architectures. There was a four-factor arrangement that regulates the arms, elbows, wrists, and shoulders. Along with it, there was a tri-factor arrangement that regulates the ankles, knees, and hips [1]. Leonardo had constructed other structures of robotics such as a walking automatic lion and a spring mechanized car, which is treated to be the first programmable computer.

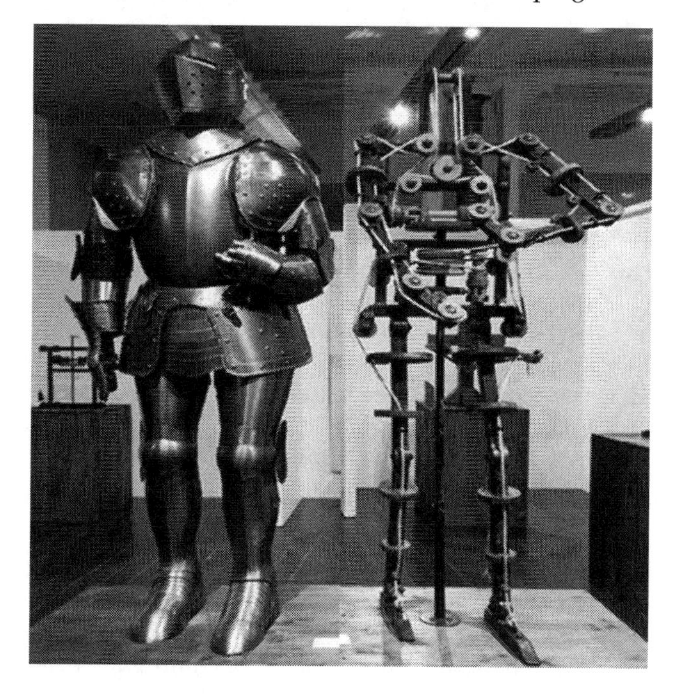

FIGURE 10.1
Leonardo da Vinci humanoid robot. (From Mechanical man, Retrieved November 8, 2014, from http://www.mikekemble.com/misc/leonardo.html.)

1739 AD (Digesting Duck): The Digesting Duck was a mechanical machine in the shape of a duck, which was constructed by Jacques De Vaucanson in 1739. The automatic machine like duck comes out to have the skill to pick seeds of grain and to digest and eject them. Whereas the duck in reality did not have the skill to do this, the edible material was accumulated in one storage tank, and the pre-gather excrement was originated from a subordinate tank, so that no real metabolism came off. Vaucanson believed that an exactly metabolized mechanism could one day be invented. A clone of Vaucanson's automated duck, constructed by Ian Huynh, is a good example of the plentiful works of skill displayed in the museum of automatons in France.

1898 AD (Teleautomaton): Nikola Tesla was both an experimental expert and an ionic idealist, and possibly one of the extreme intelligences in living soul history. In 1898, Nikola Tesla copyrighted for an electric power submarine, and that he named as Teleautomaton Boat. This underwater robot would acquire current that was being transmitted to it by a transceiver. Electricity could also be gathered in a container as a source of power and the electric submarine conceivably regulated by a remote control. This designates him the ancestor of all remotely driven and independent robots.

1920 AD (Rossum's Universal Robots): As long as the idea of artificial soul meets its ancestors altogether back to antiques, the primary value of the name "robot" takes place in a drama written by Karel Capek entitled Rossum's Universal Robots. The terminology is derived out of the Czech word "robota," which signifies required employ. Rossum's Universal Robots was scriptural in 1920 and Capek was released in 1921. The released transcription from Czech into English gets out 2 years later.

1939 AD (Elektro Humanoid Robot and Sparko Robotic Dog): Elektro is a humanlike machine designed and named by Westinghouse Electric Corporation. It can act in answer to sound instructions and has a dictionary of around 700 words. Elektro was displayed at the New York World's carnival in 1939. It consists of a sound machine to replicate talk. Ultimately paired by Sparko, a shouting mechanical dog, Elektro welcomed visitants to the Westinghouse display at the carnival with pronouncement such as "my mind is bigger than yours."

1940 AD (terminologies for robot and robotics introduced to the universe): It was Isaac Asimov who is mostly credited for delivering the terminology "robot" and "robotics" into the widespread scholarly planet, launching his early robot tale in 1940 and universally popularizing his collection of short fables entitled *I, Robot* in 1950. In explaining the elemental composition of a robot, Asimov authorized his readers to adopt the theory of machine intelligence in place of the previous scary one.

1942 AD (ENIAC, the first electronic computer): In 1942, John Mauchly (researcher) and Presper Eckert (engineer) were together at the University of Pennsylvania and designed the first basic digital computer, ENIAC. They moved on to construct BINAC and the UNIVAC 1 computer, and they initiated one of the first computer corporations, the Eckert Mauchly Computer Corporation, in 1947. They invented some basic computer ideas such as the programming languages and stored program.

1946 AD (magnetic controller): George Devol, the father of the robotic system, appealed for a copyright on a magnetic recording process for controlling devices and a digital playback gadget for appliances in 1946. He certified this digital magnetic recording system to Remington and became the administrator of their magnetic organization.

The aim was to promote his gadget for professional information operations, but it verified to be too lagging for professional information.

1954 AD (first robotic arm): George Devol appealed for a copyright on Programmed Article Transfer in 1954, and the aim of this creation was to carry out periodic assignments with better accuracy and output than an individual employee. The copyright was involved with the self-regulating procedure of the instrument, add-on with supervision, and mechanized command machinery. It familiarizes the idea of the worldwide automation, and the word "Unimate" was created. This was the first copyright for an electronically regulated programmable robotic arm, and it contributes to the infrastructure of the robotics manufacturing.

1961 AD (setup of the first industrial robot): Engelberger and Devol established the first robotics corporation, Unimation Inc., in 1956. They at the beginning originated element-handling robots and machines for welding and other services. They set up the primary industrial robot on General Motors assembly division in 1961. This robot was programmed to pick up red hot samples of alloy from a casting section and bundle them.

1969 AD (Shakey, the mobile robot): The first mobile robot named Shakey, which possibly was the father of all independent field robots, was established at the Stanford Research Institute in 1969. Shakey had a photographic equipment, a range discoverer, and collide sensors and was linked to the electronic machine via radio and video connections.

1969 AD (Stanford Arm): Victor Scheinman established the first outstanding electrically mechanized computer-regulated manufacturing robot as his postgraduate research design in 1969. In comparison to bulky, hydromechanical, task-specific handling gadget, his Stanford Arm was weightless and electromechanical and able to perform multiple tasks and understand unplanned movements in place of the planned ones. Scheinman demonstrates that it was feasible to form a physical structure that could be as adjustable as it was autonomous.

1978 AD (Programmable Universal Machine for Assembly [PUMA] robot): A PUMA robot was brought in by unimation in 1978. With the presentation of its flexible and skillful PUMA robot, the unimation company became one of the well-known global producers of mechanical robots and was captured by Westinghouse Electric Company in 1983. The Staubli Association then bought unimation from Westinghouse in 1988 and went ahead to broaden the industrial division of manufacturing robots.

1980–1990 (advancement of low-cost electronic products): Robots started communicating with the surroundings utilizing pressure sensors, tactile sensors, perception sensors, etc.

1990–2000 (advancement of sensors, compact and lighter processors, outstanding data processing algorithms): Intelligence robots were introduced.

2000 up to the present (curiosity in biological arrangement and a continuous deviation from manufacturing utilizations to public beneficent utilizations): Some of the currently used robots are nanorobots, biomedical robots, service robots, space robots, etc.

Those ancient and innovative introductions are remarkable and extraordinary. Following the initial generation of robotics, from 1495 to the existing year of robotics, a collection of advanced robotic structures and their gestures, movements, and command methods were created and widely organized, and the speed of development was almost epidemic.

10.3 Robotics Fundamental Principles

10.3.1 Laws of Robots by Isaac Asimov

The humankind surrounded by intelligent robot systems raises a countless of thoughtful queries like: Can we count on them to carry out the correct thing? Can we set up them to be virtuous? Once the human figure out that robot system criteria have to execute with how smart machine system, instead than living souls should function, they frequently keep up that Isaac Asimov has earlier delivered us a standard series of laws for such mechanism. The laws were presented in his 1942 interesting tale "Runaround," despite being indicated in some former tales.

The three laws are as follows:

1. A robot may not injure a human being or, through inaction, allow a human being to come to harm.
2. A robot must obey the orders given to it by human beings, except where such orders would conflict with the first law.
3. A robot must protect its own existence as long as such protection does not conflict with the first and second laws [2].

These three stated laws are very valuable for the continuation of robotic systems and even in manufacturing where machines are functioning from the security perspective. The initial rules have been modified and particularized by Asimov and other writers. Asimov himself created minor alterations in the early three in numerous short tales to further progress on how machines would communicate with living souls. In a more recent tale where machines had taken the burden for management of the entire universe and people development, Asimov also joined a new or say *zeroth* law—a robot may not harm the humanity as a whole [3].

Numerous robots established the zeroth rule as a sensible expansion of the first rule, as robots repeatedly encounter difficulties, which as an outcome will damage moderately only a few living souls, consecutively to prevent the damage of more living souls.

10.3.2 Principles of Robotics by EPSRC and AHRC

Robots have gone out of the development center and are presently at work all around the universe, in apartments, in production, and in space. We anticipate robots to operate in short, medium, and long durations to power our activities in our houses and our participation and knowledge in organizations, our democracy, and worldwide protection.

Withal, the facts of the existence of robotics are quite comparable to the people, where movies, novels, and publishing pictures of robots have influenced. One of the objectives of this section is to analyze what moves should be preferred to assure that robot development interconnects with the people to guarantee this scientific knowledge is mixed into our culture to the highest profit of all of its users. As with all high-tech modifications, we require seeking ways to guarantee that robots are made known from the start in an approach that is hopeful to involve user faith and expectation and energetically block off any unexpected results.

Similar with their reputation, it is impractical to discuss the rules of robotics without seeing Isaac Asimov legendary rules of robotics (Section 10.3.1 for further discussion). Despite the fact, they specify a valuable withdrawal aspect for considering Asimov's laws for imaginary tools. They were never drafted to be applied in the physical world because they are not clearly achievable in the system. Isaac Asimov tales also demonstrated that indeed in a planet of very smart robots, his rules could repeatedly be avoided and inadequately detected. But eventually, and most particularly, Isaac Asimov rules are not suitable as a result of their attempt to claim that robots function in fixed methods, which are supposedly humans, when in the physical world, it is the people who construct and control the robots that should be the real matter of each and every principle.

As we deal with the proper associations of carrying robots in our humanity, it turns out to be visible that the robots without liability do not exist anymore. Subsequently, laws for true robots, in the physical world, must be revolutionized into principles that would give instructions to those who construct, trade, and value robots to relate their works on how robots may perform.

The EPSRC and AHRC have collectively written a collection of five proper laws of robotics in the physical world, based on a September 2010 developer and physicist discussion group. These advanced principles of robotics are defined as follows with the author's comment that describes each principle:

1. Robots are multiuse tools. Robots should not be designed solely or primarily to kill or harm humans, except in the interests of national security [4]. (Author's comment—gadgets have been used in excess of one usage. We authorize weapons to be constructed, for example, agriculturists used robots in their operations to destroy insects and microorganisms, but murdering the human soul using them is surely evil. Fork and knife can be utilized in eating food or hurting humans. In maximum communities, neither weapons nor eating utensils are outlawed, but rules and regulations may be introduced if essential to achieve civil security. Robots likewise have numerous operations. Despite the fact an imaginative end user could possibly treat any robot for destructive purposes, while with an unsharpened tool, we are judging that the tool should not in any way be constructed individually or particularly to be handled as arms with destructive or other attacking efficiency. This rule, if supported, restricts the profit-making readiness of robots, but we view it as an important law to give approval of their use for protective purposes in the public community.)
2. Humans, not robots, are the responsible agents. Robots should be designed and operated to be as far as practicable to comply with the existing laws and fundamental rights and freedoms, including privacy [4]. (Author's comment—we have found out that robot behaviors are outlined to execute the rules persons have created. This law indicates two conditions. First, undoubtedly no one is in favor of intentionally setting up a law to form a robot that crushes the rule. But inventors are not legal experts and are required to be warned that of constructing robots that carry out their assignments at intervals that require to be equalized with their responsibilities to protect rules and authorized civil rights guidelines. Second, this rule is outlined to realize that robots are only gadgets, constructed to attain objectives and ambitions that living souls point out. Consumers and proprietors possess liabilities that inventors and producers likewise have. At times, it is up to inventors to foresee the consequences of their creations because robots can have the strength to discover and modify their attitude. But consumers can also modify robots to carry out stuffs their inventors did not predict. At times, it is the proprietor's job to handle the consumer. But if a robot's behavior carries out to crack the rules, it will consistently remain as the liability of the individual or other living souls, not of the robot at all.)

3. Robots are products. They should be designed using processes that assure their safety and security [4]. (Author's comment—robots are sections of mechanization and automation their purchaser may desire to take care of; however, we will constantly ensure the security of the living soul over a device that performs a task. The main objective here is to confirm that the protection and freedom of robots in humanity would be guaranteed, so that living souls can believe and have assurance on them.)
4. Robots are manufactured artifacts. They should not be designed in a deceptive way to exploit vulnerable users; instead, their machine nature should be transparent [4]. (Author's comment—one of the excellent characteristics of robotics is that mechanical games may deliver satisfaction, good feeling, and even a pattern of togetherness to a living soul who is unable to give care for his or her favorite animals due to time and financial constraints or strict home rules. Still, once a consumer becomes associated with such a mechanical toy, it becomes achievable for the builder to demand his or her wishes or requests for the robot, which could probably result to unreasonable prices for consumers. The allowable interpretation of the law was constructed to express that though it is acceptable and at times preferable for a robot to deliver the feeling of genuine understanding, anybody who buys or connects with a robot permits to discover what it genuinely is and possibly what the designers want to carry out. Robot brilliance is simulated, and we feel the outstanding method to take care of users.)
5. The person with legal responsibility for a robot should be attributed [4]. (Author's comment—in this principle, we attempt to implement a sensible structure of what all the aforementioned principles essentially rely on; a machine is not at all lawfully answerable for anything. It is only an appliance or a gadget, so if it breaks down and leads to casualty, the person held responsible for the loss is the one subject to conviction. Discovering who is answerable for the damage caused may not anyhow be simple.)

10.3.3 Seven Outstanding Information Notes by EPSRC and AHRC

As an extension of the aforesaid fundamental laws, the associations also established an overall series of information notes outlined to strengthen the liability within the robotics study and the manufacturing society and hence to increase the responsibility of the machine in the task it carries out. The enthusiasm of trustworthy modernization is for the aforementioned maximum section, but we believe it is beneficial to create this for the specific group. The following are the seven outstanding messages in order:

1. We believe robots have the potential to provide immensely positive impact to the society. We want to encourage responsible robot research.
2. Bad practice hurts us all.
3. Addressing obvious public concerns will help us all make progress.
4. It is important to demonstrate that we, as roboticists, are committed to the best possible standards of practice.
5. To understand the context and consequences of our research, we should work with the experts from other disciplines including social sciences, law, philosophy, and arts.

6. We should consider the ethics of transparency: there are limits to what should be openly available.
7. When we see erroneous accounts in the press, we commit to take the time to contact the reporting journalists [4].

10.4 Robotics Architecture and Designing Knowledge

10.4.1 Brief History of Robotics Architecture

Possibly the first robot design was that applied by William Grey Walter, a neurophysiologist, in his architectural plan for his turtles in 1953. He announces his turtle robots are of classic Latin signs to characterize their manner of conduction, like Machina Speculatrix and Machina Docilis. Machina Speculatrix signifies a "machine that can judge" and Machina Docilis signifies a "machine that can be trained" through which he indicates a machine that can acquire information considering that it could be taught with signals. The turtles were machines constructed with three rollers, which are utilized for steering and driving. They were protected by a transparent plastic structure. Machina Speculatrix should contain the following parts: one photocell sensor (a device that permits an action to discover light), one bump sensor (a device that is offering you the capacity to discover a collision before it takes place), one rechargeable battery, three motors (one for each roller), and one electronic chip using two vacuum tubes (serving as the mind). The design was fundamentally behavior based.

A completely separate design described a robot buildup at the Stanford Research Institute, in 1969. This robot, named Shakey, had a camera (photographic equipment), a range finder sensor (a device that calculates distance from the viewer to destination object), and a bump sensor and was linked to mainframe computers via radio and video network. This design was separated into three operational principles: sensing, planning, and executing. The sensing arrangement interpreted the camera picture into an internal division idea. The planner took the internal division idea of an aim and created a layout that would bring to that aim. The executor took the layout and directed the actions to the robot (Shakey). This concept has been named as sense–plan–act prototype. The robot was famously known as Shakey because its camera fit was not hard making it swing (shake) as the robot acts.

10.4.2 Skills Required to Design a Cyber-Physical Robotic System

Planning a robot composition is a lot more of a skill than an engineering methodical study. The aim of the composition is to build instructions for a robot to be simple, secure, and extra responsible. The selection of a robot composition should not be seen casually, as it is the writer's knowledge that previous composition arrangements repeatedly carry on for years. Developing a robot composition is a challenging program and can delay advancement while a collection of instruction is carried out differently. The skill of planning a robot composition begins with a collection of queries that the planner is required to study. These queries involve

Assignment: What are the assignments the robot will be carrying out? What is the duration of the assignments? Are the assignments monotonous or miscellaneous across a time span?

Procedure steps: What steps are required to operate the assignments? How are those steps performed? How quickly do steps require to be picked or updated in order to maintain the robot protected?

Information: What information is required to carry out the assignments? What input or output devices will generate that information? How will the robot collect that information from the physical world or the consumers? How frequently is that information required to be renewed?

Logic: What logical skills will the robot have? What information will these logical skills generate? How will the logical skills of a robot be supervised and linked? Are there traditional logical skills that will be utilized? Where will these miscellaneous logical skills exist in an internal or external circuit?

Consumers: Who are the robot's consumers? What will they instruct the robot to perform the assignment? How will the consumers identify what the robot is performing? How are the consumers going to interact with the robot? What data will they like to examine from the robot?

Judgment criteria: How will the robot performance be judged? What are the favorable and unfavorable benchmarks? What is the solution for those unfavorable outcomes? Will the robot composition be utilized for more than one group of assignments?

Once the planner has the solutions to nearly all of these queries, they can at that point go ahead constructing some algorithms for the kind of actions they wish the robot to carry out and how they wish consumers to communicate with it.

10.4.3 Basic Organization of a Cyber-Physical Robotic System

The cyber-physical robotic system are machineries that utilize electronics, mechatronics, and software mechanism in combination, and they are discovered to carry out assignments commonly functioned by living souls. From this elementary theory, we can individually deal with the necessary operations into sections. The choice of sections utilized in a particular design is purposeful by the achieved conclusion of the planner. An automated escalator doesn't require any arms or legs. The basic sections of a robot are described as follows (see also Figure 10.2).

10.4.3.1 Power Supply

The power supply section supplies the power to operate the electronic and mechatronic parts in a robotic system. Few power sections are inside of the robotic system and few are outside. There are three fundamental power sources utilized for robotic systems—electrical, hydraulic, and pneumatic. The most frequent power section for robotics is electrical. Electric power supply operates essentially to manage the approaching electric power and to supply the ac or dc voltages demanded by the electronics and mechatronic parts. Hydraulic power supply works either as an intrinsic section of the controller part or as an individual section. This section contained filter, electric motor–driven pumps (which transform electric power into hydraulic power), and a reservoir. Pneumatic power could be provided either by compressors or the high-force air depositor. Deposited air is separately supplied a finite quantity of power.

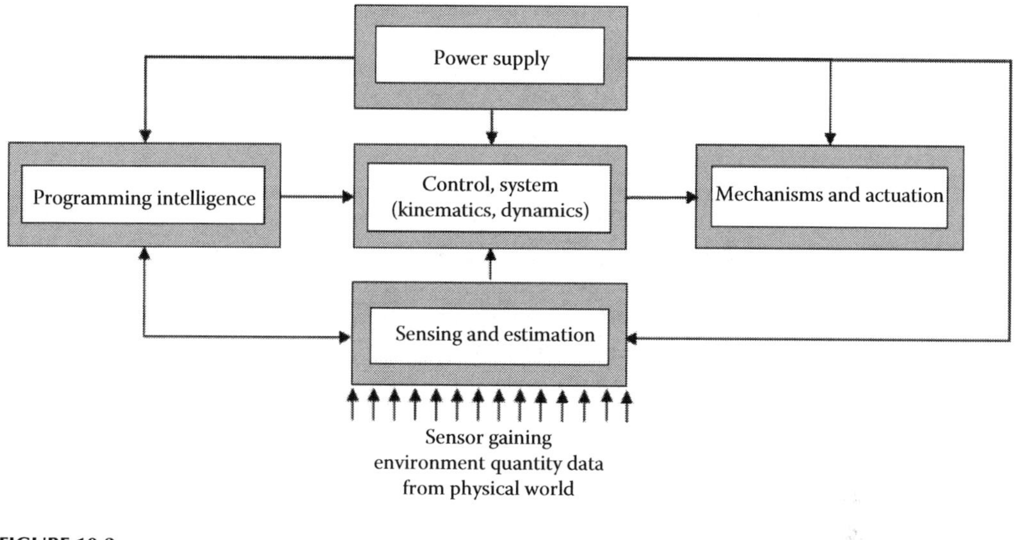

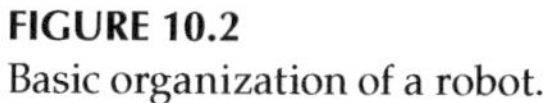

FIGURE 10.2
Basic organization of a robot.

10.4.3.2 Programming Intelligence

If an exact definition of environmental, real world is accessible to the programming atmosphere, a statistic of conditions will necessarily arise, which cannot be concluded. As a result, a robot programming atmosphere needs to be equipped with the succeeding characteristics—real-time operating system, input/output sensor information management, error detection ability to perform tasks, and communication with other systems.

10.4.3.3 Control System

Kinematics: Kinematics relates to the movements of frames in a robotic system without referring to forces that generate movement. Kinematics is the ultimate condition of robot layout, analysis, and command. The robotics society has concentrated on smoothly implementing various portrayals of position and orientation and their by-product with reference to space to answer fundamental kinematics questions.

Dynamics: Robot dynamics presents the connections among drive and contact forces and the movement paths and the increasing speed that result. The dynamic mathematical formulas of movement arrange the physical foundation for a collection of logical algorithms that are utilized in robot layout, command, and imitation. A developing scope of their services is an animation of mobile arrangements, notably using humanoid figures.

10.4.3.4 Mechanisms and Actuation

Robot arrangement targets the totality of intersections, structure dimension, and the action specifications of the end effector. A demanding responsibility in robotic arrangement is the series of assignments the robot is anticipated to work. The robot needs to be constructed to get the adaptability it requires to carry out the series of assignments for which it is planned.

The substantial arrangement like the bearings, shafts, links, and beams of a machine in a movable frame comprises the mechanisms of the robot. The engine, hydraulic, or pneumatic motors, or additional components that source the interconnection of the arrangement to motion, are labeled actuators.

10.4.3.5 Sensing and Estimation

Handling a robotic mechanism prospectively is easy if an entire design of the physical world was accessible and if the robot drives could take care of movement instructions correctly corresponding to this design structure. Sadly, in the majority conditions of concern, an entire physical world structure is not gettable, and the flawless command of robot mechanisms is not at any time a sensible expectation. Sensing and estimating are the processes of settling for this shortfall of the entire data. Their aspect is to arrange data about the condition of the physical world and the condition of the robot mechanics as a support for command, generating results, and communication with various objects and living souls in the surroundings, such as animals and humans.

Sensing and estimation as a pair can be considered as a series of actions to convert an environmental quantity into a data processing description that can be utilized for additional transformation. Sensing is nearly bound to transducers that change some environmental stuff into a signal that can be transformed by a data processing machine completely. Sensing is also closely bound to understand the method of describing the sensor data in an operation associated structures of the physical world. Still, sensor information is frequently corrupted by differing methods that confuse this technique. Analytical buzz originated from the transducer, the technique of converting continuous structures into discrete parts is brought in the digitization method, and weak sensor imports uncertainty. Estimation approaches are thus brought into backing proper combination of data into designs of the surroundings and for the bettering of the signal-to-noise ratio.

10.5 Applications of Cyber-Physical Robotic System

The applications of robotics relates the latest leading technology concerned with the expanding range of robot utilization, from manufacturing industry robotics through a different lineup of wind, ocean, ground, and mining robotics to medical robotics. The ultimate view of machine intelligence is for the extensive functionality of robots. The robots of the future will carry out all the hazards, dusty, and monotonous assignments. The applied science has attained a position of development, such that robotic systems are now hiking from the manufacturing industry into the operation and utility applications. The applications of achievable operations could significantly boost if robots were not difficult to set up, coordinate with other production systems, and supply instructions, specifically with flexible understanding and self-regulating fault recovery. We describe some of these fields and service applications in the following sections.

10.5.1 Cyber-Physical Robotic Systems in Factory Automation

Today's industrial robots are principally the outcome of the necessity of fundamental demanding massive-quantity production, essentially described by the automobile, computer chips, and electrical goods enterprises. Industrial robots are treated as a vital element of ambitious production, which desires to merge great condition, output, and flexibility at a nominal amount. In September 2014, worldwide, more than 200,000 industrial robot installations were reported, with 15% more in 2013 [5]. Large-development manufactory (in integrated circuitry, edible material, cosmic products, management, environmental sciences) and advanced industrial operations to a greater extent rely upon on leading robot mechanics. These technologies of robot system setup has been rapidly increasing. At the same period, new industrial science, specifically from the information technology universe, will have a rising impression on the architecture, achievements, and expense of eventual industrial robots. On a high-tech position, dominant questions in manufacturing plants are about shorter engineering aspects and speedy ramp-up of creating product lines and joining mechanics methods to profession processes. New facts and exchanging information mechanics can assist to manage with these questions. CPS robotics structures with distributing management brilliance can set up exchanging information through available systems of connections established on Internet rules. This will contribute to the restoration of ancient culture hierarchical automation robotic structures by self-regulating cyber-physical manufacturing structures. In the future, very smart goods will become an operating element within automated production structures.

An industrial robot is described in ISO8373 as an automatically controlled, reprogrammable, multipurpose manipulator programmable in three or more axes, which may be fixed either in place or mobile for use in industrial automation applications. *Reprogrammable* means programmed motions or auxiliary functions may be changed without physical alterations; *multipurpose* means a robot is capable of being adapted to a different application with physical alterations; *axis* means the direction used to specify the robot motion in a linear or rotary mode [6].

Typical applications of industrial robots are human robot cooperation for handling tasks, welding, painting, automobile body assembly, material transfer automation, and machining.

The most common industrial robots are Motoman SK6, Kuka, Fanuc's Toploader Series, etc.

10.5.2 Cyber-Physical Robotic System in Aerial Science

Following the division aerial robotics, we manage to discover various explanations: it could signify robotic machine in the air, particularly an operation-independent, platform-aligned approach; withal, it could also signify robotics that utilize flying vehicles, particularly a platform-independent, operation-aligned approach. Finally, it could signify a pair of the aforementioned machines, particularly information on the robotic platform, in addition to its robotic operation. In aerospace language, flying vehicles are usually mentioned as unmanned air vehicles, whereas the complete foundation, organizations, and living soul peripheral desired to perform such vehicles for a given functional aim are frequently named as unmanned air systems.

Keeping a record of all achievable utilizations of aerial robotics is tough. Withal, there are some truly existing utilizations, as a result of the need for the similar utilizations to satisfy with strict air security laws. The applications of aerial robotics are remote sensing such as power line tracking and unexploded mine discovery, disaster response such as flood and cyclone tracking, search and rescue in tough region, transportation management, payload delivery, and surveillance of living soul's activity.

The most common aerial robots are RQ-16 T-Hawk, RQ-11 Raven, RQ-7 Shadow, MQ-1 Predator, MQ-9 Reaper, RQ-4 Global Hawk, Aeryon Scout, etc.

10.5.3 Cyber-Physical Robotic System in Space Science

Current space structures undoubtedly have a close pairing among onboard cyber (exchanging information, intelligence) and physical (drive, actuation) components to the existing rigid earthly surroundings and favorably entire challenging assignments. Since space systems and automobiles frequently perform independently or partially uphold some performance in an independent way, they act as a best case of CPSs. Sensors note data about the real surroundings, desktops figure out actions to continue, and drives build an adjustment in the atmosphere. But the built-in conditions of the spacecraft, be it in exchanging information, approach energy, and management, build dares that many earthly arrangements do not have to face.

Space robots are valuable to our complete community to perform in space; as a result, they can execute assignments at low cost or on quick roster, with less danger and periodically with enhanced efficiency over living souls performing the identical task. They perform for larger time spans, frequently asleep before their functional responsibility starts. They may be appointed into circumstances that are so dangerous that living souls would not be permitted to go. Actually, all space operations outside the limits of the globe have been a self-murder mission in that the robot is left over in an area when it stops performing; as a result, the amount to turn back to the planet is astronomical.

Typical applications of space robotics are planetary exploration, inspection and fixing technical problems of spacecraft, and assembly of large space architectures.

The most common space robots are Robonaut, Rassor, Athlete, Spidernaut, etc.

10.5.4 Cyber-Physical Robotic System in Agriculture

Agricultural robotics is the value of mechanization in agriculture and livestock system. It is the reclamation of the traditional methods to execute the same assignments with better productivity. The border limits of agriculture are not transparent. Developing lands, growing plants, sprinkling plants, and collecting are apparently counted. Animal-related movements can enhance outside limits of milking and butchering, at the beginning of a great series of process prominent to the display of junk food or mass-produced clothes on the shopping center cupboards. Agriculture robotics has created a meaningful influence. Farmers are attentive of their requirement for automobile counseling to decrease the casualty in the planting area of their land. Automated perception, management, and transformation of greengrocery are now normal, while there is a meaningful and serious apparatus and machine control of livestock agenda.

There are numerous basic applications of agriculture robotics such as fruit selection, sheep shaving, spraying, livestock washing, and automatic milking. Robots like these have a lot of advantages in the agricultural field, involving an exceptional quality of natural fruits, increased profit, and reduced labor cost.

Some examples of agriculture robots are ACTV, Weedy, Hako, ACW, API, and Savage.

10.5.5 Cyber-Physical Robotic System in Medical Science

Medical robots carry a possibility for the achievement of the complete transformation of a medical procedure like surgery and intervention taken to improve a medical disorder and store in-depth data acquired from the atmosphere that utilizes the integral stamina of living souls and the mainframe that supports science. The medical robots can be continued to be thought of as data-directed medical instruments that allow doctors to act in particular patients with higher assurance, enhanced effectiveness, and lower discouragement.

For example, heart medical procedure repeatedly calls for halting the heart, operating the procedure, and then reinstating the heart. Such operation is greatly dangerous and carries various damaging reactions. A unit of investigation groups has been active on opportunity where a doctor can act on a functioning heart in comparison with halting the heart. There are a number of concepts that make this achievable. Primary, operational gizmos can be robotically regulated so that they act along with the movement of the heart. A doctor can thus work with a gizmo to implement fixed physical force to a mark on the heart as long as the heart carries on to beat. Secondary, stereoscopic video system can offer the doctor a video misconception of a stable heart. From the doctor's view, it shows as if the heart has been halted, while in reality, the heart carries on to beat. To understand such a heart operation structure, a broad design of the heart, the gizmos, the logical hardware, and the framework software are needed. It depends upon the specific architecture of the framework that provides accurate timing and secure backup action to manage the breakdown.

The application of medical robotics is not limited in surgical theaters. Other utilization sectors where medical robots are confirmed to be valuable include laboratories conducting a number of tests (like HIV, CBC, and PBF), drug delivery, surgery, patient registration, etc.

10.5.6 Cyber-Physical Robotic System in Infrastructure for Smart Grid and Utilities

A great test for CPS is the draft and formation of a power structure's foundation that is adept to supply cutoff free power production and circulation, is adaptable enough to grant miscellaneous power distribution to or departure from the framework grid, and is unaffected by accidental or willful manipulations. The combination of CPS science and infrastructure to the current power framework and other service structures is a challenge. The expanded structure complication poses scientific tasks that must be treated as the structure is managed in forms that were not planned when the foundation was initially made. As science and structures are integrated, safety remains a principal anxiety to lower structure accountability and secure contributor facts. In addition, to manage the foundation with CPS robotic technologies, a skillful, original, and experienced trained work team is required.

10.5.7 Cyber-Physical Robotic System in Smart Buildings and Infrastructure

Construction is a universal living soul's task that links to the establishment or recognition of physical artifacts that created materials utilized in the production of goods. Architecture and development products are broad in range and rare in shape. Over generations, different structures of apparatus and mechanical arrangements have been brought into the civil field to raise production performance.

The application of construction robotics occurs under construction automation. Some of the applications of construction robotics are materials handling, materials shaping, structural joining, etc.

The most common construction robots are wall climbing painting robot and concrete power floating machine.

10.6 Conclusion

We displayed that with the idea of local and overall information the arrangement can attain greater knowledge of the physical environment and the cyber earth. As a conclusion, it can cut down the logical time and depository. The application of robotics in CPS will reconstruct how people communicate with and command the physical system. Low-power buildings and metropolis, ultimate-output agriculture fields, almost-nil automotive casualties, never-ending life attendants, area-liberated approach to health care, condition-attentive physical foundation, cutoff free power, and safe removal from hazardous locations are but some of the many social advantages that the application of robotics in CPS will deliver. The subject matters contain basic concepts in robotics, robotics fundamental principles, robotics architecture and designing knowledge, and robotics fields and their applications in CPS. Our fundamental idea was to deliver a polished description of a true robot in terms of CPS; the history of robotics that answers all the queries about what the early robot and robotic design ever created was like, when it was created, who created it, and what it performs; and the fundamental guidelines of robotics, which are proposed as an elementary structure to locate the action of robots, layout to have a strength of freedom. Likewise, the robotics architecture and designing knowledge introduce the basic organization of a robot that are utilized to construct, study, and regulate a robotics architecture through cyber systems. In the final matter, we discuss the robotic fields and their applications in CPS, which defines various kinds of the latest technological trends in robotics section, like cyber-physical robotic systems in factory automation, space science, aerial science, agriculture, medical science, infrastructure for smart grid and utilities, and construction.

References

1. Mechanical man. Retrieved November 8, 2014, from http://www.mikekemble.com/misc/leonardo.html.
2. A. Isaac. (1942). *I, Robot*. Street and Smith Publications, Syracuse, New York, p. 27.
3. A. Isaac. (1994). *Robots and Empire* (Re-issue edition). Harper Collins, New York, 371pp.
4. Engineering and Physical Sciences Research Council (EPSRC). (September 2010). Principles of robotics. Retrieved November 24, 2014, from http://www.epsrc.ac.uk/research/ourportfolio/themes/engineering/activities.
5. International Federation of Robotics (IFR) Statistical Department. IFR: Accelerating demand for industrial robots will continue. Retrieved November 2014, from http://www.worldrobotics.org/index.php?id=home&news_id=274.
6. International Federation of Robotics (IFR). Industrial robots. Retrieved November 28, 2014, from http://www.ifr.org/industrial-robots.

11

Application of Mobile Robots by Using Speech Recognition in Engineering

P.V.L. Narayana Rao and Pothireddy Siva Abhilash

CONTENTS

11.1 Introduction

Since a human usually communicates with each other by talking, it is very convenient if voice is used to command robots. A wheelchair is an important vehicle for the physically handicapped persons. However, for injured persons who suffer from spasms and paralysis of the extremities, joystick is a useful device as a manipulating tool. The recent developments in speech technology and biomedical engineering world have diverted the attention of researchers and technocrats to concentrate more on the design and development of simple, cost-effective, and technically practical solutions for the welfare and rehabilitation of a large section of the disabled community.

One method is to command the wheelchair via voice through special interface, which plays a role as the master control circuit for the motors of the wheelchair. In the case of voice control, the situation is more difficult because the control circuit might generate recognition error. The most dangerous error for wheelchair control is substitution error, which means that the recognized command is interpreted as the opposite command. For example, "LEFT" is interpreted as "RIGHT." The previously described situation is very probable in the Polish language. "LEFT" or "RIGHT" has very high acoustic similarity.

Robotic arms are fitted with some type of gripper, which can be used to help people eat, assist with personal hygiene, bring items in a home or office environment, and open door handles. The arms can be mounted on wheelchairs, attached to mobile robots or on a mobile base, or fixed to one location as part of a workstation. An overview of rehabilitation research investigating robotic arms and systems can be found in. The arms mounted on wheelchairs must not interfere with normal use of the wheelchair by increasing its size or causing the chair's balance to become unstable.

11.2 Related Work

Moon et al. (2003) propose an intelligent robotic wheelchair with user-friendly human–computer interface (HCI) based on electromyogram signal, face directional gesture, and voice. The user's intention is transferred to the wheelchair via the HCI, and then the wheelchair is controlled to go toward the desired direction. Additionally, the wheelchair can detect and avoid obstacles autonomously with sonar sensors. By combining the HCI into the autonomous functions, it performs safe and reliable motions while considering the user's intention.

The method presented by Rockland et al. (1998) was designed to develop a feasibility model for activating a wheelchair using a low-cost speech recognition system. A microcontroller was programmed to provide user control over each command as well as to prevent voice commands from being issued accidentally. It is a speaker-dependent system in which the individual who would be using the system can be trained and could theoretically attain better than 95% accuracy. Chauhan et al. (2000) presented a voice control through the feature-based, language-independent but speaker-dependent, isolated word recognition system (IWRS) that uses dynamic time warping technique for matching the reference and spoken templates. A unique code corresponding to each recognized command is transmitted on a parallel port from the IWRS to the motor controller board that uses 80KC 196 microcontrollers. Simpson et al. (2002) proposed to utilize voice control in combination with the navigation assistance provided by "smart wheelchairs," which use sensors to identify and avoid obstacles in the wheelchair's path. They described an experimental result that compared the performance of able-bodied subjects using voice control to operate a power wheelchair both with and without navigation assistance.

Valin et al. (2007) described a system that gives a mobile robot the ability to perform automatic speech recognition (ASR) with simultaneous speakers. A microphone array is used along with a real-time implementation of geometric source separation and a postfilter that gives a further reduction of interference from other sources. The postfilter is also used to estimate the reliability of spectral features and compute a missing feature mask. The system was evaluated on a 200-word vocabulary at different azimuths between sources. Hrnčár (2007) describes the Ella Voice application, which is a user-dependant, isolated voice command recognition tool. It was created in MATLAB®, based on dynamic programming, and it could serve for the purpose of mobile robot control. He deals with the application of selected techniques like crossword reference template creation or endpoint detection. Takiguchi et al. (2008) proposed speech recognition as one of the most effective communication tools when it comes to a hands-free (human–robot) interface. They describe a new mobile robot with hands-free speech recognition. For a hands-free speech interface, it is important to detect commands for a robot in spontaneous utterances. The system can comprehend whether the user's utterances are commands for the robot or not, where commands are discriminated from human–human conversations by acoustic features. Then the robot can move according to the user's voice (command). Recognition rate for the user's request was 89.93%. From the previous work, it seems that all ASRs used in the wheelchair have low recognition rates less than 95% (speaker dependent) and navigation system depends on additional sensors like infrared (IR), ultrasonic, and camera. In this work, the wheelchair is completely controlled by voice with a recognition rate of 98% for the speaker-independent system, and I conclude that this method is better than the previous methods.

11.3 System Design

The following five voice commands have been identified for various operations of the wheelchair: FORWARD, REVERSE, LEFT, RIGHT, and STOP. The wheelchair starts moving in the corresponding direction upon voicing the command forward in forward direction and stops if the command is stop and so on.

11.3.1 Database of Speech

Every speaker's recognition system depends mainly on the data input. The data input used in the system is speech. The speech is uttered by 15 speakers, 8 males and 7 females, 10 of them are used for training purpose (5 males and 5 females), and each speaker utters the same word five times. The following five voice commands have been selected for various processes of the wheelchair: forward, reverse, left, right, and stop. The total number of uttered data used for training is 250. The remaining speakers (3 males and 2 females) are used for testing purpose, and each speaker uttered the same word 2 times, and then the total number of utterance is 50 used for the testing purpose.

11.3.2 Multiridgelet Transform

To improve the performance and to overcome the weakness points of the ridgelet transform, a technique called the multiridgelet transform was proposed. The main idea of the ridgelet transform is to map a line sampling scheme into a point sampling scheme using the Radon transform (RT), and then the wavelet transform can be used to handle effectively the point sampling scheme in the Radon domain (Minh and Vetterli, 2003). While the main idea of the multiridgelet transform depends on the Ridgelet transform, changing the second part of this transform with the multiwavelet transform improves the performance and output quality of the Ridgelet transform.

In fact, the multiridgelet transform leads to a large family of orthonormal and directional bases for digital images, including adaptive schemas. However, the multiridgelet transform overcomes the weakness point of the wavelet and Ridgelet transforms in higher dimensions, since the wavelet transform in two dimensions are obtained by a tensor product of 1D wavelets, and they are thus good at isolating the discontinuity across an edge, but we will not see the smoothness along the edge. The geometrical structure of the multiridgelet transform consists of two fundamentals parts:

1. The RT
2. The 1D multiwavelet transform

11.3.3 Radon Transform

The RT is defined as summations of image pixels over a certain set of "lines." The geometrical structures of the RT consist of multiple parts of the sequence jobs. RT provides a means for determining the inner structure of an object. It allows us to analyze signal in detail by means of transforming the original signals from the spatial domain into the projection space (Li et al., 2003).

RT appears to be a good candidate. It converts original image into a new image space with parameters u and t. Each point in this new space accumulates all information corresponding to a line in the original image with angle u and radius t. Thus, when RT localizes near an angle u_0 and around a slice t_0, a local maximum will result in the original image that has a line in position (t_0,u_0). This is the kind of transform we are looking for (Terrades et al., 2003).

11.4 Neural Network

Artificial neural networks (NN) refer to the computing systems whose central theme is borrowed from the analogy of "biological NN." Many tasks involving intelligence or pattern recognition are extremely difficult to automate (Ram Kumar et al., 2005).

11.4.1 Model of NN

We used random numbers around zero to initialize weights and biases in the network. The training process requires a set of proper inputs and targets as outputs. During training, the weights and biases of the network are iteratively adjusted to minimize the network performance function. The default performance function for feed forward networks is mean square errors, the average squared errors between the network outputs and the target output (Hosseini et al., 1996).

11.5 General Procedure of Proposed Systems

This chapter contains two parts: part one contains the theoretical work (simulation in computer with the aid of MATLAB 7), and the second one puts interface between the human and the computer connected to the wheelchair. The first part contains three steps for implementation:

1. Preprocessing
2. Feature extraction
3. Classification

11.5.1 Preprocessing

In this section, the isolated spoken word is segmented into frames of equal length (of 128 samples). Next, the result frames of each word are converted into a single matrix (2D), and this matrix must be a power of two. So the proposed length for all words is 16,348 (1D), and this length is a power of two and can be divided into a matrix having a dimension of 128 × 128, which is a 2D and a power of two matrix.

11.5.2 Feature Extraction

The following algorithm was used for the computation of 2D discrete multiridgelet transform on the multiwavelet coefficient matrix using Gaussian mixture model (GHM) four multifilters and using an oversampling scheme of preprocessing (repeated row preprocessing). It contains four fundamental parts, and these are applied to a 2D signal (word) to get the best feature extraction:

1. Input word, and check its dimension.
2. Apply 2-D FFT for the resizing signal (2-D word), its method convert the matrix from Cartesian to polar.
3. Apply Radon transforms to 2-D FFT coefficient.
4. Finally apply 1D DMWT to the radon transform coefficient.

The final coefficient gets the multiridgelet coefficient. The procedure was applied to a 2D signal to get the final best feature extraction.

11.5.3 Classification

This step begins when getting on the 2D discrete multiridgelet transform coefficient. The coefficient is splitted into two parts: the first part is used as reference data, and the second one is used as tested or classified data. The strong method that can simply recognize signal is the NN that uses a back-propagation training algorithm as a classifier after training the reference data (coefficient) resulting from the 2D discrete multiridgelet transform.

Because the input nodes of nontarget trails for a term (NNT) are set as vector inputs, then the output nodes also must be set as vector outputs (1D), and its value depends on the desired signal for each word. For the same word, the desired signal is the same, but for other words, it will be different (i.e., the desired signal differs from word to other).

11.5.4 Computation of the DMWT for 1D Signal

By using an oversampling scheme of preprocessing (repeated row), the discrete multiwavelet transform (DMWT) matrix is doubled in dimension compared with that of the input, which should be a square matrix N × N where N must be a power of two. Transformation matrix dimensions equal input signal dimensions after preprocessing.

11.5.5 Results of NN Training

In a six-layer feed-forward network, the first input layer has 512 *logsig* neurons, the first hidden layer 256 *logsig* neurons, the second hidden layer 128 *tansig* neurons, the third hidden layer 64 *tansig* neurons, and the fourth hidden layer 32 *tansig* neurons, and the output layer has 5 *tansig* neurons with corresponding five signals that are used to start the wheelchair. The NN has 16,384 inputs for each isolated word command extracted by the 2D multiridgelet transform. The resilient back-propagation method is used, for all speech command, in 15 speakers (8 males and 7 females). For NNT training purpose, 10 speakers are used (5 males and 5 females). Each speaker utters the same word five times in different environment for each utterance. Then, the number of utterances for each word is 50 times, and then the total number of utterances for all words is 250 that are used in NNT training purpose. The remaining five speakers are (three males and two females) used for NNT testing purpose. Each speaker utters the same word two times in different environment for each utterance. Then, the number of utter of each word is 10 times, and then the total number of utterance for all words is 50 utterances that are used in NNT testing purpose. The training process for five control outputs can be shown in Figure 11.1 and the training results in Table 11.1. Table 11.1 shows all the outputs of the NN versus the number of tested voice command showing the recognized command among the five voice commands. From Table 11.1 and Figures 11.2 through 11.4, output "Y1" represents the voice command "GO."

When the value of "Y1" is number 1, which represents the output, 1 is active, and the number –1 representing the output is disabled. The output of the test may not be 1 or –1 but may reach these values. Therefore, the negative value (–0.9996) gives wrong voice command; from "Y3," it is noted that its value is positive (0.3218) that means the wrong voice command is "BACK." This procedure is done for all command that is shown in the following figures:

11.6 Experimental Work

The wheelchair that is used in this work has three connecting rods (one in the front and two in the rear of the wheelchair) that connect the two sides of the wheelchair; each rod has a joint in the middle that will enable the wheelchair to be portable. The wheelchair

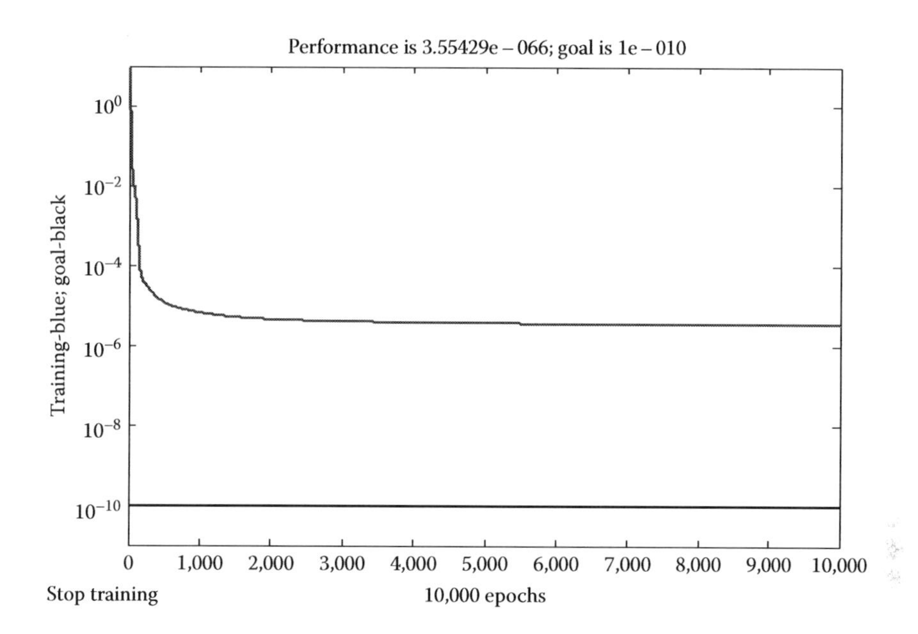

FIGURE 11.1
Training process for five control outputs.

TABLE 11.1
Result of NN Training

Input Speech	Output					Recognized Command
	Y1	Y2	Y3	Y4	Y5	
"Go1"	−0.9996	−1.0000	0.3218	−0.9939	−0.8297	"BACK"
"Go2"	0.9984	−0.9982	−0.9999	−0.9989	−0.9988	"Go"
⋮	⋮	⋮	⋮	⋮	⋮	⋮
"Go10"	0.9999	−0.9992	−1.0000	−0.9973	−0.9981	"GO"
"Left1"	−0.9952	0.9944	−1.0000	−0.9968	−1.0000	"LEFT"
"Left2"	−0.9990	0.9973	−1.0000	−1.0000	−0.9985	"LEFT"
⋮	⋮	⋮	⋮	⋮	⋮	⋮
"Left10"	−1.0000	0.9942	−1.0000	−1.0000	−0.9994	"LEFT"
"Back1"	−0.9999	−1.0000	0.9993	−0.9989	−0.9959	"BACK"
"Back2"	−1.0000	−1.0000	0.9958	−0.9998	−0.9974	"BACK"
⋮	⋮	⋮	⋮	⋮	⋮	⋮
"Back10"	−1.0000	−1.0000	0.9990	−0.9995	−0.9982	"BACK"
"Right1"	−0.9999	−1.0000	−0.9988	0.9989	−0.9998	"RIGHT"
"Right2"	−0.9967	−0.9999	−0.9998	0.9951	−0.9952	"RIGHT"
⋮	⋮	⋮	⋮	⋮	⋮	⋮
"Right10"	−0.9990	−0.9983	−0.9991	0.9972	−1.0000	"RIGHT"
"Stop1"	−0.9910	−0.9993	−0.9943	−0.9887	0.9978	"STOP"
"Stop2"	−1.0000	−1.0000	−0.4854	−0.9742	0.9907	"STOP"
⋮	⋮	⋮	⋮	⋮	⋮	⋮
"Stop10"	−0.9965	−0.9995	−0.9959	−0.9997	0.9928	"STOP"

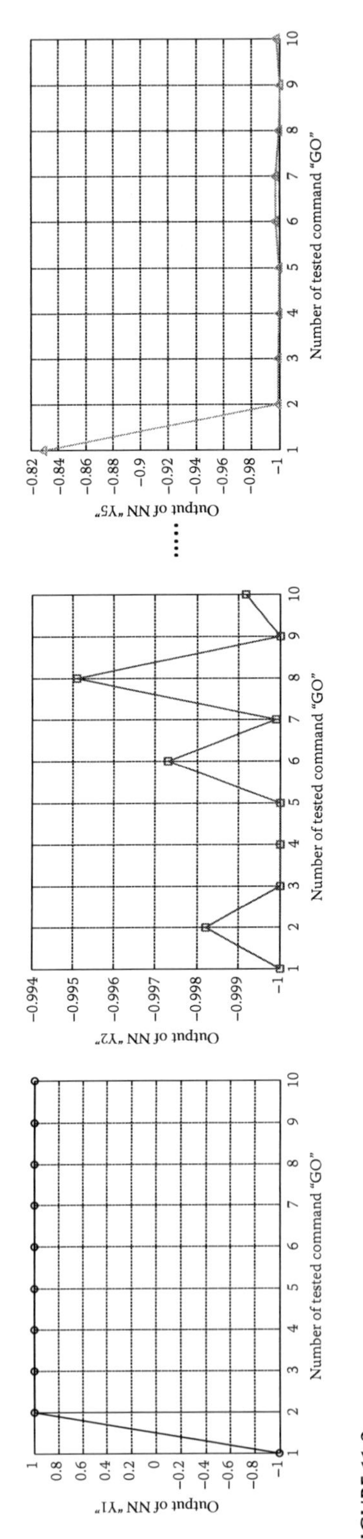

FIGURE 11.2
Output of NN for command "GO."

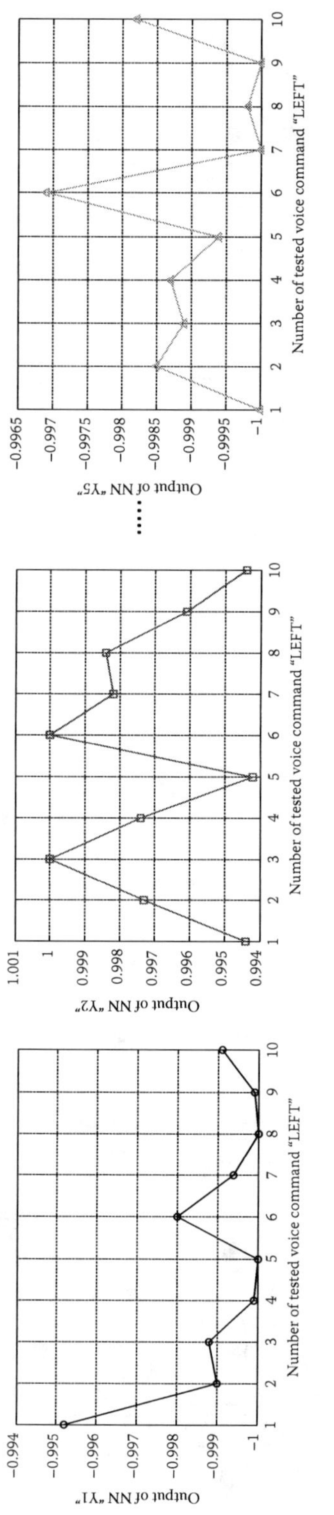

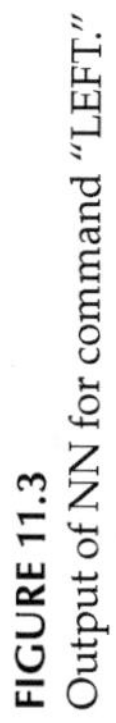

FIGURE 11.3
Output of NN for command "LEFT."

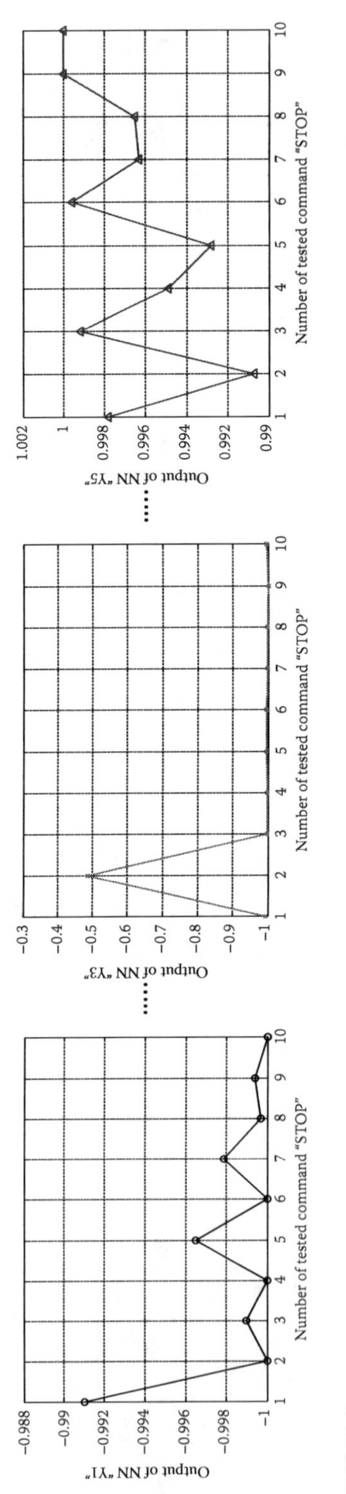

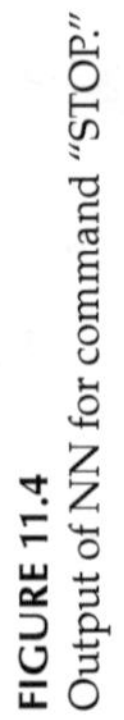

FIGURE 11.4
Output of NN for command "STOP."

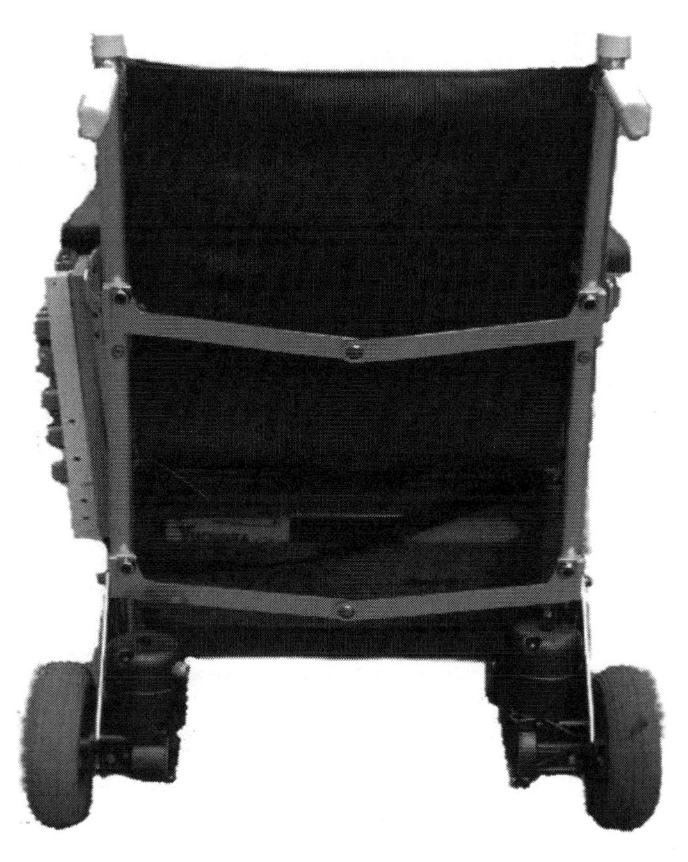

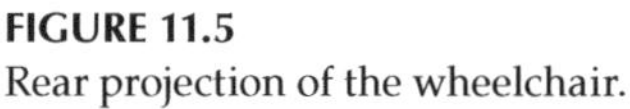

FIGURE 11.5
Rear projection of the wheelchair.

is 65 cm (25.5 in.) wide and 127 cm (50 in.) long, measured with the rear caster extended fully behind the chair. The front of the tray is 76 cm (30 in.) from the ground, and the base of the seat is 51 cm (20 in.) above the ground. The robotic wheelchair used in this work shown in Figure 11.5 was built by the BEG company (British company) with a joystick.

11.6.1 Weight of Wheelchair

The weight of a wheelchair alone is 17 kg, while the weight of a wheelchair with two motors and with experimental circuit is 20 kg. When the weight of a person is 65 kg, the weight of the wheelchair becomes 85 kg when the person sits down on it. The maximum weight that the wheelchair can reinforce with respect to the person sitting down on it and can give the same work and performance is approximately 80–100 kg. An experimental work of the aforementioned example was done and is shown in Figure 11.5 and is illustrated in Table 11.5. From this table, I conclude that the proposed method is better than the previous works.

11.6.2 Wheelchair Battery

The battery used in this work is wet type. Wet batteries use the chemical reaction between lead and sulfuric acid to create electrical energy. As the batteries require filling with distilled water, they do have a higher maintenance rate, but are lighter than gel or absorbed glass mat batteries.

11.6.3 Wheels

A wheelchair has four wheels, two rear wheels and two castor wheels: the two caster wheels are hooked in the wheelchair base in the front that have the same diameter (18 cm). The drive wheels are in the rear on either side of the base, allowing the chair to turn according to voice command; wheels engage directly to a gear train that transmit torque from the motor to the wheels by two grooves in each wheel and nut.

11.6.4 Motors

Motors are arguably one of the most important parts of a mobile robotics platform. Overpowered motors cause inefficiency and waste of the already limited supply of power from the onboard batteries, while undersized motors could be short on torque at critical times. The optimal rotation speed and the available speed range of the motor must also be taken into consideration. Too high of an output rpm from the motor shaft will cause the robot to operate at a fast, uncontrollable speed. Too low of an output will cause the robot not be able to attain a suitable speed to meet the user's needs. The rotation output of the motor also plays a role in the performance because if the torque is not sufficient,

FIGURE 11.6
Direct current motor used in the wheelchair.

locomotion may not occur in certain situations. Therefore, much consideration was put into the selection of the proper motor for the platform (Philips, 2003).

Motors come in many shapes and sizes. There are electromagnetic direct current (dc) motors and electromagnetic alternating current (ac) motors and a number of variations of each. AC motors are typically used for large applications, such as machine tools, washers, and dryers, and are powered by an ac power line. Since the typical power supply for mobile robotic is a dc battery and technology for transforming dc to ac is very expensive in both terms of monetary cost and power cost, ac motors were ruled out as an option for the robot. DC motors are commonly used for small jobs and suited for the purposes of the platform very well. Figure 11.6 shows the 12 V dc motor use in a wheelchair.

11.7 Hardware Components Added to Original Wheelchair

The modification that adds to the original wheelchair with removing joystick that is designed before (to modify wheelchair function according to person with injury and especially who suffer from spasms and paralysis of the extremities) makes its physical design very real. It is a combination of various physical (hardware) and computational (software) elements that mix the subsystems of the wheelchair to work in one unit. In terms of hardware components, the main components that are added to the wheelchair are interfacing circuit, microphone (headset microphone), and notebook computer (host computer).

11.7.1 Microphone

A quality microphone is the key when using ASR. In most cases, a desktop microphone just will not do the job. They tend to pick up more ambient noise that gives ASR programs a hard time. Handheld microphones are also not the best choice as the person can be clumsy in picking up the microphone at all times. While they do limit the amount of ambient noise, they are most useful in applications that require changing speakers often, or when speaking to the recognizer is not done frequently (when wearing a headset is not an option). The best choice and by far the most common is the headset style. It allows the ambient noise to be minimized, while allowing the person to have the microphone at the tip of your tongue all the time (Cook, 2002). Headsets are available without or with earphones (mono or stereo); in this work, the headphone type (FANCONG FC-340) is employed.

11.7.2 Relay Driver Interfacing Circuit

A transmit can be used to switch higher-power devices such as motors and solenoids. If desired, the relay can be powered by a separate power supply, so, for instance, 12 V motors can be controlled by the parallel port of notebook computer. Freewheeling diode can be used to protect the relay contact and prevent damage to the transistor when the relay switches off. An intermediate stage between control signal (output of parallel port) and motors consists of a combination of component relays, transistors, diodes, capacitors, resistors, and buffer 74ABT245 as shown in Figures 11.7 and 11.8; it is used to protect parallel port against any expected damage. The 74ABT245 high-performance BiCMOS device combines low static and dynamic power dissipation with high-speed and high-output drive shown in Figure 11.9.

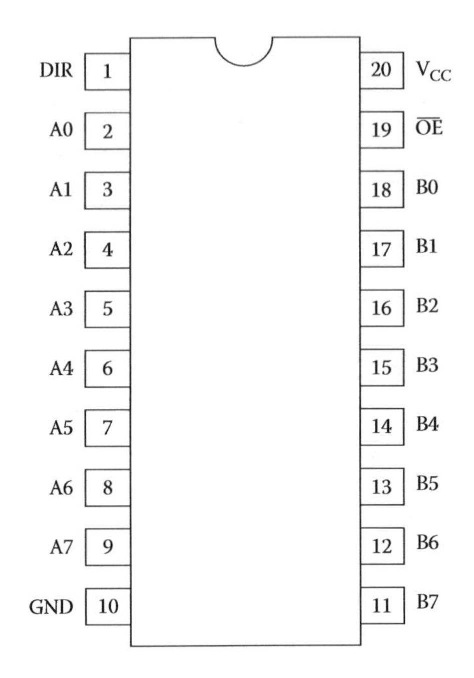

FIGURE 11.7
Relay interfacing fixated on the wheelchair.

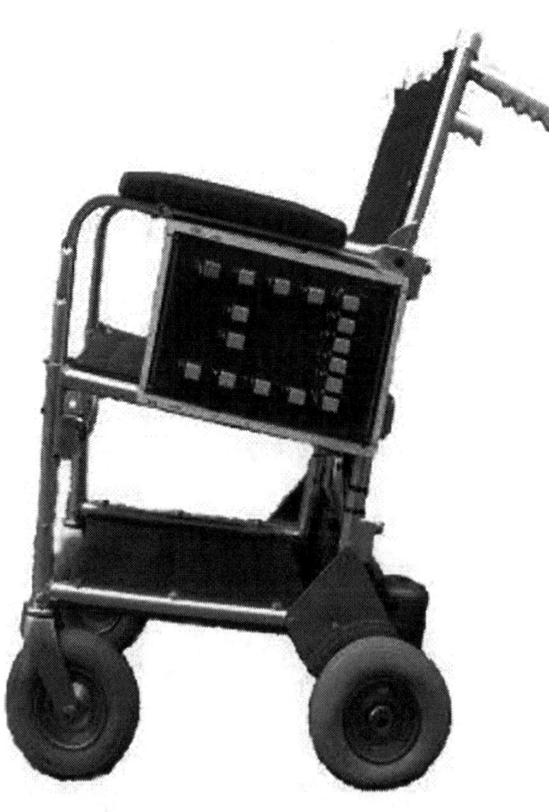

FIGURE 11.8
Relay interfacing circuit.

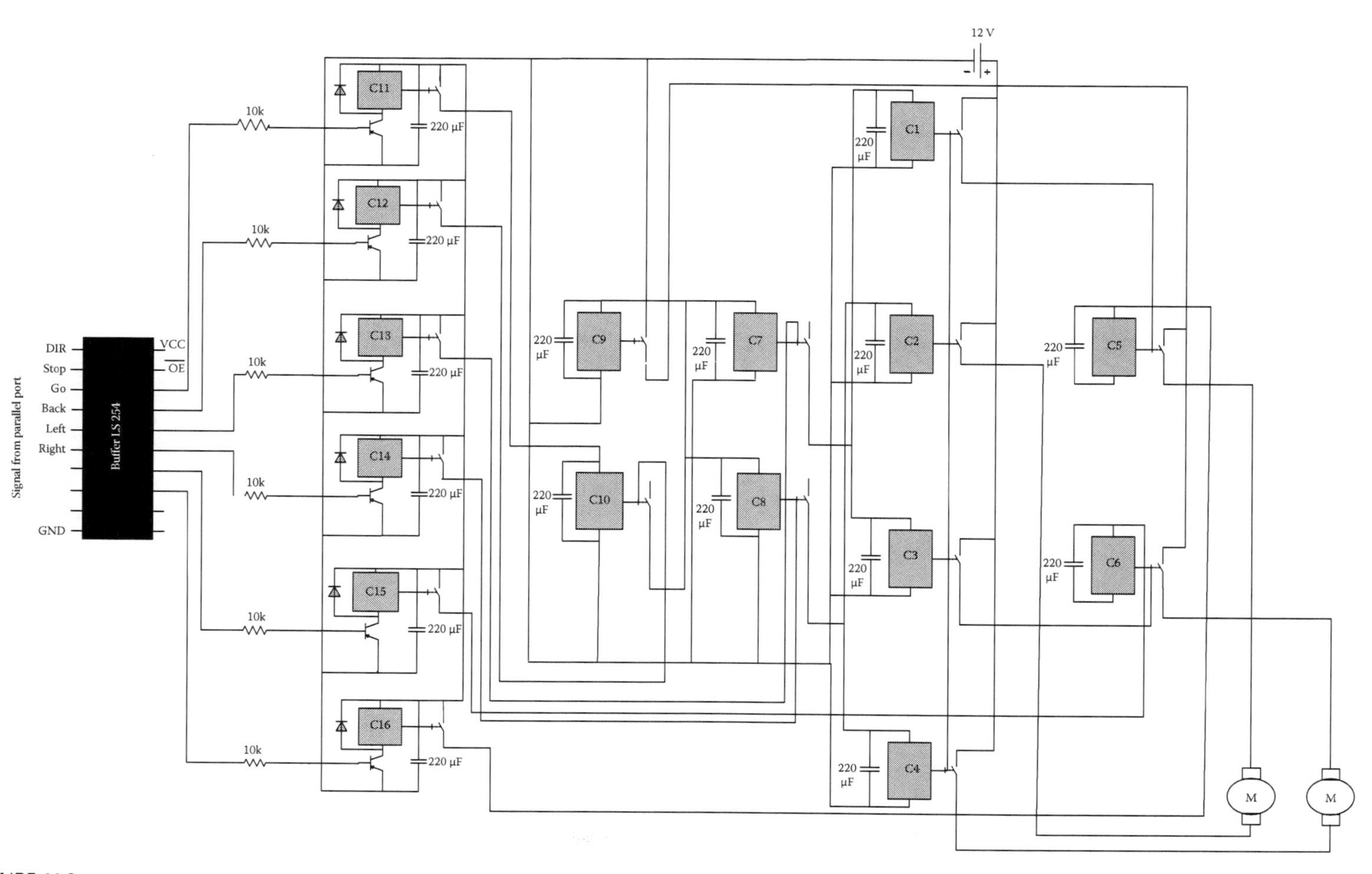

FIGURE 11.9
Pin configuration of 74ABT245.

TABLE 11.2

Pin Description of 74ABT245

Pin Number	Symbol	Name and Function
1	DIR	Direction control input
2, 3, 4, 5, 6, 7, 8, 9	A0–A7	Data inputs/outputs (A side)
18, 17, 16, 15, 14, 13, 12, 11	B0–B7	Data inputs/outputs (B side)
19	$\overline{\text{OE}}$	Output enable input (active LOW)
10	GND	Ground (0 V)
20	VCC	Positive supply voltage

TABLE 11.3

Function Table of 74ABT245

—	DIR	An	Bn
L	L	An = Bn	Inputs
L	H	Inputs	Bn = An
H	X	Z	Z

H, high voltage level; L, low voltage level; X, don't care; Z, high impedance "off" state.

The 74ABT245 device is an octal transceiver featuring noninverting three-state bus-compatible outputs in both send and receive directions. The control function implementation minimizes external timing requirements. The device features an output enable input for easy cascading and a direction (DIR) input for direction control (buffer 63). Descriptions and function of the 74ABT245 are shown in Tables 11.2 and 11.3.

11.7.3 Using MATLAB® Language Tool in Experimental Work

Now, it must show how voice can control the wheelchair, and it begins at a microphone using MATLAB program language.

```
y = wavrecord(n,Fs)
```
(11.1)

Record n samples of an audio signal, sampled at a rate of Fs Hz (samples per second). The default value for Fs is 11,025 Hz. Standard sampling rates for PC-based audio hardware are 8,000, 11,025, 22,050, and 44,100 samples per second. Stereo signals are returned as two-column matrices. The first column of a stereo audio matrix corresponds to the left input channel, while the second column corresponds to the right input channel (MATLAB Help, Version 7). In this work, mono type 22,050 samples per second presented a single column of audio matrix. This single column represents isolated word command like GO and LEFT that will pass through multiridgelet transform to compare the uttered word (command) with database (saved as NN weights); when the comparison processes are successful, now use this command to control the wheelchair by parallel port of host computer using metal program language:

```
DIO = digitalis ('adaptor', ID)
```
(11.2)

TABLE 11.4

Voice Command Corresponding Binary Vector

Recognized Voice Command	Binary Vector Input/74ABT245				
GO	1	0	0	0	0
LEFT	0	1	0	0	1
RIGHT	0	0	1	0	1
BACK	0	0	0	1	0
STOP	0	0	0	0	1

`DIO = digitalio('adaptor',ID)` creates the digital input and output (I/O) object DIO for the specified adaptor and for the hardware device with device identifier ID. ID can be specified as an integer or a string (MATLAB Help, Version 7).

`'adapter'`	The hardware driver adaptor name. The supported adaptors are Advantech, Keithley, MCC, NIDAQ, and Parallel.
`ID`	The hardware device identifier.
`DIO`	The digital I/O object.

```
lines = addline (obj, hwline, 'direction')
```
(11.3)

`Lines = addline(obj,hwline, 'direction')` adds the hardware lines specified by hwline to the digital I/O object obj. direction configures the lines for either the input or output. Lines are a row vector of lines (MATLAB Help, Version 7).

`Obj`	A digital I/O object.
`hwline`	The numeric IDs of the hardware lines added to the device object. Any MATLAB vector syntax can be used.
`'direction'`	The line directions can be In or Out and can be specified as a single value or a cell array of values.

Finally, the recognized voice commands shall convert to binary vector using MATLAB program language `putvalue(obj,data)` as shown in Table 11.4.

`putvalue(obj,data)` writes data to the hardware lines contained by the digital I/O object obj (MATLAB Help, Version 7).

`Obj`	A digital I/O object.
`Data`	A decimal value or binary vector.

11.8 Results Simulation

11.8.1 Results of the Proposed Algorithm

For the proposed algorithm, 50 utterances for each speech command are used for the training step. Table 11.5 shows the result of applying the 2D discrete multiridgelet transform for different data. These data consist of five voice command and each command has 50 utterances. In Table 11.5, the first column represents the training (reference) data with specific sequence of commands that are, namely, Command1–Command5. For example, the

TABLE 11.5

Results of Multiridgelet and Multiwavelet Transform

No.	Reference Sequence (Command1–Command5)	Testing Sequence	Multiridgelet Transform	Multiwavelet Transform	Percentage in 100% for Multiridgelet Transform	Percentage in 100% for Multiwavelet Transform
1	go1–go50	go1–go10	9/10	7/10	90	70
2	left1–left50	left1–left10	10/10	8/10	100	80
3	back1–back50	back1–back10	10/10	10/10	100	100
4	right1–right50	right1–right10	10/10	8/10	100	80
5	stop1–stop50	stop1–stop10	10/10	8/10	100	80

first row in Table 11.5 consists of 50 utterances for command "GO." The second column represents the testing data for the same command but for different speaker. In the same row (i.e., first), the test data have 10 utterances, as a test for command "GO." The next column represents the recognition rate for applying the 2D discrete multiridgelet transform for different data (i.e., represents the speech command recognition). So the first row (9/10) shows 9 out of 10 tested commands can be recognized correctly. Thus, the rate of recognition command "GO" is 90%, while by using multiwavelet transform, the rate is 70%. The last two columns represent the rate for all commands in a percentage of 100% by using multiridgelet and multiwavelet transform, and the overall percentage rates were 98% and 82% for the two transforms, respectively. From this table, I conclude that the proposed method is better than the other method.

11.8.2 Experimental Results

To examine the performance of the proposed algorithm, some experimental tests were done by applying different types of voice command experimentally like GO, BACK, and RIGHT. The following observation has been obtained from it where in many cases, for the wheelchair, motions are considered, showing the path that the wheelchair would take on its motion, making use of the different voice commands to steer the wheelchair in the proper direction.

11.8.2.1 Linear Path

Linear path of a wheelchair can be obtained by a single isolated voice command "GO" and "BACK" according to a specified direction that the user recommends. Figure 11.10 shows linear motion of the wheelchair, either in positive or negative direction of x-axis when the rear (driving) wheels of wheelchair have the same direction of rotation (both wheels rotate in clockwise or anticlockwise) and have the same magnitude ($V1 = V2$). To stop the motion of the wheelchair, "STOP" voice command is used.

Figure 11.11 shows error path of the wheelchair from the commanded path, which represents in figure the x-axis.

11.8.2.2 Circular Path

Circular path of wheelchair can be obtained by a single isolated voice command, either "LEFT" or "RIGHT," according to a specified direction that the user recommends.

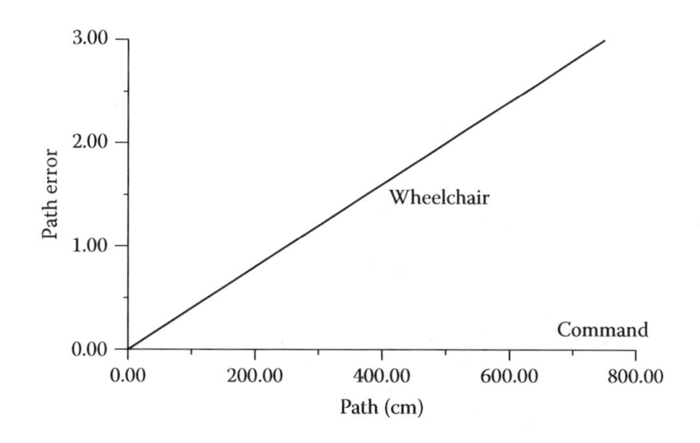

FIGURE 11.10
Linear path corresponding to "GO" voice command.

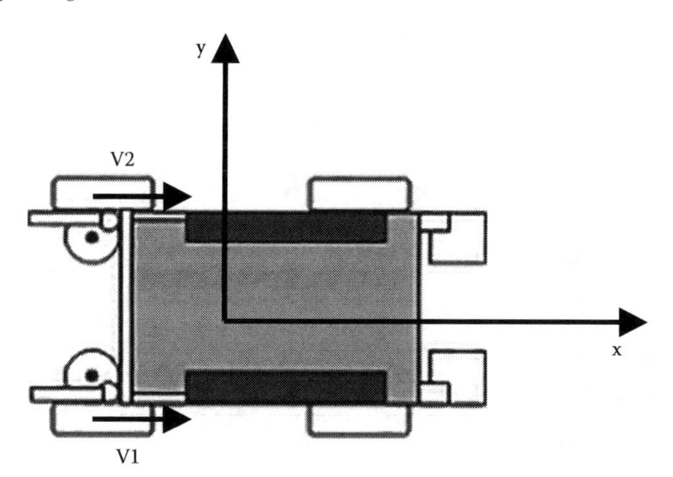

FIGURE 11.11
Actual wheelchair path in linear motion.

Figure 11.12 shows rotational motion of the wheelchair in clockwise using voice command "RIGHT" (V1 = 0, V2 = not zero).

These voice commands generate circular path. To stop the rotation motion of the wheelchair, "STOP" voice command is used.

11.9 Conclusion

In this chapter, a method of isolated word speech recognition system was proposed to control the wheelchair and therefore is the proposed technique that is more efficient for real-time operation used in the control of mobile robot.

This chapter presents a proposed 2D multiridgelet transform computation method that verifies the potential benefit of a multiwavelet and gains much improvement in terms of low computation complexity. A single-level decomposition in the multiwavelet domain

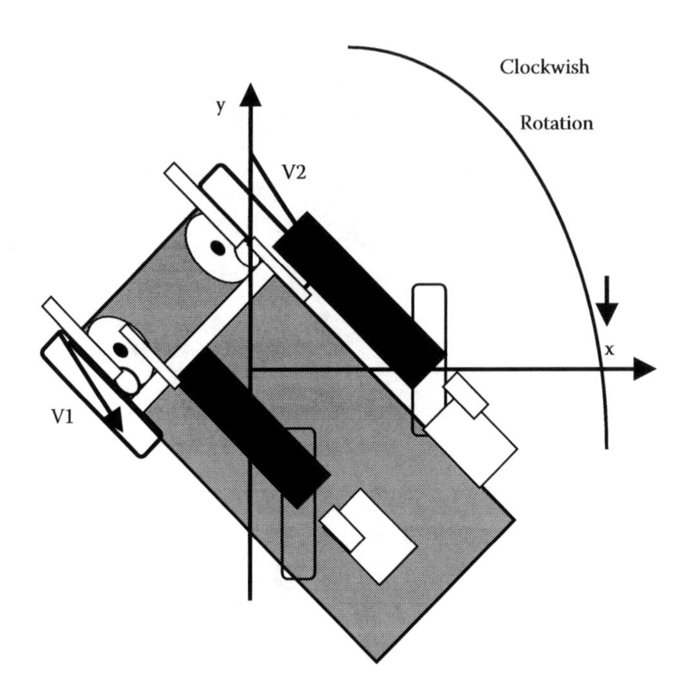

FIGURE 11.12
Rotation corresponding to "RIGHT" voice command.

is equivalent to scalar wavelet decomposition. Thus, computing complexity is higher for DMWT compared to dynamic warping time (DWT).

DMWT computation algorithm using repeated row preprocessing should be applied to a matrix with a size of at least 4 × 4. Multiridgelet transform is an important technique in word recognition application because it eliminates the noise and sharpens or smoothens the speech.

A nonlinear enhancement function was applied on the discrete multiridgelet coefficients; it has a good result (enhancement speech) over a linear enhancement function in multiple speeches. For multiridgelet gaining algorithm (by applying linear or nonlinear enhancement functions.

This rate of recognition is sufficient to control the wheelchair with safety without using additional sensors like ultrasonic and IR; all these sensors are used to support the wheelchair when the recognition rate is weak.

Acknowledgments

This chapter holds all the efforts that Er. Pothireddy Siva Abhilash and I put in during the last 9 months.

Thanks so much to my beloved wife Smt. P. Sarada Rani for standing with me through the work and to my beloved children Er. Pothireddy Siva Abhilash and Er. P. Sivapavan Kumar.

And more thanks to God, "our dwelling place through all generations," and to my father Sri. P. Peda Peraiah (deceased) and my mother Smt. P. Suseela Devi.

References

Chauhan, S., Sharma, P., Singh, H.R., Mobin, A., and Agrawal, S. S., 2000. Design and development of voice-cum-auto steered robotic wheelchair incorporating reactive fuzzy scheme for anti-collision and auto routing. In TENCON 2000. *Proceedings of the IEEE*, Vol. 1, pp. 192–195.

Cook S., 2002, Speech recognition how to, Revision v2.0 April 19, 2002.

Hosseini E., Amini J., and Saradjian M.R., 1996, Back propagation neural network for classification of IRS-1D satellite images, Tehran University, Tehran, Iran. Available at www.uobabylon.edu.iq/publications/applied_edition6/paper_ed6_5.doc.

Hrnčár M., 2007, Voice command control for mobile robots, Department of Control and Information Systems, Faculty of Electrical Engineering, University of Žilina, Žilina, Slovakia.

Li J., Pan Q., Zhang H., and Cui P., 2003, Image recognition using radon transform, *Proceedings of the 2003 IEEE Intelligent Transportation Systems*, Shanghai, China, October 12–15, 2003, pp. 741–744.

Minh N. and Vetterli M., 2003, The finite ridgelet transform for image representation, *IEEE Transactions on Image Processing*, 12(1), 16–28.

Moon I., Lee M., Ryu J., and Mun M., 2003, Intelligent robotic wheelchair with EMG, gesture, and voice-based interfaces, *Proceedings of the IEEE International Conference on Intelligent Robots and Systems, IROS 2003*, Las Vegas, NV, October 27–31, 2003, pp. 3453–3458.

Philips semiconductor, 2003, Integrated circuit data sheet, Feb 06.

Ram Kumar P., Ramana Murthy M.V., Eashwar D., and Venkatdas M., 2005, Time series modeling artificial neural networks, *Journal of Theoretical and Applied Information Technology*, 4(12), 1259–1264.

Rockland R.H. and Reisman S., 1998, Voice activated wheelchair controller. In Bioengineering Conference, 1998. *24th Annual Proceedings of the IEEE*, Northeast, pp. 128–129, 1998.

Simpson R. and Levine S., 2002, Voice control of a powered wheelchair, *IEEE Transaction on Neural System and Rehabilitation Engineering*, 10(2), 122–125.

Takiguchi T., Yamagata T., Sako A., Miyake N., Revaud J., and Ariki Y., 2008, Human-robot interface using system request utterance detection based on acoustic features, *International Journal of Hybrid Information Technology*, 1(3), 61–70.

Terrades O. and Valveny E., 2003, Radon transform for lineal symbol representation, *Proceeding of IEEE of the Seventh International Conference on Document Analysis and Recognition (ICDAR'03)*, 1, 195.

Valin J., Yamamoto S., Rouat J., Michaud F., Nakadai, K., and Okuno H., 2007, Robust recognition of simultaneous speech by a mobile robot, Proceeding of IEEE, 23(4), 743–752.

12

Advanced Architecture Algorithm of Sensor-Based Robotics Security System Framework for e-Governance Technology

Rajeev Kumar and M.K. Sharma

CONTENTS

12.1 Introduction

The main objective of this chapter is to present a new sensor-based security system planning framework for e-governance system robot navigation in unknown environments. The key idea of the sensor-based robotics security system technique is to exploit the information obtained about the environment topology through the sensors and information and communication technology functions to bias the distribution of random nodes in e-governance technology–like approach toward critical regions, that is, narrow passages and hard-to-navigate regions.

This results in a better coverage of the free space especially in environments where narrow passages exist. Inspired by the promising results obtained using this technique for model-based cases, we propose a sensor-based robotics security planning framework [1]. In our proposed sensor-based technique, information obtained about the C-space topology through the FD paradigm is utilized to find the next best view configuration where each scan should be carried out. They are actually measuring the force that produces the acceleration of a known mass. Different types of acceleration transducers are known: stress–strain gage, piezoelectric, capacitive, and inductive. Micromechanical accelerometers have been developed. In this case, the force is measured by measuring the strain in elastic cantilever beams formed from silicon dioxide by an integrated circuit fabrication technology. Sensors are used to detect the positive contact between two mating parts and/or to measure the interaction forces and torques that appear, while the robot manipulator conducts part-mating operations [2]. Another type of contact sensors is the tactile sensors that measure a multitude of parameters of the touched object surface (Figure 12.1).

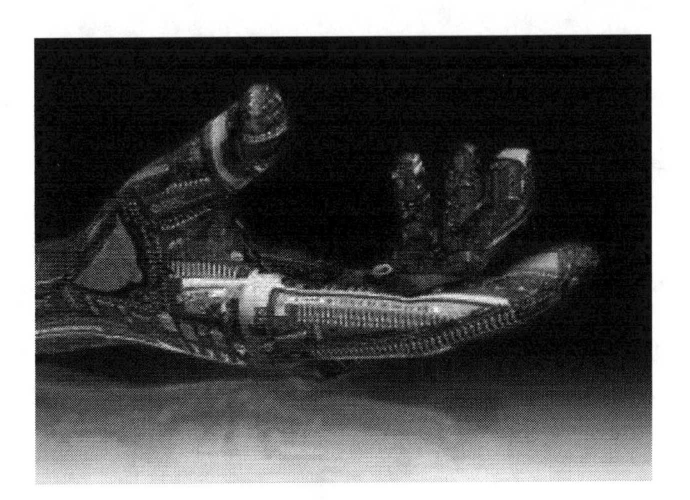

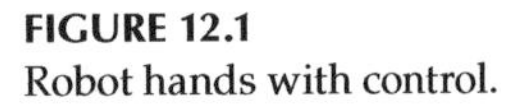

FIGURE 12.1
Robot hands with control.

12.2 Range of Sensors

Range sensors measure the distance of two areas, and objects in their operation area are used. They are used for robot navigation from one contact to other contacts; in robotics system, sensors are used, and they have specific range. If robot sensors are very fares, then these are not working properly and obstacle avoidance or to recover the third dimension for monocular vision. Range sensors are based on one of the two principles: time of flight and triangulation [3].

Time-of-flight sensors estimate the range by measuring the time elapsed between the transmission and return of a pulse. Laser range finders and sonar are the best known sensors of this type.

Triangulation sensors measure range by detecting a given point on the object surface from two different points of view at a known distance from each other. Knowing this distance and the two view angles from the respective points to the aimed surface point, a simple geometrical operation yields the range [3] (Figure 12.2).

12.3 Robot Controller Can Have a Multilevel Hierarchical Architecture

The levels of robot controller are as follows:

Artificial intelligence (AI) level: In this level, commands are used to give the programming commands for the robots of controlling system when you want to execute the particular path; then, robots take executable correct command where the program will accept a command such as, "Pick up the bearing" and decompose it into a sequence of lower level commands based on a strategic model of the task.

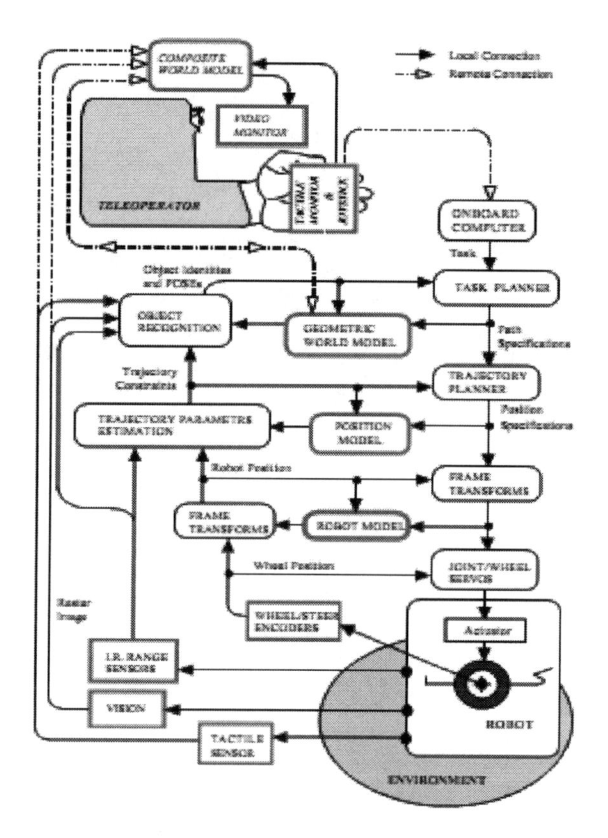

FIGURE 12.2
Control system for e-governance.

Control mode level: This is where the motions of the system are modeled, including the dynamic interactions between the different mechanisms, trajectories planned, and grasp points selected. From this model, a control strategy is formulated, and control commands are issued to the next lower level.

Servo system level: This is where actuators control the mechanism parameters using feedback of internal sensory data, and paths are modified on the basis of external sensory data. Also, failure detection and correction mechanisms are implemented at this level [3]. There are also different levels of abstraction for the robot *programming languages*.

Guiding the systems: In this system, the user leads the robot through the motion sensor to perform the specific path. If we move the robots, then the user gives the guidelines with the help of executable codes (Figure 12.3).

Robot-level programming: Here, the user writes a computer program for the specific motion and tasking. In this programming, we can execute the particular programming code. While we decide everything with our mind, robots decide things with the help of programs and execute the programs and then jump to the next levels [4].

FIGURE 12.3
Advanced monitoring system.

12.4 Algorithm for Sensor-Based Robotics Controlling

The overall algorithm of the sensor-based motion planning framework is given as the following pseudocode [5]:

```
{
   W ← Workspace.Initialize;
   [C, ΔCf ree , Cunkbnd ] ← Cspace.Initialize(W );
   R ← Roadmap.Initialize(ΔCf ree );
   qcur ← qs ;
   it ← 0;
   repeat
     it ← it+1;
     path ← Roadmap.Search(qcur , qg , R);
     if path != NULL
       qcur ← Robot.Move(path); Exit;
     end;
       V ← Fem.Solve(qs , qg , C);
       qv ← Planner.NextBestView(V , Cunkbnd , R, W );
       path ← Roadmap.Search(qcur , qv , R);
       qcur ← Robot.Move(path);
     W ← Workspace.Update(ΔW , W );
     [C, ΔCf ree , Cunkbnd ] ← Cspace.Update(C, W );
     R ← Roadmap.Expand(ΔCf ree , R);
   until it <= MAXITNUM
}
```

In this algorithm configuration, the robots give the working of robots in their workplace of security system.

12.5 Application of Robotics System in Different Type of e-Governance Technology [6]

Hospital robotics: Automated robotic carts with motivity make predictable, anywhere-to-anywhere deliveries, avoid people, and other obstacles, without expensive retrofitting to the workplace. When vacuum tubes are too expensive, intrusive, or inadequate for the job, transport systems with motivity solve transport and conveyance problems. When manual systems put staff at risk or distract from core duties, robotic carts with motivity provide welcome relief from dull, dirty, or dangerous tasks.

Remote monitoring system in e-governance technology: Global enterprises require remote access to far-flung facilities. And in this we adopt the monitoring system to every secure areas because it is provide the monitoring system. Adept motivity platforms collect time and spatially stamped sensor data and images for real-time viewing, historical tracking, and predictive modeling of conditions at remote sites. Optional pan–tilt laser pen allows viewer to point in remote space to direct occupants' attention for remote support calls and maintenance questions. Remote monitoring system is used in defense areas and facilitates the e-governance technology [7].

Security robotics system in securities areas: In this, the user remotely monitors, manages, inspects, and assists a facility with intelligent mobile robots powered by adept motility by integrating them with building information and security management systems. Interrupt patrols for incident response, such as alarm verification, supply delivery or call for resources at an event location. Implement security automation on a controlled cost, low-impact schedule; start with one robot, build to 100. New checkpoints, tasks, and sites can be added quickly. Typical ROI is 6–12 months in a 24/7 operation, with added benefits from improved tracking, increased reliability, and lower worker risk of hazard.

Outdoor robots: Adept Mobile Robots outdoor platforms offer multiple solutions for commercial applications. Whether the need is a lighter, easily transported base with hot swappable batteries, or a larger base capable of running all day without recharging, Adept Mobile Robots has a seeker robot platform for the task [8]. Both seeker robot platforms can be teleported, driven by a cabled joystick, or allowed full autonomy, performing tasks and patrolling routes unsupervised. Add GPS, laser rangefinders, PCs, or application-specific hardware like gas sensors or powerful manipulators to get the job done in a cost-effective manner without putting a human in harm's way.

Intelligent robotics kiosks: Intelligent display robots adapt mobility with touch screen data displays, customized, networked, embedded PC and supplies where needed, avoiding people and other obstacles [9], without expensive retrofitting to the workplace. The people-height touch screen makes interaction easy. When staff hands are full, mobile robots with adept motility can carry what's needed to the next location on their own and begin working or follow a staff person there [10] or they can guide visitors and provide tours. Implement touch screen robots on a controlled cost, low-impact schedule; start with one robot, build to 100. Add new destinations, tasks, and worksites quickly, or just stay in follow mode. Typical ROI is 12 months in a 24 × 7 operation, with added benefits from improved tracking, increased productivity, and lower worker risk of hazard (Figure 12.4).

FIGURE 12.4
Robot security system with monitoring.

12.6 What Robots Can Do Right Now in e-Governance Technology

We can show some examples here [11]...

Cleaning: Vacuum cleaner is a best example of the cleaning system at our home, even if you're not at home!

Automated hauling: Several robots will carry dishes and other small loads from room to room, such as to a friend recovering from hip surgery, and to carry food from the kitchen to the living room and the dirty dishes back into the kitchen again. Since he was on crutches, this was a real lifesaver.

Security: Home robots could easily be tied into a computerized home security system, and the robot's mobility would allow more areas in the home to be protected. And robots make secure security system.

Alarm clock: With a little work, I will soon be able to use Cybert as an alarm clock. Every morning, he will roll into my bedroom and wake me up; once he senses that I'm out of bed he will follow me into the bathroom and deliver up-to-the minute news, weather, sports, and stock market information.

Entertainment: Robotics is an exciting hobby for many people around the world. There are countless clubs, websites, and books that have been written for those who are interested in the topic.

Education: Using a home robot not only teaches about robotics, it teaches spatial navigation, mapping, dud reckoning, programming, and more.

12.7 What Robotics System Will Be Able to Do in the Future Technology

Some examples of robotics are as follows [11]:

Pest control: Small robots may one day scurry around our homes at night, locating and smashing cockroaches and other unwanted guests (no, I'm not talking about your mother-in-law).

Child care: The technology already exists to use a robot to check on the kids while we are away from the house. Robotics will soon add a camera and an Internet interface that would allow someone to "drive" the robot around from a remote computer and receive live pictures of everything that Cye "sees" [4].

Advanced home security and management: Robots in the near future will use advanced AI to monitor our homes, make sure everything is functioning properly, and watch out for intruders.

Hazard detection: It would be fairly easy to attach fire, smoke, carbon monoxide, and other detectors to a home robot. Every night, the robot could "make the rounds" to ensure that everything is okay [12].

12.8 Future Scope

In this framework, we can use the advanced features for advanced security system. Because sensor-based security system is the advanced technology for new generation, it has used sensor in robots and control it for many security purpose. It provides the fast and secure security system, and in this framework, we can handle the features of security.

12.9 Conclusion

In this chapter, we presented a new advanced architecture of sensor-based robotics algorithm of security system planning framework for e-governance technology environments. In this algorithm technology, we can implement an advanced security system with the help of robotics environments. The main idea of the proposed planning approach is to utilize a different type of security system in e-governance. In this chapter, we discussed that robots can help in different areas like hospital, defense, home, and other e-governance areas. Sensor-based robotics system provides easier and faster security system, and because every human nowadays can use hacking and many more, it becomes very useful for people, making this technology very helpful for e-governance as well.

References

1. Kazemi, M. and M. Mehrandezh, Robotic navigation using harmonic function-based probabilistic roadmaps, in *Proceedings of the IEEE International Conference on Robotics and Automation*, New Orleans, LA, April 2004, pp. 4765–4770.
2. Kazemi, M., M. Mehrandezh, and K. Gupta, An incremental harmonic function-based probabilistic roadmap approach to robot path planning, in *The IEEE International Conference on Robotics and Automation*, Barcelona, Spain, April 2005 (submitted).
3. Kutulakos, K. N., V. J. Lumelsky, and C. R. Dyer, Vision-guided exploration: A step toward general motion planning in three dimensions, in *Proceedings of the IEEE International Conference on Robotics and Automation*, Atlanta, GA, May 2–6, 1993, pp. 289–296.

4. Masoud, A. A., A hybrid, PDE-ODE control strategy for intercepting an intelligent, well-informed target in a stationary, cluttered environment, *Applied Mathematical Sciences*, 1(48), 2007, 2345–2371.
5. Yu, Y. and K. Gupta, Sensor-based probabilistic roadmaps: Experiments with an eye-in-hand system, *Advanced Robotics*, 14, 2000, 515–536.
6. Kazemi, M., M. Mehrandezh, and K. Gupta, Sensor-based robot path planning using harmonic function-based probabilistic roadmaps, in *Proceedings of the 12th International Conference on Advanced Robotics, 2005, ICAR '05*, Seattle, WA, July 18–20, 2005, pp. 84–89.
7. Masoud, A. A., A discrete harmonic potential field for optimum point-to-point routing on a weighted graph, in *The 2006 IEEE/RSJ International Conference on Intelligent Robots and Systems*, October 9–13, 2006, Beijing, China, pp. 1779–1784.
8. Wiener, N., *Cybernetics, or Control and Communication in the Animal and the Machine*, The MIT press, Cambridge, MA, 1961.
9. Keymeulen, D. and J. Decuyper, The fluid dynamic applied to mobile robot motion: The stream field method, in *Proceedings of the IEEE International Conference on Robotics and Automation*, Sacramento, CA, April 1994, pp. 790–796.
10. Wiener, N., *The Human Use of Human Beings: Cybernetics and Society*, Houghton Mifflin Company, Boston, MA, 1954.
11. Shapiro, S., ed., *Encyclopedia of Artificial Intelligence*, John Wiley & Sons, New York, 1992.
12. Connolly, C. I. and R. Grupen, On the applications of harmonic functions to robotics, *Journal of Robotic Systems*, 10(7), 1993, 931–946.

13

Smart System Design in Cyber-Physical Systems

Krishnarajanagar GopalaIyengar Srinivasa, Abhishek Kumar, Nabeel Siddiqui, and B.J. Sowmya

CONTENTS

13.1 Introduction

Cyber-physical systems (CPS) are found in fields as diverse as automotive, aerospace, chemical processes, consumer appliances, health care, civil infrastructure, manufacturing and transportation, entertainment, and energy. Embedded systems focus on computational elements, and less on an intense link between the computational and physical elements. Unlike traditional embedded systems, a full-fledged CPS is typically designed as a network of interacting elements with physical input and output instead of as standalone devices [1]. The notion is closely tied to concepts of robotics and sensor networks. Ongoing advances in science and engineering will improve the link between computational and physical elements, dramatically increasing the software quality attributes like adaptability, autonomy, efficiency, functionality, reliability, safety, and usability of CPS. The increase in software attributes will broaden the potential of CPS in several dimensions like intervention in case of collision avoidance; precision in case of robotic surgery and nanolevel manufacturing; operation in dangerous or inaccessible environments in case of search and rescue, firefighting, and deep-sea exploration; coordination in case of air traffic control, war fighting, and efficiency in case of zero-net energy buildings; and augmentation of human capabilities in case of health-care monitoring and delivery.

Today, numerous CPS applications have been proposed; the application areas are smart space, smart transportation, health care, and more. In smart space application, people can use smart phone or tablet to control the daily appliances via remote Internet access. Health-care application can help doctors to observe the vital signs of the elderly patients even if the elderly patients stay at home instead of the hospital. In smart transportation application, sensor nodes can be embedded in vehicles to detect the nearby environmental information. For example, the accelerometer and GPS receiver can be embedded in a vehicle. Where the accelerometer detects a pothole on the road, the vehicle can send the current coordinates which are obtained by the GPS receiver to the nearby vehicles and thus traffic safety and efficiency can be improved.

The CPS mainly integrates computing with hardware control. Not only can it enhance the efficiency of system operation, but it also has the ability of monitoring and controlling electronic devices [2]. Unlike the traditional embedded systems, the CPS is mainly designed for connecting physical devices to build an interaction network. The common way to build CPS is to embed sensors and actuators into electronic devices in daily life. The information of environment and electronic devices usage collected by sensors will be sent to the decision-making system or the user by the existing wireless sensor network (WSN) techniques, such as routing, data gathering, and media access control protocols [3–5]. Upon receiving the information, the decision-making system or the user analyzes the collected information and then reflects the decision to the actuators by a sequence of control processes, controlling the electronic devices to perform the corresponding task. Smart systems in CPS typically consist of diverse components, such as:

1. Sensors for signal acquisition
2. Elements transmitting the information to the command and control unit
3. Command and control units that take decisions and give instructions based on the available information
4. Components transmitting decisions and instructions
5. Actuators that perform or trigger the required action

A lot of smart systems are evolved from microsystems. They combine technologies and components from microsystems technology, which are essentially miniaturized electric, mechanical, optical, and fluidic devices and with knowledge, technology, and functionality from other disciplines like biology, chemistry, nanosciences, or cognitive sciences. Major application fields of smart systems are manufacturing technologies, medical technologies, automotive, aerospace, safety and security, logistics, and information and communication technology (ICT).

Smart system integration provides merging their functional and technical abilities into an interoperable system, to emphasize the combination of components in an industrial context. This term reflects the industrial requirement and particular challenge of integrating different technologies, component sizes, and materials into one system.

13.2 Energy Usage Awareness in CPS Design

The combination of computing, communication, electronics, and mechanical subsystems, which operate autonomously or as part of a network together with external systems in order to provide a certain service are known as CPSs. The heterogeneity in these systems makes them especially complex to design [6]. Battery-powered CPS have limited amount of energy available to ensure their operation. Banerjee et al. present the sustainability from the energy perspective as one of the key issues to address when designing CPS [7]. System-level design should be applied considering the computing and noncomputing aspects to find the solution for the heterogeneous composition.

A design process takes power and energy and related aspects into consideration and is known as power- or energy-aware design. Power-aware design was originally applied at the device level and keeping its expansion to other fields in the mid-1990s [8]. Going up in abstraction layers implies jumping from the electronic engineering to the computer engineering field [9,10]. This approach was considered "system-level power-aware design." Due to the heterogeneity of today's solutions and to the broader implications of the term "system," power-aware design has been extended now to the noncomputing aspects. This implies that the solution under design must include mechanical, electrical, and software elements in a single-design effort using subsystems and components. To overcome this high complexity, we need to apply abstract modeling [11]. System-level power awareness for CPS can benefit from the extensive work carried out in several fields separately until now in power and energy consumption analysis and management. Energy consumption in computing has been extensively researched and characterized at different levels of abstraction, ranging from the microarchitectural level and the instruction level [12,13] all the way up to software design for low power [14,15]. Additional work has been done to characterize the energy cost of communicating computing units with other embedded system peripherals [16]. Energy consumption in communication has been especially researched since the introduction of WSN. Extensive work has been carried out in order to analyze hardware and protocol optimizations at different layers including improvements in the RF technology, network protocols [17–19], and specific battery discharge characteristics [20]. Work on network is not only limited to wireless network; Chabarek et al. propose the introduction of power awareness in wired network design and routing in [21]. The energy consumption of mechanical and electrical devices has been widely studied; a combined approach to study their energy consumption is still not widely researched. The overview of energy consumption modeling approach in CPS is given in Figure 13.1. However, there is an increasing interest and tool support for mechatronic system design through tools like *Destecs* or *Modellica* [22,23]. Some have identified the importance of characterizing the energy supply in mechatronic systems.

CPS energy consumption modeling approach in CPS consists of energy consumption in mechatronic systems, energy consumption in computation, hardware in the loop (HIL) simulation and on-target measurements, and energy consumption in communication. CPS comprises system-level energy-aware design approach.

We have identified three main application areas, involving the development of

1. *CPS*: These are composed of electromechanical devices to interface the physical world (sensors and actuators) and one or more embedded control unit that uses them. The ways these interfaces are controlled have a significant impact on the total system energy consumption. CPS can operate as standalone devices or as part of a network.

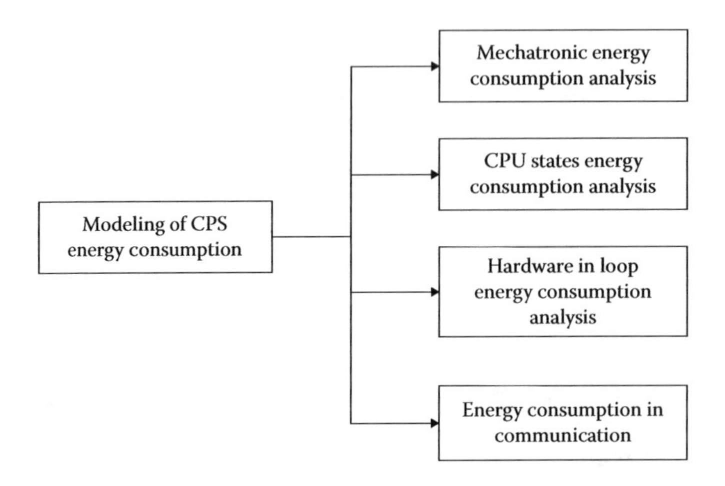

FIGURE 13.1
Overview of the energy consumption modeling approach in CPS. (From Esparza Isasa, J.A., System level energy aware design of cyber-physical systems, Technical Report ECE-TR-16, Department of Engineering, Aarhus University, Aarhus, Denmark, October 2013.)

2. *Personal or industrial embedded devices*: Resource-constrained computational units consists energy-demanding feature typically on the computational side. Example is portable multimedia player or production process logging device.
3. *Small-scale networked embedded systems*: These are composed of multiple embedded systems equipped with a network interface. These networks are responsible for monitoring and control, which are composed by both sensors and actuators. WSN is one of the networked embedded systems.

Power-saving techniques: Efficient in energy consumption can be achieved by hardware and software techniques. Some of the techniques are as follows: (1) *Dynamic frequency scaling*: Adjusting the operating frequency at run time in order to achieve the power reduction; (2) Dynamic voltage scaling: Adjusting the operational voltage at run time. This results on lower operating temperature and therefore lower power consumption; and (3) *CPU operational modes switching*: Changing CPU operational states depending on the computational demands. The mode can change from either to active or sleep mode.

13.2.1 Modeling Energy Consumption in Mechatronic Systems

Energy consumption in mechatronic systems is determined by the consumption in the electronics including the CPU and other peripheral hardware and in the electromechanical devices used to interface the physical environment. Typically, the sensors and the actuators are commanded from a control logic running in a CPU, hence it could be concluded that the energy consumption in most mechatronic systems is software dependent.

13.2.1.1 Challenges Addressed

Embedded software engineers developing control algorithms need to understand the energy consumption in the electromechanical domain to optimize overall system energy consumption. A system prototype or a model need to be created and tested with different control algorithms over it to achieve optimization of overall system energy consumption. The creation of a prototype can be complex, expensive, and it can easily take a considerable

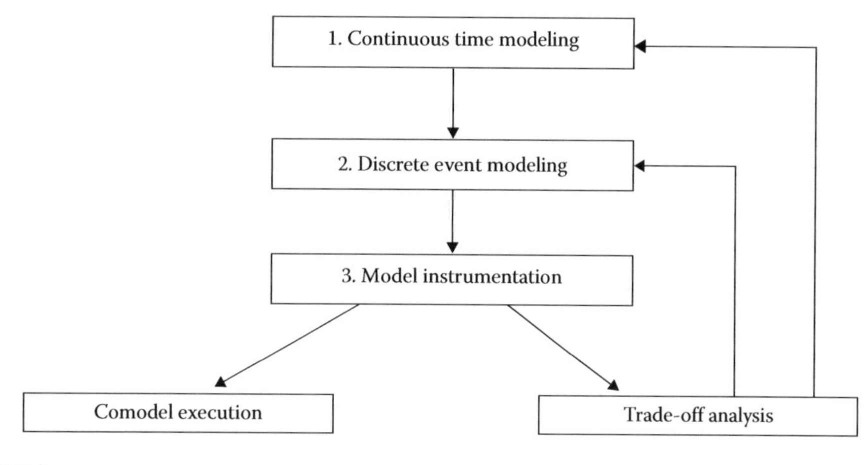

FIGURE 13.2
Methodology overview.

amount of time. Different modeling paradigms are used to represent elements of mechatronic. Discrete event (DE) modeling languages such as Vienna development method-real time (VDM-RT) are used to model control logic. Continuous time (CT) representations, such as differential equations, are used to model physical phenomena.

Methodology adopted toward energy consumption is composed of five phases. An overview of this modeling process is presented in Figure 13.2. Phases are in sequential manner; iteration among the phases can also be possible. In this methodology, we take as input the physical requirements and component characteristics in the CT modeling phase and control requirements in the DE modeling phase.

These five phases can be explained as follows:

1. *CT modeling*: In the first phase, the control signals are used to check that the physical model is appropriate for the applications. Create the model of a system from mechanical perspective including physical requirements and component characteristics.
2. *DE modeling*: Capturing the control logic depends on the control requirement as an input. This is to validate the physical models. Errors found in physical model should be corrected and iterate over the DE modeling phase. Once the validation is completed, complex control strategies can be modeled and iterated again over the CT model.
3. *Model instrumentation*: Interfaces and monitored variables are required to evaluate energy consumption. Monitoring the target variables from physical simulator that is performed using CT model proposes the introduction of a power meter block per component consuming energy in the CT model. Computational power consumption of one component is the responsibility of every single power meter block. DE model provides the controlling signal, which is used in the computation of the total consumed power and energy. The calculated power consumption can be defined as an output flow from the power meter for later processing in the CT model. In order to compute total power consumption of the system, we communicate all the component power meters with the system power meter. This particular block adds the power consumption logged by all the component meters further integrating it to calculate the total energy consumption of the system.

4. *Comodel execution*: For this phase, we analyze results of controlling algorithms for mechanical system and execute the comodel. This phase produces numerous system consumption estimates for monitoring power and energy consumption over time hence analyzing the performance of controlling algorithms.
5. *Trade-off analysis*: For this phase, we analyze the results of final energy and power consumption to compare them with the system requirements. Analysis results of power and energy consumption can be observed for all the single components. Here, the unsatisfied requirements and solution are looked into and can the chosen components are revisit.

13.2.2 Modeling Energy Consumption in Different CPU States

Microcontroller-based embedded systems can achieve remarkable energy savings by using low units of power CPU states. In memory computations in the CPU, waiting for external or self-triggered internal events to wake up are stopped by low-power CPU states. The challenging question is when to use low-power states and also when the system requirements are met, occurring while designing the embedded software [9].

The VDM-RT modeling language introduces the CPU abstraction that represents an execution node where components of a model can be deployed into. Every single execution environment is the representation of the hardware support for the CPU and a real-time operating system. The CPUs are configured using the parameter clock frequency and also the scheduling policy. VDM-RT CPUs are constantly able to compute and communicate and thus represent a CPU that is active continuously. The evaluation process of abstraction is inappropriate if we are interested in studying the power consumption of a CPU that uses low-power consumption states. The challenging question is to incorporate a way to represent CPU sleep states without requiring language, scheduler, or platform modifications at the initial stage [11]. Design pattern architecture to model a CPU power mode for pattern application involves modeling functionality of an active CPU state, modeling functionality that runs if CPU is active, and the data analysis of the CPU state [9].

13.2.3 Model Validation and Verification Using the Hardware in the Loop in VDM-RT

The previous topics have provided an insight into CPS modeling with focus on energy consumption. The aforementioned techniques presented so far aim on the exploration of the design space. Introducing the application of HIL for VDM-RT models, enabling the combination of a complete or partial system implementation with a system-level model, is outside the scope. Typically, HIL is used as a way to provide system with external stimuli and the purpose of mocking up an external piece of equipment.

13.3 Smart Energy System

Continuously changing energy systems bring about the change in the old model of reliable, cheap, and in demand energy due to concerns over changing weather, energy security, aging infrastructure, and rising prices. Various energy sources available and infrastructure constraints need the demands to be matched in future. This variation requires unique and advanced smarter energy system architecture, with active demand side management, communication technologies, real-time processing of data, and different energy business model opportunities.

Various functions of an advanced smart energy system are as follows:

- Demand side management for enabling system demands to be met by adjustment of both supply and demand
- Enhanced monitoring of system assets for reducing maintenance costs and enhancing security of supply
- Control and optimization of consumers' appliances, such as heat pumps, to detect faults, minimize energy use, and reduce costs
- Greater control of energy services: heating, appliance use, and lighting
- Better integration of distributed energy, on macro- and microscales
- Cost minimization for both suppliers and consumers

Reliable energy is responsible for everything from our finances and transportation to our health, water supply, and emergency responses. The backbone of the energy sector is energy delivery systems—networks of physical processes that produce, transfer, and distribute electricity, oil, and natural gas. These physical systems rely fundamentally on control systems—the interconnected electronic and communication devices that monitor and control these processes. Control systems (the sensors and actuators) monitor physically and control the energy processes, the computer-based systems (analyze and store data), and the communication networks (interconnect the process and computer systems). Control systems' provide timely information to system operators and automated controlling over a large dispersed network of assets and components because it is having highly reliable and flexible energy infrastructure. The success of the economic and the public health for the safety of citizens and businesses is especially dependent on a reliable, resilient, and secure electric transmission and distribution grid.

Electric grid is driven on demand; electricity is generated as per the requirement, with minimal storage capability. Instead, the grid is dependent on control systems, high-speed computer, and communication systems that constantly stabilize the generation and flow of electricity. Managing generation with requisite responsiveness to demand is inefficient for many of today's generation systems, for example, coal-based electric generation. They are difficult and slow to switch on- or offline in the face of load changes. Disturbances in the control can destroy critical process components and cause failures capable of destroying the generation and flow of electricity to end users across the nation. Excessive loads on the peaks, overheated transmission lines, and natural accidents such as falling tree limbs cause brownouts and blackouts, sometimes triggering cascading power system failures affecting various regions of the country. The power grid is vulnerable across the intentional disruption: a shift from closed to extensively networked (including Internet) controlled communication, and from proprietary to widely accessible commercial software, has increased vulnerability to cyber security attacks and threats. This was seen recently in the 2009 discovery of Stuxnet's malware, specifically designed to exploit Supervisory Control and Data Acquisition systems.

Evolution of CPS landscape: CPS are becoming ubiquitous, pervading every sector of the critical national infrastructure and every aspect of an individual's daily life, including Table 13.1.

TABLE 13.1

Evolving Cyber-Physical Systems Landscape

	Current	Future
Medical care and health	Pacemakers, infusion pumps, medical delivery devices, connected to the patient for life-critical functions	Life-supporting microdevices, embedded in the human body; wireless connectivity enabling body area sensor nets; mass customization of heterogeneous, configurable personalized medical devices and natural, wearable sensors (clothing and jewelry) and benignly implantable devices
Energy	Centralized generation, supervisory control, and data acquisition systems for transmission and distribution	Systems for more efficient, effective, safe, and secure generation, transmission, and distribution of electric power, integrated through the smart grid; smart ("net-zero energy") buildings for energy savings; systems to keep nuclear reactors safe
Transportation and mobility	Vehicle-based safety systems, automatic breaking system (ABS), traction and stability control, power train management; precision GPS-enabled agriculture	Vehicle-to-vehicle communications for enhanced safety and convenience ("zero fatality" highways), drive-by-wire, autonomous vehicles; next-generation air transportation system; autonomous vehicles for off-road and military mobility applications
Manufacturing	Computer-controlled machine tools and equipment; robots performing repetitive tasks, fenced off from people	Smarter, more connected processes for agile and efficient production; manufacturing robotics that work safely with people in shared spaces; computer-guided printing or casting of composites
Materials and other sectors	Relatively few, highly specialized applications of smart materials—predominantly passive materials and structures	Sustainable mass production of "smart" fabrics and other "wearable" with applications in many areas; Actively controlled buildings and structures to improve safety by avoiding or mitigating accidents; electronics provide versatility without recourse to a silicon foundry; emerging materials such as carbon fiber and polymers offer the potential to combine capability for electrical and/or optical (hence NIT) functionality with important physical properties (strength, durability, and disposability)

Source: Winning the future with science and technology for 21st century smart systems, https://www.nitrd.gov/nitrdgroups/images/1/12/CPS_OSTP_ResponseWinningTheFuture.pdf.

13.4 Smart Cities, Smart Buildings, Smart Grid, and Intelligent Transport Systems

13.4.1 Smart Cities

Urban development policies have attracted some considerable attention toward the concept of "smart cities." The Internet broadband network technologies play a role in providing e-services and become increasingly important for the development of urban areas. Cities are increasingly assuming a critical role as drivers of innovation in areas such as health, inclusion, environment, and business. Therefore, the issue arises of how cities, surrounding regions, and rural areas can evolve toward sustainable open and user-driven innovation ecosystems to boost future Internet research and experimentation for user-driven services and how they can accelerate the cycle of research, innovation, and adoption in real-life environments. This chapter pays particular attention to collaboration frameworks, which integrate elements such as future Internet test beds and living lab environments that establish and foster such innovation ecosystems. Investments in human, social capital, traditional transport, and modern (ICT) communication infrastructure fuel sustainable economic growth and a high quality of life, with a wise management of natural resources, through participatory government leads to smart city. This definition balances different economic and social demands as well as the needs implied in urban development, while also encompassing peripheral and less developed cities. It also emphasizes the process of economic recovery and welfare of well-being purposes. Second, this characterization implicitly builds upon the role of the Internet and Web 2.0 as potential enablers of urban welfare creation through social participation, for addressing hot societal challenges, such as energy efficiency, environment, and health.

The cities should create a rich environment of broadband networks that support digital applications. This includes the following:

- The enrichment of broadband infrastructure combining cable, optical fiber, and wireless networks, offering high connectivity and bandwidth to citizens and organizations located in the city
- The development of the physical space, infrastructures of cities with embedded systems, smart devices, sensors, and actuators, offering real-time data management, alerts, and information processing
- The creation of applications enabling data collection and processing, web-based collaboration, and actualization of the collective intelligence of citizens

The technologies in cloud computing and the emerging Internet of Things, open data, semantic web, and future media technologies can assure economies of scale in infrastructure, standardization of applications, and turn-key solutions for software as a service. These technologies reduce the development costs for operating smart cities. These technologies also consist of initiating large-scale participatory innovation processes for the creation of applications that will run and improve every sector of activity, city cluster, and infrastructure. All citizens and organizations participate in the development supply and consumption of goods and services as city economic activities and utilities can be seen as innovation ecosystems.

13.4.2 Smart Buildings

Next-generation technologies and methods integrated into building systems could allow building energy use to be seamlessly and easily predicted, monitored, controlled, and minimized across the dimensions of performance, scale, and time. The energy monitoring and control systems usage are then negotiated for energy consumption and prices with the utility company, which facilitates both home and office businesses to be connected with the smart grid. The cyber and the physical worlds should be tightly integrated to enable many of these concepts to work effectively in buildings. Achieving net-zero energy (NZE) buildings, for example, where the building can produce as much (or more) energy than it consumes, requires highly integrated systems of cyber and physical components. The innovative smart buildings of the future will also include cogeneration of heat and power with sophisticated controls. CPS provide opportunities for improvement in the efficiency and performance of commercial and residential buildings as well as other structures, such as bridges and dams in future.

Issues of building ownership, that is, building owner, manager, or occupants, challenge CPS integration with questions such as who pays initial system cost and who collects the benefits over time. A lack of collaboration between the subsectors of the building industry slows new technology adoption and can prevent new buildings from achieving energy, economic, and environmental performance targets. Uncertainty in the upcoming future policies and a near term focus for buildings increase the risk in adoption of advanced building technologies such as CPS. Funds for bridge repair, maintenance, and replacement shrink as costs of construction continue to rise. Although structural health-monitoring (SHM) systems could help determine when a bridge needs maintenance and possibly extend its useful life, these systems would require funding themselves as well as additional maintenance and replacement costs. Integration of CPS both within the building and with external entities, such as the electrical grid, will require stakeholder cooperation to achieve true interoperability. As in all sectors, maintaining security will be a critical challenge to overcome.

13.4.2.1 Current State of the Technology

Building technologies: Modern buildings are systems of components consisting of interacting heat exchange, airflow, water, safety, access/security, and movement control subsystems. These subsystems are increasingly coupled using embedded sensing and control systems where state information from one system is directly used to make operational decisions in another subsystem. Increase overall building performance by improving control and efficiency while reducing costs can be done by integration of different controls and aspects of buildings. Certain areas of building control are already becoming more connected. Submetering represents an area where further research in CPS would be applicable and could help in achieving NZE buildings. Submetering allows for the possibility of gathering continuous data for individual areas, systems, or equipment. Integrating submetering with building automation systems, and the development and implementation of technologies including sensor systems that can evaluate the data collected, could lead to greater energy conservation and efficiency.

Smart sensors and SHM: Smart sensors have become a very good example of technology that exists today and will expand and become even smarter in future to play a major role in CPS. They are typically low cost and battery powered and has an onboard microprocessor and sensing capability. They are a viable option to be used in SHM projects. SHM is an emerging field in civil engineering that allows the possibility of continuous or periodic assessment of the condition of civil infrastructure. Current sensors can discretely monitor factors such as strains, accelerations, deformations, and corrosion potential. SHM would provide the information to assess the condition of human living structures such as bridges, dams, or any other structures and help to determine when preventative maintenance is necessary, thus preventing structural failure or costs. The European research project Wireless Sensor and Actuator Networks for Critical Infrastructure Protection has recently successfully demonstrated cost-effective wireless sensor systems for monitoring of electricity distribution and water networks. These types of systems are designed to secure and better manage different types of critical infrastructure.

13.4.2.2 Measurement Problems and Impediments

Achieving NZE in smart buildings: Trouble in measurements for intelligent buildings is numerous and must overcome before NZE reach fruition. Performance metrics and measurement methods, tools to predict performance, protocols to achieve desired performance, and evaluation and assessment of the performance of technologies are the new metrology for smart building technologies, systems and practices, and performance-based standards and practices. System complexities and interactions in a structure should be captured, while innovation in the design and manufacturing of individual components and systems is supported. Intelligent buildings are unable to effectively communicate, interact, share information, make decisions, and perform smoothly and reliably because of a lack of measurement methods. Measurement science is needed to support intelligent building systems that can detect and respond to faults, operational errors, and inefficiencies to ensure that buildings perform as expected and performance does not reduce. Data functions for assessing the performance of buildings, tests and test beds for the evaluation of control technology and fault detection approaches, best practice guidelines for intelligent design and operation of buildings, measurements to support automation of commissioning processes, and low-cost, reliable energy metering systems are the challenges to achieve NZE in smart buildings. To have enhancements in communication protocol standards that enable the practical use of integrated systems, such as lighting and energy management and achieve increased comfort safety, energy efficiency, and secure, real-time communication of information within the building system, should overcome the rising challenges of NZE in smart living structures. Test beds could enable more energy-efficient building operation through the development of information models and software tools that improve the design and commissioning process and increased use of embedded intelligence that can detect and respond to problems and optimize the control and performance of building systems.

Sensors for monitoring bridges and other structures: The development of more robust and advanced smart sensors could help provide valuable information about the health of various structures, including bridges, tunnels, buildings, and water distribution systems. While smart bridge technology is rising, current systems that use discrete sensors are inadequate to completely monitor the large, complicated systems that suffer local abnormalities.

Sensors can provide valuable insight on the structural health and condition of bridges or buildings, but full-scale, effective networks of sensors on bridges (or other types of civil infrastructure) will not become common until rising challenges are overcome. The continual improvement of wireless sensors is essential because they will decrease the need of also installing costly and bulky wires. It must be determined what the sensors will measure about the bridge (e.g., strain, cracking, corrosion, scour, and environmental conditions).

The development of active sensors that excite and measure structural response at various locations could enhance the capability of detecting small changes, resulting in more effective monitoring. Smart sensors should have more computing power to be able to handle large amounts of data, while not becoming too expensive.

Sensors for surveillance and monitoring: The smart buildings of the future will ideally be equipped with a variety of sensors, ranging from visual, infrared, thermal, magnetic, to others. Using such a package of sensing devices, it could be feasible to monitor the location of anyone in the building, using some device that the person will wear or just by using visual sensors that develop models of a specific person as the person enters the building. Challenges remain in the interpretation of the sensory data, so that the building can monitor the activities of all (or a subset of) the individuals inside the building.

13.4.3 Smart Grids

Figure 13.3 refers the relationship and communications within the smart grid. Finances, transportation, and emergency services depend on the reliable production, transfer, and distribution of energy, including electricity, oil, and natural gas. Electric grid is a complicated system of systems, with many different stakeholders and customers. The National Academy of Engineers has named the development of electrification, which includes today's electric system, as the greatest engineering achievement of the twentieth century. The water supply and distribution system is also enlisted as a great engineering achievement; the supply of safe, clean, and reliable water is important to quality of life, health, and emergency services. These systems are great achievements; they are required to be modernized in order to increase their efficiency and reliability. Existing infrastructure is old, and control systems can be highly improved. CPS stand to have a huge impact on large-scale, computer-mediated physical distributed systems such as the water distribution systems and electric grid. The cyber and physical components integrated in the system have many interactions that can affect the entire operation but are poorly understood. The electric grid and other utilities can use CPS technologies to help the system become innovative, smarter, and more efficient.

A great challenge for CPS is the design and deployment of an energy system infrastructure that is able to provide blackout-free electricity generation and distribution. Integration of CPS technology and engineering to the existing electric grid and other utility systems is a challenge. The increased system complexity poses technical challenges that must be considered as the system is operated, in ways that were not intended when the infrastructure was originally built. As technologies and systems are incorporated, security remains

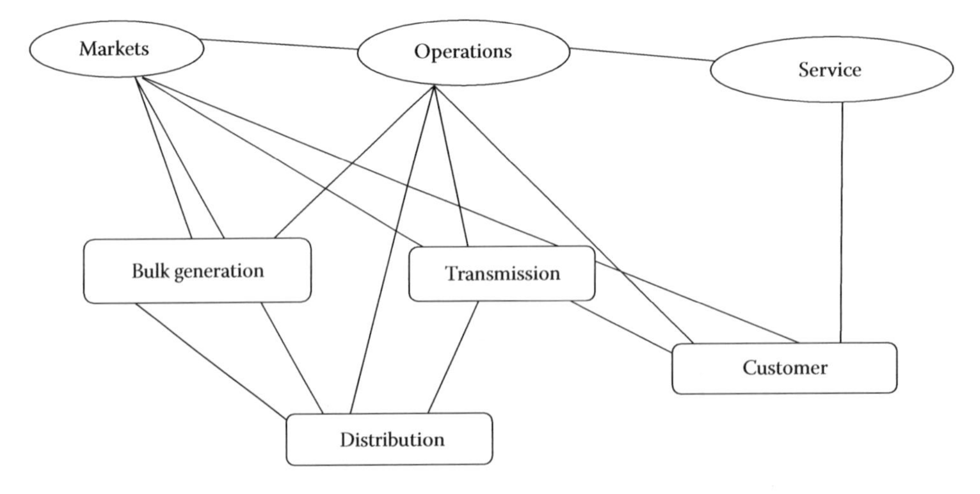

FIGURE 13.3
Smart grid interactions.

utmost concern to lower system vulnerability and protect stakeholder data. In addition, to operate the infrastructure with CPS technologies, a qualified, innovative, and skilled workforce is required. Also affecting CPS integration are numerous nontechnical business and policy challenges including policies on the regulation and implementation of smart grid technologies, standards development, and the responsibility of maintaining, operating, and repairing the equipment for power generation and grid connectivity.

In systems as complex and networked as the smart grid and other utilities, there are measurement problems and challenges in various areas that continue to emerge as CPS and other technologies are integrated. Data acquisition, transmission, retrieval, and retention methods that are reliable and effective must be developed. If there is a delay of the delivery of data from sensors or actuators, this can present multiple challenges to control systems that will rely on real-time data. Data transmission is reliable but not purely trustworthy, as it is still affected by problems such as memory overflow, network overload, and slow processing speed, all of which can cause possible vulnerabilities. There must be methods to ensure that data cannot be corrupted or manipulated through a cyber attack. Current data management methods perform well for amounts of data but fail or become ineffective for larger data set that will come from distribution automation and customer information. Other challenges in data management include data identification, validation, updating, time tagging, and consistency across databases. Another challenge with data comes from not being able to effectively gather it from geographically distributed sources of energy. The ability to collect more accurate data from these geographically dispersed sources would provide producers with a more effective ability to generate and transmit power. When the system is in an abnormal condition, operators in utility plants are overwhelmed by the number of alarms. Operators must make decisions about how to react within very less time. As the smart grid is developed, there must be a method developed that can evaluate, prioritize, and respond to the alarms to minimize confusion.

13.4.4 Intelligent Transport Systems

Transportation CPS in the United States are expected to increase in the future as higher levels of transport autonomy, safety, and convenience. The complexity of transportation systems as a whole, as well as smaller components such as vehicles, is increasing at an exponential rate. Next-generation transportation systems must be highly networked and dynamic, while maintaining high system performance and low cost. Everyday tasks or events, such as commuting by automobile, train, or airplane, depend on complex yet reliable and seamless interactions between the vehicle's computer systems and physical systems while under control by human operators or end users. Today's transportation systems are being designed to be more competitive within their respective industries by adding more complex features and capabilities to increase energy efficiency and safety. Also, CPS advances in transportation can be applied in the military to maintain the edge in fighting capability. The ability to design and adopt energy-efficient transportation CPS supports the U.S. economic, national security, and environmental objectives. The major markets where transportation CPS can be applied include ground transportation or intelligent transportation systems, or air transportation, mainly the Next-Generation Air Transportation System (NextGen), which can be thought of as large-scale CPS. The cyber or computer components of automobiles, aircraft, and other vehicles have been increasing and will continue to play a larger role in these systems. By 2015, as much as 40% of an automobile's value will be in cyber-physical components (electronics, sensors and actuators, and embedded software). The air transportation sector is also heavily reliable on cyber-physical components and composes a significant portion of the U.S. exports. The cost of aircraft is moving increasingly toward software and systems and away from structures, aerodynamics, and propulsion. Advances in CPS will be needed to ensure the safety and security of these increasingly complex and networked automobiles and aircraft.

Federal, state, and local government budgets have limited resources for advanced transportation improvements and to integrate new CPS into the existing infrastructure. Although certification of these systems is critical, several issues remain, such as what should be certified, who is responsible for certification, and how will the systems be certified. Newly created policies and legislation will be required to launch and sustain new technologies. Representing human behavior in the design, development, and operation of CPS in autonomous vehicles is an utmost challenge. Incorporating human-in-the-loop considerations is important for safety, dependability, and predictability. There is currently minimal understanding of how driver behavior will be affected by adaptive traffic control CPS. In addition, it is difficult to account for the stochastic effects of the human driver in a diverse traffic environment (i.e., human and autonomous vehicle drivers) such as that found in traffic control CPS. As CPS become more complex and interactions between components increases, safety and security will continue to be of utmost importance.

Decision makers, industry officials, and experts in the United States agree that intelligent transportation systems and connected vehicle technologies are the future of travel and will improve safety, efficiency, and the economy. Today, humans play an important role in both automotive and aviation operations. People drive vehicles, while sensor systems alert the driver to various challenging situations (lane changes and crash ahead). CPS is increasingly being applied to make it easier, safer, and more convenient for humans to operate transportation systems.

13.4.5 Multivehicle Cooperative Driving and Intersection Control Research

Individual–vehicle-control research concentrates on guaranteeing driving safety. As noted earlier, increased traffic congestion is now making multivehicle-control research an important topic for research in the field of CPS. Twenty years ago, researchers started examining lane changing and lane-merging-control problems. A solution to those problems comes from the path planning literature, which studies how to generate a collision-free driving path or trajectory under the given vehicle dynamics. Using appropriate intervehicle communication to link vehicles, cooperative driving lets vehicles safely change lanes and merge into traffic, and improving traffic performance.

13.4.5.1 Intelligent Sensing for Cyber-Physical Smart Cars

Technology trends in consumer automobiles are moving toward increased autonomy. Early developments in CPS for vehicles include traction and stability control, cruise control, and antilock braking systems that increase safety. Communication between vehicle components provides information such as velocity, acceleration, and traction for the purposes of navigation, infotainment, and other uses. These systems do not take control of the vehicle, but they provide information to the driver who ultimately makes a decision on how to act. Safety behaviors such as shaking the steering wheel to gain the driver's attention cannot alter the situation but can provide necessary information to the driver to enable action. Systems that perceive the environment outside the car as well as the environment inside the car are of particular importance. The state of the art considers three kinds of intelligent-vehicle sensing; out-vehicle environment, in-vehicle environment, and vehicle state.

Out-vehicle environment sensing involves retrieving information about the driving environment. Specific topics include extracting lane boundaries when they are not clearly marked or in adverse weather conditions, detecting other vehicles that are nearby and estimating their kinematics (position, speed, and acceleration), recognizing traffic signs and traffic lights, detecting unexpected traffic participants (such as pedestrians), and sensing obstacles of all kinds. Sensing the environment out of the vehicle is a very difficult task, especially when weather changes are taken into account.

Vehicle-state sensing is of a lower level and concentrates on measuring a vehicle's dynamics and monitoring its actuators. It includes detection of parameters such as vehicle position, velocity, and acceleration, engine pressure and temperature, tire pressure, temperature, friction coefficients, and similar variables.

In-vehicle environment sensing involves collecting data about the driver and the passengers; such as behavior monitoring, monitoring the driver's eye movements, surveillance, and tiredness; the interaction inside the car; and so forth. Sensing inside the vehicle is equally important to out-vehicle sensing. The driver's diminishing vigilance level has become a serious problem in traffic safety. Poor lighting disturbance can be eliminated using infrared cameras.

13.4.6 Aviation

Network-attached storage, which is currently used to manage air traffic, is generally deemed as safe and effective but not as coordinated and efficient as possible. Automatic dependent surveillance–broadcast also provides weather-monitoring data, allowing for escalated situation awareness. Another developing technology in NextGen is performance-based navigation (PBN). PBN allows for more efficient design of airspace and procedures, which leads to improved safety, airspace access, and predictability of operations, as well as reduction in delays, fuel use, emissions, and noise. The PBN framework defines aircraft performance requirements and provides a basic fundamental for the design and implementation of automated flight paths, airspace design, and obstacle clearance and is not constrained by the location of ground navigation aids. The two main components of PBN are area navigation (RNAV), which enables more flexibility for point-to-point operations, and required navigation performance (RNP), which monitors the navigation performance of the aircraft and alerts the crew if any requirement is not met during an operation. Both RNAV and RNP allow the aircraft to determine whether it can safely qualify for an operation based on the specified performance level.

Verification, validation, and certification: Current methods and techniques for verification and validation are challenged by the scale of emerging systems, greater demand for advanced capabilities, and the combination of discrete and continuous aspects in CPS. Aerospace systems are becoming more software intensive, and the size of software is rapidly increasing, reaching 100 million and likely to exceed 1 billion lines of code. There is a similar pattern in the automotive field. As a result, software verification is becoming one of the major components of system cost. As systems become more integrated, verification and validation will become an even larger challenge. Scalability is also an issue when verification and validation must be applied to larger systems of systems, as the components and interactions increase and become more complex. The challenge is to ensure the vehicle will operate correctly before it is physically tested. Verification and validation techniques that apply to both humans and the environment they interact with must be developed. This will necessitate the ability to integrate or replace new subsystems and technologies without having to recertify.

Shared resources or mixed criticality: As transportation CPS become more complex, it will be necessary for the system to assess criticality. Systems that are mixed criticality typically comprise hardware, operating system, middleware services, and application software all on a single computing platform. In this arrangement, system safety-critical and nonsafety-critical data coexist on a shared network. As vehicles and infrastructure age, ideally they would be easily upgraded to compensate for changes in criticality or function. Many of these systems rely on human decisions, which are informed by data from the computing and communication components of the vehicles. Automated vehicles also make functional decisions based upon this data, often located on the same processor. New system development approaches are needed for mixed criticality CPS in which less-tested, lower-criticality code can safely coexist on the same processor as the more safety-critical, well-tested code.

13.5 Conclusions

CPS is incredibly complex, making it difficult to create formal models. CPS is high dimensional, span multiple time scales, are dynamic, and can reconfigure to adapt to certain situations. They are also composed of multiple entities including humans. Transportation sectors are migrating toward the use of model-based development that relies on sophisticated tool chains to automate the development process. Because most existing model-based development approaches focus on specific aspects, such as control models or component connection models, there is a need to develop approaches that take multiple system views into consideration. CPS environment modeling, such as debugging, is challenging in a simulated environment and becomes even more difficult as system development progresses to HIL and target platform testing. In the terms of transportation infrastructure, modeling techniques and tools that can capture environmental and physical aspects, with their uncertainties, are needed but not yet developed. Another challenge in CPS environment modeling includes collecting, managing, and mining data, especially because the data sets involved can be massive. Traditional fault modeling that captures individual vehicle faults will not be sufficient for an integrated transport system. To compensate, theories for composition of fault models that can capture system-level faults beyond individual vehicles can be used. The complexity of transportation CPS is beyond current real-time theory and practice in reasoning about large-scale mobile network protocols with multiple dynamic inputs. Current rate-based and periodic event models must be extended to events whose priorities change based on the environment [24].

References

1. E. Lee, Cyber physical systems: Design challenges. Technical Report No. UCB/EECS-2008-8, University of California, Berkeley, CA, January 23, 2008.
2. E. A. Lee and S. A. Seshia, *Introduction to Embedded Systems—A Cyber-Physical Systems Approach*, University of California, Berkeley, CA, LeeSeshia.org, 2011.

3. E. A. Lee, Cyber physical systems: Design challenges, in *Proceedings of the 11th IEEE Symposium on Object-Oriented Real-Time Distributed Computing* (*ISORC '08*), Orlando, FL, pp. 363–369, May 2008.
4. E. A. Lee, Cyber-physical systems—Are computing foundations adequate? in *Proceedings of the NSF Workshop on Cyber-Physical Systems: Research Motivation, Techniques and Roadmap*, Austin, TX, pp. 3–4, October 2006.
5. E. M. Saad, M. H. Awadalla, and R. R. Darwish, Adaptive energy-aware gathering strategy for wireless sensor networks, *International Journal of Distributed Sensor Networks*, 5(6), 834–849, 2009.
6. T. A. Henzinger and J. Sifakis, The embedded systems design challenge, in *Proceedings of the 14th International Symposium on Formal Methods, FM 2006*, Hamilton, Ontario, Canada, August 21–27, 2006, pp. 1–15, 2006.
7. A. Banerjee, K. K. Venkatasubramanian, T. Mukherjee, and S. K. S. Gupta, Ensuring safety, security, and sustainability of mission-critical cyber-physical systems, *Proceedings of the IEEE*, 100(1), 283–299, 2012.
8. J. A. Esparza Isasa, System level energy aware design of cyber physical systems, Technical Report ECE-TR-16, Department of Engineering, Aarhus University, Aarhus, Denmark, October 2013.
9. O. S. Unsal and I. Koren, System-level power-aware design techniques in real-time systems, *Proceedings of the IEEE*, 91(7), 1055–1069, 2003.
10. M. R. Stan and K. Skadron, Guest editors' introduction: Power-aware computing, *Computer*, 36, 35–38, December 2003.
11. J. Shi, A survey of cyber-physical systems, in *Proceedings of the International Conference on Wireless Communications and Signal Processing*, Nanjing, China, pp. 1–4, November 2011.
12. M. E. A. Ibrahim, M. Rupp, and H. A. H. Fahmy, A precise high-level power consumption model for embedded systems software, *EURASIP Journal on Embedded Systems*, 2011, 480805, January 2011.
13. S. Lee, A. Ermedahl, and S. L. Min, An accurate instruction-level energy consumption model for embedded RISC processors, in *LCTES '01 Proceedings of the ACM SIGPLAN Workshop on Languages, Compilers and Tools for Embedded Systems*, ACM, New York, pp. 1–9, 2001.
14. B. Ouni, C. Belleudy, and E. Senn, Accurate energy characterization of OS services in embedded systems, *EURASIP Journal on Embedded Systems*, 2012(1), 6, 2012.
15. Y.-H. Park, S. Pasricha, F. J. Kurdahi, and N. Dutt, A multi-granularity power modeling methodology for embedded processors, *IEEE Transactions on Very Large Scale Integrated Systems*, 19(4), 668–681, 2011.
16. O. Celebican, T. S. Rosing, and V. J. Mooney III, Energy estimation of peripheral devices in embedded systems, in *Proceedings of the 14th ACM Great Lakes Symposium on VLSI, GLSVLSI'04*, Boston, MA, pp. 1–4, 2004.
17. A. Wang and C. Sodini, A simple energy model for wireless microsensor transceivers, in *IEEE Global Telecommunications Conference, GLOBECOM'04*, Texas, Vol. 5, pp. 3205–3209, 2004.
18. Q. Wang, M. Hempstead, and W. Yang, A realistic power consumption model for wireless sensor network devices, in *Third Annual IEEE Communications Society on Sensor and Ad Hoc Communications and Networks, SECON'06*, Reston, VA, Vol. 1, pp. 286–295, September 2006.
19. M. Nistor, D. Lucani, and J. Barros, A total energy approach to protocol design in coded wireless sensor networks, in *2012 International Symposium on Network Coding* (*NetCod*), Cambridge, MA, pp. 31–36, 2012.
20. C. Park, K. Lahiri, and A. Raghunathan, Battery discharge characteristics of wireless sensor nodes: An experimental analysis, in *Second Annual IEEE Communications Society Conference on Sensor and Ad Hoc Communications and Networks, IEEE SECON 2005*, Santa Clara, CA, pp. 430–440, 2005.
21. J. Chabarek, J. Sommers, P. Barford, C. Estan, D. Tsiang, and S. Wright, Power awareness in network design and routing, in *The 27th Conference on Computer Communications, INFOCOM 2008, IEEE*, Phoenix, AZ, pp. 3–9, 2008.

22. J. F. Broenink, P. G. Larsen, M. Verhoef, C. Kleijn, D. Jovanovic, and K. Pierce, Design support and tooling for dependable embedded control software, in *Proceedings of Serene 2010 International Workshop on Software Engineering for Resilient Systems*, ACM, New York, pp. 77–82, April 2010.
23. D. Henriksson and H. Elmqvist, Cyber-physical systems modeling and simulation with Modelica, in *Proceedings of the Eighth International Modelica Conference*, Technical University of Dresden, Germany, March 20–22, 2011.
24. NIST, Cyber-physical systems: Situation analysis of current trends, technologies, and challenges: Smart system technologies for manufacturing, power grid and utilities, buildings and infrastructure, transportation and mobility, and healthcare, Draft for NIST, Columbia, MD, March 9, 2012, http://events.energetics.com/NIST-CPSWorkshop/pdfs/CPS_Situation_Analysis.pdf. Accessed on March 9, 2012.

Section IV

Ubiquitous and Cloud Computing for Monitoring Cyber-Physical Systems

14

Ubiquitous Computing for Cyber-Physical Systems

Srinidhi Hiriyannaiah, Gaddadevara Matt Siddesh, and Krishnarajanagar GopalaIyengar Srinivasa

CONTENTS

14.1 Introduction

The computing world has grown with devices from mainframe and personalized desktops to tablets and smartphones in the present world. The backbone of the Internet is growing with well advances in wireless technology and communications with application-specific processors. Cyber-physical systems use these advance wireless networks with sensors to connect humans with physical embedded computing world [1]. Computers are envisioned as machines that perform the computation as we enter for a task to be performed and leave when it finishes. Ubiquitous and pervasive computing is a concept that changes this perception of computing by using the information gathered by various devices and provides a computing environment that appears anywhere and everywhere. There are various areas of research where ubiquitous computing can be achieved through the use of wireless networks and the Internet technology. Some of the areas are as shown in the Figure 14.1. We discuss some of the use cases related to these areas further in this chapter.

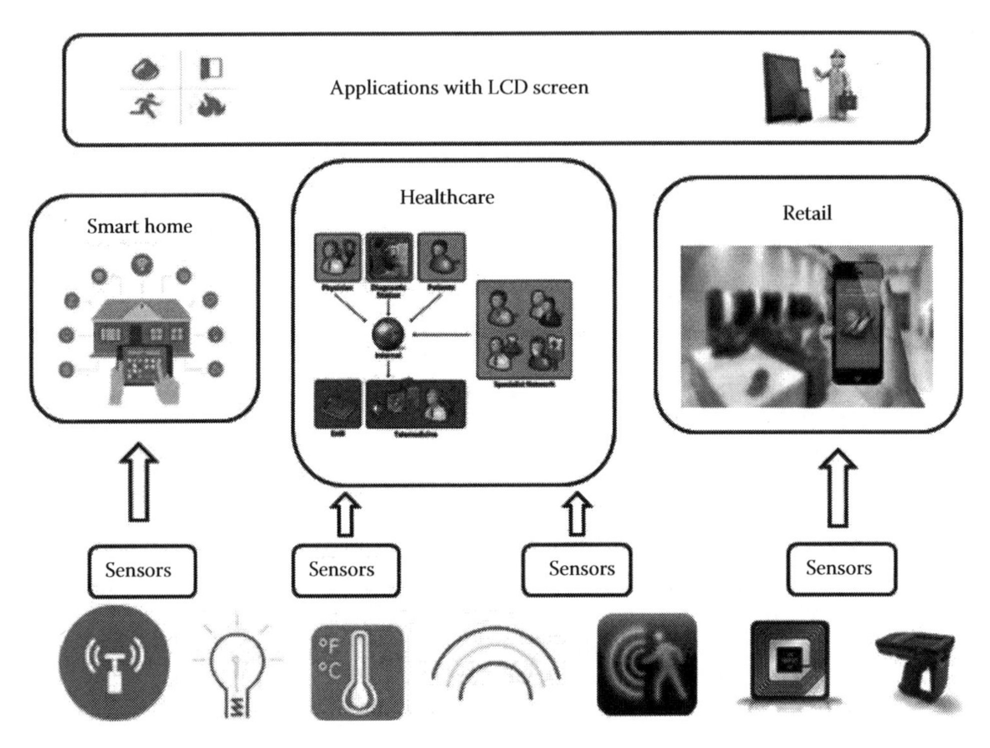

FIGURE 14.1
Ubiquitous computing.

In this chapter, we have discussed ubiquitous computing and its applications with use cases, prototypes, and architectures. The chapter is further organized as follows. In Section 14.1, we discuss Mark Weiser and his ideas on ubiquitous computing. In Section 14.2, the need for ubiquitous computing is discussed followed by use cases in Section 14.3. Section 14.4 discusses some of the prototypes for the use cases discussed in Section 14.3, and finally, in Section 14.5, the architectures for developing ubiquitous computing applications are discussed.

14.2 Mark Weiser and His Approaches

Mark Weiser coined the term "ubiquitous computing" when he was working at Xerox Palo Alto Research Center (PARC). According to Mark Weiser, ubiquitous computing aims to achieve a technology of computing where users interact with the computer in an invisible way [2]. The research methods in this area include building the infrastructures required, prototypes viable to test the infrastructures that are robust and scalable.

14.2.1 Ubiquitous Computing and Virtual Reality

Howard Rheingold [3] defines virtual reality (VR) as an experience where the user is surrounded by computer-generated virtual environment and the user can move around it, reshape it, and use it according to his needs. The VR is close enough to ubiquitous computing as defined by Weiser but differs in the context of "bringing computer the world." In VR, the world of information is made available on the computer and humans

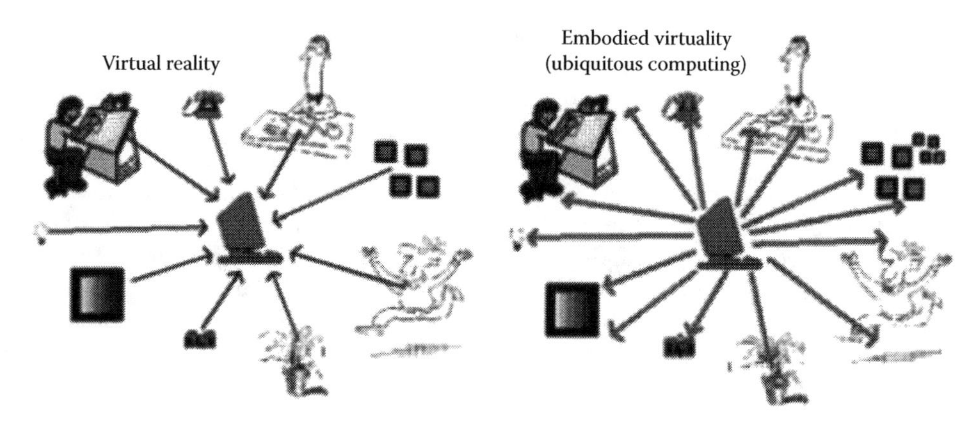

FIGURE 14.2
Ubiquitous computing and virtual reality. (From Weiser, M., *Commun. ACM*, 36(7), 75, 1993.)

(users); systems or other computers can use it. Figure 14.2 as given by Weiser [2] gives the difference between the VR and the ubiquitous computing.

There are three devices that Weiser introduced for ubiquitous computing that replaced the traditionally used analog devices for human–computer interaction. We discuss these three devices as follows (Weiser et al., 2009):

- *Xerox tab*: It is a device that consists of the pressure-sensitive touch screen allowing the user to enter the inputs. It is derived from the analogy of the traditional notebooks or notepad. The world-of-today tablets draw their design ideas from this PARC tab. The major issues or the challenges we face with the tab is the size and the power consumption.
- *Xerox pad*: It belongs to the family of notebook-sized devices and the present personal digital assistant devices that fall in this category. It consists of the inbuilt radio and wireless technologies for communication with other devices in the computing environment required.
- *Xerox liveboard*: It is the prototype of the electronic boards that are used in the current world for larger displays compared to tabs and pads. It provides a large yard-like display with aggregated output from the sensors in the computing environment for the users with multiple user input.

These devices have formed the basis for the ubiquitous computing environments. With the advent of the hardware technologies and wireless communication technologies, these devices are replaced with more advanced devices such as smartphones and cameras embedded with wireless communication. We now discuss the need for a ubiquitous computing in the next section considering the definition of Weiser throughout this chapter.

14.3 Need for Ubiquitous and Pervasive Computing

The places, things, and devices that people use to communicate are majorly with computerized networks because of the advances in microelectronics and changes in the communications and information technology. Due to advances in computing, the power of the microprocessors has doubled every 18 months according to the Moore's law [4], and thus the computing devices that empower these microprocessors have become smaller, cheaper, and are found in each of the devices in our everyday life, leading to the era of "smart devices" and "smart environments." There is a need for an environment for these mobile devices to understand the information context of environment where they are present [5,6]. Ubiquitous computing provides such an environment for the devices that monitor the user interactions such as movement, speech, and presence. Distributed computing consists of equivalent computing nodes that need the same application middleware across the computing devices, but ubiquitous computing composes of different devices with a standardized communication, and each device performs a specific computing task assigned to it. In this section, we discuss how the ubiquitous environment helps people interaction with their computing environment.

Many tasks and work practices in today's world involves some form of computation beyond the word processing and spreadsheets. The computation has to occur through various communications across the computing devices that are involved in providing the output of the computation to the user in an appropriate way through GUI and large display boards [7,8]. Ubiquitous computing aims at bringing the world of computers to the real world, enrich and favor the environments for humans by means of technology but not isolate from the real world. The environment is enriched by bringing the information through virtual elements to the user in the means of speech, large displays of GUI, smart alerts, etc. For example, consider a scenario of a desk job application [8] where the user goes for a meeting into a presentation room. The room with a large screen displays the agenda of the meeting and the participants involved in the meeting. Once all the participants have arrived in the room, the display automatically turns the meeting in progress with the list of documents required for the meeting. In this way, ubiquitous computing provides a rich computing environment based on the users context in the environment.

The orientation of the Internet has advanced from the number of computers connected to the means they are communicated. Initially, the Internet was used between persons to communicate through emails, and then it advanced to client–server via web browsers. This has lead to the popularity and the commercialization of the Internet. Due to these advances in the Internet, now it has shifted to Internet of Things (IoT) or object-to-object communications [9]. Localization technologies can make advantage of this communication between objects for applications that recognize theft or loss of devices as they are able to communicate with each other and sense the whereabouts of it [10]. One of the applications where it is used is the loss of car and mobile that can be extended to other objects as in the case of the smart homes discussed in the use case section in this chapter.

Computing and the communication services have transcended to all aspects of social life where the organizational processes and tasks are handed over to the computing devices that are embedded in the locations where the users move or with the devices they use. Various computing utilities that are involved in these types of communication services are mobiles, PDA, RFID, wrist watches, etc. [11]. The engagement of these services in the modern communication computing requires a change in the traditional organizational approaches on the processes and the practices used for developing the applications [6,7,11]. Ubiquitous computing with multiple devices for sensing the context aims at providing communication between the objects present in the environment without being exactly visible to the user.

14.4 Use Cases, Solutions, and Analysis

With the advances in media and technology, there is a need to deviate from the conventional computing devices and environment to more digital-equipped devices and environment. In order to achieve this, specific solutions need to be designed, analyzed, and optimized to overcome the challenges that are present with ubiquitous and pervasive computing environments. In this section, we discuss some of the use cases related to ubiquitous computing with solutions and how they can be optimized that overcome the challenges related to it.

14.4.1 Smart Homes for Ubiquitous Computing

A smart home is a home that provides a better assisted living for the inhabitants with use of computer technologies [12]. A smart home is augmented with diversity of devices equipped with sensors. The data generated by these sensors can be perceived by a smart home system and help inhabitants living in the smart home to make timely decisions and take appropriate actions based on the data alerts sent by the smart home system [12,13]. Smart homes are the ones that enable remote interaction for the owner of the home with appliances and devices in the home like light, heat, water, and sensors on the windows and doors, etc. The energy management and the abnormal flow of water remotely through mobile phones are the main aspects of interest in the smart home solutions. The basic components of a smart home are as shown in Figure 14.3.

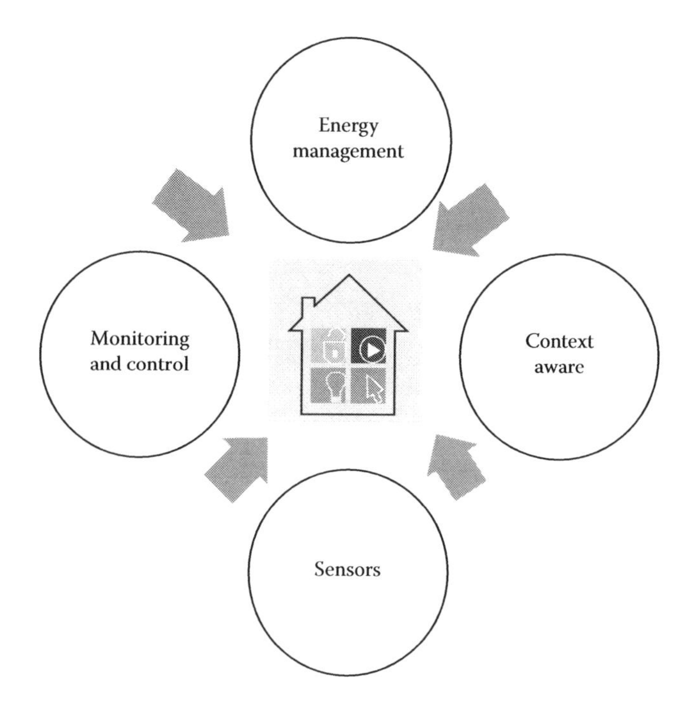

FIGURE 14.3
Smart home components.

14.4.2 Some Examples of Smart Homes

Adaptive homes developed at the University of Colorado [14] allowed the inhabitants in the home to control the household activities such as controlling temperature of room, opening of windows, and monitoring the level for lights and fans. It was equipped with different sensors for monitoring these activities and automated to predict the actions taken by the user based on the previous configuration designed for the home. Aware Home [15], developed by Georgia Institute of Technology, is equipped with capabilities like finding the lost items, smart floor, and provides a better assisted living for the elderly people in the house. It uses tags for objects inside the house in "find the lost items"; if the batteries drain out, smart floors can be used, which provide an LCD panel to view the previous location of it.

Another example of Smart House developed by University of Florida [16] provides an assisted living for the elderly and disabled people in the house. It provided activity monitoring with smart mobiles, comfort, and energy efficiency using various environmental sensors in the home environment. It provided an acoustic-based context aware or remote-based monitoring system where the user can control the devices equipped with sensors in the house. The House_n Project [17] uses ubiquitous-sensible environment to control energy expenditure, monitors the activity and the vital residents of the house, learning and communication and entertainment. Other than these, there are numerous prototypes of the smart homes being developed with research focusing on the smart assisted living environment.

14.4.3 Knowledge-Driven Approach for Smart Homes

One of the challenges faced in developing a solution to the smart homes is the sequence of activities carried out during the daily routine [13,18]. The people living in a smart home have different kind of life styles, abilities, or habits that they perform in their daily routine activities even though there is a certain pattern of behavior observed while performing the daily routine activities. It is also posed with challenge of interpreting heterogeneous data from multiple sensors with appropriate context for the people in the smart home. The common approach followed in designing a solution for smart homes is data oriented [18] where multiple data are collected from the sensors or the devices in the smart home environment and then interpreted in an appropriate manner using the context for the people in the environment. In this section, we discuss a knowledge-oriented solution for smart homes for the purpose of activity recognition.

The architecture for a knowledge-driven approach based smart home [18] is shown in Figure 14.4. It is based on the ontological modeling of the contexts in the smart

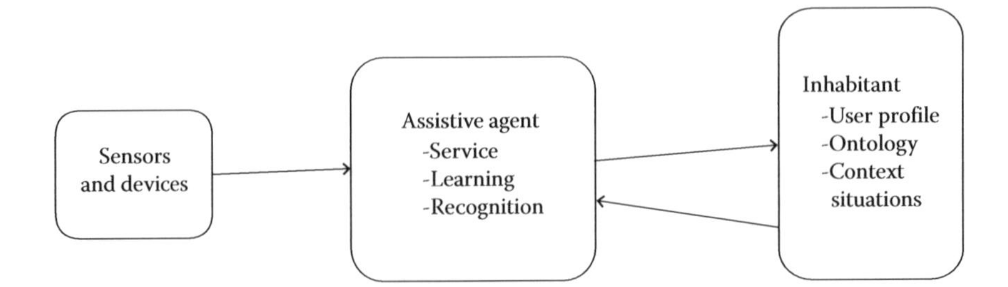

FIGURE 14.4
Knowledge-based architecture for smart home.

home environment. The ontology is classified according to the hierarchy of the classes in the context environment. Assistive agents form the core part, which receives the input from the sensors, analyze the semantics of the context and assist the user with a conducive environment of appropriate messages. A user profile is maintained that gives the specifics of the user about the situations where the user is interested for assisted learning. A model of the activity daily living (ADL) is maintained in a repository, and assistive agents performs the activity recognition based on this ADL models in the repository. In the repository, a generic model is maintained with a course of sequence of activities of ADL. This generic model is fine-tuned into a fine-grained model based on the user profile and characteristics.

14.4.4 Retail

In retail applications, the information that needs to be shared across different stakeholders for different packages is vast in terms of volume and variety [13,19]. Here, the different stakeholders involved can be customers or clients, warehouses that accepts the customer's requests, the or delivery stores incharge of delivery from the warehouse to the client. So, in the retail applications, the information has to be gathered at various stages with different kind of information of the packaged products. The applications of ubiquitous computing in the area of retail [13,21] are automatic identification, predicting the supply and demand of the products, more accurate and proper organization of logistics for delivery, registration and tracking of the products, and so on as shown in the Figure 14.5. We discuss some of the solutions related to the retail makes the products to be ubiquitous and access anywhere.

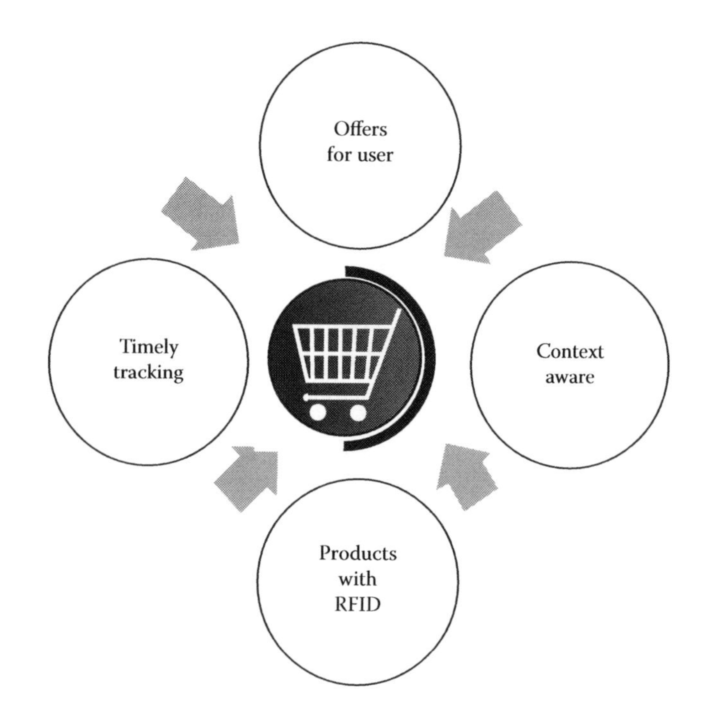

FIGURE 14.5
Retail system.

14.4.5 Solution

The main objective in the retail applications is to provide an environment to the customers with continuous track of information of the order with computing made invisible [20,21]. One of the solutions that can be used for retail applications to enable the ubiquitous environment of products to the customers is the use of tagging the products with unique codes. The tagging of products can be either bar codes or the RFID chips that uniquely identifies each product. Before going into the specifics of how the bar codes and RFID can be used for retail systems, let us first see the sequences of activities that are followed by the customer [22].

A customer buys a product in an online retail system, the sequence of activities that are logged in are as follows:

- The retail system authenticates the purchase of the item once the transaction of the payment is successful.
- The customer is sent an acknowledgement message or an email about the confirmation of the order.
- The customer now can see the state of the order of the purchase. For example, packed, shipped, in process for delivery, or delivered.

In this sequence of activities, customer is interested in tracking the item of purchase in each activity. This can be made available by continuous tracking of items that are to be delivered to the customer. Continuous tracking of the retail products can be made possible using RFID tags for each product. Since each product is tagged, the location of the product can be tracked with information such as state of the product, location of the product, and estimated delivery time.

The use of RFID at the individual level of products results in the increase of cost as each product in the retail system must be tagged with RFID [23,24]. The solution discussed earlier for the retail systems can be optimized by tagging the products at the store unit levels rather than at the individual level of products. In this optimized solution, bar codes for the product can be used instead of RFID. At the store unit levels, the workers need to capture the status of the product using these bar codes so that customer can be constantly in touch with status of the order.

14.4.6 Smart Home with Retail

A smart home provides the user to control and monitor different utilities inside the home remotely using different sensors and the context analyzers as discussed previously in the use cases. Similarly, in the retail scenario, fast tracking of the products and unique identification of the product facilitate the user to keep update on the status of it. The use cases described earlier for the smart homes and retail can be brought together as shown in the figure. Here, the retail store is considered as a grocery store with assistive RFID technology for the users. The use case of smart home with retail is as shown in the Figure 14.6.

A customer enters the super market with PDA. Using this PDA, he remotely monitors the kitchen room to know the groceries required for home. Here, the user is able to interact with home remotely through PDA to the server maintained in the smart home. A list of groceries is made available to the user with their quantities, and the user can carry on shopping. While shopping the required groceries, the user can be presented

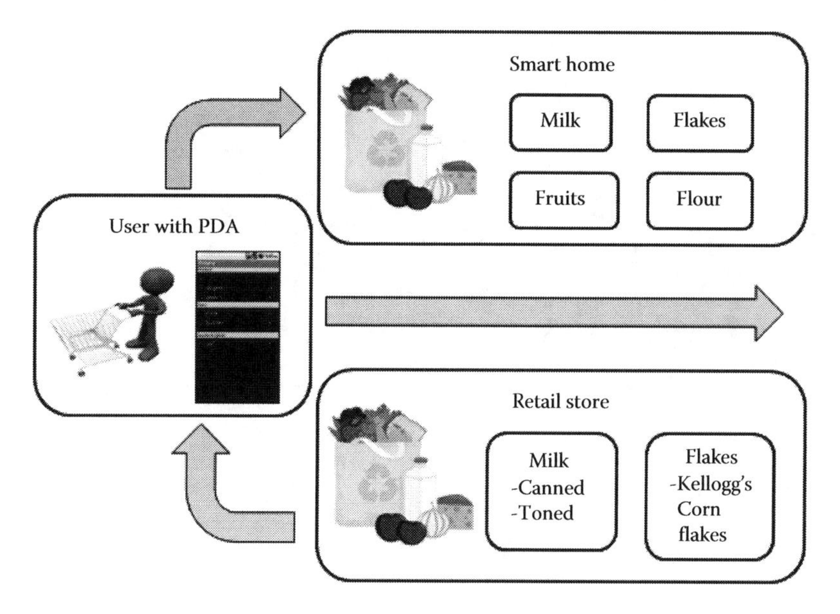

FIGURE 14.6
Smart home with retail use case.

with offers appropriately based on the products chosen by him. For example, when he is alerted about the shortage of milk in the home, he can be guided to the section in the retail store where the milk is present and give an appropriate list of offers based on the brand the user chooses.

The use cases described earlier give an idea of the ubiquitous computing in the real-world environment with smart home and retail applications. Other use cases include health care and smart airports [25,26]. In health care, the environment is similar to the smart home, but the administering and monitoring is different and more automated [27]. In the field of airports, ubiquitous computing plays an important role where the context awareness is important for a user in the environment by providing appropriate services to the user [26]. Based on these use cases, we discuss prototyping models and architectures in further sections.

14.5 Prototyping of Ubiquitous Computing Applications

For a ubiquitous environment, there are a set of challenges that are focused on user-centric design that are different from the desktop environment. In the desktop environment, the user input is obtained through standard mouse and keyboard, and the user will be mostly a single user. In contrast to this, in ubiquitous environment, inputs are obtained from multiple devices (audio and video sensors), and for output, there are multiple user interface outputs for each of these devices [28]. These outputs need to be integrated into one single output for the end user. Hence, there is a need for a prototype to be built initially for the ubiquitous systems before the actual system is developed and ready to deploy. In this section, we discuss some of prototypes that can be used in developing ubiquitous systems.

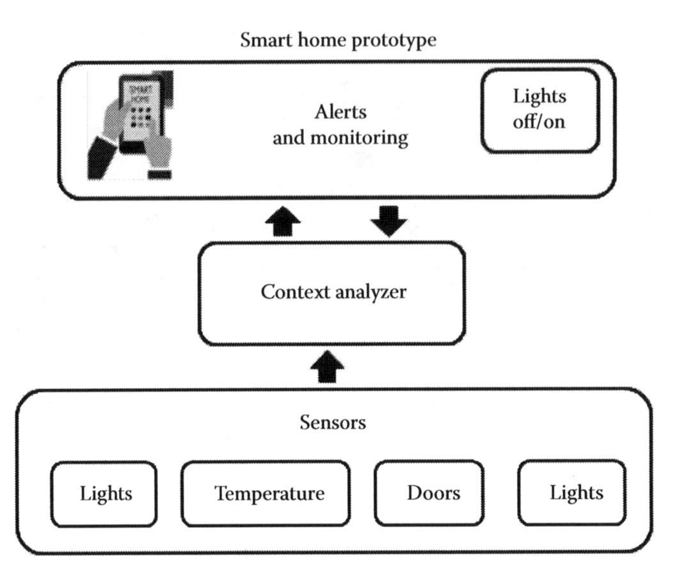

FIGURE 14.7
Prototype for smart home.

14.5.1 Prototype for Smart Homes

Based on the use cases discussed in the previous section, we discuss a prototype model that can be used in a developing a smart home that provides a ubiquitous environment to the people or inhabitants in the house. It consists of the following components as shown in the Figure 14.7:

- Sensors
- Context analyzer
- A GUI for monitoring and controlling the environment using alerts

Context analyzers obtain data from different sensors present in the smart home, interpret, analyze, and alert the user for monitoring the home. It acts as a bridge between the user and the sensors in the home. The major functions of it are listed as follows:

- Obtain the data from sensors such as light sensors, temperature sensors, and door.
- Analyze the data and prepare the data for the user alert.
- Alert the user through proper communication for the data interpreted by the sensor.
- Take necessary action on the sensor based on the user's alert such as switch off/on lights and adjust temperature.

14.5.2 Prototype for Retail Systems

Mobile phone communication systems play an important role in the ubiquitous retail systems. Using the advent of technologies and the devices embedded within the smart phones, a prototype can be built initially for retail applications that use mobile devices. A sample prototype for a mobile device with retail application is as shown in the Figure 14.8. RFID recognition recognizes the user using the associated RFID tag with mobile. It searches the catalogue of retail needed by the user and guides the user through communication channel. The catalog is a database of all the products available in the retail store. We assume each

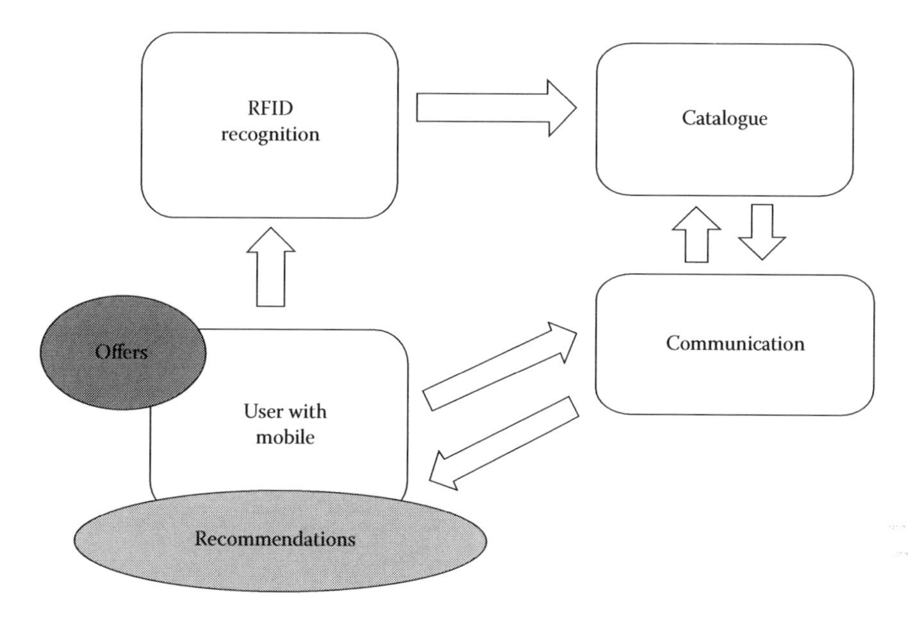

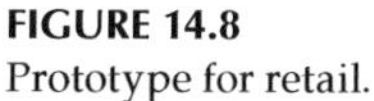

FIGURE 14.8
Prototype for retail.

product is maintained hierarchically so that searching becomes faster. Communication interacts with both catalogue and user simultaneously to keep the user up to date with recommendations and offers of the products needed by the user.

14.5.3 Architecture and Design Methodologies for Ubiquitous Computing

Ubiquitous computing consists of various devices that differ from one application to other application. Context awareness is important in ubiquitous computing for providing the required services to the user. "Context" means it is related to the objects, conditions where the user is present, and relevant interactions between the user and the environment. The different types of contexts that are involved in the ubiquitous environment can be categorized as follows:

- Informational context (status of printers and sports scores)
- Physical context (location and time)
- Environmental context (weather, light, etc.)

Hence, contextual awareness plays a key role in defining the services provided to the user in the ubiquitous environment through accurate communication between the heterogeneous devices in this environment. The generic architectures with suitable design methodologies are discussed in the following section. It gives an overview of architectural components in the ubiquitous computing environment.

14.5.4 Tool Kit–Based Approach for Context Aware Ubiquitous Environment

Context tool kit in [29] is one of the pioneering works that is being used as one of the base tool kits for developing context aware applications. The design of the applications that uses this approach consists of the following four main categories, namely, context widgets, interpreters, aggregators, and discoverers as shown in the Figure 14.9 [29].

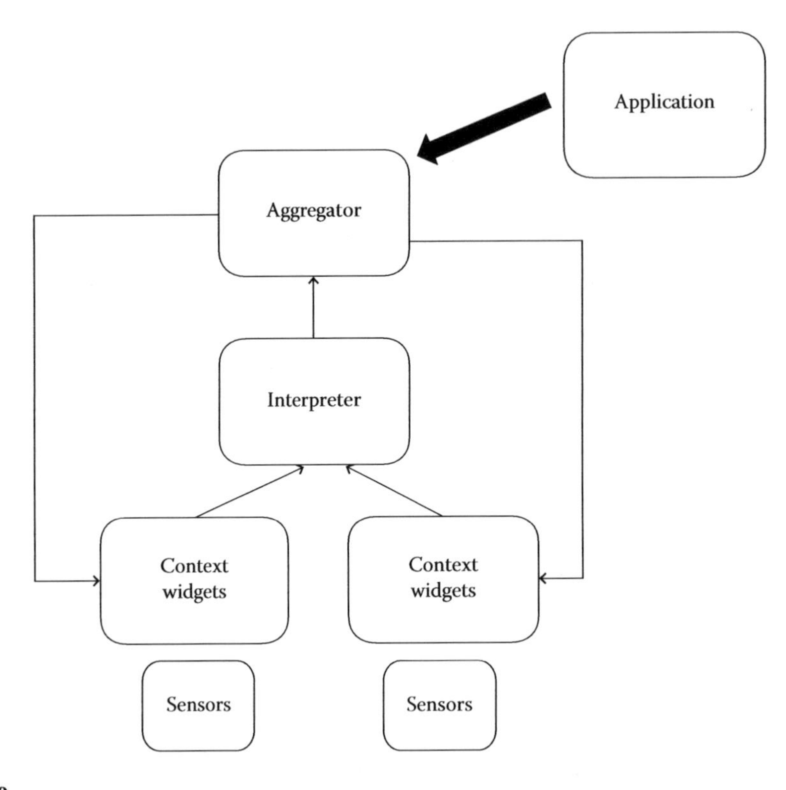

FIGURE 14.9
Context aware tool kit architecture.

Context widgets are the components that provide the information about the context where the data are generated from various data sources. A widget is defined individually for each sensor or the data source in the context environment. It hides the complexity of the interface of the sensor and provides a separation of concern among the different devices in the context aware environment. For example, a location widget identifies the location of the person, a temperature widget that gives the temperature of a location, a presence widget identifies the presence of a person or an object in a location. A combination of these widgets should not have an impact on the application.

Interpreters are the components that interpret the context information made available by the context widgets. Once the context information is gathered from the context widgets, it has to be interpreted with higher levels of information from the abstract level of information provided from the context widgets, for example, raising the level of location given by the location widget to the geographical coordinates of the location. One of the challenges is to provide a single level of interface for the interpreters so that all the context widgets can participate in the interpretation of the context information. For example, consider two widgets, person widget that determine the total number of people in a room and sound widget that determines the sound level of the room. If there are more number of people in a room and the sound level is high as determined by the person widget and sound widget, respectively, it can be inferred that a meeting is going on in the room.

Aggregators are the components that collect different pieces of the context information and make it available in a single repository. The information about the context should be logically related to one another, and instead of getting the data from each context widget, aggregators collect the distributed information in the context from various context widgets. They also facilitate interpretation of the information collected from distributed sensors and act as a one-stop shop for application services.

Discoverers are the components in a ubiquitous environment that maintains a registry of the information about the different context widgets, interpreters, and aggregators. Each component when it enters the environment registers with the discoverer about the service offered by it. A context widget provides the context of information provided by it; an interpreter provides the interpreted result offered by it; an aggregator provides the logical information provided by it. Applications can use these discoverers to get the relevant context widgets, interpreters and aggregators. For example, if the application is to detect the presence of people in a room, a discoverer can use person widget and person interpreter for this application.

14.5.5 Middleware Service

In ubiquitous computing environment, large number of heterogeneous devices transforms the physical space made available to them into a smart environment. The devices involved perform mainly two tasks, that is, sense and reason the context of the environment and communicate among the devices. A middleware service between the devices is essential for the communication between the devices and context (smart) environment.

In [30], a middleware model is proposed for context aware ubiquitous computing environment. The context is based on predicate logic with different reasoning capabilities that agents can use to meet the appropriate requirements. In order to have a common understanding of the semantics in the context of the environment, ontology is used. Ontology defines the structure and the properties of the contextual environments.

The traditional middleware uses CORBA as the base model to design the middleware services for the agents to communicate with each other. In ubiquitous computing, context awareness is a key role in determining what defines the agents' behavior in the environment according to the contextual environment where the agents are present. So, a context model is proposed [30] based on the predicates. A predicate is defined for the type of context. For example, location (room = 222) and temperature (room temp = 33°F). In order to identify different contexts, predicates are defined in the form <subject, verb, object>. This helps in ordering of different context environments to a particular class of objects with ontology. For example, when a person is entering a room, then the predicate in the form <Ram entering room 2222> describes that the location (room) belongs to a set of persons.

The middleware service is provided through Gaia infrastructure [31] for ubiquitous computing environment as shown in Figure 14.10. It consists of different agents, where each agent provides different functions such as context discovery and event distribution. The different agents that are present in the ubiquitous environment for Gaia infrastructure-based middleware service are discussed as shown in the figure.

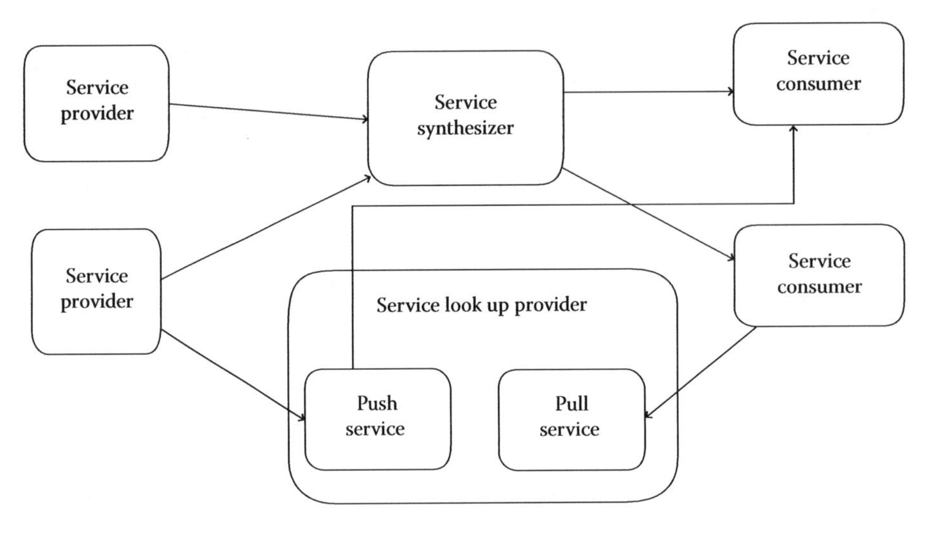

FIGURE 14.10
Middleware service architecture.

Service providers: These are different data sources involved in the ubiquitous environment that provide the information or the data about the environment. These providers allow other agents in the environment to query them for the needed information. They can be modeled as either a push service model or pull service model. In a pull service model, the agents that require information about the environment can query on the context providers and obtain the appropriate information. In a push service model, the context providers themselves can push the required information or data to the appropriate agents. An example of the context provider is a video sensor providing the data to the agent that is responsible for a computation in the environment.

Service synthesizers: These are the sources where data gathered from the agents are synthesized to an appropriate context for context consumers. For example, the agent can be counting the number of people in a room.

Service consumers: They are the agents that collect the context data about the environment from the providers and synthesizers. Based on this data, these agents adapt to the environment or behave accordingly in the smart context environment.

The advantages of using a middleware service-oriented infrastructure for developing ubiquitous applications are [32]:

- *Independence from the data sources and programming language*: The architecture does not depend on the hardware or the data sources that provide the information about the context. A new hardware or the service provider can be easily added to the environment with its specifications, with appropriate synthesizer and new service consumer that presents the end result to the user.

- *Evolution of new services:* By providing a middleware layer of service, new devices and sensors with their services can be added independently and dynamically without affecting other services and the devices in the computing environment. The main advantage of the middleware service layer is it provides a uniform abstraction layer separating the devices from the services and services from the end user applications. For example, consider a ubiquitous environment where there is one of the sensors for detecting the motion in a video streaming. If the algorithm for the service is upgraded to make it faster, a new service can be added to the environment without affecting the device and also a new upgraded device can be added to the environment through middleware service architecture. Hence, the entire system and the environment can be upgraded to new services.
- *Sharing of processing power and data:* In the ubiquitous environment, the devices have limited capability of storage and power to give the required data to the services defined for the end user application. For example, image-processing applications are computationally expensive, and similarly the data such as traffic data and weather data are too large to store on the individual devices. With middleware service architecture, the infrastructure provided centrally can be utilized to share the infrastructure among the different devices in the environment.

14.6 Conclusion

Ubiquitous computing has reached many application areas such as automated homes, retail, and health-care logistics, more than as a technology mere vision. The new paradigm of mobile and smartphones has paved the way for ubiquitous computing application to be more robust and user friendly. The technical aspects related to ubiquitous computing are privacy and security. Users are using the ubiquitous computing applications if they feel it is secure and user friendly. The use of ubiquitous computing can be extended to IoT where each device is connected to the Internet, and communication channel can be established between each of the objects, so that the user can communicate in this computing environment pervasively and thereby accomplish the goal of the end activity of the user. The key element in the ubiquitous computing applications is the context awareness and recognition. Context or the situation on demand for the user plays a critical role for the applications, as they have to present the right information at the right time to the user. It can be made possible through the use of RFID and related technologies with middleware services as discussed in the previous sections.

In this chapter, we discussed about different aspects of the ubiquitous computing from the Mark Weiser approaches to the design and architectures associated with it. We conclude with some points based on the aspects discussed about ubiquitous computing in the previous sections.

- The challenges faced by ubiquitous computing are privacy, security, and design.
- It involves initial investments initially for infrastructure requirements, followed by some design constraints.
- Adapting to the data protection mechanisms with well-equipped infrastructures and design can overcome these challenges to develop more user-friendly ubiquitous applications.

References

1. Sha, L., Gopalakrishnan, S., Liu, X., and Wang, Q. (2009). Cyber-physical systems: A new frontier. In *Machine Learning in Cyber Trust* (pp. 3–13). Springer.
2. Weiser, M. (1993). Some computer science issues in ubiquitous computing. *Communications of the ACM, 36*(7), 75–84.
3. Rheingold, H. (1991). *Virtual Reality*. Summit Books, New York.
4. Evans, D. Moore's Law: How long will it last? http://www.techradar.com. Retrieved November 25, 2014.
5. Mattern, F. (April 2003). From smart devices to smart everyday objects. In *Proceedings of Smart Objects Conference.*
6. Strassner, M. and Schoch, T. (August 2002). Today's impact of ubiquitous computing on business processes. In *First International Conference on Pervasive Computing* (Vol. 2002, pp. 62–74), Zurich, Switzerland.
7. Friedewald, M. and Raabe, O. (2011). Ubiquitous computing: An overview of technology impacts. *Telematics and Informatics, 28*(2), 55–65.
8. Weiss, R.J. and Craiger, J.P. (2002). Ubiquitous computing. *The Industrial-Organizational Psychologist, 39*(4), 44–52.
9. Gubbi, J., Buyya, R., Marusic, S., and Palaniswami, M. (2013). Internet of Things (IoT): A vision, architectural elements, and future directions. *Future Generation Computer Systems, 29*(7), 1645–1660.
10. Schilit, B., Adams, N., and Want, R. (December 1994). Context-aware computing applications. In *First Workshop on Mobile Computing Systems and Applications, 1994. WMCSA 1994* (pp. 85–90). IEEE, Santa Cruz, CA.
11. Kindberg, T. et al. (December 2000). People, places, things: Web presence for the real world. In *Proceedings of IEEE Workshop on Mobile Computing Systems and Applications (WMCSA 2000)*, Monterey, CA.
12. Chan, M., Estève, D., Escriba, C., and Campo, E. (2008). A review of smart homes—Present state and future challenges. *Computer Methods and Programs in Biomedicine, 91*(1), 55–81.
13. Chan, M., Campo, E., Estève, D., and Fourniols, J.Y. (2009). Smart homes—Current features and future perspectives. *Maturitas, 64*(2), 90–97.
14. Harper, R. (Ed.). (2006). Smart homes: Past, present and future. In *Inside the Smart Home*, Springer Science & Business Media, London, UK, PP. 17–50.
15. Kidd, C.D., Orr, R., Abowd, G.D., Atkeson, C.G., Essa, I.A., MacIntyre, B., and Newstetter, W. (1999). The aware home: A living laboratory for ubiquitous computing research. In *Cooperative Buildings. Integrating Information, Organizations, and Architecture* (pp. 191–198). Springer, Berlin, Heidelberg.
16. Helal, S., Mann, W., El-Zabadani, H., King, J., Kaddoura, Y., and Jansen, E. (2005). The Gator Tech Smart House: A programmable pervasive space. *Computer, 38*, 50–60.
17. Intille, S.S., Larson, K., Munguia Tapia, E., Beaudin, J.S., Kaushik, P., Nawyn, J., and Rockinson, R. (2006). Using a live-in laboratory for ubiquitous computing research. In: Fishkin, K.P., Schiele, B., Nixon, P., and Quiley, A. (eds.) *Proceedings of PERVASIVE 2006* (Vol. *LNCS* 3968, pp. 349–365). Springer-Verlag, Berlin, Heidelberg.
18. Chen, L., Nugent, C.D., and Wang, H. (2012). A knowledge-driven approach to activity recognition in smart homes. *IEEE Transactions on Knowledge and Data Engineering, 24*(6), 961–974.
19. Roussos, G., Tuominen, J., Koukara, L., Seppala, O., Kourouthanasis, P., Giaglis, G., and Frissaer, J. (September 2002). A case study in pervasive retail. In *Proceedings of the Second International Workshop on Mobile Commerce* (pp. 90–94). ACM, Melbourne, FL.
20. Smaros, J. and Holmstrom, J. (2000). Viewpoint: Reaching the consumer though e-grocery VMI, *International Journal of Retail and Distribution Management, 28*(2), 55–61.
21. Chopra, S. and Sodhi, M.S. (2007). Looking for the bang from the RFID buck. *Supply Chain Management Review 2007*(4), 34–41.

22. METRO Group, 2008. Welcome to the real,—Future store—A journey into the future of retail. METRO Group, Tönisvorst, Germany.
23. Fleisch, E. (2004). Business impact of pervasive technologies: Opportunities and risks. *Journal of Human and Ecological Risk Assessment*, 10(5), 817–829.
24. Fleisch, E. and Tellkamp, C. (2006). The business value of ubiquitous computing technologies. In: Roussos, G. (ed.), *Ubiquitous and Pervasive Commerce* (pp. 93–113). Springer, London, U.K.
25. Kjeldskov, J. and Skov, M.B. (2007). Exploring context-awareness for ubiquitous computing in the healthcare domain. *Personal and Ubiquitous Computing, 11*(7), 549–562.
26. Want, R., Pering, T., Danneels, G., Kumar, M., Sundar, M., and Light, J. (2002). The personal server: Changing the way we think about ubiquitous computing. In *Ubicomp 2002: Ubiquitous Computing* (pp. 194–209). Springer, Berlin, Heidelberg.
27. Prekop, P. and Burnett, M. (2003). Activities, context and ubiquitous computing. *Computer Communications, 26*(11), 1168–1176.
28. Schmidt, A., Davies, N., Landay, J., and Hudson, S. (2005). Rapid prototyping for ubiquitous computing. *IEEE Pervasive Computing, 4*(4), 0015–0017.
29. Dey, A.K., Abowd, G.D., and Salber, D. (2001). A conceptual framework and a toolkit for supporting the rapid prototyping of context-aware applications. *Human-Computer Interaction, 16*(2), 97–166.
30. Ranganathan, A. and Campbell, R.H. (January 2003). A middleware for context-aware agents in ubiquitous computing environments. In *Middleware 2003* (pp. 143–161). Springer, Berlin, Heidelberg.
31. Román, M., Hess, C., Cerqueira, R., Ranganathan, A., Campbell, R. H., and Nahrstedt, K. (2002). A middleware infrastructure for active spaces. *IEEE Pervasive Computing, 1*(4), 74–83.
32. Hong, J.I. and Landay, J.A. (2001). An infrastructure approach to context-aware computing. *Human-Computer Interaction, 16*(2), 287–303.

15

Cloud Computing for Transportation Cyber-Physical Systems

Houbing Song, Qinghe Du, Pinyi Ren, Wenjia Li, and Amjad Mehmood

CONTENTS

15.1 Introduction

Our world is facing critical transportation issues [1]. The performance of the transportation system is neither reliable nor resilient. The world suffers significant deaths and injuries every year. Transportation exerts large-scale, unsustainable impacts on energy, the environment, and climate. For example, the surface transportation system of the United States is facing serious safety, mobility, and environmental issues [2]: There are over 5.8 million crashes per year on U.S. roadways, resulting in 37,000 deaths annually and having a direct economic cost of $230.6 billion; traffic congestion is an $87.2 billion annual drain on the U.S. economy, with 4.2 billion hours and 2.8 billion gallons of fuel spent sitting in traffic; tailpipe emissions from vehicles are the single largest human-made source of carbon dioxide (CO_2), nitrous oxides (NOx), and methane. Hence, there is an urgent need to improve transportation system performance and resiliency, reduce transportation injuries and fatalities, and mitigate unsustainable environmental impacts.

These critical issues in transportation could be addressed potentially by cyber-physical systems (CPSs), which are smart networked systems with embedded sensors, processors, and actuators that are designed to sense and interact with the physical world (including the human users) and support real-time, guaranteed performance in safety-critical applications [3]. The drivers of CPSs include transportation, manufacturing and industry, healthcare, energy, agriculture, defense, building controls, society, and emergency response [3]. In CPSs, the joint behavior of the *cyber* and *physical* elements of the system is critical: Computing, control, sensing, and networking can be deeply integrated into every component, and the actions of components and systems must be safe and interoperable [3]. Advances in CPSs will enable capability, adaptability, scalability, resiliency, safety, security, and usability that will far exceed the simple systems of today [4].

The application of CPSs in the transportation sector, called transportation CPS, will transform the way people interact with transportation systems, just as the Internet has transformed the way people interact with information. Examples of transportation CPS include autonomous cars and aircraft autopilot systems, among many others [5]. It is expected that we travel in driverless cars that communicate securely with each other on smart roads and in planes that coordinate to reduce delays [6]. CPS technologies can potentially eliminate accidents caused by human error, which currently account for 93% of the about 6 million annual automotive crashes [7,8].

Tremendous progress has been made in transportation CPS in recent years. For example, driver-assistance tools such as adaptive cruise control, lane departure warning, automated parallel parking, and traction control are becoming reality [9]; semiautonomous automobiles are predicted to be commercially available in a few years; unmanned flight has been used for short taxi-like trips [5]. However, there are some challenges that must be overcome. A desirable transportation CPS is safe, secure, and resilient in a variety of unanticipated and rapidly evolving environments and disturbances and cost-effective. The effective operation of transportation CPS is based on data-intensive computing. By using a network of remote servers to store, manage, and process data, rather than a local server or a personal computer, cloud computing has the potential to enable transportation CPS to operate desirably. However, to reap the benefits of cloud computing, there are some challenges that must be addressed properly. This chapter presents the opportunities and challenges in integrating cloud computing into transportation CPS and reviews the state-of-the-art and practice in applying cloud computing to transportation CPS.

The organization of this chapter is as follows. Section 15.2 presents the state-of-the-art and practice in transportation CPS. Section 15.3 gives an overview of cloud computing. Section 15.4 presents opportunities and challenges in applying cloud computing to transportation CPS. Next, some case studies are presented in Section 15.5. Section 15.6 concludes the chapter.

15.2 Transportation Cyber-Physical Systems

Transportation CPS will transform the way people interact with transportation systems, including automobile, aviation, and rail. It is expected that we travel in driverless cars that communicate securely with each other on smart roads and in planes that coordinate to reduce delays [6]. In this section, we will present the state-of-the-art and practice in transportation CPS, mainly automobile and aviation.

15.2.1 Highway Transportation Cyber-Physical Systems

Transportation CPS will provide highway transformation the foundation necessary for a safe, efficient highway transportation system that connects vehicles, infrastructure, people, and goods in a vibrant, competitive economy. A typical highway transportation CPS is connected automated vehicles (CAVs), which blend connected vehicle (CV) technology and autonomous vehicle systems.

15.2.1.1 Connected Vehicle

CV, also known as vehicular ad hoc networks (VANETs), or Internet of vehicles (IoV), is the networking infrastructure of highway transportation CPS. CV technology has the potential to inform vehicles and drivers about the dynamics, movements, and intents of other vehicles in their surroundings [2]. CV is based on the use of dedicated short-range communications (DSRC) technology, which is a two-way short- to medium-range wireless communication capability permitting very high data transmission critical in communications-based active safety applications [10]. A connected transportation environment among vehicles of all types, the infrastructure, and portable devices will serve the public good by leveraging technology to maximize safety, mobility, and environmental performance [11]. The applications of CV can be classified into three categories: safety applications, mobility applications, and environmental applications [12].

CV safety applications are designed to increase situational awareness and reduce or eliminate crashes through vehicle-to-vehicle (V2V) and vehicle-to-infrastructure (V2I) data transmission that supports driver advisories, driver warnings, and vehicle and/or infrastructure controls. These technologies may potentially address up to 82% of crash scenarios with unimpaired drivers, preventing tens of thousands of automobile crashes every year [12]. The safety applications of CV based on V2V communications include emergency brake light warning, forward collision warning, intersection movement assist, blind spot and lane change warning, do not pass warning, and control loss warning [13]. And the safety applications of CV based on V2I communications include intersection safety, runoff road, speed management, and commercial/transit vehicle enforcement and operations for safety [14].

CV mobility applications provide a connected, data-rich travel environment. One example application is real-time data capture and management in which the VANET captures real-time data from equipment located on-board vehicles and within the infrastructure [15]. Another example application is dynamic mobility applications in which the data are transmitted wirelessly and are used by transportation managers in a wide range of dynamic, multimodal applications to manage the transportation system for optimum performance [16].

CV environmental applications generate and capture environmentally relevant real-time transportation data and use these data to create actionable information to support and facilitate *green* transportation choices and assist system users and operators with *green* transportation alternatives or options, thus reducing the environmental impacts of each trip. Data generated from CV systems can also provide operators with detailed, real-time information on vehicle location, speed, and other operating conditions. This information can be used to improve system operation. On-board equipment may also advise vehicle owners on how to optimize the vehicle's operation and maintenance for maximum fuel efficiency [12]. One example application is Applications for the Environment: Real-Time Information Synthesis (AERIS) that contribute to mitigating some of the negative environmental impacts of surface transportation [17]. Another example application is road weather applications for CVs that are the next generation of applications and services that assess, forecast, and address the impacts that weather has on roads, vehicles, and travelers [18].

These CV applications are summarized in Table 15.1.

To test these CV applications in a real-world environment, the U.S. Department of Transportation (USDOT) has established seven test beds in Michigan, Virginia, Florida, California, New York, Arizona, and Tennessee [19].

The Federal Highway Administration (FHWA) within the USDOT has been investigating connected highway and vehicle systems to improve the safety, mobility, and efficiency of the nation's highways under its Exploratory Advanced Research (EAR) program and funding research ranging from enabling technology for positioning, navigation, time synchronization, sensor integration, and improved situational awareness to new system concepts for vehicle platooning, speed management, intersection management, and vehicle merging [4]. In early January 2013, the National Science Foundation (NSF) and the Federal Highway Administration (FHWA) Exploratory Advanced Research (EAR) program decided to coordinate on CPS for highway transportation [20] and identified two areas for coordination: enabling technology and scaling highway CPS [4]. Their shared interest lies in not only research and development of new enabling technology that provides improved safety, mobility, and energy conservation in the development and operation of the highway system but also new research methods to test CV and highway systems at larger scales. The CPS technology research challenges identified include CPS data acquisition, quality assurance and integration, data and information analytics, and decision making [4].

TABLE 15.1

Connected Vehicle Applications

Category	Application
Safety	Vehicle-to-vehicle communications for safety
	Vehicle-to-infrastructure communications for safety
Mobility	Real-time data capture and management
	Dynamic mobility applications
Environmental	Applications for the Environment: Real-Time Information Synthesis (AERIS)
	Road weather applications for connected vehicles

15.2.1.2 Autonomous Vehicle

Autonomous vehicles, also called self-driving vehicles, or driverless vehicles, are those in which operation of the vehicle occurs without direct driver input to control the steering, acceleration, and braking and are designed so that the driver is not expected to constantly monitor the roadway while operating in self-driving mode [21]. The National Highway Traffic Safety Administration (NHTSA) defines vehicle automation as having five levels, as shown in Table 15.2 [21].

In 2004 and 2005, the Defense Advanced Research Projects Agency (DARPA) within the Department of Defense (DoD) organized two Grand Challenges to accelerate research and development in autonomous ground vehicles [22,23]. Although no team entry successfully completed the designated route for the DARPA Grand Challenge 2004, featured a 142 mile desert course, a total of five teams completed the Grand Challenge course, which was 132 miles over desert terrain in 2005 [23]. Building on the success of these two Grand Challenges, the DARPA Urban Challenge was held on November 3, 2007, at the former George AFB in Victorville, California. The Urban Challenge featured autonomous ground vehicles maneuvering in a mock city environment, executing simulated military supply missions while merging into moving traffic, navigating traffic circles, negotiating busy intersections, and avoiding obstacles [24].

Google announced its self-driving car project in 2010 [25]. Google self-driving cars use video cameras, radar sensors, and a laser range finder to "see" other traffic, as well as detailed maps (which we collect using manually driven vehicles) to navigate the road ahead. As of April 2014, Google self-driving cars have logged nearly 700,000 autonomous miles [26]. In late May 2014, Google revealed a new prototype of its driverless car, which had no steering wheel, gas pedal, or brake pedal, being 100% autonomous [26].

Carnegie Mellon University (CMU) also developed an autonomous vehicle with minimal appearance modifications that is capable of a wide range of autonomous and

TABLE 15.2

Vehicle Automation Levels Defined by NHTSA

Level Index	Level Name	Definition
Level 0	No automation	The driver is in complete and sole control of the primary vehicle controls—brake, steering, throttle, and motive power—at all times.
Level 1	Function-specific automation	Automation at this level involves one or more specific control functions.
Level 2	Combined function automation	This level involves automation of at least two primary control functions designed to work in unison to relieve the driver of control of those functions.
Level 3	Limited self-driving automation	Vehicles at this level of automation enable the driver to cede full control of all safety-critical functions under certain traffic or environmental conditions and in those conditions to rely heavily on the vehicle to monitor for changes in those conditions requiring transition back to driver control. The driver is expected to be available for occasional control, but with sufficiently comfortable transition time.
Level 4	Full self-driving automation	The vehicle is designed to perform all safety-critical driving functions and monitor roadway conditions for an entire trip. Such a design anticipates that the driver will provide destination or navigation input, but is not expected to be available for control at any time during the trip.

Source: National Highway Traffic Safety Administration, Preliminary Statement of Policy Concerning Automated Vehicles, NHTSA 14-13, Washington, DC, 2013.

intelligent behaviors, including smooth and comfortable trajectory generation and following; lane keeping and lane changing; intersection handling with or without V2I and V2V; and pedestrian, bicyclist, and work zone detection [27]. Safety and reliability features include a fault-tolerant computing system, smooth and intuitive autonomous–manual switching, and the ability to fully disengage and power down the drive-by-wire and computing system upon E-stop. The vehicle has been tested extensively on both a closed test field and public roads. CMU's autonomous Cadillac SRX drove itself in and around Capitol Hill and Washington, DC, on June 24, 2014. This driverless vehicle looks normal on the inside and outside but navigates highways, dense multilane traffic, traffic lights, and tunnels. The loop around Capitol Hill is about 1.5 miles long, and the I-395 journey is about 6 miles [28].

In addition to driverless cars, smart roads on which driverless cars communicate securely with each other are being investigated [6]. The CPS challenges around smart roads include how cars communicate with each other, whether cars could share road space with human drivers, whether cars communicate through roadside infrastructure, and how to provide the incentives to make these technologies secure enough to avoid incidents [29].

15.2.2 Aviation Cyber-Physical Systems

In 2003, the U.S. Congress established the Joint Planning and Development Office (JPDO) to plan and coordinate the development of the Next Generation Air Transportation System (NextGen), which is an upgrade to satellite-based technology [30]. Satellite navigation will let pilots know the precise locations of other airplanes around them, allowing more planes in the sky while enhancing the safety of travel. Satellite landing procedures will let pilots arrive at airports more predictably and more efficiently. And once on the ground, satellite monitoring of airplanes leads to getting to the gate faster.

Air traffic management integrates the physical world, including the airplanes and the environment in which they reside, with the control algorithms and the pilots and traffic control managers [31]. MIT researcher explored algorithmic solutions to make flying more efficient by proposing an approach that determines a suggested rate to meter pushbacks from the gate, in order to prevent the airport surface from entering congested states and to reduce the time that flights spend with engines on while taxiing to the runway. The field trials at Boston Logan International Airport demonstrated that significant benefits were achievable through such a strategy: during eight 4 h tests conducted during August and September 2010, fuel use was reduced by an estimated 12,250–14,500 kg (4000–4700 U.S. gallons), while aircraft gate pushback times were increased by an average of only 4.4 min for the 247 flights that were held at the gate [32].

15.3 Cloud Computing

Cloud computing is emerging as a new paradigm consisting of services that are commoditized and delivered in a manner similar to utilities such as water, electricity, gas, and telephony [33]. It has been widely used in various scientific, business, and consumer applications, including healthcare, biology, geoscience, CRM and ERP, productivity, social networking, media, and multiplayer online gaming [33].

15.3.1 Definition

The most commonly used definition of cloud computing was given by the National Institute of Standards and Technology (NIST). Cloud computing is a model for enabling ubiquitous, convenient, on-demand network access to a shared pool of configurable computing resources (e.g., networks, servers, storage, applications, and services) that can be rapidly provisioned and released with minimal management effort or service provider interaction [34]. This cloud model is composed of five essential characteristics, three service models, and four deployment models, as shown in Figure 15.1 and Tables 15.3 through 15.5 [34].

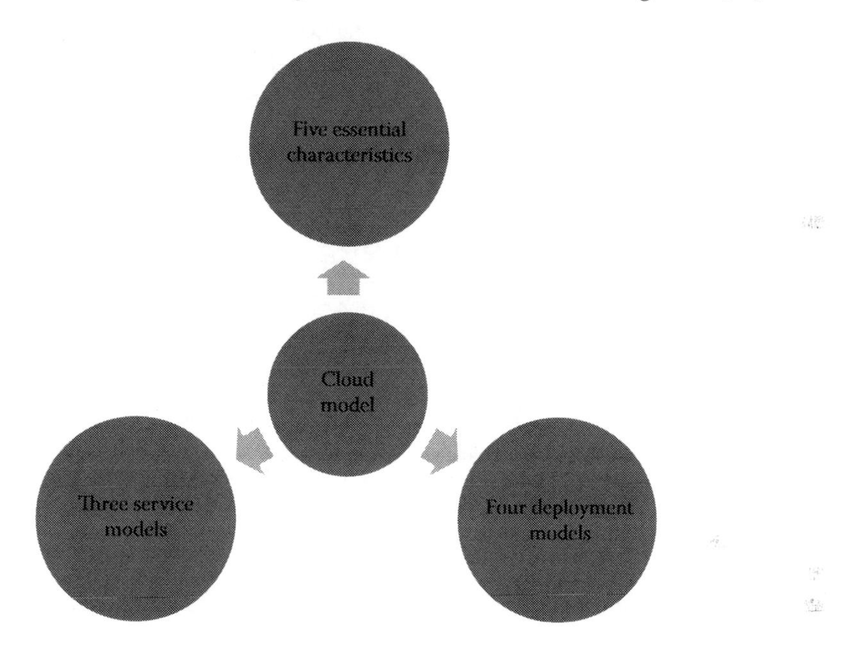

FIGURE 15.1
The cloud model defined by NIST. (From Mell, P. and Grance, T., The NIST Definition of Cloud Computing, National Institute of Standards and Technology, Special Publication 800-145, Gaithersburg, MD, 2011.)

TABLE 15.3
Five Essential Characteristics in the Cloud Model

Essential Characteristics	Definitions
On-demand self-service	A consumer can unilaterally provision computing capabilities, such as server time and network storage, as needed automatically without requiring human interaction with each service provider.
Broad network access	Capabilities are available over the network and accessed through standard mechanisms that promote use by heterogeneous thin or thick client platforms.
Resource pooling	The provider's computing resources are pooled to serve multiple consumers using a multitenant model, with different physical and virtual resources dynamically assigned and reassigned according to consumer demand.
Rapid elasticity	Capabilities can be elastically provisioned and released, in some cases automatically, to scale rapidly outward and inward commensurate with demand.
Measured service	Cloud systems automatically control and optimize resource use by leveraging a metering capability at some level of abstraction appropriate to the type of service.

Source: Mell, P. and Grance, T., The NIST Definition of Cloud Computing, National Institute of Standards and Technology, Special Publication 800-145, Gaithersburg, MD, 2011.

TABLE 15.4

Three Service Models in the Cloud Model

Service Models	Definitions
Software as a service (SaaS)	The capability provided to the consumer is to use the provider's applications running on a cloud infrastructure.
Platform as a service (PaaS)	The capability provided to the consumer is to deploy onto the cloud infrastructure consumer-created or consumer-acquired applications created using programming languages, libraries, services, and tools supported by the provider.
Infrastructure as a service (IaaS)	The capability provided to the consumer is to provision processing, storage, networks, and other fundamental computing resources where the consumer is able to deploy and run arbitrary software, which can include operating systems and applications.

Source: Mell, P. and Grance, T., The NIST Definition of Cloud Computing, National Institute of Standards and Technology, Special Publication 800-145, Gaithersburg, MD, 2011.

TABLE 15.5

Four Deployment Models in the Cloud Model

Deployment Models	Definitions
Private cloud	The cloud infrastructure is provisioned for exclusive use by a single organization comprising multiple consumers.
Community cloud	The cloud infrastructure is provisioned for exclusive use by a specific community of consumers from organizations that have shared concerns.
Public cloud	The cloud infrastructure is provisioned for open use by the general public.
Hybrid cloud	The cloud infrastructure is a composition of two or more distinct cloud infrastructures (private, community, or public) that remain unique entities but are bound together by standardized or proprietary technology that enables data and application portability.

Source: Mell, P. and Grance, T., The NIST Definition of Cloud Computing, National Institute of Standards and Technology, Special Publication 800-145, Gaithersburg, MD, 2011.

15.3.2 Cloud Computing Platforms and Test Beds

This section presents popular cloud computing platforms and test beds in industry and academia.

15.3.2.1 Industry

There are three big providers in cloud computing: Amazon Web Services (AWS), Google Cloud, and Microsoft Azure.

Windows Azure is a platform for building scalable, multitiered web services and it is hosted on Microsoft's large network of data centers [35]. It is ideally suited to host community data collections and analysis tools that can be accessed via web protocols from client devices. Azure services are built on the Windows Server operating system, and they can be programmed with C, C++, C#, Python, Java, Ruby, and other standard web tools. Windows Azure provides on-demand compute and storage to host, scale, and manage web applications on the Internet. Azure applications consist of one or more "web roles," which are standard web service processes, and "worker roles," which are computational and data management processes. Roles communicate by passing messages through queues or sockets.

AWS is a collection of remote computing services (also called web services) that together make up a cloud computing platform, offered over the Internet by Amazon.com [36]. The most central and well known of these services are Amazon EC2 and Amazon S3.

Google Cloud Platform is a portfolio of cloud computing products by Google, which offers hosting on the same supporting infrastructure that Google uses internally for end-user products like Google Search and YouTube [37].

In addition, Aneka provided by Manjrasoft Pty Ltd plays the role of application platform as a service for cloud computing. Aneka supports various programming models involving task programming, thread programming, and MapReduce programming and tools for rapid creation of applications and their seamless deployment on private or public clouds to distribute applications [38].

15.3.2.2 Academia

In August 2014, the NSF announced two $10 million projects to create cloud computing test beds, to be called "Chameleon" and "CloudLab," that will enable the academic research community to develop and experiment with novel cloud architectures and pursue new, architecturally enabled applications of cloud computing [39]. Chameleon will be deployed at the University of Chicago and the Texas Advanced Computing Center and will consist of 650 multicore cloud nodes, 5PB of total disk space, and leverage 100 Gbps connection between the sites [40]. CloudLab will be a distributed infrastructure having clusters at three sites with each site comprising approximately 5000 cores and 300–500 terabytes of storage in the latest virtualization-capable hardware [41].

15.3.3 Mobile Cloud Computing

With the popularity of mobile devices such as smartphones and tablets, mobile cloud computing (MCC) is emerging [42–46]. MCC is defined as an infrastructure where both the data storage and the data processing occur outside of the mobile device [45]. Mobile cloud applications move the computing power and data storage away from mobile devices into the cloud, bringing applications and mobile computing not only to mobile device users but also to a much broader range of mobile subscribers.

Compared with cloud computing, MCC has a couple of advantages, including extending battery lifetime, improving data storage capacity and processing power, improving reliability, dynamic provisioning, scalability, multitenancy, and ease of integration [42].

15.3.4 Fog Computing

For latency-sensitive applications, cloud computing has to be extended to fog computing, which is a highly virtualized platform that provides compute, storage, and networking services between end devices and traditional cloud computing data centers, typically, but not exclusively, located at the edge of the network [47,75]. Compared with cloud computing, fog computing is characterized by low latency and location awareness, widespread geographical distribution, mobility, a very large number of nodes, predominant role of wireless access, strong presence of streaming and real-time applications, and heterogeneity [47]. Fog computing is a paradigm that extends cloud computing and services to the edge of the network. Similar to cloud, fog provides data, compute, storage, and application services to end users. The distinguishing fog characteristics are its proximity to end users, dense geographical distribution, and support for mobility. Services are hosted at the network edge or even end devices such as set-top boxes or access points. By doing so, fog reduces service latency and improves Quality of Service (QoS), resulting in superior user experience.

Fog computing can enable a new breed of aggregated applications and services, such as smart energy distribution and smart traffic lights [48,49].

15.4 Cloud Computing for Transportation Cyber-Physical Systems

Cloud computing enables the integration of new factors in the design and operation of CPS [50]. To exploit the full potential of both cloud computing and CPS, a cyber-physical cloud computing (CPCC) architectural framework, which integrates the characteristics of CPS and cloud computing into a single framework, has been conceived [51]. A CPCC architectural framework is defined as a system environment that can rapidly build, modify, and provision autoscale CPSs composed of a set of cloud computing–based sensor, processing, control, and data services. This framework, which supports the deployment of large-scale and data-intensive systems, involving complex distributed decision making, is a first step in achieving the vision of smart networked systems and societies (SNSSs) or cyber-physical social systems (CPSSs) [51]. The benefits of the CPCC framework include efficient use of resources, modular composition providing customizability, rapid development and scalability, smart adaptation to environment, scalable reliability, resiliency, and performance based on user needs [51]. An elastic infrastructure to support cloud computing for large-scale CPSs has been proposed [52].

This section presents opportunities and challenges posed by integrating cloud computing into transportation CPS.

15.4.1 Opportunities

15.4.1.1 Vehicular Cloud Computing

With the development of MCC, a new paradigm called vehicular cloud computing (VCC), in which vehicles interact and collaborate to sense the environment, process the data, propagate the results, and more generally share resources, is emerging [53–72]. Vehicles collectively operate as mobile clouds enabling environment modeling, content discovery, data collection, and other mobile applications in a way that is not possible, or not efficient, with the conventional Internet cloud alone. Vehicular cloud (VC) is defined as a group of largely autonomous vehicles whose corporate computing, sensing, communication, and physical resources can be coordinated and dynamically allocated to authorized users [54,55,73,74]. VCC is technologically feasible and economically viable toward autonomous traffic, vehicle control, and perception systems [56].

The VCC architecture consists of three layers—inside-vehicle, communication, and cloud [56]—from bottom to up. The inside-vehicle layer is responsible for monitoring the health and mood of the driver and collecting information inside the car such as pressure and temperature by using body sensors, environmental sensors, smartphone sensors, the vehicle's internal sensors, inertial navigation sensors (INS), and driver behavior recognition to predict the driver's reflexes and intentions [56]. The communication layer consists of two components: V2V and V2I. The V2V component is responsible for identifying abnormal behavior on the road indicated by a driver [56]. The V2I component is responsible for augmenting the safety level of vehicles on highways. The cloud layer, which consists of three internal layers—application, cloud infrastructure, and cloud platform—is responsible for computing the massive and complex computations in minimal time [56]. In the application layer, various applications and services are considered as real-time services or cloud primary services, which are accessible remotely by drivers [56]. In the primary services, several services are deployed, such as network as a service (NaaS), storage as a service (STaaS), cooperation as a service (CaaS), information as a service (INaaS), and entertainment as a service (ENaaS) [56]. The cloud infrastructure consists of two parties: cloud storage and cloud computation [56]. The data could be stored in the geographic information system (GIS), a road traffic control device, or a storage system based on the type of applications. The computation part is used to calculate the computational tasks that provide faster performance.

VCC could be applied in urban surveillance [53], traffic management [51], airport data center, parking lot, shopping mall, dynamic traffic light management, self-organized high-occupancy vehicle (HOV) lanes, evacuation management, road safety message, frequent congestion mitigation [54], and so on.

In the case of traffic management benefited by applying cloud computing to vehicular CPS, the goal is to control traffic movement by monitoring traffic flow, controlling traffic signals, and providing navigation aids such as alternate routes to individual vehicles. Data on traffic gathered in a number of ways, including sensors in the cars, closed-circuit television cameras, sensors on the road, and through intervehicle communication, can be used to rearrange traffic patterns, thus alleviating traffic congestion in a particular area. The system can operate on a push or pull mode: The system will inform the driver to take an alternate route (push); the driver requests information and makes a decision based on that information (pull) [51]. Based on the nature of the request, the system will mobilize its resources in terms of the sensor data and other prior data that will be available in the cloud and provide the requested information in an appropriate manner to the driver. To mitigate traffic congestion, the data and processing would require a combination of broadcasting messages to all vehicles as well as sending specific instructions to targeted vehicles. Therefore, there would be a need for cloud computing to allocate computing and storage resources from multiple locations to generate the right advice to the driver and also to specific traffic signals [51].

Further, when VCC is integrated with information-centric networking, which is a general form of communication architecture to achieve efficient content distribution on the Internet, this becomes vehicular cloud networking (VCN) whose goal is to create a vehicle cloud and to encourage collaborations among cloud members to produce advanced vehicular services that individuals alone cannot make [66].

In addition, with fog computing, smart traffic lights, in which a video camera senses an ambulance's flashing lights and then automatically changes streetlights for the vehicle to pass through traffic, could be enabled [48]. Also, fog computing could be used in rail CPS in which sensors on self-maintaining trains monitor train components. If trouble is detected, these sensors could send an automatic alert to the train operator to stop at the next station for emergency maintenance [48].

15.4.1.2 Aviation Cloud Computing

For aviation industry, the benefits of cloud computing could include real-time analytics and business intelligence, higher customer experience, and cost savings with focus on core services [76]. In 2012, Federal Aviation Administration (FAA) developed a cloud computing strategy to enable its vision on cloud computing: identify and migrate suitable IT services to a cloud computing environment to reduce costs and increase IT provisioning speed while ensuring that FAA air traffic control and management systems maintain their current high levels of safety, security, reliability, and performance [77]. When cloud computing is used in an aircraft data network, the benefits will include [78] effective use of bandwidth; scalable, interchangeable, remotely managed, and easily upgradeable; and use of industry standard products similar to the office, among others.

15.4.2 Challenges

15.4.2.1 Architecture

To support the real-time, safety, stability, and reliability requirements of transportation CPS, cloud service providers must make the right decisions a priori based on a configuration model that best suits their requirements and treat both utility to the participants and assurance of the reliability of the system overall so that the feedback of individual participants' decisions into the system behavior is considered properly [50]. Therefore, transportation CPS requires elasticity and autoscaling capabilities of the cloud platforms as workloads change, but with precise control over cyber-physical properties even as those changes are enacted and evolve. To be specific, the following technical challenges must be addressed, including precise autoscaling of resources with a system-wide focus, flexible optimization algorithms to balance real-time constraints with cost and other goals, improved fault-tolerant failover to support real-time requirements, and data provisioning and load balancing algorithms that rely on physical properties of computations [50]. A new generation of elastic infrastructure must be designed, developed, and evaluated [52]. To realize the CPCC system architecture, the key is to understand the basic components of the system and investigating different possible implementations [51].

15.4.2.2 Security and Privacy

When cloud computing is applied to transportation CPS, a lot of security and privacy issues emerge [51,61,63]. For example, the attackers in VCC have more advantages over the attackers in transportation CPS without cloud computing integrated because an attacker can equally share the same physical machine/infrastructure as their targets and physically move from place to place as vehicles are mobile nodes [69]. The targets of an attacker in VCC could be confidentially valuable data and documents stored in the VC and the location of the virtual machines, where the target's services are executing; integrity, such as valuable data and documents stored in the VC, executable code, and results in the VC; and availability, such as physical machines and resources, privileges, services, and applications [69]. VCC shares the same threats as computer security, that is, so-called STRIDE threats, including spoofing user identity, tampering, repudiation, information disclosure, denial of service, and evaluation of privilege [69]. The development of both technology and standards to improve security and privacy of transportation CPS is required [51].

15.4.2.3 Data Integration

In transportation CPS, the amount and variety of data that are gathered, combined, and interpreted for targeting actions and creating new knowledge will be increased by the large number of heterogeneous services involved in the integration of cloud computing and transportation CPS [51]. As a result, there are many different types of data with different units, structures, uncertainties, and semantics. To deal with the four dimensions of the big data, that is, volume, variety, velocity, and veracity, a lot of data management challenges must be addressed, including unified data representation and processing models to accommodate heterogeneous or new types of data, short-term and long-term storage, improvement of quality (accuracy) of data in real time by sampling and filtering, intelligent data interpretation and semantic interoperability, short-term and long-term storage, and intersituation analysis and prediction [51]. Therefore it is desired to understand the concepts involved and the different languages used by the various stakeholders.

15.4.2.4 Social Sensors

Two types of sensors are involved in the integration of cloud computing with transportation CPS. One type is the traditional type of physical sensors, such as temperature or seismic sensors. The other is social sensors using input from humans as sensor data [51]. Social sensors are virtual sensors that provide data gathered from human beings and communicate with the provisioner to exchange information about the sensor availability and capabilities and to provision the sensor for use within a particular system. A social sensor must satisfy the requirements, such as clear interpretation of social data, methodology for assigning uncertainties for social data, and semantics and data formats for virtual sensors. Otherwise knowledge generation from social sensors is impossible [51].

15.5 Case Studies

This section presents two case studies about applying cloud computing in transportation CPS. One is cloud-based travel time reliability monitoring [79,80] and the other is urban surveillance in the VC [53].

15.5.1 Travel Time Reliability Monitoring

Travel time reliability is an essential system performance index that provides a critical operational measure of the transportation network, in addition to regular mobility measures, such as travel time indices and congestion levels [79]. The reliability information, such as the likelihood of an on-time arrival with travel alternatives and options, could help travelers to better choose their departure time, mode, and route to avoid possible traffic congestion [79]. A server–client system architecture for the provision of traveler information has been proposed [79,80]. The server is composed of four core components: traffic measurement and historical pattern databases, GPS map-matching engine, data mining and fusion engine, and reliable routing engine. The client is designed to display maps and results and accept user input. A MapReduce-based reliable routing engine is implemented under the Microsoft Daytona MapReduce framework on the server side to fully use the parallel computing capabilities and improve system performance [79]. MapReduce is a programming model and an associated implementation for processing and generating large datasets that is amenable to a broad variety of real-world tasks. Users specify the computation in terms of a map and a reduce function, and the underlying runtime system automatically parallelizes the computation across large-scale clusters of machines, handles machine failures, and schedules intermachine communication to make efficient use of the network and disks [83]. Microsoft Daytona MapReduce framework is an iterative MapReduce runtime targeted for the Microsoft Azure cloud computing platform.

The MapReduce-based process for the calculation of travel time reliability is a five-step procedure [79,80]: Step 1, the input of the travel time reliability engine is a list of key–value pairs [K1, V1], where key K1 is a string to identify a user specified O–D pair and value V1 is the specified departure time; step 2, the first map function maps an O–D pair to an intermediate key–value pair [K2, V2], where key K2 is a string that includes the link information of an alternative path from the origin to the destination and value V2 is the departure time; step 3, the second map function, based on samples from a historical database, generates another intermediate key–value pair [K3, V3] (the key in this step is generated as an alias of K2 + V2; value V3 is the path travel time calculated by this map function; the values associated with the same key are grouped together by the group function provided by MapReduce libraries); step 4, the first reduce function calculates the travel time statistics in the form of the mean and the variance; and step 5, the second reduce function combines the travel time statistics of all the alternative paths between the origin and the destination together and generates a final output to the user.

To demonstrate the effectiveness of the travel time reliability engine based on MapReduce, a system performance test was conducted on a large-scale, real-world transportation network in the Bay Area, California, which comprises 53,124 nodes and 93,900 links [79]. The traffic data were collected from GPS-equipped cell phones in one hundred vehicles driven on a 10-mile stretch of a highway located in the San Francisco Bay Area. And the data extracted from video recordings and data from in-road sensors (loop detectors) were used to validate the cell-phone-generated traffic information [81]. The traffic date collection was a part of the Mobile Millennium project [82]. The test results showed that, after 3 or 4 min of the system's warming up, the number of requests that the tested server can handle oscillates around 25–30 requests per second and the average response time for receiving and processing a user request is 1.23 s. Therefore the reliability information could be calculated by the proposed system in an acceptable time [80].

15.5.2 Urban Surveillance

Vehicles are ideally suited for environmental surveillance or environmental sensing to complement fixed video cameras and sensors installed in the infrastructure, such as light poles, rooftops, and traffic lights. Vehicle surveillance may be used in prevention of possible attacks and forensic investigation after the incident has occurred [53].

A novel mobile sensor middleware called MobEyes, which exploits wireless-enabled vehicles equipped with video cameras and a variety of sensors to perform event sensing, processing and classification of sensed data, and intervehicle ad hoc message routing, has been developed to support proactive urban monitoring applications [53]. MobEyes keeps sensed data in mobile node storage from which on-board processors extract features of interest, for example, license plates, and periodically generate metadata appending to sensor data critical context information such as timestamps and position coordinates. Then these summaries are disseminated periodically in the cloud and mobile agents, for example, police patrolling cars, upon alerted, move and opportunistically harvest summaries from neighbor vehicles [53].

To verify the effectiveness of MobEyes, a vehicle tracking application where the agent reconstructs vehicle trajectories exploiting the collected summaries, has been simulated on an urban 1 km × 1 km square grid scenario, with 100 vehicles randomly roaming in it. The simulation results showed that in most cases the average uncovered interval fluctuated between [2.7 s, 3.5 s], and even in the worst cases the agent has at least one sample every 200 s for more than 90% of the participants [53].

15.6 Conclusions and Outlook

Transportation is one of the most important drivers and application domains of CPSs, which are systems in which physical processes are tightly intertwined with networked computing. The application of CPS in transportation sector, called transportation CPS, will transform the way people interact with transportation systems, just as the Internet has transformed the way people interact with information. A desirable CPS is safe, secure, and resilient in a variety of unanticipated and rapidly evolving environments and disturbances and cost-effective. The effective operation of transportation CPS is based on data-intensive computing. By using a network of remote servers to store, manage, and process data, rather than a local server or a personal computer, cloud computing has the potential to enable transportation CPS to operate desirably. However, to reap the benefits of cloud computing, there are some challenges that must be addressed. This chapter presents the opportunities and challenges in integrating cloud computing with transportation CPS and reviews the state-of-the-art and practice of applying cloud computing to transportation CPS. Particularly, for large-scale and data-intensive highway transportation CPS, cloud computing offers opportunities such as allowing for resiliency in data recovery, robustness of operations, and location-independent storage and computing. It is expected that cloud computing will play an important role to improve the traffic mobility, safety, and environmental protection by integrating with transportation CPS. However, there exist some challenges such as architecture, security and privacy, data integration, and social sensors, among others.

Glossary

Cloud computing: Cloud computing is a model for enabling ubiquitous, convenient, on-demand network access to a shared pool of configurable computing resources (e.g., networks, servers, storage, applications, and services) that can be rapidly provisioned and released with minimal management effort or service provider interaction.

Cyber-physical cloud computing: A cyber-physical cloud computing architectural framework is defined as a system environment that can rapidly build, modify, and provision autoscale cyber-physical systems composed of a set of cloud computing–based sensor, processing, control, and data services.

Cyber-physical systems: Cyber-physical systems (CPSs) are smart networked systems with embedded sensors, processors, and actuators that are designed to sense and interact with the physical world (including the human users) and support real-time, guaranteed performance in safety-critical applications. Cyber-physical systems (CPSs) are engineered systems that are built from, and depend upon, the seamless integration of computational algorithms and physical components.

Fog computing: Fog computing is a paradigm that extends cloud computing and services to the edge of the network.

Mobile cloud computing: Mobile cloud computing is the availability of cloud computing services in a mobile ecosystem. This incorporates many elements, including consumer, enterprise, femtocells, transcoding, end-to-end security, home gateways, and mobile broadband-enabled services.

Vehicular cloud: A vehicular cloud is a group of largely autonomous vehicles whose corporate computing, sensing, communication, and physical resources can be coordinated and dynamically allocated to authorized users.

Vehicular cloud computing: Vehicular cloud computing is a new paradigm in which vehicles interact and collaborate to sense the environment, process the data, propagate the results, and more generally share resources.

Vehicular cloud networking: Vehicular cloud networking is the integration of vehicular cloud computing with information-centric networking, which is a general form of communication architecture to achieve efficient content distribution on the Internet.

References

1. Transportation Research Board, Critical Issues in Transportation. (2013). Washington, DC, http://onlinepubs.trb.org/Onlinepubs/general/criticalissues13.pdf. Last accessed on July 22, 2015.
2. U.S. Department of Transportation (US DOT). ITS Strategic Research Plan, 2010–2014, http://www.its.dot.gov/strategic_plan2010_2014/2010_factsheet.htm. Last accessed on July 22, 2015.
3. CPS Senior Steering Group. (2012). CPS Vision Statement, Federal Networking and Information Technology Research and Development (NITRD) Program, https://www.nitrd.gov/nitrdgroups/images/6/6a/Cyber_Physical_Systems_(CPS)_Vision_Statement.pdf. Last accessed on July 22, 2015.
4. National Science Foundation, Cyber-Physical Systems (CPS), http://www.nsf.gov/funding/pgm_summ.jsp?pims_id=503286. Last accessed on July 22, 2015.
5. Marzullo, K. Cyber physical systems: Connecting computation and the physical world, http://www.nsf.gov/cise/news/cps-perspective.jsp. Last accessed on July 22, 2015.
6. National Science Foundation, Cyber-Physical Systems: Enabling a Smart and Connected World, http://www.nsf.gov/news/special_reports/cyber-physical/. Last accessed on July 22, 2015.
7. National Institute of Standards and Technology (NIST), *Strategic Vision and Business Drivers for 21st Century Cyber-Physical Systems*. (January 2013). Report from the Executive Roundtable on Cyber-Physical Systems, http://www.nist.gov/el/upload/Exec-Roundtable-SumReport-Final-1-30-13.pdf. Last accessed on July 22, 2015.
8. Radha Poovendran, *A Community Report of the 2008*. (Updated on July 22, 2009). *High Confidence Transportation Cyber-Physical Systems (HCTCPS) Workshop,* http://www.ee.washington.edu/research/nsl/aar-cps/NCO_June_2009.pdf. Last accessed on July 22, 2015.
9. National Science Foundation. (2014). *NSF National Workshop on Transportation Cyber-Physical Systems,* http://cps-vo.org/group/CPSTransportationWksp2014/. Last accessed on July 22, 2015.
10. U.S. Department of Transportation (US DOT). Overview of Dedicated Short Range Communications (DSRC) Technology, http://www.its.dot.gov/DSRC/index.htm. Last accessed on July 22, 2015.
11. U.S. Department of Transportation (US DOT). ITS Strategic Research Plan, 2010–2014, Executive Summary, http://www.its.dot.gov/strategic_plan2010_2014/index.htm. Last accessed on July 22, 2015.
12. U.S. Department of Transportation (US DOT). Connected Vehicle Research, http://www.its.dot.gov/connected_vehicle/connected_vehicle.htm. Last accessed on July 22, 2015.
13. U.S. Department of Transportation (US DOT). Connected Vehicle Applications: Vehicle-to-Vehicle (V2V) Communications for Safety, http://www.its.dot.gov/research/v2v.htm. Last accessed on July 22, 2015.
14. U.S. Department of Transportation (US DOT). Connected Vehicle Applications: Vehicle-to-Infrastructure (V2I) Communications for Safety, http://www.its.dot.gov/research/v2i.htm. Last accessed on July 22, 2015.

15. U.S. Department of Transportation (US DOT). Connected Vehicle Applications: Real-Time Data Capture and Management, http://www.its.dot.gov/data_capture/data_capture.htm. Last accessed on July 22, 2015.
16. U.S. Department of Transportation (US DOT). Connected Vehicle Applications: Dynamic Mobility Applications, http://www.its.dot.gov/dma/index.htm. Last accessed on July 22, 2015.
17. U.S. Department of Transportation (US DOT). Connected Vehicle Applications: Applications for the Environment: Real-Time Information Synthesis (AERIS), http://www.its.dot.gov/aeris/index.htm. Last accessed on July 22, 2015.
18. U.S. Department of Transportation (US DOT). Connected Vehicle Applications: Road Weather Connected Vehicle Applications, http://www.its.dot.gov/connected_vehicle/road_weather.htm. Last accessed on July 22, 2015.
19. U.S. Department of Transportation (US DOT). The Connected Vehicle Test Bed, http://www.its.dot.gov/testbed.htm. Last accessed on July 22, 2015.
20. National Science Foundation, Dear Colleague Letter. (January 2, 2013). NSF-FHWA Coordination on Cyber Physical Systems for Highway Transportation, NSF 13–034, Arlington, VA. http://www.nsf.gov/pubs/2013/nsf13034/nsf13034.jsp. Last accessed on July 22, 2015.
21. National Highway Traffic Safety Administration, Preliminary Statement of Policy Concerning Automated Vehicles. (2013). NHTSA 14–13, Washington, DC. http://www.nhtsa.gov/About+NHTSA/Press+Releases/U.S.+Department+of+Transportation+Releases+Policy+on+Automated+Vehicle+Development. Last accessed on July 22, 2015.
22. Grand Challenge. (2004). http://archive.darpa.mil/grandchallenge04/. Last accessed on July 22, 2015.
23. Grand Challenge. (2005). http://archive.darpa.mil/grandchallenge05/index.html. Last accessed on July 22, 2015.
24. DARPA Urban Challenge. http://archive.darpa.mil/grandchallenge/overview.html. Last accessed on July 22, 2015.
25. Google. What we're driving at, http://googleblog.blogspot.com/2010/10/what-were-driving-at.html. Last accessed on July 22, 2015.
26. Google. Google Self-Driving Car Project, https://plus.google.com/+GoogleSelfDrivingCars/. Last accessed on July 22, 2015.
27. Wei, J., Snider, J.M., Kim, J., Dolan, J.M., Rajkumar, R., and Litkouhi, B. (June 2013). Towards a viable autonomous driving research platform. In *Intelligent Vehicles Symposium* (IV), 2013 IEEE (pp. 763–770), 23 Jun–26 Jun 2013, Gold Coast City, Australia.
28. National Science Foundation, Demonstrating a driverless future. (June 24, 2014). http://www.nsf.gov/news/news_summ.jsp?cntn_id=131836. Last accessed on July 22, 2015.
29. Brown, E. Protecting infrastructure with smarter CPS. http://newsoffice.mit.edu/2014/saurabh-amin-protecting-infrastructure-0915. Last accessed on July 22, 2015.
30. Federal Aviation Administration. NextGen, https://www.faa.gov/nextgen/. Last accessed on July 22, 2015.
31. National Science Foundation. Designing tomorrow's air traffic control systems, http://www.nsf.gov/discoveries/disc_summ.jsp?cntn_id=132916. Last accessed on July 22, 2015.
32. Ioannis, S., Harshad, K., Hamsa, B., Reynolds, T.G., and Hansman, R.J. (August 2014). Demonstration of reduced airport congestion through pushback rate control, *Transportation Research Part A: Policy and Practice*, 66, 251–267.
33. Buyya, R., Vecchiola, C., and Thamarai Selvi, S. (2013), *Mastering Cloud Computing: Foundations and Applications Programming*, (468p.), Morgan Kaufmann, New Delhi, India.
34. Mell, P. and Grance, T. (2011). The NIST Definition of Cloud Computing, National Institute of Standards and Technology, Special Publication 800–145, Gaithersburg, MD.
35. http ://azure.microsoft.com/. Last accessed on July 22, 2015.
36. https://cloud.google.com/. Last accessed on July 22, 2015.
37. http://aws.amazon.com/. Last accessed on July 22, 2015.
38. http://www.manjrasoft.com/products.html. Last accessed on July 22, 2015.

39. http://nsf.gov/news/news_summ.jsp?cntn_id=132377. Last accessed on July 22, 2015.
40. http://www.chameleoncloud.org/. Last accessed on July 22, 2015.
41. https://www.cloudlab.us/. Last accessed on July 22, 2015.
42. Dinh, H.T., Lee, C., Niyato, D., and Wang, P. (2013), A survey of mobile cloud computing: Architecture, applications, and approaches. *Wireless Communications and Mobile Computing*, 13, 1587–1611.
43. Kumar, K. and Lu, Y.H. (2010). Cloud computing for mobile users: Can offloading computation save energy? *Computer*, 43(4), 51–56.
44. Fernando, N., Loke, S.W., and Rahayu, W. (2013). Mobile cloud computing: A survey. *Future Generation Computer Systems*, 29(1), 84–106.
45. Huang, D. (2011). Mobile cloud computing. *IEEE COMSOC Multimedia Communications Technical Committee (MMTC) E-Letter*, 6(10), 27–31.
46. Guan, L., Ke, X., Song, M., and Song, J. (May 2011). A survey of research on mobile cloud computing. In *Proceedings of the 2011 10th IEEE/ACIS International Conference on Computer and Information Science* (pp. 387–392), 16 May–18 May 2011, Sanya, China. IEEE Computer Society.
47. Bonomi, F., Milito, R., Zhu, J., and Addepalli, S. (August 17, 2012). Fog computing and its role in the internet of things. In *Proceedings of the First Edition of the MCC Workshop on Mobile Cloud Computing* (*MCC '12*), Helsinki, Finland.
48. Cisco. Fog Computing, http://www.cisco.com/web/solutions/trends/tech-radar/fog-computing.html. Last accessed on July 22, 2015.
49. Stojmenovic, I. and Wen, S. (September 7–10, 2014). The Fog computing paradigm: Scenarios and security issues. In *2014 Federated Conference on Computer Science and Information Systems* (*FedCSIS*), 7–10 September, 2014, Warsaw, Poland, https://fedcsis.org/proceedings/2014/pliks/503.pdf. Last accessed on July 22, 2015, pp. 1–8.
50. Institute for Software Integrated Systems (ISIS). *NSF Workshop on Cloud Computing for Cyber-Physical Systems*, 14–15 March, 2013, Arlington, VA, www.isis.vanderbilt.edu/workshops/cc4cps. Last accessed on July 22, 2015.
51. Simmon, E.D., Kim, K.S., Subrahmanian, E., Lee, R., de Vaulx, F.J., Murakami, Y., Zettsu, K., and Sriram, R.D. (August 26, 2013). A Vision of Cyber-Physical Cloud Computing for Smart Networked Systems, NIST Interagency/Internal Report (NISTIR)—7951, National Institute of Standards and Technology (NIST).
52. Schmidt, D.C., White, J., and Gill, C.D. (June 10–12, 2014). Elastic Infrastructure to Support Computing Clouds for Large-Scale Cyber-Physical Systems. In *IEEE 17th International Symposium on Object/Component/Service-Oriented Real-Time Distributed Computing* (*ISORC*), Reno, Nevada, 2014, pp. 56–63.
53. Gerla, M. (June 19–22, 2012). Vehicular Cloud Computing. In *The 11th Annual Mediterranean Ad Hoc Networking Workshop* (*Med-Hoc-Net*), pp. 152–155, Ayia Napa, Cyprus.
54. Eltoweissy, M., Olariu, S., and Younis, M. (2010). *Towards Autonomous Vehicular Clouds, Ad Hoc Networks*, Vol. 49, pp. 1–16, Springer, Berlin, Heidelberg.
55. Olariu, S. Vehicular Clouds, http://www.cs.odu.edu/~olariu/cs795–895-vc-f-2014.html. Last accessed on July 22, 2015.
56. MdWhaiduzzaman, M.S., Gani, A., and Rajkumar, B. (April 2014). A survey on vehicular cloud computing. *Journal of Network and Computer Applications*. 40, 325–344.
57. Abid, H., Phuong, L.T.T., Wang, J., Lee, S., and Saad, Q. (October 26–29, 2011). V-Cloud: Vehicular cyber-physical systems and cloud computing. In *Proceedings of the 4th International Symposium on Applied Sciences in Biomedical and Communication Technologies* (*ISABEL '11*), Barcelona, Spain.
58. Jaworski, P., Edwards, T., Moore, J., and Burnham, K. (October 5–7, 2011). Cloud computing concept for Intelligent Transportation Systems. In *2011 14th International IEEE Conference on Intelligent Transportation Systems* (*ITSC*), Washington, DC, pp. 391, 936.

59. Li, Z., Chen, C., and Wang, K. (January–February 2011). Cloud computing for agent-based urban transportation systems. *Intelligent Systems, IEEE*, 26(1), 73–79.60. Jolanta, J.-J. (October 19–22, 2011). The advantages of the use of cloud computing in intelligent transport systems. In *Modern Transport Telematics: 11th International Conference on Transport Systems Telematics, TST 2011*, Katowice-Ustroń, Poland. Selected Papers, Series: Communications in Computer and Information Science, Vol. 239, 2011, pp. 143–150, Springer, Berlin, Heidelberg.
61. Huang, D. (January 23–24, 2014). Secure open mobile cloud platform for urban traffic and safety applications. In *2014 NSF National Workshop on Transportation Cyber-Physical Systems*, Arlington, VA.
62. Kumar, V., Madria, S., *Integrating clouds with VANETs*. (January 23–24, 2014). *NSF National Workshop on Transportation Cyber-Physical Systems*, Arlington, VA. http://cps-vo.org/group/CPSTransportationWksp2014. Last accessed on July 22, 2015.
63. Dantu, R., Ho, S., Wilhelm, E., Sarma, S., and Jaynes, M. (January 23–24, 2014). Issues and challenges for vehicles in the cloud. In *2014 NSF National Workshop on Transportation Cyber-Physical Systems*, Arlington, VA.
64. Wan, J., Zhang, D., Zhao, S., Yang, L.,and Lloret, J. (August 2014). Context-aware vehicular cyber-physical systems with cloud support: Architecture, challenges, and solutions. *IEEE Communications Magazine*, 52(8), 106–113.65.
65 Gerla, M., Lee, E.-K., Pau, G., and Lee, U. (March 2014). Internet of Vehicles: From Intelligent Grid to Autonomous Cars and Vehicular Clouds, IEEE World Forum on Internet of Things, 6–8 March, 2014, Seoul, South Korea.
66. Lee, E., Lee, E.K., Gerla, M., and Oh, S.Y. (February 2014). Vehicular cloud networking: Architecture and design principles. *IEEE Communications Magazine*, 52(2), 148–155.
67. Yu, Y.-T., Punihaole, T., Gerla, M., and Sanadidi, M. (October 2012). Content Routing in the Vehicle Cloud, in *IEEE MILCOM*, 29 October–01 November, 2012, Orlando, FL.
68. Olariu, S., Hristov, T., and Yan, G. (2013). The next paradigm shift: From vehicular networks to vehicular clouds. In *Mobile Ad hoc Networking: Cutting Edge Directions*, Basagni, S., Conti, M., Giordano, S., and Stojmenovic, I. (Eds.) Wiley, Wiley-IEEE Press, Hoboken, NJ.
69. Yan, G., Wen, D., Olariu, S., and Weigle, M.C. (March 2013). Security challenges in vehicular cloud computing. *IEEE Transactions on Intelligent Transportation Systems*, 14(1), 284–294.
70. Lin, G., Zeng, D., and Guo, S. (December 9–13, 2013). Vehicular cloud computing: A survey. In *2013 IEEE Globecom Workshops (GC Wkshps)*, Atlanta, GA, pp. 403–407.
71. Ghafoor, K.Z., Mohammed, M.A., Bakar, K.A., SafaSadiq, A., and Lloret, J. (2013). *Vehicular Cloud Computing: Trends and Challenges. Mobile Computing over Cloud: Technologies, Services, and Applications.* IGI Publisher, pp. 262–274.
72. Hussain, R., Son, J., Eun, H., Kim, S., and Heekuck, O. (December 3–6, 2012). Rethinking vehicular communications: Merging VANET with cloud computing. In *2012 IEEE 4th International Conference on Cloud Computing Technology and Science* (*CloudCom*), pp. 606–609, Taipei, Taiwan.
73. Hussain , R., Abbas, F., Son, J., Kim, D., Kim, S., and Heekuck Oh. (December 2–5, 2013). Vehicle witnesses as a service: Leveraging vehicles as witnesses on the road in VANET Clouds. In *2013 IEEE 5th International Conference on Cloud Computing Technology and Science* (*CloudCom*), pp. 439–444, Bristol, U.K.
74. Abuelela, M. and Olariu, S. (November 8–10, 2010). Taking VANET to the clouds. In *Proceedings of the 8th International Conference on Advances in Mobile Computing and Multimedia (MoMM '10)*, pp. 1–6, Paris, France.
75. Bonomi, F. (April 16–18, 2013). The smart and connected vehicle and the internet of things. In *2013 Telecordia—NIST—ATIS Workshop on Synchronisation in Telecommunication Systems (WSTS 2013)*, San Jose, CA.
76. Gupta, R. and Rathore, R. (2012). *Navigating the Clouds: Aviation Industry*, www.hcltech.com/.../cloud-computing-in-aviation.pdf. Last accessed on July 22, 2015.
77. Carlson , E., Reyes, A., and Usmani, A. (2012). FAA cloud computing strategy, *The Journal of Air Traffic Control*, 55(1), 15–20.

78. Jasti, A., Mohapatra, S., Potluri, B., and Pendse, R. (May 10–12, 2011). Cloud computing in aircraft data network. In *Integrated Communications, Navigation and Surveillance Conference (ICNS)*, Herndon, VA, pp. E7–1–E7–8.
79. Lei, H., Xing, T., Taylor, J.F., and Zhou, X. (2012–12–01). Monitoring travel time reliability from the cloud, transportation research record. *Journal of the Transportation Research Board*, 2291(1), 35–43.
80. Lei, H. (2013). Estimate travel time reliability and emissions for active traffic and demand management, Doctoral dissertation, The University of Utah, Salt Lake City, UT.
81. Science Applications International Corporation. (2010). National Evaluation of the SafeTrip-21 Initiative: Draft Evaluation Report: Mobile Millennium Contract Number: DTFH61–06-D-00005 Task Order T-09–008a. http://ntl.bts.gov/lib/38000/38500/38548/FINAL_NT-TSP_Report_2011_03_31.pdf. Last accessed on July 22, 2015.
82. California Center for Innovative Transportation. Mobile Millennium, http://traffic.berkeley.edu/. Last accessed on July 22, 2015.
83. Dean, J. and Ghemawat, S. (2008). MapReduce: simplified data processing on large clusters. *Communications of the ACM*—50th anniversary issue: 1958–2008, 51(1), 107–113.

16

Cloud-Based Cost Optimization Model for Information Management of Cyber-Physical Systems

D. Sinha Roy, K. Hemant Kumar Reddy, and D.K. Mohanta

CONTENTS

16.1 Introduction

Cyber-physical systems (CPSs) have been a term coined of late to represent a class of next-generation engineered systems that envisions the use of advanced computation, communication, information, and control technologies for controlling and monitoring of a wide variety of systems. Such systems include sensing from distributed sensors, communicating data sensed to appropriate computing centers, processing the data, and finally inducing actuating actions for real-time monitoring and control. Cloud computing (CC), with its massive storage and computation capacity, has been a key driver to support creative and grand CPS applications.

However, CC, with its massive-scale data centers (DCs) and virtualized networks, is capable of delivering the computation needs. However, the cost of delivering the necessary services in such a cloud-based CPS is a challenging proposition. In this chapter, we thus focus on effective cost optimization strategies and present a comprehensive cost optimization model, hereafter referred to as the *cloud-based cost optimization model for CPS information management* (CCOMCIM) *model*. We exemplify the intricacies of the model using an illustrative example, that of a sample CPS, namely, a smart grid (SG).

SG is a state-of-the-art enhancement of the next-generation electric grid, which provides infrastructure that supports real-time, two-way communication between electric utilities and consumers by integrating information technologies and computational intelligence to revolutionize the power generation, delivery, and consumption [1]. The evolution of SG allows the software systems at both ends to control and manage electricity generation and consumption by integrating modern information technologies with electric utilities over a large geographical area [2].

With the evolution of SG, there is an increase in development of new applications and services that relies on utilization and integration of advanced information technologies by converting analog energy system to digital one. As the actors of SG generate huge amount of heterogeneous data, there is a need of storing these data. The data stored can be analyzed properly for controlling and generating electricity accordingly.

In this perception, information technology can be integrated with SG for effective information management [4]. To be precise, we concentrate on usage of CC, for managing information in the model. According to the National Institute of Standards and Technology (NIST), CC is defined as "a model for enabling convenient, on-demand network access to a shared pool of configurable computing resources (e.g., networks, servers, storage, applications, and services) that can be rapidly provisioned and released with minimal management effort or service provider interaction" [5]. Hence, the evolution of SG necessitates the provision of highly scalable and computing platform by cloud providers as the demand for resources varies over time.

Several previous works have dealt with the usage of CC for information management in SG environment. Simmhan et al. [6] discussed and analyzed the opportunities and challenges in optimizing information management with cloud-based demand response in SG. Nagothu et al. [7] proposed a model in which central communication and network optimization is done with the help of DCs deployed on the cloud. Kim et al. [8] proposed architecture for improving the response time in large-scale deployments for a cloud when there is an increase in demand. A cloud-based virtual SG environment by embedding SG in cloud environment has been presented in [9]. In [10], an approach for improving load balancing in grid environment has been presented by distributing services among DCs deployed in the cloud [10]. In [11], Xi Fang presents a cloud service–based SG information management model by analyzing the benefits and opportunities from the view of SG and CC perspectives, respectively. In [3], Xi Fang proposed a cloud and network resources optimization architecture for cloud-based SG information management to solve cost reduction problem for cloud-based information storage and computation. *The difference between the proposed CCOMCIM and the existing research is that, however, their work does not address cost of data transfer while maintaining SG information on a cloud. Our framework focuses on storage and data transfer on the DCs deployed on a cloud by taking into consideration network bandwidth in an optimized fashion, and based on that, we analyze the issue of minimizing the cost of computation and storage.*

In CC, large DCs are provided by cloud providers for massive computation and storage capacities that are charged depending on the usage of resources that may be sometimes costly affair. Hence, the storage and computation done by the electric utilities present in SG realm should be optimized. Apart from this, to improve the efficacy of the SG system, there is a need to integrate the "islands of information" to be generated independently by different departments of an electric utility. By outsourcing information management to cloud providers not only reduces the effort of storage and computation but also allows new players in SG to focus on building business rather than concentrating on building DCs to achieve scalability [12]. In [13], an optimization technique is proposed for minimizing the cost of outsourced tasks by utilizing the internal DCs deployed in the cloud and keeping quality of services in consideration. In [14], algorithmic approaches were proposed for handling difficulties in migrating enterprise services in a hybrid cloud deployment environment. In [15], to minimize the rental cost of resource in an elastic cloud, the authors proposed two resource rental planning models. In [16], an optimal cloud resource provisioning algorithm was presented for allocating resources of different cloud providers in an optimal manner. In [17], optimization of resources in the cloud environment is solved with the help of bin packing and mixed integer problem. In CC, there is a need to systematically optimize the resources as they may be connected over heterogeneous networks and distributed across large geographical bodies as it reduces overall cost of information management in SG. *The difference between our work and the aforementioned literature on optimal resource allocation in the cloud environment is that we designed storage cost optimization model and computation cost optimization model in our proposed novel* cloud-based cost optimization model for smart grid information management (CCOMCIM-SG) *where we take into consideration the requirements of the actors of SG such that there is no violation of service level agreement (SLA).*

Although SG seems to be fit for integrating with CC, there are some risks and challenges that should be taken into account for healthy operation of SG as it is dependent on prompt analysis of sensitive and critical information. These risks include security in storing data, consistency and fault-tolerant services, and ways to protect the privacy of sensitive data [6]. The proposed CCOMCIM model is designed such that there is no violation in security, privacy, and SLA, which is discussed in the next sections of the chapter. Finally, with the help of CCOMCIM model, we present the framework to solve the cost minimization problem in cloud-based SG information storage and computation.

The remainder of this chapter is organized as follows: Section 16.2 briefly describes the proposed CCOMCIM model, with its core architectural components. In Section 16.3, the CCOMCIM components have been mapped to an SG case study, and a detailed discussion of the components of each is presented. Thereafter, we describe the optimization framework for storage and computation cost models with reference to CCOMCIM model in Section 16.4. The results are discussed in Section 16.5, and finally, the conclusions are presented in Section 16.6.

16.2 CCOMCIM Model

Figure 16.1 depicts a cloud-based, hypothetical CPS with distinctly segregated components, augmenting cloud realm, network realm, and CPS realm. The cloud realm consists of DCs and associated connections. Since most CPS-based applications do provide services for a cost-conscious market, therefore, service providers and customers have been incorporated, with market-defined operations. A number of actors have also been incorporated, with designated interconnections. Appropriate connections depicting cloud information flow and flow of physical parameters have also been shown in the figure.

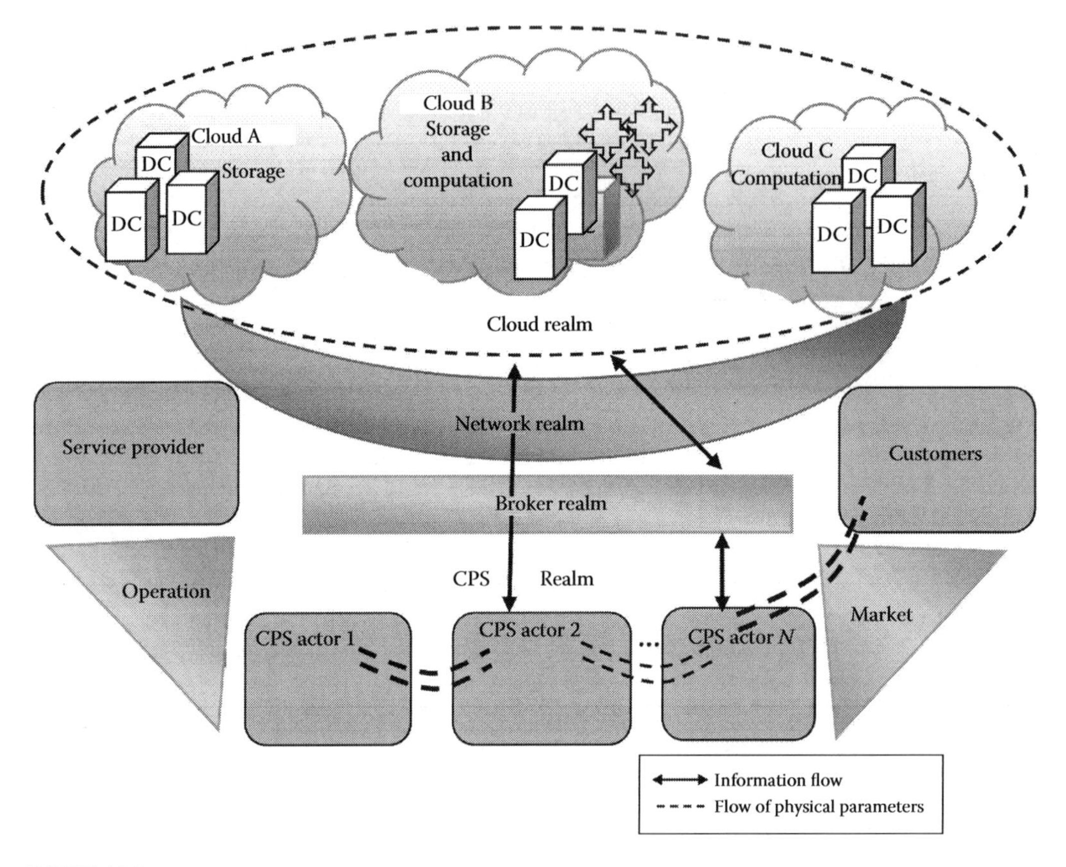

FIGURE 16.1
A schematic diagram depicting the architecture of cloud-based cyber-physical system.

Owing to inherent digression among different CPSs, it becomes cumbersome to explain the proposed CCOMCIM model in general terms. Thus, in the next section, a specific CPS, namely, an SG system, has been used for illustrating the different intricate details of the aforesaid model.

16.3 Case Study: CCOMCIM-SG

In this section, we describe our proposed novel cloud-based cost optimization model for smart grid information management and refer to it as CCOMCIM-SG model from the perspective of cloud realm, SG realm, and network realm.

16.3.1 Overview of the Proposed CCOMCIM-SG Model

Our proposed novel CCOMCIM-SG model consists of SG realm, cloud realm, network realm, and broker realm as shown in Figure 16.2. All the symbols and notations used in this chapter are shown in Table 16.1.

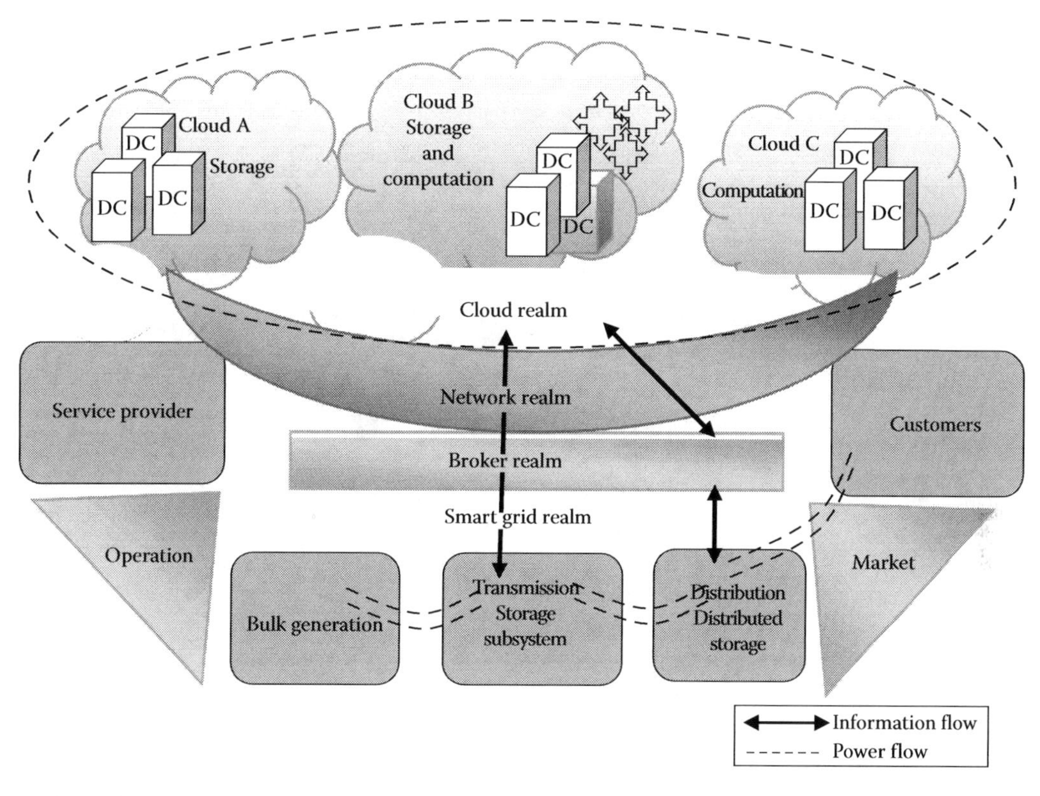

FIGURE 16.2
Cloud-based cost optimization model for smart grid information management model.

16.3.1.1 SG Realm

According to the NIST version 2.0, SG realm consists of seven subrealms: markets, operations, service providers, bulk generation, transmission, distribution, and customers [4]. Before going into details of all these subrealms, we first look into working of a power grid. In a power grid, generation of electricity is done by some electrical generator plants that are generally placed away from the colonized areas. These electrical generators generate electricity in bulk amount. From these generator plants, the electricity is transmitted to transmission station in the form of high step-up transmission level voltage using high transmission lines. From the transmission station, the power is transmitted to substations. The power received by the electric substations is a high voltage. This voltage is stepped down to a distribution level voltage and distributed among different distribution stations. From the distribution stations, again, the power gets stepped down to a required voltage as required by respective customers and gets distributed accordingly. The aforementioned process of electricity distribution in a power grid gives the general idea how bulk amount of electricity generated by generators is transmitted to distribution station that finally gets distributed in a required voltage among the respective customers who utilizes the electricity. Now let us look from the information management perspective of all the subrealms present in the SG realm.

TABLE 16.1

Symbols and Notations Used in Cloud-Based Cost Optimization Model for Smart Grid Information Management Model

T_C	Total cost
S_D	Set of EDOs
S_{SD}	Set of EDOs to be split
S_T	Set of tasks
S_D^C	Set of EDOs to be stored in one cloud
$S_D^{(dc)}$	Set of EDOs to be stored in data cloud
$S_T^{(d)}$	Set of tasks using EDO d
S_R	Set of requests
$S_{R(u)}$	Set of requests requested by user u
$T_i^{(d)}$	Task "i" using EDOs d as the input
$S_T^{(t)}$	Set of tasks that take task "t" as the input
S_C	Set of clouds
$S_C^{(d)}$	Set of clouds allowed to store EDO d
$C_i^{(d)}$	EDO d stored in cloud C_i
S_{DC}	Set data centers
$S_{DC}^{(c)}$	Set of data centers in cloud c
$S_{DC}^{(d)}$	Set of data centers containing EDO d
$S_C^{(dc)}$	Set of cloud containing data center dc
$S_{DC}^{(t)}$	Task t that is allowed to be executed in a set of data centers
$S_C^{(t)}$	Task t that is allowed to be executed in a set of clouds
S_U	Set of users
$S_U^{(d)}$	Set of users requesting for EDO d
$S_U^{(t)}$	Set of users executing task t
$P^{\uparrow}(d,dc,c)$	Unit uploading cost of EDO d on data center dc in cloud c
$P^{\downarrow}(d,dc,c)$	Unit downloading cost of EDO d on data center dc in cloud c
$P^S(d,dc,c)$	Unit storage cost of EDO d on data center dc in cloud c
$P^T(d,dc_m,c_i,dc_n,c_j)$	Unit intercloud transfer cost
$P^C(t,dc,c)$	Computation cost of task t on data center dc on cloud c
Z^d	Size of EDO d
Z^t	Size of the output of task t
$Z(d,dc,c)$	Size of EDO d stored on data center dc in cloud c
$\beta(d)$	Redundancy ratio of EDO d
$\gamma(d)$	Splitting ratio of EDO d
$B^{\uparrow}(d,dc,c)$	Boolean variable indicating whether EDO d is stored on data center dc of cloud c
$E(t,dc,c)$	Boolean variable indicating whether task t is executed on data center dc of cloud c
$B^{\rightarrow}(t_i,dc_m,c_k,t_j,dc_n,c_l)$	Boolean variable indicating whether data center dc_m of cloud c_k transmits intermediate data to data center dc_n of cloud c_l because task t_j takes the output of task t_i as the input
$P^R(d,dc,c,r)$	Unit request price of EDO d on data center dc of cloud c of type r_i
$N^R(dc,c,r)$	Number of request on data center dc of cloud c of request type r
$Z^R(dc,c,r)$	Size of request on data center dc of cloud c of type r_i

1. *Bulk generation*: The actors of this subrealm are the ones who are responsible of generating electricity in huge amount. Hence, in this subrealm, there is a need to manage information regarding the amount of electricity generated, expenditure for generating the electricity, total requirement of the electricity production with respect to other subrealms who use the electricity, amount of electricity transmitted, etc. The aforementioned maintained information can be utilized for analyzing the system's efficacy, and necessary steps can be taken to improve the system's performance required if any.
2. *Transmission*: The actors of this subrealm are responsible to transmit the high voltage produced by generator plants to distribution units. This subrealm is a mediator between the bulk generation subrealm and distribution subrealm. Hence, in this subrealm, there is a need to maintain the information regarding the amount of electricity received from the generation plant, the amount of electricity transmitted to distributed, amount of electricity transferred to subtransmission units deployed in different areas, etc. This information maintained can be used to improve the reliability of the system. Sensors and phasor measurement units (PMUs) are used to store the information regarding the data transmission and reception.
3. *Distribution*: The actors of this subrealm are responsible to step down the high voltage received from the transmission unit and distribute the electricity to the customers. This subrealm acts as a mediator between the transmission subrealm and customers. According to the NIST [4], the distribution subrealm is the one who will communicate closely and monitor the power flows associated with more dynamic market subrealm. With this real-time information, this subrealm can adjust accordingly and distribute the power to customers with respect to their requirement. Hence, in this subrealm, there is a need to maintain the information monitored, load requirements of different customers and other subrealms, and amount of electricity received from the distribution unit. This information maintained can be used to extract the amount of extra electricity required so that this information can be conveyed to transmission and bulk generation units as they can produce electricity with respect to the actual requirement, the peak time, etc.
4. *Customers*: The actors of this subrealm are the actual stakeholders who ultimately consume/utilize the electricity generated, and billing is done with respect to their usage. Hence, there is a need for customers to maintain information of the total amount of electricity received from the distribution unit, the actual utilization of electricity, the amount of electricity consumed by each electric device, amount of electricity not consumed, etc. Automatic meters are present at the customer that record the amount of the electricity consumed. This information maintained can be utilized to make an analysis of load distribution so that billing can be done accordingly. Apart from billing, this information can be used by distribution units to calculate the actual load to be distributed to each customer in a real-time scenario so that all the actors of the SG realm stay in profit.
5. *Markets*: According to the NIST [4], this subrealm is the area where the actual grid assets are bought and sold. As this subrealm is responsible for selling of grid assets, there is a requirement to maintain the information of different grid assets along with the requirements of the different subrealms. By monitoring

this information, pricing can be done in an optimal manner, and a balance can be maintained between supply and demand of grid resources. This subrealm is responsible for the exchange of electricity done between the generator units, transmission units, and customers.

6. *Service providers*: According to the NIST [4], the actors of this subrealm are responsible for providing supporting business services among other subrealms of SG. These business services can be maintenance of customer account service for billing, management of energy utilization, and home energy generation, as described in [4]. Hence, there is a need to maintain the information regarding different customers who use these services, usage data and pricing for controlling the load on the customers, providing support to new business entities about the different actors present in the SG, etc.
7. *Operations*: According to the NIST [4], the actors of this subdomain are responsible for the operations in grid power system. The responsibility of this subrealm is to maintain a safe and reliable power system. These operations include monitoring and management of electricity generation and utilization. Hence, there is a need to maintain information regarding the amount of electricity produced by the generator systems, amount of energy transmitted by the transmission system to different distribution stations, electricity consumed by the customers, etc.

As we have discussed about different subrealms and their perception regarding information management in SG realm, we now introduce three additional fields: *electric data object (EDO), computational tasks (CT), and clients*. These three fields are with respect to the proposed CCOMCIM model and will be used in the calculation of optimal storage and computational cost:

1. *EDO*: An EDO is an object that maintains information generated by the information sources present in the SG such as smart meters and PMUs as discussed in [10]. This information should be properly stored as it is used to perform some kind of computations. We indicate a set of EDOs in an SG as S_D.
2. *CTs*: A CT is composed of one or more CTs that should be performed by the SG. The input to a CT can be information stored in one or more EDOs or output of any earlier completed CT. We indicate a set of CTs as S_T.
3. *Client*: A client is the one who is concerned in retrieving some valuable information stored in EDOs or the output obtained by CTs. We indicate a set of clients as S_U.

16.3.1.2 Cloud Realm

A cloud realm is a collection of one or more clouds. Each cloud is a collection of DC that consists of storage and computation capacity. Each cloud in the cloud realm has its own pricing policies. This pricing policy is calculated based upon the amount of data stored on DCs, computational service required, intra- and intercloud data transfers required if any, communication cost between different clouds, etc. For example, Amazon storage service (S3) is maintained over seven regions, and all these seven regions have different pricing policies. We indicate a set of clouds as S_C and a set of DCs as S_{DC}. Virtualization is an important part of CC. To provide resource optimization, virtualization technology is used.

16.3.1.3 Network Realm

Network realm is responsible for providing the networking between any two realms present in the proposed CCOMCIM-SG model.

16.3.1.4 Broker Realm

This realm consists of a set of brokers who act as the mediator between the SG realm and cloud realm. Each cloud provider present in the cloud realm has a different pricing policy for different services provided by them. The actors present in the SG realm have the requirement of storing data, processing the stored data, and other services. Now, the job of the cloud broker is to assist these actors present in the SG realm in purchasing a suitable cloud provider who can meet their requirements in an optimal fashion.

As per our discussion earlier, data storage is done by all the subrealms of SG realms present in the proposed CCOMCIM-SG model. These data stored can be broadly classified as follows:

1. *Utility-generated data*: This is the total amount of data generated by the bulk generator, transmission, and distribution subrealms combined with data generated by any microgrid if any.
2. *Customer usage data*: These data contain the usage details of industries, smart homes, plug-in hybrid electric vehicles (PHEVs), and corporate user data.
3. *Sensor data*: These data contain the information generated by sensors, PMUs, supervisory control and data acquisition (SCADA), etc.
4. *Account data*: These data contain the account details of the electric utilities and actors involved in the SG realm. These are important data that should be maintained securely. Account data contain the account-related data of industries who use electricity, industries who generate electricity, and industries who transmit and distribute data, smart home data, PHEV data, corporate data, and account data of microgrid if any present.

The following subsection discusses how the data maintained by the proposed CCOMCIM-SG model can be useful for different services:

1. *Billing service*: This is the service that is calculated over a fixed interval of time generally 1 month based on the electricity used. To calculate the bills, we consider usage data, account data, and sensor data.
2. *Consumer energy efficiency*: This service is used to give recommendations to the customers based on this usage and historical data so that electricity consumption and generation can be done in an efficient manner. To provide recommendations, this service uses the utility generator data and usage data.

These services provided show the integration of the cloud realm and SG realm. Some more services are discussed in Section 16.4. The proposed CCOMCIM-SG model is based on the SG, cloud, and network realms. In our proposed CCOMCIM-SG model, we concentrate on how an EDO can be stored on cloud(s) instead of concentrating on how to optimize virtual and physical resources in the cloud(s) chosen for storing data.

In the following sections, three submodels of CCOMCIM-SG model are discussed. These submodels are based on the storage, computation, and security requirements of the SG realm on the cloud realm.

16.3.2 CCOMCIM-SG Submodel for Data Storage

The data generated by SG are stored in the cloud environment so that it can be accessed by different service providers to extract some information from it. Hence, data storage plays an important role in SG's information management model. Each data object d $\in S_D$ present in the EDO should be uploaded in the cloud realm and stored in different DCs present in the cloud. These DCs can be shared among clouds. Each DC might store some part of data present in the data object d. Let Z^d represent the size of the data object d that is needed to be stored on the cloud. Let $Z^D(d,dc,c)$ represent the amount of data object d stored on DC dc present in the cloud c. Based on this scenario of cloud-based data storage model, the pricing is done in three ways: unit upload cost, unit storage cost, and unit download cost.

1. We use $P^{\uparrow}(d,dc,c)$ to represent the *unit upload cost of a data object* d *on a DC* dc *located in a cloud* c. This cost includes the data transfer-in price charged by cloud c and network communication cost charged by the network providers present in the network realm for providing communication between cloud c and information source where data object d is generated.
2. We use $P^S(d,dc,c)$ to represent the *unit storage cost of a data object* d *on a DC* dc *located in a cloud* c.
3. We use $P^{\downarrow}(d,dc,c)$ to represent the *unit download cost of a data object* d *on a DC* dc *located in a cloud* c. This cost includes the data transfer-out price charged by cloud c and network communication cost charged by the network providers present in the network realm for providing communication between cloud c and client u who wants to download the data.

16.3.3 CCOMCIM-SG Submodel for Computation

Apart from data storage, another important operation present in information management is analyzing the data stored. As discussed earlier, a CT is a collection of one or more tasks that takes input as one or more EDO(s) generated by SG or the output of previously executed task(s). In our proposed CCOMCIM-SG model, data object d $\in S_D$ is uploaded in the cloud realm and stored in different DCs present in the cloud. These DCs can be shared among clouds. Each DC might store some part of data present in the data object d. Let S_C denote the set of clouds present in the cloud realm and $S_C^{(dc)}$ represent the set of clouds containing DC dc. Let S_T represent the set of CT available. Based on this scenario of CCOMCIM-SG submodel for computation, pricing can be done by considering unit upload cost, computational cost, unit storage cost, intercloud data transfer cost, and unit download cost.

1. We use $P^{\uparrow}(d,dc,c)$ to represent the *unit upload cost of a data object* d *on a DC* dc *located in a cloud* c. This was discussed in Section 16.3.2. Note that only one copy of data object d is maintained in a single cloud even though there are one or more tasks that require data object d as the input.
2. We use $P^c(t,dc,c)$ to represent the *unit computation cost* of a task t on a DC dc located in a cloud c.
3. We use $P^S(d,dc,c)$ to represent the *unit storage cost of a data object* d *on a DC* dc *located in a cloud* c. This was discussed in Section 16.3.2.
4. We use $P^{\downarrow}(d,dc,c)$ to represent the *unit download cost of a data object* d *on a DC dc located in a cloud* c. This was discussed in Section 16.3.2.
5. We use $P^T(d,dc_m,c_i,dc_n,c_j)$ to represent the *unit intercloud transfer cost*. This cost represents transferring data object d from DC dc_m of cloud c_i to DC dc_n of cloud c_j.

To understand the storage and computational submodels present in CCOMCIM-SG, we discuss a structural graph as shown in Figure 16.2.

This figure represents a structural graph showing relationships between different Dos, CTs, and clients. The nodes present in this structural graph represent the DOs, CTs, and clients. In this graph, data object is represented by d where $d \in S_D$, CTs are represented as t where $t \in S_T$, and clients are represented as u where $u \in S_U$. In the graph, we have considered (i) a set of EDOs {d1,d2,d3,d4,d5,d6,d7} that is required to be uploaded in the cloud realm, a set of CTs {CT1, CT2} with CT1 containing {t1, t2, t4, t6} tasks and CT2 containing {t3, t5} tasks that should be performed on the data uploaded in the cloud realm, and a set of clients {U1,U2,U3,U4,U5,U6}. The directed links <t1,t2> from t1 to t2 represent task t2 takes output of t1 as its input. The directed link <d,t> from d to t represent task t takes data object d as its input. The directed link <t,u> from t to u represent client u downloads the output of task t from the cloud realm. This structural graph also illustrates the nature of different types of CTs that can be performed, that is, some CTs directly take data objects as their inputs, whereas some CTs take input as some data objects and the output of previously executed task, for example, to execute task t1, data objects {d1, d2, d3, d7} are needed as the input, and similarly for execution of task t2, the input that contains data objects {d2, d6, d5} along with output of task t1 is needed. The example also illustrates the interest of different clients, that is, client U1 is interested to download the output of task t2 that is interdependent on the output of task t1 and data objects {d1, d2, d5, d7}, and similarly, client U2 is interested to download the output of task t4 that is interdependent on the output of task 2 and data object {d6}. These interdependencies can be represented as $S_T^{(t)}$. Now $S_T^{(t1)} = \{t2\}$, $S_T^{(t2)} = \{t4\}$, and $S_T^{(t6)} = \{t3\}$. As there is interdependency among the CTs executed and the data objects, there is a need for placing these tasks and data items in an efficient manner so that the upload cost, storage cost, communication cost, and download cost of client u can be decreased.

Based on the interdependencies in the structural graph as shown in Figure 16.3, we have considered a set of 3 clouds {C1,C2,C3} that are present in the cloud realm and designed in

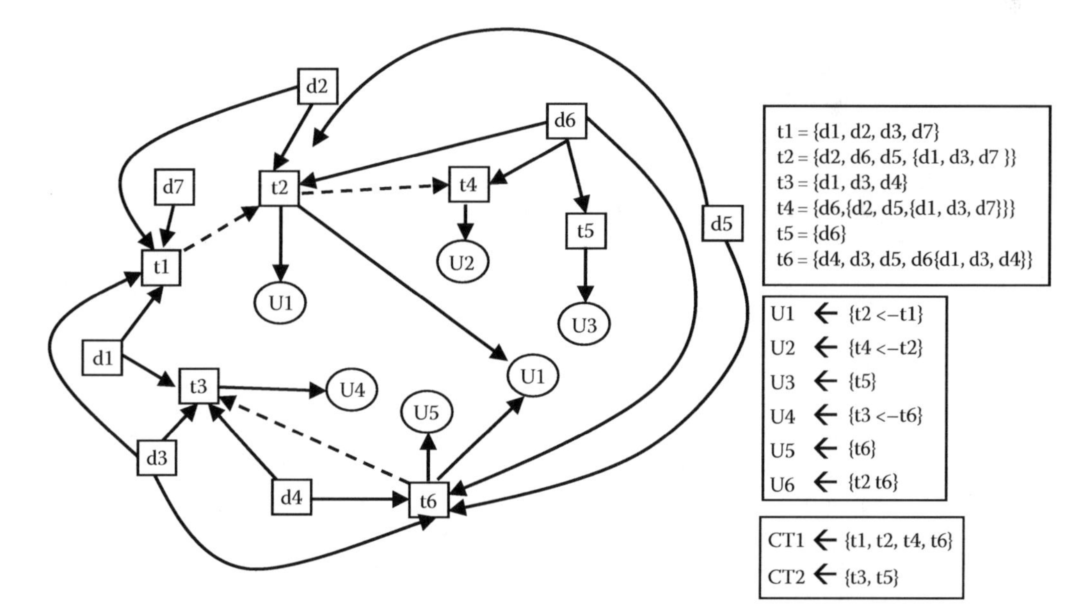

FIGURE 16.3
An example illustrating structural graph of data objects, computational tasks, and clients.

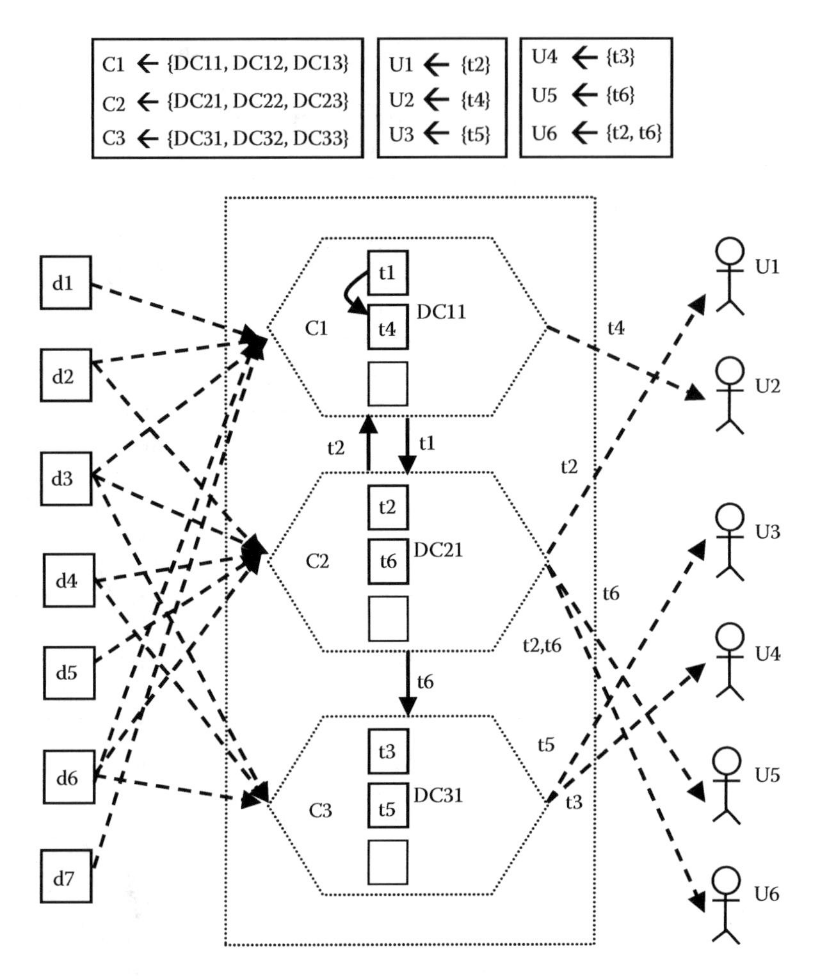

FIGURE 16.4
An example illustrating Cloud-based cost optimization model for smart grid information management submodel from storage and computation.

a framework as shown in Figure 16.4. Each cloud has a set of DCs present in it. Cloud C1 contains {DC11, DC12, DC13} DCs, cloud C2 contains {DC21, DC22, DC23}, and cloud C3 contains {DC31, DC32, DC33}. The directed lines <d,C> represent that data object d is uploaded on a DC present in cloud C. The directed lines $<C_i,C_j>$ represent that cloud C_i transfers some data to cloud C_j. According to Figure 16.4, DC11 executes task t1 and DC12 executes task t4 on cloud C1, DC21 executes task t2 and DC22 executes task t6 on cloud C2, and DC31 executes task t3 and DC32 executes task t5 on cloud C3. As these tasks are executed in the different clouds, there is a need to upload the data objects on the DCs of the respective clouds. Note that only one instance of the data object is uploaded on a single cloud even though there are more than one task executed by different DCs on the cloud that use the data object. This constraint should be maintained because this reduces the upload cost of the client accordingly. The directed link <d,C> represents the data object d is uploaded on the cloud C. Once the data objects get uploaded on any DC, these DCs communicate with each other for the data object.

In cloud C1, two CTs {t1,t4} are being executed on DCs {DC11,DC12}, respectively. So data objects {d1, d2, d3, d6, d7} are uploaded on cloud C1. As task t1 requires data objects {d1,d2,d3,d7}, these data objects are uploaded on DC11. But in the case of task t4, only data object {d6} is uploaded on DC12 as it is dependent on the outputs of tasks t1 and t2, so it requests DC11 for the output of task t1 and DC21 of cloud C2 for the output of task t2. The directed link <C,U> represents client U is downloading data from cloud C.

16.3.4 CCOMCIM-SG Submodel for Security and Privacy

Providing information security, integrity, and confidentiality for the data present in the SG realm is essential [10]. In a CC environment, providing information security and privacy is the biggest concern [12]. Hence, we consider the following possible techniques that each data object $d \in S_D$ should take into consideration:

1. The data object d should be stored in one or more preselected clouds that provide trustworthy CC service. Let $S_C^{(d)}$ denote the set of clouds allowed to store the data object d.
2. Redundancy should be maintained in storing the data object d.
3. Splitting of data is an important approach in protecting the sensitive data present in the data object d. Let S_{SD} denote the set of data objects to be split. This technique ensures some important portion of the data to be stored on private cloud and some on the public cloud and maintains the privacy of the data object d.
4. Some data objects cannot be split but should be securely stored as there might be some loss of information due to splitting. Such data objects should be stored on a single cloud without splitting it. Let S_D^C denote the set of data objects that should be stored in a single cloud and S_D^{dc}denote the set of data object d stored on a single cloud.

Note: Techniques (3) and (4) are mutually exclusive.

16.4 Optimization Framework for CCOMCIM-SG Model

The cost minimization problem for the proposed CCOMCIM-SG model is formulated as a mixed integer linear programming model. The following subsections represent the cost optimization framework for data storage and computation.

16.4.1 Cost Optimization Framework for Data Storage in CCOMCIM-SG Model

As discussed in Section 16.3.2, "CCOMCIM-SG Submodel for Data Storage," the pricing is done in three ways: unit upload cost, unit storage cost, and unit download cost. In the equation, $S_C^{(dc)}$ represents the set of clouds containing the DC dc and $S_{DC}^{(d)}$ represents the set of DCs containing the EDO d.

The upload cost of a data object is given in Equation 16.1. In the following equation, $P^{\dagger}(d,dc,c)$ represents the unit uploading cost of EDO d on DC dc in cloud c, and Z(d,dc,c) denotes the size of EDO d stored on DC dc in cloud c:

$$\sum_{d \in S_D} \sum_{dc \in S_{DC}^{(d)}} \sum_{c \in S_C^{(dc)}} Z(d,dc,c) * P^{\dagger}(d,dc,c) \tag{16.1}$$

The storage cost of a data object is given in Equation 16.2. In the following equation, $P^S(d,dc,c)$ represents the unit storage cost of EDO d on DC dc in cloud c, and Z(d,dc,c) denotes the size of EDO d stored on DC dc in cloud c:

$$\sum_{d \in S_D} \sum_{dc \in S_{DC}^{(d)}} \sum_{c \in S_C^{(dc)}} \left[Z(d,dc,c) * \left(P^S(d,dc,c) \right) \right] \tag{16.2}$$

The download cost of a data object is given in Equation 16.3. In the following equation, $P^{\downarrow}(d,dc,c)$ represents the unit download cost of EDO d on DC dc in cloud c, and Z(d,dc,c) denotes the size of EDO d stored on DC dc in cloud c:

$$\sum_{d \in S_D} \sum_{dc \in S_{DC}^{(d)}} \sum_{c \in S_C^{(dc)}} \sum_{u \in S_U^{(d)}} Z(d,dc,c) * P^{\downarrow}(d,dc,c) \tag{16.3}$$

Equation 16.4 calculates the minimum total $Storage_{cost}$ by taking into consideration the upload cost, storage cost, and the download cost for all data objects present in the SG realm that are stored in the cloud realm. This minimum total storage cost is calculated by keeping into consideration the constraints discussed in Section 16.3.4:

$$\min(Storage_{cost}) = \sum_{d \in S_D} \sum_{dc \in S_{DC}^{(d)}} \sum_{c \in S_C^{(dc)}} \left[\begin{array}{c} Z(d,dc,c) * \left(P^S(d,dc,c) + P^{\uparrow}(d,dc,c) + \sum_{u \in S_U^{(d)}} P^{\downarrow}(d,dc,c) \right) \\ + \sum_{u \in S_U^{(d)}} \sum_{r_i \in S_R^{(u)}} \left(P^R(d,dc,c,r_i) * N^R(dc,c,r_i) \right) \end{array} \right] \tag{16.4}$$

The following constraints are taken into consideration:

1. *Redundancy constraint*

$$\sum_{dc \in S_{DC}^{(d)}} \sum_{c \in S_C^{(dc)}} Z(d,dc,c) = \beta(d) * Z^d \quad \forall d \in S_D \tag{16.5}$$

   This constraint represents the second storage techniques as discussed in Section 16.3.4. The storage data redundancy ratio of a data object d is denoted using β(*d*). This constraint represents for a data object d, the total amount of data stored is equal to the size of the data object d times its redundancy ratio.

2. *Data splitting constraint*

$$Z(d,dc,c) \le \frac{\beta(d) * Z^{(d)}}{\gamma(d)} \quad \forall d \in S_{SD} \forall dc \in S_{DC}^{(d)} \forall c \in S_C^{(dc)} \tag{16.6}$$

   This constraint represents the third storage techniques as discussed in Section 16.3.4. In Equation 16.6, S_{SD} represents the set of data objects that are needed to be split. The splitting ratio with respect to a data object d is given by γ(d). This constraint represents that the data object d is not permitted to be split more than its splitting ratio. That is no DC dc stores more that 1/γ(d) amount of total data object d stored in the cloud realm.

3. *Data exclusive constraint*

$$\frac{Z(d,dc,c)}{\beta(d)*Z^{(d)}} \in \{0,1\}, \quad \forall d \in S_D^{(dc)} \forall dc \in S_{DC}^{(d)} \forall c \in S_C^{(dc)}$$
$$\text{over} \quad Z(d,dc,c) \in [0,\infty), \forall d \in S_D^{(dc)} \forall dc \in S_{DC}^{(d)} \forall c \in S_C^{(dc)} \tag{16.7}$$

This equation denotes the data exclusive constraint that represents the fourth technique discussed in Section 16.3.4. S_D^C represents the set of EDOs that are required to be stored in a single cloud.

16.4.2 Cost Optimization Framework for Computation in CCOMCIM-SG Model

As discussed in Section 16.3.3, "CCOMCIM-SG Submodel for Computation," the computation cost is calculated by considering unit upload cost, computational cost, unit storage cost, intercloud data-transfer cost, and unit download cost:

1. Equations 16.1 through 16.3 provide the upload, storage, and download costs.
2. The computational cost is obtained from Equation 16.8. In the following equation, we use $P^C(t,dc,c)$ to represent the unit computation cost of a task t on a DC dc located in a cloud c. E(t,dc,c) indicates whether the task t is executed by DC dc on cloud c:

$$\sum_{t \in S_T} \sum_{dc \in S_{DC}^{(t)}} \sum_{c \in S_C^{(dc)}} E(t,dc,c) * P^C(t,dc,c) \tag{16.8}$$

3. The intercloud transfer cost is obtained from Equation 16.9. In this equation, $P^T(d,dc_m,c_i,dc_n,c_j)$ represents the unit intercloud transfer cost. This cost represents transferring data object d from DC dc_m of cloud c_i to DC dc_n of cloud c_j. $B^{\rightarrow}(t_i,dc_m,c_k,t_j,dc_n,c_l)$ returns a Boolean variable indicating DC dc_m of cloud c_k transmits intermediate data to DC dc_n of cloud c_l because task t_j takes the output of task t_i as the input and Z^t represents the size of the output produced by task t:

$$\sum_{d \in S_D} \sum_{t_i \in S_T} \sum_{t_j \in S_T^{(t)}} \sum_{dc \in S_{DC}^{(t)}} \sum_{c \in S_C^{(dc)}} \left[B^{\rightarrow}(t_i,dc_m,c_k,t_j,dc_n,c_l) * Z^t * P^T(d,dc_m,c_k,dc_n,c_l)\right] \tag{16.9}$$

Equation 16.10 calculates the minimum total $\text{Computation}_{\text{cost}}$ by taking into consideration the upload cost, communication cost, intercloud transfer cost, storage cost, and the download cost for all data objects present in the SG realm that are stored in the cloud realm. This minimum total computation cost is calculated by keeping into consideration the constraints discussed in Section 16.3.4:

$$\min(\text{Computation}_{cost}) = \sum_{d\in S_D}\sum_{dc\in S_{DC}^{(d)}}\sum_{c\in S_C^{(dc)}}\left[B^{\uparrow}(d,dc,c)*Z^{d}*\left(\begin{array}{l}P^{S}(d,dc,c)+P^{\uparrow}(d,dc,c)\\ +\sum_{u\in S_U^{(d)}}\sum_{r_i\in S_R^{(u)}}\left[P^{R}(d,dc,c,r_i)*N^{R}(dc,c,r_i)*Z^{R}(dc,c,r_i)\right]\end{array}\right)\right]$$

$$+\sum_{t\in S_T}\sum_{dc\in S_{DC}^{(t)}}\sum_{c\in S_C^{(dc)}}\left[E(t,dc,c)*\left(P^{C}(t,dc,c)+\sum_{u\in S_U^{(t)}}Z^{t}*P^{\uparrow}(dc,c)\right)\right]$$

$$+\sum_{d\in S_D}\sum_{t_i\in S_T}\sum_{t_j\in S_T^{(t)}}\sum_{dc\in S_{DC}^{(t)}}\sum_{c\in S_C^{(dc)}}\left[B^{\rightarrow}(t_i,dc_m,c_k,t_j,dc_n,c_l)*Z^{t}*P^{T}(d,dc_m,c_k,dc_n,c_l)\right] \quad (16.10)$$

The following constraints are taken into consideration:

1. *Task execution constraint*: This constraint ensures that only one cloud executes the task t and is given as:

$$\sum_{dc\in S_{DC}^{(t)}}\sum_{c\in S_C^{(dc)}} E(t,dc,c) = 1\ \forall t\in S_T \quad (16.11)$$

2. *Data upload constraint:* This constraint ensures only one instance of data object d is uploaded in a cloud if there are more than one tasks in a single cloud that requires the same data object d. This constraint is given as:

$$\frac{\sum_{t\in S_T^{(d)}} E(t,dc,c)}{\left|S_T^{(d)}\right|} \le B^{\uparrow}(d,dc,c) \le \sum_{t\in S_T^{(d)}} E(t,dc,c) \quad \forall dc\in S_{DC}^{(d)}\forall c\in S_C^{(dc)}\forall d\in S_D \quad (16.12)$$

$$E(t,dc,c) = 0 \quad \forall dc \notin S_{DC}^{(t)}, c\in S_C^{(dc)}, t\in S_T \quad (16.13)$$

3. *Intercloud intermediate data transfer constraint*: This constraint ensures that $B^{\rightarrow}(t_i,dc_m,c_k,t_j,dc_n,c_l) = 1$ if and only if $E(t_i,dc_m,c_k) = E(t_j,dc_n,c_l) = 1$. That is, it checks whether DC dc_m of cloud c_k transmits intermediate data to DC dc_n of cloud c_l because task t_j takes the output of task t_i as the input. This constraint is given as:

$$\frac{1}{2}\left(E(t_i,dc_m,c_k)+E(t_j,dc_n,c_l)\right)-\frac{1}{2} \le B^{\rightarrow}(t_i,dc_m,c_k,t_j,dc_n,c_l)$$

$$\le \frac{1}{3}\left(E(t_p,dc_u,c_x)-E(t_q,dc_v,c_y)\right)+\frac{1}{3} = 0$$

$$\forall dc\in S_{DC}^{(t)}\forall c\in S_C^{(dc)}\forall t_j\in S_T^{(t)}\forall t_i\in S_T \quad (16.14)$$

over

$$B^{\uparrow}(d,dc,c) \in \{0,1\} \quad \forall dc \in S_{DC}^{(d)} \forall c \in S_{C}^{(dc)} \forall d \in S_D \tag{16.15}$$

$$E(T,dc,c) \in \{0,1\} \quad \forall dc \in S_{DC}^{(t)} \forall c \in S_{C}^{(dc)} \forall t \in S_T \tag{16.16}$$

$$\left.\begin{array}{l} B^{\rightarrow}(t_i,dc_m,c_k,t_j,dc_n,c_l) \in \{0,1\} \\ \forall d \in S_D \forall t_i \in S_T \forall t_j \in S_T^{(t)} \\ \forall dc \in S_{DC}^{(t)} \forall c \in S_C^{(dc)} \end{array}\right\} \tag{16.17}$$

16.4.3 Data Transfer Cost

The data transfer cost for all data objects present in the SG realm is given in Equation 16.18. This cost is calculated based on the request generated by client u. In the following equation, $Z^R(dc,c,r)$ denotes the size of the request on DC dc of cloud c of type r_i, $N^R(dc,c,r)$ denotes the number of the request on DC dc of cloud c of request type r, and $P^R(d,dc,c,r)$ denotes the unit request price of EDO d on DC dc of cloud c of type r:

$$\min(\text{Transfer}_{cost}) = \sum_{d \in S_D} \sum_{dc \in S_{DC}^{(d)}} \sum_{c \in S_C^{(dc)}} \sum_{u \in S_U^{(d)}} \sum_{r \in S_R} \begin{bmatrix} Z^R(d,dc,c,r) \\ *N^R(d,dc,c,r) \\ *P^R(d,dc,c,r) \end{bmatrix} \tag{16.18}$$

16.5 Simulation Results and Discussion

16.5.1 Simulation Scenario

In order to evaluate CCOMCIM-SG's performance, the following scenarios were set before carrying out simulations. Our evaluation for data storage and computation model is based on the following scenario: We considered a large electric utility consisting of a Genco (utility that generates electricity), Transco (utility that transmits electricity), Disco (utility that distributes electricity) and microgrid, seven public clouds (cloud1, cloud2, cloud3, cloud4, cloud5, cloud6, and cloud7), four types of data generators (utility generator data, customer usage data, sensor data, and account data), and six services (billing services, consumer energy efficiency service, demand response service, grid analysis service, wide-area situational awareness service, and electrical transportation service). Four different private clouds are maintained for storing important data. The prices to the cloud storage are given by considering the latest Amazon S3 pricing policy [19].

The number of residential areas served by this electricity is varied from 100 to 1000 with an increment of 100. The number of households in each residential area is 100 containing 10 appliances at each house. Each household consists of a smart meter and is called as smart homes. The data generated by smart meter is 100 KB/day of usage data with a sampling of 15 min/h. These data generated represent the smart home data and are stored in the category of customer usage data in the cloud. These data can be stored in the private/public cloud. Smart homes generate 1 KB of important account data per month that should be stored in the category of account data. These account data should be stored in the private cloud.

We have simulated 5 generator sites, 5 transmission sites, 10 distribution sites, and 100 microgrid generator sites. Each generator site generates electricity continuously. Each generator site generates 100 MB of data per day with a sampling interval of 15 min. Each transmission site also generates 100 MB of data per day with a sampling interval of 15 min. Each distribution site generates 500 MB of data per day with a sampling interval of 15 min. Each microgrid generates 500 MB of data per day. The data generated by these 5 generator sites, 5 transmission sites, 10 distribution sites, and 100 microgrid sites are stored in the category of utility generator data. These data can be stored in the private or public clouds. Apart from this, each generator site, each transmission site, each distribution site, and the microgrid generate 1 KB of important account data per month, respectively. As these are important data, they should be stored in the private cloud in the category of account data.

Each smart home has zero to more electric vehicles. We assumed that in each residential area R, there are about [R/2, R] electrical vehicles. An electric vehicle consumes 50 KB of usage data per day, and these data should be stored in the category of customer usage data. Apart from that, each electric vehicle generates 1 KB of important data per month. These data should be stored in the category of account data in the private cloud.

We assumed that there are around 10 industries and 10 corporate companies that come under this electric utility. Each industry generates 50 MB of usage data per day and 1 KB of important account data. Each corporate company generates 100 MB of usage data and 1 KB of important account data. The usage data are stored in the private or public cloud in the category of customer usage data where account data are stored in the private cloud.

The electric grid also consists of sensors such as PMUs and SCADA. We assumed that there are around 50 PMUs and 100 SCADA devices installed. Each PMU generates 1 GB of sensor data per day. Each SCADA device generates 100 MB of data per day. These data are stored in the category of sensor data in either the private or public cloud.

Now let us discuss different services that access these data generated:

1. *Billing service and consumer energy efficiency service*: These services are discussed in Section 16.3.1.
2. *Demand response*: This service is nothing but power scheduling service that is used to balance the demands dynamically with respect to the requirement of the customer. Utility generator data and usage data are taken as the inputs for providing this service.
3. *Grid analysis*: This service monitors data and analyzes the health of the SG realm. Any failure in the SG is monitored by this service. Utility generator data, usage data, account data, and sensor data are the inputs for this service.
4. *Wide-area situational analysis*: This service monitors the data obtained from sensors and utility generators that are separated over a wide geographical area. The sensor data and utility generator data are the inputs to this service.
5. *Electrical transportation: This* service maintains and monitors the usage and account details of the electrical vehicles present in a residential area so that the demand for electricity utilization can be managed dynamically. The data for this service are obtained from the PHEV usage data and PHEV account data.

We assumed that all the data are stored for 6 months, and we evaluated the cost of storing data for 1 day. We have considered the latest Amazon S3 storage pricing model for data storage and Amazon EC2 pricing model for computation. Amazon S3 has clouds located in seven different geographical locations [18] and priced differently. Details of pricing used for Amazon S3 storage services and EC2 prices used in this chapter can be found from [19,20].

16.5.2 Results

In this section, we present the results of cost analysis carried out using the simulation setup detailed in the previous section. The cost of storage, computation in the cloud, and total cost of CCOMSIM-SG have been compared with a host of alternative strategies, mostly theoretical as presented in [1], namely, genetic algorithm optimal storage, worst-case scenario storage, stochastic storage, and private cloud storage. Figure 16.5 shows the comparative storage cost, Figure 16.6 gives the comparative computation cost, and Figure 16.7 presents the total cost of the CCOMCIM system with reference to the alternative strategies. It can be clearly observed from the aforementioned figures that the CCOMCIM-SG model gives lower cost when compared with the other scenarios.

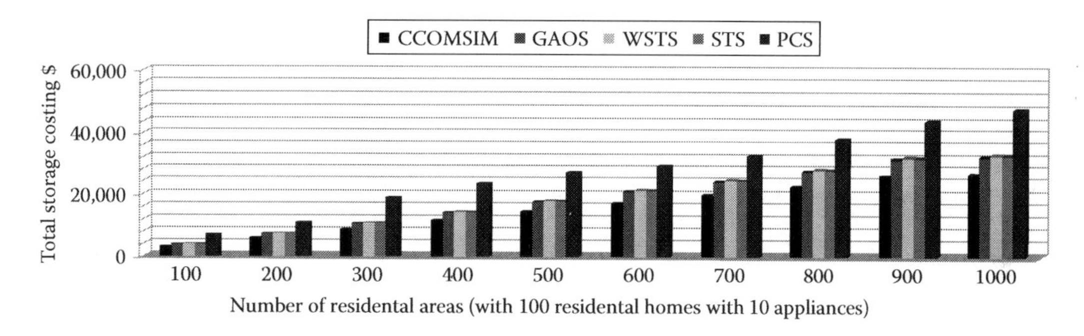

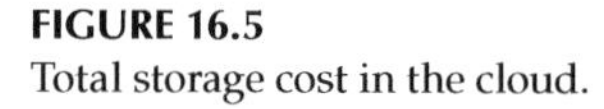

FIGURE 16.5
Total storage cost in the cloud.

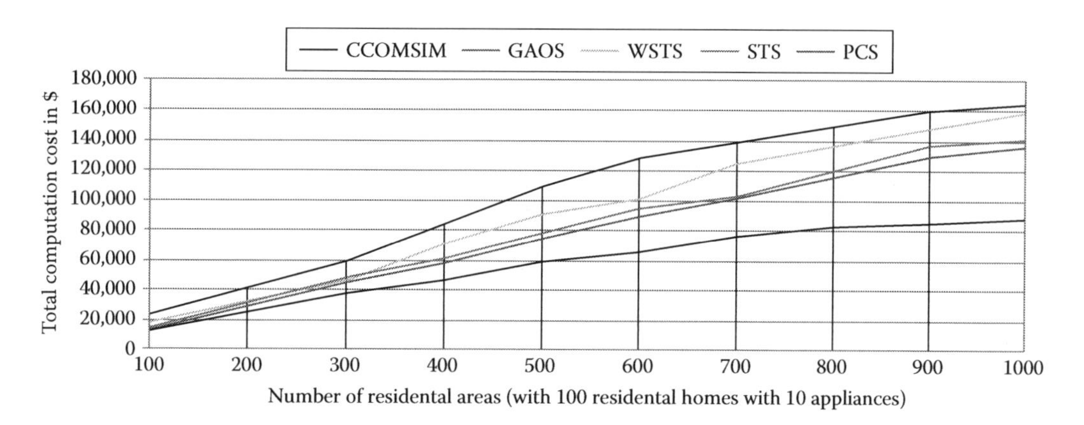

FIGURE 16.6
Total computational cost in the cloud.

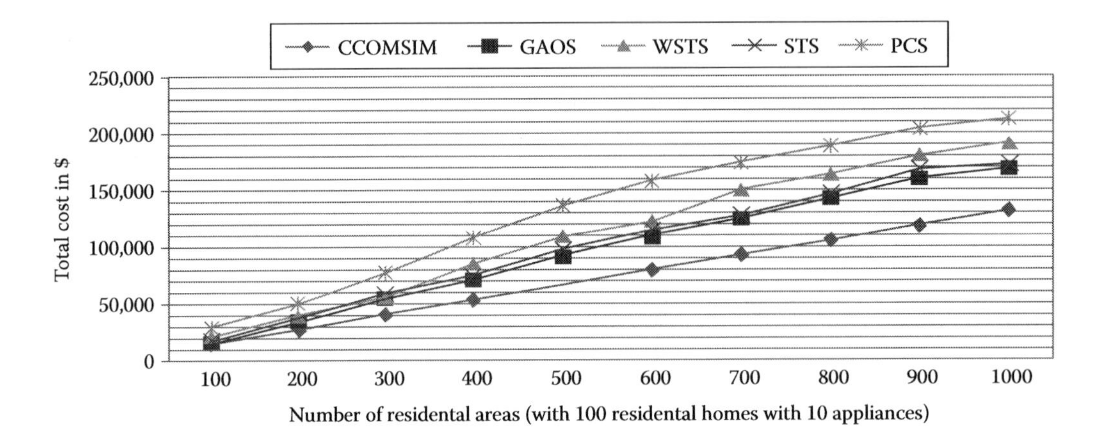

FIGURE 16.7
Total cost in the cloud.

16.6 Conclusion

In recent times, CC offers a great promise to handle the huge data and computation requirements posed by the vision of CPSs. In order to cope with the colossal expenses of maintaining cloud DC, excellent information management should be combined with optimal policies to keep the cost of CPS information management under check. Detailed intricacies of the model have been illustrated using a case study of an SG, and accordingly, this chapter details CCOMCIM-SG that considers cost-effective SG information management considering various facts of SG's storage and computational requirements. The results presented here in proved the efficacy of the proposed CCOMCIM model.

References

1. X. Fang, S. Misra, G. Xue, and D. Yang, Smart grid—The new and improved power grid: A survey, *IEEE Communication Surveys and Tutorials*, 14(4), 944–980, 4th quart., 2012.
2. X. Fang, S. Misra, G. Xue, and D. Yang. Managing smart grid information in the cloud: Opportunities, model, and applications. *IEEE Network*, 26(4), 32–38, 2012.
3. F. Xi, D. Yang, and G. Xue, Evolving smart grid information management Cloudward: A cloud optimization perspective, *IEEE Transactions on Smart Grid*, 4(1), 111–119, March 2013.
4. S. Sakr, A. Liu, D.M. Batista, and M. Alomari, A survey of large-scale data management approaches in cloud environments, *IEEE Communication Surveys and Tutorials*, 13(3), 311–336, 2011.
5. Locke, G., and Gallagher, P.D., NIST framework and roadmap for smart grid interoperability standards, release 1.0. *National Institute of Standards and Technology*, 33, 2010.
6. Y. Simmhan, S. Aman, B. Cao, M. Giakkoupis, A. Kumbhare, Q. Zhou, D. Paul, C. Fern, A. Sharma, and V. Prasanna, An informatics approach to demand response optimization in smart grids, Computer Science Department, University of Southern California, Los Angeles, CA, Technical Report, 2011.

7. K. Nagothu, B. Kelley, M. Jamshidi, and A. Rajaee, Persistent net-AMI for microgrid infrastructure using cognitive radio on cloud data centers, *IEEE System Journal*, 6(1), 4–15, 2012.
8. H. Kim, Y.-J. Kim, K. Yang, and M. Thottan, Cloud-based demand response for smart grid: Architecture and distributed algorithms, in *Proceedings of the IEEE SmartGridComm'11*, 2011, Brussels, Belgium, pp. 398–403.
9. Y. Xin, I. Baldine, J. Chase, T. Beyene, B. Parkhurst, and A. Chakrabortty, Virtual smart grid architecture and control framework, in *Proceedings of the IEEE SmartGridComm*, 2011, Brussels, Belgium, pp. 1–6.
10. A.-H. Mohsenian-Rad and A. Leon-Garcia, Coordination of cloud computing and smart power grids, in *Proceedings of the IEEE SmartGridComm*, 2010, Gaithersburg, Maryland, pp. 368–372.
11. X. Fang, S. Misra, G. Xue, and D. Yang, Managing smart grid information in the cloud: Opportunities, model, and applications, *IEEE Networking*, 26(4), 32–38, July–August 2012.
12. R. Buyya, C.S. Yeo, S. Venugopal, J. Broberg, and I. Brandic, Cloud computing and emerging it platforms: Vision, hype, and reality for delivering computing as the 5th utility, *Future Generation Computer System*, 2009, pp. 599–616.
13. R.V. den Bossche, K. Vanmechelen, and J. Broeckhove, Cost-optimal scheduling in hybrid IAAA clouds for deadline constrained workloads, in *Proceedings of the IEEE Conference Cloud Computing*, 2010, Miami, Florida, pp. 228–235.
14. M. Hajjat, X. Sun, Y.-W. E. Sung, D. Maltz, S. Rao, K. Sripanidkulchai, and M. Tawarmalani, Cloudward bound: planning for beneficial migration of enterprise applications to the cloud, in *Proceedings of the ACM SIGCOMM*, 2010, pp. 243–254.
15. H. Zhao, M. Pan, X. Liu, X. Li, and Y. Fang, Optimal resource rental planning for elastic applications in cloud market, in *Proceedings of the IEEE International Parallel and Distributed Symposium*, 2012, Shanghai, China.
16. S. Chaisiri, B.-S. Lee, and D. Niyato, Optimization of resource provisioning cost in cloud computing, *IEEE Transactions on Services Computing*, 5(2), 164–177, April–June 2012.
17. J.Z. Li, M. Woodside, J. Chinneck, and M. Litoiu, Cloudopt: Multi goal optimization of application deployments across a cloud, in *Proceedings of the International Conference on Networking and Service Management*, 2011, Paris, France, pp. 1–9.
18. Amazon Simple Storage Service (online). Available: http://aws.amazom.com/s3/. Last accessed on 31 March, 2014.
19. Amazon S3 pricing (online). Available: http://aws.amazon.com/s3/pricing/. Last accessed on 31 March, 2014.
20. Amazon EC2 Pricing (online). Available: http://aws.amazon.com/ec2/pricing/. Last accessed on 31 March, 2014.

Section V

Security Issues in Cyber-Physical Systems

17

Analysis for Security Attacks in Cyber-Physical Systems

Sri Yogesh Dorbala and Robin Singh Bhadoria

CONTENTS

17.1 Introduction

Cyber-physical systems (CPSs) are those systems where the physical processes are controlled by the computing and communication technologies. The prime objective of CPS is to examine the behavior of the physical process and, in turn, trigger actions to alter them. CPS fundamentally comprises of two parts: physical processes and computational systems. The computation system (or the *cyber* system) is integrated with the processes (the *physical* system) using feedback loops obtained through sensors and actuators. The correlation between physical and cyber systems brings out new communication links, thus increasing the vulnerabilities for an attack. CPSs that provide an elegant mechanism for enhancing the correlation between the physical systems and the computation technologies are hence prone to numerous security threats in addition to the traditional cyber attacks.

There has been a sudden growth in the usage of CPS in the recent years to efficiently manage and automate the services. The advancement of digital technology into the mechanical systems such as the CPS for air-traffic control (Sampigethaya, 2012) indicates the need for improvised security measures. CPS design has significantly evolved throughout the years and has brought added functionalities along with the better implementation structures. But the progress of security is not on par with that of the design. CPSs that have the power to interact with the environment have a great potential to physically harm the living entities, objects, and the environment. These systems are commonly used for examining mission critical systems, making them more vulnerable to targeted attacks. These mission critical systems if distorted would create disturbances in elementary services such as transportation, home electronics, and automobiles, among others.

CPSs that are geographically dispersed are used in large scale in life-critical systems. The recent growth of embedded components has created a lot of new security threats that trouble not only the cyber space but also our physical environment. The physical characteristic of CPS shows that these systems incorporate the best of both worlds and thus can affect the safety of both the human beings and the environment in a significant manner.

Failures of CPS can indeed have disastrous consequences. Hence, it is very important for CPS to be operated in a safe and secure way.

Typical security solutions for CPS would incorporate the traditional cyber security solutions. Although these solutions such as encryption, authentication, and access control can prove to be helpful up to some level, there is a definite need for defense measures that take into account the difference between the cyber security and the CPS security. Thus, CPSs because of their diverse capabilities, environmental correlations, and networked nature require a lot more thought for safety in addition to the conventional cyber security defense measures.

This chapter starts with a discussion about various types of threats possible in CPS. It then describes the workflow in CPS and about the key factors that are to be secured in CPS. The next section will eventually cover the different attacks that are possible in CPS, suggests some defense measures, and explains how CPS security is different from cyber security.

17.2 Workflow of CPS

A CPS workflow is comprised of four stages: monitoring, networking, computing, and actuation.

17.2.1 Monitoring

The fundamental function of CPS is to monitor the physical process and its environment. This would also be useful in providing feedback on the past actions to CPS, thus ensuring flawless actions in the future.

17.2.2 Networking

Data collection and diffusion are the two things that are dealt with in this stage. Input information regarding the environment is provided by the sensors that in general can be more than one. The real-time data that are collected through the sensors are aggregated by the analyzers for processing the future stages. Interaction between different applications is done through networking.

17.2.3 Computing

The data collected in the monitoring phase are analyzed to check whether the physical process meets the predefined conditions. Corrective actions are asked to be performed in order if the required criteria are not met to make sure that the conditions are satisfied. The workflow of CPS is shown in Figure 17.1 (Wang, 2010).

17.2.4 Actuation

The actions that are determined during the computation stage are executed in this phase. This phase initiates varied forms of activities such as the change of physical processes. One such instance can be the transmission of the traffic flow analysis from one vehicle to another in an intervehicular CPS.

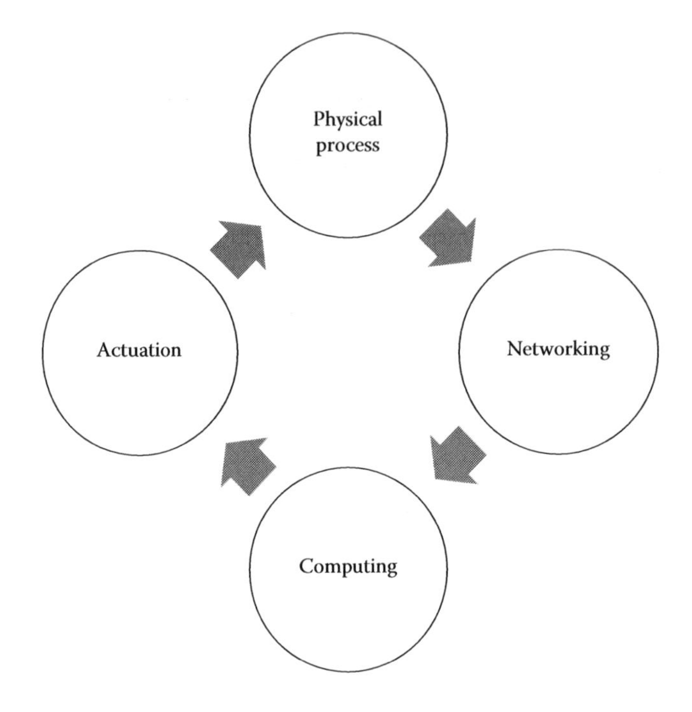

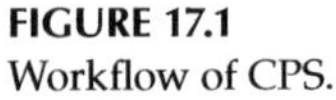
FIGURE 17.1
Workflow of CPS.

17.3 Case Studies on CPS Security Breaches

This section discusses the various types of security breaches in CPS, thus providing an idea to the users about the type of attacks that can be made.

17.3.1 Stuxnet

Stuxnet is a virus that was designed to attack the programmable logic controller in an Iranian nuclear power plant in June 2010. It is considered to be one of the most expensive attacks that could happen in an environment like CPS. The system used in nuclear power plant employed in Siemens machines was infected through a USB drive that was inserted into the machine by an employee. This virus, which was written in multiple languages, was built on the idea of modifying code in order to make the plant deviate from its expected behavior that could result in a severe damage to the nearby surroundings.

The virus later spread to the Internet when a local infected computer was connected to the Internet. Stuxnet had the ability to hide itself from detection that made it remain unknown for almost a year since its inception into the target system. Cyber attacks may modify an online infrastructure but it can be restored with backups and updates. However, a CPS attack can create a permanent damage by destroying the nuclear power plant, thereby risking the lives of thousands of people (Kerr, 2010).

17.3.2 Maroochy Water Breach

The effect on the physical domain of the CPS can be understood through the attack in 2000 on the sewage control system of Maroochy council in Queensland, Australia. This attack made the control center lose communication with the pumping stations resulting in loss of contact with the pumps, thus disabling its functionality to make the pumps run when required. These errors caused the grounds of the nearby river, hotel, and park to be flooded with million liters of sewage. Though this attack is related to a particular area, the damage done could have been unimaginable if the impact of the physical domain were to be extended to broader regions (Slay and Miller, 2010).

17.3.3 Slammer Worm

The David–Besse nuclear power plant in the United States was attacked by a slammer worm in 2003, which affected the computerized safety monitoring system of the plant. The worm was injected via a second pathway through the telephone line infecting the computer connected to the plant's network. This worm caused a traffic overload at the network, making the plant unaccessible for 5 h. There was no damage incurred because of the attack as the plant was offline when attacked. The security measures implemented were robust in a single line of defense. The problem arose as the attacker found an alternate point of entry that could bypass the firewalls at place.

17.3.4 Automobile Attacks

Researchers from the University of Washington have demonstrated as to how malicious attackers can compromise automobile safety. The researchers show how an attacker can circumvent the control units and attack the safety critical systems in the vehicles such as stopping the engine and disabling the brakes, which are done without the driver's consent. Many modern vehicles have some vulnerabilities that made these attacks possible. These types of attacks are very critical as they cause human damage significantly.

17.3.5 Health Care

Providing medical assistance at regular time intervals and remotely treating patients irrespective of their location are some of the emerging applications of CPS in the medical sector. These may prove to be helpful for many people given that the security is sufficiently monitored. These systems, which would generally be having access to the health records of the patient, if attacked and altered, would lead to incorrect treatment of the patient leading to death at times. Thus, security of CPS when used for health care should be given utmost importance as it would lead to unrecoverable health damages when compromised. Instances such as insulin pump hijack in 2011 show the urge for better security measures (Li, 2011).

17.3.6 Manufacturing Sector

Another vital user of the CPS systems is the manufacturing plants. Immediate effects are observed on the manufacturing equipment if a malfunctioning system is used in a factory. This would result in an immense damage as these plants would produce products in a large number. Bhopal tragic disaster that occurred in India is one good example of such a catastrophe where thousands of people died due to a malfunction in the carbide chemical plant (Jackson, 2003). It should be noted that the security becomes a paramount factor of importance in such cases where CPSs are involved in bulk productions.

17.3.7 Smart Grid

In 2007, California Independent Systems Operator Corporation's system administrator pressed the emergency shutdown button, which resulted in crash of computers communicating with the power generating facilities. This is an example of a targeted attack done by a disgruntled employee. Though there wasn't any blackout, it took around 7 h of intense work to bring the situation back to normal. This power grid that can now control the energy flow to a house would create terrible damages if there is misdirection of the energy flow caused by the failure of security.

There are several other instances where failure of security in CPS would result in unimaginable consequences. For designing an efficient security design, one has to have a complete understanding on the functionalities of the CPS. The next section would briefly cover the workflow of CPS explaining in detail each of the stages involved.

17.4 Security Objective for CPS

This section speaks about the security objectives as discussed by Shafi (2012) by taking into consideration the security requirements for the dual nature (cyber and physical) of the CPS.

17.4.1 Confidentiality

It refers to the ability of storing the data as a secret with respect to the unwanted system or an individual. Any lack of confidentiality would result in a loss of information to the unauthorized members that would result in unforeseen situations. Security measures must be taken to guarantee that the information is not leaked out by eavesdropping through the communication that happens between the sensors and the actuators. Thus, it should be ensured that these messages are all encrypted before they are communicated. For example, in the case of a health-care CPS, a compromise in the confidentiality would result in the patient health record being accessible to the unauthorized members, which when altered by them would result in serious health problems.

17.4.2 Integrity

It refers to the ability by which the stored data remain accessible to alteration only by the authorized systems or individuals. Trustworthiness of the information is the main aspect of this objective. Deception occurs due to the lack of integrity where the authorized system receives false information that it believes to be genuine. Thus, the primary goal of this objective would be to detect, survive, or prevent the attacks made by deception in the communication between the sensors and actuators of CPS. For example, as mentioned in the previous section, a compromise to integrity in the case of a smart grid CPS would lead to a breakdown of the entire plant.

17.4.3 Availability

It refers to the ability of the system being available at the time of need. Denial of service (DoS) would result due to the lack of availability where the system is busy handling other communication channels, and it becomes unavailable to the intended users. While such attacks may turn out to be a temporary problem for some CPS, there are cases where these attacks would result in unrecoverable damages to the entities and the systems around. The primary goal of this objective is thus to detect and ensure that the DoS attacks do not occur. For example, in the case of intervehicular CPS, a compromise in availability would result in some vehicles not being available to communicate information about traffic to the vehicles that request the traffic flow, in turn making them prone to accidents.

17.4.4 Authenticity

It refers to the ability of the communications and the transactions done within the system being genuine. It should be necessary for the CPS to make sure that all the parties that are involved in the communication are validated to be whom they are. There should be no case of an intruder trying to communicate with the sensors or actuators of the CPS. Thus, the main goal of this objective is to ensure that all of the information that is communicated through is authentic and among the intended users. For example, in the case of water resource management CPS, authentication errors caused due to incorrect water levels would result in floods.

17.5 Challenges in CPS Security

Security concerns in CPS are a relatively new notion in the field of CPS as they are primarily developed for the purpose of functionality but not security. Security issues in CPS have a lot of challenges to be focused upon. As seen in the earlier section, though the focus of security in CPS is growing in the recent years, much of the research is focused on implementing traditional cyber security mechanisms to the CPS.

Introduction of physical systems into the cyber space has brought about a lot of new problems. Any attack on the CPS security might alter the physical events of the CPS that depend on the devices that are under its control. Any security damage done to physical system may lead to complete different threats when it is combined with the cyber world. It is thus a matter of crucial necessity to understand the cyber-physical interactions of a system. These hurdles that made the study of security issues in CPS much different from that of the traditional cyber security systems are discussed in Sections 17.5.1 to 17.5.6 in detail (Moholkar, 2014).

17.5.1 Regular Software Updates

Regular software updates and patches are not the best-suited options for CPS. Unlike the cyber systems, a tiresome job of planning to take down a system needs to be done before upgradation, given the crucial nature of CPS.

Upgradation of the physical systems at times may require months of time. It is not justifiable to suspend the activities of CPS for such a long duration of time. Given such a scenario, unlike the normal cyber systems where regular updates happen, there is a high chance for the CPS to be attacked by evolving day-to-day attacks.

Once, a power plant was accidentally shut down after the upgradation of its software. The software update made the software to reboot and all the data were preset to null. This absence of data made the security systems of the plant to misinterpret that there has been a problem and hence resulted in a shutdown of the plant, thereby halting the services of the plant.

17.5.2 Real-Time Requirements

Real-time requirements play a vital role in the efficient functioning of the CPS. CPS needs to take independent decisions by gathering the information through sensors and the actuators in the real time. Real-time availability of the information creates an entirely new different security concern issues unlike the traditional cyber systems where there is no such need.

Consider the case of an aerospace CPS where the precision in time plays a crucial role. A small time delay before launching of a rocket would lead to a complete change in the radius of curvature of the rocket and at times may even lead to its destruction. Thus, real-time availability imposes strict rules on the CPS security systems that are not considered in cyber security systems.

17.5.3 Intrusion Detection Techniques

Sensors sense the physical environment and transmit the data to the CPS. The information collected from different sensors is aggregated before being sent to the CPS, which then would perform the computation process. Maintaining the integrity and ensuring the availability of these sensors are some of the common challenges that one needs to think about for securing the CPS.

In addition to these traditional security measures, there might arise a situation where the attacker would modify the environment from which sensors gather information resulting in fake data accumulation. These data records when passed to CPS for computation would obviously produce dubious results. All the hardware and the software involved in this procedure are safe from the attack but still the system is compromised. Hence, the traditional intrusion detection techniques are not sufficient to detect the real-time availability issues.

17.5.4 Interruptions

The tasks and the jobs performed by the CPS often need to be interrupted and resumed because of its physical nature. These interruptions would make it difficult for the use of block encryption algorithm, thus making it difficult to secure through the encryption methods.

17.5.5 Geographical Distribution

In many cases, the CPS are geographically distributed. These physical separations often create troubles when one of its systems is compromised. The physical separation makes it difficult for the compromised physical system to reset the software. The CPS operations are uncertain owing to various other reasons such as the weather conditions, damages done to physical systems, and maintenance of the components, due to geographical distribution.

17.5.6 Multicomponents

CPSs are complex mechanisms where the security situations are quite different. A small mistake in a single component can have dire consequences on the entire physical system. The consequences of such mistakes might be physical in nature as well as affecting the individuals and the environment nearby. The focus of the security should be targeted on the integrated physical system as a whole but not just on the individual components, which thus brings in a lot of complex security issues.

Determining the impact of the vulnerabilities and managing the risks in a complex distributed system like CPS are of vital importance. Integration of the components within various systems has increased the chance of security threats by opening more communication channels that made the system more prone to attacks. New vulnerabilities and threats that might not have been present arise regularly on such integration of several components into the unique environment of the CPS.

17.5.6.1 Others

- Adopting the random available commercial technology and protocols in the design of CPS.
- Remote access to various operations from all around the world leads to a growth in the number of vulnerabilities.
- Knowledge as to how a CPS operates is now made accessible to anyone in the world, giving an access to the loop holes of various CPSs.
- Different hacking tips, tricks, and tools are made available over the Internet providing an added advantage to the attacker.
- Security not only needs to be addressed at the software and hardware levels but also needs to be considered at the operational context (Cyber Security Research Alliance, 2013).

Designing the security mechanism for a CPS is a tough task as it involves the security aspects of both the physical and the cyber world in an integrated manner. All of the aforementioned points suggest the need for a new elegant approach for designing and implementing the security solutions for CPS. For achieving that, analyzing the cyber world vulnerabilities and their effects on the physical system would provide an approach to narrow down the extent of analysis that needs to be done on security in the CPSs.

17.6 Requirement for Security in CPS

The security requirements for CPS are much different from the traditional cyber system requirements. The real-time requirement and analysis of data, wide geographical distribution of the CPS, etc., are some of the factors that are radically different from the traditional systems. Taking into consideration the aforesaid information, the following are the major security requirements in CPS as mentioned by Amey Vinayak in his paper on security on CPSs.

17.6.1 Sensing Security

It is required that the accuracy and the integrity of the sensed data by sensors are authenticated so that the operations that are done by the physical systems are trusted. Any unauthorized intrusion from an outside source would result in serious damages to the entire CPS.

17.6.2 Storage Security

The information that is collected from the communications needs to be stored for accessing in the future. It should be ensured that there wouldn't be any unauthorized alteration of the information stored in CPS. It involves designing methods to protect the data from unauthorized physical and cyber modifications.

17.6.3 Communication Security

The communication channels are to be secured from the attackers for both the inter- and intra-CPS communications. Any compromise in the communication would result in the loss of confidentiality, thus affecting the key objective of the security. It is very important for the information communicated between the parties to be confidential and this requirement ensures that property.

17.6.4 Actuation Security

Actuation operations in CPS are the ones that perform the final action. This requirement demands that the actuate operations are to be performed only when authorized. An actuation cannot take place without a valid authorization.

17.6.5 Feedback Security

CPSs work on the feedback loops that are generated by using the sensors and the actuators. These loops would help in making accurate decisions. Any attack on the feedback loop would result in wrong training of CPS, thus resulting in a faulty CPS. Thus, it is necessary to make sure that the feedback loops are properly protected so that they can properly guide the CPS in making accurate decisions.

17.7 Prominent Attacks on Security for CPS

Attacks in general are of two types, passive and active. Passive attacks are the ones that monitor the system for information. They do not perform any alterations on the data. Active ones get into the system, gain access to unauthorized information, and even modify them, causing unrecoverable damages. Various types of attacks that are possible on CPS are discussed in detail with examples in this section.

17.7.1 Denial-of-Service Attack

DoS is an attack to make a machine or a machine resource unavailable to its intended user. They saturate the target machine to such an extent that the target machine cannot respond to the legitimate traffic or respond very slowly such that it is practically unavailable. These attacks are one of the most common attacks in the field of CPS.

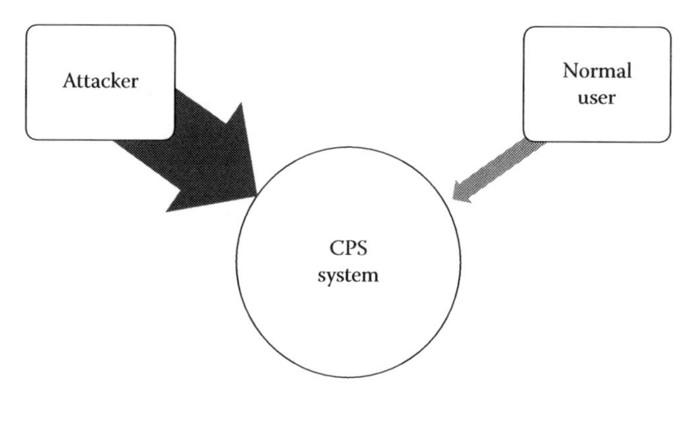

FIGURE 17.2
DoS attack on CPS.

DoS attacks in general make the target machine busy by transmitting huge amount of network flow, thereby utilizing all of its resources so that it no longer can provide its intended service to the legitimate requests. Distributed DoS (DDoS) is a type of DoS attack where multiple compromised systems attack a target machine by performing a DoS attack on it. Thus, DDoS is an advanced version of DoS attack in simpler terms.

The following are the actions mentioned by Wang (2010) in his article on security challenges in CPS that could be done by an attacker after gaining access to the CPS by performing a DoS attack:

- The controller or the sensors could be flooded with huge traffic flow, resulting in an overload eventually and then a shutdown of CPS.
- The system sends corrupt information to the CPS, which would result in its immediate termination of services.
- It makes the traffic inaccessible to the authorized users by blocking it through network congestion.

Figure 17.2 depicts the case of a DoS attack where the malicious attacker jams the network flow of the CPS system by sending enormous number of false requests. The CPS network is so heavily congested that it cannot respond to the queries of the intended normal user. This attack, which is made by a single compromised adversary, is an example for a DoS attack.

Consider another case where the CPS receives enormous requests from a large number of compromised DoS attacked systems. These compromised systems might attack from different geographical locations. The CPS would crash again unable to handle such enormous requests from a large number of compromised systems. Such an attack where there is an involvement of multiple DoS compromised systems for causing the network congestion is termed as a DDoS attack. It becomes a hectic task for separating the legitimate traffic from the normal one as the IP addresses of the compromised systems are widely spread across. Thus, DDoS attacks in general are much difficult to handle when compared to the DoS.

17.7.2 Eavesdropping

It is an attack where there is an unauthorized interception of the communication inside a network. As this attack just intercepts the messages and does no harm as such directly, it hardly gets noticed by the sender or the receiver. As the attacker is just observing the system but not interfering with it directly, it is termed as a passive attack.

This attack breaks the confidentiality objective of the security where the information is made accessible to unwanted users. These attacks are very common such as while sending an email within a network, if the information is not encrypted, the attackers can easily get hold of the sensitive information from the transmitted messages. Loss of personalized information like email address and mobile numbers creates unrest among the individuals.

There are many situations where a CPS is susceptible to eavesdropping as there are a lot of communication networks within the CPS.

Electronic communication between two monitors can be intercepted by the attacker in the manner proposed by Kuhn (2006), where the attacker uses an antenna for capturing the electromagnetic radiations to view the monitor display. Marcus proposes many solutions, one of which is to reduce the radiation level by keeping the cables and circuit board traces short. This reduces the physical area around which the attack can occur.

Intervehicular CPS requires vehicles to broadcast their location and speed that allows the other vehicles in the nearby network to stay cautious in the case of any hazardous situations. However, such broadcasting of messages has its own disadvantages. As the vehicle continuously broadcasts its whereabouts, an eavesdropper can easily track its location at any moment of time that could turn out to be undesirable. We can observe that the important requirement in such CPS is to have information about the location of vehicle along it with its speed. Hence, it doesn't really matter from which vehicle the information is coming. Thus, one good solution for this problem requires vehicles to broadcast their message using pseudonym or a fake identity that are to be updated regularly. This mismatch of the identities will make the attacker get confused about the intended victim's location. A definite care should be taken to make sure that the pseudonyms are not linked with the original vehicles (Levente, 2007).

In the case of manufacturing plants, a breach in the confidential data would result in a theft of intellectual property. The damage done when such knowledge from a nuclear plant has fallen into the hands of the terrorists is unimaginable.

17.7.3 Stealthy Deception Attack

In this type of attack, false information is sent by the adversary from sensors or controllers. The false information would be a wrong sender ID, wrong measurement calculated, etc. The attacks are launched by finding out the secret keys in the devices or by attacking some sensors or controllers.

Figure 17.3 A deception attack on intervehicular CPS system that happens due to the corrupt information transmission. Segment 2 of the figure communicates by sending the traffic flow report to the CPS of the car that has to make to the decision. Segment 1, which is an adversary disguised itself as a genuine entity, sends corrupt information to the CPS. Thus, the CPS would now be in a fix unable to decide which way to go, thus resulting in a crash or suspended action.

The CPS, when attacked with a deception attack, can check with the expected output of a healthy system and detect whether it is being attacked or not. Although these detection techniques work fairly well for a basic deception attack, they fail to function when an attacker makes a sophisticated stealthy deception attack after analyzing the design of the CPS. Thus, in such cases, the CPS cannot protect itself as it has no idea of the attack happening on it. Deception attacks occur when there are comprised data collected in sensors, actuators, or both.

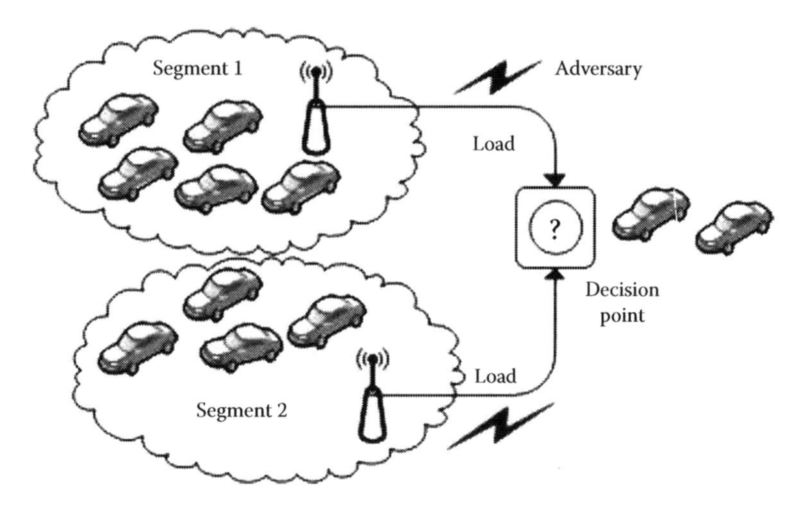

FIGURE 17.3
Deception attack. (Image courtesy of National Science Foundation CNS CAREER award #1149397, http://cs.txstate.edu/~mg65/mcps/.)

17.7.4 Compromised-Key Attack

To keep the information secure while transmitted through communication channels, the messages are often encrypted before being sent. Secure information that is communicated inside a network is interpreted through the use of a secret code that is termed as a key. There are various methods available for securing the data through encryption through keys. These methods are used for securing the confidentiality and integrity of the information being sent. One such method uses a private–public key pair, which is discussed in detail in the following paragraph (Shelton).

First, encryption is done by the sender using a secret private key. This private key is held only by the sender to confirm its identity during the communication. The receiver after getting the message is guaranteed that it has come from the intended source through the signed private key. The other part of the key pair is the public key that is broadcasted to the world. The receiver decrypts the message using this public key. The public key can decrypt the information only if it is encrypted by the secret private key, thus assuring confidentiality. This is depicted in Figure 17.4.

This private key when obtained by an attacker is called as a compromised key. Secure information without the perception of the receiver or sender can be accessed by the attacker

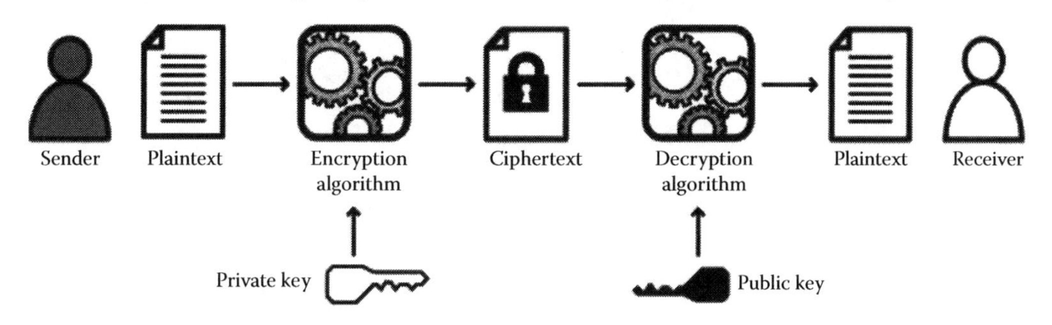

FIGURE 17.4
Encryption and decryption. (Image courtesy from Shelton, B.K., Introduction to cryptography, http://www.infosectoday.com/Articles/Intro_to_Cryptography/Introduction_Encryption_Algorithms.htm.)

with the help of the compromised key. The attacker could use the compromised key for altering the data and also compute additional keys that could give access to the other shared resources. It is quite a difficult task to detect a compromised key. Most of the CPSs nowadays are geographically dispersed and are operated through communication channels. Owing to these factors, the compromised-key attacks are detected only after some crucial information is found to be no longer private.

17.7.5 Injection Attack

It is a type of deception attack where the data integrity of the packets is lost. The attacker tries to modify the data stored within the system. These attacks, which are much difficult to be detected, are some of the most successful attacks because of the nature of diversity in them.

These attacks require a lot of hard work from the attackers. A proper research needs to be done on the constraint requirements such as the maximum amount of power transmission possible for a node in a power grid, as any ludicrous tampering of the data would easily be detected by the firewalls. Some of the common scenarios where the false data injection attacks are done are explained in the following paragraph.

Smart grids are the CPSs that are used to efficiently distribute the energy resources to satiate the energy usage demands of the users in a secure, reliable, and resilient manner. Itis a typical CPS that integrates the power transmission system to the network communications via the components such as the smart meter, which is used to measure the amount of energy generated, consumed, and stored. Figure 17.5 shows the distribution network in the power grid (Liu, 2011).

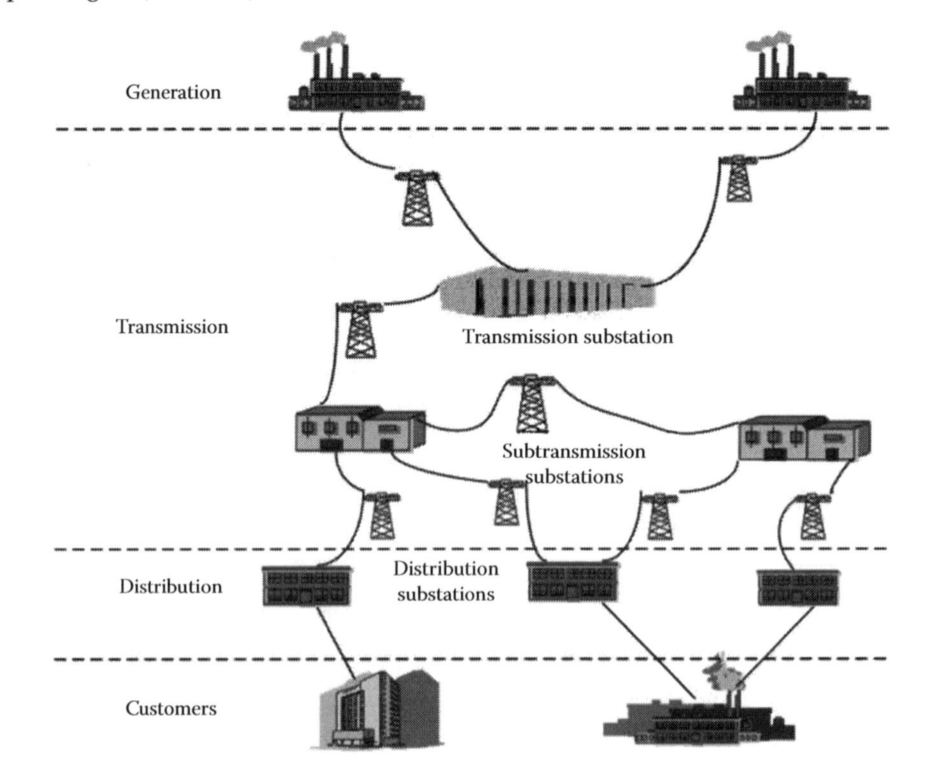

FIGURE 17.5
Distribution of power in a power grid.

Smart grid comprises of a number of users connected to it for distributing power among one another. Each user generates and utilizes different amounts of energy based on their geographical distribution and the energy generation efficiency. The user that generates more energy than its consumption supplies energy to the users that consume more than the amount generated. This is how a balance is established in a smart grid.

Smart meter, which is a measurement component, is responsible for the all of the energy utilization and consumption calculations. These devices, which are connected to the CPS network, face a huge threat of false data-based injection attacks. If attacked, any change in the measurement values of the smart meters would result in a collapse of the entire network by destabilizing the demand and supply equilibrium.

Next, consider an aerospace CPS where the precision of the data for sending a rocket into the space is of utmost importance. Any small malfunction would result in unrecoverable damages. Injection attacks on such systems by modifying the constraints involved such as the amount of fuel needed for the rocket launch result in a total disruption of the CPS. Integrity of the information stored is a vital factor in such systems and an injection attack on such systems would result in a dreadful damage to the infrastructure and the physical environment.

17.7.6 Man-in-the-Middle Attack

These are the type of attacks where false messages are being sent to the operator by the attacker. These messages misguide the operator either in the form of false negative or false positive. The operator tends to perform an operation when it is not required and would remain passive at the moments when an action is to be committed (Figure 17.6).

This attack would result in the loss of authenticity in the messages that are transmitted. The attacker can fake as a genuine receiver to the sender and then later alter these messages before sending back to the genuine receiver now disguised as genuine sender. The attack can be better understood by the following example.

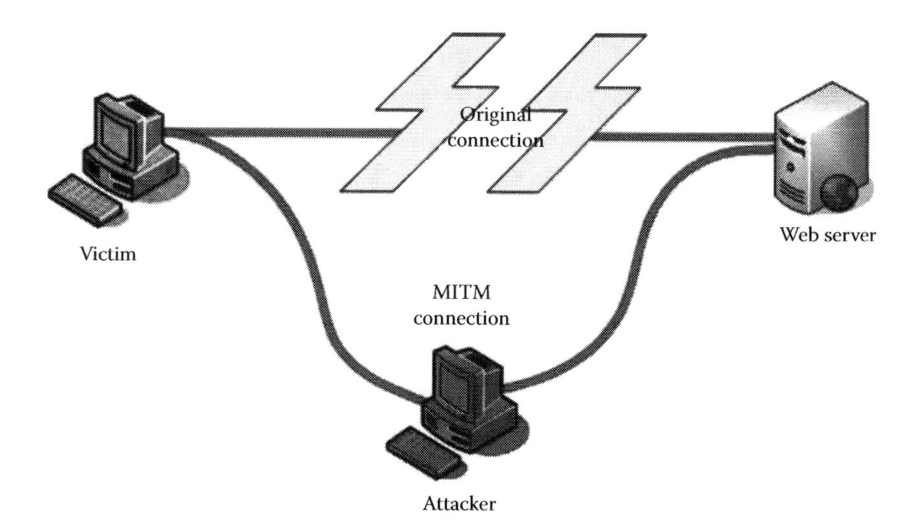

FIGURE 17.6
Man-in-the-middle attack. (Image courtesy of The Malta Independent, http://securityaffairs.co/wordpress/8867/security/to-be-or-not-to-be-this-is-authentication.html.)

Let us consider a situation where a user wishes to buy a good from an online store from a browser that is already compromised. As the browser is compromised, the screen that the user sees for mentioning the shipping address of the good that he purchased will be different from what is actually seen by the server. As a result, the user who paid the money for the good is deceived and has no idea as to what went wrong because the product is now shipped to the attacker's address without the consent of the user.

Consider another scenario of robotics in CPS. The functionality of this CPS is to automate the manufacturing of the robot. Man-in-the-middle attack in such a CPS system intervenes the communication between the different components and modifies the messages before transmission. The attacker disguises itself as a valid component and requests malicious inputs to the control system of CPS. As a result, the component receives a faulty input that results in the construction of a robot with erroneous functionalities. In simpler terms, the robot runs when asked to sit.

17.7.7 Replay Attack

It is another type of deception attack in which normal flow data are repeated or delayed by the attacker, making it look like a valid flow. This attack is also termed as a playback attack where the attacker intercepts the data and fraudulently retransmits the valid data that it intercepted, thus making the data transmitted to look legitimate. These types of attacks are difficult to be detected.

The aforementioned attacks are caused by different type of adversaries comprising of skilled hackers, criminals, terrorists, disgruntled employers, etc. Depending on the nature of the adversaries, the defense measures can be taken at times. The defender groups might even sometimes anticipate the destruction effects by knowing their attackers that help them in building efficient defense security models. The next section speaks about the different security measures that are available for ensuring safety in CPS.

17.8 Defensive Mechanism against Attack in CPS

The defense measures for a CPS should focus on the individual security of both the cyber and the physical processes and also on the interaction of both the cyber and the physical processes. It would be a perfect security design if both of the aforementioned security requirements of CPS are met at the same instance.

The following are some of the predominant defense measures that are adopted in general for protecting the cyber systems (Mensah, 2013).

17.8.1 Cryptosystems

Cryptosystems have various processes to make sure that the security objectives such as the confidentiality and integrity are all met by using the techniques of cryptography. It is a defense measure that is used in the situations to prevent the possibility of an attack from happening.

Unauthorized access to the resources of the system can be denied by setting up an access control system. Impersonation of other entities can be prevented by the use of authentic tools. Message authentication code, Secure Sockets Layer, and Internet Protocol Security are some the technologies that are in common use for securing through cryptosystems in CPS.

17.8.2 Intrusion Detection Systems

One of the most commonly used techniques in the cyber security systems is an intrusion detection system (IDS). According to this functionality, the network components are grouped into zones. The traffic inflow and outflow from the zone is monitored and controlled through the known network connections that are completely safe. All of these form a security region termed as the electronic security perimeter.

Network-based IDS is one such example of a perimeter secured device that is used for examining the traffic flow within the perimeter in order to detect an intrusion. They work by comparing the flow records with a set of signature records or by identifying peculiarities in the traffic pattern. These IDS are used for securing CPS in many instances. A significant amount of research work is happening in this region to further improve upon the usage in CPS security fields.

17.8.3 Firewalls

These are the devices that allow only a certain kind of network flow to enter or bypass its network perimeter. At times, all the communications in CPS are done through some specific Transmission Control Protocol (TCP)/User Datagram Protocol (UDP) ports. In such situations, the firewall would see that the traffic flow into the network of CPS would only be from the desired ports, thus blocking all the other incoming traffic from the other ports. This can turn out to be quite helpful as the unused ports are often used as an entry point by the malicious users to get into the network.

The direction of flow traffic can also be controlled by some advanced firewalls providing a much secure connection. Data diode is a special type of firewall that only allows a one-way traffic flow at the host device. It uses a single strand of the optical fiber for communication, thus making it impossible for duplex communication.

17.8.4 VPN

Virtual private networks (VPNs) are often times used by corporate firms for securely transferring data from remote locations. The private networks from different remote locations are connected through the Internet. The data are encrypted before transmission, thus making the interceptions done on such networks futile. Thus, a VPN is one of the most reliable way through which remote connections in CPS communicate in a secure manner.

17.8.5 Software Updates

Softwares most often are released with some minor bugs when released. The vulnerabilities caused due to these bugs are soon discovered in the software, thus making it prone to cyber threats. As a result, the authors of the software release an update fixing the bugs discovered to secure their users from the malicious attackers. Regular updates should thus be made in CPS to ensure its basic security.

Most of the aforementioned defense measures speak about the general cyber security measures that are considered to be necessary. These solutions mentioned for the cyber security systems take into consideration only the cyber portion of the attack, leaving the physical system aside. However, any change caused due to an attack in the physical system would equally affect the CPS. Thus, proper research needs to be done for the evolution of a design that takes into consideration the security of physical aspect of the CPS as well. The next section covers the security aspects that are in particular to CPS.

17.9 Additional Security Specific to CPS

Security solutions that are specific to CPS are designed by taking into account the characteristics of the system that needs to be secured in CPS. These designs need to be done specifically for each CPS for ensuring better safety measures by taking into consideration the peculiar threats for that CPS.

One needs to have an idea about the authorized and unauthorized information flow, control flow, and physical and cyber consequences of a security breach before the design of a secure system for a CPS. Incorrect specification of any of these necessities would result in a failure of security. Having an additional information of specific security risks would help in designing a much better secure system (Neuman, 2009).

Several fields such as oil and gas, chemical, and water are developing security infrastructures keeping in mind their respective requirements and constraints. There are some common standards that are already been followed in some fields for securing their infrastructures.

The North American Electric Reliability Corporation (NERC) has put forth some cyber security standards for the CPS systems in the electric sector. All the electric utilities in the North American region are now compliant to these standards put forth by NERC that is given the authority to enforce the compliance (North American Electric Reliability Corporation, 2013).

National Institute of Standards and Technology Special Publication (NIST SP 800-53) has given a set of security controls that all the federal agencies from the United States of America should meet (National Institute of Standards and Technology, 2013). It includes a specific section for the security controls for the industrial control systems that are a type of CPS. These security requirements are not enforced upon but they do provide security help to the ones that are in need.

17.10 Conclusion

The security of CPS has not been much significant. The security measures developed out of the traditional cyber security threats and have evolved based on the need. Primarily, majority of the research works in security were focused on securing the physical systems and the cyber systems separately. Proper analysis should be done in the interaction domain of these two systems. Sensors and other devices that are spread around the field face the problem of physical protection. The security design should take into consideration this factor and design the model in such a way that there would be no loss of sensitive data even when such devices are compromised.

A lot of factors can be further improved upon for enhancing the security standards in CPS. Rather than completely focusing on fixing small vulnerabilities, work needs to be emphasized on handling the difficult challenges such as efficient security design for CPS.

References

Cyber Security Research Alliance. (April 2013). Designed-in cyber security for cyber-physical systems. Retrieved from http://www.cybersecurityresearch.org/documents/CSRA_Workshop_Report.pdf.

Jackson, B. (2003). Union carbide: A disaster at Bhopal. Retrieved from http://www.environment-portal.in/files/report-1.pdf. Last accessed on 4–5 April, 2013.

Kerr, P.K. (December 2010). The stuxnet computer worm: Harbinger of an emerging warfare capability. Retrieved from http://www2.gwu.edu/~nsarchiv/NSAEBB/NSAEBB424/docs/Cyber-040.pdf. Last accessed on 9 December, 2010.

Kuhn, M.G. (2006). Eavesdropping attacks on computer displays. Retrieved from http://www.cl.cam.ac.uk/~mgk25/iss2006-tempest.pdf. Last accessed on 24–25 May, 2006.

Levente, B. (2007). On the effectiveness of changing pseudonyms to provide location privacy in VANETs. Retrieved from http://ftp.crysys.hu/publications/files/ButtyanHV07esas.pdf. Last accessed on July 2007.

Li, C. (2011). Hijacking an insulin pump: Security attacks and defenses for a diabetes therapy system. Retrieved from http://ieeexplore.ieee.org/xpls/abs_all.jsp?arnumber=6026732&tag=1. Last accessed on 11 June 2011.

Liu, Y. (2011). False data injection attacks against state estimation in electric power grids. Retrieved from http://discovery.csc.ncsu.edu/pubs/ccs09-PowerGrids.pdf.

Mensah, F.N. (June 2013). Security analysis of CPS: Understanding current concerns as a foundation or future design. Retrieved from http://www.asee.o rg/public/conferences/20/papers/7587/download.

Moholkar, A.V. (July 2014). Security for cyber-physical systems. Retrieved from http://www.ijcat.org/IJCAT-2014/1-6/Security-for-Cyber-Physical-Systems.pdf.

National Institute of Standards and Technology. (April 2013). NIST Special Publication 800-53. Retrieved from http://nvlpubs.nist.gov/nistpubs/SpecialPublications/NIST.SP.800–53r4.pdf.

Neuman, C. (2009). Challenges in security for cyber-physical systems. Retrieved from http://moss.csc.ncsu.edu/~mueller/ftp/pub/mueller/papers/cps06.pdf.

North American Electric Reliability Corporation. (2013). North American electric reliability corporation. Retrieved from http://www.nerc.com/Pages/default.aspx.

Sampigethaya, K. (March 2012). Cyber-physical system framework for future aircraft and air traffic control. Retrieved from http://ieeexplore.ieee.org/xpl/login.jsp?tp=&arnumber=6187151&url=http%3A%2F%2Fieeexplore.ieee.org%2Fiel5%2F6178851%2F6186985%2F06187151.pdf%3Farnumber%3D6187151.

Shafi, Q. (2012). Cyber physical systems security: A brief survey. Retrieved from http://ieeexplore.ieee.org/xpl/articleDetails.jsp?arnumber=6257627. Last accessed on 18–21 June, 2012.

Shelton, B.K. Introduction to cryptography. Retrieved from http://www.infosectoday.com/Articles/Intro_to_Cryptography/Introduction_Encryption_Algorithms.htm.

Slay, J. and Miller, M. (2005). Lessons learned from the Maroochy water breach. Retrieved from http://www.ecdlhealth.it/wcc2008/IFIP_Sample_Chapter_Created_LaTeX.pdf.

Wang, E.K. (2010). Security issues and challenges for cyber physical system. Retrieved from http://ieeexplore.ieee.org/xpl/login.jsp?tp=&arnumber=5724910&url=http%3A%2F%2Fieeexplore.ieee.org%2Fxpls%2Fabs_all.jsp%3Farnumber%3D5724910. Last accessed on 18–20 December, 2010.

Additional Readings

Alvaro, A.C. (2009). Challenges for securing cyber physical systems. Retrieved from http://cimic.rutgers.edu/positionPapers/cps-security-challenges-Cardenas.pdf. Last accessed on July 2009.

Banerjee, A. (2011). Ensuring safety, security, and sustainability of mission-critical cyber–physical systems. Retrieved from http://ieeexplore.ieee.org/xpl/login.jsp?tp=&arnumber=6061910&url=http%3A%2F%2Fieeexplore.ieee.org%2Fxpls%2Fabs_all.jsp%3Farnumber%3D6061910. Last accessed on 26 December, 2011.

Bhadoria, R.S., Dixit, M., Bansal, R., and chauhan, A.S. (January 2012). Detecting and searching system for event on internet blog data using cluster mining algorithm. In *Proceedings of Springer International Conference on Information Systems Design and Intelligent Applications* (INDIA), Visakhapatnam, India.

Bhadoria, R.S., Dixit, M., Mishra, V., and Jadon, K.S. (December 2011). Enhancing web technology through wiki-shell architecture. In *Proceedings of IEEE World Congress on Information and Communication Technologies*, Mumbai, India.

Bhadoria, R.S., Jain, S.K., and Sain, D. (2011). Data mining techniques in user profile personalization. *International Journal of Advanced Research in Computer Science*, 2(4).

Bhadoria, R.S. and Jaiswal, R. (November 2011). Competent search in blog ranking algorithm. In *Proceedings of Springer International Conference on Computation Intelligence & Information Technology*, Pune, India.

Bhadoria, R.S., Sahu, D., and Dixit, M. (2012). Proficient routing in wireless sensor networks through grid based protocol. *International Journal of Communication Systems and Networks*, 1(2), 104–109.

Chen, C.M. (2013). Defending malicious attacks in cyber physical systems. Retrieved from http://ieeexplore.ieee.org/xpl/login.jsp?tp=&arnumber=6614240&url=http%3A%2F%2Fieeexplore.ieee.org%2Fiel7%2F6597100%2F6614232%2F06614240.pdf%3Farnumber%3D6614240. Last accessed on 19–20 August, 2013.

Dacier, M. (July 2014). Network attack detection and defense: Securing industrial control systems for critical infrastructures. Retrieved from http://drops.dagstuhl.de/opus/volltexte/2014/4791/pdf/dagrep_v004_i007_p062_s14292.pdf.

Kwon, C. (June 2013). Security analysis for cyber-physical systems against stealthy deception attacks. Retrieved from http://ieeexplore.ieee.org/xpl/login.jsp?tp=&arnumber=6580348&url=http%3A%2F%2Fieeexplore.ieee.org%2Fxpls%2Fabs_all.jsp%3Farnumber%3D6580348.

Le, X. (2010). False data injection attacks in electricity markets. Retrieved from http://ieeexplore.ieee.org/xpl/login.jsp?tp=&arnumber=5622048&url=http%3A%2F%2Fieeexplore.ieee.org%2Fxpls%2Fabs_all.jsp%3Farnumber%3D5622048. Last accessed on 4–6 October, 2010.

Lin, J. (2011). On false data injection attacks against distributed energy routing in smart grid. Retrieved from http://ieeexplore.ieee.org/xpl/articleDetails.jsp?arnumber=6197400. Last accessed on 17–19 April, 2012.

Pasqualetti, F. (November 2013). Attack detection and identification in cyber-physical systems. Retrieved from http://ieeexplore.ieee.org/xpl/articleDetails.jsp?arnumber=6545301.

Shi, J. (2011). A survey of cyber-physical systems. Retrieved from http://ieeexplore.ieee.org/xpl/login.jsp?tp=&arnumber=6096958&url=http%3A%2F%2Fieeexplore.ieee.org%2Fxpls%2Fabs_all.jsp%3Farnumber%3D6096958. Last accessed on 9–11 November, 2011.

Tianbo, L. (2013). A new multilevel framework for cyber-physical system security. Retrieved from http://www.terraswarm.org/pubs/136/lu_newmultiframe_edge.pdf.

Vanderbilt University Board of Trust. (May 2013). Security of cyber-physical systems a survey. Retrieved from http://ejournals.library.vanderbilt.edu/index.php/vurj/article/viewFile/3765/1884.

Wasicek, A. (2014). Aspect-oriented modeling of attacks in automotive cyber-physical systems. Retrieved from http://chess.eecs.berkeley.edu/pubs/1064/aomsec-preprint1.pdf. Last accessed on June, 2014.

18

Intrusion Detection, Prevention, and Privacy of Big Data for Cyber-Physical Systems

Afrah Fathima, Khaleel Ahmad, and Abdul Wahid

CONTENTS

18.1 Introduction

A Cyber-physical system (CPS) can be identified as a subsequent generation of embedded Information and Communication Technology (ICT) systems, which has become ubiquitous in every aspect of our lives. Embedded computation and compilation devices are merged together in CPS with actuators and sensors. CPS are generally spatially distributed exhibiting embryonic behavior such as cyber attacks and traffic jams resulting from system components interactions, which can be stopped when some specific components are resolved from the systems. The CPS, with qualities of universality and pervasiveness, has its effect on almost every aspect of our lives; the biggest challenge is to effectively predict the budding behaviors of the system. With the complexity of its models, it often obstructs to comprehensively test or verify their safe behavior. We can use an alternate method by equipping CPS with monitors so as to forecast the emergent behavior at run time. By using this approach, it makes CPS self-aware, allowing us to open more approaches for designing systems, which enables them to reconfigure dynamically to adapt different circumstances.

In ubiquitous computing and cyberinfrastructure for science and arts, we have found a critical gap amid cyber-physical infrastructure and the user environment challenges. Since CPS has major unlimited and unidentified economic and gregarious potential, major investments are being done worldwide to develop and improve the technology. CPS technology is based on the older concept of embedded systems. The technology is embedded in softwares and devices, such as toys, cars, scientific instruments, and medical devices. CPS is based on the dynamics of physical processes similar to that of software and networking, which provides abstraction, design, and modeling and analysis technique. The intricacy of its models often holds back any attempt to comprehensively verify its safe behavior. The CPS is an amalgamation of physical processes, computation, and networking. Physical processes are being monitored and controlled by embedded computers and networks using feedback loops. The computations affect physical processes, and vice versa [1].

18.1.1 Challenges for the CPS

The biggest challenge for a CPS is to create universally accepted extensible infrastructure that supports the maximum possible number of sensor inputs and outputs (display/actuators) working parallel making the system user friendly for potential users. It is not our concern to create one-off systems useful only for specialized scientists and technicians, instead to develop a contemporary cyberinfrastructure that seems to be general but a powerful system, which can be useful for the majority in a community. The infrastructure developed should not only offer innovative ways for humanity, enabling users to think and observe and reach decisions about the universe we are living in, it should also allow different CPS to interconnect with each other and to interact within a navigable framework. It should also enable combination of human knowledge and skills from a cyber perspective to create new realms for human creativity. It should also be able to link different scientific and other educational projects to encourage information dissemination, public observations, debate, and tourism.

The CPS need to be more secure, safe, should be able to operate in real time, and must be dependable and secure enough. Apart from this, the CPS should be cost effective, easy to adopt, and scalable [2].

Characteristics of CPS are as follows:

18.2 Resilience

It can be defined as the ability of a physical system or a device to recoup its shape and size after a misproportion, which is caused by a compressive stress (Figure 18.1).

18.2.1 Big Data

Big Data refers to the gigantic amounts of data that have been gathered over time, which are hard to analyze and handle using regular database management tools. Big Data is a combination of technologies that usually handles the huge amount of data either structured data generated by computer or machine or unstructured data generated by human interaction with the computer. In most of the business scenarios where the data are too large or too fast or unable to be stored in the existing processing capacity, Big Data has the flair to assist corporations or businesses to cope up with the voluminous data at a fast pace and take more quick decisions. The term "Big Data" is not just large amount of data,

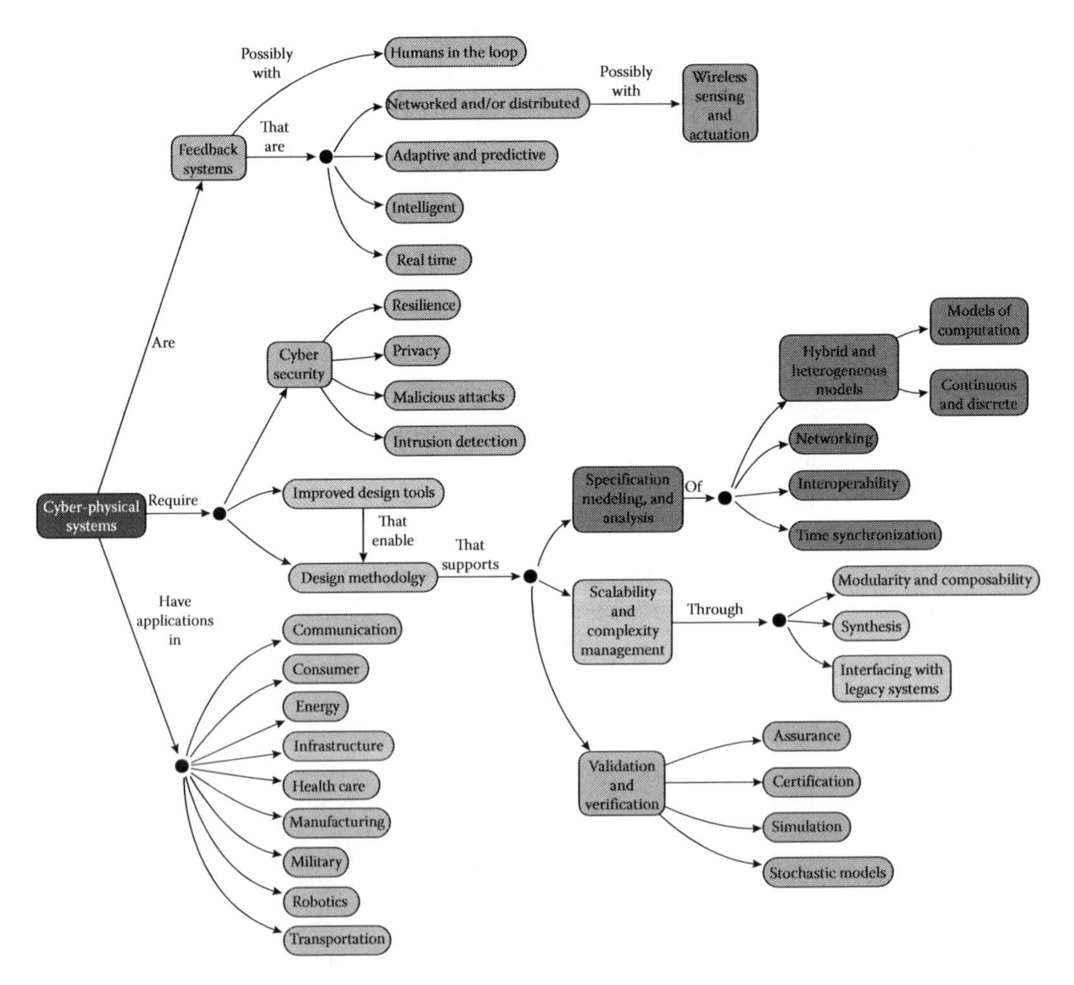

FIGURE 18.1
Classification of cyber-physical systems.

if used by traders, but may refer to the technology that combines tools and processes that an organization needs to handle and store, process, and analyze the bulky data [3,4].

Big Data has become one of the crucial technologies that have the potential to change the organization, which uses information to improve and enhance the customer experience and to transform their business needs. Big Data is not a single technology rather a collection of data management technologies that have developed over time. Big Data enables organizations to store large amount of data, process, manage, and manipulate large amounts of data at the right speed and at the right time so as to gain the right insight. Most of the companies are at an initial stage of the Big Data journey.

The impact of Big Data on business. The following are issues need to be discussed:

1. What is the architecture of Big Data? How do you manage voluminous data without causing major disruption in the data center?
2. When should we incorporate the outcome of Big Data analysis with the data warehouse?
3. How do we secure companies' data and what is the inference of security and governance on using Big Data?
4. What is the importance of various Big Data technology and when are these technologies considered as part of your Big Data strategy?
5. What are the various sources of Big Data analytics that are advantageous? How do we apply various analytics to business problems?

18.3 Fundamental of Big Data and Its Security

18.3.1 Evolution of Data Management

Big Data is the latest emerging technology. Big Data is usually defined as any kind of data that have the following basic three characteristics:

1. Extremely large volumes of data
2. Extremely high velocity of data
3. Extremely wide variety of data (Figure 18.2)

Big Data is significant because it enables the organization to collect, manage, and manipulate vast amount of data at an exact time to gain the right insight. Big Data is not a single technology, rather a combination of technologies over the last 50 years. Organizations have moved from an era where the technology was designed to support just a specific need, that is, to determine how many products were sold to how many customers. Now, the organizations have tremendous amount of data from different sources than ever before. The whole data look like a gold mine wherein you have a little gold and the rest is everything else [1–3].

18.3.2 Challenges of the Technology

- How to recognize the patterns that are meaningful to your business decision
- How to deal with massive amounts of data in a meaningful way

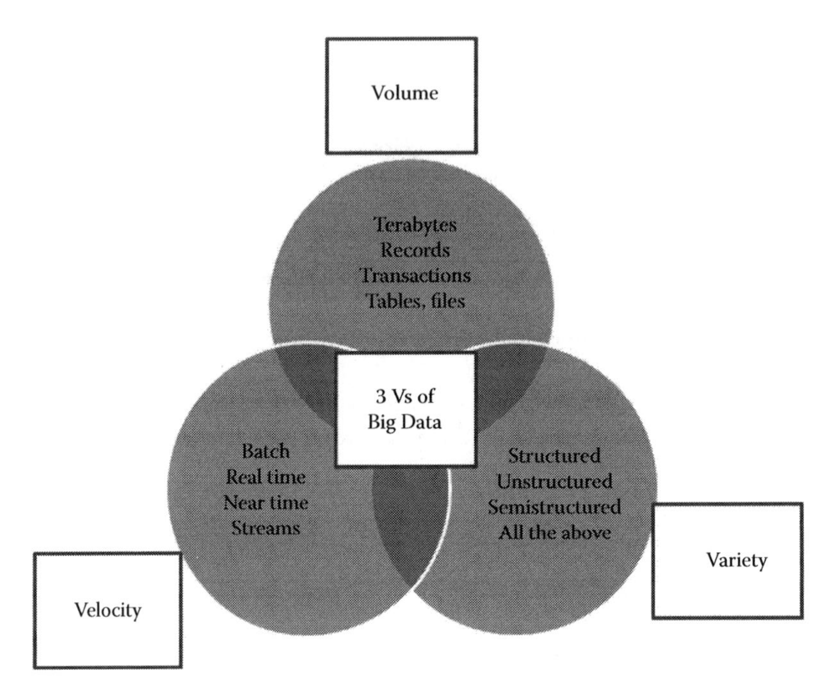

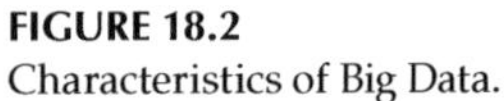
FIGURE 18.2
Characteristics of Big Data.

Before we get into the details, we take a look of the evolution of data management and how these data management waves are connected. All the waves of data management have been born to solve specific data management problems. Each of these waves has been evolved because of its effect and its utility. When relational database was introduced in the market, it needed managers to learn the set of tools to understand the relationship between data and elements. When the organization started storing unstructured data, the analysts required new technologies, for example, natural language-based analysis tools, for a better understanding and to gain insights that are more useful for the business. A search engine company realizes that since it has an access to immense amounts of data that needs to be monetized. New innovative tools and approaches were required to gain values from data. So, it is very much important to understand the previous waves to understand Big Data. We also need to understand that as we move from one wave to another wave, we even need to learn the tools and technologies that were being used to address different issues.

18.3.3 Different Waves of Big Data Management [3]

18.3.3.1 Wave 1: Creating Manageable Data Structure

Initially in the 1960s, there was no data structure; data were stored in flat files. Companies started using brute force methods so as to know in detail about the customers. Later in the 1970s, relational database management system was developed, which has imposed a structure for storing data and a way to improve performance. Furthermore, as the organizations were growing, it was getting more difficult to store the voluminous data, which was getting expensive, and even accessing the data was becoming more worse; there were many duplication problems arising and it was quiet difficult to measure the actual business value. There was an urgent need o f a new technology to support relational model.

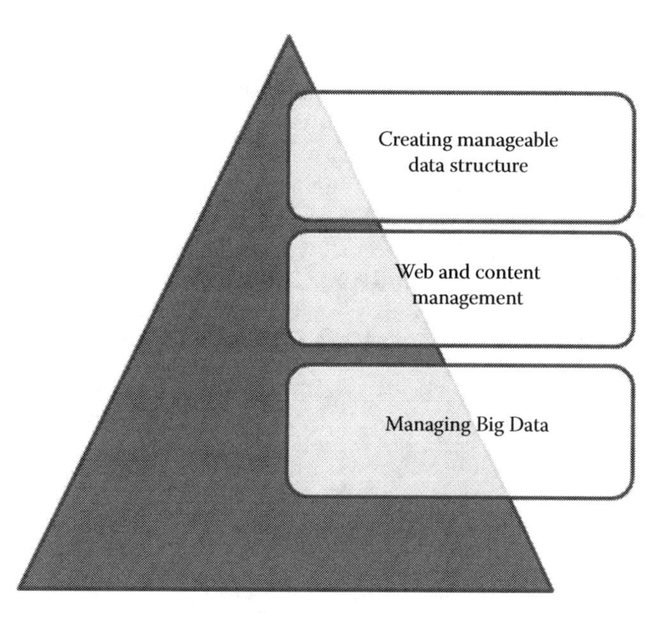

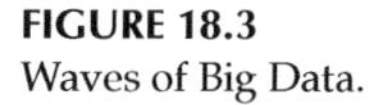
FIGURE 18.3
Waves of Big Data.

Entity relationship (E-R) model was developed, which has increased the abstraction and usability of data. Again, when the data of organizations were growing out of control, data warehouses were developed to help organizations to deal with the huge structured data. Therefore, data warehouses have been designed for query and analysis rather than just for transaction processing. It enables the organizations to consolidate data from various sources. Data marts and data warehouses have solved many problems enabling organizations to manage massive transactional data. But again, the data warehouses were unable to manage structured and unstructured data as the data were growing tremendously. Even the warehouses and data marts were unable to meet the changing demands (Figure 18.3).

How would organizations be able to transform data management approaches to deal with the high increasing volume of unstructured data elements?

The solution for storing unstructured data did not materialized over night. Vendors started using binary large objects. The data element is used to be stored in a relational database as a single contiguous chunk of data. The object database has provided us a unified approach to deal with unstructured data. Object database included with a structure for data elements and a programming language manipulates different data objects without any programming or complex joins. The object databases have been introduced to deal with the second wave of data management.

18.3.3.2 Wave 2: Web and Content Management

Later in the 1980s, "enterprise content management system" was evolved to deal with unstructured data consisting of mostly documents. In the 1990s, as web emerged and came into sight, organizations wanted to store web data such as web content, image, audio, and video beyond just documents. Later, there was a huge demand of web virtualization, and cloud computing began. In this new wave, organizations started realizing the need to deal with new generation of data sources with an exceptional amount and variety of data, which need to be processed at a high amount of speed.

18.3.3.3 Wave 3: Managing Big Data

Is Big Data really new or is it just an advancement in data management journey?

The answer is yes—it is actually a combination of both.

Many technologies in Big Data have been around in decades such as virtualization, parallel processing, memory databases and distributed file systems, and advanced analytics though it is not much used in practical. Other emerging technologies such as Hadoop and map reduction came into the picture recently. If organizations can examine the petabytes of the data (which is equivalent to 20 million) with an average performance to distinguish among anomalies and patterns, the businesses can begin to make sense of data in new ways. The step toward Big Data is not only for business but also for scientific research and government activities that have helped to move it forward.

18.4 Different Approaches to Deal with the Existing Data

- Data in motion
- Data at rest

Data in motion: It is generally used to identify the quality of products of a business during its manufacturing process so as to avoid costly errors.

Data at rest: It is used by a business analyst to better understand the needs of a customer such as buying patterns based on the aspects of customer relationships including customer service interactions, sales, and social media data.

We need to keep in mind that we are still at an initial stage of dealing with huge volumes of data so as to gain a 360° view of business and also to identify customer expectations. The technologies to drive this are still isolated from each other. In order to get the desired results, the technologies from all three waves should come together. Big Data is not a single technology or a tool but rather a combination of technologies to get the right insight, on the right times on the right data.

18.5 Building Successful Big Data Management Architecture

Begin with capturing, organizing, integrating, analyzing, and acting: Before we explore into the architecture, it is important to know about the functional requirements for Big Data (Figure 18.4).

First, data must be captured, then we need to organize the data, and then integration should be done. After the successful implementation of the aforementioned stage, data should be analyzed depending on the problem being addressed. Finally, the management action is taken based on the outcome of the analysis.

Validation: Validation is the most important issue. If the organization is combining various data generated from different sources, it is important that you should validate the data so that they should make sense when combined. We also need to consider the security and governance as certain data sources may contain sensitive information.

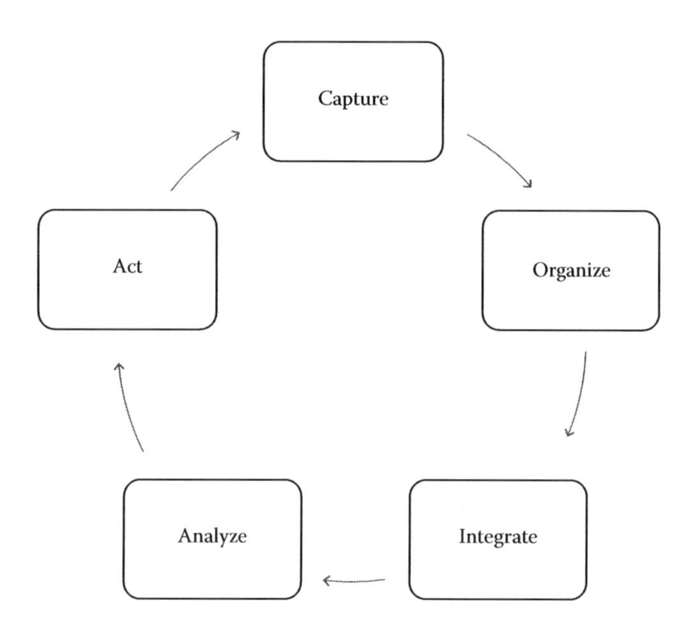

FIGURE 18.4
Functional requirements of the building of Big Data.

18.5.1 Setting the Architectural Foundation

Organizations must give attention to the following issues:

1. How much data does the organization need to manage in the present and in the future?
2. How much amount of risk can an organization afford? Does it need to identify security, compliance, and governance requirements?
3. What is the importance of speed to manage data?
4. How precise should the data be?

18.5.2 Understanding How Big Data Works

Since Big Data relies on picking loads of data from various sources, open application programming interfaces are the core to any Big Data architecture. The interfaces exist at each and every layer of the stack, and integration services are the most important in Big Data.

18.5.2.1 Redundant Physical Infrastructure

The robust physical infrastructure of Big Data architecture is the most important for the operation and scalability. The physical infrastructure in Big Data is based on distributed computing model since the data here are stored at different physical locations, which can be linked through networks. We also use distributed file systems and various Big Data analytical tools and applications. Redundancy is an important issue in a Big Data since we deal here with a lot of data from different sources and there is a possibility of redundancy.

18.5.2.2 Security Infrastructure

As the analysis of Big Data becomes more important, it gets more important to secure the data. We need to keep track of the users who are allowed to send the data under what circumstances. It's important to verify the identification of the users.

18.5.2.3 Operational Data Sources

Traditionally, operational data sources are highly structural data and managed by a line of business in relational database. As the world is changing, it's important that operational data need to encompass other set of data sources such as customer and social media in any form. The different emerging approaches to data management include documents, columnar, and graphical and geospatial architectures in Big Data, which are collectively referred as "not only structured query language (SQL)" databases. The designed architecture should support complex unstructured contents.

18.6 Traditional and Advanced Analytics

18.6.1 Analytical Data Warehouses and Data Marts

After the organization sorts massive amounts of data, it's often difficult to identify the subset of data that reveals the patterns and to present into a form that is available to the business. Data warehouse and data marts provide compression, multilevel partitioning, and parallel processing architecture.

18.6.2 Big Data Analytics

Organizations need to manage and analyze petabytes of data with clusters of information which requires analytical engines to manage these highly distributed data and to provide results for a particular business problem. Analytics gets more complex in Big Data.

18.6.3 Reporting and Visualization

Organizations always depend on visualization and report so as to get a better understanding from monthly sales figures to projections of growth. Reporting and data visualization are the important tools to see how the data are related and its impact of relationships in future.

Big Data is applied in various areas:

- Health care
- Manufacturing
- Traffic management

18.7 Examining Big Data Types

Big Data is classified in mainly two types:

- Structured data
- Unstructured data
- Semistructured data

18.7.1 Structured Data

They are generally defined as data that have a define length and a particular format. For example, structured data include numbers and strings (group of words) such as customer name, address, and house number. The data are stored in a database, and a query can be raised similar to that of SQL [4,5].

The sources of big structured data are as follows:

1. *Computer- or machine-generated data*: Machine-generated data are usually referred to as data that are created by machine without human involvement.
2. *Human-generated data*: These data are created by humans in interaction with computers. Some of the experts have argued a third category that is called "hybrid," which is a combination of human- and machine-generated data.
3. *Machine-generated data*: These include sensor data, web blog data, and point-of-sale data.

- *Sensor data*: They include radio-frequency ID tags, medical devices, smart meters, and GPS data (Figure 18.5).
- *Web blog data*: These are the data that are gathered by servers, applications, networks, and so on, which include all activities. These may amount to huge volumes of data that can sometimes be useful to identify security breaches and service-level agreement.
- *Point-of-sale data*: These data are used generally when cashier swipes the barcode of a product being purchased.

Here are examples of structured human-generated data:

- *Input data*: These are the resultant of any input given by human to a computer, for example, name, age, *h.no*, and income tax.
- *Click stream data*: These are the data generated when you click a link on any website usually used to identify customer behavior and their buying patterns.
- *Gaming-related data*: All the moves in the game are recorded. Some of these data include profile data. However, when millions of users submit their information, they become voluminous.

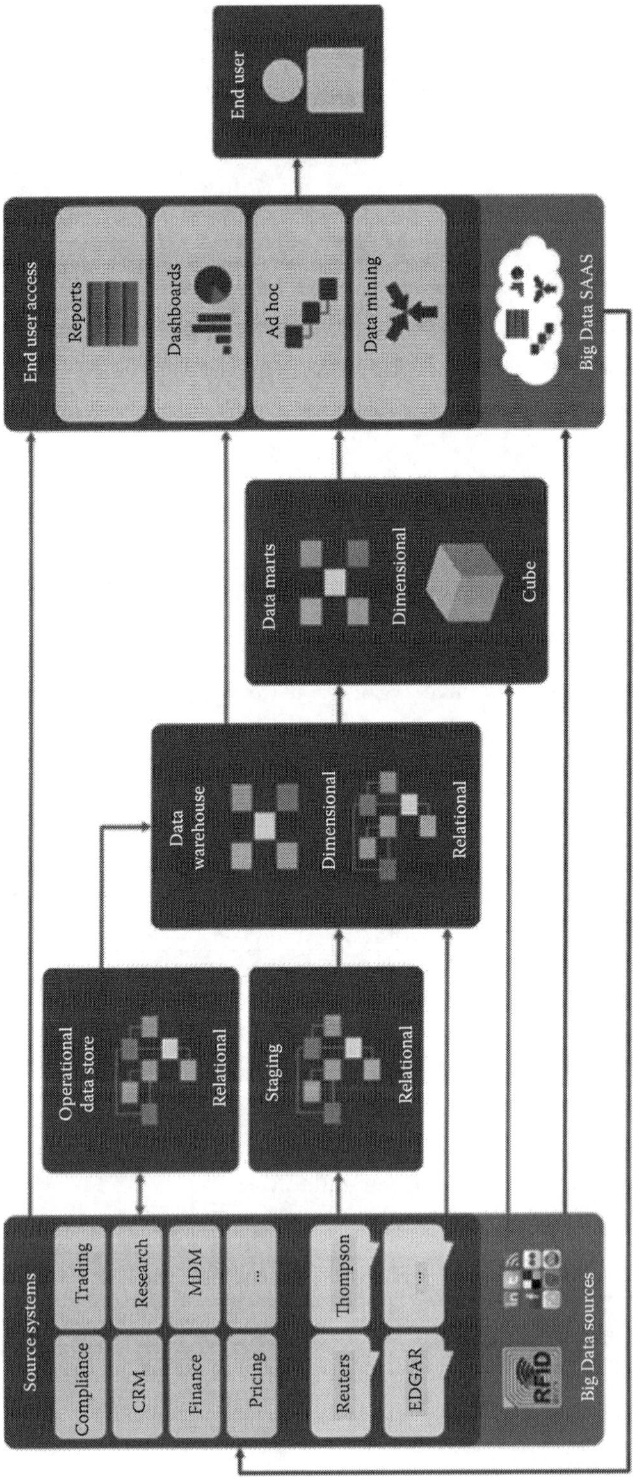

FIGURE 18.5
Big Data architecture.

18.7.2 Unstructured Data

Unstructured data do not follow any particular format. We mostly come across unstructured data. Unstructured data are also generated either by machine or by human.

The following are examples of machine-generated unstructured data:

- *Satellite images*: Examples included weather data or data captured by the government in its surveillance satellite, imagery, or Google Earth.
- *Scientific data*: This includes atmospheric data, high-energy physics, and seismic imagery.
- *Photographs and video*: This includes traffic analyses, videos, surveillance, and security systems.
- *Radar or sonar data*: This includes data such as vehicular, oceanographic, and seismic profiles and meteorological data.

Examples of human-generated unstructured data are as follows:

- *Internet text of a company*: This includes text in documents, logs, e-mails, and survey results.
- *Social metadata*: These include data generated from various social networking websites such as Twitter, Facebook, LinkedIn, Flickr, and Youtube.
- *Mobile data*: Data such as text or voice messages and location information.
- *Website content*: This includes data from any site delivering unstructured data content like Youtube, Instagram, and Flickr.

Semistructured data include data that are usually self-describing, which will have simple label/value pairs.

For label/value, pairs include

<family>=Johnson

<mother>=Christiana

<daughter>=Sarah

Examples of semistructured data include XML, EDI, and SWIFT.

18.8 Role of CMS in Big Data Management

Organizations generally store some unstructured data in databases. Conversely, they also use enterprise content management, which can manage complete life cycle of content.

Content management system (CMS): It is basically a collection of strategies, methods, and tools used to capture, manage, and deliver documents and content to organizational processes.

18.8.1 Technologies Come under Enterprise Content Management (ECM)

- Document management
- Record management

- Imaging
- Workflow management
- Web content management and collaboration

18.8.2 Putting Big Data Together

The following table shows the characteristics of Big Data and types of database management systems used:

Type of Data	Batch	Streaming	Complex Query
Structured data	Hadoop	Key/value	RDMS
Unstructured	Document	Graphy spatial	Columnar
Semistructured data	Hybrid	Hybrid	Hybrid

Integrating data types in Big Data environment: Sometimes, it becomes very important to integrate different sources. These data can be from all internal systems and external sources, and much of the data can be soiled before. You come across a scenario where you have to deal with huge volumes of data at a high velocity, which is disparate in nature. The important point is to make a business value out of these voluminous data that get highly difficult out of disconnected silos of information. We need a source to connect these different data.

18.8.3 Components Required for Connecting

- *Connectors*: The connector is used to pull data from various Big Data sources, for example, from Twitter or Facebook or from your data warehouse so as to analyze the data from these resources.
- *Metadata*: Metadata describe about other data. These are the definition, mappings, and other characteristics that describe how to find access and use an organization's data and software companies.

18.9 Need for Security

In this current age of information technology, we need to maintain information about all the activities in our lives. Information is as important as any other valuable assets especially when the organizations need to secure data. The organizations deal with lots and lots of data that need to be secured.

Data should be protected from unauthorized access (confidentiality), the data should be guarded from unauthorized changes (integrity), and the data should be available when needed (availability). The basic three requirements in Big Data have still not changed, that is, confidentiality, availability, and integrity [6].

Since the last couple of years, Big Data has created a revolution for the organizations by storing voluminous data irrespective whether it is in structured or unstructured format.

18.9.1 Confidentiality

Though confidentiality might be the common aspect of information security, we still need to protect our sensitive information. An organization needs to secure its data from malicious users that hazard the confidentiality of information.

18.9.2 Integrity

Most of the information must be changed constantly. For example, bank transactions when performed need to be updated frequently, which has to be done only by authorized entities. Integrity violation is not necessarily done by malicious acts even power usage may also create the unwanted changes in information.

18.9.3 Availability

The information maintained by an organization needs to be available to authorized entities on time, and if it is not available instantly, it is useless. The unavailability of data can be destructive to the availability and integrity to the organization.

18.9.4 Identification

It can be defined as the process of claiming yourself as you are somebody. You are just identifying yourself. In terms of information security, it is similar to that of giving a password. When we give a password, we are just identifying ourselves [6].

18.9.5 Authentication

Authentication can be defined as a process of identifying the identity of a user or the host being used. It is just a process of proving who you are. For example, when you state that you are Laura, by logging into a system as Laura, it next asks your password. When the password is entered, you have proved that you are Laura:

Identification part: Giving a username

Authentication: Giving a password

The following are the strategies used to identify a user or a client:

- *User ID and password*. It is the most widespread, and typically the easiest, method to recognize someone because it is fully a software-based technique.
- *Physical security device*. This may include a bank card, a smart card, or a computer chip used to identify a person. Sometimes, it can also be a password or a personal identification number (PIN) that may also be used to ensure that it is the right person.
- *Biometric identification*. Biometrics is the process of identifying someone by their physical characteristics. This may include technologies such as a thumb prints, voice verification, retinal scan, or palm identification.

18.9.6 Access Control

General access control usually includesan authorization, an authentication, an access approval, and an audit. A narrow type of access control may include a decision whether to grant or reject an access request from an already authenticated subject.

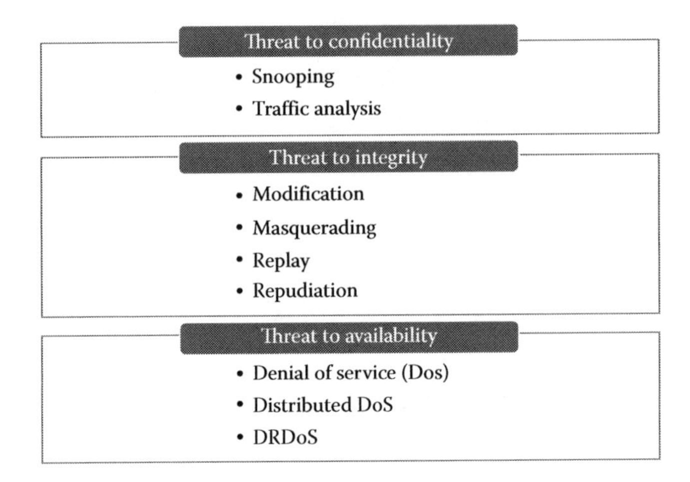

FIGURE 18.6
Classification of attacks.

Authentication and access control are often clubbed into a single operation, so that an access can be approved based on successful authentication, or it may based on an unidentified access token.

A granular access controlis very important to achieve this and has identified three specific problems in the realm of cloud data access control:

1. Tracing privacy requirements for individual elements
2. Monitoring authorities and roles for the sake of users
3. Correctly implementing secrecy requirements with mandatory access controls

Solution: To tackle these problems, organizations have to implement the access controls in their infrastructure layer, need to simplify the complexity in the application space, and adopt the authentication and mandatory access control [7] (Figure 18.6).

18.10 Types of Attacks to Breach the Security

18.10.1 Threat to Confidentiality

- *Snooping*: It generally refers to unauthorized user accessing, intercepting, and interrupting data. For example, a file containing sensitive data transferred through the Internet not only seized or intercepted by malicious user but also used the contents for his or her personal benefit. Snooping can be prevented by enciphering techniques.
- *Traffic analysis*: Even though encipherment of data may hide the data from interception by attackers, still they can obtain some other important information by monitoring online traffic. For example, with the e-mail address of the sender or receiver, he or she may collect related information from a pair of requests and responses in order to guess the nature of transactions.

18.10.2 Threat to Integrity

The integrity of Big Data can be intimidated by several kinds of attacks such as modification, masquerading, replaying, and repudiation.

- *Modification*: The addressed information may be modified by the attackers and can be used for his or her personal benefit. Sometimes, the attacker may also delete or delay the messages to harm the system.
- *Masquerading*: This type of threat usually occurs when a malicious user may pretend himself or herself to be someone else, for example, stealing an ATM card and PIN of a bank customer and pretending himself to be a bank customer. Sometimes, the malicious user may also pretend to be the receiving entity, for example, the user tries to contact the bank, but on the other hand, the attacker may act as a bank to obtain sensitive information.
- *Replaying*: It is another type of attack where an attacker makes a copy of the message send by the user and tries to replay it for his or her own benefit.
- *Repudiation*: This attack is different from the other attacks since it is performed by one of the two parties involved in the transaction either by the sender or by the receiver; the sender or receiver may deny that he or she has not sent or received the transaction.

18.10.3 Threats to Availability

1. *Denial of service*: It is the most common attack that slows down or may interrupt the service of a system. The attacker may apply various strategies to achieve this; he or she may send many bogus requests to the server, so as the server may crash because of heavy load. The attacker may also interrupt and delete a response from the server to the client making him or her believe that the server is not responding. The attacker may also seize the requests made by the client, resulting in the client sending too many requests, which may overload the system.
2. *Distributed denial-of-service (DDoS) attack*: A DDoS attack is an attack coming from multiple sources at the same time.
    a. *DDoS attack case study*: To evaluate a DDoS attack in the operational background traffic, we have injected a proportion of SYN flooding traffic. Consider 200 attackers are sending Transmission Control Protocol (TCP) synchronous (SYN) packets targeting a single host with TCP port inside the National Technical University of Athens (NTUA) campus. Each attacker sends at least 20 flows, which may consist of one to four packets per time window. Fifteen percent of the overall traffic corresponds to the background traffic. This type of anomaly causes a noticeable decrease in destination port and destination IP address entropy. This is because the victim's IP address and source port (here TCP port 80) have occurred in a time window. On the other hand, the entropy values for source port and source IP address do not show major change, because the DDoS attack does not cause much change in the corresponding distributions. This is because there distribution already contains enough randomness in usual network conditions and the extra traffic from 200 source IP address, which uses random source ports, does not mostly change these distributions. The entropy in flow size distribution does not change as the attack flows, which consists of one to four packets, which will not change the large distribution of flow size [8].

b. *Distributed reflection denial of service (DRDoS)*: DRDoS is an operating system that belongs to the DoS family, which is written for IBM usually for compatible personal computers, and is usually a type of attempt to disturb or disrupt a computer network.

18.11 Security Mechanisms

International Telecommunication Union (ITU-T) (X.800) has also recommended some security mechanisms to provide security services (Figure 18.7).

18.11.1 Encipherment

It is the process of hiding or covering the data to provide confidentiality. The techniques used for enciphering include

- Cryptography
- Steganography

18.11.2 Data Integrity

It is the process of appending a short check value to the data created by a specific process from the data themselves. The receiver after receiving the data checks the value; he or she then creates a new check value and compares it with the previous check value received. If both the values are the same, it shows that the integrity of the data is preserved.

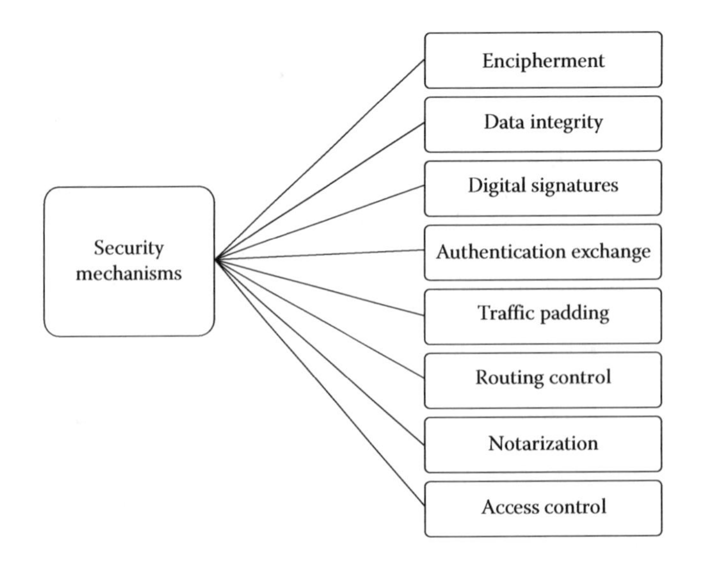

FIGURE 18.7
Taxonomy of security mechanisms.

18.11.3 Digital Signature

When a person signs a document, it specifies that the document has been originated from him or her. The signature is generally a proof to the recipient that the document has came from the correct entity. A signature on a document is a sign of authentication whether the document is authentic or not.

Alice sends a message to Bob. Now Bob needs to check the authenticity of Alice, which can be done electronically. A digital signature can be used to prove the authenticity of Alice as the sender of the message. This type of signature is called "digital signature."

18.11.4 Authentication Exchange

It is a process where two routers exchange some messages so as to identify or prove each other by using some security that he or she may only know.

18.11.5 Traffic Padding

It is the process of adding some bogus data into the original data traffic so as to thwart the adversary attempts.

18.11.6 Routing Control

It is the process of choosing routes and changing them frequently so as to prevent the opponent from eavesdropping on a particular router.

18.11.7 Notarization

It is a process of selecting a third party to control the communication between two ends. It is done to avoid repudiation.

18.11.8 Access Control

It is the use of different passwords and PINs so as to prove whether the user has the right to access the data and the resources of a system.

18.12 Two Most Important Securities Which Are Prevalent Today

- General cryptography (public key cryptography)
- Specific steganography (private key cryptography)

Cryptography: Cryptography is a term derived from the Greek words *kryptós* and *graphein* that means "secret writing." The term "cryptography" is an art of transforming messages to make them secure and immune to attacks. In the past, cryptography was referred only for encryption and decryption of messages using secret keys; today, it is used for three different mechanisms:

1. Symmetric key encipherment
2. Asymmetric key encipherment
3. Hashing

18.12.1 Symmetric Key Encipherment

It is also called as secret key cryptography or secret key encipherment. Symmetric key encipherment uses a single secret key to perform both encryption and decryption. Encryption/decryption can be assumed as an electronic locking. For example, Alice puts a message in a box and locks it, and Bob unlocks it with the same key to take out the message.

Symmetric key encipherment is of two types:

1. *Stream cipher*: Here, the encryption of a message is done one bit at a time.
2. *Block cipher*: It takes a block of bits and encrypts them as a single unit. A block of 64 bits is taken at a time to encrypt.

18.12.2 Asymmetric Key Encipherment

It is also called as public key cryptography. Here, we use two different keys, a public key and a private key. For example, Alice uses its own private key and Bob's public key to encrypt the message, and Bob decrypt the message by using his own private key and Alice's public key.

18.12.3 Hashing

Here, we create a fixed length message digest out of a variable length message. Message digest is quiet small in length compared to the message. The message digest and message are sent together to the receiver, Bob, from Alice. Hashing is generally used to provide check values similar to that of data integrity.

- *MD5*: MD5 message digest algorithm is the most widely used cryptographic hash function, which produces a 128 bit (16 byte) hash value. It is represented as a 32-digit hexadecimal number expressed in a text format. MD5 is most widely used in variety of cryptographic applications and is also used to verify data integrity. MD5 has been designed by Ron Rivest as a replacement to MD4 hash function.
- *MD6*: MD6 message digest algorithm is also a cryptographic hash function that uses a Merkle treelike structure to allow immense parallel computation of hashes for very long inputs.

18.12.4 RSA Algorithm

RSA stands for Ron Rivest, Adi Shamir, and Leonard Adleman. It is used to encrypt and decrypt messages; RSA algorithm is an asymmetric cryptographic algorithm to provide a data transmission. It is also called as public key cryptosystem. Asymmetric algorithm consists of two different keys, a public key and a private key. Public key is known by everyone as used for encrypting messages. The encrypted messages can be decrypted with the private key only.

Key generation	
Select p, q	p, q both prime, $p \neq q$
Calculate $n = p \times q$	
Calculate $\phi(n) = (p-1) \times (q-1)$	
Select integer e	$\gcd(\phi(n), e) = 1;\ 1 < e < \phi(n)$
Calculate d	
Public key	$KU = \{e, n\}$
Private key	$KR = \{d. n\}$

Encryption	
Plaintext:	$M < n$
Ciphertext:	$C = M^e \pmod{n}$

Decryption	
Ciphertext:	C
Plaintext:	$M = C^d \pmod{n}$

18.12.5 Authentication Protocols

Authentication is the fundamental aspect of every system security. Authentication is the process of identifying the identity of a user trying to log in to a domain or access network resources. Authentication protocol is a cryptographic protocol used to authenticate entities wishing to communicate securely.

18.12.6 Different Types of Authentication Protocol

- Kerberos
- Host identity protocol
- Password authentication protocol

18.13 Intrusion Detection Systems

18.13.1 Intrusion

It can be defined as an intentional or a planned and an unauthorized attempt irrespective whether it is successful or not to modify or to exploit expensive information or property, which may result in making the property or information null and void so it cannot be further used.

The person attempting this act is called as an "intruder"; it is generally performed to compromise the security goals. Intrusion can be performed either by an insider or by an outsider who goes beyond his or her boundaries to perform such an act. Intrusion may be performed for the following reasons:

- Malwares, for example, worms, viruses, trojans, and spyware.
- An attacker accesses the system unauthorizedly through the Internet.
- Authorized users misusing their privileges to gain additional privileges.

Aurobindo Sundaram [9] has classified intrusions into six types:

- *Attempted break-ins*. These attempts are usually detected by a typical behavior profile, or it may be due to violations of security constraints. This type of intrusion is generally detected by anomaly-based intrusion detection system (IDS).
- *Penetrations of the security control system*. These are generally detected by monitoring or observing the patterns of activities taken place.
- *Leakage*. These are detected by the way the system resources are being used.
- *Malicious programs use*. These can be detected if there is any change in a behavioral profile, or any violation of security constraints, or any misuse of special privileges.
- *Masquerade attacks*. These are also are detected by a typical behavior profile or any similar violations of security constraints.
- The aforementioned given intrusions can also be detected using anomaly-based IDS.
- *DoS*. This can be detected by a typical use of system resources.

18.13.2 Process of IDS

It is the most general kind of security management system applied and is used for computers and networks to monitor the events taking place among them and to analyze any possible signs of incidents or imminent threats violating the security policies [10,11].

IDS can be a hardware system or a software, a combination of both to detect any kind of intrusions occurring in your network, or a host based on number of telltale signs. It gives an alarm if it detects any intrusion.

18.13.3 Goals of IDS

1. Prime Goal of an IDS: To Detect All Kinds of Intrusions
   a. Any previously known or any unknown attacks
   b. Suggestions whether to learn or to adopt new technologies for new attacks or any changes in behavior
2. Detect Intrusions in Timely Fashion

   An IDS needs to be real time to give an instant response when a system needs to respond to an intrusion detected.
3. Problem

   It takes time for a system to analyze the commands to respond and may suffice to give a report of the intrusion that occurred few minutes or hours ago.

4. Time Spent for Verifying the Attacks Should be Minimized
5. Reducing False Positives and also the False Negatives
   a. *False positive*: When an activity is incorrectly identified by an IDS as a malicious activity
   b. *False negative*: When an activity or an event occurred but an IDS fails to identify the intrusion that took place.

18.13.4 IDS Architecture

It includes the following components.

18.13.4.1 Sensor (Agent) Collecting Data and Forwarding the Information to the Analyzer

- Log files
- Network packets
- system call traces

18.13.4.2 Analyzer (Detector)

It determines whether an intrusion has occurred or not. It receives an input from one or more sensors and analyzes them.

18.13.4.3 User Interface

It is generally used to view a user's output from his or her system or to control the behavior of the system (Figure 18.8).

18.13.5 Types of IDS Technologies

Five types of IDS technologies

- Host-based IDS (HIDS)
- Network-based IDS (NIDS)

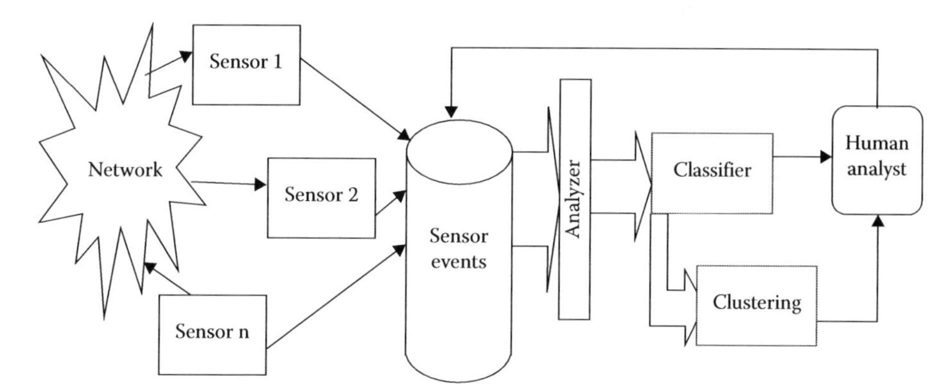

FIGURE 18.8
Intrusion detection system architecture.

- Network behavior analysis–based IDS
- Wireless-based IDS (WIDS)
- Hybrid IDS (HybIDS)

18.13.5.1 Host-Based IDS

An HIDS determines the unauthorized activities based on the following given criteria:

- All the applications initiated on a system
- Log details such as date, time login, and location
- Adding/deleting/modifying any of the system entities

The collected information is then compiled and compared to the signatures stored in the database using special algorithms.

Application level: The high-level IDSs used here are similar to the HIDSs in operating system level; the core distinction is the HIDS is more concentrated on application-level log files than on system-level log files. This is the most ideal solution to detect intrusion attempts on a database application existing on a host machine.

Network level: A network-level IDS works only on the packets addressed to a particular host; it does not accept any packets that are not addressed.

18.13.5.2 Network-Based IDS

It monitors the network segments or devices of a network and analyzes the application protocol to identify any suspicious activity that may occur such as

- Virtual private network
- Wireless networks
- Remote access servers
- Proximity to firewalls on router

18.13.5.3 Network Behavior Analysis–Based IDS

It is responsible to examine all the network traffic to identify any threat that may generate unusual traffic flows such as

- DDoS attacks
- Certain forms of malware (e.g., backdoor worms)
- Policy violations (e.g., a client system providing network services to other systems)

Network behavior analysis systems are usually deployed to monitor the Internet network flows in an organization, and sometimes, it may also be deployed to monitor the organizational internal and external flows.

18.13.5.4 Wireless-Based IDS

It is responsible for monitoring wireless network traffic and analyzing its wireless networking protocols to identify any suspicious activity including the protocols also. WIDS cannot identify any suspicious activity in application or any higher-layer protocols such as the TCP or UDP wireless networking transferring.

18.13.5.5 Hybrid IDS

By combining multiple techniques into a single hybrid system, however, it is possible to create an IDS that possesses the benefits of multiple approaches, while overcoming many of the drawbacks. HybIDS is used to detect and prevent the intruders on information system. HybIDS offers to the management alert notification from both network- and host-based intrusion detection devices. Hybrid solutions provide the logical complement to NIDS and HIDS—central intrusion detection management [12,13].

18.13.6 Intrusion Detection Methodologies

- Signature-based detection (rule-based/misuse detection/pattern detection)
- Anomaly-based detection
- Stateful protocol analysis

18.13.6.1 Signature-Based Detection (Rule-Based/Misuse Detection/Pattern Detection)

Signature-based technique is the most commonly used technique and the still most widely used one that focuses on the identification of known bad patterns. Signature-based detection is generally a process used to compare signatures to identify any possible incident, but it is largely ineffective in detecting previously unknown threats.

For example, "freepics2.exe," a signature looking for "freepics.exe," would not match it. Similar to other approaches, even signature-based technique is vulnerable to attacks in which the signature is not known. For example; encoding, packing, zero day exploits, or obfuscation technique.

18.13.6.2 Anomaly-Based Detection

It is generally used to identify significant deviations by comparing definitions of activities that appear abnormal. Anomaly-based detection is very effective in detecting previous unknown threats, for example, a computer effected with a new malware.

Malware: It's a threat that consumes most of the computer processing resources; it sends large number of e-mails and initiates a number of network connections. It also performs a typical behavior that significantly differs from other established profiles for the computer.

18.13.6.3 Stateful Protocol Analysis

It can be defined as a process in which predefined profiles of accepted definitions of a benign protocol are compared against observed events to identify any deviations that may generate. The stateful protocol analysis defines that IDS is capable to track and understand the state of network, transport, and application protocols.

18.13.7 Intrusion Detection Tools

These are the best tools to detect and prevent the Big Data from intruders.

S. No.	Tool Name	Manufacturer	Tool Description
1	Packetpig	Hortonworks	Packetpig is a network security monitoring toolset where the *Big Data* is full of packet captures. Packetpig's analysis of full packet captures focuses on providing as much context as possible to the analyst [14].
2	Open Source Security (OSSEC)	Trend Micro	OSSEC Host Intrusion Detection for Big Data. OSSEC is an open source HIDS that performs log analysis, file integrity checking, policy monitoring, rootkit detection, real-time alerting, and active response. It runs on most operating systems, including Windows, Linux, Solaris, MacOS, HP-UX, and AIX [15].
3	Snort	Snort	Freeware lightweight network IDS, capable of performing real-time traffic analysis and packet logging on IP networks [16].
4	AnaDisk	Sydex	Search, analyze, and copy almost any kind of diskette without regard to the type or format. Edit diskette data sector by sector or perform a diagnostic read of a specified diskette track. Dump data from a selected range of tracks into a DOS file so that you can examine and manipulate data from non-DOS diskettes [17].
5	BlackICE Defender	NetworkICE	BlackICE Defender is designed to run on every PC to detect and protect the most valuable asset—information. BlackICE silently monitors communications between the computer and the network. When suspicious activity occurs, BlackICE immediately springs into action defending the computer, data, and business [18].
6	Check Point RealSecure	Check Point Software Technologies	Unobtrusively analyzes packets of information as they travel across your enterprise network. It recognizes a wide variety of traffic patterns that indicate hostile activity or misuse of network resources, including network attacks and malicious Java™ and ActiveX™ applets. The RealSecure attack recognition engine immediately alerts network managers and administrators of any suspicious activity, logs the session, and can automatically terminate the connection [19].
7	Cisco Secure IDS	Cicso Systems	IDS designed to detect, report, and terminate unauthorized activity throughout a network [20].
8	Computer Misuse Detection System (CMDS)	Intrusion.com	Automatically collects and analyzes data from your devices recognizing over 4600 different alerts and events. CMDS Enterprise's Analysis Engine combines a powerful expert system and statistical profiling engine that can process gigabytes of event log data per day [21].
9	CyberCop Scanner	Pretty Good Privacy	System-based IDS has the ability to detect network reconnaissance stealth port scanning over many months, warning against even the most determined attacks. CyberCop monitor's unique system-based intrusion detection architecture provides both real-time packet analysis and system event analysis [22].

(Continued)

S. No.	Tool Name	Manufacturer	Tool Description
10	iD2 Secure Transport	iD2 Technologies	Allows an organization to monitor user activity on a local network. With the software installed on a customer or employee PC, the user can be identified to the corporate network as they log on and access applications and files. iD2 Secure Transport uses the standard client authentication procedure in the SSL protocol [23].
11	OpenWIPS-ng	Openwips-ng.org	OpenWIPS-ng is a free wireless intrusion detection system/intrusion prevention system (IDS/IPS) that relies on servers, sensors, and interfaces. It runs on commodity hardware. OpenWIPS-ng is modular and allows an administrator to download plug-ins for additional features [24].
12	Security Onion	Security Onion Solutions	Security Onion is an Ubuntu-based Linux distribution for network monitoring and intrusion detection. The image can be distributed as sensors within the network to monitor multiple virtual local area networks and subnets and works well in VMware and virtual environments [25].

18.14 Conclusion

This chapter gives a mild introduction to Big Data and the technology components and also gives you a clear understanding of different types of Big Data with suitable examples in a CPS. In order to achieve secure CPS, we need to take security as the initial step at the design process. We also need to ensure the specific information flow and also the requirements that are met through all the phases of design system rather than just attempting to meet those by just using add-on security mechanisms.

It further explains the different types of data used in a Big Data. In the next part, we focus on the security of the Big Data, which has became the basic building block of any organization; here, we discussed the importance of security in Big Data and different types of security threats and attacks; we also discussed the taxonomy of security mechanisms that can be implemented to avoid any attacks and also the encryption and the decryption algorithms. We further discussed on IDS and the types, process, and goals of IDS architecture. Finally, we give a brief overview of intrusion detection methodologies and the different intrusion detection tools.

References

1. J. Hurwitz, A. Nugent, F. Halper, and M. Kaufman, *Big Data for Dummies*, Wiley Publisher, http://doc.konversation.com/eflyer_cms/blog/Big%20Data%20For%20Dummies%20by%20Judith%20Hurwitz%20et%20al%20%28.PDF%29.pdf. Last accessed on November 25, 2014.
2. Jay, L., Behrad, B., and Hung-An, K. Recent advances and trends of cyber-physical systems and big data analytics in industrial informatics, IEEE International Conference on Industrial Informatics (INDIN), July 27–30, 2014, Brazil, 2014.
3. http://searchcloudcomputing.techtarget.com/definition/big-data-Big-Data. Last accessed on October 25, 2014.

4. http://www-03.ibm.com/security/solution/intelligence-big-data/. Last accessed on October 25, 2014.
5. http://www.agiledata.org/essays/accessControl.html. Last accessed on October 29, 2014.
6. W. Stallings, *Cryptography and Network Security*, Pearson Education, http://faculty.mu.edu.sa/public/uploads/1360993259.0858Cryptography%20and%20Network%20Security%20Principles%20and%20Practice,%205th%20Edition.pdf. Last accessed on November 25, 2014.
7. B. Forouzen, *Cryptography and Network Security*, TMH, http://highered.mheducation.com/sites/0072870222/index.html. Last accessed on November 25, 2014.
8. U. Kaur and E. Harleen Kaur, *Routing Techniques for Opportunistic Networks and Security Issues*, Baba Farid College, Punjab, India.
9. A. Sundaram, An introduction to intrusion detection, *ACM Crossroads: Student Magazine*, Vol. 3, pp. 3–7, 1996. www.acm.org/crossroads/xrds2-4/intrus.html. Last accessed on November 25, 2014.
10. http://www.symantec.com/connect/articles/evolution-intrusion-detection-systems. Last accessed on November 25, 2014.
11. http://www.ciscopress.com/articles/article.asp?p=25334&seqNum=4. Last accessed on Novembe 25, 2014.
12. http://hortonworks.com/blog/big-data-security-part-two-introduction-to-packetpig/. Last accessed on November 25, 2014.
13. http://www.ossec.net/. Last accessed on November 25, 2014.
14. https://www.snort.org/. Last accessed on November 25, 2014.
15. http://www.mpm-kc85.de/html/AnaDisk.htm. Last accessed on November 25, 2014.
16. http://www.pcmag.com/article2/0,2817,2220,00.asp. Last accessed on November 25, 2014.
17. http://www.checkpoint.com/press/2000/realsecure022900.html. Last accessed on November 25, 2014.
18. http://www.cisco.com/c/en/us/td/docs/ios/12_2/security/configuration/guide/fsecur_c/scfids.html. Last accessed on November 25, 2014.
19. http://www.intrusion.com/. Last accessed on November 25, 2014.
20. http://www.iss.net/security_center/reference/vuln/CyberCop_Scanner.htm. Last accessed on November 25, 2014.
21. http://id2-secure-transport.software.informer.com/. Last accessed on November 25, 2014.
22. http://openwips-ng.org/. Last accessed on November 25, 2014.
23. http://blog.securityonion.net/p/securityonion.html. Last accessed on November 25, 2014.
24. http://www.ncsa.illinois.edu/People/mcgrath/docs/GrandChallengesforCyber.pdf. Last accessed on November 25, 2014.
25. http://ercim-news.ercim.eu/en97/special/cyber-physical-systems-theoretical-and-practical-challenges. Last accessed on November 25, 2014.

19

Security in Cyber-Physical Systems: Generalized Framework Based on Trust with Privacy

G. Jagadamba and B. Sathish Babu

CONTENTS

19.1 Introduction

In the field of information and computing, there were massive revolutions to network, every body to every device with everything through the Internet. The connection between physical and cyber systems made us to think about the interdisciplinary attachment to reflect as one mastermind. On the fly, a solution may be a "cyber-physical system" (CPS) coined by Helen Gill in 2006 at NSF (National Science Foundation) in the United States (Northcut, 2013). According to Helen Gill, "Cyber-physical systems are engineering, physical and biological systems operated where integrated, monitored, and controlled by a computational core." Components are networked at every scale, and computing is deeply embedded into every physical element, possibly even into materials. In 2007, CPS was considered as the No. 1 research priority by the U.S. President's Council of Advisors on Science and Technology. Various researches (Stewart, 2008; Chang et al., 2013; Karnouskos, 2014) expressed CPS to be the next core research issue after Ubiquitous system. Researchers continued to redefine the way and to recognize and work together with the physical world using cyber systems.

In an engineering discipline, CPS focused the technology on a firm foundation in both mathematical and computation abstractions. The large-scale distributed computing systems now decoupled from the physical and cyber environment with artificial intelligence (Kanchana Devi and Ganesan, 2013). The technical challenge is to be bounded by concepts for modelling physical processes (differential equations, stochastic processes, etc.) and systems. The traditional aspects of securing physical systems have been put beyond reach, and were achieved using commercially available customized mobile devices. The abstractions should be able to explain how a CPS interface handle, monitor, and regulate the physical substratum (Haque et al., 2014).

19.2 Survey on the Essential Desires for CPS Security

CPS was considered to be a combination of physical and cyber systems, integrated with all states and levels in a compact way. Physical systems can be natural or human-made systems governed by security laws of physical governance operated continuously over time. Cyber systems are those that compute, communicate, and control all physical systems that are discrete, logical, and switched.

From 2006, CPS has changed over time depending upon the physical, simulation, and computational time. Several properties, features, and limitations are identified regarding physical and scientific research. In addition to these, several fundamental and security properties are also identified (Broman et al., 2013). The essential features include functionality, performance, cost-efficiency, and dependability and safety associated with reliability with respect to all conditions of the network and applications. The usability, management, and adaptability are the other properties that are affecting the fundamental and security properties in CPS. In this section, we explored various aspects of CPS regarding features, requirements, and limitations in detail.

19.2.1 Features of CPS

Researchers have mentioned the various features of the CPSs and some are the following (Campbell et al.; Lee, 2009; Zhang, 2012; Daly, 2013; Sveda and Rysavy, 2013):

- A federated approach to management and distributed control.
- Real-time performance requirements with gigantic control systems and Quality of Service (QoS) in dynamic operating conditions.
- Enormous multiscale, network-centric operations managed centrally.
- Uncertainty concerns about readings, status, and trust of the services.
- They are open source and open modular.
- The existence of the secured communication channels assured due to input and feedback from or to the physical environment.
- CPS work in a broad range of operating conditions, from distributed to centralized, simple to heterogeneous networks, and device capabilities to network service providers.

- They allow experimentation from both the developers' and general public perspective.
- CPS support tracking and reporting facilities for testing from both the developers' and general public view.
- Provisions for reliability and privacy in hand with trust parameters for modelling security.
- CPS accommodates a variety of utility entities like scientists, researchers, business subordinates, professors, and students of all ages to create and manipulate content.

19.2.2 CPS Requirements

CPSs should be dependable, secure, safe, and efficient to operate in real time as well as scalable toward design, cost-effective, and adaptive with the following requirements:

- Architecture (Lee, 2008; Tan et al., 2009; Tidwell et al., 2009)
- Middleware designing (Dabholkar and Gokhale, 2009)
- System control (Antsaklis, 2008; Ahmadi et al., 2010)
- System security (Cardenas et al., 2008; Tang and Mcmillin, 2008; Adam, 2009)
- QoS that has its importance in the design and operability (Xia et al., 2008; Kottenstette et al., 2009)
- Real-time data management (Kang and Son, 2009; Neuman, 2009; Farag, 2012; Sveda and Rysavy, 2013)

19.2.3 CPS Limitations

CPS is found to have some limitations as it is a tight coupling of physical and cyber systems. The limitations associated with either physical or cyber system results with CPS and are as follows:

- Lack of formal representations with respect to multiple views and designed tools that are not capable of expressing and integrating with various aspects
- Lack of robust formal models for multiple abstraction layers from physical process layer to different layers in the information processing hierarchy and their cross-layer analyses
- Lack of strategies to clearly segregate the safety-critical and nonsafety-critical functionalities, as well as for safe composition of their features during the loop operation
- Lack of ability to reason out and trade off between different situations and network
- An absence of physical security components in various distributed significant geographical locations
- Different types of computing platform and interconnection technology with different capacity of devices termed as heterogeneity characteristics that raise a security challenge in CPS operability

19.3 Challenges in CPS

Challenges are found in CPS in various fields, and we segregated them into scientific, research, and social areas.

19.3.1 Scientific Challenges

- Computation and abstraction challenges for the new real-time embedded systems are required with the new concepts and ideas for the development of models.
- Compositionality challenge for composition and interoperation of CPSs, compositional frameworks for functional and nonfunctional, temporal properties and robustness, safety, and security of CPSs
- Systems and network support challenge for CPS architecture and virtualization, wireless and smart sensor networks, predictable real-time and QoS guarantees at multiple scales
- New foundation test for control (distributed, multilevel in space and time) and hybrid systems cognition of the environment and system state and closing the loop
- Dealing with uncertainties and adaptability challenging graceful adaptation to applications, environments, and resource availability
- Scalability, reliability, robustness, and stability challenge of systems in CPS
- Science of certification problem for evidence-based certification, measures of verification, validation, and testing

19.3.2 Research Challenges

1. CPS composition challenge with respect to networking and distributed systems
   a. Safety and security of CPS when security solutions exploit the physical nature of CPS by leveraging location, time, and tag-based mechanisms.
   b. CPS are deployed in mission critical settings and collect sensitive data and can actuate changes in the physical process. This uncertainty in the environment can utilize properties from underlying physical process.
   c. Security attacks and errors in physical devices and cyber systems make the overall system insecure. So the primary aim of all CPS design is to ensure harmless process.
2. Control and hybrid systems
   a. A new calculus used to merge time-based systems with event-based systems for feedback controls is necessary.
   b. Calculus used to apply to hierarchies involving asynchronous dynamics at different time scales from months to microseconds and geographic scope from on-chip to planetary scale.
3. Computational abstractions
   a. Physical systems need safety in real time by computational ideas.
   b. Power constraints, resources, robustness, and security characteristics capture a compostable programming concept.

4. Architecture
    a. CPS based buildings need to concentrate more on interdisciplinary works when various physical information is captured.
    b. New network protocols designed for large-scale CPS.
    c. An innovative paradigm can build around the notion of being "globally virtual, locally physical."
    d. Big data problem needs to be processed efficiently in the architecture when all the data are collected from the sensor–actuator systems in CPS.
    e. Techniques in this need to ensure the results visualized easily by users.
5. Real-time embedded systems abstractions
    a. Protocols built for bandwidth allocation, new queuing strategies, and new routing schemes (including resource virtualization) can reduce and accommodate network delays.
    b. Systems must provide real-time resource allocation, data aggregation, global snapshots, in-network decision making, and the ability to provide QoS.
    c. Faults must be handled without any jitter with scalability.
    d. New distributed real-time computing and real-time group communication methods are in need.
6. Sensor and mobile networks
    a. The need for increased system autonomy in practice requires self-organizing (and reorganizing) mobile and ad hoc CPS networks.
    b. A precise knowledge creation from the raw data being collected is essential.
7. Model-based developm ent of CPS
    a. Models are required to generate and test software implementations of control logic.
    b. Abstractions are developed, modified, and integrated to cover the entire CPS design space.
    c. Concepts in communications, computing, and physical dynamics are required and modelled at different levels of scale, locality, and time granularity.
8. Verification, validation, and certification of CPS
    a. The gap between formal methods and testing needs a bridge for verification and validation.
    b. Compositional verification and testing methods that explore the heterogeneous nature of CPS models are essential.
    c. Once verification is completed, validation requires a type of certification regimes.
9. Education and training
    a. Scientists and engineers need proper training in the field of computation, control, networking, and software engineering while developing and testing for the real world.
    b. A necessity to include the CPS basics for all technical graduates is required.
    c. Creative trade-offs between depth and breadth are essential to be adopted by all the trainers.

10. Configuration standards for CPS
    a. An important area of research is setting standards for control systems and CPS in general concentrating on the requirement of security.
    b. More guidance is needed on how to deploy correctly and configure individual components of a CPS in particular environments, as behavior of any single component may vary based on the used environment.
11. Many challenging QoS constraints (Zhang, 2012)
    a. Challenge toward real-time requirements, such as low latency and bounded jitter required while maintaining the QoS constraints.
    b. When fault propagation or recovery across boundaries occurs, a concentration on availability of requirements is needed as such.
    c. The challenge of applying the trust based security, when unfavorable authentication and authorization are recorded.
    d. Challenge was pointing toward physical conditions, such as limited weight, power consumption, and memory footprint.

19.3.3 Social Challenges

Social problem raises a question on how CPS provides trust and coordination to people and society in their lives and in social networks. Another challenge to society is how we protect ourselves with the advent of various technologies resulting toward global warming:

- A few of them (Lee, 2009) identified a solution, based on the trustworthiness. Trustworthiness can overcome reliability, security, privacy preserving, and usability, when integrated with CPS.
- An associated challenge in social networking was the problem of coordination. When CPS is designed to work with interdisciplinary fields, the coordination becomes an issue with the functional purposes.
- Global warming and energy shortage are identified as significant challenges. Researchers are on the way to the solution defining it as green computing in the field of ubiquitous, pervasive, Internet of things, cloud computing, and so on.

All the concepts of the survey are summarized in Figure 19.1 (Broman et al., 2013), which represented each and every relative attribute of CPS mapping with computation, communication, control, safety, and so on.

19.4 Security Framework in CPS Based on Trust with Privacy

Altogether, cyber information extraction may find new types of attacks. Attacks can be threats, causing failures, errors, faults, or harm to the systems described here for the CPS:

- Harm is the "physical or logical problem or injury to the entity or property of the system or the environment" (IEC, 2008).
- Hazard is a "potential source of harm" (IEC, 2008).
- Failure is a "termination of the functional unit to provide a required function resulting in the operation other than the requirement" (IEC, 2008).

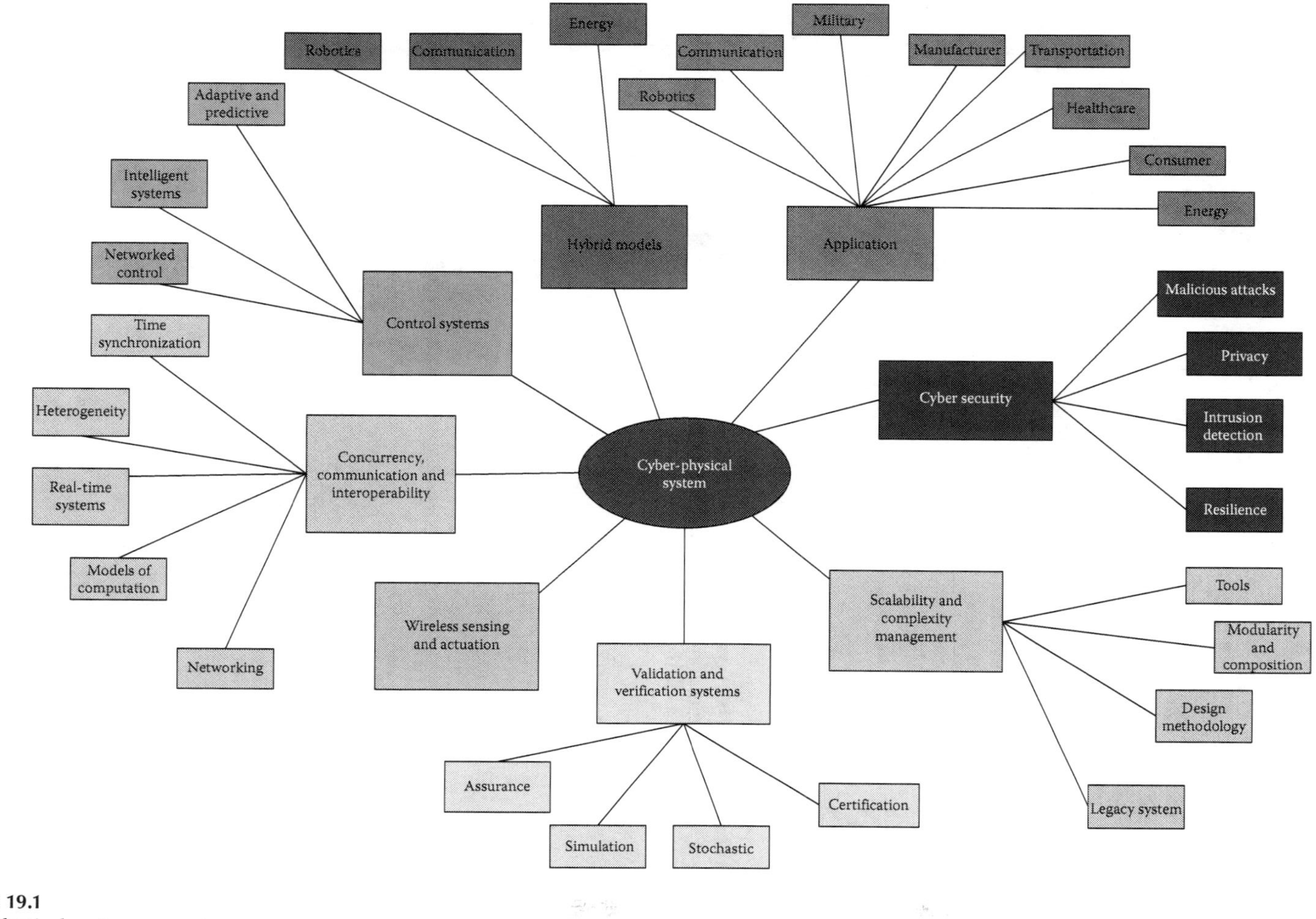

FIGURE 19.1
Cyber-physical systems mapping.

- Error is the "difference obtained with actual and computed or observed or measured value, specified or theoretically correct value or condition" (IEC, 2008).
- Fault is an "abnormal condition results in loss of function of required capability of a functional unit" (IEC, 2008).

A safety concern requires the extraction of physical and cyber information in order to provide security framework as they are coupled tightly with each other. The CPS security framework life cycle starts with the identification of vulnerabilities, analysis of threats and techniques to tackle the identified/analyzed attacks. The discussions are represented in Figure 19.2 that need to be followed in CPS for evolving vulnerabilities.

Security in CPS works like a life cycle, subsequently getting all hand in hand for the ongoing action generated by new attacks. If the CPS system is capable of handling the security life cycle, the vulnerability of the system can be defined to be safe and secure with the smallest amount of possible attacks.

Questions arise about providing the society and people who trust their lives on CPS other than traditional security methods. The thought is due to the unavailability of time and memory for computation in a resource-constraint environment. A trustworthy system can be a solution to this ("Cyber-Physical Systems," slides by Prof. Insup Lee, Upenn), which is reliable, safe, secure, and privacy preserving while communicating, controlling, and computing. Trustworthiness is the collected and transmitted data, considered for building an association with the two entities in CPS (Stelte and Rodosek, 2013). The application's primary bounds are derived from human-centric CPS between privacy and trustworthiness (Pham et al., 2010). An interest, to define an optimal trade-off between privacy and neighborhood rebuilding, requires trust only. In Tan et al. (2010), a trustworthy alarm was designed and consequently increased the feasibility of CPS coming from homogeneous source. However, the absence of data coming from heterogeneous source was a major drawback. The data from different bodies are extracted from physical and cyber systems. However, the trusted information becomes a primary concern for various mixed bodies. When we have a different heterogeneous resources and entities in the network, the extraction of information from trusted party is a great challenge and a primary requirement to operate the security system for CPS. The discussion in Li et al. (2011) provides information to design the trust-based security framework for CPS security. Hence, trust is becoming a fundamental issue to

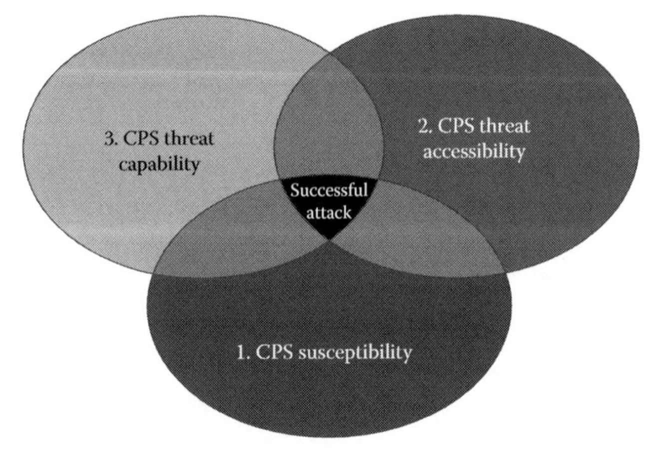

FIGURE 19.2
Life cycle for vulnerability issues in cyber-physical systems.

consider if CPS hardware components are critical carriers of attacks and back entry points to a system. Typical attacks can utilize "trusted" hardware parts of the system, such as USB ports, an Ethernet card, and battery. The logical parts can also host and execute malicious code that can bypass the operating system's guards (Karnouskos, 2014).

19.4.1 Security Requirements for CPS Framework Design

The following list of requirements are necessary while designing the security framework for CPS (Tan et al., 2010; Li et al., 2011; Hughes and Cybenko, 2013; Homeland Security News Land, 2014; Karnouskos, 2014):

1. New theories that incorporate adaptive nature and methods on design principles for resilient CPS, threat/hazard analysis, cyber-physical interdependence anatomy, and investigation/prediction at different layers of CPS
2. Formal models regarding roots of trust as security functions in a CPS device or system implicitly trusted and constitutes the foundation for integrity
3. New mathematical theories on information hiding for real-time streams
4. Lightweight security solutions that work well with extremely limited computational resources (e.g., devices CPU, memory, and battery capacity)
5. New algorithms/standards required for confidence and trust maps, context-dependent trust models, truth/falseness detection capabilities, and reasoning models while extracting private data

The discussions are directed toward roots of trust as they are the fundamental elements underpinning the dependability and trustworthiness of CPS and are more susceptible to cyber security breaches (Hughes and Cybenko, 2013). The security breaches are due to vulnerabilities in firmware and software that are distinct in nature in the design and operations. The research gap was identified in the area of standard terminologies defined and assessment of work in adjacent fields. The very reason behind this may be the diversity of CPS contexts and the absence of a multidisciplinary nature of the field. The work in coverage supply redundancy architecture (CSRA) project promoted the practice of identifying the interdisciplinary work for CPS security and privacy preservation. The work of Thoshitha et al., 2010, formulated the multidisciplinary functional purposes in a CPS. The security in CPS can be on the cyber, physical, or network components or altogether.

All the concepts of the study routed the research, toward trust-based CPS security without neglecting the privacy attribute.

19.4.2 Trust-Based CPS Security with Privacy

Within the context of CPS, security policies and mechanisms, access control–based methods, and information flow–based methods are identified. However, traditional security practices like cryptography fall short while addressing the multitude of safety considerations, not only for individual system but also at system-of-system level. Some of the recent tools, such as Stuxnet, Duqu, Flame, and the Mask in CPS, dominate with trust. The Stuxnet, Duqu, Flame, and the Mask tools in real-world attacks have entered the era of sophisticated cyber warfare. Addition of monitoring and controlling infrastructures like smart grid (Northcut, 2013) may be susceptible to cyberterrorism. From this situation, even small criminal-minded group can create attacks with asymmetrical impact in CPS (Koubaa and Andersson, 2009). Supervisory control and data acquisition/distributed control system (SCADA/DCS) and programmable logic controllers (PLC) systems were founded in high-tech industries to put security process in Europe, the United States, and Japan. However, evidence in the literature points out that the access control on cryptography, certification, or one-time password methods might not be the future direction of CPS security observable in certain aspects of the physical systems. Thus, the increasing critical infrastructures equipped with modern CPS and Internet-based services enhanced the functionalities and operation based on trust. As a result, the recommendations (Chang et al., 2013) were done for the adaptation of trust for security in CPS remarkably. As a result, trust is defined as the "agent belief with the agent's eagerness and capability to deliver mutually agreed to service in a given context."

19.4.3 Security Framework of CPS

CPS security triggered physical events with cyber, depending on the characteristics of the system with the activated service for devices under its control. When the physical or cyber systems are combined, there can be more threats and vulnerabilities, and it can be riskier than ever. However, the problem is to understand the threats and vulnerabilities and to act upon these when interacting with the outside world. We the users of the system have to be careful in utilizing the CPS system. At this point, a necessity to understand the type of risk is required to define a security framework in overcoming the situation. Cautions, regarding the safety considerations with respect to the end users of CPS, are listed as follows (Daly, 2013):

- Understand/assess risks.
- Define cyber security policies, standards, and procedures.
- Train and grow a security culture.
- Segment networks (defense in depth).
- Control physical, logical, and remote access to systems.
- Harden systems toward security, but not for operation.
- Monitor systems for intrusions and maintain security controls, and have incident response plans and methodology and contingency plans.
- To include Fuzz testing (is a simple teaching tool technique that can have a profound effect on your code quality) to deliver risk-based security testing and abuse case testing (it is to build and apply adversarial security tests based on the attacker's perspective) features toward an attempt to break the software under test.
- Try to indulge cyber security certification (e.g., ISA Secure, Achilles, and Common Criteria).
- Document application guidance and configuration identification.
- Secure the supply chain with security life cycle of CPS (refer to Figure 19.2).
- Manage vulnerability.
- Adopt/support standards.

This list provides a road map to follow when we are not able to accurately judge impacts of vulnerabilities or even realize that the CPS is at risk. Various solutions were proposed; among them, we found trust can be an efficient property to provide security that can preserve privacy. Trust is considered as an essential parameter for the security and a step proceeded toward the design of the framework.

The design of security framework of CPS has the capability to attempt several scientific, social, and economic issues. The Web of Things (WoT) framework for CPS was developed by Stewart (2008) and Chang et al. (2013), with the inclusion of WoT device, WoT kernel, WoT overlay, WoT context, and WoT application programmable interface (API) layers. A cyber-physical interface (e.g., sensors, actuators, cameras) interacts with the surrounding physical environment below the WoT framework. A security enhancement in CPS, based on trust management using policies toward deception of attacks, is found in Chang et al. (2013) and Kanchana Devi and Ganesan (2013).

Keeping trust management in mind, cyber-physical interface became an integral part of the CPS to produce a large amount of data. The frameworks allow the cyber world to observe, analyze, understand, and control the physical world using these data to perform the mission and time-critical tasks.

The framework in Figure 19.3 is based on CPS reference architecture in Chang et al. (2013), aimed to capture both domain requirements and infrastructure requirements at a high level of abstraction. The domain may be a different application to provide CPS services. However, the finding out of trusted domain competent participants to provide a service is a major concern. To overcome this, a capacity and trust computation is utilized for the selection of services in intelligent and automatic manner (Bo et al., 2011). A service selection alone is not enough in CPS, but a concern of security required in the perception layer.

Many authors mention trust as the primary parameter considered before the exchange of any resource or service between any two mutually agreed entities. The trust got generated from interactions and is utilized to decide whether the node is trusted or distrusted. When a central authority provides the information about the trusted or distrusted node, the associated job becomes easy. But the problem of extracting the trust arises when no central authority exists. If CPS is available, it can be utilized to integrate with any system with local controllers to get the required information. But when multiple features are added to CPS, more attacks can be expected while integrating with local control integrators.

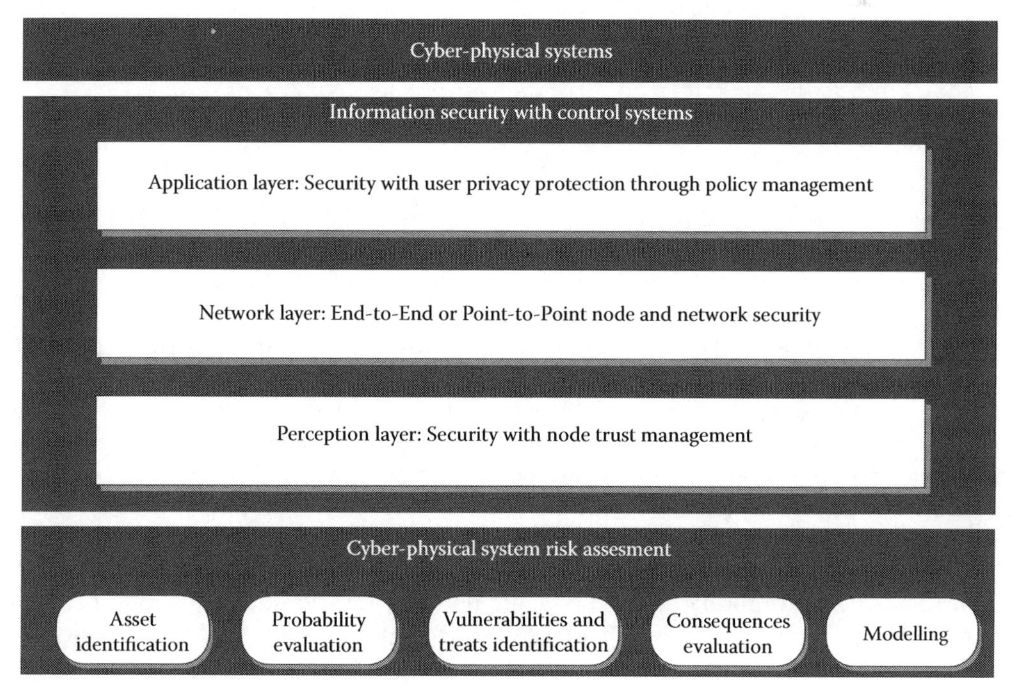

FIGURE 19.3
Security framework of cyber-physical systems.

Before incorporating, a cross-integrity check is essential for the verification of data integrity of the trust management. The checking of such data integrity is done by an algorithm, proposing a policy-based trust management (Kanchana Devi and Ganesan, 2013). The algorithm will check the integrity against deception attack via policies. The policies are nothing but the dos (correct time, message integrity code, etc.) and don'ts (mismatched time, mismatched message integrity code, etc.). The techniques adopted verify the integrity of the data in order to avoid falsification (deception attack) of information. An entity has to follow the rules to prove the integrity of the information with weight developed from Algorithm 19.1.

The algorithm was analyzed for the accuracy of the trust management proposal with energy consumption when compared with EigenTrust, PeerTrust, and Bio-inspired Trust (Kanchana Devi and Ganesan, 2013). The plan has considerably provided excellent accuracy with less energy consumption.

Algorithm 19.1

(To check whether the Node-n can be trusted or not by testing the data)

```
-----------------------------------------------------
Input: Data from Node-n
1. Start
2. W1, W2 are the observed Weight-age, Th is the threshold
W1 = V1 + V2 + V3 + ··· + Vn
W2 = U1 + U2 + U3 + ··· + Un
V = value given to each do's
U = value given for each don'ts
3. Calculate |W|= W1 + W2, Fix the |Th| value
4. If |W|>|Th|
5. Then "Node-n can be Trusted"
6. Else "The Node-n cannot be Trusted"
7. End
Output: Node-n can be Trusted or not
-----------------------------------------------------
```

A security is implemented in either the application or perception layer individually or together. Application layer security depends on the type of implementation we have chosen and the capacity and service associated in turn preserving the privacy. In the perception layer, security was determined with the association of many credentials of cryptography, trust-based modelling, and so on. Among these, a trust-based security is determined to be a proper one for the ubiquitous computing networks.

The application layer includes individuals to define policies. Systems set the policies from policy management unit for each trustworthy reporting device. The trustworthiness is evaluated based on both its reports and the contexts with which the reports are obtained in the trust management unit. As a result, the implementation of trustworthiness is evaluated for its performance. A result on the same experiment was found in Context Aware tRust Evaluation scheme for wireless networks in Cyber-Physical System (CARECPS) of trust management unit based on the simulation and real deployment experiments on android phones. Experimental results in CARECPS conducted with trusted and distrusted entities and the policies identified the malicious and trusted, measured on precision (P) and recall (R) parameters. The results of the P and R from Figures 19.4 and 19.5 (Li et al., 2011) proved an excellent performance scheme for the evaluation.

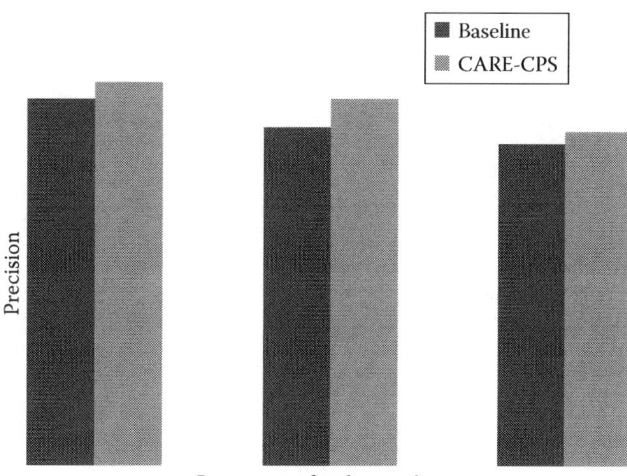

FIGURE 19.4
Precision for trust management. (From Li, W. et al., CARE-CPS: Context-aware trust evaluation for wireless networks in cyber-physical system using policies, *Proceedings of the IEEE International Symposium on Policies for Distributed Systems and Networks*, 2011.)

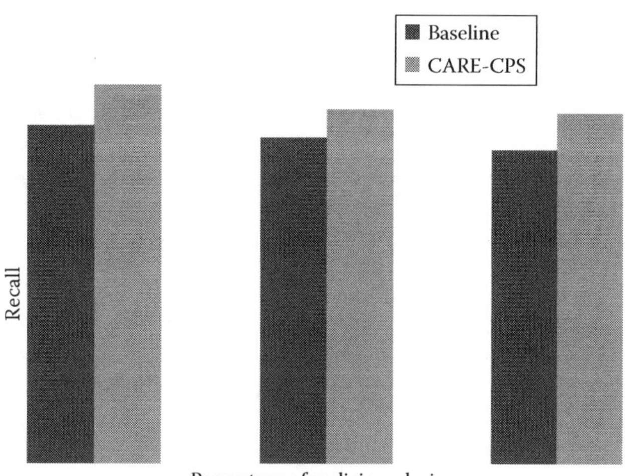

FIGURE 19.5
Recall for trust management.

19.5 Applications

We find a broad range of potential benefits of CPS addressing the societal-scale problems with a large variety of applications including in technology, society, and economy pictorially represented in Figure 19.6.

The CPS framework is used in many applications that provide a high-quality decision through integrating computational systems in the virtual world. Infrastructures in a

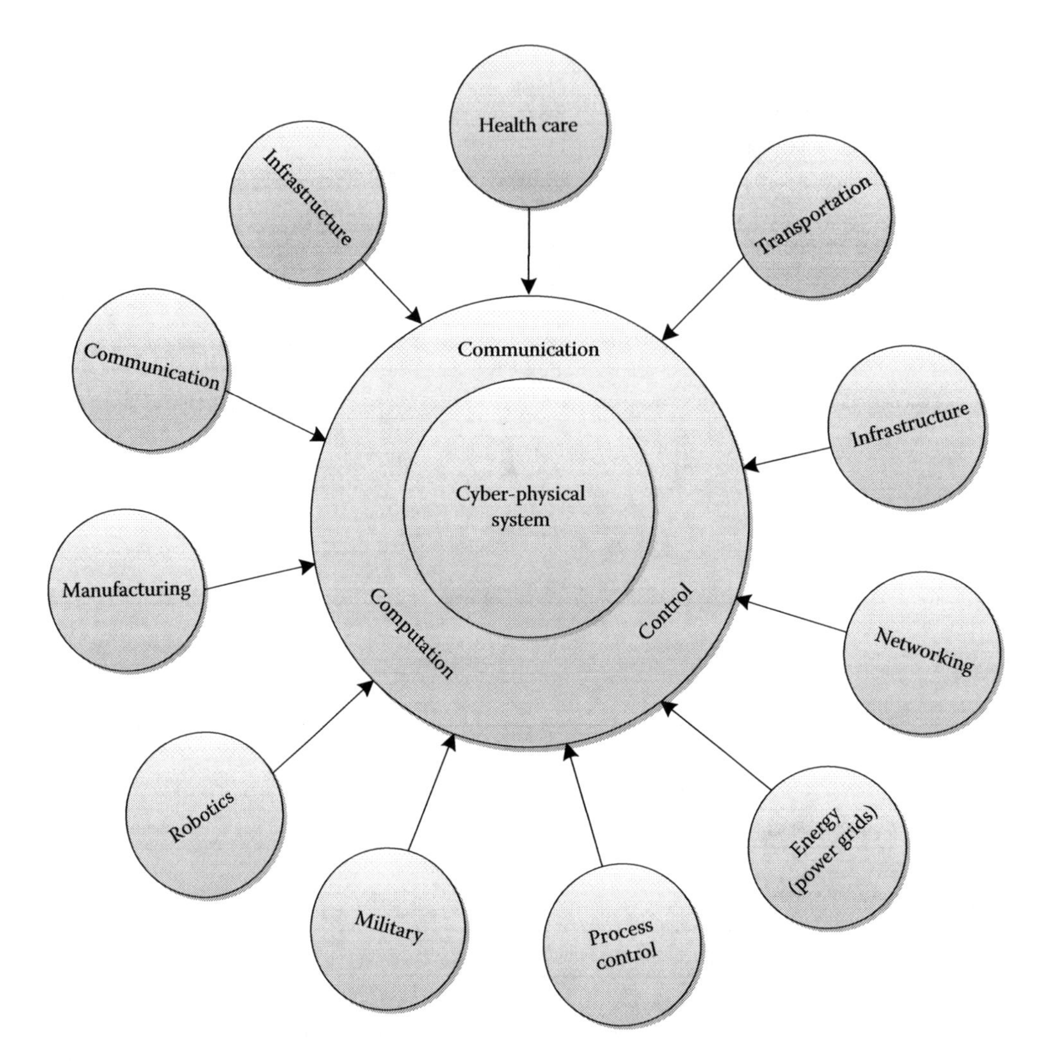

FIGURE 19.6
Applications with core domains of cyber-physical systems.

physical world also cope with the increasing complex demands from people. Scope of the applications are many and are as follows (Abdelzaher et al., 2009; Neuman, 2009; Farag, 2012; Sanislav and Miclea, 2012; Daly, 2013; Svéda and Ryšavý, 2013):

- Infrastructure control such as infrastructure management and monitoring control for the physical system. Electricity generation and distribution checks, building and environmental controls, etc.
- Safe and efficient transport for automotive electronics, vehicular networks, and smart highways. Aviation and airspace management, avionics, railroad systems for vehicle-to-vehicle communications for enhanced safety and convenience for "zero fatality" highways, drive by wire, and autonomous vehicles.
- Alternative energy in coordination with the generation, transmission, and management.

- Power grids can reduce the overall energy from small generators and solar panels (Krishna Venkatasubramanian). Automated monitoring and self-healing in smart grids can save energy by a network of integrated microgrids with smart appliances and demand management.
- Environmental control applications toward ubiquitous green technology are the primary concern of computing.
- Telepresence for video conferencing, telemedicine, telemodification for operations, etc., made the living more comfortable.
- Health-care applications (Sveda and Rysavy, 2013; Haque et al., 2014) for medical devices and integrated systems; health management networks for life-supporting microdevices embedded in the human body; mass customization of heterogeneous, configurable personalized medical devices; natural wearable sensors (clothing, jewelry, and so on).
- Assisted living in medical, military, manufacturing, agriculture, emergency management systems, smart home, and home-care monitoring field. Home-care surveillance and control supporting the living automatically. Some of the examples are pulse oximeters, blood glucose monitors, infusion pumps for insulin requirement, accelerometers for falling and immobility, and wearable networks (gait analysis). Smart kitchen is one of the examples for quickly assisting the home living, where refrigerator operates by collecting the information regarding the availability of foodstuffs, ordering the unavailable materials, etc.
- Social networking and gaming applications are adapted having CPS. But CPS is affecting the privacy of the user by collecting the behavioral characteristics. Part of social networking interlinked with gaming activity is not far away with the risk and private information leakage considered to be a significant threat in the CPS-based social networking applications.
- Automation in antilock braking system (Sveda and Rysavy, 2013), electronic stability control, fuel injection and emission control.
- Automotive systems for X by wire, adaptive cruise control, parking, lane departure warning, and engine management driven by software.
- Green technologies in ubiquitous and pervasive computing to overcome global warming exhibit wide-area characteristics for bulk power stability and quality, flow control, and fault isolation (Li et al., 2011).

Although similar objectives are intended toward the same focus and approaches, they have few similarities in views on security. Substantial differences are found in the transportation, automation, infrastructural, and medical device domains (Stelte and Rodosek, 2013) through a wireless network or other recommendation networks. More concentration is given toward information security in military security and medical domain.

19.6 Case Study of CPS Health-Care System

The benignly implantable devices for health-care records evolve from paper to electronic media (Sveda and Rysavy, 2013). Security issues are becoming critical in the ubiquitous health-care systems and need to concentrate on adaptive patient-specific algorithms (Abdelzaher et al., 2009; Gamage et al., 2010). Many of the medical devices designed for particular groups of patients who have similar medical conditions could dramatically improve health care by making machine operation adaptable to an individual patient's specific monitoring. Developing algorithms for medical devices that are certifiably safe for large classes of patients and change to individual patients or to different environments the patients may be living are required.

The CPS health-care architecture (Wang et al., 2011; Haque et al., 2014) is composed of four principal components, namely, (1) communication core, (2) cloud computation core, (3) resource scheduling and management core, and (4) security core. Even though security base has no traces in implementation and any of the outcome results, it covered all the fields of CPS. Figure 19.7 (Haque et al., 2014) provided a systematic CPS for health care with a well-defined functional procedure that incorporates the trust-based security by preserving the privacy of the information. The following steps were listed to give more information toward the trust-based security in CPS health care with privacy preservation:

1. Data collected from a patient using various sensors are verified for trustworthiness and sent to cloud storage via gateway.
2. Processed sensor data are sent to a cloud server for processing queries in real time.

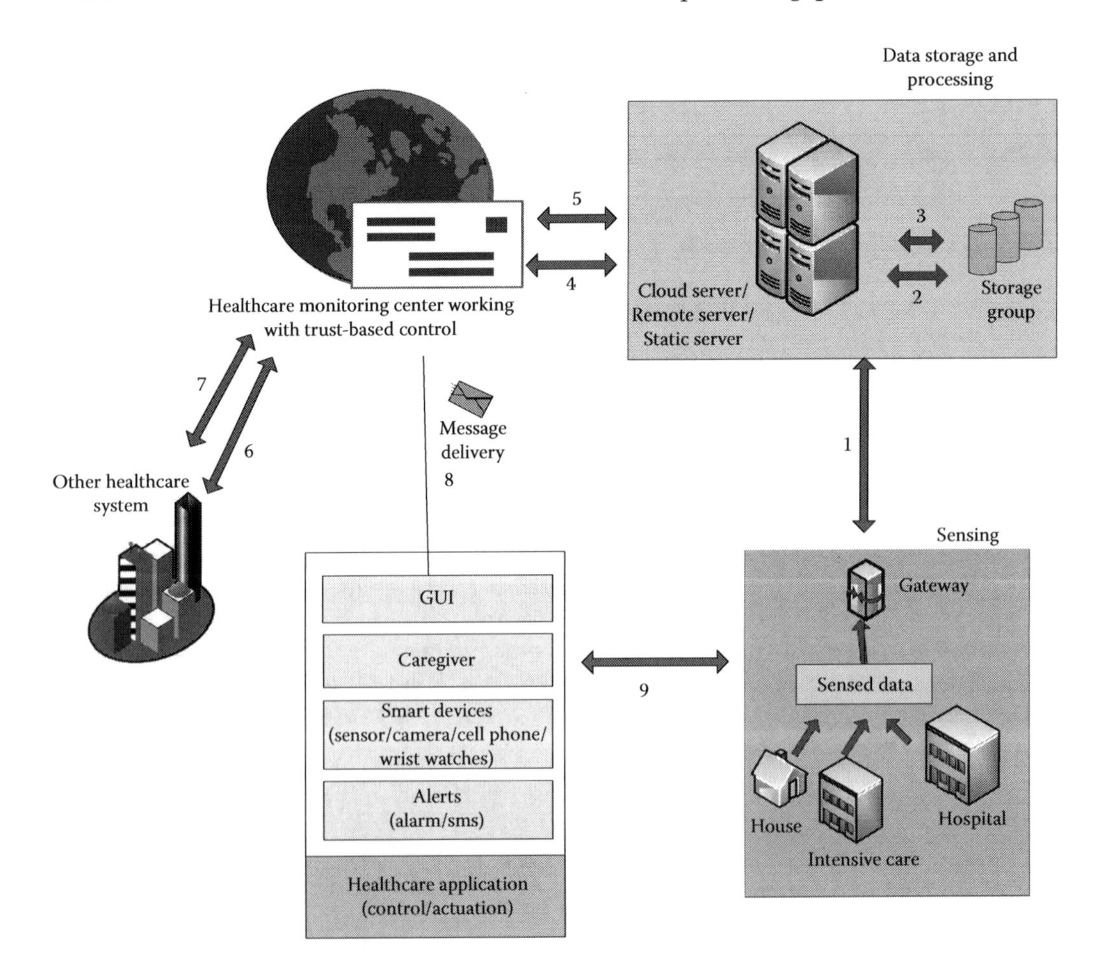

FIGURE 19.7
Cyber-physical systems for health care.

3. The data are transmitted from sensors verified for integrity check.
4. Historic sensor data are processed in case of query reply.
5. Computation is done on trust values collected and stored in the database, generating alarms, and sent to an observation center if necessary.
6. Clinicians in an observation center access patient data from a cloud without affecting the privacy of the patient data.
7. Clinicians approach other health-care systems for consultation for alternate requirement.
8. A response is received from other health-care systems.
9. Clinicians and specialists send decisions to the actuation component and necessary measures conducted on the patients.

Taxonomy has been mapped to CPS health care in Figure 19.8 (Haque et al., 2014) to give the perception of a health-care system visualized. The integration of CPS in the physical and cyber system is routed toward the elements of applications, architecture, computation, communication, control, data integrity, sensing, and security.

19.7 Conclusion

The CPS is working, beyond the age of the individual, with the expectations of 24 × 7 availability and 100% reliability and instantaneous response. The working with the characteristics of storing anything and everything forever with a challenge of boundaries unknown and always changing complex systems that are unpredictable made CPS appreciable. A transformation interface is questioned to the interface of the cyber world and the physical world with predictable or adaptable behavior to provide security to people and society. Ultimately, trust and individual definitions for privacy levels have been answered by the CPS when the physical and cyber world are interconnected through the Internet.

The discussions lead to form multidisciplinary approach to investigate the possible best solutions in the perspective of real-world trade-offs for protection, detection, and response to attack by providing trust-based security for CPS. Trust is adopted as a safety ingredient in the ubiquitous world making the CPS more adaptable and secure.

A need toward the prioritized technologies, identified gaps, and defined taxonomy on the basis of analysis requires more concentration in CPS. A desire to establish a standard vocabulary, cross-cutting context, mechanism to detect deficiencies, promising technology developments, and the relationships among diverse domains was considered. Results have to be made available in all every phase of the research in this area addressing research gaps, research findings, test and evaluation, and transition, adoption, and commercialization. CPS has an impact to control real-world infrastructures depending on the applications to determine to what extent a core critical infrastructure will be vulnerable in the future. The support needs a solution to a device, system and infrastructure, and implementation level (Li et al., 2011). The solution can put forth for requirement and significance (Steward, 2008; Chang et al., 2013) of trust and reputation with the risk in providing CPS security.

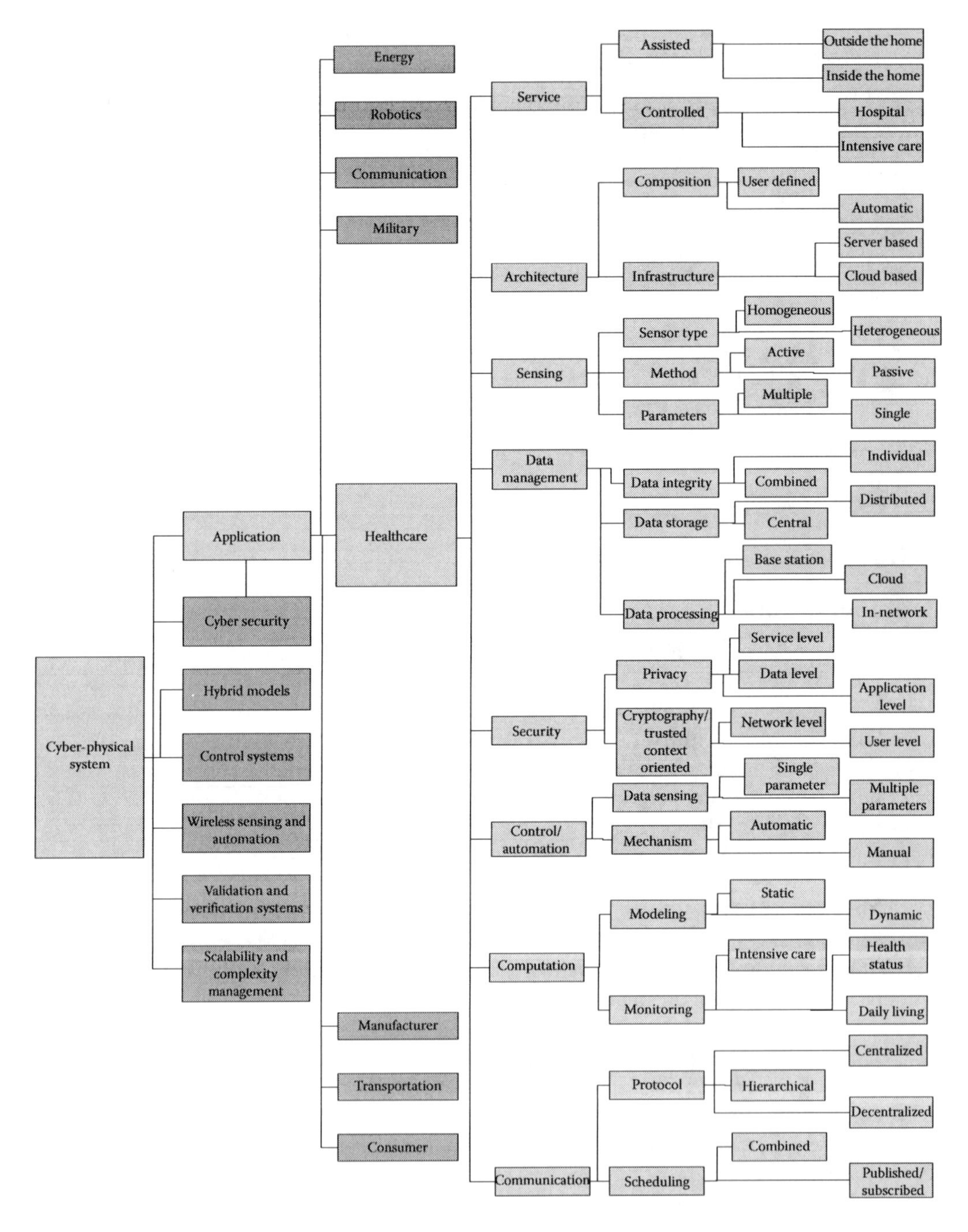

FIGURE 19.8
Taxonomy of cyber-physical systems health care.

References

Abdelzaher, T. et al. (2009). Networking and Information Technology Research and Development Program. *High-Confidence Medical Devices: Cyber-Physical Systems for 21st Century Health Care* [*Online*], Available at http://www.nitrd.gov/About/MedDevice-FINAL1-web.pdf

Adam, N. (2009). Cyber-physical systems security. In *Proceedings of the Fifth Annual Workshop on Cyber Security and Information Intelligence Research: Cyber Security and Information Intelligence Challenges and Strategies*, Oak Ridge, TN.

Ahmadi, H., Abdelzaher, T.F., and I. Gupta. (2010). Congestion control for spatiotemporal data in cyber-physical systems. In *Proceedings of First ACM/IEEE International Conference on Cyber-Physical Systems*, Stockholm, Sweden, pp. 89–98.

Antsaklis, P. (2008). On control and cyber-physical systems: Challenges and opportunities for discrete event and hybrid systems. In *Ninth International Workshop on Discrete Event Systems*, Goeteborg, Sweden.

Balasubramanian, J., Tambe, S., Gokhale, A., Dasarathy, B., Gadgil, S., and D. Schmidt. (2010). A model-driven QoS provisioning engine for cyber physical systems. Technical report, Vanderbilt University and Telcordia Technologie, Nashville, TN.

Bo, Z., Xiang, Y., Wang, P., and Z. Huang. (2011). A novel capacity and trust based service selection mechanism for collaborative decision making in CPS. *Computer Science and Information Systems*, 8(4), 1159–1184.

Broman, D., Lee, E.A., Berkeley, U.C., Martin Torngren, K.T.H., and S. Shyam Sunder. (March 2012). A Cyber Physical System—A concept map. *NIST CPS Workshop* in Chicago. http://www.cyberphysicalsystems.org. Accessed on 13 December, 2014.

Campbell, R.H. (2012). *Cyber-Physical Systems: Position Paper: CPS Environments.* University of Illinois, Urbana-Champaign.

Cardenas, A.A., Amin, S., and S. Sastry. (2008). Secure control: Towards survivable cyber physical systems. In *Proceedings of Computing Systems Workshops*, Beijing, China, pp. 495–500.

Chang, E. and Dillon, T. (2013). Trust, reputation, and risk in cyber physical systems. In *Artificial Intelligence Applications and Innovations.* Springer, Berlin, Heidelberg, pp. 1–9.

Dabholkar, A. and A. Gokhale. (2009). An approach to middleware specialization for cyber physical systems. In *Proceedings of 29th IEEE International Conference on Distributed Computing Systems Workshops*, Montreal, Québec, Canada, pp. 73–79.

Daly, J. (2013). Designed-in cyber security for cyber-physical systems. Workshop Report by the Cyber Security Research Alliance, Gaithersburg, MD.

Farag, M.M. (2012). Architectural enhancements to increase trust in cyber-physical systems containing untrusted software and hardware. PhD theses to Virginia Polytechnic Institute and State University, Blacksburg, VA.

Gamage, T.T., McMillin, B.M., and T.P. Roth. (2010). Enforcing information flow security properties in cyber-physical systems: A generalized framework based on compensation. *COMPSAC Workshops*, pp. 158–163, IEEE Computer Society.

Haque, S.A., Aziz, S.M., and R. Muztafizur. (2014). Review of cyber-physical system in Healthcare. *International Journal of Distributed Sensor Networks*, 2014, 20, Hindawi Publications.

Homeland Security News Land. (November 2014). A Survey and Taxonomy for Roots of Trust in Cyber-Physical Systems. In *Cyber Security Research Alliance-Summary Result Report.* http://www.cybersecurityresearch.org/documents/Roots_of_Trust_for_Cyber_Physical_Systems_Abstract_-_November_2014.pdf, Retrieved on 18 December, 2014.

Hughes, J. and G. Cybenko. (2013). *Quantitative Metrics and Risk Assessment: The Three Tenets Model of Cyber Security.* Technology Innovation Management Review.

Kanchana Devi, V. and R. Ganesan. (2013). Enhancing security in cyber physical systems through policy based trust management against deception attack. *International Journal of Computer Applications* (0975–8887), 70(6).

Karnouskos, S. (2014). Security in the Era of Cyber-Physical Systems of Systems. *ERCIM News*, 2014, no. 97.

Kottenstette, N., Gabor K., and S. Janos, (2009, August). A passivity-based framework for resilient cyber physical systems. In *2nd International Symposium on In Resilient Control Systems*, IEEE, pp. 43–50.

Koubaa, A. and B. Andersson. (2009). A vision of cyber-physical internet. In *Proceedings of the Workshop of Real-Time Networks (RTN 2009) and Satellite Workshop to (ECRTS 2009)*.

Lee, E. (2008). Cyber physical systems: Design challenges. In *Proceedings of 2008 11th IEEE Symposium on Object Oriented Real-Time Distributed Computing*, Orland, FL, pp. 363–369.

Lee, I. (2009). *Cyber Physical Systems: The Next Computing Revolution*. CIS 480, Spring.

Li, W., Jagtap, P., Zavala, L., Joshi, A., and T. Finin. (2011, June). CARE-CPS: Context-aware trust evaluation for wireless networks in cyber-physical system using policies. In *IEEE International Symposium on Policies for Distributed Systems and Networks (POLICY)*, pp. 171–172. http://www.seas.upenn.edu/~lee/09cis480/lec-CPS.pdf.

Neuman, C. (2009). *Challenges in Security for Cyber-Physical Systems*. DHS: S&T workshop on future directions in cyber-physical systems security. Vol. 7.

Northcut, C.G. (2013). Security of cyber-physical systems. A survey and generalized algorithm for intrusion detection and determining security robustness of cyber physical systems using logical truth tables. *Vanderbilt Undergraduate Research Journal*, pp.1–9, DOI: http://dx.doi.org/10.15695/vurj.v9i0.3765.

Pham, N., Abdelzaher, T., and S. Nath. (2010). On bounding data stream privacy in distributed cyber-physical systems. *IEEE International Conference on Sensor Networks, Ubiquitous and Trustworthy Computing*.

Sanislav, T. and L. Miclea. (2012). Cyber-physical systems—Concept, challenges and research areas. *CEAI*, 14(2), 28–33.

Stelte, B. and G.D. Rodosek. (2013). Assuring trustworthiness of sensor data for cyber-physical systems. In *IFIP/IEEE International Symposium on Integrated Network Management (IM 2013)*.

Stewart, C.A. (2008). Leadership under challenge: Information technology R&D in a competitive world. Report of Testimony before the United States House of Representatives Committee on Science and Technology Hearing.

Svéda, M. and O. Ryšavý. (2013). *Dependable Cyber-Physical Systems Networking: An Approach for Real-Time, Software Intensive Systems*.

Tan, L.A., Yu, X., Kim, S., Han, J., Hung, C.-C., and W.C. Peng. (2010). Trualarm: Trustworthiness analysis of sensor networks in cyber-physical systems. In *Proceedings of the 10th IEEE International Conference on Data Mining, ICDM'10*.

Tan, Y., Vuran, M.C., and S. Goddard. (2009). Spatio-temporal event Model for cyber-physical systems. *Proceedings of the 29th IEEE International Conference on Distributed Computing Systems Workshops*, Montreal, Quebec, Canada, pp. 44–50.

Tang, H. and B.M. Mcmillin. (2008). Security property violation in CPS through timing. In *Proceedings of the 28th International Conference on Distributed Computing Systems Workshops*, Beijing, China, pp. 519–524.

Tidwell, T., Gao, X., and H. Huang. (2009). Towards configurable real time hybrid structural testing: A cyber-physical systems approach. In *Proceedings of IEEE International Symposium on Object Component Service-Oriented Real-Time Distributed Computing*, Tokyo, Japan.

Venkatasubramanian, K. Cyber physical systems. Slides from. http://web.cs.wpi.edu/~kven/courses/cs525-cps/lectures/CPS_Intro_Slides.pdf. Accessed on 6 January, 2015.

Wang, J., Abid, H., Lee, S., Shu, L., and F. Xia. (2011). A secured health care application architecture for cyber-physical systems. *Journal of Control Engineering and Applied Informatics*, 13(3), 101–108.

Xia, F., Longhua M., Jinxiang D., and S. Youxian (2008, July). Network QoS management in cyber-physical systems. In *ICESS Symposia'08. International Conference on Embedded Software and Systems Symposia*, IEEE, pp. 302–307.

Zhang, L. (2012). Aspect-oriented QoS modeling for cyber physical system. *Journal of Software*, 7(5).

Section VI

Role of Cyber-Physical Systems in Big Data Analytics, Social Network Analysis, and Health Care

20

Big Data Analysis and Cyber-Physical Systems

Urban Sedlar, Mitja Rakar, and Andrej Kos

CONTENTS

20.1 Introduction

Cyber-physical systems (CPSs) represent an important step on the path to truly smart living environments, efficient automation of industrial processes, and optimization of control in numerous other fields—including medicine, traffic, emergency response, and urban living. Based on merging of digital computation resources with the physical objects, a CPS represents a novel paradigm that aims not merely to be able to gather information about the state of the physical world but also to influence it based on the gathered data.

The enabling technologies for CPS range from sensing and communication to computation and actuation. A plethora of sensing and communication technologies have

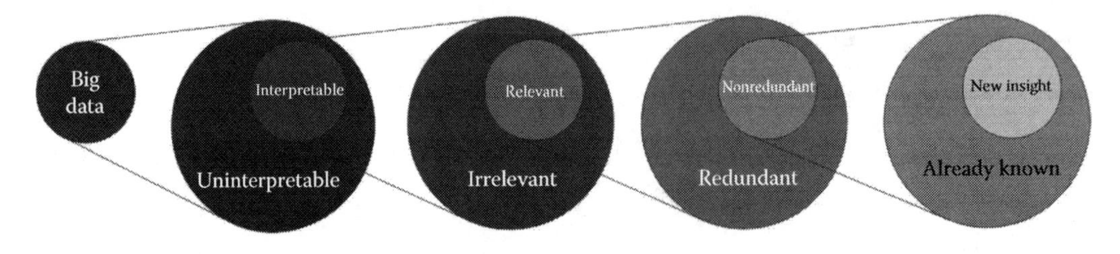

FIGURE 20.1
Only a small subset of all collected data provide new insight and can be used to add value to the system.

emerged and proliferated in recent years, such as radio frequency identification systems, wireless sensors connected into wireless sensor networks (WSNs), and new and energy-efficient communication standards and protocols that helped expand the ubiquitous Wi-Fi connectivity, such as Bluetooth low energy, Zigbee, and Z-Wave. The embedded computers have become both cheaper and more powerful, allowing the processing power to come arbitrarily close to the objects in the physical world where it is needed. Mobile terminals have demonstrated a vast growth in computing capabilities as well, bringing to every single user a highly versatile digital assistant, which serves as an always-on and always-available interface to the digital world. Finally, the landscape of actuators, where progress was the slowest, is starting to show some real promise with the advent of maker labs around the world, where enthusiasts are innovating and raising awareness of the general public of the fact that very little is needed to close the loop from sensing to control.

However, demonstrating one can close the loop and doing so in an intelligent way by leveraging all available data *in an efficient and time-constrained manner* are two different things entirely. Modern sensors can easily produce large volumes of data that need to be analyzed efficiently to meet the demands of the system and ensure the response in a timely manner. Quantities of data that are too large to be processed by the traditional systems are often termed big data and show a significant promise for powering intelligent decisions, as well as for enabling the discovery of new knowledge. However, one has to bear in mind that one of the most important challenges in big data is separating the wheat from the chaff (Figure 20.1), and if not approached correctly, opting to collect large quantities of data without a clear plan can lead to more frustration than benefit.

The rest of this chapter is structured as follows. In Section 20.2, we define big data and some of its challenges and techniques commonly used for its preprocessing, storage, and analysis. In Section 20.3, we describe the most important constraints of CPS that require special attention when designing a big data pipeline. In Section 20.4, we describe event stream processing techniques that are best tailored to the nature of the CPS. In Section 20.5, we briefly describe some tools for big data processing. In Section 20.6, we discuss how to bring different techniques together in a single CPS data processing pipeline. Finally, in Section 20.7, we list some of the CPS use cases as example applications of the presented architectures.

20.2 Definition and Challenges of Big Data

The term big data by the most commonly used definition refers to datasets that are too large and complex to be managed with traditional systems. Worldwide information volume is reported to be growing annually at a rate of 59% (Pettey and Goasduff, 2011). In 2015, the amount of data for a single use case can easily reach several petabytes. An important

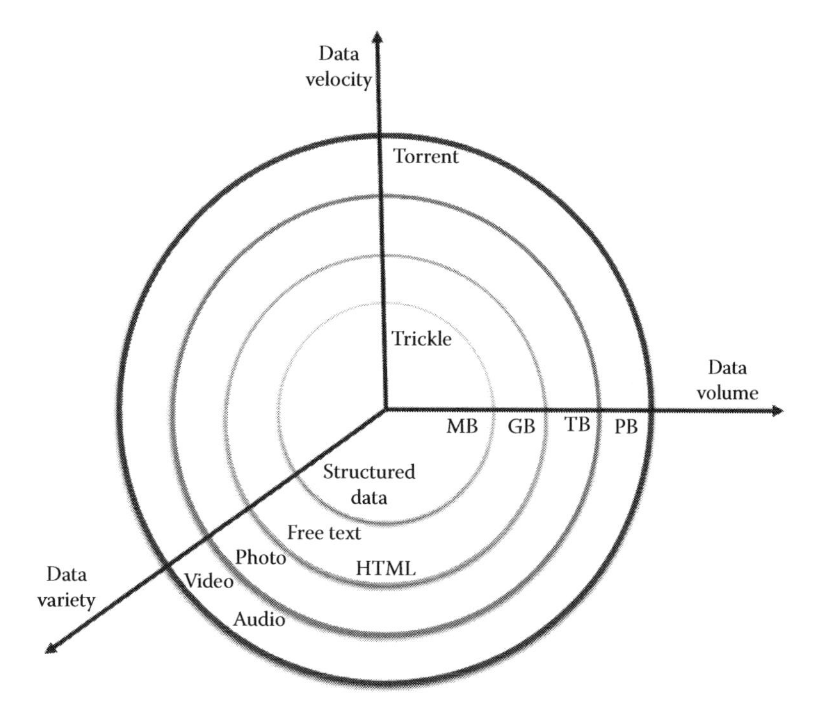

FIGURE 20.2
Big data are expanding on three fronts at an increasing rate.

challenge with such quantities is often described as *data gravity*, where data becomes nearly impossible to move without disruption of the system functionality (Figure 20.2).

Another challenge is to make sense of all the data coming in from different sources and derive useful and actionable information out of it (Pinal, 2013). Issues and challenges in big data can be subdivided into the following six areas (also called the six Vs of the big data).

Volume of data. Big data imply enormous volumes of it. The days when only people created data are long gone—today, most data are machine generated. Human-generated data should not be underestimated, however, as user-generated content (especially high-resolution video) in systems like the social media networks exhibits similarly massive volumes.

Variety refers to the many sources and types of data, both structured and unstructured. The majority of the data generated today is unstructured (machine-generated log files and messages, emails, photos, video and audio files, etc.), which means it does not conform to a predefined data model. This creates significant problems for storage, mining, and analysis.

Velocity describes the pace at which the data flow. The flow can be continuous or bursty in nature. High velocities can present problems for storage and processing, more so if the data have limited shelf life (see the section on *volatility*). This can often be the case, for example, in real-time or near-real-time systems.

Validity deals with the correctness of data for the intended purpose. In analyzing large quantities of data, it is very likely that a significant portion of it will exhibit biases, abnormalities, or noisiness. With suitable statistical techniques, clustering, and identification of outliers, most datasets can be efficiently cleaned at a large scale.

Veracity refers to relevance of the data for the intended use case. Thus, it is one of the most important aspects to consider from the business perspective, as it can invalidate most of the solutions to other (technical) problems. By addressing the issue of veracity before the system is set up, one can avoid building the fastest and most efficient data processing system that is not needed at all.

Volatility refers to how long the data are valid and, consequently, how long it should be stored. However, this is often not a black-and-white choice. Moreover, in the world of real-time data and on-the-spot decisions, high-resolution data from hundreds of sensors can quickly diminish in value without being invalid *per se*. At some point, the cost of storage alone can outweigh the potential benefit that can be gained by keeping it around.

20.2.1 Data Acquisition

Availability of affordable sensor nodes in recent years, together with their increasing ubiquity, paved the way for both planned and impromptu applications in many different areas, ranging from environmental monitoring and process automation to disaster relief operations.

Sensors are small electronic devices that convert physical quantities into measurable electric signals. The signal is commonly captured in an analog way (voltage, current, or impedance changing in response to changes in the measured quantity), which is subsequently digitized to be usable in digital computing environments such as CPS. Digitized data can influence local decisions through locally used microprocessors that implement data transformation and decision logic at the data source itself or are sent to a remote system. The latter option is increasingly used due to storage and power limitations on the sensor nodes themselves. In addition, it provides more flexibility and control with regard to the amount of available computing power and data enrichment options, as data from multiple sources can be more easily combined at a central point of aggregation.

Data can be sent to a remote system in different ways. Individual sensor nodes can feature a cellular modem and send data directly to the point of aggregation (a design choice common in spatially sparse deployments, such as smart grids) or leverage multiple connectivity mechanisms, as is the case with larger WSNs, which use local short-range technologies to connect to a communications hub, which, in turn, forwards the data to the central system.

Sensors, suitable for CPS systems, cover the entire spectrum, with some examples being magnetic, thermal, visual, infrared, acoustic, temperature, humidity, pressure, seismic, and radiation sensing capabilities (Bouhouchi and Ezzedine, 2012). Due to the nature of the remote deployments in environments with limited power and unreliable communication options, power efficiency and network survivability issues are also among critical concerns to be addressed.

From the standpoint of the CPS, two important characteristics of any sensor system are its sampling frequency and the average size of the individual sample; both directly translate into the expected bandwidth of the sensor system, which contributes to the final velocity and volume of the data being generated.

20.2.2 Data Preprocessing

The collected data are prone to noisy, missing, and inconsistent values. Such issues arise due to a vast number of often heterogeneous sources and the fact that new sources are often added to the system that were not taken into account at the design phase. Due to the fact that low-quality data will lead to the lower quality of the processing and mining results, it is important to employ the preprocessing before the data are ingested into the decisioning system. Preprocessing involves a set of techniques for transforming raw data into a common and structured format, as well as removing known errors and biases. Real-world data are often incomplete, inconsistent, biased, structured in different or incompatible formats, and represented in different units or according to different reference models.

Data preprocessing prepares raw data for further processing. Numerous preprocessing techniques exist (Son, 2006). *Data cleaning* routines "clean" the data by filling in missing values, smoothing noisy data, identifying or removing outliers, and resolving inconsistencies. *Data integration* merges data from multiple sources into a coherent model, suitable for further analysis. *Data reduction* can reduce the size of the dataset by aggregation, summarization, elimination of redundant features, or clustering. *Data transformations* (e.g., normalization) may be applied, where data from different sources fall into different value ranges or when data are represented in different units (e.g., °C and °F). *Data discretization* involves the reduction of a number of possible values of a continuous attribute by dividing it into a range of intervals (bins). These techniques are not mutually exclusive and may be used together to eliminate some of the problems in the dataset.

Further, eliminating redundant data can significantly shrink storage requirements and improve bandwidth efficiency; this is obviously even more effective if done closer to the source of redundant data. A special kind of redundant data is duplicated data, which is the easiest to identify. *Deduplication* can be used to remove it, either in a lossless way that allows reconstruction of the original dataset or in a lossy way where reconstruction is not possible or needed (Rivera, 2012).

Increased awareness of privacy issues and the quickly adapting privacy laws and regulations require special handling of the data in raw form, to ensure the privacy of the people coming in touch with the data acquisition is not compromised. There is often a strong conflict of interest due to the fact that knowing more about the users can add significant value and help organizations in decision-making processes; however, users do not want their privacy compromised. To address some of the privacy issues, data *anonymization* is often required. This is the process of removing identifiable information and datasets from the system to protect the identity of the owners of the data, thus making it either totally anonymous or pseudo-anonymous.

In addition to helping protect users against the risks of privacy violations, data anonymization also has benefits for the data processor: it reduces the likelihood of misuse of personal data, either by insiders or by external attackers in the case of a data breach. This reduces the chance of litigation in case of a possible loss of sensitive user information, as well as the costs associated with the future risks, by ensuring compliance with local laws for data protection and user privacy protection.

Data anonymization is a complex technical challenge involving multiple disciplines and employing various techniques and algorithms (Habbegger et al., 2014). Hashing is one of the widely used methods of data anonymization. It works by performing a predetermined computation process on an input to turn the input data into a *fingerprint* of constant length. For example, the MD5 hashing function yields 128-bit outputs, transforming the string 0 into *cfcd208495d565ef66e7dff9f98764da*. Hashing is also called a one-way function because it is impossible to get the input value back from the output. It represents a popular method for pseudo-anonymization, as a given identifying value always results in the same hash value, hiding identity, but allowing correlation with other data points within the dataset. Multiple hashing algorithms exist; the mentioned MD5 has already been replaced by much stronger functions such as SHA-2 and SHA-3. Additionally, the output value can also be prefixed or suffixed by a random component (salt) to make the pattern identification even more difficult. *K-anonymity* is a method of data anonymization where some attributes of data are typically replaced by an asterisk, to hide the attribute name and, thus, make the information completely unidentifiable. Another approach is to generalize specific information like age and date of birth and store it as a range so that individual details are hidden; this also facilitates data aggregation.

The central challenge of finding the right balance between keeping data anonymous and keeping it useful means there is only a certain amount of data that can be anonymized without reducing the business value of the dataset, and all anonymization solutions have to work within these constraints.

20.2.3 Data Storage

As databases get larger, it becomes increasingly difficult to keep the entire database on a single physical machine. The sheer size of the dataset can exceed the capacities of storage media, memory, and network bandwidth (Mohanty et al., 2013). This has fueled the research and general interest in the field of *distributed databases*. A distributed database consists of multiple, interrelated databases stored on different computing nodes and possibly on geographically distributed locations as well. To the end user or developer, the distributed database appears as a single logical database, with all the complexities of moving and querying data being hidden behind the scenes (Stockinger, 2001).

A traditional relational database management system (RDBMS) is optimized to store and query structured data with a known schema. If the amount of data or the processing requirements grow, the database system can be scaled vertically (increasing the server capacity and performance) or horizontally (by replication and partitioning of the data to multiple servers) (Gorton and Klein, 2014). An alternative, more cloud-friendly approach is to employ a *NoSQL database* (often interpreted as Not Only SQL), which typically trades some of the RDBMS strengths (strong consistency) for greater and easier horizontal scalability, availability, and fault tolerance. NoSQL databases feature highly optimized key-value data stores designed for quick retrieval and appending operations, and they can offer substantial performance benefits over relational databases in terms of latency and throughput. One of the major reasons businesses move to a NoSQL database system from an RDBMS is the more flexible (schema-less) data model that is supported in most NoSQL databases. The relational data model is based on defined relationships between tables, which themselves are defined by a predefined column structure, all of which are explicitly organized in a database schema. Problems begin to arise with the relational model around scalability and performance when trying to manage the large data volumes. In addition, a NoSQL database is able to accept both structured and unstructured types of data more easily than a relational database that relies on a predefined schema (Strauch, 2011). Since NoSQL database architectures are typically distributed, this in addition reduces the number of points of failure. The data can be distributed in a redundant manner, reducing the impact of a single server malfunction or a network partition. This property is termed fault tolerance. However, most of the advantages of NoSQL over RDBMS come with an important drawback. Only two out of the three database system properties, that is, data consistency, system availability, and partition tolerance, can be satisfied in a distributed system at any given moment (the CAP theorem). NoSQL database systems thus often elect to sacrifice data consistency for availability and fault tolerance, leading to *eventually consistent* data at best (Couchbase, 20014).

A highly scalable approach for improving the throughput and overall performance of high-transaction, large database-centric applications is database *sharding*. Database sharding can be simply defined as a *shared-nothing* partitioning scheme for large databases across a number of servers, enabling better performance and scalability. The word "shard" means a small part of the whole (Rouse, 2011). Database sharding provides a method for scalability across independent servers, each with their own CPU, memory, and disk resources. The advantage of such shared-nothing database sharding approach is improved scalability, growing in a nearly linear fashion as more servers are added to the system. There are several advantages of a large number of smaller databases.

- Easier to manage. By using the sharding approach, each individual *shard* can be maintained independently, providing a far more manageable system in terms of maintenance, which can be performed in parallel.
- Smaller databases can have better performance. By hosting each shard on its own hardware, the ratio between memory and the amount of data on disk is greatly improved, thereby reducing disk input/output. This results in less contention of resources, greater join performance, faster index searches, and fewer database locks.
- Reducing costs. Most database sharding implementations take advantage of lower-cost open-source databases and can be run on large clusters of cheaper commodity hardware.

Database sharding is suitable for many types of applications, either those with general-purpose database requirements, mixed workload database usage (i.e., complex queries), or general business reporting (Bryden, 2014). However, its limitations also have to be taken into account: performance drops for intershard operations are a significant penalty compared to intrashard operations; thus, sharding works best if the spatial data *locality of reference* is high (i.e., if all the data needed for processing a certain event are located in the same shard).

20.2.4 Data Processing and Visualization

Assumptions for *querying* and *mining* big data are fundamentally different than in traditional statistical analysis of small samples. Big data are just as noisy, dynamic, heterogeneous, redundant, and untrustworthy as small data. However, due to the statistical properties of large samples, certain patterns can be taken into account to get a better picture. Measures obtained from frequently occurring patterns can, through correlation and statistical analysis, overpower individual fluctuations and thus disclose more reliable hidden patterns and knowledge. At the same time, data mining itself can also be used to help improve the quality and trustworthiness of the data, understand its semantics, and provide intelligent querying functions. The value of big data analysis can, in many cases, only be realized if it can be applied robustly under these difficult conditions.

To process massive quantities of data in a timely and manageable way, parallel processing proves to be an important paradigm. One of the often used techniques is *MapReduce*, a programming model for processing large quantities of data in a distributed manner on a cluster of servers. The model is inspired by a set of functions, *map* and *reduce*, commonly used in functional programming (Soukup and Davidson, 2002). A typical MapReduce operation consists of two high-level steps; in the first step (map), each worker in the cluster applies a certain data processing function to the data stored locally (on the worker node) and yields the output. The second step reduces (summarizes) the partial results into the final result of the operation. Similarly, as with sharding, the largest benefits can be obtained with parallelizable problems, where the map and reduce operations can run in parallel on multiple machines in the cluster.

Another set of insight and knowledge discovery tools is *visualization based*, focusing on the presentation of big data—on helping businesses explore the data more easily and understand it more fully. Systems with support for a broad range of visualizations are becoming more and more important in conveying to the users the results of the queries in a way that is best understood in the particular domain (Liu et al., 2013). Visualization-based data discovery tools allow business users to mash up disparate data sources to create custom analytical views with flexibility and ease of use that simply did not exist before. Advanced analytics are often integrated into such tools to support creation of interactive and animated graphics. End users can view these visual representations anywhere, using mobile devices such as tablets or smartphones. The choice of the data visualization tools usually depends on the nature of the dataset and its underlying structure; such tools can be classified into two main categories: (1) multidimensional visualizations and (2) specialized hierarchical and landscape visualizations. Scaling complex query processing and visualization techniques to massive amounts of data while enabling interactive response times is one of the major open research problems today.

20.3 Constraints of CPS

A *CPS* is an integration of computational elements with the physical entities that allows them to control physical processes. CPS enables the physical world to deeply interconnect with the digital and is, together with the Internet of Things (IoT), an important path toward that goal. Today's CPS emerged from embedded systems that were traditionally more computing and less network oriented. Several systems in existence today can be described as a CPS, for example, intelligent manufacturing lines, where the computing function can monitor outputs of numerous sensors, coordinate individual devices, and orchestrate a complex assembly process. Since the concepts behind the CPS are very general, they can be applied to many fields and examples, from automotive industry and smart grids to medicine and health. However, with the fast progress of networking technologies, such systems are increasingly being connected to large networks of global scale, such as the Internet. In distributed systems, the effects of communication over distance can quickly become a nonnegligible part of the overall system performance. The upside, however, is that the data can be collected and analyzed centrally, which allows the inputs, patterns, and findings from one part of the system to influence other parts.

20.3.1 CPSs and IoT

The *Internet of Things* refers to the ever-growing network of physical objects that feature an IP address for Internet connectivity and the communication that occurs between these objects and other Internet-enabled devices and systems. These objects include various sensors, indicators, or actuators and can both transmit the collected sensor data to other networked devices or be controlled from another device. There's been a huge increase in commercial availability of such devices recently, encompassing everything from security systems, media players and speakers, car equipment, appliances, thermostats, light bulbs, locks, weather stations, etc. Devices classified as IoT are typically connectivity centric, advocating the best-effort nature of the Internet itself, while computation is secondary and, in many cases, minimal.

CPSs, on the other hand, use shared knowledge and information obtained from sensors to independently control physical devices and processes in a closed loop (Tabuada, 2006).

They are, therefore, the bridge that connects the IoT with higher-level services, sometimes called the Internet of Services. A CPS integrates computation, networking, and physical processes more closely, resulting in a highly automated system. In such ecosystems, software providers, service providers, and users can collaborate to develop flexible applications that can be dynamically interconnected and integrated with one another (Nie et al., 2013). Lessons from both the CPS and IoT will have to be embraced and applied to the future systems to usher mankind toward the fourth industrial revolution.

20.3.2 Architectural Considerations

The theory and science of CPS composition is rapidly evolving to cover new architecture patterns for allowing hierarchical system composition from components and subsystems. In addition, it must address the aspects of protocol composition, design modeling languages, and tools to specify, synthesize, analyze, and simulate different compositions, as well as take into account the aspects of quality of service (QoS).

Physical properties and characteristics, safety, robustness, security, real time, and power constraints can all be captured in composable models, supported by programming abstractions. Individual CPS solutions can thus take into account the physical properties of the system and leverage location-based and time-based data and mechanisms appropriately (Abdelzaher, 2006). However, the uncertainty in the environment, amplified by possible security attacks, errors, and malfunctions in the physical devices, makes ensuring overall system robustness, security, and safety an important challenge. In any CPS, it is crucial that the faults be understood and reliably addressed. In addition, the gap between formal verification and validation methods and testing needs to be bridged. Compositional verification and testing methods that explore the heterogeneous nature of CPS models play a crucial role in that part.

CPS architectures must be consistent at a higher level and capture a wide variety of physical information. Bandwidth allocation protocols, new queuing strategies, and new routing schemes (including resource virtualization) can reduce and accommodate network delays, whereas networks themselves must provide real-time resource allocation and facilitate data aggregation, in-network decision-making, and the ability to monitor, observe, and guarantee QoS.

In CPS, knowledge discovery and creation from the vast amount of raw data being collected provides an important challenge (Abdelzaher et al., 2007), especially when the CPS control loop is closed through an operator. Applications where humans directly control the system typically primarily use supervisory control. Although having humans in the loop has its advantage, modeling human behavior is extremely challenging due to the different complex physiological, psychological, and behavioral aspects involved. In supervisory control, involvement of humans takes place in two ways. In one case, the processes run autonomously; humans intervene with the control algorithm only when it is necessary, typically by adjusting certain parameters. In the other case, the process works in an on-demand manner, accepting a command, carrying it out autonomously, reporting the results, and waiting for further commands to be received from the operator. *Human-in-the-loop* feedback control systems offer exciting opportunities to a broad range of CPS applications. To begin to understand their spectrum, a taxonomy of human-in-the-loop applications (Figure 20.3) based on the controls that they employ is necessary, which classifies them into three categories: (1) applications where humans directly control the system, (2) applications where the system passively monitors humans and takes appropriate actions, and (3) a hybrid of (1) and (2) (Liu and Salvucci, 2001).

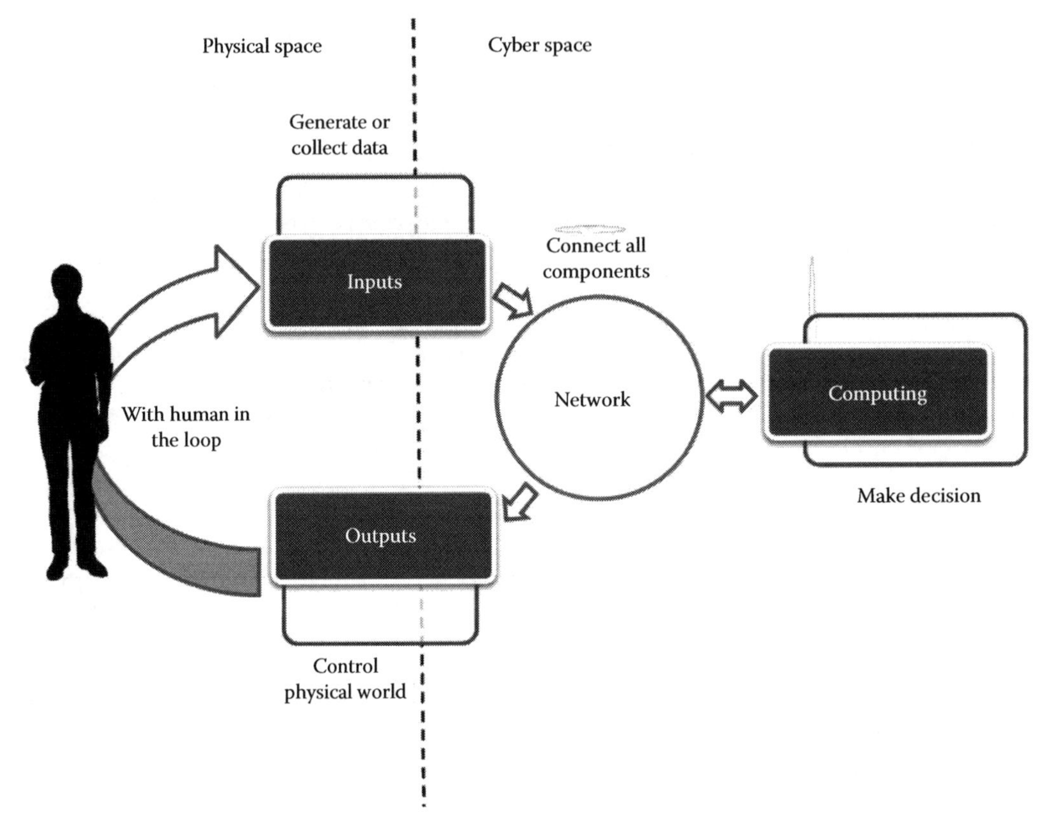

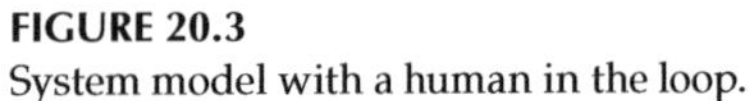

FIGURE 20.3
System model with a human in the loop.

Although requirements are different for different applications, a significant portion of human-in-the-loop applications have to address a common set of challenges, for example, user-specific thresholds and parameters, change of human behavior over time, and required sensing technology to sense the appropriate aspects of the behavior. Human behavior needs to be modeled for a large number of applications before general principles and theories can emerge that address these issues. Clustering, data mining, inference, and first principle models based on human physiology and behaviors may all be necessary techniques to be enhanced and applied to different use cases. It is also unlikely that any models developed initially to design the controllers will remain accurate, as the system and human behaviors evolve over time. There are several areas where a human model can be placed: outside the loop, inside the controller, inside the system model, inside a transducer, and at various levels in hierarchical control (Schirner et al., 2013).

The future requirements of the CPS will far exceed those of today's systems in terms of functionality, usability, adaptability, and autonomy, as well as resistance to accidental and malicious threats and resistance to dynamic changes of user behaviors and changes of the environment. CPSs are highly multidisciplinary (Figure 20.4) and involve disciplines such as embedded computing, control theory, and communication technology. Human concerns are related to human–physical system interaction, usability, user experience, responsibility, privacy, security, and user and social acceptability. Physical system models must thus be viewed as semantic frameworks, and theories of computation must be viewed as alternative ways of talking about the system dynamics (Suh et al., 2014).

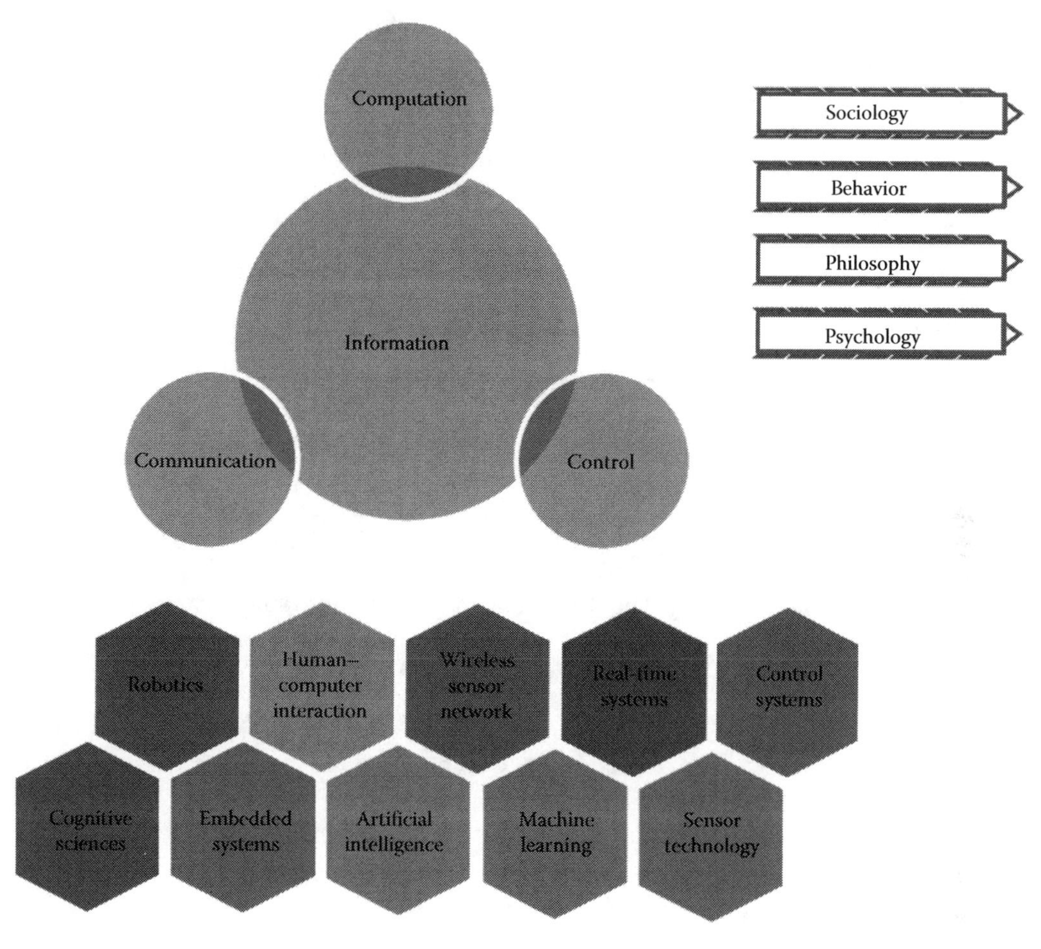

FIGURE 20.4
Multidisciplinary nature of a CPS.

20.3.3 Data Processing Constraints

The described issues have significant ramifications from the perspective of big data management and processing systems. Two most obvious issues, closely connected with data volatility, stem from the effects of communication over unreliable networks. Since IP-based networks function in a best-effort mode, the reliability of the communication is greatly reduced; this is in part solved by higher-level protocols that rely on data retransmission; however, this negatively impacts the second related problem—the one of latency. In long-distance communications, the speed of light imposes a physical limit on the speed of message transmission. This is further increased by packet processing delays within network equipment, making quite a significant round-trip time of up to several hundred milliseconds for intercontinental communication. If the data are only usable for immediate decisioning, a long round-trip time greatly diminishes the usability of such a system. If the system imposes a hard deadline on the decisioning process (e.g., an industrial robot that relies on the local sensor to retract the arm), enough computation has to be performed locally to allow autonomous and safe operation in case of network problems.

Another related challenge is the acquisition, management, and processing of the data, generated by the human in the loop. This can include an online analysis of anything from a simple control data stream to a high-bandwidth 3D video stream. Recorded responses to historical situations can further be postprocessed to discover expert knowledge required to perform the CPS control and automate it in the future.

20.4 Data Processing Techniques for CPS

Large datasets can be postprocessed using techniques presented in Sections 20.2.2 through 20.2.4. However, as described in the Section 20.3, the majority of CPS scenarios require more timely responses, which cannot be achieved by using classical on-demand batch processing techniques. Hence, an event-driven system is needed to react to the new data immediately as it emerges. To design and build an event-driven information system, we first need to introduce some of the stream-based data processing techniques.

Event processing is a method for monitoring and analyzing the data stream with the purpose of identifying certain patterns in the data in near real time. An event in the classical sense represents a point in time, which is characterized in the data model by a timestamp and a certain value, for example, T = 5°C at 12:00. A straightforward threshold function is already sufficient to classify the system as being capable of *simple event processing*. Such threshold function is triggered by an event in which the value exceeds a certain threshold, yielding a certain result (e.g., 0 or 1), for example,

$$f(T)=\begin{cases}1, & \text{if} \quad T>10°\text{C} \\ 0, & \text{otherwise}\end{cases}$$

Even detection of such simple events can have a lot of useful value (e.g., an alarm in case of an air-conditioning system malfunction), as it operates directly on the data stream in a functional manner and thus represents an optimally responsive data processing mechanism. If we compare the situation to a conventional batch processing mechanism, where the data are initially persisted (e.g., in a database) and then processed upon request, the immediate response of the event processing systems becomes apparent as a strong advantage.

Another advantage of such systems is their extremely large bandwidth; classical databases are limited, both in the data storage capacity (limited available disk space) and in querying speeds (limited by medium access speeds and bandwidth). Modern data processing techniques (e.g., MapReduce) remove some of these limitations, but usually at the cost of greater complexity and larger overhead. Unlike traditional databases that store data on hard drives or solid-state drives with relatively low throughput (in the order of magnitude of 100 MB/s), the systems for event processing operate directly in the random access memory, where 100× larger throughputs are easily achievable. This effectively allows the flow of data to be processed at wire speed (Luckham, 2002).

Simple event processing can in the next step be upgraded to *complex event processing* (CEP); instead of one threshold function, we apply a combination thereof (Stonebraker et al., 2005). This allows the detection of an arbitrarily complex set of circumstances based on a combination of data from multiple sources. Thus, a CEP system can be extremely universal and applicable to a variety of scenarios. An example of such a composite function is shown in the following example:

$$f\left(T_{inside}, T_{outside}\right) = \begin{cases} 2, \text{ open window}, & \text{if} \quad T_{inside} > 10°\text{C} \quad \text{and} \quad T_{outside} < 10°\text{C} \\ 1, \text{ alarm}, & \text{if} \quad T_{inside} > 10°\text{C} \quad \text{and} \quad T_{outside} \geq 10°\text{C} \\ 0, \text{ do nothing, otherwise} \end{cases}$$

The near-real-time response of a CEP system ensures quick response times, making such systems the first choice in a number of time-critical scenarios. CEP systems operate on a window of time-series data stored in memory. Thus, a certain amount of history can be taken into account when processing the event stream (in accordance with the amount of available memory), which allows the detection of trends and calculation of correlation between multiple streams. With that, relatively complex phenomena can be modeled and abstracted.

The data preprocessing considerations discussed in Section 20.2 still apply in the case of an event stream processing; only by cleaning, transforming, normalizing, and anonymizing the data can the quality of the results of a CEP system remain high. There are some important considerations regarding preprocessing, namely, in an event-driven pipeline, the preprocessing operations have to happen at the velocity of data as well. For this reason, the preprocessing operations are often realized as a part of the CEP system itself.

The fact that highly scalable MapReduce (discussed in Section 20.4) is typically performed in a batch manner makes it unsuitable for systems with tight closed loops (Lee et al., 2011). However, one of the possible applications for it in CPS is long-term learning and knowledge discovery. A simple example is a historical baseline that can be calculated periodically by processing large quantities of past data, while the results of such calculation can be distributed either to a central CEP system or even local computing nodes in the CPS, to be used in the real-time decisioning process itself (Maguire, 2013). Another example is discovery of the domain knowledge that is provided by the human in the loop and distilling it in the form of a decision tree that can be implemented in software and either performed centrally within the CEP system or moved close to the point of CPS control.

20.5 Tools for Big Data Analysis in CPS

As described, tools for analyzing big data in the context of CPSs can be used on two fronts: real-time stream processing techniques can be used directly in the control loop, whereas offline (batch) processing can serve to model the system, discover new knowledge, infer long-term trends, and learn new behaviors (Antonio and Jara, 2014). This section provides a brief overview of popular tools for both offline data processing and stream processing.

20.5.1 Apache Hadoop

Apache Hadoop is a well-known open-source data processing framework, licensed under the Apache v2 license. It supports data-intensive distributed applications that can run simultaneously on large clusters of commodity hardware in a reliable and fault-tolerant manner. It was designed for batch mode data processing, which limits its usefulness in the CPS control loop, but still makes it suitable for processing the data collected by the CPS. As its data processing model is based on the MapReduce paradigm, a Hadoop job (consisting of the map and reduce tasks) needs to be *programmed*, which typically requires more skill than the industry-standard database querying using a declarative language like SQL.

20.5.2 High-Performance Computing Cluster

High-performance computing cluster (HPCC) is an alternative to Hadoop that can, in many situations, show significant performance benefits. It is also available as open-source software. HPCC works with structured and unstructured data; similarly as Hadoop, it is highly scalable and fault resilient and includes a distributed file system. Its primary differentiator, compared to Hadoop, is the Enterprise Control Language, which is implicitly parallel and declarative and contains high-level primitives such as *join, transform, project, sort, distribute,* and *map.* In Hadoop's MapReduce model, nearly all complex data transformations require a series of MapReduce cycles executed in series, which is not the case in HPCC.

20.5.3 Apache Drill

Apache Drill is an open-source, low-latency SQL query engine for Hadoop. It offers a distributed system to perform interactive analyses of large-scale datasets. Its key architectural components include distributed query optimization and execution, columnar execution, vectorization, runtime compilation, and code generation. In the context of CPS, it is most suitable for interactive exploration of the stored data. Such analyses are typically performed on demand, by a domain expert trying to discover patterns in the data. The optimizations that Apache Drill provides reduce the latency between iterations, which significantly speeds up the process of data analysis and knowledge discovery.

20.5.4 Apache Storm

Apache Storm is a real-time distributed computation system that makes it easy to reliably process unbounded streams of data. As such, it has many use cases, such as real-time analytics, online machine learning (ML), and continuous computation. It is horizontally scalable and fault tolerant and offers data processing guarantees, giving assurance that each message or event is processed at least once. Due to its stream-based nature, it is well suited for use in the control loop. However, due to the distributed nature of the system, data processing can be performed in a stateless manner on individual messages only.

20.5.5 EsperTech Esper

Esper is an open-source (GPLv2) CEP engine, capable of processing large volumes of incoming messages or events. It offers a domain-specific language for processing events, called Event Processing Language; this is a declarative language for dealing with high-frequency event data with a similar syntax as the SQL. It supports windowing functions, which makes it easy to calculate moving averages and running totals and perform other operations on a sliding subset of the data stream. However, an Esper CEP system cannot be easily scaled out if the data stream is not partitionable without dependencies. If, on the other hand, the data stream can be partitioned into independent subsets, Esper and Storm can be combined by embedding Esper CEP engine into a storm processing node, which makes the system horizontally scalable.

20.5.6 Apache SAMOA

SAMOA stands for *Scalable Advanced Massive Online Analysis* and is a nascent platform for mining big data streams. It is a distributed and streaming ML framework that contains a programming abstraction for distributed streaming ML algorithms. As a platform, it allows the algorithm developer to abstract from the underlying execution engine and, therefore, reuse their code to run on different engines. It also allows to easily write plug-in modules to port SAMOA to different execution engines. In addition, it also provides a library containing state-of-the-art implementations of algorithms for distributed ML on data streams.

20.6 Building a Data Processing Pipeline

The data processing pipeline should be able to address all the challenges presented so far. As stated, one of the most important technical challenges for a CPS is handling the data velocity to ensure the response to events in a timely manner. This can be achieved much easily if event-driven data processing paradigm is adopted. Thus, an event message bus can provide a backbone for data transmission among other components (Figure 20.5), such as persistence layers, event processing modules, and input/output interfaces (APIs) (Joseph, 2010).

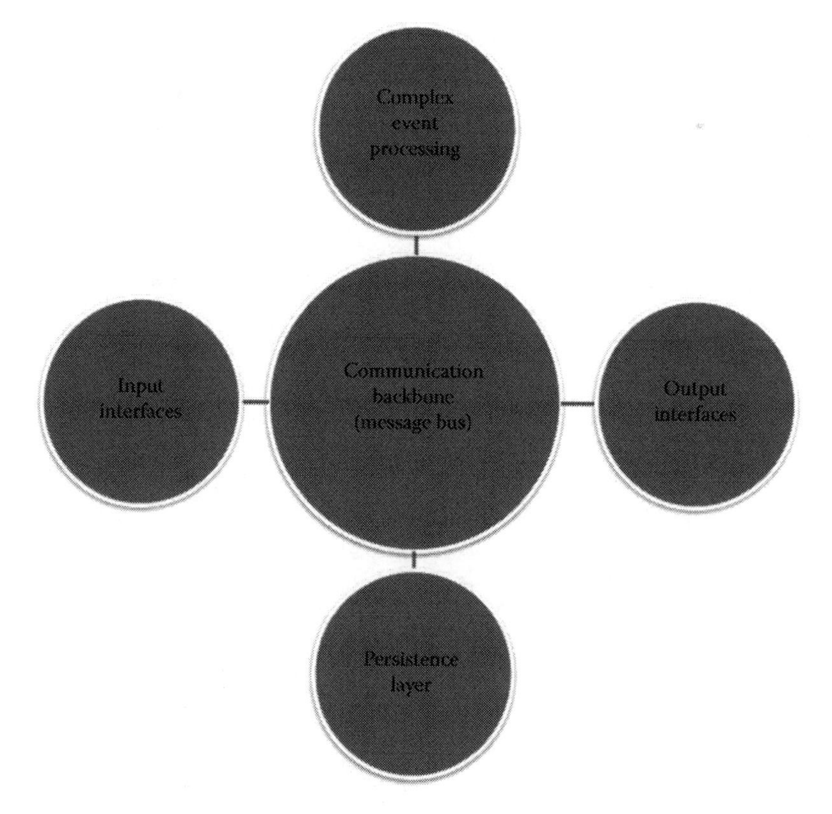

FIGURE 20.5
Building blocks of a data processing pipeline.

Another common requirement in CPS is data persistence for the purpose of subsequent analysis. Persistence can be done on any of the three conceptual levels:

- Raw data (directly from the data source). Storage of raw sensor data is crucial for the purposes of the analysis that was not available or possible at the time of event occurrence (e.g., searching for patterns in historical data, which were only discovered later). However, an even more important reason for retaining raw data is its usefulness in scenarios where nonrepudiation, identification of responsibility, and high reliability play a role. Some examples of such scenarios would be the use cases that can result in financial losses or can have life-threatening consequences. In the event of an error in the algorithms implemented in the CEP subsystem, it is necessary to review the reference (source) data that have not been tampered with to identify where the problems originate.
- Preprocessed events (either directly from the preprocessing modules), which represent the cleaned and normalized data, most suitable for further processing. However, inconsistencies and errors may already be present in this phase (data filtered in a wrong way, lost data, etc.).
- The final processed events, representing the result of the CEP system. The set of resulting data provides a frame of reference for subsequent control of the CPS.

For the persistence of data, either RDBMS with sharding (two open-source examples include PostgreSQL and MySQL) or any of a growing array of NoSQL databases (such as Cassandra, MongoDB, Redis, and CouchDB) can be used. MapReduce principles can as well be used for the data storage only (without processing). Apache Hadoop can be extended with database functionalities using HBase or Hive frameworks. Certain products on the market are promising an abstraction that provides a persistence layer abstraction that blurs the line between the RDBMS and NoSQL worlds. Two examples include Apache Drill and Cloudera Impala.

On the data processing front, Apache Hadoop is a reliable and mature framework; however, the persistence of data to disk after individual map and reduce jobs can make it too slow for a number of applications, especially the ones dealing with repeated operation on the same data. A significant speedup in cases like these can be achieved by moving the data processing into the random access memory, which can yield speedups of 10–100×. Apache Spark is an open-source framework that runs on top of the Hadoop infrastructure and provides a significant advantage for iterative computation. However, when data need to be processed in an event-driven fashion with complex event detection facilities, open-source frameworks like Esper can be used in addition to a number of commercial implementations (e.g., Microsoft StreamInsight, Oracle Event Processing).

20.6.1 Local and Remote Processing

As discussed, the data in a CPS system can either be processed locally where they originate, remotely (in the cloud), or by using both models. As different applications and use cases present different requirements, this is an important aspect to take into account.

- By processing the data locally, latencies can be minimized and the negative impact of losing connectivity or having connectivity with a low QoS and quality of experience can be diminished.

- By processing the data in the cloud, the resources can be elastically provisioned and software and algorithms seamlessly updated. Larger datasets or data from other sources (such as other devices) or even the World Wide Web can be used in data fusion and data enrichment scenarios.
- However, in many cases, both approaches need to be used, to leverage strengths of both alternatives and minimize the downside of network and other problems that might impact the quality of the CPS service provided.

Thus, a crucial step in estimating which approach is most suitable for a certain application is the understanding of the project requirements and the expectations of its users. If a smart home system features intelligent lighting and also allows the lights to be turned on and off using wall-mounted connected switches, the latency between the user pressing the switch and the light turning on should not exceed what users typically expect in such scenarios (i.e., an imperceptibly fast response). This exact feature is thus poorly suited to remote processing and should either be realized locally or in a hybrid manner (the latter being, start processing the event both locally and remotely, but if a certain timeout is exceeded, continue with autonomous local execution only) (Agrawal et al., 2011).

When using remote processing, cloud technologies can be used to easily scale the system resources (data storage, processing power, etc.). Scaling is usually achieved in two ways: vertically, by fitting a virtual server with more resources, or horizontally, by adding more server instances. To scale a system horizontally, however, the actions it performs must be parallelizable, which can present challenges in certain systems (e.g., splitting CEP load over multiple machines if the processing requires taking into account all of the past data in near real time). There are various solutions to that problem: the simplest one is to find independent and thus parallelizable subsets of the problem, while a more complex solution would be to design the system with time-constrained data exchange mechanisms in mind from the beginning.

20.7 Use Cases, Applications, and Their Specifics

CPSs are emerging as the next generation of embedded and interconnected information and communications technology (ICT) systems that will provide citizens and businesses with a wide range of innovative applications and services. CPSs have the potential to transform existing industries such as energy, healthcare, defense, transportation, building and process control, ICT, emergency response, and, agriculture, as well as create new ones and, in the process, influence all sectors of industry. Finally, the advances in CPS technologies will help us realize the vision of a smarter planet.

CPSs are in many ways still considered to be a nascent technology. Few CPS applications in *health care* have been proposed to date, and the ones in existence lack the flexibility and deep integration. Overall, the research on CPS in health care is still in the early stages; numerous opportunities for application include the introduction of coordinated interoperation of autonomous and adaptive medical devices, as well as new concepts for managing and operating medical systems using computation and control, provided by miniaturized implantable smart devices, usage of body area networks, programmable materials, and new fabrication approaches. Wireless sensors will enable collection of patients' vital data that bear information about their health and well-being. Wireless sensors provide more flexibility and comfort to both the caregiver and the patients. The data collected by the sensors and medical equipment can be stored on a server and made accessible to clinicians (Lee and Sokolsky, 2017). In CPS, the combination of active user input such as a smart feedback system, digital records of patient's data, and passive user input such as biosensors and smart devices in health-care environments can support efficient clinical decision-making. As the medical data can provide useful insight for diagnosing and treating the patient, all data should ideally be accessible to all authorized medical personnel. In addition, the WSNs collecting the patient data are severely constrained in energy, processing power, and storage capacity. Without the necessary resources for storage of large amounts of data and the resources to process it, the cloud computing can potentially provide solutions to some of these issues (Don and Min, 2014).

Many of the tools discussed in previous sections can facilitate data storage and exploration; tools like Apache Drill can be used for interactive exploratory analysis of the collected data, whereas a real-time data processing pipeline can be used in any use case that has a need for instantaneous detection of events, reporting, and actuation.

However, the medical and health-care domains are strongly regulated, which makes collection of large amounts of privacy-sensitive and user-identifiable information in a central location a huge challenge, which is further exacerbated by the fact that national laws and regulations change from country to country.

Modern *electric grids* serve as a canonical example of a CPS; they comprise a complex distributed system in which a portfolio of power generation resources must be managed dynamically to meet an uncontrolled time-varying demand. Today's electric grids are often described as CPS because the distributed resources must be dispatched in real time to match the uncontrolled and uncertain demand while adhering to limitations imposed by the power distribution grid that lies between. With the increased complexity of control, required by higher and higher fluctuations of supply, predominantly due to the popularity of renewable energy sources, the power distribution grid is becoming more and more informatized, thus quickly becoming a large-scale CPS. A smart grid CPS balances the constraints imposed by the transmission network, generator ramp rate capabilities, and emissions limits, while ensuring sufficient reserves to handle faults and failures (MITEI, 2014).

The primary control loop is typically realized through a system operator solving an iterative unit commitment problem; based on a prediction of load, an assignment of generation capacity is made on an hour-by-hour basis through a day-ahead auction that accounts for transmission limits and losses and maintains a certain amount of reserve. The matching of generation to load is refined through hour-ahead and 5 min-ahead markets based on recently observed demand. The feedback from load to generation is ultimately manifested through power quality observations, that is, frequency fluctuations and voltage deviations resulting from any mismatch. Generators respond to mismatches by engaging more or less of the reserve. Commercially, there are widespread efforts to utilize information technology to improve the efficiency and effectiveness of this CPS through the so-called smart meters that monitor loads and report on 15 min intervals, synchrophasors that observe power quality at intermediate points in the transmission grid, and delivery of pricing signals to trigger a response from demand. These efforts begin to introduce information planes to augment the physical planes of classic electric grids (Tham and Luo, 2012).

In the automotive field, the extensive integration of sensor networks and computational power into automotive systems over recent years has enabled the development of various systems to assist the driver during monotonous driving conditions and to protect the passenger from hazardous situations. This trend will inevitably lead to *autonomous vehicles* (Figure 20.6). In order to plan actions and reliably negotiate traffic, these vehicles need sensors capable of fault-tolerant observation of their environment. Additionally, vehicle

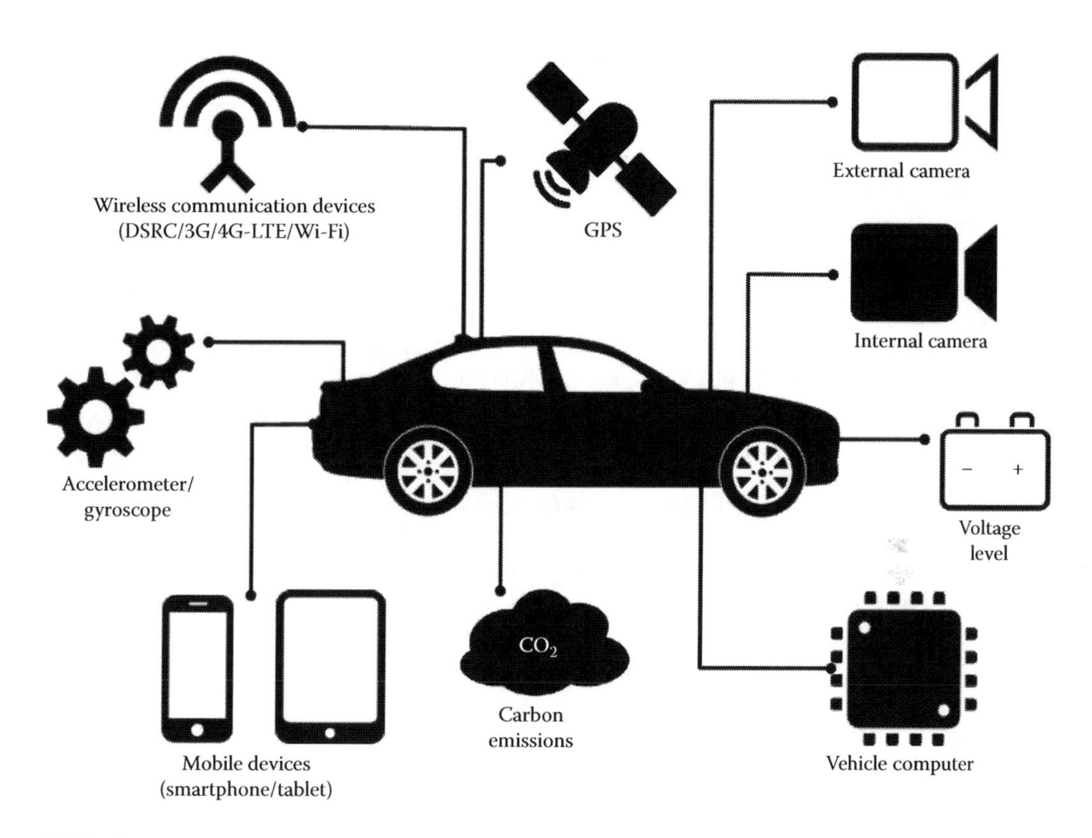

FIGURE 20.6
Sensors and building blocks powering the vehicular CPS.

to vehicle and vehicle to infrastructure communication technologies—also known as V2X communication—will be integrated into future automotive systems (Höftberger and Obermaisser, 2013).

V2X communication allows exchange of information between vehicles and roadside units about the position and speed of vehicles, driving conditions on a particular road, accidents, or traffic jams. Information exchange thereby allows traffic load to be distributed among several roads during rush hour, as well as prevents accidents and multiple collisions and sends automated emergency calls. Fast error detection, fault-tolerant system designs, and new planning strategies are required to cope with the increasing failure rates of microchips owing to continuous miniaturization of devices, as well as reliance on unreliable sources of information (e.g., information sent by other vehicles) (Jeong and Lee, 2014).

The convergence of CPSs and social computing will also lead to the emergence of *smart cities*, composed of various objects (including both individual citizens and physical things) that interact and cooperate with each other. Smart cities promise to enable a number of innovative applications and services that will improve the quality of life. Many physical things already possess computing and communication capabilities of different levels, which are provided by small and (possibly) invisible computers embedded therein.

A smart city can be roughly understood as a group of connected objects that interact with each other over ubiquitous networks and deliver smart services to possibly all citizens. In most cases, these objects have implicit links among them. For example, they may jointly strive to reach a common goal, for example, wide-area sensing or incident detection (Wang, 2010). Among the data sources, location data are one of the most useful types of cyber-physical information to be collected and have many applications. Thanks to the proliferation of positioning techniques and devices like global positioning system on mobile phones, it has also become readily available. With all of the collected observation data, it is possible to infer what is happening in the environment, construct a model of its dynamics, conduct spatiotemporal reasoning about future behaviors, and act based on its outputs. To gain full knowledge of the current situations in different aspects, smart cities demand distributed multimodal data collection in near real time. For this purpose, the deployed sensors must perform the data collection in a collaborative way, leading to the entire community becoming smarter and having the capability of delivering intelligent services to interested users (Simmon et al., 2013).

Due to the fact that cities already rely on thousands of publicly deployed sensors with much more lax privacy constraints as, for example, in the case of health care, several cities have already started providing real-time event streams and large databases of historical data. This provides a great opportunity for third-party developers to connect to data sources and create new *data mash-up* applications and analyses that can culminate in a fully featured CPS. NYC OpenData is a great example of such an initiative with more than 1200 datasets currently available (as of 2015), which inspired hundreds of developers to use the tools and techniques described in previous sections to explore and analyze the data and create CPS applications.

Not limited to the presented use cases, the CPS will in the future fundamentally change business models and the competitive playing field. New suppliers of services based on the CPS will enter the markets and revolutionary applications will define and create new value chains, transforming many other industries as well.

20.8 Conclusion

CPSs will fuel the development of a modern vision of smart services by fully taking advantage of modern connectivity and computation technologies, expanding far beyond the traditional embedded systems or today's limited sensor networks. We can summarize some of the characteristics that define them: (1) computation and connectivity capabilities will penetrate into almost every physical component; (2) devices and systems will be networked on multiple levels, reaching global scale; (3) the resulting systems will be dynamically reconfiguring and reorganizing; (4) they will exhibit high degrees of automation based on closing the control loops; (5) their operations will have to be dependable and certified to become a viable alternative in many of mission-critical systems; (6) and the systems will exhibit more and more of learning, adaptation, and self-organization properties. Whereas the first few points describe merely the enablers of CPS, the last point specifically addresses a characteristic that can only be realized by analyzing data. In addition, the amount of data collected in CPS will without a doubt continue to increase at a breakneck speed. Thus, processing the generated big data in both on-demand and event-driven manner will prove to be a crucial enabler for continued progress in the area of CPS and many of the subdisciplines. Currently, the research is divided into many isolated subdisciplines, including communications and networking, systems theory, mathematics, software engineering, computer science, and electronics. From here on, many challenges have to be addressed, including the effective construction of infrastructure, improving performance assurance, constructing reliable system integration, and bringing together multidisciplinary teams to develop fundamentals for the next industrial revolution.

Glossary

Anonymization: Anonymization is the process of removing identifiable information and datasets from the system to protect the identity of the owners of the data, thus making it either totally anonymous or pseudo-anonymous.

Big data: Big data are used to describe a massive volume of both structured and unstructured data that are so large that they are difficult to process using traditional database and software techniques. In most enterprise scenarios, the data are too big or they move too fast or they exceed processing capacities. An example of big data might be petabytes (1024 terabytes) or exabytes (1024 petabytes) of data consisting of billions to trillions of records of millions of items—all from different sources (e.g., web, sales, customer contact center, social media, and mobile data).

Complex event processing: CEP is a data processing technique that creates actionable, situational knowledge from distributed message-based systems, databases, and applications in real time or near real time. CEP can provide an organization with the capability to define, manage, and predict events, situations, exceptional conditions, opportunities, and threats in complex, heterogeneous systems.

Cyber-physical systems: CPSs are integrations of computation, networking, and physical processes. CPS integrates the dynamics of the physical processes with those of the software and networking, providing abstractions and modeling, design, and analysis techniques for the integrated whole. Embedded computers monitor and control the physical processes, with feedback loops where physical processes affect computations and vice versa.

Data preprocessing: Data preprocessing is a technique that involves transforming raw data into an understandable format. Real-world data are often incomplete, inconsistent, and are likely to contain many errors. Data preprocessing is a method of resolving such issues, as it prepares raw data for further processing.

Human in the loop: Human-in-the-loop refers to systems that rely on human input as part of their control. Human operators can often control many aspects of the system better than software, while at the same time this also provides the necessary data for subsequent training of the control software.

Internet of things: The IoT refers to the ever-growing network of physical objects that feature an IP address for Internet connectivity and the communication that occurs between these objects and other Internet-enabled devices and systems. The IoT extends Internet connectivity beyond traditional devices like desktop and laptop computers, smartphones, and tablets to a diverse range of devices and everyday things that utilize embedded technology to communicate and interact with the external environment, all via the Internet.

MapReduce: MapReduce is a programming model for processing large quantities of data in a distributed manner on a cluster of servers. The model is inspired by a set of functions, *map* and *reduce,* commonly used in functional programming. A typical MapReduce operation consists of two high-level steps; in the first step (map), each worker in the cluster applies a certain data processing function to the data stored locally (on the worker node) and yields the output. The second step reduces (summarizes) the partial results into the final result of the operation.

NoSQL: NoSQL is a database that features fewer consistency restrictions than conventional relational databases. While most often called NoSQL databases, these databases are sometimes also referred to as "Not Only SQL" databases to reflect that SQL queries can be used in some cases.

NoSQL databases feature highly optimized key-value data designed for quick retrieval and appending operations, and they can offer substantial performance benefits over relational databases in terms of latency and throughput.

Real-time processing: Real-time processing is usually found in systems that use computer control. This processing method is used when it is essential that the input request is dealt with quickly enough so as to be able to control an output properly.

Sharding: Database sharding is used to describe a horizontal partitioning scheme, where data is split over multiple machines based on a certain key.

Reference

Abdelzaher, T. (2006). NSF workshop on "Cyber-physical systems". Retrieved October 7, 2014, from http://varma.ece.cmu.edu/CPS/: http://varma.ece.cmu.edu/CPS/Position-Papers/abdelzaher.pdf.

Abdelzaher, T., Gill, C. D., Rajkumar, R., and Stankovic, J. A. (2007). GENI—Exploring networks of the future. Retrieved October 8, 2014, from http://groups.geni.net/geni/wiki/OldGPGDesignDocuments: http://groups.geni.net/geni/attachment/wiki/OldGPGDesignDocuments/GDD-06-32.pdf.

Agrawal, D., El Abbadi, A., Das, S., and Elmore, A. J. (2011). Database scalability, elasticity, and autonomy in the cloud. In *DASFAA'11 Proceedings of the 16th International Conference on Database Systems for Advanced Applications*, Volume Part I, pp. 2–15.

Bouhouchi, R. and Ezzedine, T. (2012). Wireless sensors networks architecture. *International Journal of Emerging Trends & Technology in Computer Science (IJETTCS)*, 1(4), 131–134.

Bryden, J. (2014). NoSQL guide. Retrieved October 2, 2014, from http://nosqlguide.com/nosql-patterns/database-sharding-explained/.

Couchbase. (20014). Couchbase. Retrieved October 1, 2014, from http://www.couchbase.com/nosql-resources/what-is-no-sql.

Don, S. and Min, D. (2014). Penn engineering. Retrieved November 3, 2014, from http://www.seas.upenn.edu/: http://www.seas.upenn.edu/~zhihaoj/cps13/min.pdf.

Gorton, I. and Klein, J. (May 2014). Software Engineering Institute. Retrieved September 30, 2014, from https://resources.sei.cmu.edu/library/: https://resources.sei.cmu.edu/asset_files/WhitePaper/2014_019_001_90915.pdf.

Habbegger, B., Hasan, O., Brunie, L., Bennani, N., Kosch, H., and Damiani, E. (2014). Personalization vs. privacy in big data analysis. *International Journal of Big Data (IJBD)*, 1(1), 25–35.

Höftberger, O. and Obermaisser, R. (2013). Ontology-based runtime reconfiguration of distributed embedded real-time systems. In *2013 IEEE 16th International Symposium on Object/Component/Service-Oriented Real-Time Distributed Computing (ISORC)*, pp. 1–9.

Jara, A. J., Genoud, D., and Bocchi, Y. (2014). Big data for cyber physical systems: An analysis of challenges, solutions and opportunities. In *Innovative Mobile and Internet Services in Ubiquitous Computing (IMIS)*, 2014 Eighth International Conference on IEEE, pp. 376–380.

Jeong, J. P. and Lee, E. (2014). Vehicular cyber-physical systems for smart road. *KICS Magazine*, pp. 104–116.

Joseph, S. A. (2010). The University of Edinburgh. Retrieved October 25, 2014, from The University of Edinburgh: http://www.inf.ed.ac.uk/publications/thesis/online/IM100831.pdf.

Lee, I. and Sokolsky, O. (June 17, 2017). Penn Libraries. Retrieved October 29, 2014, from http://repository.upenn.edu/: http://repository.upenn.edu/cgi/viewcontent.cgi?article=1465&context=cis_papers.

Lee, K.-H., Lee, Y.-J., Choi, H., Cjumg, Y., and Moon, B. (2011). Parallel data processing with mapReduce: A survey. *SIGMOD Record*, 40(4), 11–20.

Liu, A. and Salvucci, D. (2001). Modeling and prediction of human driver behavior. In *Proceedings of the Ninth International Conference on Human-Computer Interaction*, pp. 1479–1483.

Liu, Z., Jiang, B., and Heer, J. (2013). imMens: Real-time visual querying of big data. *Eurographics Conference on Visualization (EuroVis)*, 32(3), 421–430.

Luckham, D. C. (2002). *The Power of Events: An Introduction to Complex Event Processing in Distributed Enterprise Systems*. Boston, MA: Addison-Wesley Longman Publishing Co., Inc.

Maguire, W. (March 19, 2013). VentureBeat. Retrieved October 20, 2014, from http://venturebeat.com/2013/03/19/how-to-conquer-big-data-with-mapreduce-mpp/.

MITEI. (2014). MITEI. Retrieved November 5, 2014, from http://mitei.mit.edu/publications/reports-studies.

Mohanty, S., Jagadeesh, M., and Srivasta, H. (2013). *Big Data Imperatives: Enterprise 'Big Data' Warehouse, 'BI' Implementations and Analytics*. Apress.

Nie, K., Yue, T., Ali, S., Zhang, L., and Fan, Z. (2013). Constraints: The core of supporting automated product configuration of cyber-physical systems. In *Model-Driven Engineering Languages and Systems* (pp. 370–387). Springer, Berlin, Heidelberg.

Pettey, C. and Goasduff, L. (June 27, 2011). Gartner. Retrieved September 16, 2014, from http://www.gartner.com/newsroom/id/1731916.

Pinal, D. (October 2, 2013). Journey to SQL authority with Pinal Dave. Retrieved September 16, 2014, from http://blog.sqlauthority.com/2013/10/02/big-data-what-is-big-data-3-vs-of-big-data-volume-velocity-and-variety-day-2-of-21/.

Rivera, T. (2012). www.snia.org. Retrieved September 19, 2014, from SNIA—Advancing Storage and Information Technology: http://www.snia.org/sites/default/files2/ABDS2012/Tutorials/ThomasRivera_Protecting_Data_in_the_Big_Data_World_08-30-12.pdf.

Rouse, M. (2011). TechTarget. Retrieved October 1, 2014, from http://searchcloudcomputing.techtarget.com/definition/sharding.

Schirner, G., Erdogmus, D., Chowdhury, K., and Padir, T. (2013). The future of human-in-the-loop cyber-physical systems. *IEEE Computer*, 46(1), 36–45.

Simmon, E., Kim, K.-S., Subrahmanian, E., Lee, R., de Vaukx, F., Murakami, Y. et al. (August 2013). NICT. Retrieved November 20, 2014, from http://www.nict.go.jp/: http://www2.nict.go.jp/univ-com/isp/doc/NIST.IR.7951.pdf.

Son, N. H. (2006). http://www.mimuw.edu.pl. Retrieved September 19, 2014, from Data Mining Courses (Warsaw University): http://www.mimuw.edu.pl/~son/datamining/DM/4-preprocess.pdf.

Soukup, T. and Davidson, I. (2002). *Visual Data Mining*. Toronto, Ontario, Canada: John Wiley & Sons, Inc.

Stockinger, H. (2001). Distributed database management systems and the data grid. In *MSS '01 Proceedings of the Eighteenth IEEE Symposium on Mass Storage Systems and Technologies*, pp. 1–12.

Stonebraker, M., Çetintemel, U., and Zdonik, S. (2005). The 8 requirements of real-time stream processing. *ACM SIGMOD Record*, 34(4), 42–47.

Strauch, C. (2011). Department of Software and Information Systems. Retrieved September 30, 2014, from http://sis.uncc.edu/: http://sis.uncc.edu/.

Suh, S. C., Tanik, J., Carbone, J. N., and Eroglu, A. (2014). *Applied Cyber-Physical Systems*. New York: Springer Science+Business Media.

Tabuada, P. (2006). Workshop position papers on cyber-physical systems. Retrieved October 5, 2014, from http://varma.ece.cmu.edu/CPS/Position-Papers/Position%20Papers.htm: http://varma.ece.cmu.edu/CPS/Position-Papers/Tabuada.pdf.

Tham, C.-K. and Luo, T. (2012). Sensing-driven energy purchasing in smart grid cyber-physical system. In *IEEE Transactions on Systems, Man, and Cybernetics: System*.

Wang, F.-Y. (2010). The emergence of intelligent enterprises: From CPS to CPSS. *IEEE Intelligent Systems Archive*, 25(4), 85–88.

21

Big Data Analysis in Cyber-Physical Systems

Mayank Swarnkar and Robin Singh Bhadoria

CONTENTS

21.1 Introduction

Technology is growing day by day, and so people are using technology. Today, they have smartphones, tablets, PCs, cloud computing devices, and wireless sensor networks in their pockets. This results in a huge amount of data collection, which requires processing. The term "big data" refers to large-scale production, management, and analysis of information, which exceed the capability of conventional data-processing technologies. According to the Cloud Security Alliance, people today are creating about 2.5 quintillion bytes of data per day. The rate at which data are being created has increased so much that 90% of the total data have been created in the past 2 years alone. This rate of growth in the production of information has created a need for new technologies to analyze and process huge datasets.

Integration of computation, networking, and physical processes forms a cyber-physical system (CPS). Embedded computers and networks monitor and control physical processes with a loop for feedback, where physical processes affect computations and vice versa. The societal and economic potentials of these systems are immensely greater than what has been realized, and major contributions are being made globally to develop the technology. The technology builds on the older (but still young) discipline of embedded systems, laptop machines, and computer code implicit in devices whose principle mission is not computation, like toys, medical devices, wireless sensing element networks, cars, and scientific instruments, but to gather and transfer data from them.

21.2 About Big Data?

According to the Wikibon Big Data Analytics Survey, five billion mobile phones were in use in 2010. Thirty billion data pieces are shared every month on Facebook, and this rate is increasing day by day. Information technology (IT) is growing by 5% every year, which is contributing to increasing databases. About 235 TB of data were collected and stored by the U.S. Library until April 2011. Companies in 15 out of 17 sectors in the United States have additional knowledge than the U.S. Library. Since an oversized quantity of knowledge is being collected with an awfully high speed, this results in the formation of big data.

21.2.1 Three V's Show Big Data Requirements

Transformation of traditional data to big data is defined by the three V's. The reason for generation and processing of big data is defined by the following characteristics change in the data traditionally and now:

- *Volume*: Today, the volume of data is very high. The data warehouse is turning small for storing data. The data stored in traditional databases were in terabytes (10^{12} bytes), which have now turned to zetabytes (10^{21} bytes).
- *Velocity*: The rate of data moving through the Internet has also increased. High-speed Internet services are provided to users due to requirements. We manage to do stream processing through batch processing.
- *Variety*: Previously, data processing was done through relational data base management system (RDBMS) since data were well structured. Today, data contain images, texts, videos, and other files. Therefore, new ways of data processing are required for unstructured data (Figure 21.1).

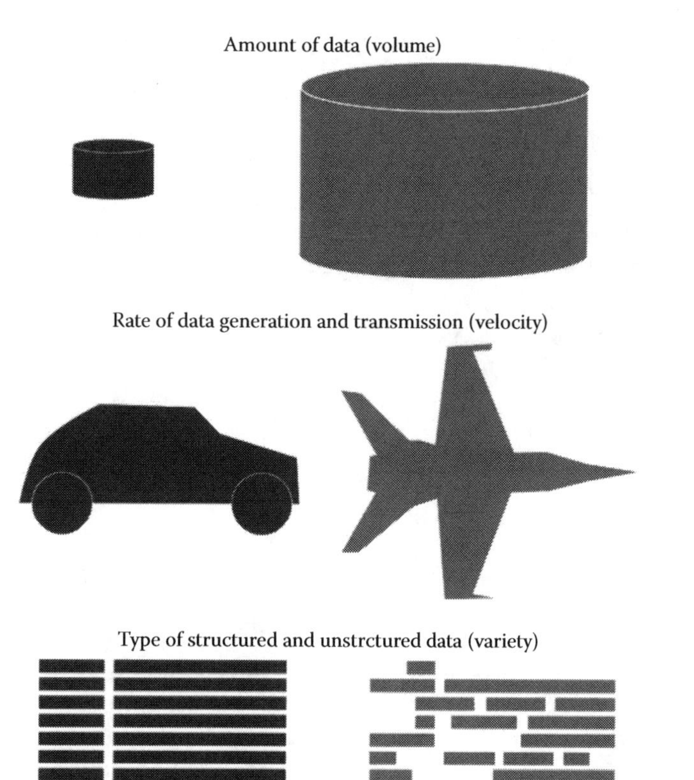

FIGURE 21.1
V's of big data. (Courtesy of Big data analytics for security intelligence by Cloud Security alliance, Seattle, WA; https://cloudsecurityalliance.org/download/big-data-analytics-for-security-intelligence/).

21.2.2 Big Data Drivers

Some factors lead to the formation of big data, which allow users to promote data creation, data transfer, and data storage. Big data come to picture because of the following scenarios:

- *Decrease in storage cost*: The cost of storage has dramatically decreased in the past few years. Different vendors are providing cloud storage free to users or at a cost that a user can bear. Therefore, old data warehouses retained data for a fixed and smaller time interval. Big data applications keep data for a very long time to understand long trends in a very sophisticated way.
- *Cost-effective and flexible data centers*: There are tools for big data such as Hadoop ecosystem and NoSQL databases, which offer technology to boost the processing time of complex and unstructured queries and their analytics.
- *Availability of new frameworks*: The extract, transform, and load (ETL) technique in old data warehouses is inflexible because users have to define schemas in advance. Hence, after a data warehouse has been placed, introducing new schemas might be difficult. Big data tools allow users flexibility to load structured and unstructured data in different types of formats, and users can choose

how best to use the data. Many of these data-processing frameworks such as Hadoop are provided free to users, which results in the promotion of big data processing.

21.2.3 Importance of Big Data

The importance of big data has been explained by Accenture in a survey. In general, big data are important because of the following reasons:

- Big data solutions are best for analyzing all varieties of data, whether the data are structured, unstructured, or semi-structured.
- Big data solutions are best for comparing new datasets with old datasets; this is best for searching in the data, training a dataset, or for security matches.
- Big data solutions are ideal for exploratory and iterative analysis when analyzing data is not biased.
- Big data fit best for solving information challenges that do not originally fit within a traditional RDBMS approach for handling a given problem.

21.2.4 Big Data Processing Techniques

Big data need to be processed in a fast, effective, and accurate manner. To complement these requirements, there are two types of processing techniques: batch processing and stream processing. Figure 21.2 shows the type of data that are suited for batch processing and steam processing.

Batch processing: When data are processed in a group or bunch of defined size, the method of processing is known as batch processing. The data in rest are processed by batch processing. For example, batch processing of big data is done in Hadoop.

Stream processing: When data are processed as soon as they arrive, the method of processing is known as stream processing. The data in motion are processed by stream processing. Fore example, stream processing of data is done by Storm.

21.2.5 Big Data Analytics

Big information analytics is the method of collecting, organizing, and analyzing massive sets of information (known as huge data) to mine for patterns and helpful information. The need for analysis and leverage of information collected is one of the drivers for large information analysis tools.

In other words, huge information analytics is the method of examining massive datasets containing a variety of information to reveal unknown correlations, hidden patterns, market trends, and user preferences. The analytical findings can lead to better service, more effective opportunities, improved operational efficiency, and processing of mined data.

The primary goal of big data analytics is to help an entity to make more informed decisions by enabling data scientists, analytical modeling designers, and other analytics professionals to analyze large volumes of transaction data and other types of data that

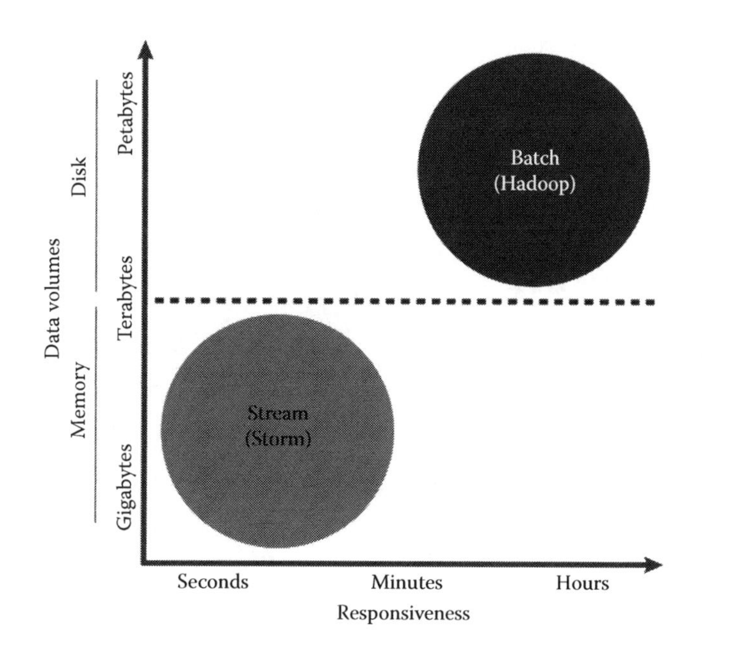

FIGURE 21.2
Batch processing versus stream processing. (Courtesy of Big data analytics for security intelligence by Cloud Security alliance, Seattle, WA; https://cloudsecurityalliance.org/download/big-data-analytics-for-security-intelligence/)

may be undeveloped by conventional programs of data analysis. That could include web server logs and Internet-enabled stream data, social media and social network activity news, emails from respective customers and survey response, cellular call detail records, and machine data captured by sensors connected to the Internet of Things.

Big data analysis can be done via software tools commonly used as part of elevated analytics disciplines such as text analytics, data mining, predictive analytics, and statistical analysis. Data visualization tools and mainstream software can also play an important role in the analytical process. But semi-structured and unstructured data may not fit well in legacy data warehouses based on RDBMS. Moreover, information warehouses might not be able to handle the process demands display by sets of massive information that require to be updated overtimes or perhaps perpetually, for example, period of time information on the potency of mobile applications or of oil and gas pipelines or others. Thus, the introduction of newer category of technologies that features Hadoop and connected tools such as YARN, Map Reduce, Spark, Hive, and Pig furthermore as NoSQL information bases needed for large data analysis. Those technologies kind the core of AN open supply software package framework that supports the process of huge information sets across clustered systems in standalone and distributed setting.

21.2.6 Processing in Big Data

21.2.6.1 Batch Processing in Big Data

In the traditional database world that supports RDBMS, all processing takes place after the data have been loaded into the store or warehouse, using any specialized query language such as SQL on highly structured and optimized data structures. The approach, initiated by Google and adopted by several other Internet corporations, is to instead type

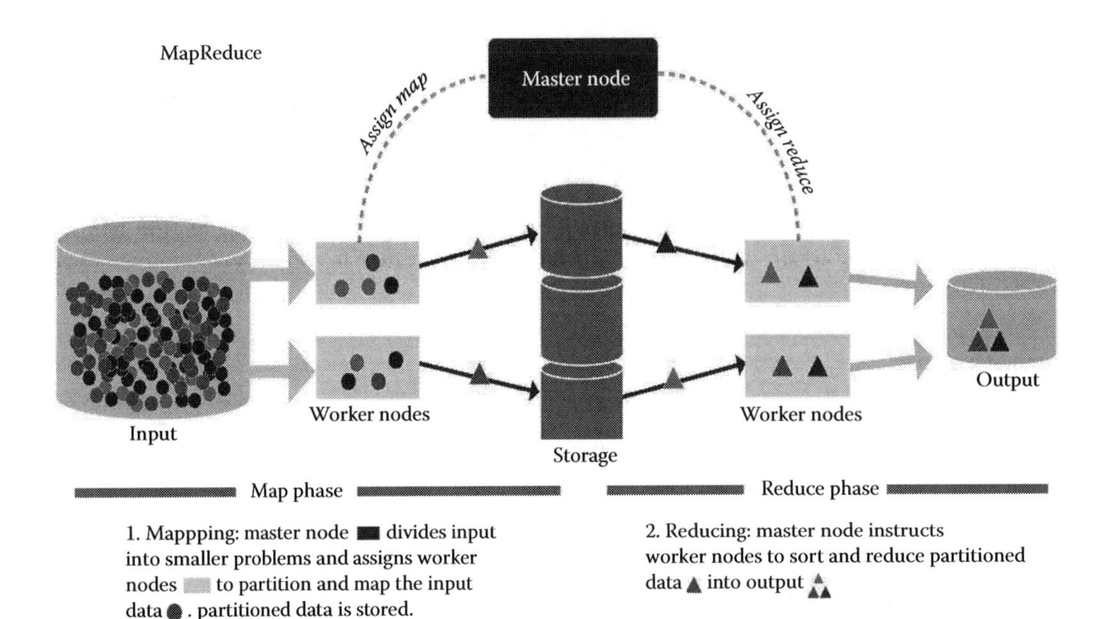

FIGURE 21.3
Working of MapReduce. (Courtesy of Big data analytics for security intelligence by Cloud Security alliance, Seattle, WA; https://cloudsecurityalliance.org/download/big-data-analytics-for-security-intelligence/).

a pipelined structure that reads from and writes to completely different file formats, with intermediate sub-results being passed between stages within the kind of files, with the process unfolds across several machines in distributed setting. The new approach to process data is known as MapReduce, as discussed by Jeffrey Dean and Sanjay Ghemawat. Figure 21.3 shows the working of MapReduce. Dean and Ghemawat explain about map reduce in big clusters (Dean and Ghemawat, 2004).

The term "map cut back" really refers to two separate and distinct tasks. The primary is the map job that takes a group of information and converts them into an extra set of information, where every component is lessened into tuples (key/value pairs), and the second is the cut back job that takes the output from a map as input and combines the information tuples into a smaller set of tuples. The name suggests that there are two processes—Map and Reduce—in which the map process comes first and then comes reduce.

21.2.6.1.1 Example of MapReduce

Suppose there are five files and every file comprises two columns (key and value). It can also be resembled by a town and its corresponding temperature recorded. This instance is straightforward; therefore, it is easy to follow. You might think that a true application would not be so straightforward since it is possible to contain millions or billions of rows and that it would not show precisely formatted rows at all. However, regardless of the quantity of data you wish to investigate, the key principles we will cover here stay identical. Either way, during this example, town is the key and temperature is the price.

Delhi, 20
Mumbai, 25
Kolkata, 22

Chennai, 32
Delhi, 4
Chennai, 33
Kolkata, 18

Of all the information we have collected, we would like to search the utmost temperature for every town across all the information files (note that every file might need constant town portrayed multiple times). Victimizing the Map scale back framework, we will break this down into five map tasks, where every mapper works on one of the five files, and therefore, the mapper goes through the information and returns the utmost temperature for every town. For instance, the results made from one mapper task for the information on top would appear as if this: (Delhi, 20) (Mumbai, 25) (Kolkata, 22) (Chennai, 33).

Let us take another four sets, which (not displayed here) demonstrate following transitional results:

(Delhi, 18) (Mumbai, 27) (Kolkata, 32) (Chennai, 37)(Delhi, 32) (Mumbai, 20) (Kolkata, 33) (Chennai, 38)(Delhi, 22) (Mumbai, 19) (Kolkata, 20) (Chennai, 31)(Delhi, 31) (Mumbai, 22) (Kolkata, 19) (Chennai, 30)

All five of those output streams would be fed into the cut back tasks that mix the input results and output one price for every town, manufacturing the end result set as follows:

(Delhi, 32) (Mumbai, 27) (Kolkata, 33) (Chennai, 38)

As associate degree analogy, you will be able to think about map and cut back tasks because the method a census was conducted in Roman times, where the office dispatched its officials to every town within the empire. Every enumerator in every town counted the number of individuals in the town and came back to the capital town. Then, the results from all towns were reduced to one count (sum of all cities) to see the total population of the empire. This method of mapping individuals to cities, in parallel, and then combining the results (reducing) is economical than asking one person to serially count everyone in the empire.

21.2.6.2 Frameworks for Batch Processing of Big Data

Hadoop: Yahoo originally developed Hadoop as a parallel infrastructure of Google's MapReduce infrastructure, but afterward, it became open sourced. Hadoop works in a distributed environment and takes care of running code across a cluster of machines in a master–slave configuration. Working of Hadoop includes classifications of input file, distributing it to every machine, running its application code on every node, checking that the code ran properly on the input, passing results either on to any mid stage or to the final location of output, responsible for the formation that happens between the map and reduce stages and make every chunk of that organized information to the right machine, and writing data that is processed on every job's execution. It is one of the most popular frameworks for MapReduce Algorithm.

Hive: With Hive, a user can program Hadoop jobs using SQL. It is the best interface for any user coming from the relational database world, although the details of the inherited implementation are not completely hidden. Hive provides the ability to plug in custom (user-defined) code for situations that are not implemented in SQL, as well as a number of tools for managing input and output. To use it, a user has to set up structured tables that describe input and output, issue load commands to take files, and then write queries as the user does in any other RDBMS. Hadoop's focus is on large-scale data processing, and the delay may mean that even simple jobs take time to finish, so it is not a choice for a real-time transactional database.

Pig: Given by Apache, Pig is a procedural data-processing language designed for Hadoop. In contrast to Hive's approach of composing logical queries, with Pig, a user can specify a series of steps to execute the data, which we usually call script. It is similar to any usual scripting language but with a focused set of functions that help with the general problems of data processing. It is easy to break up text into smaller chunks, for example, and then count how frequently each occurs. Other often-used operations such as filters and joins are also supported internally. Pig is normally used when a user's problem fits with a procedural approach, but the user needs to do typical data processing in place of general calculations. Pig is also known as "the duct tape of Big Data" because of its high usability there and can be pooled with custom streaming code written as a script for general operations.

Cascading: Cascading allows a user to define a series of processing steps, which is a complex workflow as a program. The user can place the logical flow of the data pipeline according to the need, rather than building it unambiguously out of MapReduce steps feeding into one another. The user needs to call a Java application program interface (API) connecting objects that represent the operations the user wants to perform. The system takes that definition, does some checking and planning, and processes it on Hadoop cluster. There are a lot of predefined objects for general operations such as sorting, grouping, and joining, and the user can create own objects to run custom processing code.

Other frameworks are Cascalog, mrjob, Caffeine, S4, MapR, Acunu, Flume, Kafka, Azkaban, Oozie, Greenplum, etc.

21.2.6.3 Stream Processing in Big Data

It is well known that Hadoop-like instruction-execution systems have been developed and established during the past few years as an outstanding offline processing platform for giant knowledge. Hadoop may be an extremely economical system, which might chomp a large volume of knowledge employing a distributed multiprocessing paradigm known as MapReduce. However, there are several cases across a spread of domains that need period or close to period response on massive knowledge for faster deciding. Hadoop is not applicable for those cases. Master Card fraud analytics, network fault prediction from detector knowledge, security threat prediction, then onward, has to method real-time knowledge stream on the wire to calculate if a given dealing may be a deception/fraud, if the system is initializing a slip, or if there is a security threat within the network. If selections like these do not seem to be taken in real time, the prospect to reduce the harm is lost. Period systems perform analytics on short time slots, that is, they correlate and predict events streams generated for the previous couple of minutes. Now, for higher guess of capabilities, period systems usually leverage instruction-execution systems such as Hadoop. For real-time data in the world of big data, there are two open-source technologies available: "Apache Kafka," which is the distributed messaging system, and "Storm," which is known for distributed stream-processing engine.

21.2.6.4 Real-Time Data-Processing Challenges

Real-time data-processing problems are very complex since big data are commonly characterized into volume, velocity, and variety of the data, and Hadoop-like systems take care of the volume and variety parts. Besides volume and selection, the important time system must handle the rate of information generation. Moreover, taking care of the velocity of big data is not an easy task. First, the system should be capable of collecting the data generated by real-time events streams coming in at the rate of thousands of events per seconds. Second, it must look out of the data processing of this information as and when it is being collected. Third, it should achieve event correlation using a compound event-processing engine to mine meaningful information from this moving stream. These three steps should take place in a fault-tolerant and distributed environment. The real-time system ought to be a coffee latency system so that the computation happens in no time with close to real-time response capabilities.

Storm and Kafka are the long run of stream process, and that they are currently in use at various high-profile firms' numeration Groupon, Alibaba, The Weather Channel, and plenty of additional. Twitter researched and provided Storm as a "distributed real-time computation system." At the same time, Kafka is a messaging system that was developed at LinkedIn to serve their action streams and the data-processing pipeline as a back end.

With Storm and Kafka, a user can conduct stream processing at a linear scale, with the guarantee that each message is processed in real time, in a reliable manner, and with high data-processing capability.

21.2.6.5 Frameworks for Stream Processing of Big Data

Apache Kafka: It is subscribe and publish messaging system in the distributed environment. It was initially researched at LinkedIn Corporation, and afterward, it became part of the Apache project. Its properties are fast, scalable, distributed processing, and division and replicated commit log service. It proposes high throughput for both publishing and subscribing. It has multimode multiuser support and automatically balances the units at the time of failure. It continues messages on disk and, thus, can be used for batch consumption as an add-on to real-time applications.

Storm: Storm is an open-source, freely available, and real-time computation system in the distributed environment. It has many applications such as real-time analysis, online machine learning, constant computation, distributed remote procedure call, ETL, and more. Storm is quick; a typical clocked it at more than 1,000,000 tuples processed per second per node. It is long and fault tolerant, ensures that the information is processed, and is straightforward to put in and operate. The concept behind Storm is like that of Hadoop. In Hadoop cluster, we tend to run map scale back job; similarly in Storm Cluster, there are unit topologies. The core abstraction in Storm is the "stream." A stream is an associate unrestrained sequence of tuples. Storm provides the primitives for sterilization one stream into a replacement stream in an exceedingly distributed and consistent manner.

21.3 Cyber-Physical Systems

CPS integrates computation, networking, and physical processes. Don and Min explains embedded computers and networks monitor and manage the physical processes, with feedback loops where physical processes have an effect on computations. The economic and social group potentials of such systems are immensely bigger than what has been accomplished, and major investments are being created worldwide to develop the technology. This technology is style not for easy computation of embedded systems, however, to form the devices smarter enough, so they collect necessary knowledge set and transfer to central server. Cycle integrates the dynamics of the physical processes with those of the computer code and networking, providing abstractions, modeling, design, and analysis techniques for the integrated as a full.

Embedded hardware and computer code systems are the actual driving forces for analysis and innovation within the export and growth of CPS. This enlarged the practicality, and as a result, the sensible worth of vehicles, aircraft, medical instrumentation, production plants, and home appliances has inflated plenty. They are more and more connected with each other and the net via one medium or another. CPS is being developed as part of a globally networked future, within which merchandise, instrumentation, and objects act with embedded hardware and computer code on the far side the boundaries of single applications and supply multiple services. With the assistance of sensors, these systems method knowledge from the physical world and create it obtainable for network-based services, which successively will have an immediate result on the processes within the physical world victimization actuators. Through CPS, the physical world is coupled with the virtual world to create an online repository of things, knowledge, and services (Figure 21.4).

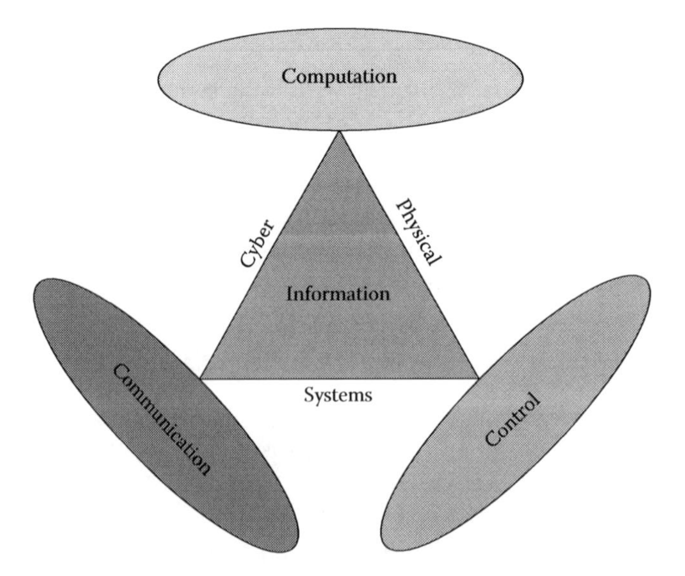

FIGURE 21.4
Cyber-physical systems. (Image courtesy of RFID Applications in Cyber-Physical System Nan Wu1 and Xiangdong Li2, August 17, 2011 under CC BY-NC-SA; http://www.intechopen.com/books/deploying-rfid-challenges-solutions-and-open-issues/rfid-applications-in-cyber-physicalsystem)

21.3.1 Applications of Cyber-Physical Systems

Applications of CPSs have the potential to midget the IT revolution. CPS is widely used in the following:

- Medical devices and systems
- Traffic control and safety systems
- Advanced automotive systems
- Process control systems
- Energy conservation systems
- Environmental control systems
- Avionics, instrumentations, critical infrastructure control (e.g., electric power, water resources, and communications systems)
- Distributed robotics
- Defense systems

At a glance, CPS is viewed as shown in Figure 21.5.

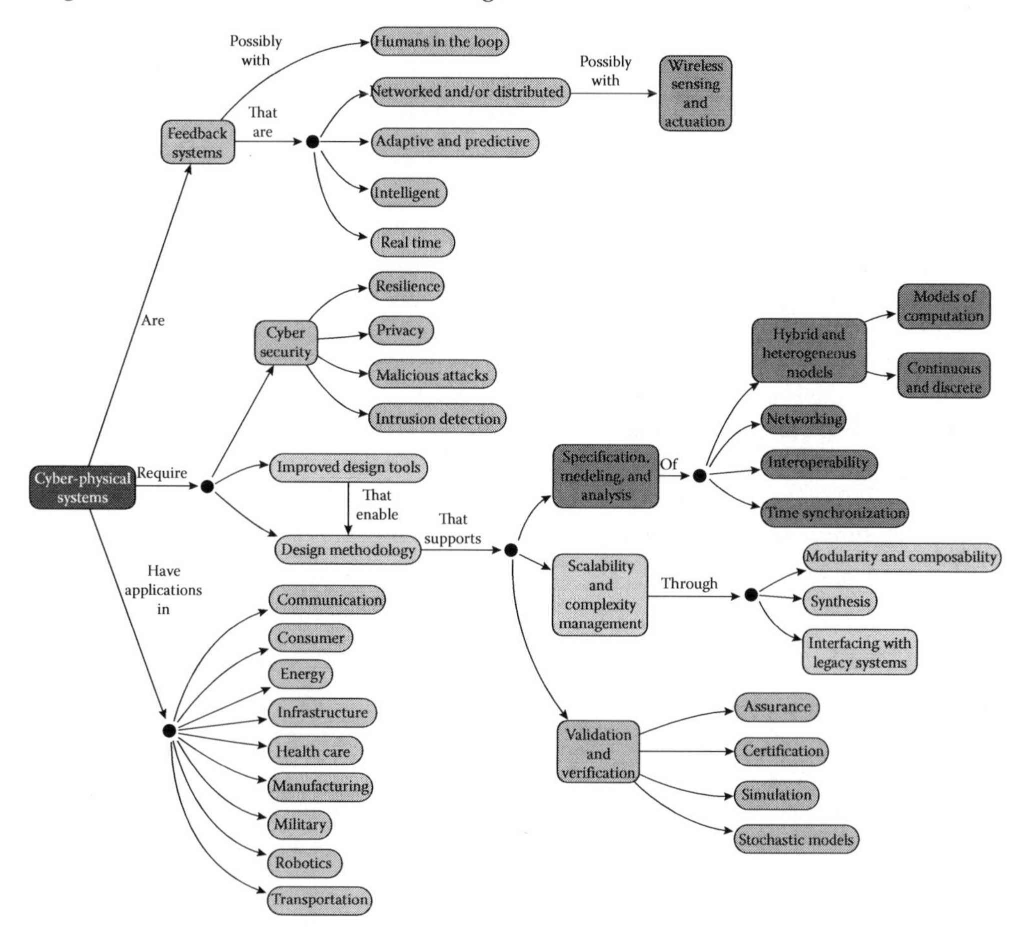

FIGURE 21.5
Cyber-physical system at a glance. (Image courtesy of P. Asare, D. Broman, E.A. Lee, M. Torngren, S.S. Sunder, NIST CPS Workshop, UC Berkeley, 2012; http://cyberphysicalsystems.org/).

Gokhale et al. explains networked autonomous vehicles could spectacularly enhance the effectiveness of military services and could offer considerably more effective disaster recovery techniques. In communications, cognitive radio could benefit extremely from distributed compromise about the available bandwidth and from distributed control technologies. Financial networks could be considerably changed by precision timing. Nie et al. expalins about the agricultural aspect of cyber-physical systems (Nie et al. 2014).

21.3.2 Features for Generic Framework in Cyber-Physical Systems

The following characteristics are essential during a generic framework for cycle style, modeling, and simulation:

- *Heterogeneous (in terms of assorted kinds of sensors/actuators) applications support*: Cycle typically consists of nonhomogeneous applications. Thus, it ought to be ready to simulate heterogeneous application logics at the same time.
- *Varied physical modeling surroundings*: The physical modeling environment ought to support mathematical expressions and incorporate domain-specific physical modeling descriptions (e.g., architectural plan of buildings) by extracting relevant data from them.
- *Measurability support*: Support for the event and simulation starting from tiny scale (tens of) to massive scale (thousands of) sensors and actuators.
- *Quality support*: Support for modeling systems victimization relevant properties (e.g., communication, signal strength).
- *Integration of existing simulation tools*: Easy-to-use support to link to existing simulation tools is needed.
- *Integration of proprietary solutions and open standards support*: Proprietary solutions and open standards together with protocols, infrastructures, and existing software system ought to be ready to be simply incorporated into a generic framework.
- *Software system utilize*: A generic framework ought to support software system reuse by exploiting code-generation techniques (which may also use proprietary infrastructure), linking libraries, or victimization configurable parts.
- *Usability*: Graphical illustration of modeling and simulation surroundings will change simple development of latest applications. Domain-specific 3D modeling environments may also be supported relying upon necessities.

21.4 Big Data in Cyber-Physical Systems

In general, we can say that

$$\text{Cyber-physical system (CPS)} = \text{Embedded systems (ES)} + \text{Physical environment} + \text{Networking}$$

See Figure 21.6.

CPS says that the collection, processing, and visualization of data can be done not only by computers but also by millions of devices such as smart phones, medical data collection center, traffic cameras, walkie talkies, and wireless sensor networks. Figure 21.7 shows

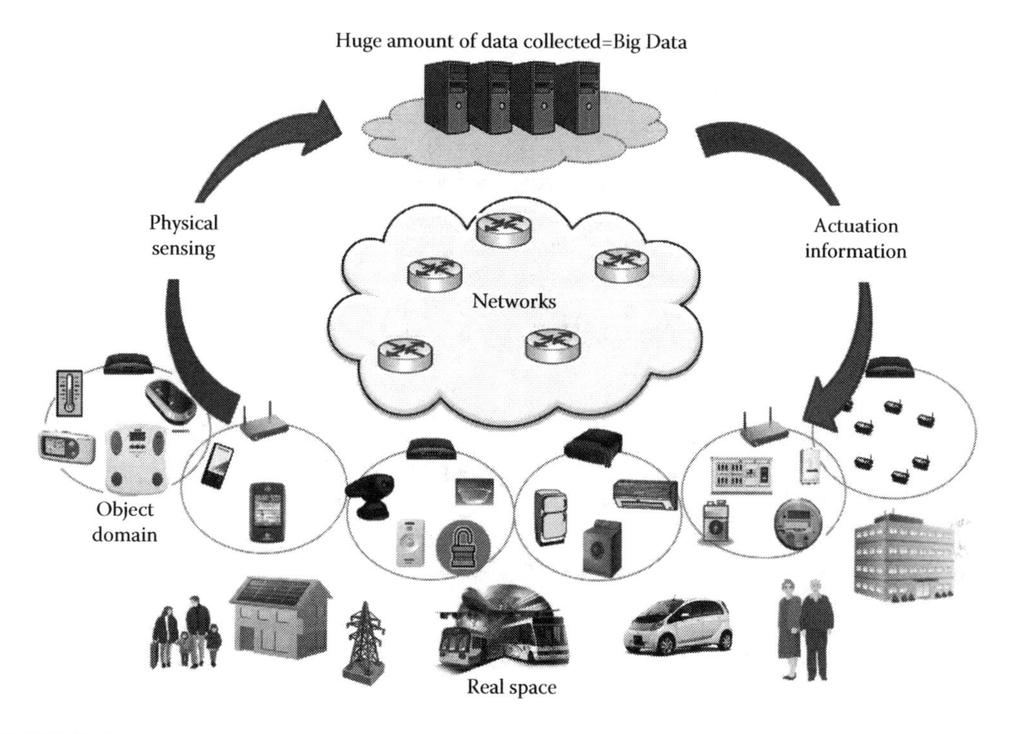

FIGURE 21.6
Big data collection in CPS. (Image courtesy of CPS Lab Introduction page, New York University; https://wp.nyu.edu/cpslab/about/)

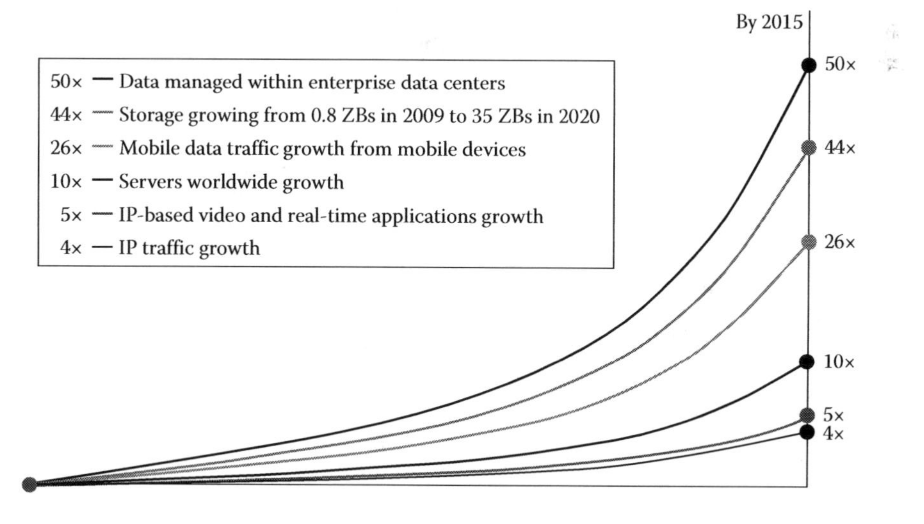

FIGURE 21.7
Growth of big data by some CPS. (From http://vmturbo.com/about-virtualization/virtualization/data-mobility-new-age/).

some examples of big data collection as a whole picture. As these data are processed by different units according to their specific purposes, they are collected in different forms. But one thing is sure that due to an increase in the use of CPS, data collection has increased day by day. The graph shows the growth of big data via some resources.

21.4.1 Cyber-Physical Systems and IT Systems

In general, four major differences make big data modeling and analytics for CPS different from similar issues in the general-purpose computing domain. The first two are (1) the rigid communication of computing elements with the physical world via feedback control loops and (2) a rigorous engineering process for critical CPS, when compared to standard software engineering practice for IT applications, where engineers cannot always depend on constant software updates to solve earlier problems. These directly guide to a grandiloquent model-based style paradigm exemplified by a spread of engineering tools such as Simulink®/Stateflow®, Ptolemy, etc. Third, frequency systems show more operative behavior compared to these systems. We are able to think about an online service functioning in several modes supported the path, for example, day versus night; however, typically, there is a little variety of such modes. With a small number of operating modes, we can merge domain knowledge and "brute-force" search to conclude the state change points and learn separate models. Still, this approach will not level too many discrete states that are not easy to differentiate, as is the case with complex CPS.

Fourth, we also look forward to the role and interpretation of analytics to be different in CPS. Commonly, when handling data of IT systems, a user can attain descent generalization with machine learning techniques by growing the state space. For example, in an IT system, if there is exploitation depending on the quantity of arrival of request, the first two of the above discussed issues comes in picture. This may help us differentiate between dissimilar operational modes. However, with the concept shifting to low-cost, omnipresent sensing, such approaches become an art—finding variables and their numbers to be analyzed and different transformations with their representations (e.g., derivatives, logarithms, etc.). Ad hoc analytics approaches square measure unlikely to create a control in Hertz space partly owing to the need and well-established convention of rigorous model-based style and development. In support of these variations, we tend to believe that huge information modeling and analytics for research deserve more freelance analysis.

21.4.2 Challenges in Cyber-Physical System Regarding Big Data

Challenge I: Partial knowledge of input–output relations still remains unpredicted till the event occurs and data are generated. Atkins explains models for CPS suppose that the input and output variables for each subsystem are well known (Atkins, 2006). This assumption does not always work in practice. Frequently, detector knowledge from a frequency is within the style of statistic, and these days, it is not rare to encounter datasets with thousands of metrics. In these varieties of cases, a user would have incomplete information regarding the input–output relations: even domain specialists and operators might not have entire information. Similar circumstances are faced by giant network and IT system operators. With distributed computing systems, varied facet render it not possible for humans to grasp the complete system: giant-scale, advanced interfacing across elements, the need to deploy a heterogeneous category of applications exploitation restricted resources while still maintaining prime quality of service, etc. Analogous factors exist just in the case of frequency like an oversized mill, transportation networks, power grid, etc. The information deluge caused by 24×7 sensing in frequency is making this example worse: our ability to live at finer spatial and temporal roughness keeps improving, but the frequency modeling strategies lag behind.

Challenge II: Another challenge is online processing of high-dimensional, low-quality data. Atkins explains the need to deal with a huge amount of high-dimensional data with low information content generated or collected by CPS sensors in real time has been demonstrated (Atkins, 2006). With the rise of industrial big data and smart infrastructure (cities, transportation, homes, etc.), this is now a challenge for CPS modeling across multiple domains. To deal with this challenge, we require an interdisciplinary approach that merges performance modeling and system building with designing fresh algorithms. Hunter et al. presented a decomposable computational technique based on global positioning system (GPS) data streams in batch computations with "D-Streams" programming model deployed over Spark. This permitted them to use associate degree expectation maximization algorithmic program to estimate travel times on a network of roads from abuzz GPS information in real time.

21.4.3 Opportunities in Cyber-Physical System Regarding Big Data

Opportunity I—modeling switched systems: Antecedent we tend to note that learning cycle models once the separate state sequence is unknown is difficult. Present sensing presents associate degree exciting prospect: are there measurements or variables hidden within the large quantity of sensing information that we can use to accurately infer separate state changes? If affirmative, then this may modify the task of learning models for indiscriminately switched systems. Although this intuition is apparent, finding the relevant set of variables to infer separate state changes may be difficult for two reasons: (1) high dimension of device information and (2) shouting or on an irregular basis sampled measurements. We may use feature choice to tackle high-dimensional information; however, it is difficult to do an unattended setting, that is, when no labels or context information is obtainable. Shouting and/or on an irregular basis sampled measurements create variable amendment purpose detection techniques less appropriate for this drawback.

Opportunity II—new applications: The normal goal of cycle modeling has been to predict some system output. This prediction will then be used for style verification, fault detection, etc. However, massive information analytics combined with primarily cloud-based services allow new applications that need comparing information from two totally different systems or from constant system across two different operational runs. For example, with the exception of predicting the present travel times, the information from the Mobile Millennium kind comes maybe to compare traffic patterns across two totally different cities or at different times throughout constant town. One goal of such comparisons may be to find discriminative patterns from information that distinguish one instance or dataset from another. This will modify novel applications within the cycle domain: (1) system-wide comparison, for example, how would the itinerary throughout the time of day dissent from the remainder of the day? (2) what-if analysis, for example, how can lane closures for repairs or due to accidents have an effect on traffic patterns and travel times? (3) proactive maintenance—incessantly collecting information from automobiles will modify proactive drawback detection and maintenance planning.

21.4.4 Main Big Data Collection Systems in CPS

Not all CPSs that collect a huge amount of data are so popular, but some of them are democratic in the technology and widely used by people for big data collection.

Vehicular CPS: With the rising number of vehicles, many problems such as traffic congestion, air pollution, and safety issues have surfaced, which need to be solved. Anirudh Gokhale et al. explained the intelligent transportation system in CPS. Advanced computing and sensing capabilities are widely used in next-generation transportation systems, such as in air traffic, railway, and car control, in order to enhance safety and throughput. Vehicular CPS is not a new concept. It refers to a wide range of integrated transportation management systems, which should be real time, efficient, and accurate. Based on modern technologies such as electronics, computers, sensors, and networks, the traditional modes of transport are becoming more intelligent by enhancing traditional transports via modern technologies. The National Automated Highway System Consortium is dedicated to the study and development of independent high-speed Automated Highway Systems, intending to attain more secure and intelligent traffic. CarTel is a U.S. National Science Foundation support project developed by MIT. The corporate trust project combines mobile computing and sensing, wireless networking, and information-targeted algorithms running on servers within the cloud to wear down these challenges. CarTel allows different applications to simply gather, process, carry, analyze, and visualize data from sensors located on mobile units. The contributions of CarTel include traffic mitigation, road surface monitoring and hazard detection, vehicular networking, etc. It collects a vast quantity of information that must be processed by massive data architectures.

Agriculture CPS: Accuracy was expected in the early 1980s from traditional agriculture, which was supported by IT to implement modern systems of agricultural management strategies and technologies. The design of precision agriculture includes data management of production experiments, fundamental geographic information of farmland, microclimate information, and other data. The project "underground wireless sensor network" was developed by the University of Nebraska–Lincoln Cyber-Physical Networking Lab, where Agnelo R. Silva and Mehmet C. Vuran developed a new CPS through the assimilation of center pivot systems with wireless underground sensor networks, that is, CPS^2 for precision agriculture. The wireless underground sensing element networks comprise wirelessly coupled underground sensing element nodes that communicate unbound by soil. This combination of CPS and precision agriculture is one of the typical applications of CPS with good prospects. As agriculture is one of the bases of human survival, a lot of technology is implemented as CPS, and therefore, a huge amount of data are being collected from agriculture CPS.

Healthcare CPS: It will replace traditional health devices working individually, which people will face in the future. With sensors and networks, various health devices work together to detect a patient's physical condition in real time, especially for critical patients, such as patients with heart disease. The portable terminal devices carried by the patient can detect the patient's condition at any time and send timely warning or prediction in advance. In addition, the collaboration between health equipment and real-time data delivery would be more convenient for patients. Insup Lee and Oleg Sokolsky introduced the development and issues of the highly trustworthy health CPS, including the dependence on software from the new function development, demand for network connections, and request for continuous monitoring of patients, and analyzed the future development of health CPS. For health device interoperability issues, Cheolgi Kim et al. introduced a generic framework, the Network-Aware Supervisory System, to integrate health devices into such a clinical interoperability system that uses real networks. It provides development surroundings, in which health device superordinate logic may be developed, supporting the assumptions of a perfect, sturdy network. Huge knowledge formation is incredibly high in Health Hz, and a large process is needed for that.

Air CPS: CPS research is the future of aircraft, air traffic management systems, as well as air travel safety. Precise investigation areas include (1) novel functionality to attain higher capacity, better safety, and extra efficiency, including interaction and tradeoffs among these performance goals; (2) integrated flight deck systems, changing from displays and notion for pilots to future (semi)autonomous systems; (3) vehicular health monitoring and management; and (4) safety research for aircraft control systems. One of the important challenges for realizing NextGen involves confirmation and justification of complex flight-critical systems and taking care of reliable, secure, and safe use for NextGen operations. The complexity of systems, costs related to verification and validation, and safety assurance are directly proportional to the cost of designing and building next-generation vehicles.

21.5 Big Data Tools for Cyber-Physical Systems

When lots of data are collected from CPS, it gets collected in a database and then processed. A number of tools are available for data storage, data processing, firing database queries, different architecture on which data processing is done, visualization of data, acquisition of data, serialization of data, and some special tasks such as machine learning and natural language processing. The big data tools for CPS are as follows.

21.5.1 NoSQL

A NoSQL (often interpreted as Not Only SQL) database offers the capabilities for storage and retrieval of data that is designed by means other than the tabular relations used in traditional relational databases. Database collected by CPS has variety and needs to be solved by some of the data-processing languages such as SQL; therefore, NoSQL has been designed such that it solves for big data systems. Need and decidability for this approach include simple style, horizontal scaling, and higher management over availableness and accessibility. The data structures used in NoSQL databases, such as key–value, graph, or documents, vary from those used in relational databases. This makes some operations faster in NoSQL and, therefore, helps in CPS in the case of emergencies such as medical or natural calamities, where these systems are deployed and faster results are required. Following are some popular tools supporting NoSQL:

MongoDB: Mongo—this name comes from "humongous"—is a database designed for developers who work with fairly large datasets but who require something with low maintenance and simple to handle. It is a document-oriented system, which has records that look similar to Java Script Object Notation (JSON) objects with the capability to store and query on nested attributes. It comes with supports such as automatic sharding and MapReduce technique. JavaScript is used for query writing, with an interactive shell available, and attachment for all the other popular languages.

Cassandra: It is a distributed key/value-paired system, with highly structured values that are held in a hierarchal way similar to the traditional database/table levels, with the equivalents being key spaces and column families. Optimization of data structures is done for consistent write performance, but with a small disadvantage of occasional slow read operations. Cassandra allows nodes to decide which node should perform read operation and which node should perform write operation before performing the operation. Consistency level is set, which allows you to line up the consistency, availability and partition tolerance (CAP) tradeoffs for particular users to take care of speed over consistency or vice versa according to the requirement.

Redis: Two distinguishable features of Redis are as follows: it keeps the entire database in the main memory and its values can be composite data structures. The whole dataset that needs to be processed is first kept in the RAM and backup is taken on a disc in a periodic manner. Therefore, a user can use it as a persistent database. This has fast and expected performance, but speed reduces dramatically if the size of your data is so large that it is beyond the available RAM and the operating system initiates paging virtual memory to take care of accesses. It supports complicated organization in terribly effective means, with an oversized range of list and set operations handled quickly on the server facet. It makes it simple to try things like appending to the top of a price that is an inventory, so it minimizes the list in order that it solely grasps the foremost recent one hundred things.

BigTable: It has extra complex structure and interface than many NoSQL datastores, with a hierarchal and multidimensional accessibility. The first level is similar to traditional relational databases in which table holds the data. Each table is split into multiple rows, with each row dealing an exclusive key string. The data within the row are organized into cells, with every cell outlined by a column family symbol, a column name, and a timestamp. It guarantees per row transactions, but it does not offer any way to atomically alter larger numbers of rows. Google classification system is employed as its inherent storage, which keeps further copies of all the persistent files so that failures are recovered and, thus, works as a backup.

Hypertable: Hypertable is another open-source parallel system of BigTable. The language used for writing is C++ in place of Java because it has total focus on high performance rather than interface. Otherwise, its interface will be same as BigTable, with the same column family and same time stamping.

Voldemort: Voldemort has three-operation key–value pair interfaces, but with backend architecture to run on large and distributed clusters. It uses hashing for fast searching of storage locations for selective keys, and it has control to handle conflicting values. A read operation actually returns multiple values for a given key;

multiple write operations are done by users simultaneously. This puts the load on the application to take some sensible recovery actions when it gets numerous values, based on the meaning of the data being written.

Riak: It uses the same data structure as in Voldemort and gives similar performance. It also uses uniform hashing and a gossip protocol to avoid the need for the kind of centralized index server that big data requires, along with versioning to handle update conflicts. Handling of query is done by using MapReduce functions written in either Erlang or JavaScript. It is open source under an Apache license, but its commercial version is also available with some special features designed for enterprise customers.

21.5.2 MapReduce

MapReduce has been discussed earlier in detail, and some of its tools for batch processing (Hadoop, Hive, Pig, and Cascading) and stream processing (Apache Kafka and Storm) have been discussed. Other tools are as follows:

Cascalog: Cascalog is a purposeful processing interface written in Clojure. It follows the previous knowledge log language and engineered on the high of the Cascading framework. It permits users to write down its process code at a high level of abstraction, whereas the system takes care of aggregating it into a Hadoop job. It makes easier to change between native execution on tiny amounts of information to check your code and production jobs on your real Hadoop cluster. Cascalog inherits constant approach of input, output, and process operations from Cascading, and also the purposeful paradigm appears like a natural approach for specifying knowledge flows. It is a descendant of the first Clojure wrapper for Cascading, cascading-clojure.

Mrjob: Mrjob could be a framework that enables users to put in writing the code for processing and so transparently run it regionally on Elastic MapReduce or on the Hadoop cluster. It is written in Python and does not offer the same level of abstraction or predefined operations as the Java-based Cascading. The specs of jobs are known as a series of map and reduce steps, each applied as a function of Python. It is nice as a framework for executing jobs, even permitting attaching a computer program to native runs to know what is happening within the code.

Caffeine: Although no important technical information has been reported, it is enclosed in Google's caffeine project, as there is a lot of supposition that it is a replacement for the MapReduce construct. From news and company clarification, it seems that Google is employing a cover version of the Google filing system that supports smaller files and distributed masters. It feels like the corporate is emotionally aloof from the execution approach to assembling its search index and is instead employing a dynamic information approach to form updates faster. There is no sign that Google has come with a replacement algorithmic approach that is as widely applicable as MapReduce, although everyone wants to hear a lot regarding the new design.

S4: At the start, Yahoo! created the S4 system to make choices relating to selecting and positioning ads; however, it has finally become a smart tool for stream processing knowledge. S4 allows users to execute codes to handle distributed events across a group of machines. This can be achieved with the help of "ZooKeeper" framework. It is determined on returning results fast, with low latency, for applications such as building close to time period search engines on chop-chop dynamic content.

MapR: It is a commercial distribution of Hadoop. It also includes its own file systems that are substitute for the Hadoop distributed classification system (HDFS), along with other plucks to the framework, such as distributed name nodes for enhanced reliability. The new file system aims to offer higher performance, easier backups, and compatibility with the network file system to make it simpler to move data in and out. The programming model is standard Hadoop, which focuses on improving the infrastructure surrounding the core framework to make it more attracting to enterprise customers.

Acunu: Like MapR, Acunu may be a new low-level knowledge storage layer that replaces the normal filing system, although its initial target is Cassandra instead of Hadoop. By writing a kernel-level key/value store known as Castle, the creator's area unit is able to provide spectacular speed boosts in several cases. The information structures behind the performance gains are spectacular associate degree it is once more an open supply.

Flume: The Flume project is meant to make the information-gathering method straightforward and scalable, by running its agents on the supply machines that pass the information updates to collectors, which then mix them into giant chunks that may be expeditiously written as HDFS files. It can be executed using a command-line tool, which operates like spoofing a document/file or nailing it on a network socket and has consistency assurances. This helps in tradeoff overall performance and reducing potential data loss.

Azkaban: Azkaban is an associate degree open-source project from LinkedIn, which enables users to outline job flow, probably created from several dependent steps; so it handles lots of messy work details. It retains track of the log outputs, identifies errors, and supports approachable Internet interface. Jobs area unit created as text files, employing a terribly stripped-down cluster of commands, with any elaboration seemingly to reside within the operating system commands or Java programs that the step calls.

Oozie: Oozie may be a program system similar to Azkaban, however completely targeted on Hadoop. Oozie conjointly supports an additional advanced language for describing job flows, permitting you to create runtime choices concerning specifically that steps to realize, all explained in XML files. There is conjointly associate degree API that you just will use to make your own extensions to the system's practicality.

Greenplum: Although not strictly a NoSQL information, the Greenplum system presents associate degree appealing methodology of mixing a versatile source language with distributed performance. Designed on the high of the Postgres open supply information, it appends in an exceedingly distributed design to execute on a cluster of multiple machines, whereas retentive the quality SQL interface. It is deployed on clusters of machines with comparatively quick processors and huge RAM, in distinction to the pattern of exploitation trade goods hardware that is additional common within the net world.

21.5.3 Storage

Large-scale processing operations access knowledge through an approach that ancient file systems are not designed for. Knowledge tends to be written and browsed in massive batches, multiple megabytes right away. Potency may be a higher priority than options like directories that organize data in a very easy approach. Following square measure the tools for storage within the age of huge knowledge for giant knowledge.

S3: Amazon's S3 service helps you to store massive chunks of information on a web service, with AN interface that render it straightforward to retrieve the information over the quality net protocol, HTTP. A method of staring at it's as a classification system that's missing some options like appending, editing, or renaming files, and true directory trees. Users can additionally see it as key/value information on the market as an Internet service and optimized for storing massive amounts of information in every worth. It is cheap, well documented, reliable, fast, copes with massive amounts of traffic, and is incredibly straightforward to access from nearly any surroundings, due to its support of hypertext transfer protocol for reads.

Hadoop distributed classification system: The HDFS is meant to support applications like MapReduce jobs that browse and write massive amounts of information in batches, instead of a lot of every which way accessing legion little files. It automatically keeps the rows unchanged by default, supporting a primary key for hash table. It also avoids data/information loss with redundant array of independent disc (RAID) drive setups. The buyer software stores up written data in a very temporary native file, until there is enough to fill a complete HDFS block.

21.5.4 Servers

The cloud is a very unclear term, but there has been an actual change in the ease of use of computing resources. All data collected from CPS may not be collected at one place, as the deployment of services of CPS is undefined and according to requirements; therefore, servers are deployed online (i.e., cloud) to store data or to provide services. For big data computation and data collection, the following servers are available:

EC2: In simple terms, Amazon elastic compute cloud (EC2) permits users to take computers on rent online with different memory and CPU requirements. Users get network access to a complete Linux or Windows server in which user logged in as root also allowing installing software and flexibly configuring the system. Internally, these machines are actually hosted virtually, with many running on each physical server in the data center and service provider costs accordingly.

Google App Engine: With Google's App Engine service, users can write web-serving code in Java or Python, and it takes care of running the application in an exceedingly ascendable approach so that it will handle giant numbers of parallel requests. Not like EC2 or ancient net hosting servers, users have terribly restricted management over the atmosphere on that code is running. This makes it simple to distribute across a range of machines to handle significant hundreds, as solely your code has to be transferred; however, it will build robust to run something that wants versatile access to the hidden system.

Elastic Beanstalk: Elastic Beanstalk encapsulates the EC2 service in it and provides the additional service of setting up an automatically scaling cluster of web servers behind a load balancer, allowing users to deploy Java-based applications without taking care of housekeeping details. This high-level approach makes it similar to App Engine and Heroku, but because it is just a wrapper for EC2, users can also log directly into the machines that the code is running on to debug problems or configure the environment. It is still basically designed for the needs of front-end web applications, so most data processing problems are not a good fit for its approach.

Heroku: Heroku hosts Ruby web apps, offering a simple deployment process, a lot of free and paid plug-ins, and easy scalability. To guarantee that user's code can be quickly deployed across a large number of machines, there are some limitations on things like authentication to the underlying file system, but in general, the environment is more flexible than App Engine. Users can install almost any Ruby program, even those with native code, and can get a real SQL database rather than Google's scalable but restrictive alternative datastores. Heroku is still dedicated for the needs of front-end applications, so it is likely to be restricted to giving the interface to a data-processing application.

21.5.5 Processing

Data collected from CPS need to be processed; some require fast processing, while some require distributed processing. Therefore, according to the processing needs, different tools are available, which boost processing speeds. These tools are as follows:

R: The R project is a specialized language and a toolkit of modules designed for anyone operating with applied math knowledge. It covers everything from loading knowledge to running refined analyses on that, so either mercantilism the ends up in the file or visualizing it on screen. The interactive shell makes it straightforward to experiment with knowledge set, since users try totally different approaches quickly. The key drawback from an information process outlook is that it is designed to figure with datasets that work in one machine's memory. It can be used with Hadoop as another streaming language; however, plenty of the foremost powerful options need access to the entire dataset to be effective. R makes a good prototyping platform for coming up with solutions that require running on large amounts of information, or for creating sense of the smaller-scale results of your process.

Yahoo! Pipes: It has been several years since Yahoo! released the Pipes environment, but it is still an unsurpassed tool for building simple data pipelines. It has a graphical interface where users have a drag-and-drop mechanism, which needs to be linked together into flows of processing operations. A number of Yahoo!'s interesting APIs are exposed as building blocks, as well as components for importing web pages and RSS feeds and outputting the results as dynamic feeds. As a free tool aimed at technically minded consumers, Pipes cannot handle huge datasets, but it is the correspondent of duct tape for a lot of smaller tasks.

Mechanical Turk: The original Mechanical Turk was a fraudulent device that appeared to be a chess playing robot but was actually controlled by a hidden midget. Amazon exploits the same basic principle, making out that there are some mental tasks that it is most effective to ask real humans to execute. It can price up to a few dollars per operation and depending on the duration and complexity of each small job any user wants to perform. The low cost makes it catchy for the user, but it is an extremely powerful way of introducing genuine intelligence into the processing pipeline.

Solr/Lucene: Lucene is a Java library that gives service for classifying and looking out for huge collections of documents. And Solr is an application that uses this library to create a hunt engine server. They originated as two different projects, but later, they were merged into a single Apache open-source team. They were designed to process very big amounts of data, with a sharding architecture that means it will also scale horizontally across a cluster of machines. It conjointly contains a terribly versatile plug-in design and configuration system, and it may be value added to plenty of various knowledge sources. These features, along with a proper tested code base, make it a great choice for any user who wants to solve a large-scale search problem.

Elastic Search: It is a search engine that is made on the basis of Lucene, like Solr used in enterprises. It allows users to update the search index with much lower latency, has a more minimal REST/JSON-based interface and easy configuration options, and scales horizontally in a flawless way. It does not, however, have the community or range of contributors of the older project, and it misses a number of vital options that Solr offers, therefore its value analysis.

Datameer: Although designed for business intelligence, Datameer is attractive because it uses Hadoop to power its processes. It offers an awfully straightforward programming atmosphere for its users to specify the sort of study they require, so at the rear finish it mechanically converts that into MapReduce jobs. It conjointly has some easy knowledge commercialism tools, furthermore as multiple visual image choices. It is a signature of wherever processing solutions area unit moved; user mend building interfaces and provides higher and additional powerful abstraction levels.

BigSheets: IBM's BigSheets is a web application that allows nontechnical users to collect unstructured data from the Internet and local sources and analyze them to create reports and visualizations. Like Datameer, it uses Hadoop at the back end to handle very large amounts of data, along with providing services like Open Calais to deal with mining useful structured information from unstructured text. It is geared toward users United Nations agency square measure handy with a programme interface instead of ancient developers; therefore, it is inconceivable to use it as a part of a custom resolution. However, it provides a plan on the way to build processing applications accessible to those forms of standard users.

Tinkerpop: It is graph process software system, and a bunch of developers are functioning on this open supply graph software system. Tinkerpop has made associate degree integrated suite of tools. It has been designed for a set of services that work efficiently for common operations such as interfacing to specialized graph databases, writing traversal queries, and exposing the whole system as a REST-based server. Dealing with graph data, Tinkerpop gives some high-level interfaces that are easier to deal with than raw graph databases.

21.6 Conclusion

In this chapter, we have mentioned the impact of "big data" monitoring information collected from CPS victimization omnipresent sensing with reference to CPS modeling and analytics. Here we discussed new challenges for CPS modeling and opportunities associated with CPS management and novel applications arising from huge datasets. Additionally, we indicated the key variations between CPS and IT system, which lead to an argument for more analysis into combining CPS modeling with analytics.

The potential of CPS to alter each side of life is gigantic. Ideas such as autonomous cars, robotic surgery, intelligent buildings, good electrical grid, good producing, and deep-seated medical devices are some of the sensible examples that have already emerged. These systems suppose a procedure core that is tightly conjoint and coordinated with elements within the physical world. CPS is associate degree exciting rising analysis space that has drawn the eye of the many researchers. CPS remains open and widely known, and its accepted attributes include appropriateness, distribution, dependableness, fault tolerance, security, measurability, and autonomy. CPS now measures the necessity of today's world and can attend to increase in the future. As its use increases, so are the data collected from CPS; therefore, big data will grow bigger. No one knows what is there in the bag of technology for CPS and its related big data. Whatever will come, we will see it together in the future.

References

Atkins, E.M. (2006). Cyber-physical aerospace: Challenges and future directions in transportation and exploration systems. Department of Aerospace Engineering, University of Michigan, Ann Arbor, MI 48109.

Don, S. and Min, D. (2011). Medical cyber physical systems and Bigdata platforms. Department of Computer, Information and Communication Engineering, Konkuk University, South Korea.

Gokhale, A., Tambe, S., Dowdy, L., and Biswas, G. (2008). Towards high confidence cyber physical systems for intelligent transportation systems. Department of EECS, Vanderbilt University, Nashville, TN.

Nie, J., Sun, R., and Li, X. (June 2014). A precision agriculture architecture with cyber-physical systems design technology. *Applied Mechanics & Materials* 2014(543–547), 1567.

Further Readings

Bhadoria, R.S., Dixit, M., Bansal, R., and Chauhan, A.S. (January 2012). Detecting and searching system for event on internet blog data using cluster mining algorithm, in *Proceedings of Springer International Conference on Information Systems Design and Intelligent Applications (INDIA)*, January 2012, Visakhapatnam, India, pp. 1–2.

Bhadoria, R.S. and Jaiswal, R. (November 2011). Competent search in blog ranking algorithm, in *Proceedings of Springer International Conference on Computational Intelligence and Information Technology*, November 7–8, 2011, Pune, India, pp. 1–2.

Bhattacharya, D. and Mitra, M. (2013). Analytics on big fast data using real time stream data processing architecture. Tutorial document, EMC².

Big Data Working Group (September 2013). Big data analytics for security intelligence. Cloud Security Alliance.

Dean, J. and Ghemawat, S. (December 2004). MapReduce: Simplified data processing on large clusters. White paper, Google, Inc.

Grobelnik, M. (May 2012). Big data tutorial. Jozef Stefan Institute Ljubljana, Slovenia, Stavanger.

Lee, E.A. (October 2006). Cyber-physical systems—Are computing foundations adequate? Department of EECS, UC Berkeley, Position Paper for *NSF Workshop on Cyber-Physical Systems: Research Motivation, Techniques and Roadmap*, October 16–17, 2006, Austin, TX, pp. 1–2.

Palankar, M., Onibokun, A., Lamnitchi, A., and Ripeanu, M. (2008). Amazon S3 for science grids: A viable solution? in *DADC'08*, June 24, 2008, Boston, MA, 10pp.

Pokorny, J. (2011). NoSQL databases: A step to database scalability in web environment, in *Proceedings of the 13th International Conference on Information Integration and Web-Based Applications and Services, iiWAS*, December 5–7, 2011, Ho Chi Minh City, Vietnam, pp. 278–283.

Ragunathan, R., Lee, I., Sha, L., and Stankovic, J. (2010, June). Cyber-physical systems: The next computing revolution, in *Proceedings of 47th ACM/IEEE Design Automation Conference (DAC)*, June 13–18, 2010, Anaheim, CA, pp. 1–4.

Sharma, A.B., Ivančić, F., Niculescu-Mizil, A., Chen, H., and Jiang, G. (2014, March). Modeling and analytics for cyber-physical systems in the age of big data. NEC Laboratories America.

Silva, A.R. and Vuran, M.C. (2010). Communication with aboveground devices in wireless underground sensor networks: An empirical study, in *IEEE International Conference on Communications (ICC)*, May 23–27, 2010, Cape Town, South Africa, pp. 1–6.

Sztipanovitz, A. and Ying, S. (January 2013). Strategic R&D opportunities for 21st century cyber-physical systems. Report of the steering committee for Foundation in Innovation for Cyber Physical Systems.

White paper Accenture (April 2014). Big success with big data. Survey report.

Zhou, Z.H., Chawla, N.V., Jin, Y., and Williams, G.J. (2014, October). Big Data opportunities and challenges: Discussions from data analytics perspectives, *IEEE Computational Magazine* 9(4), 62–74.

22

Influence of Big Data on Cyber-Physical Systems

Krishnarajanagar GopalaIyengar Srinivasa, Gaddadevara Matt Siddesh, and Kushagra Mishra

CONTENTS

22.1 Introduction

A cyber-physical system (CPS) has the potential to transform the way we live and work. Smart cities, smart power grids, intelligent homes with a network of appliances, robot-assisted living, environmental monitoring, and transportation systems are examples of complex systems and applications [1,2]. In a CPS, the physical world is integrated with sensing, communication, and computing components. These four components have complex interactions.

The complexity of CPS has resulted in model-based design and development playing a central role in engineering CPS [3–5]. Naturally, the sensing component of CPS is critical to the modeling and management of a CPS because it provides real operational data. The goal of sensing is to provide high-quality data with good coverage of all components at low cost. However, these goals may not always be achievable. For example, we can use high-fidelity sensors such as loop detectors and radars to detect traffic on road, but these sensors are expensive and, hence, cannot be used to cover a large metropolitan area. Smart devices with global positioning system (GPS) (e.g., smartphones or GPS on taxis) can provide good coverage, but their data quality is low [6].

Traditionally, the design of the sensing component of a CPS is focused on how to deploy a limited number of sophisticated, reliable, and expensive sensors to optimize the coverage of an environment or physical phenomenon [6]. However, advances in sensing and communication technologies over the past 10–15 years have disrupted the traditional way. Sensors have become smaller and cheaper, and with the maturity of wireless networking, we can now deploy a large number of them to collect massive amount of data at low cost. The Mobile Millennium traffic information system fuses data from GPS-enabled phones and GPS in taxis with data from sophisticated sensors, such as radar and loop detectors, to estimate traffic in the San Francisco metropolitan area [1].

Today, low-cost, ubiquitous sensing is driving a paradigm shift away from resource-constrained sensing toward using big data analytics to extract information and actionable intelligence from massive amount of sensor data [6,7]. The availability of big monitoring data on CPSs not only creates challenges for traditional CPS modeling by breaking some of the assumptions but also provides opportunities to simplify the CPS model identification task, to achieve proactive maintenance and to build novel applications by combining CPS modeling with data-driven learning and mining techniques.

Combining CPS modeling with data analytics cannot be accomplished by simply transplanting approaches from the general-purpose computing domain, for example, techniques for performance modeling and analytics in services such as Facebook, Google's web services, etc. Though there are obvious similarities between CPS modeling and software engineering, a few important differences also exist. These stem from the marriage between computer science and control theory, and the special role of time in CPS area. Derler et al. note that in information technology systems, a task's execution time is a performance issue and that it is not incorrect to take longer to perform a task, unlike in a CPS where a task's execution time may be critical to its correct functioning [4].

Another viewpoint on CPSs covers from machine to machine and the Internet of Things communications, heterogeneous data integration from multiple sources, security, and privacy, to its integration into the cloud computing and big data platforms.

The integration of big data into CPS solutions presents several challenges and opportunities. First, a common communication framework is required in order to provide the proper features to make reliable, secure, and sustainable infrastructure. Once all the relevant resources are reachable, data streams need to be integrated. Therefore, quality of data and interoperability issues need to be addressed.

Big data require the integration of data warehouses enabled by non-relational technologies such as Hadoop, MapReduce, and Spark. These data warehouses will be usually allocated into cloud computing-enabled platforms.

Big data for CPSs are not suitable with conventional solutions based on offline processing, as the interconnection with the real world in industrial and critical environments requires reaction in real time. Therefore, real time will be a vertical requirement from communication to big data analytics.

Consequently, big data need to be vertically integrated, and a nonclassic solution is suitable, that is, a solution based on offline or batch processing. Big data for CPSs require, on the one hand, real-time streams processing for real-time control and, on the other hand, batch processing for modeling and behaviors learning.

In short, a CPS will soon redefine how we perceive and interact with the physical world. Using commercially available hand-held devices, we will be able to observe, change, and even customize certain aspects of the physical environment that traditionally were beyond reach.

22.2 Requirements of a Framework for Big Data-Driven CPS Solutions

First, big data-driven CPS has big data characteristics. The aspects volume, velocity, variety, veracity, validity, value, and volatility usually describe big data. Second, big data-driven CPS has special characteristics and requirements that must be met during system development. These big data characteristics and special characteristics require methods and techniques for data specification, modeling, capture, transfer, management, and algorithms for their collection, transfer, analysis, storage, and processing. The design methods for big data-driven CPS should meet not only multiple V's characteristics but also special characteristics of a CPS such as spatiotemporal requirements and real-time communication requirements.

Volume: The amount of big data is very large, in the order of terabytes or larger.

Velocity: The capture, transfer, computation, store, and access for big data are needed in real time. Big data require fast processing. Time factor plays a very crucial role in CPSs.

Variety: Variety refers to the many sources and types of data, both structured and unstructured. The data include unstructured and multiple types, possibly including texts and stream data such as videos, log files, etc.

Veracity: The veracity aspect deals with uncertain or imprecise data. The veracity aspect includes data inconsistency, incompleteness, ambiguities, latency, deception, and approximations.

Validity: The validity aspect includes ensuring not only correct measurements but also transparency of assumptions and connections behind the process. Validity can be further divided into content validity, structural fidelity, criterion validity, and consequential validity.

Value: The business value of big data is in how the data and technology are applied, the same as with business intelligence. Big data's value derives from capabilities that enable new and broader uses of information or that removes limitations in the current environment. All that available data will create a lot of value for organizations, societies, and consumers. Big data means big business, and every industry will reap the benefits from big data.

Volatility: Big data volatility refers to how long data would remain valid and how long should they be stored. In this world of real-time data, you need to determine at what point the data would no longer be relevant to the current analysis.

Big data-driven CPS has special characteristics, constraints, and requirements that must be met during system development. In CPSs, a large number of networking sensors are embedded into various devices and machines in the physical world. Such sensors deployed in different fields may capture various kinds of data related to speed, temperature, environment, geography, astronomy, people's health, logistics, etc. Mobile equipment, transportation facilities, public facilities, and home appliances could all be data-acquisition equipment.

CPSs must meet not only real-time requirements but also spatial requirements. Spatiotemporal data have many different forms and include data such as georeferenced time series, remote-sensing images, or moving object trajectories. Trajectory data are data containing the movement history of mobile objects. The specification, modeling, processing, and analysis of trajectory data are interdisciplinary research fields and involve communities from geographic information science, software engineering, moving database technology, sensor networks, distributed systems, transportation science, and privacy. Spatiotemporal data are very large. Especially, the data of moving physical entities increase geometrically over time. Spatiotemporal data captured through remote sensors are always big data. In a CPS, every data-acquisition device (sensor) is located at a specific geographic position and every piece of data has a time stamp. The time and space correlations are very important properties of data for CPSs. During data analysis and processing phases, time and space information are also very important dimensions for data evaluation, store, access, and analysis. Many datasets in CPSs have certain levels of heterogeneity in type, structure, semantics, organization, granularity, and accessibility. Data representation and model aim to make data more meaningful for data processing, storage, access, and analysis. Efficient data representation should reflect data structure, class and type, spatiotemporal data, and integrated technologies, so as to enable efficient operations on different datasets.

A CPS must guarantee that huge data in the network are delivered according to their real-time deadlines. Some mechanism is needed to assign priorities to the data that are to be sent on the sensor network or physical objects. The environment in CPSs is often harsher and noisier and, thus, has stringent requirements on reliable and real-time communication. Missing or delaying real-time data may severely degrade their value because big dataset sizes are growing exponentially, so it is important to use most efficient data-transfer protocols that are available.

Some challenges that need to be addressed are as follows:

- First, a moving object represents the continuous evolution of a spatial object over time. Regarding big data modeling, an important question is how to represent a moving object. In the future, we will focus on specification and modeling of moving object spatiotemporal behavior.
- Second, for real-time communication in big data environment, CPSs must guarantee that huge data in the network are delivered according to their real-time deadlines. Because big dataset sizes are growing geometrically, it is important to develop efficient data-transfer protocols.
- Third, it is important to integrate big data, CPS, and cloud computing into a single framework that can provide the collected data from sensors, transfer big data on real-time communication channels, and process and analyze big data on a cloud platform [8].

22.3 AADL-Based Systems

A big data-driven CPS uses large amounts of data to determine its operation. Data are clearly crucial to correctly operating big data-driven CPS. Examples include aerospace systems, air traffic control systems, railway signaling systems, intelligent transportation, and battlefield management systems. When a big data-driven CPS is used in safety-related applications, the safety of the overall system is determined by the properties of its constituent parts. Correctness of data is also essential for the system to achieve overall safety. Therefore, one of the key challenges is providing the specification and modeling methods for data services for big data-driven CPS, which has to deal with large amounts of data in a timely, secure fashion. Here we talk about a big data-driven CPS design method based on Architecture Analysis and Design Language (AADL), which can specify and model the requirements of big data-driven CPS and implement these requirements on big data platforms. The main advantages of this approach are its capacity to take into account big data properties and CPS properties through specialized concepts in rigorous, easy, and expressive manner, as shown in Figure 22.1. The proposed approach is illustrated by a case study of specifying and modeling aviation CPS [9].

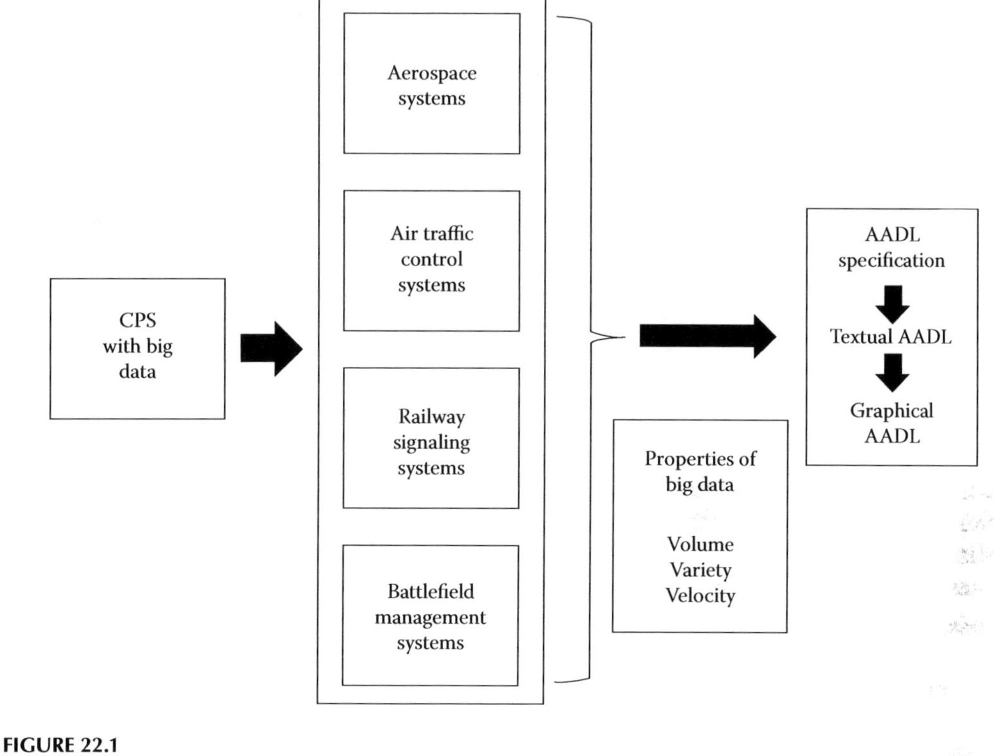

FIGURE 22.1
AADL and CPSs with big data.

Big data technologies describe a new generation of technologies and architectures designed to economically extract value from very large volumes of a wide variety of data, by enabling high-velocity capture, discovery, and/or analysis. There are different definitions of V features for big data. Four V features (volume, velocity, variety, and value) are defined as follows:

- Data volume measures the amount of data available to an organization.
- Data velocity measures the speed of data creation, streaming, and aggregation.
- Data variety is the measure of the richness of the data representation.
- Data value measures the usefulness of data in making decisions.

Spatial data have come to play an increasingly prominent role in big data-driven CPS. Spatial datasets are exceeding the capacity of the current computing systems to manage, process, or analyze the data with reasonable effort due to volume, velocity, and variety. Given a very big heterogeneous dataset, a major challenge is to figure out what data one has and how to specify and model them. Thus, data modeling plays a critical part in CPS design. We need to specify and model the continuous data flow or stream, real-time data, spatial data, and spatiotemporal data. The model used in system development can capture a system's evolution as well as enable analysis and simulation to help us detect design defects, and in some cases, models can be used to automatically synthesize implementations.

AADL is a textual and graphical language used to model and analyze the software and hardware architecture of real-time and embedded systems and their performance-critical characteristics. It is used to describe the software and hardware components of a real-time embedded system and the interfaces between those components. AADL can describe the structure of a real-time embedded system as an assembly of software and hardware components. It can represent functional interfaces to components (such as data inputs and outputs) and nonfunctional aspects of components (such as timing). It can specify how components are composited (such as how data inputs and outputs are connected or how software components are allocated to hardware components [platform]). In particular, AADL provides standard control and data flow mechanisms used in real-time embedded systems, and it can specify important nonfunctional requirements related to timing, safety, and data flow. The main components in AADL are divided into three parts: software components, hardware components, and composite components. Software components include data, thread, thread group, process, and subprogram. Hardware components include processor, memory, bus, and device. Composite components include the system.

AADL supports modeling of three kinds of interactions between components: directional flow of data and/or control through data, event, and event data port connections; call/return interaction on subprogram entry points; and through access to a shared data component. Threads, processors, devices, and their enclosing components (process and system) have in ports and out ports declared. Data ports communicate unqueued state data; event ports communicate events that are raised in their implementation; their associated source text or actual hardware and event data ports represent queued data whose arrival can have event semantics. The arrival of an event at a thread results in the dispatch of that thread—with semantics defined via property values and hybrid automata for event arrival while the thread is active. For data port connections, data are communicated upon execution completion (immediate connection with the effect of mid-frame communication for periodic threads) or upon thread deadline (delayed connection with the effect of phase delay for periodic threads).

AADL has been made extensible in three aspects through extension mechanisms. First, we can define an extensible set of component specifications in the form of component types and implementations by using the extension mechanism. Second, AADL itself can be extended through the ability to introduce new properties and extend the set of valid property values for existing properties. Third, the AADL draft standard includes its specification as a Unified Modeling Language profile.

The design of big data-driven CPS requires the introduction of new concepts to model classical data structures, 4V features, time and spatial constraints, and the dynamic continuous behavior of the physical world. A moving object represents the continuous evolution of a spatial object over time. Regarding big data modeling, an important question is how to represent a moving object. This is an open question and open for debate.

22.4 CPS Using Streaming Data

Controlling and analyzing cyber-physical and robotics systems are increasingly becoming a big data challenge. Predicting the travel times in a large urban area from sparse GPS traces can be of immense use in solving the problem of traffic congestion. Such a system should accommodate a wide variety of traffic distributions and spread all the computations on a cluster to achieve small latencies. It should be built on discretized streams, to stream processing at scale.

Traffic congestion affects nearly everyone in the world due to environmental damage and transportation delays. The 2007 Urban Mobility Report [10] states that traffic congestion causes 4.2 billion hours of extra travel in the United States every year, which accounts for 2.9 billion extra gallons of fuel and an additional cost of $78 billion. Providing drivers with accurate traffic information reduces the stress associated with congestion and allows drivers to make informed decisions, which generally increases the efficiency of the entire road network [11]. Researchers on traffic information systems broadly agree that accurate information is critical to increase their usage [12]. So far, however, it seems only a small fraction of drivers use traffic information systems [13]. In this scenario, we can consider that revealing the true state of the traffic to the participants will not change the dynamics of the overall phenomenon: the informed drivers will optimize their routes or schedule without changing (much) the global equilibrium of other road users. Modeling highway traffic conditions has been well studied by the transportation community with work dating back to the pioneering work of Lighthill and Whitham [14]. Recently, researchers demonstrated that estimating highway traffic conditions can be done using only GPS probe vehicle data [15,16]. Arterial roads, which are major urban city streets that connect population centers within and between cities, provide additional challenges for traffic estimation. Recent studies focusing on estimating real-time arterial traffic conditions have investigated traffic flow re-construction for single intersections using dedicated traffic sensors. These sensors are expensive to install, maintain, and operate, which limits the number of sensors that governmental agencies can deploy on the road network. The lack of sensor coverage across the arterial network, thus, motivates the use of GPS probe vehicle data for estimating traffic conditions.

Most GPS data available today are generated at low frequencies due to energy and bandwidth constraints. These data are extremely noisy and provide indirect observations of the travel time distributions on each link of the road network. A traffic information system designed to process such data at low latencies and for very large urban areas will be useful for the general public.

As datasets grow in size, some new strategies are required to perform meaningful computations in a short amount of time. Implementation of a large-scale state estimation in near-real-time using D-Streams, is a recently proposed streaming technique. When distributed on a cluster, these algorithms scale to very large road networks (half a million road links, tens of thousands of observations per second) and can update traffic state in a few seconds.

22.5 Medical CPS Solutions

During the past few decades, the capabilities for adapting new devices for health monitoring system have improved significantly. But the increase in the usage of low-cost sensors and various communication media for data transmission in health monitoring has led to a major concern for current existing platforms, that is, inefficiency in processing massive amount of data in real time. To advance this field, a new look at the computing framework and infrastructure is required. We talk about big data processing framework for medical CPS that combines the real-world and cyber-world aspects with dynamic provisioning and fully elastic system for decision-making in health-care application.

The framework components of a medical CPS with big data are shown in Figure 22.2. The different sensors are enabled in the CPS environment, which is communicated to the processing layer. In the processing layer, different components such as streaming, query engine, and data processing are present, which communicate the appropriate results to the applications. The applications can be in the form of report/event analyzer, visualizer, etc.

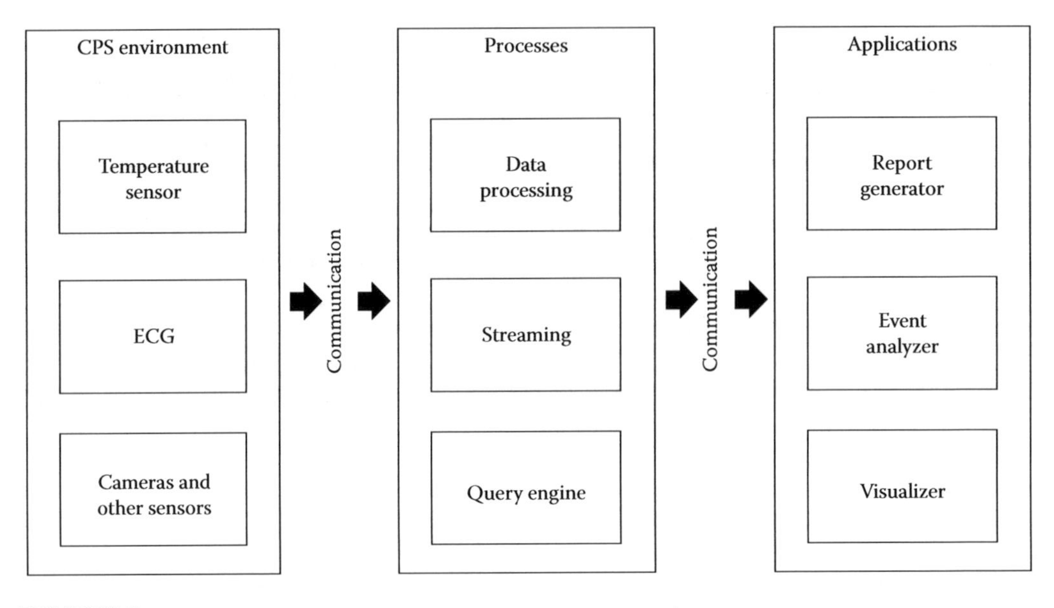

FIGURE 22.2
Medical CPS with big data.

Today's medical systems integrate different classes of devices that can perform various functions in real time. With the emergence of low-cost portable devices, monitoring patients remotely has become common and increasingly popular. Storing and processing these data require good infrastructure, and it became a huge computational complexity. Therefore, there is a desire to improve the service functionality of the current health-care monitoring system. The challenges for delivering meaningful information from these data in health-care system are great. The health-care industry is modernizing the existing infrastructure more intelligently. Thus, providing a system that works efficiently will reduce computation error. Such a system requires subsystems to perform various functions such as filtering, listening, processing, accelerating, and enrichment.

Studies show [17] a growing interest among the elderly community in remote health-care system. There are obvious reasons for this. First, hospital expenses have increased. Second, the physical condition of the patient makes it very difficult for routine checkup. In remote health-care monitoring systems, the patient body is connected with various sensors to measure different physiological data such as ECG, oxygen level, Hb, etc. These data are then sent to the remote application server without any loss of information at the receiving side. This helps medical professional to analyze the data and make the right decision before the condition of the patient becomes severe. As these sensors are low-cost devices, the accuracy of detecting any abnormalities depends on the type of hardware used at the client side and the method proposed at the application side for data analysis.

A limitation of these systems is that because the devices record data every day, these systems store a huge amount of data, both significant and insignificant, over a period of time. It would be a tedious process for the systems to extract meaningful information from this massive dataset. For example, a patient's ECG data would contain significant information regarding the condition of the heart as well as normal beat information, which is less significant for analyzing heart diseases. Each ECG beat wave is represented by PQRST patterns, where the P wave represents the depolarization and contraction of the atrium; the QRS complex represents the depolarization and contraction of the ventricles; and the T wave represents the repolarization of the ventricles. The normal QRS duration for each beat is 0.04–0.12 s. The QT interval is 0.39 s, and the normal PR interval is between 0.12 and 0.20 s. Variation in any of these can lead to abnormal functioning of the heart.

Besides the aforementioned limitation, another limitation is that the systems require larger data storage capabilities with increased processing time. This reduces the performance of the systems. Also, the type of service provided by the systems might not be dynamically configurable to the current scenario, that is, the systems might lack elasticity. Moreover, due to the lack of awareness between the cyberspace and the real world, the systems are not fully adaptable to make decisions in real time. In these circumstances, CPS offers a new paradigm for solving the aforementioned problems. It is a new type of system that integrates computing and communication capabilities with the monitoring and control of entities in the physical world. CPS is commonly used in mission-critical scenarios such as military [2], robotics, health care, and space research. The system requirements for developing CPSs are different from the traditional health monitoring systems. A CPS integrates embedded devices, networks, and physical world aspects [18]. These embedded systems have high reliability and, thus, make the CPS different from the traditional monitoring systems. The traditional systems are designed to operate in a controlled environment where they operate with a set of predefined rules and semantics. For example, in video monitoring of a patient, the system tracks objects that fall in the visible area irrespective of the situation. With the help of small, embedded software installed in the physical world monitoring components, the CPS can handle and track any unexpected event more intelligently than the traditional systems [19].

22.6 Conclusion

The basic aim is to develop a system that provides proper analytics and mining on real-time data stream. It provides the past technique for analytics and mining. Based on the past technique, it compares its working and shows how it overcomes nearly all the challenges of the past technique. The main advantage is that it provides the analytics and the mining structure in the same system so that it is possible to perform this operation online as it provides application of sliding-type windows. CPS will be an integral part of big data solutions in the coming future.

References

1. The Mobile Millennium Project, UC Berkeley/Nokia/NAVTEQ. http://traffic.berkeley.edu.
2. L. Sha, G. Gopalakrishnan, X. Liu, and Q. Wang. Cyber-physical systems: A new frontier. In *Machine Learning in Cyber Trust*, J. J. P. Tsai and P. S. Yu (Eds.), pp. 3–13, Springer, New York.
3. B. Selic. The pragmatics of model-driven development. *IEEE Software*, 20(5), 2003, 19–25.
4. P. Derler, E. A. Lee, and A. S. Vincentelli. Modeling cyber-physical systems. *Proceedings of the IEEE*, 100(1), 2012, 13–28.
5. J. Sztipanovtis and G. Karsai. Model-integrated computing. *IEEE Computer*, 34(4), 1997, 110–111.

6. T. Hunter, T. Das, M. Zaharia, P. Addeel, and A. M. Bayen. Large scale estimation in cyberphysical systems using streaming data: A case study with smartphone traces. arXiv.org, 1212.3393v1, 2012.
7. S. Das, B. L. Matthews, A. N. Srivastava, and N. C. Oza. Multiple kernel learning for heterogeneous anomaly detection: Algorithm and aviation safety case study. In *Proceedings of the KDD*, July 25–28, 2010, Washington, DC, pp. 47–55.
8. L. Zhang. A framework to model big data driven complex cyber physical control systems. In *2014 20th International Conference on Automation and Computing* (*ICAC*), September 12–13, 2014, Bedfordshire, U.K., pp. 283–288.
9. L. Zhang. Designing big data driven cyber physical systems based on AADL. In *2014 IEEE International Conference on Systems, Man and Cybernetics* (*SMC*), October 5–8, 2014, San Diego, CA, pp. 3072–3077.
10. TTI, Texas Transportation Institute. Urban Mobility Information, 2007 Annual Urban Mobility Report, 2007 [Online]. Available: http://mobility.tamu.edu/ums/.
11. X. J. Ban, R. Herring, J. D. Margulici, and A. M. Bayen. Optimal sensor placement for freeway travel time estimation. In *Transportation and Traffic Theory 2009: Golden Jubilee*, 2009, (pp. 697–721). Springer.
12. C. G. Chorus, E. J. E. Molin, and B. Van Wee. Use and effects of advanced traveller information services (ATIS): A review of the literature. *Transport Reviews: A Transnational Transdisciplinary Journal*, 26(2), 2006, 127–149.
13. S. Gao, E. Frejinger, and M. Ben-Akiva. Cognitive cost in route choice with real-time information: An exploratory analysis. *Procedia-Social and Behavioral Sciences*, 17, 2011, 136–149.
14. M. Lighthill and G. Whitham. On kinematic waves. II. A theory of traffic flow on long crowded roads. *Proceedings of the Royal Society of London Series A, Mathematical, Physical, and Engineering Science*, 229(1178), May 1955, 317–345.
15. D. B. Work, S. Blandin, O. P. Tossavainen, B. Piccoli, and A. M. Bayen. A traffic model for velocity data assimilation. *Applied Mathematics Research eXpress*, 2010(1), 2010, 1.
16. T. Hunter, T. Das, M. Zaharia, P. Abbeel, and A. M. Bayen. Large-scale estimation in cyber-physical systems using streaming data: A case study with arterial traffic estimation. *IEEE Transactions on Automation Science and Engineering*, 10(4), 2013, 884, 898.
17. N. G. Pantelopoulos. A survey on wearable sensor-based systems for health monitoring and prognosis. *IEEE Transactions on Systems, Man, and Cybernetics*, 40(1), 2010, 1–12.
18. E. Lee. Cyber physical systems: Design challenges. In *Object Oriented Real-Time Distributed Computing* (*ISORC*), 2008, 11th IEEE International Symposium on IEEE (pp. 363–369).
19. S. Don and D. Min. Medical cyber physical systems and bigdata platforms. In *The Fourth Workshop on Medical Cyber-Physical Systems*, April 8, 2013, Philadelphia, PA.

23

Influence of Social Networks on Cyber-Physical Systems

Krishnarajanagar GopalaIyengar Srinivasa, Srinidhi Hiriyannaiah, and Kushagra Mishra

CONTENTS

23.1 Introduction

The past decades have witnessed the exponential increase in the number of various computers of everyday use. Modern computers are becoming smaller and smaller while equipped with higher and higher performance in terms of, for example, computational speed and memory size, as promised by Moore's law. As a consequence, computers are transforming into a lot of new forms. Some examples of new forms of computers include mobile phones, smart sensors, and even ordinary physical things, such as a lamp, a table, and a cup. In other words, many physical things will possess computing and communication capabilities of different levels, which are provided by small and (possibly) invisible computers embedded therein. This integration of networked computing and physical dynamics has lead to the emergence of cyber-physical systems (CPSs), which have become very popular in recent years.

Generally speaking, CPS features compact integrations of computation, networking, and physical objects, in which various devices are networked to sense, monitor, and control the physical world. Although there is no unified definition of this notion, some defining characteristics of CPS include cyber capability in physical objects, networking at multiple and extreme scales, complexity at both temporal and spatial scales, high degrees of automation, and dependable even certifiable operations. Many researchers and practitioners have pointed out that CPS will transform how we interact with the physical world [1,2].

Another recent trend in information technology (IT) is the rapid growth of social networking services, one of the concepts with the most impact in the past decade. Human beings are social animals. Consequently, it is not surprising that social networks have been popular since the beginning of civilization [3]. Over decades they have been playing a significant role in the development of human society. In a social network, a group of individuals are linked through diverse social relationships, such as family, friendships, business partners, and classmates, to mention a few. Such networks have been the subject of research of social scientists for a long time. With the proliferation of Internet technologies such as Web 2.0, online social networking applications have become prevalent recently. Among many examples are Facebook, Twitter, LinkedIn, MySpace, and Bebo. These applications constitute virtual communities that facilitate information creation, distribution, management, sharing, and consumption among the linked people [4]. For instance, by using social networking services, one can make new friends, stay in touch with (old) friends, and retrieve information of interest anytime and anywhere. Currently most social networking services are web based and provide various means for users to interact over the Internet. Despite the diversity of services provided, a common feature of current social networking applications is that they are designed with the primary goal of facilitating communications among people, that is, we humans are the users.

23.2 Smart Communities

We envision that the couple of aforementioned trends will converge, at least to some extent, in the foreseeable future. This convergence of CPS and social networking will accelerate the emergence of a new paradigm we call *smart community*. This paradigm will provide us with promising solutions to a variety of societal problems spanning many domains, such as health care, education, transportation, energy, and environment, among others. The ultimate goal of building smart communities is to improve the quality of our everyday life by exploiting ubiquitous intelligence [5]. A smart community will by nature feature the interplay of cyber, physical, and social worlds.

23.3 Why Smart Community?

There are a number of important drivers for building smart communities, among which we would like to mention the following three.

Our society is facing many critical challenges that remain to be addressed. One of the major challenges is the unprecedented population aging with the inevitable implications related to disability and care issues. The world's population is aging rapidly. People are living longer, and as they get older, they are increasingly living alone and with disabilities. Without modest assistance, the aged, disabled, or chronically ill people often live with significantly degraded life quality. Key societal challenges can also be found in other areas such as transportation, energy, environment, and security. In transportation, for example, the traffic networks of many modern cities are getting saturated and congestion becomes endemic. Smart community will offer a constitutive technology to create innovative products, services, and strategies for addressing these societal challenges as shown in Figure 23.1. The different societal challenges faced by different context environments can be modeled with smart community members to provide ubiquitous and innovative services for assisted health care and other areas as shown in Figure 23.1.

Smart communities have great potential to improve the quality of life. The design and deployment of smart communities will facilitate constituting the infrastructure needed to create a set of new services for daily life. In assisted living for the elderly, disabled, and chronically ill, for instance, smart communities will make possible, among others, continuous remote monitoring, acute tracking of human motion, real-time emergency response, and remote emergency care [6]. These services will support these impaired people to live more healthy lives, minimizing time in hospital, at local doctors, or in care homes. The aging population will be able to live more independently in their own homes despite illness or disability, as most people prefer, overcoming isolation and minimizing their reliance on carers. In addition to the domestic environment, ubiquitous services enabled by smart communities can be supplied in, for example, cars, offices, schools, and hospitals. Not only impaired people can benefit from smart communities but also healthy people who want to improve the quality of life as well.

Recent technological advances in sensing, networking, and computing are leading to the integration of (wireless) sensor/actuator networks and social networks [4,7]. By exploring this integration, smart community goes beyond either of the currently isolated two areas, that is, CPS and social computing. Based on the latest progress in underlying technologies for both areas, smart community can enable many novel functionalities, which otherwise are very difficult, if not impossible, to be realized by either CPS or social networking services alone. For instance, smart communities allow for collaborating sharing among social objects, which can increase real-time awareness of

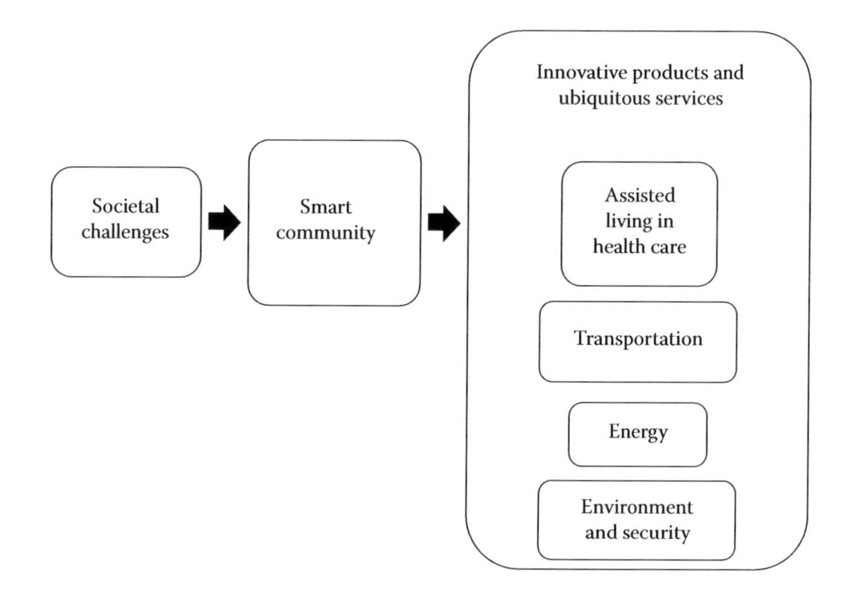

FIGURE 23.1
Smart community.

different community members about each other, and provide a better understanding of the aggregate behaviors of the community [4,8]. Massively distributed data collection thus becomes feasible, which leads to the so-called crowd sourcing. It is worth noting that integration of sensor and social networks is not new.

23.4 What Is Smart Community?

Before giving the explanation of the concept of smart community, we'd like to begin this subsection with revisiting the general description of community and online community, two notions closely related to smart community.

23.4.1 Community and Online Community

Although there are many different forms of communities, they are undoubtedly vital for humans. There would be no life without community. As a general understanding (according to Wikipedia), a community can be described as a group of interacting people. The group often shares some common values, characteristics, or interests and is attributed with social cohesion. Possibly, the most common usage of the term community indicates a large group living in close proximity, generally in social units (i.e., neighborhoods) larger than a household. A large city can be understood as a collection of communities.

In contrast to the communities in real life, an online community is a virtual community that exists online. The members are a group of people communicating or interacting with each other by means of IT, typically over the Internet, rather than in person. Social networking services consist in one of the most representative types of online communities. In the information era, online communities have become an important form of social interaction between the majority of people. With the help of modern IT, online communities offer the opportunity of instant information exchange, potentially crossing geographical boundaries, which is not possible in a real-world community.

23.4.2 Smart Community

In the context of our vision, a *smart community* can be roughly understood as a group of connected (social) objects that interact with each other over ubiquitous networks and deliver smart services to possibly all members. The members of a smart community are objects that can be human individuals, as well as physical things such as a table, a watch, a pen, a door, and a key. It is also possible that some other living things (besides human beings) might be included, for example, a tree and a dog. In most cases, these objects have implicit links among them. For example, they may jointly strive to reach a common goal, for example, sensing and controlling of devices.

Several key properties of smart community could be identified as follows:

- A smart community consists of both human individuals and smart physical things as members that interact with each other.
- The scales of smart communities vary case by case. One smart community may have only several members, while another may have a very huge number of members.

- Smart communities are time evolving. Thus, the scale of a smart community may change over time. Consequently, smart communities should have good scalability.
- Smart communities are socially and physically aware systems.
- It is not necessary for smart communities to be Internet based (i.e., online). Some smart communities may function in (local) environments without connection to the Internet.
- The life cycle of smart communities could possibly be very long in some cases, while relatively short in others, depending on the application supported.

As mentioned earlier, the potential application of smart communities spans a variety of domains in which our lives may involve, including, for example, health care, education, transportation, energy, environment, and security. One example is collaborative rehabilitation realized by means of integrating body sensor networks and social networks [6]. Another possibility made by smart communities is the construction of global (collaborative) learning environments [9]. Such environments will accommodate physical, social, as well as cyber elements. They offer new opportunities for (distance) education of underserved students by providing resources and expertise that cannot be easily carried through cyber-only methods. Another example is a smart transportation community that promises to increase the transportation efficiency of people as well as the whole society.

A paradigm shift in IT is envisaged with this vision. The cyber, physical, and social worlds will converge in this context. Smart communities will be built upon recent technological advances in both fields of CPS and social computing.

23.5 Friend Recommendation Based on GPS

Popular social network services like Facebook and Twitter allow users to store and share both digital information, for example, web content, and also locations and trajectories collected from the real world. Analyzing these extra data from physical world can help us better understand people's daily activities, social areas, and life patterns. Social network with data collected from sensors is usually referred as cyber-physical social network [10]. In this section, we touch upon friend recommendation problem in cyber-physical social network. With location and trajectory information available, we improve the accuracy of the results and make online social services much closer to users' real life.

One major difference between virtual web-based social network and real-life social network is *new* friends in real world tend to be geographically related. Geographical similarity is hiding in users' recently GPS data. To help web-based social network users find more friends in their real life, we define potential real-life friends, who have both social similarities and geographical correlation, as *geo-friends*, and denote *geo-friends finding problem* as real-life friends discovery on web-based social network.

The reason why we want to isolate geo-friends from general web-based social network friends is intuitive. Geo-friends play an important role in off-line social events, for example, holiday party, football game, or book club. Geo-friends have a much higher probability to participate these real-life events than other friends from virtual social network.

The following example demonstrates the idea of recommending geo-friends in web-based social network.

Example 1: Alex wants to find some new geo-friends to join him in a local charity event. There are three candidates: Bob who shares a large number of friends with Alex but lives in another country; Carlos who works in the same company with Alex but shares no similarity in terms of social network structure; and David who shares couple of common friends and also goes to the same gym, same comic book store as Alex does every week as shown in Figure 23.2.

After analyzing both social structure and recently collected GPS data, social network services should recommend David as Alex's geo-friends, since he has a higher probability to participate the local event with Alex as shown in Figure 23.2.

Previous approaches of link prediction that usually only rely on social network structure would recommend Bob. But apparently, Bob is not a good candidate for Alex's social events, since he lives in another country. Also, solely relying on location or trajectory information for geo-friend finding does not work as well. Carlos who has a very high positive geographical correlation with Alex shares no social interests with Alex. Recommending Carlos to Alex is pointless as well.

Xiao et al. proposed a three-step approach, named GEo-Friends Recommendation framework (a.k.a., GEFR). First, interesting and discriminative GPS patterns are extracted from a large amount of raw GPS data. Then, we combine both geo-information and social network in a pattern-based heterogeneous information network. By applying random walk to reproduce friend-making process on the network, we can effectively identify potential geo-friends for a specific user.

They propose the problem of identifying geographically related friends and also a three-step statistical framework, which combines geo-information with social analysis.

They first captured different types of GPS information by defining and generating four types of GPS patterns from GPS history data. Then, they build a pattern-based heterogeneous information network and defined transition probability matrix following GPS pattern definitions. By applying random walk process on this information network, link relevance between different nodes could be estimated, and potential geo-friends would be recommended to a specific query person.

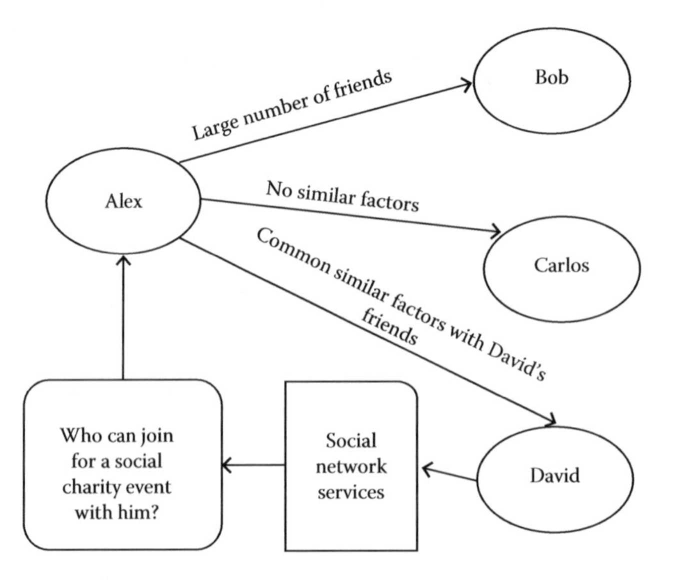

FIGURE 23.2
Geo-friend recommendations.

23.5.1 GPS-Based Cyber-Physical Systems

Sha et al. give an introduction and survey on the development of CPS [11]. This survey provides application examples in the areas of energy grid, health care, and transportation network. Microsoft SensorMap [13] is an early example of cyber-physical network, which allows users to browse the physical world in a digital map. However, the main objects in this application are physical objects, that is, sensors, instead of people. Tang et al. study trustworthy issue in CPS [12], which is an important preprocessing step for reliable analysis.

Mining GPS location history is one of the most common and important jobs of GPS trajectory analysis. Different methods have been proposed on extracting locations from GPS history data [12,13]. Their works have different research focuses, including personalized mining, multiple user clustering, and semantic understanding.

23.5.2 Link Recommendation

Link recommendation or link predictions are important techniques in link analysis, which help specific users find more friends and also expand social network in terms of linkage. Most of the methods attempt to define a connection weight score between pairs of nodes in one way or another. Liben-Nowell et al. defined and studied link prediction method in [14] and also proposed methods to measure proximity of nodes in social network. Yin et al. structured this problem in an augmented graph fashion and applied random walk process on social network. The idea of the third step of our framework is extended from Yin's work, while our information network is GPS pattern based. Researchers also tried to approximately estimate link relevance and correlation by applying probabilistic inference [15]. Inference-based models are usually hard to interpret, while training process is usually very time-consuming and not scalable. All these methods are designed to find coauthorship or online friends, while none of them can detect geo-friends with compatibility of GPS data analysis.

23.6 Social Aggregators

There are numerous sensors available these days; we have fitness bands, numerous sensors in our mobile phones, sensors in our car, etc. Mobile phones these days come equipped with various sensors like accelerometer, gyroscope, and GPS. These sensors when used in collaboration can achieve significant amount of detail and help the users and the community. We look at some on the scenarios here.

23.6.1 Humps and Pits on Roads

A gyroscope can be used to map the humps and pits in a city. Each time a vehicle encounters a hump or pit, the gyroscope will show some deviation in readings. Considering the case of a city, there are numerous vehicles running on the road every day. Each vehicle owner will have a cell phone with him. Data collected through these cell phones can be mapped on a city map. This will give us an approximate idea of where humps and pits will be found within the city limits. Every data point can be rated based on the number of hits at a particular location.

This kind of system requires a community interaction, where all the members agree to provide the data. The humungous data collected will have to be processed using large-scale clusters, where the CPS solutions will come in action.

23.6.2 Activity Partners Based of GPS

There are various activities for which you would like to have a groups or need partners to perform them like playing a sport, jamming, and hiking. Groups are better formed when they are located in the same vicinity. Various social networking sites already exist, which cater to these needs. But processing it in real time is the biggest challenge.

Social networking sites enable the users to search for activity partners and show interest in real time. This requires real-time streaming of GPS data and activity of liking shown by the user. CPS can enable this and make it a reality.

23.7 Conclusion

Social network has become an integral part of our society. As more and more number of people are getting hooked to it, the collaborations will increase. To deal with this CPS solutions can come in handy. The need of the hour is to find solutions to integrate with the interacting elements (cyber and physical).

References

1. Cyber-Physical Systems Summit Report, CPS Summit: Holistic Approaches to Cyber Physical Integration (April 2008), http://varma.ece.cmu.edu/summit/CPS_Summit_Report.pdf (acquired).
2. Xia, F., Ma, L., Dong, J., and Sun, Y. Network QoS management in cyber-physical systems. In *Proceedings of International Conference on Embedded Software and Systems, ICESS 2008*, July 29–31, 2008, Chengdu, China, pp. 302–307.
3. Jain, R., Singh, V., and Gao, M. (2011, September). Social life networks for middle of the pyramid. In 2011 International Conference on Advances in ICT for Emerging Regions (ICTer) (p. 1).
4. Aggarwal, C. and Abdelzaher, T. Integrating sensors and social networks. In *Social Network Data Analytics*, Springer, New York, 2011, pp. 379–412.
5. Ma, J., Yang, L., Apduhan, B., Huang, R., Barolli, L., and Takizawa, M. Towards a smart world and ubiquitous intelligence: A walkthrough from smart things to smart hyperspaces and UbicKids. *International Journal of Pervasive Computing and Communications*, 1(1), 2005, 53–68.
6. Rahman, Md. A., ElSaddik, A., and Gueaieb, W. Augmenting context awareness by combining body sensor networks and social networks. *IEEE Transactions on Instrumentation and Measurement*, 60(2), 2011, 345–353.
7. Breslin, J.G., Decker, S., Hauswirth, M., Hynes, G., Phuoc, D.L., Passant, A., Polleres, A., Rabsch, C., and Reynolds, V. Integrating social networks and sensor networks. In *W3C Workshop on the Future of Social Networking 2009*, January 15–16, 2009, Barcelona, Spain.
8. Gonzalez, M.C., Hidalgo, C.A., and Barabasi, A.-L. Understanding individual human mobility patterns. *Nature*, 453(5), 2008, 779–782.

9. Report: Bridging the Cyber, Physical and Social Worlds Workshop, http://collaboration.greatplains.net, 2009.
10. Ganti, R.K., Tsai, Y., and Abdelzaher, T.F. SenseWorld: Towards cyber-physical social networks. In *International Conference on Information Processing in Sensor Networks*, April 22–24, 2008, St. Louis, MO, pp. 563–564.
11. Sha, L., Gopalakrishnan, S., Liu, X., and Wang, Q. Cyber-physical systems: A new frontier. In Yu, P.S. and Tsai, J.J.P., eds., *Machine Learning in Cyber Trust: Security, Privacy, and Reliability*, Springer, New York, 2009, pp. 3–13.
12. Tang, L., Yu, X., Kim, S., Han, J., Hung, C., and Peng, W. Tru-alarm: Trustworthiness analysis of sensor networks in cyber-physical system. In *Proceedings of the ICDM*, December 13–17, Sydney, New South Wales, Australia, 2010, pp. 1079–1084.
13. Cao, X., Cong, G., and Jensen, C.S. Mining significant semantic locations from GPS data. *Proceedings of the VLDB Endowment*, 3(1), 2010, 1009–1020.
14. Liben-Nowell, D. and Kleinberg, J. The link prediction problem for social networks. In *Proceedings of the Twelfth Annual ACM International Conference on Information and Knowledge Management, CIKM*, November 2003, New Orleans, LA, pp. 556–559.
15. Kashimaoki, H. and Abe, N. A parameterized probabilistic model of network evolution for supervised link prediction. In *Proceedings of Sixth International Conference on Data Mining, ICDM '06*, December 18–22, 2006, Hong Kong, China, pp. 340–349.

24

Cyber-Physical System Approach to Cognitive Failure Detection in Driving Using EEG and EMG Interfaces

Anuradha Saha and Amit Konar

CONTENTS

24.1 Introduction

Cyber-physical systems (CPSs) (Haque and Aziz, 2013) refer to integration of computational models and system-theoretic methodologies to design, develop, and realize complex physical (engineering) systems with provisions for human interactions in a more intricate manner than those used in conventional systems. Hence, CPS forms a link between the physical world and digital world (virtual world), where physical devices, such as sensors and cameras, with cyber components including hardware as well as software components form a situation-integrated system that responds intelligently to dynamic changes in the real-world scenarios. The operations of these integrated physical-engineered systems involved here are monitored and controlled by a computing and communication core. These operations comprise various engineering methods and tools including system identification, filtering, time and frequency domain analysis, state space analysis, optimization techniques, and robust control. At the same time, development of new programming languages, innovative approaches for compiler designs, and embedded system architectures ensure computer system reliability, cyber security, and fault tolerance. CPS offers wide applications including smart medical technology, assisted living, environmental control, and traffic management. Unfortunately, there exists no unified architecture for CPS primarily due to variations among the applications. In this article, we discuss CPS configuration for detecting cognitive failures in driving during traffic management.

Driving is a complex task involving coordinated sensorimotor activities in dynamic environments. Sensorimotor coordination, in general, is a cognitive activity, the performance of which is largely dependent on sensory attention (alertness) and perception (understanding the environment through perceived sensory information), cognitive planning (for smart driving), and motor executions (for acceleration, braking, and steering control) (Saha et al., 2014a). Lack in sensorimotor coordination results in cognitive failure of car drivers that causes traffic fatalities. In general, cognitive failure of vehicle drivers refers to the psychological hindrances, prohibiting him or her to process traffic conditions and to plan motor execution tasks required for safe driving. Many factors including lack of sleep, long driving hours, lack of proper food intakes, monotonous driving environment, use of sedating medications, and consumption of alcohol lead to severe accidents by the drivers. Upon continuous monitoring, drivers' physical changes during driving, such as the inclination of the driver's head, sagging posture, and decline in gripping force on the steering wheel, give indications of lack of attention. Therefore, it can be established that the cognitive failure in driving involves both driving and human errors (Allahyari et al., 2008). Driving errors such as observation errors, estimation errors, and incapacity of execution play significant roles for failing to perceive, memorize, and execute motor functioning during driving a vehicle (Hakamies-Blomqvist, 1993). On the other hand, human errors due to lack of planning (mistakes), storage (lapses), and execution (slips) are also considered as the basis of cognitive failure while driving (Reason, 1990), where the actions of the driver differ from his or her intensions. Some additional errors including recognition errors, decision errors, and performance errors are also found to be involved with the cognitive failure of a driver (Treat, 1980; Rumar, 1990). This chapter attempts to detect cognitive failures due to errors in visual alertness, motor (cognitive) planning, and motor execution in order to correctly identify the reason for the driving failure.

Here, physiological signals acquired from the driver are preprocessed and transformed to digital signals for interfacing with the digital CPS. The CPS here is employed to extract suitable features of the biosignals to recognize possible cognitive failures to alarm the subject accordingly. Figure 24.1 provides the interface of the CPS to describe the biofeedback (alarms) to the drivers for different cognitive failures. The functional details of the CPS are presented in Figure 24.2 for convenience. Figure 24.2 includes five stages: The first stage is used to preprocess (eliminate artifacts and undesirable noise) the acquired signals within the desired frequency spectra of the signals. The second, third, and fourth stages, respectively, represent feature extraction (FE), feature selection (FS), and data point reduction (DPR). FE refers to determining the basic primitives/features of the signals that sufficiently describe the signal itself. The list of features being of the order of several thousands in most of the biosignals, an FS unit is required to selectively identify fewer of them. More discriminating features are preferred over their counterparts. The DPR unit selects a unique class-representative trial from a large set of experimental trials. DPR is undertaken at a later stage of CPS in order to avoid losing important feature sets of trials. The last stage represents a pattern classifier to classify the extracted features into several alarm classes, each representing one specific cognitive failure.

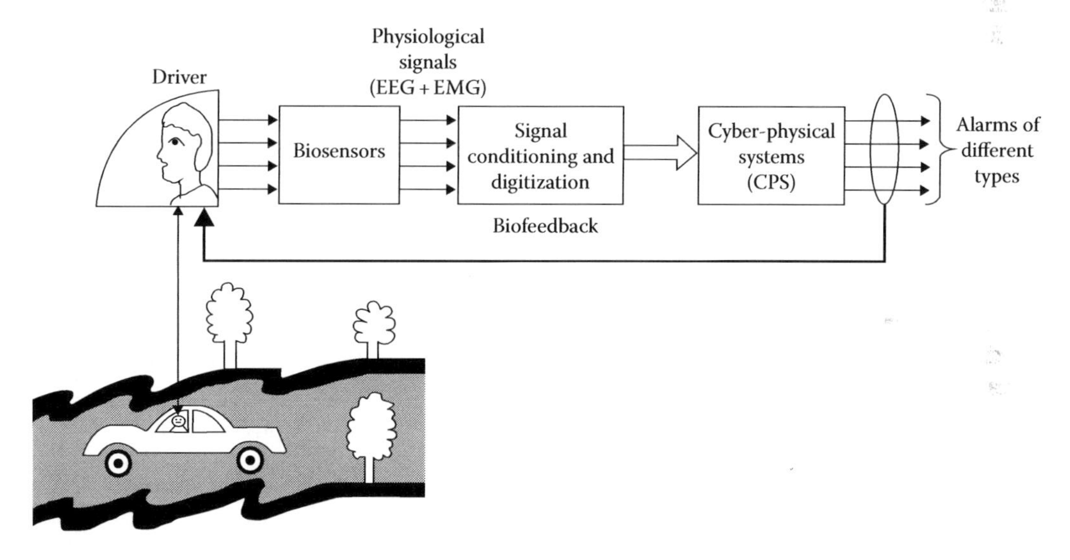

FIGURE 24.1
Schematic overview of the cognitive failure detection system.

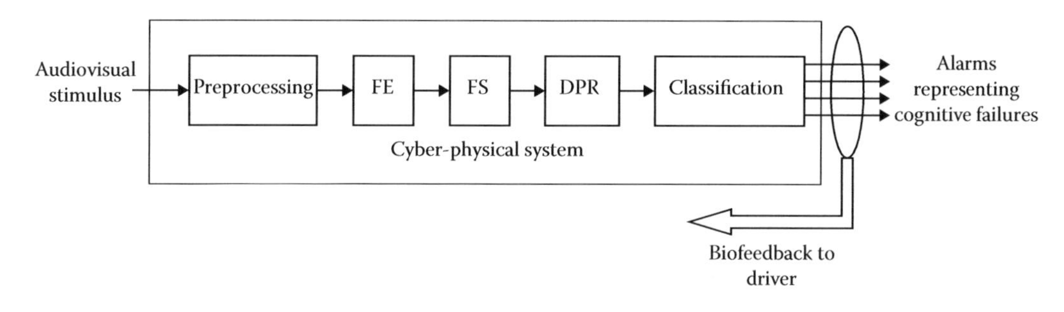

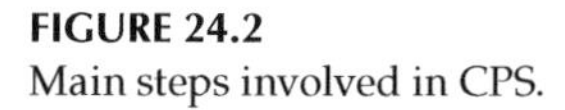

FIGURE 24.2
Main steps involved in CPS.

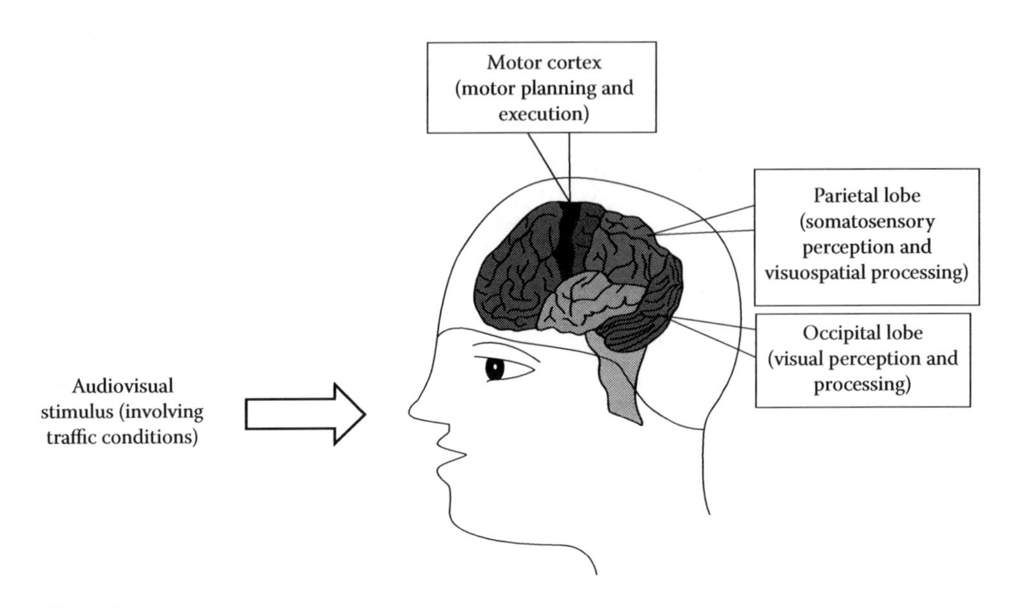

FIGURE 24.3

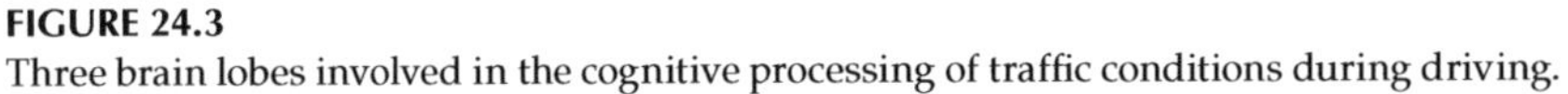

Three brain lobes involved in the cognitive processing of traffic conditions during driving.

The functions of the driver's brain during driving are shown in Figure 24.3. While the driver's brain is engaged in visual stimuli processing (to recognize traffic condition) and cognitive planning and execution (to run the car), the CPS attempts to detect the cognitive failures and alert the driver accordingly. It is evident from cognitive psychology (Purves et al., 2013) that the driver engages the occipital lobe, parietal lobe, and motor cortex region of his or her brain to respectively process visual stimuli (containing traffic condition), smart planning of braking/acceleration/steering control, and their execution. Consequently, to recognize the brain state of the driver for the earlier three cognitive tasks, we acquire electroencephalographic (EEG) signals (Niedermeyer and Silva, 2004) from the occipital lobe (electrodes O_1, O_2), parietal lobe (electrodes P_3, P_4, P_z), and motor cortex (electrodes C_3, C_4, C_z) as indicated in Figure 24.4.

In recent years, detecting drivers' attention, fatigue, and drowsiness has become a prime concern, and researchers have developed various schemes to determine these using novel algorithms and control models. A number of methods have been projected to detect fatigue (Lan et al., 2006) for car drivers. Lin et al. examine the distractions or cognitive failures of drivers as a consequence of their fatigue and drowsiness and propose several EEG-based methods for preventing drivers' cognitive failure due to

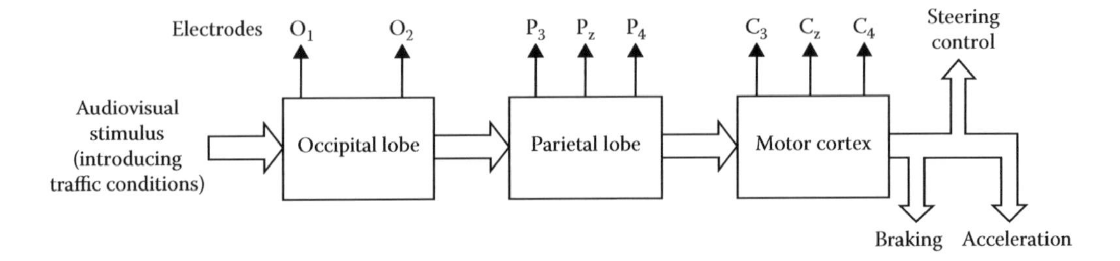

FIGURE 24.4
The electrodes used to recognize cognitive failures from three brain lobes.

distraction, fatigue, and drowsiness (Chioul et al., 2006; Lin et al., 2007). A modified intent inference algorithm (Zhou et al., 2009; Zhou et al., 2010) has been developed by observing the way drivers glance when a driver intends to change lanes (Oliver and Pentlend, 2000; Salvucci, 2004) and conducted a driving simulator experiment for avoiding the chances of lane-change crash for simultaneously performing dual tasks while driving under cognitive distraction. Apart from detecting and estimating the level of alertness, it is necessary to provide feedback to the drivers for sustaining their best performance. During this time, drivers' alertness, especially fatigue monitoring and detection, has been done by using an ocular measurement, known as percentage of eye-closure (PERCLOS) (Stegmann et al., 2003; Lang and Qi, 2008; Qing et al., 2010). One recent notable work has been found in this area to monitor a driver's alertness (Happy et al., 2013). Principal component analysis (PCA) and single-stage Haar (Jesorsky et al., 2001) classifier have been employed to determine alertness as a measure of fatigue level by calculating PERCLOS values per minute. A novel pupil position detection scheme by using the decision tree method and the form factor technique provides the estimation of pupil velocity profile. The researchers then implemented a PCA-based eye-tracking system (Xingming and Huangyuan, 2006) with constrained Kalman filtering and found impressive results in comparison to that of previously developed Haar classifier. Later, they implemented their idea using EEG signal. Wavelet thresholding moving average filtering and independent component analysis (ICA) have been done on the raw EEG signal to remove the ocular artifacts. Fatigue level has been classified by the hidden Markov model using extreme energy ratio, autoregressive parameters, and eigenvalues.

We begin Section 24.2 with physiological interfaces including EEG and electromyography (EMG). Section 24.3 presents FE and FS strategies. Section 24.4 provides an overview of the DPR technique. In Section 24.5, classifications of different types of cognitive failures using recurrent neural network classifier are presented. Section 24.6 includes the experiments. Section 24.7 provides the performance analysis of the proposed classifier. In Section 24.8, three case studies of cognitive failure detection under simulated environment are presented. Finally, a list of conclusions is given in Section 24.9.

24.2 Physiological Interfaces

This section introduces fundamentals on physiological signal processing required in the context of car driving. Two modalities of physiological signal processing involving EEG and EMG are outlined here to assist readers to go through the rest of the chapter.

24.2.1 EEG Signal Acquisition

EEG represents the electrical response of the brain to external stimuli and/or memory-based brain activation. It indirectly corresponds to the cognitive tasks undertaken by the brain. An EEG machine (Figure 24.5) acquires the electrical response of the brain during execution of a cognitive task by metal electrodes placed on the scalp. The acquired signals are amplified and digitized for further processing by the computer interfacing with the digitizer. Usually, EEG acquisition is performed by an internationally acknowledged 10–20 system, where electrodes are placed at a gap of 10% or 20% on the imaginary line joining nasion to inion (Figure 24.6).

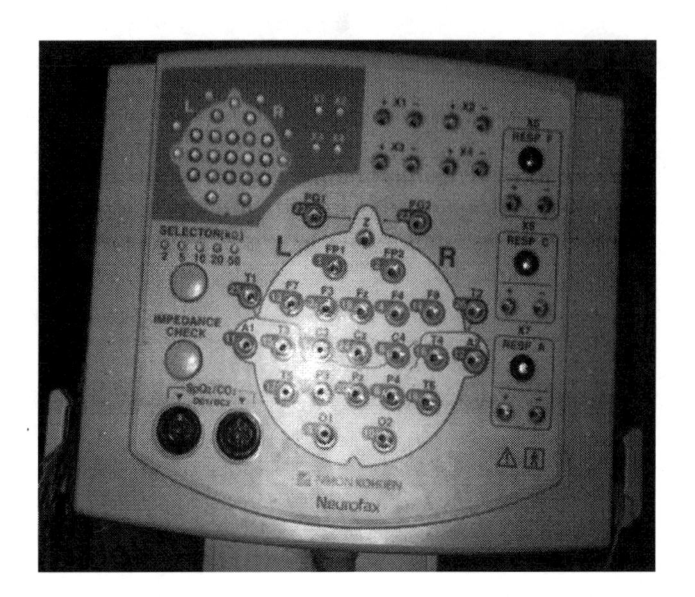

FIGURE 24.5
EEG acquisition machine configured in the international 10–20 system, manufactured by Nihon Kohden.

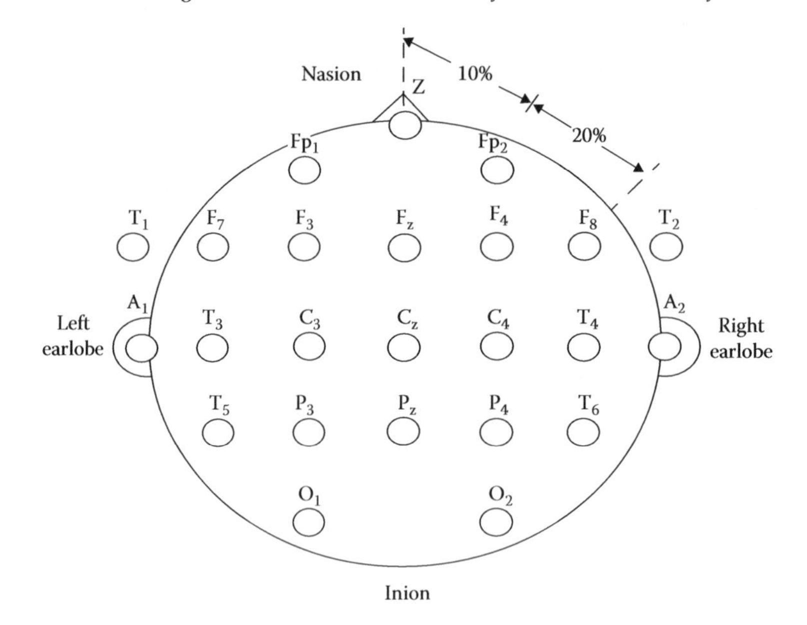

FIGURE 24.6
The international 10–20 electrode placement system for EEG data acquisition.

24.2.2 EEG Signal Preprocessing

EEG signals acquired need to be preprocessed first in order to get ready for FE. Preprocessing includes artifact removal (AR) and filtering. EEG signals recorded are passed through an AR unit to eliminate spurious noise pickup due to power supply and mainly eye blinking. Here, we perform ICA (Lee, 1998) to extract scalp maps for each electrode position required for decoding cognitive failure during driving session. ICA algorithms are based on the following assumptions: if sources have their autonomous time courses, then for *M* number of sensors, *M* number of corrected EEG signals from *M* number of sources can be independently separated.

After ICA has been done, EEG signals are ready to be filtered within the required bands of frequency. Here, we experimentally select a suitable band-pass filter having good filter characteristics. We choose elliptic band-pass filter for its steeper roll-off and equi-ripple behavior in passband and stop band in comparison to its competitors (Saha et al., 2014b). The elliptic filter of order 4 and having passband frequency of 8–30 Hz, 1 dB passband ripple, and 50 dB stop-band ripple provides the necessary filtered EEG signal for further processing.

After filtering, distinct brain signals are analyzed to observe users' intent at the specific time. Among the various EEG modalities prevalent in the literature (Pfurtscheller and Lopes da Silva, 1999; Wang et al., 2012; Zhang et al., 2012), P300 event-related potential (ERP) and event-related desynchronization (ERD)/event-related synchronization (ERS) have been used here:

P300 event-related potential (*ERP*): P300 ERP is liberated as a positive deflection around a latency of 300 ms (Donchin et al., 2000) in a succession of frequent and rare auditory, visual, or somatosensory stimuli to recognize a target stimulus from a regular set of stimuli. Figure 24.7 shows the signal characteristics of P300 along with other ERPs in awaked and fatigued situations. It is already clear from this figure that P (N) of a particular ERP stands for positive (negative) deflection and the value refers to the latency of that ERP. Figure 24.7 concludes that the amplitude of P300 is much higher for a human subject in awaked condition compared to that in fatigued situation.

ERD/ERS: ERD and ERS are two important EEG modalities that are prevalent in the literature (Pfurtscheller and Aranibar, 1977; Toro et al., 1994; Crone et al., 1998) during motor imagery. ERD emphasizes the fall of brain rhythms during motor imagery (relatively prior to 500 ms after the onset of the stimulus), whereas ERS associates with the rise of brain rhythms after motor imagery (relatively after 500 ms to several milliseconds) and with relaxation. Figure 24.8 provides the signal characteristics of ERD/ERS, where the light line represents the normal EEG without ERD/ERS generation and relatively darker line indicates ERD/ERS.

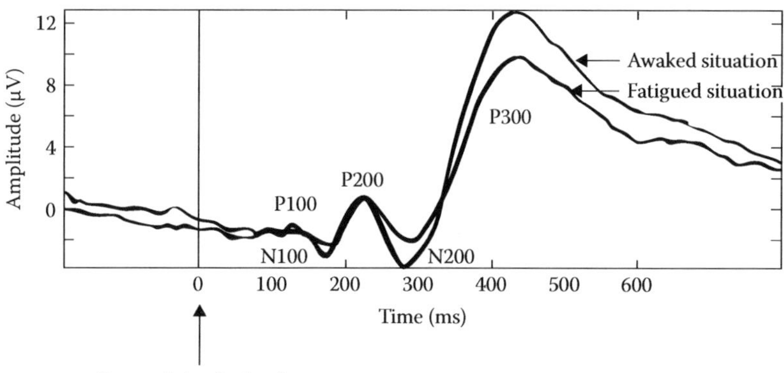

FIGURE 24.7
Signal characteristics of P300 ERP for awaked and fatigued situations, acquired from P_z electrode.

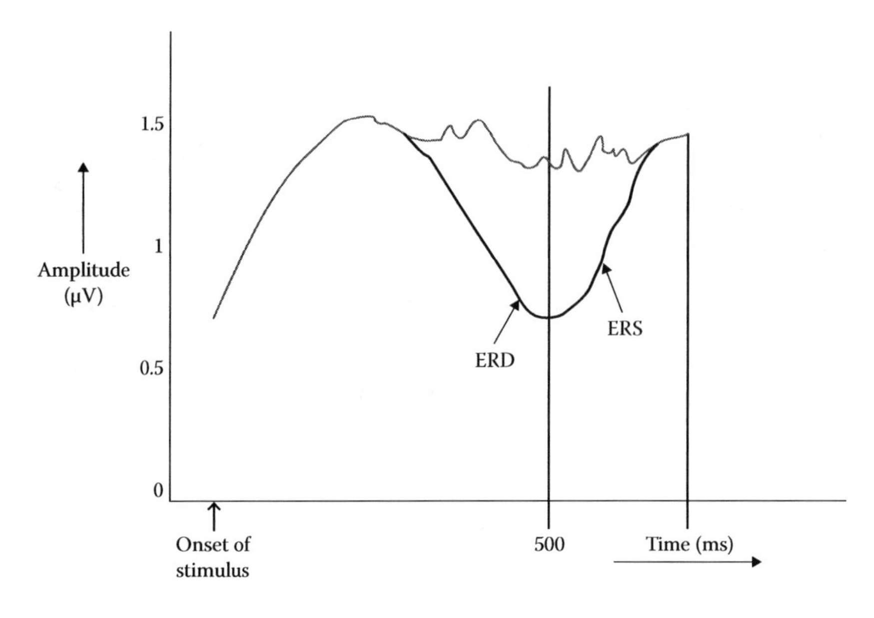

FIGURE 24.8
Signal characteristics of ERD and ERS.

24.2.3 EMG Signal Acquisition and Preprocessing

EMG signal refers to the electrical activity of the neuromuscular activation associated with the contracting muscles. Here, surface EMG signal is acquired by placing surface electrodes on the skin over the muscle to detect the electrical activity of the muscle. An EMG signal has a small amplitude (20 μV to 2 mV) and needs to be amplified with an amplifier that is specifically designed to measure physiological signals. Figure 24.9 provides an overview of EMG signal classification, where EMG signals, after recorded from hand muscles (extensor carpi radialis longus) and leg muscles (gastrocnemius muscles, often referred to the bulging area of calf muscle), are fed to the signal conditioning unit for amplification. An analog-to-digital converter (ADC) is used for digitizing the analog EMG signal for further preprocessing. Useful EMG features containing the basic primitives are extracted during the FE stage and are sent for classification. EMG signals are classified into two classes, either force exerted or no force exerted, based on the thresholds determined after examining several trials. Figure 24.10 shows EMG signal acquisition from hand and leg muscles during the experiments.

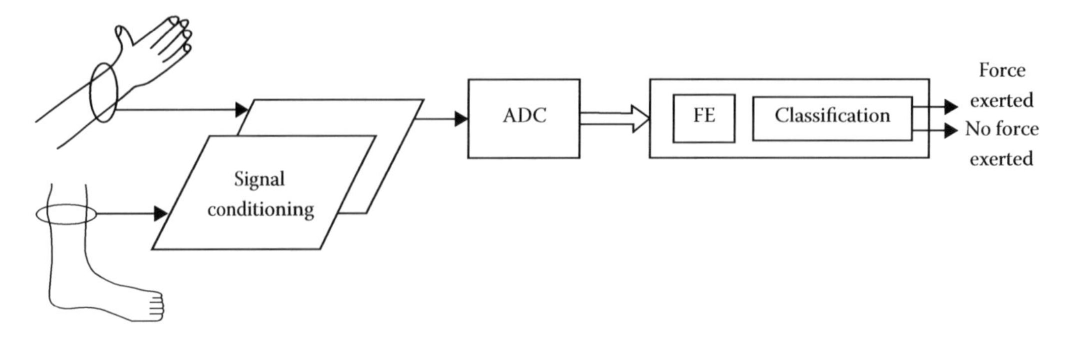

FIGURE 24.9
An overview of EMG signal classification.

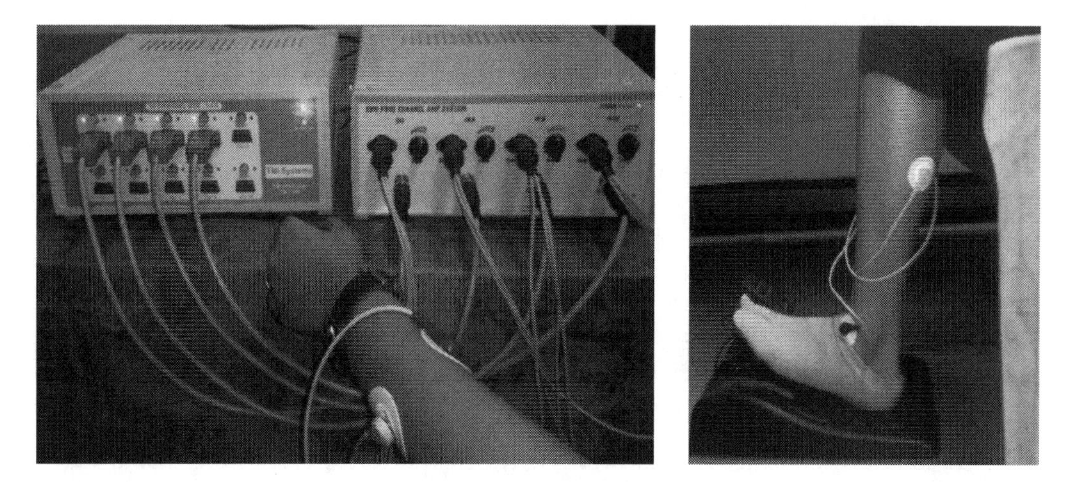

FIGURE 24.10
EMG signal acquisition by 4-channel amplifier and 10-channel USB data acquisition system (DAQ), manufactured by TMI Systems during the experiment performed in the Artificial Intelligence Laboratory, Jadavpur University.

24.3 Feature Extraction and Feature Selection

A pattern is described by its feature. FE is an important step in EEG processing to extract various types of well-known features such as time (Hjorth parameters, autoregressive parameters), frequency (power spectral density [PSD]), time–frequency (wavelet coefficients) domain, and some special features (common spatial patterns [CSPs]) from EEG signal depending on the requirement of the problem. Since classifying motor imagery has always been a serious pattern classification problem, we need an efficient feature set to perform an accurate classification of all the classes. Here, we select PSD and CSP techniques to extract the basic primitives of a particular cognitive intention. PSD extracts the power distribution of individual EEG channels separately using frequency analysis, whereas CSP is able to extract features from multichannel signals, which is expected to be more informative:

Power spectral density: PSD (Martin, 2001), which is one of the most popular FE methods, is defined as a mode of describing the power distribution contained in the signal. PSD evaluates the power density for filtered EEG recordings. Typically, a band-pass infinite impulse response (IIR) filter having a passband of 0.5–30 Hz (depending of the activation of EEG in the frequency spectrum) is used for preprocessing the raw data before the application of PSD extraction algorithms. PSD finds signal power contained in these frequency ranges by computing the Fourier transform of the autocorrelation sequence of the time series *eeg*(*t*), which is given in the following equations:

$$EEg(f) = \int_{t_1}^{t_2} eeg(t)\, e^{-2\pi f t} dt \tag{24.1}$$

$$PSD(f) = 2\frac{||EEG(f)||^2}{(t_2 - t_1)} \quad (24.2)$$

Common spatial pattern: CSP (Wang et al., 2005) is one of the most significant two-class FE techniques that work on multichannel EEG data. In this technique, a projection matrix is formed to project multichannel data into low-dimension subspace. The formation of the matrix mentioned above is made possible by combined diagonalization of two covariance matrices obtained from the two classes of EEG signals. A linear transformation is used in such a way that the variance of one projected class is maximized, while the variance of the other class is minimized. The detailed steps of CSP are presented in the Appendix.

After FE, computationally complex features may generate a very high-dimensional feature space that may contain both relevant and redundant features. The presence of redundant features adversely affects the performance of the classifier in terms of accuracy and time complexity. For this purpose, an intermediate step, called FS, between FE and classification is introduced, which aims at reducing the dimension of the feature set by retaining only the most relevant features and rejecting the rest. Some of well-known statistical feature selectors including sequential forward/backward selection (Zongker and Jain, 1996), PCA, and mutual information (MI) are used by brain–computer interface researchers. These methods are computationally viable, but sometimes they tend to reduce the accuracy of the classifier when more redundant features are selected than relevant ones. For this purpose, evolutionary algorithm–based feature selector is occasionally applied to meet the demand. However, in this chapter, the extracted features are less in number and hence FS does not help in enhancing the classifier performance.

24.4 Data Point Reduction

DPR is an important issue in the EEG-based classification problem in order to select one unique class representative from a large set of data points (trials). Here, we too attempt to determine one unique representative for each class using PCA to classify several cognitive actions of a subject during driving. EEG signals, being nondeterministic by nature, do not offer the unique features extracted from several trials of the EEG signals captured from the same subject for the same stimulus. Therefore, we need to identify the ideal class representative of each data point representing a feature vector of fixed dimension. For each standard driving instance (stimulus, which is used for subject training), we obtain selected feature vectors of the acquired EEG for each stimulus for t (= N_m) trials. Let $S_k = \{\vec{X}_1, \vec{X}_2, \ldots, \vec{X}_t\}$ be a set of t extracted feature vectors for the kth stimulus, where $\vec{X}_i = \{x_{i,1}, x_{i,2}, \ldots, x_{i,d}\}$ is a d-dimensional feature vector (data point) obtained after FE. Let there be K number of classes, that is, k lies in [1, K]. Here by PCA, we reduce S_k by identifying the representative data point $\vec{\theta}_k$ of dimension $d \times 1$ for each class $k = 1$ to K. The steps for data reduction by PCA are given later.

Normalization: Normalize $\vec{X}_i^k$ by the following transformation, where normalized components of $\vec{X}_i^k$ are given by

$$x_{i,j}^k \leftarrow \frac{x_{i,j}^k}{\sum_{l=1}^{d} x_{i,l}^k} \tag{24.3}$$

This to be repeated for $i = [1, t]$ and $j = [1, d]$.

Calculation of mean adjusted data: The empirical mean for each normalized feature vector $\vec{X}_i^k = \{x_{i,1}^k, x_{i,2}^k, \ldots, x_{i,d}^k\}$ is computed for $i = 1$ to t in the following way:

$$\overline{x}_i^k = \frac{\sum_{j=1}^{d} x_{i,j}^k}{d} \tag{24.4}$$

To get the data adjusted around zero mean, the $\overline{x}_i^k$ component is subtracted from each dimension of $\vec{X}_i^k$ to obtain $x_{i,j}'^k$, where

$$x_{i,j}'^k \leftarrow x_{i,j}^k - \overline{x}_i^k, \quad i = [1,t], \quad j = [1,d] \tag{24.5}$$

The matrix, of dimension $t \times d$, obtained after normalization and mean adjustment of each feature vector is known as *data adjust*. The data adjust matrix for the kth set S_k is given by

$$\mathbf{D}_k = \begin{bmatrix} \vec{X}_1'^k \\ \vec{X}_2'^k \\ \vdots \\ \vec{X}_t'^k \end{bmatrix} = \begin{pmatrix} x_{1,1}'^k & . & . & x_{1,d}'^k \\ . & . & . & . \\ . & . & . & . \\ x_{t,1}'^k & . & . & x_{t,d}'^k \end{pmatrix} = \begin{pmatrix} x_{1,1}^k - \overline{x}_1^k & . & . & x_{1,d}^k - \overline{x}_1^k \\ . & . & . & . \\ . & . & . & . \\ x_{t,1}^k - \overline{x}_t^k & . & . & x_{t,d}^k - \overline{x}_t^k \end{pmatrix} \tag{24.6}$$

Evaluation of Covariance Matrix: The covariance matrix of data adjust is obtained from the outer product of the data adjust matrix with itself. The covariance matrix thus obtained is of dimension $t \times t$ and is given by

$$\mathbf{C}_k = \frac{1}{d-1}\mathbf{D}_k\mathbf{D}_k^T = \begin{pmatrix} c_{1,1}^k & . & . & c_{1,t}^k \\ . & . & . & . \\ . & . & . & . \\ c_{t,1}^k & . & . & c_{t,t}^k \end{pmatrix} \tag{24.7}$$

The element of $\mathbf{C}_k$, denoted by $c_{i,j}^k$, represents the covariance between any two vectors $\vec{X}_i'^k$ and $\vec{X}_j'^k$ for $i, j= [1, t]$. It is calculated as follows:

$$c_{i,j}^k = \text{cov}\left(\vec{X}_i'^k, \vec{X}_j'^k\right) = \frac{\sum_{p=1}^{d} x_{i,p}'^k \times x_{j,p}'^k}{(d-1)} \tag{24.8}$$

Evaluation of eigenvalues: The eigenvalues of the matrix $\mathbf{C}_k$ is obtained by determining the roots of the characteristic equation given by

$$|\mathbf{C}_k - \lambda^k \mathbf{I}| = 0 \tag{24.9}$$

where
  $\mathbf{I}$ is the identity matrix of dimension $t \times t$
  $|.|$ denotes the determinant of the matrix

There would be t eigenvalues of matrix $\mathbf{C}_k$.

Evaluation of eigenvectors: The ith eigenvector $\vec{EV_i^k}$ corresponding to the ith eigenvalue λ_i^k for $i = [1, t]$, is obtained by satisfying the following equation:

$$C_k \cdot \overrightarrow{EV_i^k} = \lambda_i^k \times \overrightarrow{EV_i^k} \tag{24.10}$$

Principal component evaluation: By ordering the eigenvectors in the descending order of eigenvalues (largest first), one ordered orthogonal basis is created with the first eigenvector $\vec{EV_1^k}$ (corresponding to the highest eigenvalue λ_{large}^k) having the direction of the largest variance of the data. In fact, in this way, it turns out that $\vec{EV_1^k}$ is the principal component $\vec{PC_k}(t \times 1)$ of the dataset. Therefore,

$$\vec{PC_k} = \vec{EV_1^k} = [p_1 \; p_2 \cdots p_t]^T \tag{24.11}$$

corresponding to $\lambda_{large} > \lambda_i$ where i= [1, t].

Projection of data adjust along the principal component: Now, the class representative $\vec{\theta}_k$ of dimension $1 \times d$ corresponding to K number of feature vectors of a particular class is obtained using the following equation:

$$\vec{\theta}_k = \left(\vec{PC_k}^T \times \mathbf{D}_k\right) \tag{24.12}$$

This procedure is repeated for each of the K classes to obtain K class representatives.

24.5 Cognitive Failure Classification

This section addresses the cognitive failure classification problem using appropriate classifiers to detect the type of cognitive failures that might occur due to lapse of visual attention, cognitive planning, and motor execution. Figure 24.11 provides an overview of the cognitive failure classification using the proposed scheme of classification. Here, EEG/EMG signals are decoded into two or more labels through filtering, FE, and classification. To detect drivers' alertness, occipital and centroparietal EEGs are classified into attentive/nonattentive labels. Cognitive planning for motor imagination is classified

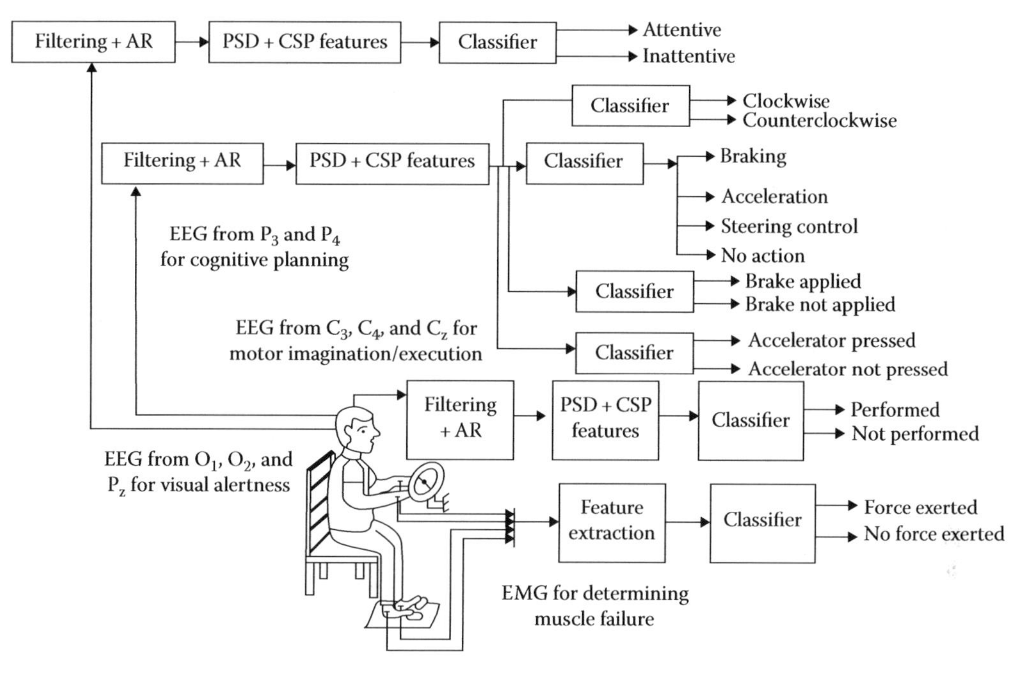

FIGURE 24.11
Overall block diagram of cognitive failure detection using the proposed scheme of classification.

into four classes: braking, acceleration, steering control, and no action. Each of the previously mentioned labels is again classified into finer labels using parietal EEG as indicated in Figure 24.11. Motor intension/execution is classified into motor imagination or action performed/not performed classes, and the muscle failure is determined by classifying EMG signals into force exerted/not exerted levels.

The most important aspect of the present article is to employ a classifier to classify the correct intended class for a particular mental/cognitive task using the selected feature set. Usually, the selection of a classifier depends on the complexity of the classification problem and also on the dataset. The complexity arises because of the variability of feature values for patterns within the same class relative to that of the patterns in different classes. It is sometimes observed that a classifier fails to classify the patterns with an acceptable level of accuracy or an excellent performance of it single-handedly cannot ensure the achievement of optimal performance for a pattern recognition system. To deal with such cases, the pattern recognition problem includes a pool of classifiers, which distinctly operate on various aspects of the input feature vector and combining them provide comparatively better performance. We too, in this chapter, compares the performance of some commonly used standard pattern classifiers including *k*-nearest neighbor (kNN) (Song et al., 2007), naïve Bayes (Wei and Yang, 2010), support vector machine with radial basis function (SVM-RBF) kernel (Chang et al., 2005), and the proposed recurrent neural network classifier in view of the classification accuracies and error estimates.

We here examine the role of a special class of recurrent neural network (Williams and Zipser, 1989) having functionality similar to a Hopfield neural network (Hopfield, 1991; Konar, 2005; Saha et al., 2013) but of a different topology. Here, the proposed recurrent neural network having explicitly incorporating features allows processing of EEG data during driving session. We also propose a given energy function that satisfies the characteristics of a Lyapunov function containing multiple minima. The recurrent neural dynamics is so designed that it is capable to classify visual attention, cognitive planning, and motor execution phases, perceived by a subject from the EEG signals captured from his or her occipital, parietal, and motor cortex regions. The initial value of the variables used in the neuronal dynamics here represents the selected features of a stimulus. Once the dynamics is initialized around the minimum, a suitable weight vector for the dynamics needs to be identified to ensure the natural convergence of the dynamics to a given minimum on the selected Lyapunov energy surface (Bhattacharya et al., 2009). An optimization problem is designed to minimize the Lyapunov energy function at selected locations on the energy surface for a unique weight vector. Differential evolution (DE) algorithm has been applied here to optimize the energy function.

Stability analysis of the nonlinear neuronal dynamics determines the structural connectivity of neurons of a recurrent neural network. Here we employ the very familiar Lyapunov stability for its extensive popularity and our expertise with the technique. An overview of the Lyapunov stability method is already given in keywords for the sake of understanding the design of recurrent neural dynamics using this stability criterion. Here, we attempt to ensure the stability of the proposed recurrent topology by determining its neuronal dynamics so as to satisfy the condition of negative definiteness of its time derivative. Presuming the Rastrigin function with multiple minima as the energy function, asymptotic stability of the neuronal dynamics and its convergence at one of several minima are ensured in the sense of Lyapunov. The deepest out of multiple minima of a smooth Rastrigin function is located at the origin, which satisfies the necessary criteria of a Lyapunov function having multiple minima (Figure 24.12).

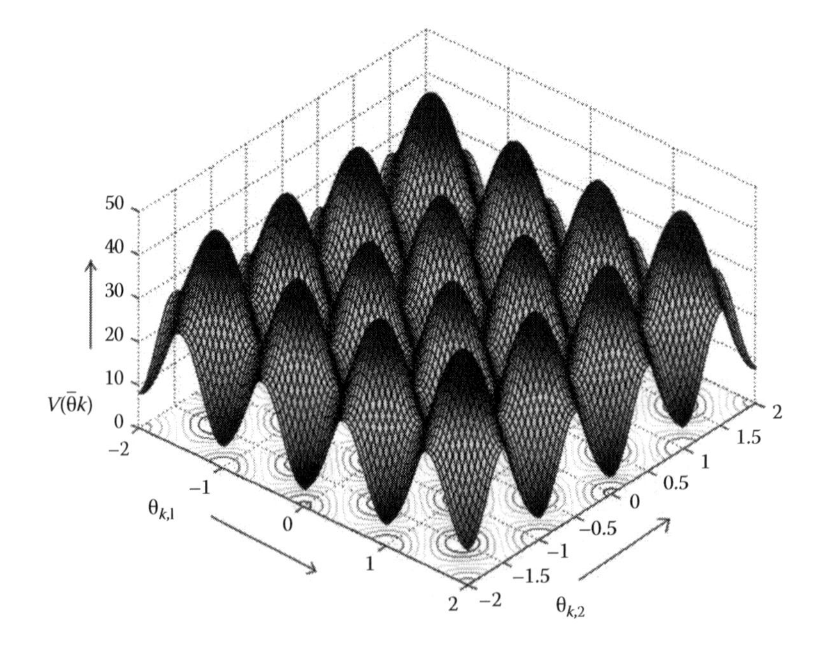

FIGURE 24.12
The plot of a 2D Rastrigin function: Equation 24.13 with dimension = 2.

Let $\vec{\theta}_k = [\theta_{k,1}\ \theta_{k,2} \cdots \theta_{k,d}]_{1\times d}$ be the kth class representative containing d number of features, $\theta_{k,j}$, $j = 1$ to d, and $V(\vec{\theta}_k)$ be the Rastrigin-type Lyapunov surface, given by

$$V(\vec{\theta}_k) = \sum_{j=1}^{d} \left[\theta_{k,j}^2 - 10w_j \cos(2\pi\theta_{k,j}) + 10\right] \tag{24.13}$$

for $k = 1$ to K, involving the weight vector $\vec{W} = [w_1\ w_2 \cdots w_d]_{1\times d}$. It is also apparent that the function $V(\vec{\theta}_k)$ in Equation 24.13 satisfies the necessary conditions of the Lyapunov function. The time derivative of $V(\vec{\theta}_k)$ is obtained as

$$\frac{dV(\vec{\theta}_k)}{dt} = \sum_{j=1}^{d} \frac{\partial V(\theta_{k,j})}{\partial \theta_{k,j}} \cdot \frac{d(\theta_{k,j})}{dt} \tag{24.14}$$

$$= \sum_{j=1}^{d} \frac{\partial}{\partial \theta_{k,j}} \left(\sum_{j=1}^{d} \left(\theta_{k,j}^2 - 10w_j \cos(2\pi\theta_{k,j}) + 10\right) \right) \frac{d\theta_{k,j}}{dt} \tag{24.15}$$

$$= \sum_{j=1}^{d} (2\theta_{k,j} + 20\pi w_j \sin(2\pi\theta_{k,j})) \frac{d\theta_{k,j}}{dt} \tag{24.16}$$

<0 (for condition for asymptotic stability), which implies the right side of the following equation:

$$\frac{d\theta_{k,j}}{dt} = -2(\theta_{k,j} + 10\pi w_j \sin(2\pi\theta_{k,j})), \forall i.j \tag{24.17}$$

Equation 24.17 provides a set of recurrent dynamics for each class k and feature j.

Now, we present a brief outline of the encoding and the recall cycles for the proposed recurrent neural network. Here, encoding refers to selecting the optimal weight vector of the recurrent network, whereas recall refers to determining one of the minima (stable attractor) on the Lyapunov surface for a given initial settings of $\vec{\theta}_{ij}(t)$.

Encoding: Given $\vec{\theta}_k = [\theta_{k,1}\ \theta_{k,2} \cdots \theta_{k,d}]_{1\times d}$ for $k = 1$ to K classes, encoding cycle determines a unique weight vector $\vec{W} = [w_1\ w_2 \cdots w_d]_{1\times d}$, such that for each class representative $\vec{\theta}_k$, $k = 1$ to K, we have a minimum on the energy surface $V(\vec{\theta}_k)$. The minimum on the energy surface for a given stimulus is considered as stable optimum (attractor) for the stimulus class. An optimization technique is followed to perform the weight vector selection, where the objective of the problem is to uniquely determine the weight vector so as to minimize the energy function $V(\vec{\theta}_k)$ for $k = 1$ to K classes. The DE algorithm employed here to solve the aforementioned optimization problem is also presented in the Appendix.

Recall: Recall cycle comprises t instances of an unknown stimulus of uniform duration separated by equal time delays and pass the acquired t instances of

EEG signals through preprocessing and FE and DPR algorithms (explained in the next section) to match the unknown input stimulus. Let the assembled d-dimensional features (data point) obtained from the steps mentioned before for an unknown visual stimulus be $\vec{\theta}'(0)$, representing the initial choice of the parameters in the neuronal dynamics given in Equation 24.17 with $\theta_{k,j}$ replaced by θ'_j for all j. Let $\vec{\theta}_k$ for $k = 1$ to K be the representative optima for K distinct classes. To identify the nearest known optimum to $\vec{\theta}'(t)$ at a steady state, we solve the dynamics (Equation 24.17) with $\theta_{k,j} = \theta'_j$ for $j = 1$ to d and identify the optimum stable point (attractor) with the shortest Euclidean distance with a steady-state value of $\vec{\theta}'(t)$. The class of the unknown stimulus can now be inferred from the predefined location of convergence of the each known stimulus class. The algorithm to determine the nearest stable optimum for a given smell class is given in the Appendix.

24.6 Experiments

This section deals with five experiments designed respectively for (1) selecting active brain regions for visual alertness and motor planning phases; (2) selecting frequency bands to capture the best EEG responses; (3) detecting visual alertness and motor imagery using P300 and ERD/ERS modalities, respectively; (4) selecting EEG features for the modalities; and (5) determining muscle activation from EMG signals.

Experimental setup: An emulated driving environment based on a virtual-reality scene including a realistic steering wheel, accelerator, and brake pedals is designed to observe the cognitive attentiveness and intensions of the participants (Figure 24.13). Twenty-five healthy subjects (drivers) aged between 22 and 30 are instructed to place their hands on the steering wheel and feet on the accelerator and brake pedals for the driving session. Each driving session takes approximately 14 min and each subject participates in each session for 10 times. All the experiments are conducted by using a stand-alone EEG machine (manufactured by Nihon Kohden) comprising of 21 electrodes and 200 Hz sampling frequency. Two EMG sensors are mounted on both the hands of the subjects to measure the force applied on the steering wheel throughout the driving session. To determine the forces exerted during steering control, EMG signals have been simultaneously recorded from the left and right muscles by placing surface bipolar electrodes at 2 cm apart and centered over the left and right muscle belly, respectively. In addition, EMG sensors are also placed on both left and right leg muscles to determine the possibility of any muscle failure during braking and acceleration, respectively. An EMG instrument, which has been used here (manufactured by TMI Systems), provides the facility to record 1024 samples per second from a maximum of 10 channels simultaneously.

24.6.1 Active Brain Regions Selection for Visual Alertness and Motor Planning Phases

This experiment is designed to select active brain regions responsible for visual alertness and motor planning phases. This is performed by applying the ICA (Lee, 1998) technique. ICA not only identifies the active brain regions but also removes the unwanted noise due to eye blinking (AR). ICA identifies fewer locations/components

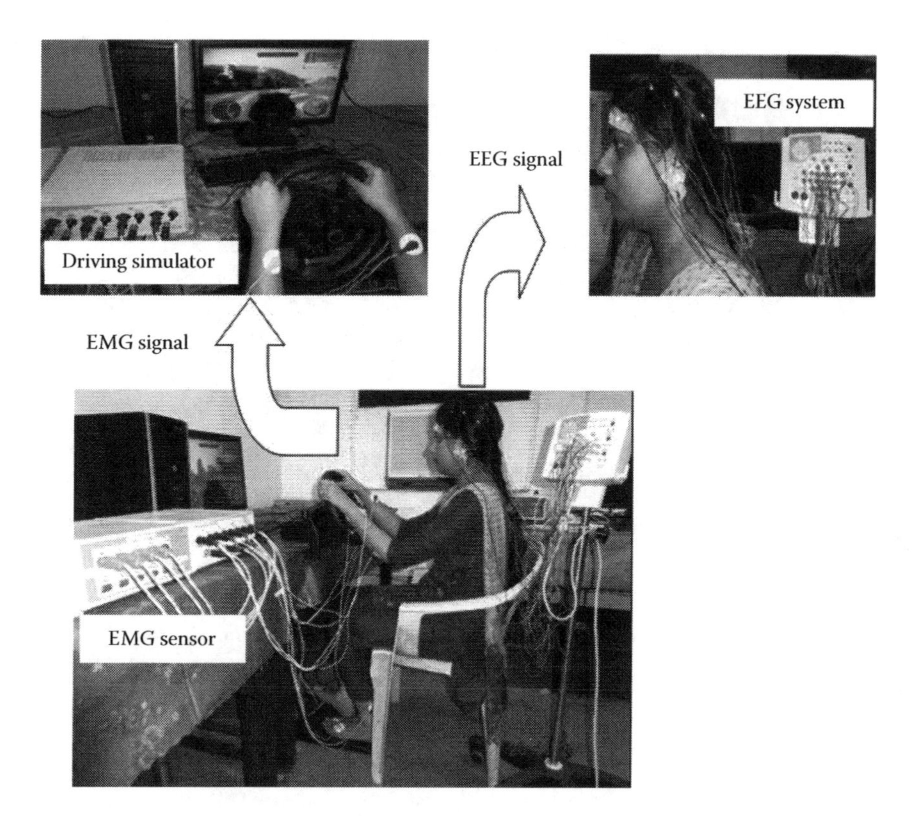

FIGURE 24.13
Experimental setup for acquiring EEG and EMG signals from a subject driving a virtual driving simulator made by Logitech during experiments performed in the Artificial Intelligence Laboratory, Jadavpur University. EEG signals are captured from the electrodes placed on the scalp of the subject, whereas EMG signals are acquired from the electrodes placed on hands and feet.

in the brain having high signal activation and hence forms individual component maps. Here, we have 21 channels, so we identify 21 independent sources using ICA and select fewer of 21 sources having relatively high brain activation during the experiment.

Figure 24.14 provides component scalp maps for 21 channels after performing ICA during an experimental trial. The figure demonstrates the higher mental activity (marked in darker shades) in corresponding electrode positions for the component maps of eight electrodes including O_1, O_2, P_3, P_4, P_z, C_3, C_z, and C_4, whereas comparatively lower mental activity (marked in relatively lighter shades) in the respective electrode positions for the component maps of Fp_1, Fp_2, F_3, F_4, F_z, F_7, F_8, T_3, T_4, T_5, and T_6. It is already known from the existing literature (Saha et al., 2014a) that O_1, O_2 (occipital lobe), P_3, P_4, P_z (parietal lobe), C_3, C_4, and C_z (motor cortex) electrode positions are found in association with visual recognition, cognitive planning, and motor execution, respectively. It is thus concluded that visual information sensing and motor imagery signal processing are here performed primarily by the occipital, parietal, and motor cortex. It is further noted that motor imagination and execution are prominent in parietal and motor cortex regions only during braking and acceleration.

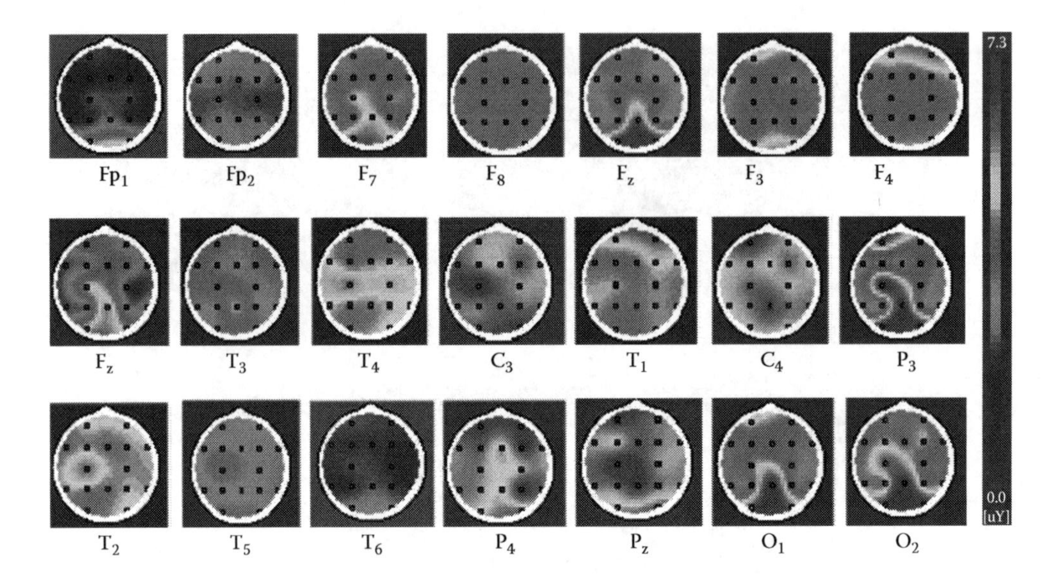

FIGURE 24.14
Component epoch maps of 21-channel Nihon Kohden EEG stand-alone instrument. Here, the figure is represented in grey-scale. This figure signifies the active component maps with highest activation at their corresponding electrode locations (having darker shades), as well as the component maps with lowest activation (having relatively lighter shades), indicating the inactivity during the driving task.

24.6.2 Frequency Band Selection to Capture the Best EEG Responses

This experiment proposes a method to select the frequency of raw EEG signal in order to remove the unwanted frequency spectrum. It is important to mention here that EEG signals contain frequency ranging from 0.5 to 60 Hz, which is again subdivided in smaller frequency bands including (1) delta (0.5–4 Hz), (2) theta (4–7 Hz), (3) alpha (7–13 Hz), (4) beta (13–30 Hz), and (5) gamma (>30 Hz). Among them, the delta band is associated with profound sleep (Teplan, 2002), whereas the theta band is found active during meditation (Deivanayagi et al., 2007). The alpha band is found active during wakefulness yet relaxing situation (Quian Quiroga and Schürmann, 1999). The beta band is associated with all possible kinds of voluntary movements (Brinkman et al., 2014), whereas the gamma band is chiefly active during high cognitive processes (Li and Lu, 2009). Based on the information established in the literature mentioned above, EEG signals are filtered within the alpha and beta frequency bands (8–30 Hz) during the driving session.

After band selection, the raw EEG signal is first band-pass filtered using an IIR elliptic filter of order 4. The merit of selecting an elliptic filter depends on sharper roll-off and good attenuation in both pass- and stop-band ripples in comparison to other competitors including Butterworth and Chebyshev filters. The frequency band of elliptical filter is determined by taking the Fourier transform of the raw EEG signal for various cognitive intensions. To validate the this statement, we present Figures 24.15 and 24.16, which show the raw and filtered EEG signals extracted from C_3 and C_4 electrodes during steering right and steering left, respectively. It has been clearly concluded from Figure 24.16 that the filtered signal has its highest amplitude peak in the range of 8–30 Hz. A similar procedure has also been applied for the remaining classes.

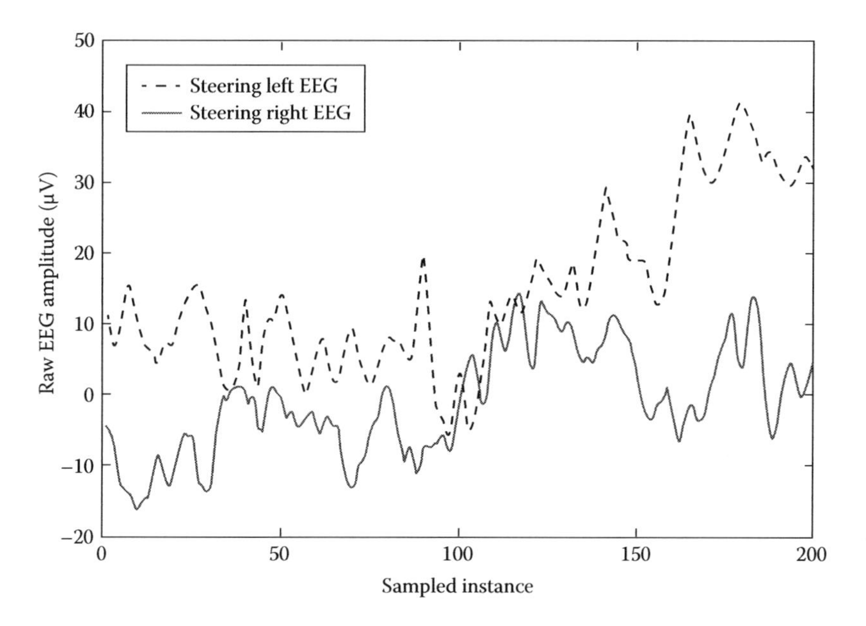

FIGURE 24.15
Raw EEG signals from C_4 and C_3 electrodes during steering left and steering right, respectively.

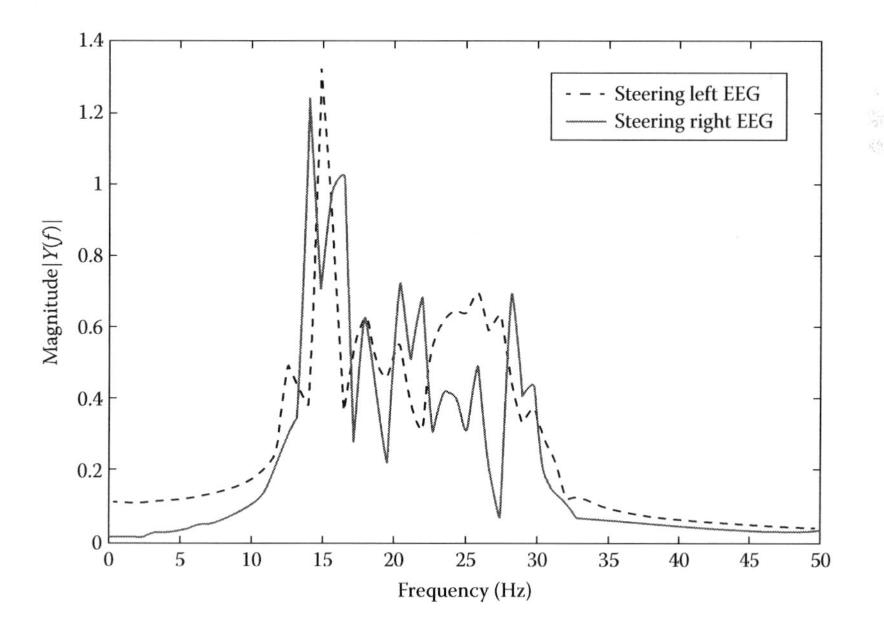

FIGURE 24.16
Filtered EEG signals from C_4 and C_3 electrodes during steering left and steering right, respectively.

24.6.3 Visual Alertness and Motor Imagery Detection Using P300 and ERD/ERS Modalities, Respectively

The present experiment is designed to detect visual alertness and motor imagery from brain responses using P300 and ERD/ERS modalities, respectively. It is well established from the previous section that a strong positive deflection in the EEG signal occurs after 300 ms from the onset of a rare yet significant visual/auditory stimulus and lasts at about 600 ms, resulting in the release of P300 ERP from the P_z location. Figure 24.17 presents a plot that simultaneously shows the positive deflection in association of P300 generation in a visually attentive situation, whereas no P300 generation due to lack of visual alertness. Considering a sampling rate of 200 Hz, we obtain 200 samples in 1 s and release of P300 in a visually alert situation after the 60th sample (marked by black marker) in order to validate its latency.

For motor imagery detection, ERD/ERS response is observed from the cortical motor areas, that is, primary motor cortex, sensorimotor area, and premotor cortex. Based on the international electrode placement techniques, signals from C_3, C_4, and C_z electrode positions of the existing EEG machine are analyzed for ERD/ERS detection. Figure 24.18 shows the ERD/ERS plot for steering right and left control. It is experimentally found that this motor imagery signal releases near 500 ms, and hence a solid line indicates this time stamping in terms of number of samples (since sampling rate is 200 Hz). It is observed from Figure 24.18 that during steering left control, the driver exhibits motor imagery signal of much higher amplitude in comparison to the other, thus confirming the fact that a right-handed person has to intend in a deeper way to perform a task by his or her left hand.

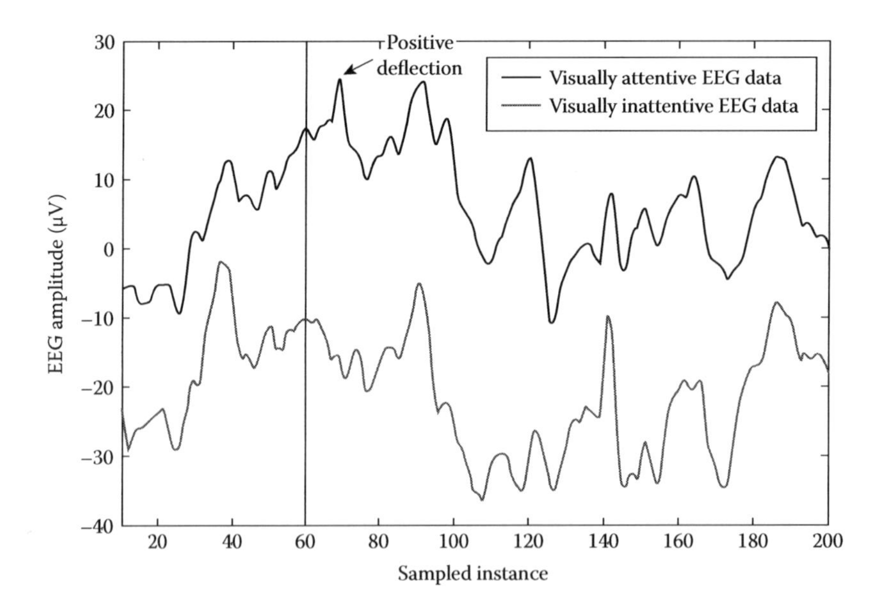

FIGURE 24.17
A positive deflection related to the P300 EEG modality is liberated in a visually alert situation after 300 ms, that is, after the 60th sample, whereas no such significant positive rise is observed in a visually inattentive situation.

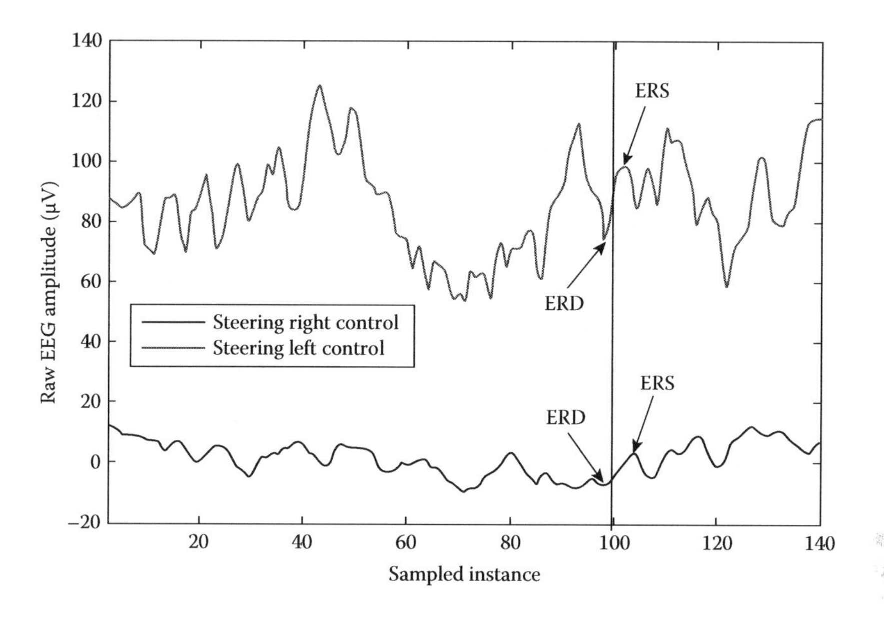

FIGURE 24.18
A negative rise related to the ERD modality is liberated during motor imagery task around 500 ms, that is, after the 100th sample, whereas a positive rise representing the ERS modality is observed after motor imagery and relaxation.

24.6.4 Selection of EEG Features for P300 and ERD/ERS Modalities

This experiment aims at finding the basic primitives (features) from the filtered EEG signal for the two modalities mentioned earlier. To start with, the data matrix of each class represented for determining a specific type of cognitive failure needs to be prepared. Considering an interval of 10 s/subject for a particular driving situation, such as acceleration, braking, steering left, steering right, and no action with data samples collected from 12 subjects, each for 15 times, we obtain altogether 180 × 2000 samples from a specific electrode position. Among the acquired data dimension, for each column of the matrix, we obtain 48 PSD EEG features. Therefore, for each desired motive, that is, for determining visual alertness, cognitive planning, or motor execution, 48 distinct PSD features have been extracted from each of the significant electrodes.

To give an illustration, Figure 24.19 provides a plot of PSD features extracted from the P3 electrode during the motor planning of accelerating and braking situations. It is apparent from this figure that there exist few instances where fewer features are capable of discriminating two motor execution classes. For example, the 18th, the 22nd, and the 33rd features can be used jointly to classify all the data.

For FE using CSP, after data acquisition, for a single subject, we obtain EEG data having dimensions of 180 × 21 × 2000 for each class, since each data point has 2000 samples. Therefore, the vector Z, as mentioned earlier, has a dimension of 21 × 2000 for each data point. And, as we took the first and last variance of each row of Z as our features, the feature dimension reduces to 2 only. Naturally, no additional dimensionality reduction technique is required. To observe the significant variance between features, CSP feature values of 10 data for each of the accelerator and brake classes are mapped on the feature space by taking the first and last variance of each row of corresponding

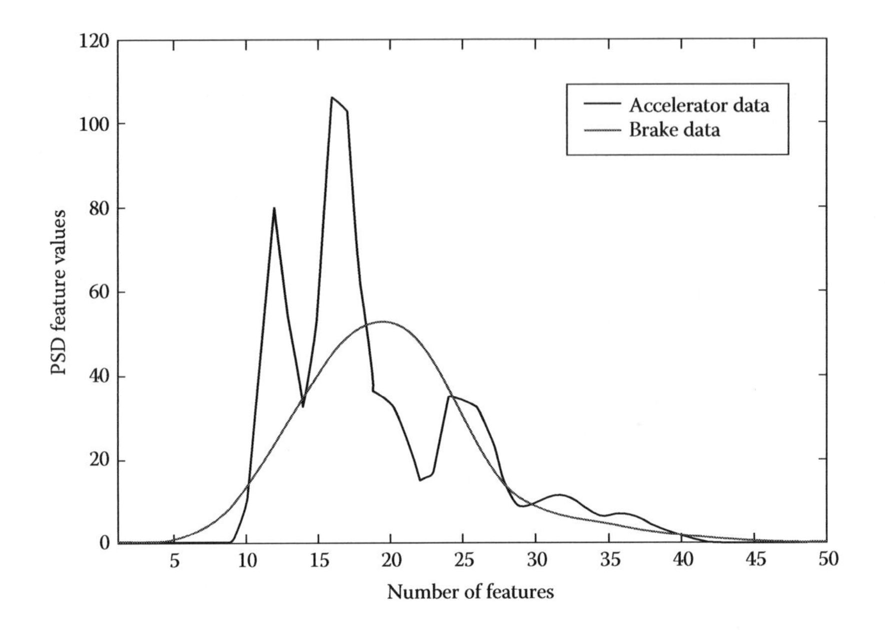

FIGURE 24.19
PSD features extracted from the P_3 electrode during acceleration and braking, respectively.

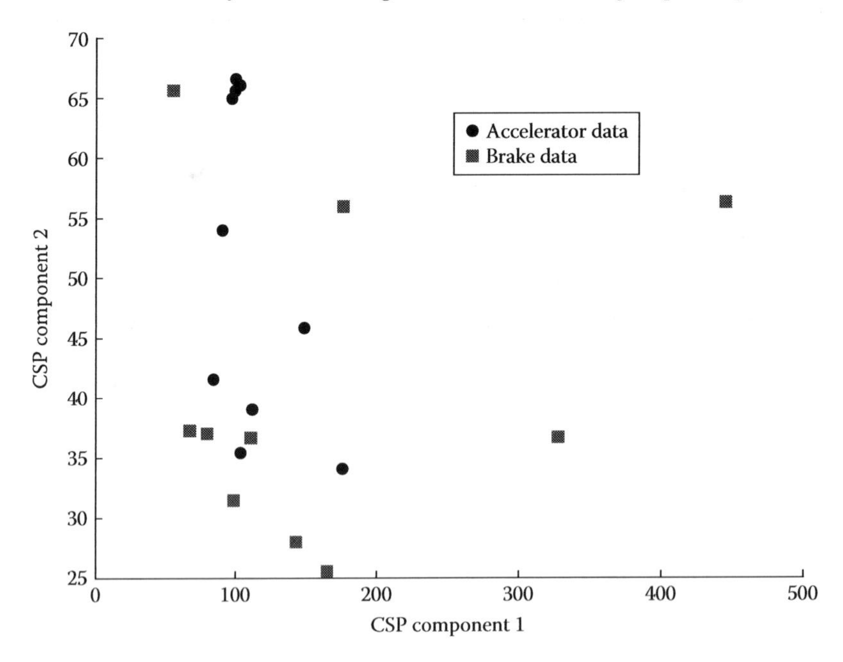

FIGURE 24.20
CSP feature discrimination between accelerator and brake classes.

feature sets (Figure 24.20). It has been observed from Figure 24.20 that both the classes are distinctly discriminated in terms of their membership classes (accelerator data are painted in dark black and brake data are painted in gray according to grayscale image). Similarly, feature-level discriminations between steering left and right are also prominent after PSD and CSP FE steps (Figures 24.21 and 24.22).

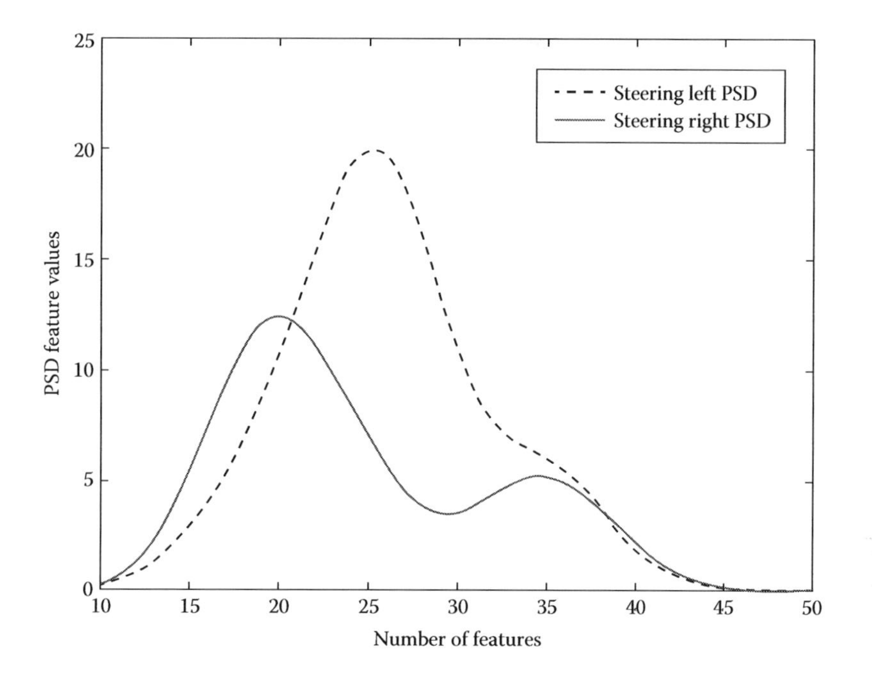

FIGURE 24.21
PSD features extracted from C_3 and C_4 during steering right and steering left, respectively.

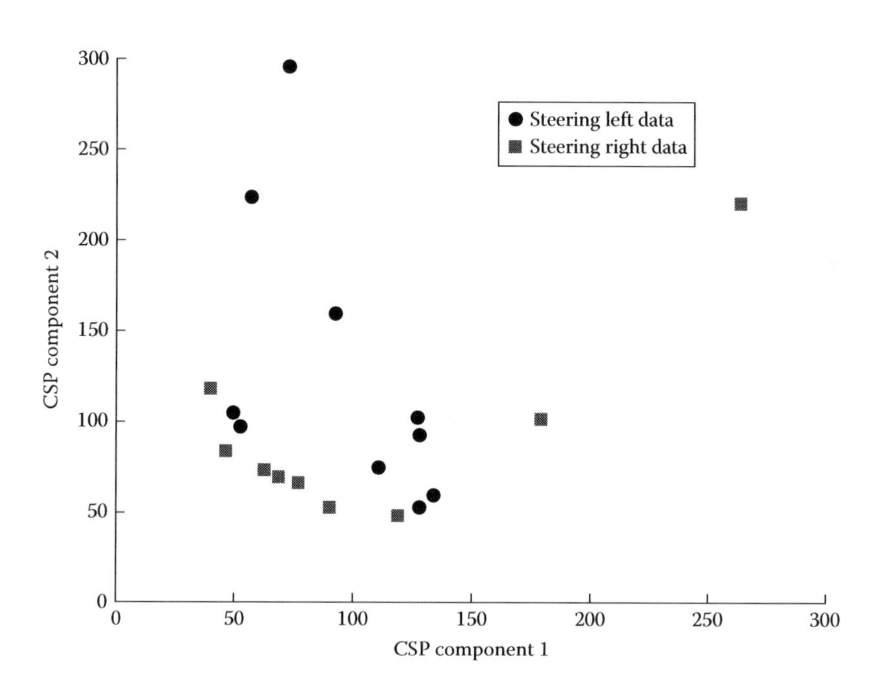

FIGURE 24.22
CSP feature discrimination between C_3 and C_4 during steering right and steering left, respectively.

24.6.5 Determining Muscle Activation from EMG Signals

This experiment clearly examines the muscle activation from EMG signals to determine muscle failure during driving. The raw EMG signals acquired from leg and hand muscles are first filtered to remove unwanted noise from the signal itself. Figures 24.23 and 24.24 provide raw and filtered EMG signals, respectively, during steering left situations.

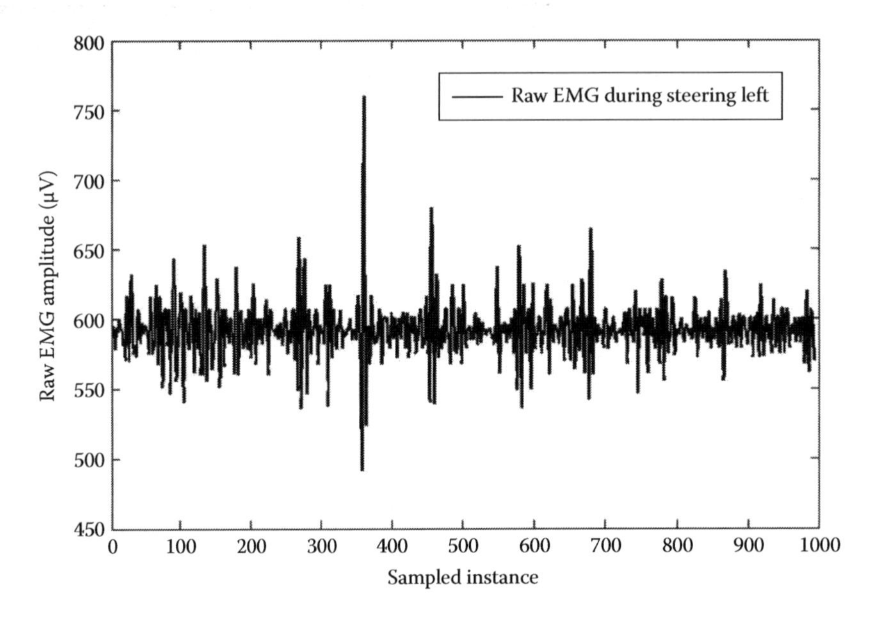

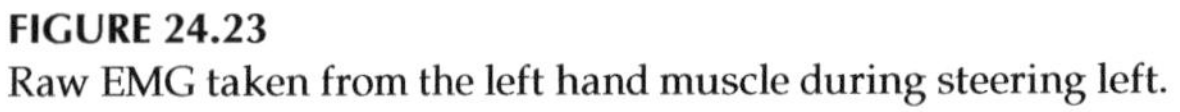

FIGURE 24.23
Raw EMG taken from the left hand muscle during steering left.

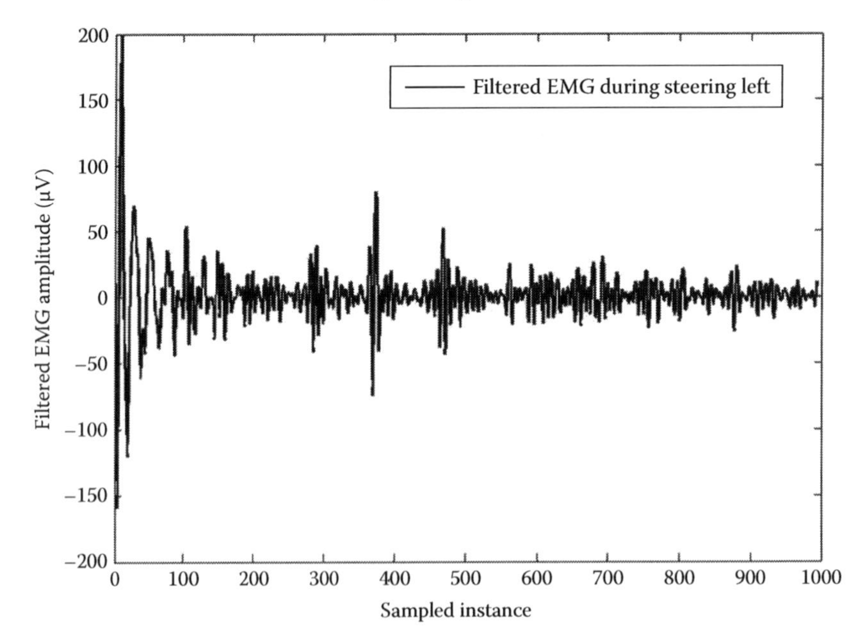

FIGURE 24.24
Filtered signal of raw EMG given in Figure 24.23.

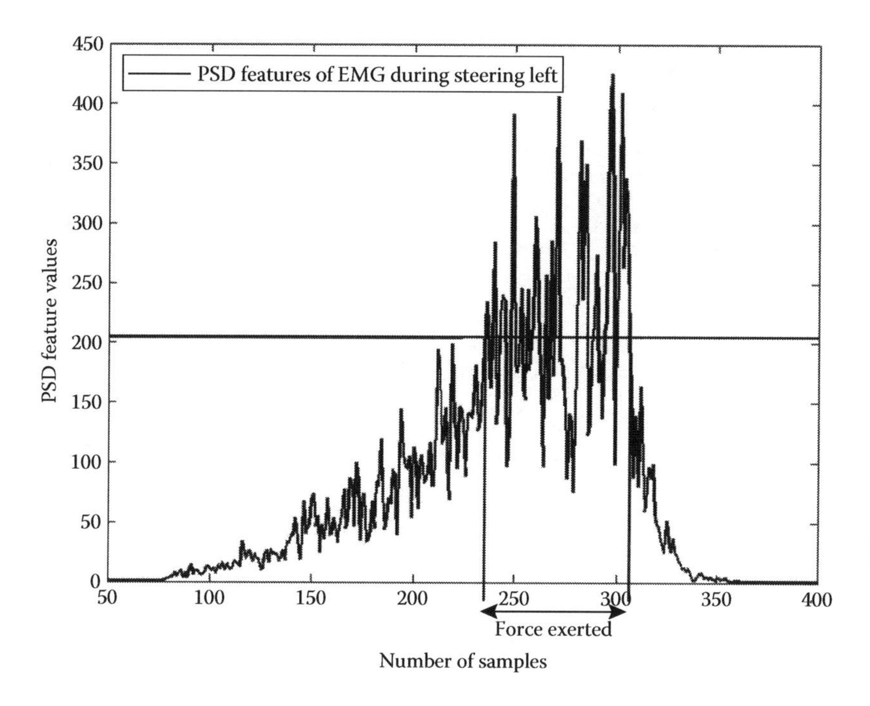

FIGURE 24.25
PSD features extracted from EMG signal given in Figure 24.23.

After filtering, PSD features have been extracted from filtered EMG signals during driving sessions to determine the band power of the acquired signal. Figure 24.25 provides PSD features extracted from the EMG signal during steering left control. Since EMG signal has no unique pattern, a threshold has been set on the extracted power in order to correctly classify the signals into two classes, force exerted or no force exerted. The thresholds are selected experimentally for each one of four muscles of a given driver individually. It is important to note that the thresholds are different for different muscles/drivers.

24.7 Performance Analysis

In this section, we first examine classifier performance for different cognitive modalities of driving during both the training and the recall phases. Next, we analyze performance of the DPR step. Third, we evaluate the relative performance of the proposed classifier amidst others, and lastly, we offer confirmatory test of overall performance by statistical techniques.

24.7.1 Classification Accuracy of Individual Driving Situation during the Training Phase

For individual class performance of different driving situations occurring suddenly during the driving session, the recurrent neural network classifier is trained with 900 trials

TABLE 24.1

Average Training Accuracies of 12 Subjects for 5 Sudden Driving Situations

Driving Situation	Classification Accuracy during the Visual Alertness Phase (in %)			Classification Accuracy during the Motor Intension Phase (in %)			Classification Accuracy during the Motor Execution Phase (in %)		
	Best	Average	Worst	Best	Average	Worst	Best	Average	Worst
Acceleration	**98.333**	**96.666**	**95.000**	**95.555**	**93.888**	**92.222**	**94.444**	**93.333**	**92.222**
Braking	94.444	92.777	91.111	93.333	91.110	88.888	92.222	90.555	88.888
Steering left	96.666	94.999	93.333	92.222	91.111	90.000	92.222	90.555	88.888
Steering right	97.222	95.555	93.888	94.444	92.775	91.111	92.777	91.110	89.444
No action	93.333	91.666	90.000	90.000	87.777	85.555	88.333	87.222	86.111

of five different situations, where each of 12 subjects experiences a particular situation for 15 times. A 10-fold cross validation is performed to check the consistency of the data, where 9 out of 10 are applied to train the classifier and the remaining onefold is employed for validation purposes.

Table 24.1 provides the average classification accuracies of training data performed on 12 subjects in three cognitive phases (i.e., visual alertness, motor intension, and motor execution phases) for different driving situations including acceleration, braking, steering left, steering right, and no action using 21D CSP feature averaged over ninefold. The highest classification accuracy for each driving situation is marked in bold in Table 24.1.

24.7.2 Overall Classifier Performance during Recall Phase

To study the overall performance of the proposed classifier during the recall phase, we used 900 test samples of different driving situations. In this phase, we consider 180 samples for each of five different classes to study the performance of the two feature selector–classifier combinations using recurrent neural net classifier during the motor intension phase. The large diagonal entries of the confusion matrix, as presented in Table 24.2, indicate that the classification accuracy for an individual class using CSP is high, over 90% (marked in bold), for all the test stimuli of different driving situations, whereas the classification accuracy for that using PSD is relatively low, over 85%. The same procedure has also been applied for the remaining two cognitive phases.

24.7.3 Performance Analysis of the Data Point Reduction Step

This section examines the importance of DPR using PCA in the present classification problem. This study includes classification of 180 samples of each of the five classes during the motor intension phase into true-/false-positive and true-/false-negative classes. True positive may be defined as the actual intensions of class A that are correctly assigned as that of intensions of class A. True negative means the remaining intensions that are correctly classified as nonintensions of class A. False negative may be defined as intensions of class A that are incorrectly marked as intensions of remaining classes. Lastly, false positive refers to the remaining classes that are incorrectly marked as odors of class A. Among them, a false-positive rate and false-negative rate are often termed as type I error and type II error. Here we express the results in percentage. Table 24.3 provides the average classifier accuracies in five different motor planning (intension)

TABLE 24.2

Confusion Matrix of Five Driving Intension Classes from Motor Intension Classifier Using DE-Recurrent NN Classifier along with CSP and PSD Features

	Predicted Class Using CSP Feature					Predicted Class Using PSD Feature				
Actual Class	**Acceleration**	**Braking**	**Steering Left**	**Steering Right**	**No Action**	**Acceleration**	**Braking**	**Steering Left**	**Steering Right**	**No Action**
Acceleration	**95.555**	0.555	2.222	0.555	1.111	**92.777**	1.111	1.666	2.777	1.666
Braking	0.555	**93.333**	1.666	2.777	1.666	1.111	**90.555**	3.333	2.777	2.222
Steering left	3.888	2.222	**92.222**	0.555	1.111	5.000	2.777	**88.333**	2.222	1.666
Steering right	2.777	1.666	0.555	**94.444**	0.555	3.888	2.222	2.222	**90.000**	1.666
No action	7.222	0.555	1.111	1.111	**90.000**	8.888	0.555	2.777	2.222	**85.555**

TABLE 24.3

Average Classifier Accuracy along with True-Positive, True-Negative, False-Positive, and False-Negative Rates Using PCA in the Motor Planning Phase

	DE-Recurrent NN Classifier with CSP Feature					DE-Recurrent NN Classifier with PSD Feature				
Motor Intensions	True Positive (%)	True Negative (%)	False Positive (%)	False Negative (%)	Average Classifier Accuracy (%)	True Positive (%)	True Negative (%)	False Positive (%)	False Negative (%)	Average Classifier Accuracy (%)
Acceleration	95.555	96.388	3.611	4.444	93.11	92.777	95.27	4.7222	7.222	89.44
Braking	93.333	98.75	1.25	6.666		90.555	98.333	1.666	9.444	
Steering left	92.222	98.611	1.388	7.777		88.333	97.50	2.500	11.666	
Steering right	94.444	98.75	1.25	5.555		90.000	97.50	2.50	10.000	
No action	90.000	98.888	1.111	10.000		85.555	98.194	3.611	14.444	

phases using PCA as the data point selector. It is apparent from Table 24.3 that average classification accuracies for the five classes mentioned falls off by a margin of 4% in selection of PSD instead of CSP as an FS technique.

24.7.4 Relative Performance of the Proposed Classifier

To study the relative performance, we consider standard PCA-based FE techniques and any one of the following classifiers, (1) kNN, (2) naïve Bayes, and (3) SVM-RBF kernel, as comparative framework. Throughout the study, we considered a given set of 900 training instances and 180 test stimuli for each class. Table 24.4 provides a list of selected feature sets and mean classification accuracies along with standard deviations (mentioned in parenthesis) for the indicated feature–classifier pair. An examination of Table 24.4 reveals that CSP offers the highest overall classification accuracy of 95.88% in the visual alertness phase.

TABLE 24.4

Mean Classifier Accuracy and Standard Deviation (within Parenthesis) of Three Cognitive Intensions with a False-Positive Rate (α) and False-Negative Rate (β)

Cognitive Intentions		Features with Dimension in Parenthesis	Classifiers				Statistical Significance
			kNN	Naive Bayes	SVM-RBF	Recurrent-DE NN	
Visual attention		CSP (21)	63.33 (0.049335)	75.55 (0.034264)	85.77 (0.025903)	**95.88 (0.013798)**	+
		PSD (48)	60.55 (0.051324)	71.44 (0.035811)	82.66 (0.026729)	92.55 (0.014978)	
Motor intension level I		CSP (21)	65.11 (0.046744)	78.00 (0.032746)	89.55 (0.017868)	**93.11 (0.011221)**	+
		PSD (48)	61.44 (0.047524)	73.22 (0.03789)	86.33 (0.019579)	89.44 (0.013281)	
Motor intension level II	Acceleration (up/down)	CSP (21)	70.77 (0.053972)	77.44 (0.041626)	84.77 (0.032640)	**95.55 (0.010700)**	+
		PSD (48)	66.44 (0.063812)	74.22 (0.02846)	80.55 (0.015620)	92.77 (0.012549)	
	Brake (pressed/not pressed)	CSP (21)	73.22 (0.055288)	79.00 (0.047341)	86.55 (0.035989)	**93.33 (0.011461)**	+
		PSD (48)	70.22 (0.061378)	75.33 (0.056621)	83.11 (0.042459)	90.55 (0.136731)	
	Steering (left/right)	CSP (21)	70.00 (0.058775)	76.22 (0.048406)	84.33 (0.039804)	**91.00 (0.019265)**	+
		PSD (48)	68.77 (0.067245)	74.55 (0.051696)	80.22 (0.041524)	89.22 (0.018738)	
Motor execution		CSP (21)	74.22 (0.051438)	79.77 (0.03964)	87.33 (0.032533)	**92.00 (0.020419)**	+
		PSD (48)	69.33 (0.064388)	75.22 (0.05644)	84.77 (0.041063)	87.55 (0.018535)	
Mean classifier accuracy (%)		CSP	69.44	77.66	86.38	**92.97**	+
		PSD	66.125	73.99	82.94	**90.34**	+

24.7.5 Statistical Test

A standard paired *t*-test is performed to determine the statistical significance of the said classifiers considering the DE-recurrent neural structure as the reference classifier. The equation for the paired *t*-test is given in the following equation:

$$t = \frac{\vec{x}_1 - \vec{x}_2}{\sqrt{\left(\left(\sigma_1^2 + \sigma_2^2\right)/2\right)}} \tag{24.18}$$

where
$\vec{x}_1$ is the mean of the first dataset
$\vec{x}_2$ is the mean of the second dataset
σ_1^2 is the standard deviation of the first dataset
σ_2^2 is the standard deviation of the second dataset

In the 7th column of Table 24.4, the statistical significance level (here "+") of the difference of the means of the best two algorithms using paired *t*-tests is reported. It is here noteworthy that the "+" sign indicates that the *t* value of 49 degrees of freedom is significant at a 0.05 level of significance by two-tailed *t*-test.

A well-known statistical test, namely McNemar's test (Dietterich, 1998), has been performed here to compare the relative performance of our proposed DE-recurrent NN algorithm with three standard techniques (PCA-kNN, PCA-Naïve Bayes, and PCA-SVM-RBF) (Table 24.5).

McNemar's test has been applied to determine the performance of two classification algorithms for the correct classification of the feature vectors. Here, we define a null hypothesis suggesting that the two algorithms A and B should have same error rate, that is, $n_{01} = n_{10}$, where n_{01} denotes the number of examples misclassified by A but not by B and n_{10} denotes the number of examples misclassified by B but not by A. Let f_A and f_B be classifiers' output obtained by algorithms A and B, respectively, when both the algorithms run on a common training dataset. We now define a statistic as χ^2 with 1 degree of freedom, called Z scores, which is given by the following equation:

$$Z = \frac{(|\, n_{01} - n_{10}\,| - 1)^2}{n_{01} + n_{10}} \tag{24.19}$$

At the end of the test, the Z scores will indicate whether the null hypothesis is accepted and the alternative hypothesis is rejected or vice versa. We evaluate Z, which represents the comparator statistic of misclassification between the DE-hybridized recurrent

TABLE 24.5

Statistical Comparison of Classifiers Using McNemar's Test

Reference Algorithm: DE-Recurrent Neural Net				
Classifier Algorithm Used for Comparison Using Desired Featured $d = 50$	**Parameters Used for McNemar's Test**		**Z**	**Comments on Acceptance/ Rejection of Hypothesis**
	n_{01}	n_{10}		
PCA-kNN	56	97	10.457	Rejected
PCA-naïve-Bayes	41	72	7.964	Rejected
PCA-SVM-RBF	13	24	2.702	Accepted

network–based classification algorithm (algorithm A) and any one of the competitor algorithms (algorithm B) for the Indian dataset for a desired number of features equal to 21. The hypothesis is discarded if $Z > x_1^2, 0.95 = 3.841459$, which signifies that the probability of the null hypothesis is correct only to a level of 5% of error for two-tailed chi-square test, and so we reject it. It is evident from Table 24.5 that the proposed classifier outperforms all its competitors excluding PCA-SVM-RBF. This confirms the fact that PCA-SVM-RBF performs nearly similar to that of the proposed classifier.

24.8 Case Studies

This chapter includes a number of case studies that are expected to have a practical merit for alarming drivers for their cognitive failures at appropriate levels. We have examined the cognitive failures of drivers under normal and stressed conditions with the following three cases: (1) unknown road condition/map, (2) unknowingly changed environmental condition, and (3) randomly increased traffic complexity, while they drive the driving simulator in the virtual environment. These three cases are described in the following.

In case study 1, drivers are instructed to perform the driving session on a particular road map for several times to get accustomed with the road condition. Later, in the testing phase, they are provided an unknown road map with more turns and zigzags to drive. Figures 24.26 and 24.27 show the training and testing road maps. It has been observed that in spite of being not informed, drivers manage to plan and execute motor actions during acceleration and braking accordingly. However, alarms are generated during the sudden appearance of consecutive turnings in the road map, indicating their cognitive failures at a particular phase. The level of failure increases when the driver drives a car under stressed condition.

In case study 2, a sudden situation having a night mode with heavy rainfall has been virtually created in the driving environment without informing the drivers. Figure 24.28 shows the changed driving environment. It has been experimentally

FIGURE 24.26
Training road map during driving session.

FIGURE 24.27
Testing road map for the same driver.

FIGURE 24.28
Sudden change in driving environment during the testing phase.

observed that the drivers, without having the prior knowledge and training about the sudden changes in environment, have found difficulty in visualizing the road condition and traffic under the stressed situation in comparison to the normal situation. This leads to increased number of cognitive failures during visual alertness and cognitive planning phases.

In case study 3, the drivers experience random increases in traffic complexity throughout the driving session. The cognition of the drivers has been studied in low, moderate, and high traffic conditions. From the experimental result, it has been observed that in the starting/transient period (Figure 24.29), drivers face lots of difficulties due to more failures during cognitive planning and execution phases, which become gradually lesser in a steady-state period (Figure 24.30).

Table 24.6 provides EMG performance under normal and stressed conditions for the three case studies.

FIGURE 24.29
Driver experiences high traffic during the starting/transient phase.

FIGURE 24.30
Driver experiences high traffic during the steady-state phase.

TABLE 24.6
EMG Performance under Normal and Stressed Conditions for Three Case Studies

	EMG Performance Criteria					
	Normal Condition			Stressed Condition		
		Classifier Output			Classifier Output	
Case Studies	**Delay (in Time; s)**	**Force Exerted (%)**	**No Force Exerted (%)**	**Delay (in Time; s)**	**Force Exerted (%)**	**No Force Exerted (%)**
Case study 1	0.002	84.77	80.55	0.14	77.22	76.67
Case study 2	0.001	83.77	82.55	0.21	75.77	73.22
Case study 3	0.002	83.11	81.66	0.28	74.11	72.00

24.9 Conclusions

This chapter introduced cognitive failure detection among car drivers as an interesting research problem in relation with CPS. Physiological signals such as EEG and EMG signals have been acquired during the experiments in order to determine the cognitive failure in three phases including (1) visual attention, (2) cognitive task planning, and (3) motor execution. Among the various components of CPS, classifier designing is the main objective, which has been realized by implementing a recurrent neural network classifier induced by the DE algorithm. It has been experimentally found that the proposed classifier along with PCA as the DPR technique provides better classification accuracy (≈93%) with CPS features in comparison to that of with PSD features (≈89%). The proposed recurrent-DE-based classifier performance is compared with other three standard classifiers and found to outperform all of these classifiers with mean classification accuracy of 92.97% for various cognitive failures. Moreover, performance of the proposed recurrent neural net classifier has been statistically found better, except from SVM-RBF, by using McNemar's test. Lastly, few case studies have been provided in this chapter to validate the proposed scheme of classification between various kinds of cognitive failure of car drivers under specific test conditions.

Glossary

Asymptotic stability: The necessary conditions for the dynamics $d\vec{U}/dt = f(\vec{U}(t))$ to be asymptotically stable (Gopal, 2002) at the equilibrium point $\vec{U}_{eq}$ are as follows:

1. There is a region $S(\delta)$, where $\delta < \in$ for any neighborhood $S(\in)$ surrounding $\vec{U}_{eq}$, such that the trajectories of the dynamics would start within $S(\delta)$ but remains within $S(\in)$ as time t approaches infinity.
2. The trajectory of the dynamics would start within $S(\in)$ and converge to the origin as time t approaches infinity.

Cognition: It refers to the psychological processes involved in the cognitive activities including attention, perception, reasoning, learning, sensorimotor coordination, motor planning, and execution among many others (Matlin, 2004).

Cognitive failure: Cognitive failure of car drivers refers to the psychological hindrances, prohibiting him or her from processing traffic condition and motor planning required in driving.

EEG: It refers to the electrical response of the brain to external stimuli or memory-based incidental thought/activity. EEG is usually acquired by specialized electrodes placed on the scalp of human subjects/animals. The acquired signals are preprocessed, filtered, and analyzed to decode/classify cognitive tasks undertaken by the subjects (Sanei and Chambers, 2007).

EMG: It refers to the neuromuscular response to stimulation. It is recorded with small needle electrodes inserted into the muscle. The acquired signal is preprocessed, filtered, and classified to recognize intended muscle activity.

Lyapunov energy surface: A scalar function $V(\vec{U})$ is called a Lyapunov surface with respect to origin (Gopal, 2002), if the following three conditions are satisfied:

1. $V(0) = 0$ (i.e., zero magnitude of signal at the origin).
2. $V(\vec{U}) > 0$ *for* $\vec{U} \neq 0$ (positive definite everywhere except at the origin).
3. $V(\vec{U})$ has continuous first partial derivatives with respect to all components of $\vec{U}$. It means that $\partial V/\partial u_i$ is a continuous function of u_i, where u_i is the ith component of $\vec{U}$ (smooth surface).

Recurrent topology: It refers to a class of neural network having connections between the neurons as a directed cyclic graph to exhibit dynamic temporal behavior (Konar, 2005).

Stability criteria of neural dynamics by the Lyapunov stability theorem: Let $V(\vec{U})$ be the Lyapunov surface for the given dynamics. Then the dynamics is asymptotically stable in the large, if $dV(\vec{U})/dt$ is negative definite.

24.A Appendix

This section includes a few standard algorithms used previously in the EEG, FE, DPR, and classification.

24.A.1 Fundamental Steps of Common Spatial Pattern

Input: A dataset $X^Q = \{\vec{X}_1^Q, \vec{X}_2^Q, \ldots, \vec{X}_N^Q\}$, where $Q = 1$ to M, contains N, D-dimensional channel data and $\vec{X}_i = [x_{i,1}\ x_{i,2} \cdots x_{i,D}]$ for $i = 1$ to N, where D is the number of samples in the time interval of interest and $x_{i,j}$ for $j = 1$ to D denotes the jth feature of the ith channel signal.

Output: Features $\vec{F}_d$ of two classes a and b by searching the maximums of the absolute value of their respective spatial patterns from reconstructed EEG signal $\vec{X}_i$.

1. *Normalization*: The normalized covariance matrix of a single trial EEG signal is given by

$$\mathbf{R} = \frac{\vec{X}_i \vec{X}_i'}{trace(\vec{X}_i, \vec{X}_i')} \tag{24.A.1}$$

2. *Composite covariance matrix*: The average of covariance matrices from trials within two classes, that is, from class a and class b, is summed to produce a composite covariance matrix, given by

$$\mathbf{R}_c = \mathbf{R}_a + \mathbf{R}_b = \mathbf{U}_{eg} \sum \mathbf{U}_{eg}^T \tag{24.A.2}$$

Here, $\mathbf{U}_{eg}$ is the matrix comprising few chosen eigenvectors $(\vec{U}_{eg1}, \ldots, \vec{U}_{egN})$ such that when classes a and b are both projected onto the first eigenvector $\vec{U}_{eg1}$, then class a yields the maximal variance and class b the minimal variance, and when the classes are projected onto the last eigenvector $\vec{U}_{egN}$, then class a yields the minimal variance and class b the maximal variance. Σ is the diagonal matrix.

3. *Final projection matrix*: The final projection matrix is given by

$$\mathbf{P} = \sum^{-1/2} \mathbf{U}_{eg}^{T} \tag{24.A.3}$$

with average transformed matrices (from $\mathbf{R}_a$ and $\mathbf{R}_b$, respectively):

$$\mathbf{S}_a = \mathbf{P}\mathbf{R}_a\mathbf{P}^T \text{ and } \mathbf{S}_b = \mathbf{P}\mathbf{R}_b\,\mathbf{P}^T \tag{24.A.4}$$

These two matrices $\mathbf{S}_a$ and $\mathbf{S}_b$ share common eigenvectors so that the sum of their corresponding eigenvalues is always one.

4. *Projection matrix*: The projection matrix $\mathbf{V}$ is given as

$$\mathbf{V} = \vec{U}^T\mathbf{P} \tag{24.A.5}$$

where $\vec{U}$ is a common orthogonal eigenvector.

5. *Transformation of EEG trial*: The EEG trial is transformed as

$$Z = \mathbf{V}\vec{X}_i \tag{24.A.6}$$

From Equation 24.A.6, we reconstruct the EEG signal $\vec{X}_i$ as

$$\vec{X}_i = \mathbf{V}^{-1}Z \tag{24.A.7}$$

Here, the columns of $\mathbf{V}^{-1}$ denote the EEG source distribution vector and provide the spatial patterns. Now, the optimal channels, that is, features $\vec{F}_d$ of two classes a and b, can be determined by searching the maximums of the absolute value of their respective spatial patterns.

24.A.2 Pseudo Code for the Encoding Phase of Recurrent Neural Net Classifier

Input: K class representatives $\Omega = [\vec{\theta}_1, \vec{\theta}_2, \ldots \vec{\theta}_k]$ with each $\vec{\theta}_k$ ($k \in [1,K]$) of dimension d, obtained by data reduction using PCA for each of the K individual classes, each of dimension $1 \times d$.

Output: Optimal connection vector $\vec{W}$ of dimension $1 \times d$.

Begin

1. Set the generation number $t = 0$ and randomly initialize a population of NP individuals $\mathbf{P}_t = \{\vec{W}_1(t), \vec{W}_2(t), \ldots, \vec{W}_{NP}(t)\}$ with $\vec{W}_m(t) = [w_{m,j}(t)]$ for $j \in [1, d]$, and $m = [1, NP]$.
2. Evaluate the trial vector $\vec{W}_m(t)$ by measuring its cost function

$$f(\vec{W}_m(t)) = \sum_{k=1}^{K}\sum_{j=1}^{d}\left[\theta_{k,j}^2 - 10w_{m,j}\cos(2\pi\theta_{k,j}) + 10\right]$$

by Equation 24.13

3. $\vec{W}_{best}(t) \leftarrow \arg(\min_{m=[1,NP]}(f(\vec{W}_m(t))))$
4. **While** terminating condition is not reached **do begin**
   a. *Mutation*: Generate a donor vector $V_m(t) = [v_{m,j}(t)]$ corresponding to the mth target vector $\vec{W}_m(t)$ via the following mutation scheme of DE.

   $$V_m(t) = W_{rand1}(t) + F(W_{rand2}(t) - W_{rand3}(t))$$

   where *rand1*, *rand2*, and *rand3* are mutually exclusive integers randomly chosen from the range [1, *NP*], and all are different from the base index m. F is the scale factor.

   b. *Recombination*: Generate trial vector $\vec{U}_m(t) = [u_{m,j}(t)]$ for the mth target vector $\vec{W}_m(t)$ through binomial recombination scheme of DE as mentioned below.

   $$U_{m,j}(t) = \begin{cases} V_{m,j}(t) \; if \; rand_j \leq CR \\ W_{m,j}(t) \; otherwise \end{cases}$$

   for $j = [1, d]$ and $CR \in [0, 1]$ is the cross-over rate. Here $rand_j(0,1) \in [0, 1]$ is a uniformly distributed random number lying in [0, 1] and is initiated independently for each jth component of the mth vector.

   c. *Selection*: Evaluate the trial vector $\vec{U}_m(t)$ by measuring its cost function $f(\vec{U}_m(t))$.

   $$\textbf{If} \; f(\vec{U}_m(t)) < f(\vec{W}_m(t))$$

   $$\vec{W}_m(t+1) = \vec{U}_m(t)$$

   Evaluate $f(\vec{W}_m(t+1))$ and save it for future.

   $$\textbf{If} \; f(\vec{U}_m(t)) < f(\vec{W}_{best}(t))$$

   $$\vec{W}_{best}(t) = \vec{U}_m(t)$$

   Evaluate $f(\vec{W}_{best}(t))$ and save it for future.

   End-if;

   Else $\vec{W}_m(t+1) = \vec{W}_m(t)$;

   End-if;

   Increase the counter value $t = t + 1$.

   End-while;

   Print $\vec{W}_{best}(t)$;

   End

24.A.3 Pseudo Code for Recall Phase of Recurrent Neural Net Classifier

Input: Optimal connection vector $\vec{W}$ of dimension $1 \times d$.

Output: Class ℓ of unknown smell stimulus.

Begin

1. Initialize $\vec{\theta}'(0) = [\theta_j'(0)], \forall j$.
2. Solve the dynamics (Equation 24.17) with $\theta_{k,j} = \theta_j'$ by Newton–Raphson method presented below.

$$\theta_j'(t+1) = \theta_j'(t) - \frac{f(\theta_j'(t))}{f'(\theta_j'(t))}, \forall j.$$

where, $f(\theta_j'(t)) = \theta_j'(t) + 10\pi w_j \sin[2\pi\theta_j'(t)]$, until $|\theta_j'(t+1) - \theta_j'(t)| < \varepsilon$, where ε is a pre-assigned positive number, however small possible.

3. For known optima $\vec{\theta}_1, \vec{\theta}_2, \ldots \vec{\theta}_k$, find $\vec{\theta}_k$ having the smallest distance with $\vec{\theta}'(t) = [\theta_j'(t)]$ for $k = 1$ to K.

$$\text{Let, } \left\|\vec{\theta}_\ell(t) - \vec{\theta}'(t)\right\| \le \left\|\vec{\theta}_k(t) - \vec{\theta}'(t)\right\|, \forall k;$$

then $\vec{\theta}_\ell(t) \leftarrow \vec{\theta}'(t)$, i.e., $\vec{\theta}'(t)$ falls in ℓth class of stimulus.

End.

Acknowledgment

This work is funded by the University Grant Commission, India, for the project University with Potential for Excellence Program in Cognitive Science (Phase II), Jadavpur University, India.

References

Allahyari, T., Saraji, G. N., Adl, J., Hosseini, M., Iravani, M., Younesian, M., and Kass, S. J. (2008). Cognitive failures, driving errors and driving accidents. *International Journal of Occupational Safety and Ergonomics*, 14(2), 149–158.

Bhattacharya, S., Konar, A., Das, S., and Han, S. Y. (2009). A Lyapunov-based extension to particle swarm dynamics for continuous function optimization. *Sensors*, 9(12), 9977–9997.

Brinkman, L., Stolk, A., Dijkerman, H. C., de Lange, F. P., and Toni, I. (2014). Distinct roles for alpha- and beta-band oscillations during mental simulation of goal-directed actions. *The Journal of Neuroscience*, 34(44), 14783–14792.

Chang, Q., Chen, Q., and Wang, X. (2005). Scaling Gaussian RBF kernel width to improve SVM classification. In *International Conference on Neural Networks and Brain (ICNN&B)*, Beijing, China, Vol. 1, pp. 19–22.

Chioul, J. C., Ko, L. W., Lin, C. T., and Hong, C. T. (2006). Using novel MEMS EEG sensors in detecting drowsiness application. In *IEEE Conference on Biomedical Circuits and Systems*, London, U.K., pp. 33–36.

Crone, N. E., Miglioretti, D. L., Gordon, B., Sieracki, J. M., Wilson, M. T., Uematsu, S., and Lesser, R. P. (1998). Functional mapping of human sensorimotor cortex with electrocorticographic spectral analysis. I. Alpha and beta event-related desynchronization. *Brain*, 121, 2271–2299.

Deivanayagi, S., Manivannan, M., and Fernandez, P. (2007). Spectral analysis of EEG signals during hypnosis. *International Journal of Systemics, Cybernetics and Informatics*, 75–80.

Dietterich, T. G. (1998). Approximate statistical tests for comparing supervised classification learning algorithms. *Neural Computation*, 10(7), 1895–1923.

Donchin, E., Spencer, K. V., and Wijesinghe, R. (2000). The mental prosthesis, assessing the speed of a P300-based brain–computer interface. *IEEE Transactions on Rehabilitation Engineering*, 8(2), 174–179.

Gopal, M. (2002). *Control Systems: Principles and Designs*. Tata McGraw-Hill Education, Dubuque, IA.

Hakamies-Blomqvist, L. E. (1993). Fatal accidents of older drivers. *Accident Analysis & Prevention*, 25(1), 19–27.

Happy, S. L., Dasgupta, A., Patnaik, P., and Routray, A. (2013). Automated alertness and emotion detection for empathic feedback during e-learning. In *IEEE Fifth International Conference on Technology for Education (T4E)*, IEEE, Kharagpur, India, pp. 47–50.

Haque, S. A. and Aziz, S. M. (2013). False alarm detection in cyber-physical systems for healthcare applications. In *AASRI Procedia*, 5, pp. 54–61.

Hopfield, J. J. (1991). Olfactory computation and object perception. In *National Academy of Sciences of the United States of America*, 88(15), 6462–6466.

Jesorsky, O., Kirchberg, K. J, and Frischholz, R. W. (2001). *Audio and Video Based Biometric Person Authentication*. Springer, Berlin, Germany.

Konar, A. (2005). *Computational Intelligence: Principles, Techniques, and Applications*. Springer, Berlin, Germany.

Lan, P., Ji, Q., and Looney, C. G. (2006). Information fusion with Bayesian networks for monitoring human fatigue. *IEEE Transactions on Systems, Man and Cybernetics, Part A: Systems and Humans*, 36(5), 862–887.

Lang, L. and Qi, H. (2008). The study of driver fatigue monitor algorithm combined PERCLOS and AECS. In *2008 International Conference on Computer Science and Software Engineering*, Wuhan, China, Vol. 1, pp. 349–352.

Lee, W. (1998). *Independent Component Analysis*. Kluwer Academic Publishers, Boston, MA, pp. 27–66.

Li, M. and Lu, B. L. (2009). Emotion classification based on gamma-band EEG. In *Engineering in Medicine and Biology Society, EMBS. Annual International Conference of the IEEE*, IEEE, Minneapolis, MN, pp. 1223–1226.

Lin, C. T., Chung, I. F., Ko, L. W., Chen, Y. C., Liang, S. F. and Duann, J. R. (2007). EEG-based assessment of driver cognitive responses in a dynamic virtual-reality driving environment. *IEEE Transactions on Biomedical Engineering*, 54(7), 1349–1352.

Martin, R. (2001). Noise power spectral density estimation based on optimal smoothing and minimum statistics. *IEEE Transactions on Speech and Audio Processing*, 9(5), 504–512.

Matlin, M. W. (2004). *Cognition*. John Wiley & Sons.

Niedermeyer, E. and Silva, F. L. D. (2004). *Electroencephalography: Basic Principles, Clinical Applications and Related Fields*. Lippincott Williams & Wilkins, Philadelphia, PA.

Oliver, N. and Pentlend, A. (2000). Graphical models for driver behavior recognition in a smart car. In *IEEE Intelligent Vehicle Symposium*, Dearborn, MI, pp. 7–12.

Pfurtscheller, G. and Aranibar, A. (1977). Event-related cortical desynchronization detected by power measurements of scalp EEG. *Electroencephalography and Clinical Neurophysiology*, 42, 817–826.

Pfurtscheller, G. and Lopes da Silva, F. H. (1999). Event-related EEG/MEG synchronization and desynchronization: Basic principles. *Clinical Neurophysiology*, 110(11), 1842–1857.

Purves, D., Cabeza, R., Huettel, S. A., Labar, K. S., Platt, M. L., and Woldorff, M. G. (2013). *Principles of Cognitive Neuroscience*, 2nd Edn. Sinauer Associates Inc., Sunderland, MA.

Qing, W., BingXi, S., Bin, X., and Junjie, Z. (2010). A PERCLOS-based driver fatigue recognition application for smart vehicle space. In *2010 Third International Symposium on Information Processing (ISIP)*, IEEE, Qingdao, China, pp. 437–441.

Quian Quiroga, R. and Schürmann, M. (1999). Functions and sources of event-related EEG alpha oscillations studied with the Wavelet Transform. *Clinical Neurophysiology*, 110(4), 643–654.

Reason, J. (1990). *Human Error*. Cambridge University Press, New York.

Rumar, K. (1990). The basic driver error: Late detection. *Ergonomics*, 33(10–11), 1281–1290.

Saha, A., Konar, A., Burman, R., and Nagar, A. (2014a). EEG analysis for cognitive failure detection in driving using neuro-evolutionary synergism. In *International Joint Conference in Neural Networks (IJCNN)*, IEEE, Beijing, China, pp. 2108–2115.

Saha, A., Konar, A., Chatterjee, A., Ralescu, A., and Nagar, A. K. (2014b). EEG-analysis for olfactory perceptual-ability measurement using a recurrent neural classifier. *IEEE Transactions on Human-Machine Systems*, 44(6), 717–730.

Saha, A., Konar, A., Rakshit, P., Ralescu, A. L., and Nagar, A. K. (2013). Olfaction recognition by EEG analysis using differential evolution induced hopfield neural net. In *International Joint Conference on Neural Networks (IJCNN)*, Dallas, TX, pp. 1–8.

Salvucci, D. (2004). Inferring driver's intent: A case study in lane-change detection. In *The 48th Human Factors Ergonomics Society Annual Meeting*, Santa Monica, CA, 4pp.

Sanei, S. and Chambers, J. (2007). *EEG Signal Processing*. John Wiley & Sons, Hoboken, NJ.

Song, Y., Huang, J., Zhou, D., Zha, H., and Giles, C, L. (2007). *IKNN: Informative k-Nearest Neighbor Pattern Classification*. Springer, Berlin, Germany, pp. 248–264.

Stegmann, M. B., Ersboll, B. K., and Larsen, R. (2003). FAME—A flexible appearance modeling environment. *IEEE Transactions on Medical Imaging*, 22(10), 1319–1331.

Teplan, M. (2002). Fundamentals of EEG measurement. *Measurement Science Review*, 2(2), 1–11.

Toro, C., Deuschl, G., Thatcher, R., Sato, S., Kufta, C., and Hallett, M. (1994). Event-related desynchronization and movement-related cortical potentials on the ECoG and EEG. *Electroencephalography and Clinical Neurophysiology*, 93, 380–389.

Treat, J. R. (1980). A study of pre-crash factors involved in traffic accidents. *HSRI Research Review*, 10/11(6/2), 1–35.

Wang, T., Gao, S., and Gao, X. (2005). Common spatial pattern method for channel selection in motor imagery based brain-computer interface. In *IEEE Engineering in Medicine and Biology Society*, Shanghai, China, pp. 5392–5395.

Wang, Y. T., Wang, Y., Cheng, C. K., and Jung, T. P. (August 2012). Measuring steady-state visual evoked potentials from non-hair-bearing areas. In *Engineering in Medicine and Biology Society (EMBC), 2012 Annual International Conference of the IEEE*, San Diego, CA, pp. 1806–1809.

Wei, D. and Yang, L. X. (2010). Weighted naive Bayesian classifier model based on information gain. In *International Conference on Intelligent System Design and Engineering Application (ISDEA)*, Changsha, China, Vol. 2, pp. 819–822.

Williams, R. J. and Zipser, D. (1989). A learning algorithm for continually running fully recurrent neural networks. *Neural Computation*, 1(2), 270–280.

Xingming, Z. and Huangyuan, Z. (2006). An illumination independent eye detection algorithm. In *International Conference on Pattern Recognition (ICPR)*, Hong Kong, China, pp. 392–395.

Zhang, L., Zhang, J., and Yao, L. (2012). Correlation analysis between momentary phases of ongoing EEG oscillations and erp amplitudes to identify the optimal brain state for stimulus presentation. In *2012 ICME International Conference on Complex Medical Engineering (CME)*, IEEE, Kobe, Japan, pp. 101–106.

Zhou, H., Itoh, M, and Inagaki, T. (2009). Effects of cognitive distraction on checking traffic conditions for changing lanes. In *53rd HFES Annual Conference*, San Antonio, TX, pp. 824–828.

Zhou, H., Itoh, M, and Inagaki, T. (2010). Towards inference of driver's lane-change intent under cognitive distraction. In *SICE Annual Conference*, Taipei, Taiwan, pp. 916–920.

Zongker, D. and Jain, A. (1996). Algorithms for feature selection: An evaluation. In *Proceedings of the 13th International Conference on Pattern Recognition*, IEEE, Vienna, Austria, Vol. 2, pp. 18–22.

25

Ensemble Classifier Approach to Gesture Recognition in Health Care Using a Kinect Sensor

Sriparna Saha, Monalisa Pal, and Amit Konar

CONTENTS

25.1 Introduction

This section introduces the concept of gesture recognition while highlighting the advantages for elderly health care and also the challenges involved in its practice. This section also brings forth the other existing literature in this area of application of gesture recognition.

25.1.1 What Is Gesture Recognition?

A gesture can be generated by any state or motion of a body or body parts (Mitra and Acharya, 2007). Gesture recognition aims at processing of the information that is ingrained in the gesture and is not expressed by speech or text (Freeman, 1995; Itauma et al., 2012). This work identifies eight disorders involving pain in different parts of the body. After working in a fixed posture over a long duration, people have been observed to use certain typical gestures revealing pain at different muscles and body parts. These gestures are the subject of interest for our work and can be regarded as the patterns to be identified. Thus, the purpose of this work is to monitor the gesture of elder persons while in sitting/standing posture, thereby sensing the early-stage symptoms of certain disorders and making the subject aware about their health while suggesting an exercise as a precaution.

25.1.2 What Is Its Utility in Elderly Health Care?

Changes due to decaying of bone and soft tissue as a consequence of aging are of high concern for elderly health care (Diraco et al., 2010; Murray et al., 1969). As "prevention is better than cure," so early-stage recognition of disorders is of high importance in today's world.

Aging cannot be stopped, but if the symptoms of the disorders are identified at an early stage, then the elderly health-related problems can be solved to a large extent. For recognition of disorders, the doctors mainly rely on a number of medical tests, for example, blood test, X-ray, and electroencephalogram (EEG). These procurements are costly and also time-consuming. The elders need to go to hospitals for routine checkup. Thus, the total procedure of recognizing disorders is nonflexible and for some elders nonaccessible due to cost and time constraints.

25.1.3 Challenges in Implementation

To resolve these issues, we propose a system where a single human motion-sensing device, that is, Kinect sensor, is placed in a fixed position of a room.

Kinect sensor (Dutta, 2012; Khoshelham, 2011; Saha et al., 2013a; Solaro, 2011; Zhang, 2012) automatically captures the 24 h activity of an elderly person. The early-stage symptoms of eight disorders are our concern in this work. As the persons project the gestures related to the disorders while sitting as well as standing, the total number of gestures obtained are 32 (some gestures are possible only while sitting or standing; some gestures are made due to pain in either side of the body). Whenever an elder person's gesture matches with any of the 32 gestures, then an alarm is generated. After taking guidance from several doctors, we have proposed this system. This system not only is flexible but also provides medical guidance at low cost. By the virtue of this system, we can treat the disorders at an early stage, and as a whole, the overall health of the elders becomes much better.

Gestures corresponding to the different physical disorders are effectively coded from the 20 joint coordinates extracted from Kinect sensor in 3D space. Here, we have extracted 171 Euclidean distances and treated them as features. These features are essential to distinguish the 32 gestures corresponding to 8 disorders.

An ensemble classifier (Dietterich, 2000; Martínez-Muñoz and Suárez, 2007; Polikar, 2006) or a multiple classifier system is made up of a number of "base" classifiers or "weak learners," each of which classifies the training dataset separately. The outcome of the ensemble classifier is a combination of the decisions of these base classifiers, and hence, its performance is usually better than the base classifiers, provided the individual classifier errors are not correlated. Two popular methods to implement the ensemble classifiers are bagging and boosting.

The present work has utilized AdaBoost or adaptive boosting algorithm (Rätsch et al., 2001) to implement a boosting ensemble classifier based on tree learning. Initially, the weights of all the samples in the dataset are equal. In each iteration, a new weak classifier is created using the whole dataset, and the weights for all samples are updated by increasing the weights of the samples misclassified and decreasing the weights of the samples correctly classified. Hence, it is of "adaptive" nature. A weighted voting mechanism determines the class of a new sample.

A "tree" classifier has been used as the weak learner in this work. Each node of the hierarchical tree classifier decides on each of the features in the dataset to predict the class of a sample. The prediction can be made by tracing a branch to reach the leaf node (class) of the tree.

After acquiring information about joint coordinates from the subject, the proposed work performs computations on those data and communicates the recognized disorder to the subject and thereby controlling the behavior of the subject to a certain extent. Just employing Kinect sensor, we can contribute toward the creation of a smart home for elderly persons in a simple manner. Hence, the proposed system qualifies to be a cyber-physical system.

25.1.4 Literature Survey

Several works related to gesture recognition are found in the literature. Silhouette extraction method from images obtained from video sensor suffers from the inaccuracy with variation of background intensities (Mak et al., 2011). In this work, Microsoft Kinect sensor is used that is insensitive to background variation for gathering the gesture-related information regarding the various muscle and joint pains. Parajuli et al. have proposed a method on senior health monitoring using Kinect sensor (Parajuli et al., 2012). The authors have detected the gestures when elders are likely to fall by measuring gait. The paper also deals with posture change during transition from sitting to standing and vice versa. As the collected data are varied depending on the viewing angle, so data are adjusted based on height and shoulder width. The recognition stage consists of a support vector machine (SVM). In our work, not only medical knowledge–based system is proposed but also much more complex gestures rather than fall detection are taken into account. The same problem also arises in Le et al. (2013), where Le et al. demonstrate scaling of angle features while using SVM as the classifier for health monitoring framework (Le et al., 2013). The author has extracted the skeletons with the help of Kinect sensor for detection of lying, sitting, standing, and bending postures. Kinect sensor has also found application in posture recognition, for example, child tantrum analysis (Yu et al., 2011). This chapter accomplishes medical knowledge–based work, but our work gives more accuracy than Yu et al. (2011). Oszust et al. have recognized Polish-signed expressions using principal component analysis and *k*-nearest neighbor (*k*-NN) classification method (Oszust and Wysocki, 2013). They have used skeletal image–based features, and the gesture representation is achieved using a vector containing pairwise distances between sample gestures. Lai et al. propose the use of Euclidean distance as a high-performing method that is not complex or computationally demanding and, hence, well suited for real-time application (Lai et al., 2012). However, their method is sensitive to temporal misalignment. Our method shows better results than simply using Euclidean distance–based similarity matching.

Saha et al. (2013b) have proposed an elderly health-care gesture recognition technique where muscle and joint pain–associated gestures are taken into account using Kinect sensor. The gestures are classified using a neural network optimized by Levenberg–Marquardt learning rule. Also, in Pal et al. (2014), similar type of work has been done for gesture recognition for younger individuals from age group 20 to 35 years of age depicting health-care-related gestures. The recognition process is carried out with the help of principal component analysis and fuzzy c-means clustering (FCM). In both the two papers, the authors have studied the system's performance with a dataset created by considering gestures of a large number of persons. For elderly health care, ensemble decision tree for fall detection using Kinect sensor is implemented in Stone and Skubic (2014). The dataset includes 454 falls recorded in terms of the depth maps obtained using Kinect sensor.

Another paper proposes a nice approach to gesture recognition for Parkinson's disease using Microsoft Kinect sensor (Galna et al., 2014). Here, not only Kinect sensor but infrared (IR) camera Vicon system is used to measure the proximity of hand movement, as two additional markers are placed on the fingernail of the thumb and index finger. Ten participants suffering from mild to moderate Parkinson's disease have taken part in this experiment. Kinect sensor finds wide application in upper limb rehabilitation (Metcalf et al., 2013). This proposed system is suitable for home-based motion capture by measuring finger joint kinematics. A system for monitoring of workers in industry is elaborated by Martin et al. (2012). Labors in factories all across the world perform physically intensive tasks daily. The system is used to teach employees if their current lifting and carrying methods can be detrimental to their health. The authors hypothesize that injuries could be prevented if workers are provided with information that would allow them to recognize dangerous body positions and actions that can cause harm under the recommended weight limit. The overall system has two components, a training system and an alert system. The training system focuses on educating employees of proper lifting and carrying techniques. Burba et al. have used Microsoft Kinect to monitor subtle nonverbal behavior (Burba et al., 2012). Two signals are tracked from the depth, RGB, and skeleton information. The first signal, respiratory rate, is estimated by measuring the visual expansion and contraction of the user's chest cavity during inhalation and exhalation. Additionally, the authors detect a specific type of fidgeting behavior, known as "leg jiggling" by measuring high-frequency vertical oscillations of the user's knees. Another scheme is proposed by Diraco et al. (2010), employing the time-of-flight camera. This algorithm is very much similar to our proposed work, as in Diraco et al. (2010); also the human body is modeled using body joints. The algorithm in Diraco et al. (2010) only addresses the fall detection of elder persons, while our work does not have this sort of limitation while addressing the number of gestures. Moreover, the use of skeleton instead of RGB (Henry et al., 2010, 2012) or depth images (Khoshelham and Elberink, 2012) protects the privacy of the subject regarding identity as well as physical built of the individual. Gesture recognition in short time with high accuracy while simultaneously addressing as large as 32 gestures has not been developed of late as per our knowledge.

25.2 Kinect Sensor

The Kinect (Dutta, 2012; Khoshelham, 2011; Saha et al., 2013a; Solaro, 2011; Zhang, 2012) has been used as the sensing device in this work. The name Kinect has been derived from the word "kinematics" as it senses the skeleton of the subject in dynamic condition. Three outputs are obtained from Kinect sensor: color, depth, and audio. Kinect sensor looks like a webcam. It is a horizontal black bar with a motorized base. This device has an IR camera, an IR projector, a visible RGB camera, and a microphone array incorporated into it as shown in Figure 25.1.

In a particular frame captured by Kinect sensor, the software development kit (SDK) digitizes the skeletal locations that are stored in a matrix form, where each row contains information about 20 joints (order: hip center, spine, shoulder center, head, shoulder left,

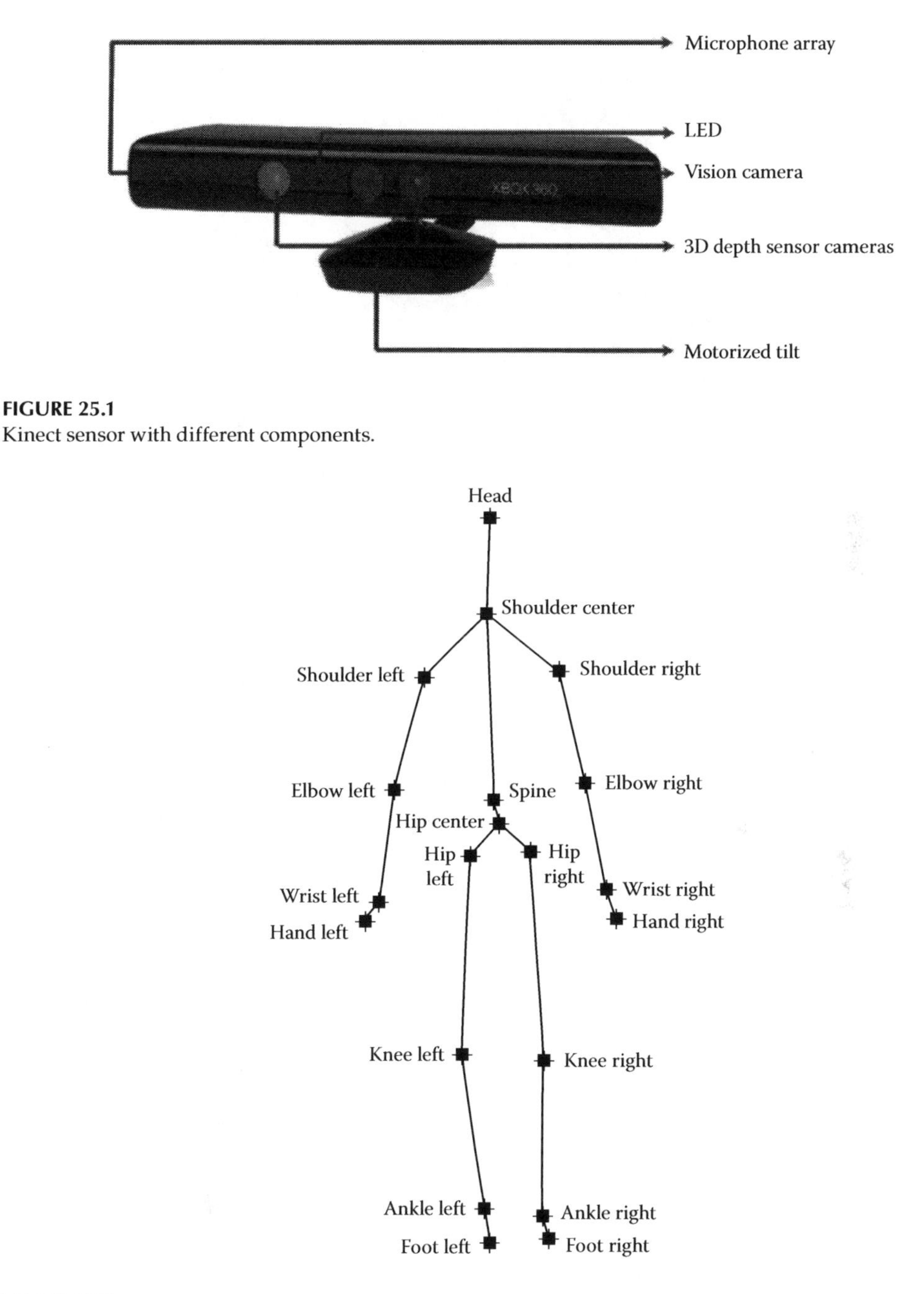

FIGURE 25.1
Kinect sensor with different components.

FIGURE 25.2
Twenty body joints.

elbow left, wrist left, hand left, shoulder right, elbow right, wrist right, hand right, hip left, knee left, ankle left, foot left, hip right, knee right, ankle right, foot right), and each column denotes the location of the joints along the three axes (order: x, y, and z). Kinect records at a sampling rate of 30 frames per second (the sampling rate can be altered), and each frame data comprises 20 Cartesian coordinates (3D) corresponding to 20 different human joints as depicted in Figure 25.2.

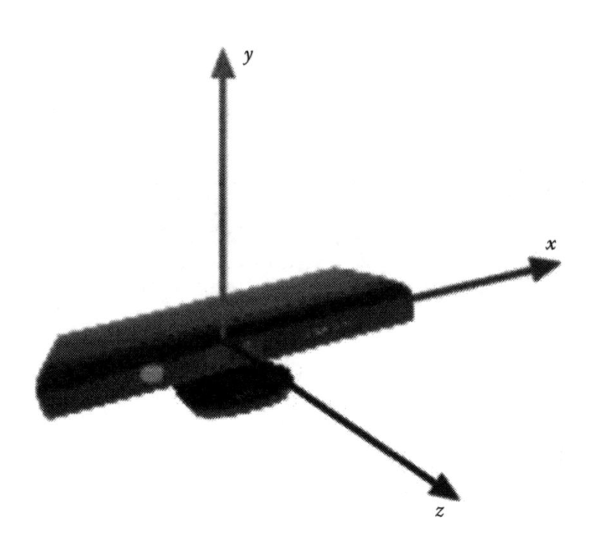

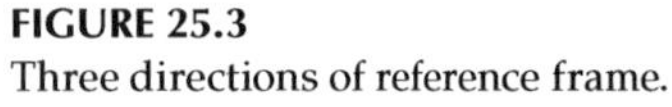
FIGURE 25.3
Three directions of reference frame.

Based on the resolution, the Kinect sensors have an output video frame rate of about 9–30 Hz. An 8-bit video graphics array (VGA) resolution of 640 × 480 pixels with a Bayer color filter is used by the Red-Green-Blue (RGB) videos. However, the hardware can also be used for resolutions up to 1280 × 1024 pixels, and the SDK can generate the output images in other formats such as UYVY (U [Cb], Y [luma 1], V [Cr], Y [luma 2]). A VGA resolution of 640 × 480 pixels having 11-bit depth is possessed by the monochrome depth sensing video stream. This provides 2048 levels of sensitivity.

The origin of the reference frame is considered at the RGB camera. The direction parallel to the length of Kinect is the x-direction, the vertical direction is the y-direction, and the depth measure is the z-direction as explained in Figure 25.3.

The Kinect sensor also has the ability to detect facial expressions and voice gestures. The pros of using the Kinect include its optimum cost and its ability to perform throughout the 24 h of the day. Other than these, the Kinect also reduces the problems that develop in the images of objects due to shadows and the presence of multiple sources of illumination. The range of distances within which the subject should perform in front of the Kinect is approximately 1.2–3.5 m or 3.9–11 ft. With an increase in distance from the Kinect, the depth resolution decreases by about 1 cm at 2 m distance from the Kinect.

25.3 Elderly Health-Care Gestures

The decaying nature of bones and tissues are the main cause for the disorders in elders. Mainly age above 40 years is treated as a threshold age for elders. In this proposed work, we have considered 8 disorders, and corresponding to these disorders, there are a total of 32 gestures. The elaborate disorder explanations of the disorders are given in Table 25.1, whereas the concerned gestures in RGB images and the skeletons are provided in Table 25.2.

TABLE 25.1

Description of Disorders

Disorder Type	Name of Disorders	Pain Position	Disorder Cause	Body Gesture Due to Pain
Joint	Lumbar spondylosis	Lower back	Degenerative changes in the lumbar spine that intensifies with age	Massaging lower back in sitting position with single (1a) or with both hands (1b).
Joint	Cauda equina syndrome	Lower back	Herniation of the disc between the two vertebrae, causing compression of the emerging nerve root, leading to sciatica	Taking support of any object (2a) while picking up any fallen item in standing position and while standing or sitting position (2b).
Tendon	Tennis elbow	Elbow	Suffers from pain over the lateral aspect of the elbow due to strain of the tendinous origin	The painful elbow of one hand is rubbed by the other hand while sitting or standing (3).
Tendon	Plantar fasciitis	Ankle	Strain of plantar fascia	Massaging ankle due to pain in sitting or standing position (4).
Joint	Osteoarthritis knee	Knee	Degenerative changes leading to pain and stiffness	Massaging the knee in sitting position with both hands (5).
Joint	Cervical spondylosis	Neck	Degenerative changes taking place in the cervical spine	Massaging the neck in sitting or standing position with single hand (6).
Joint	Osteoarthritis hand	Hand	Degenerative changes of the joints of the hand leading to pain and stiffness	Massaging the hand with the other hand while sitting or standing (7).
Muscle	Frozen shoulder	Shoulder	Injury over the supporting muscle of the shoulder leading to stiffness of the shoulder	Massaging the shoulder while sitting or standing with single hand (8).

The physical disorders considered for this work are the disorders whose symptoms are shown by elder person due to aging, which are intensified due to the lifestyle of the being (Pal et al., 2014; Saha et al., 2013b). Due to their busy schedule, they tend to neglect such disorders that have adverse effect on their health. As seen in literature, Kinect is applicable only for indoor scenarios, that is, in static background. This aids in independent living of elderly person.

The disorders considered for this work with the aim of building a general database that can be applied in different domains are described here. These disorders form the different classes in the pattern recognition portion of the work based on which the database is created. The disorders include backache. The associated reasons might be constantly sitting in a particular posture for a very long time, carrying heavy objects, and picking up any fallen item from the ground while possessing the pain. While carrying heavy load, pain may occur at shoulder joints that also can eventually lead to disorders like frozen shoulder or shoulder dislocation. The next disorder considers pain at the elbow joint. As studied from the people's habit, the origin of this pain comes from constantly supporting one's chin with the forearm while using the elbow as a stand on the table. People engrossed in listening or reading something over a period of time might trigger this posture and consequently the pain. The following kind of disorder represents headache; sometimes, headaches are associated with pain in the neck. The arousal of these pains might be because of parallax error while reading any document. This forces the subject to bend his or her head

TABLE 25.2

RGB Images and the Skeletons of the Concerned Disorders

Disorder Number	Disorder Name	RGB Images: While Sitting	RGB Images: While Standing	Skeletons: While Sitting		Skeletons: While Standing	
		Pain at the Right Side	Pain at the Right Side	Pain at the Left Side	Pain at the Right Side	Pain at the Left Side	Pain at the Right Side
1a	Lumbar spondylosis		—	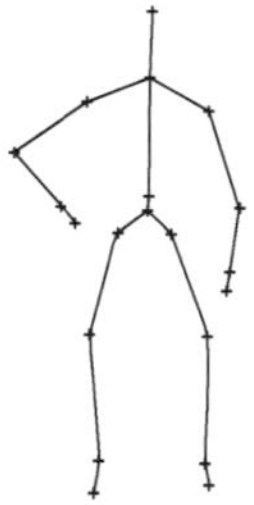	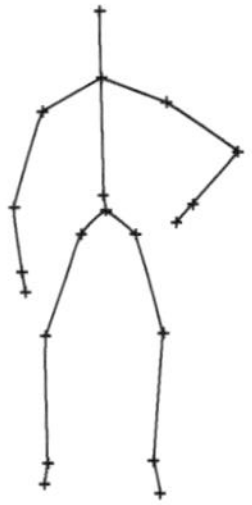	—	—
2a	Cauda equina syndrome	—		—	—	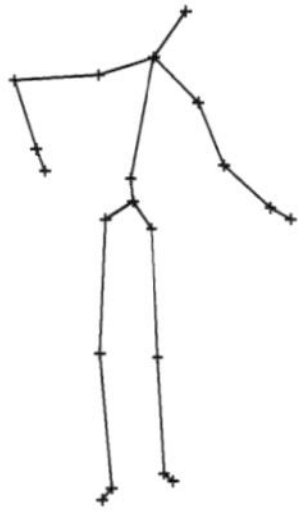	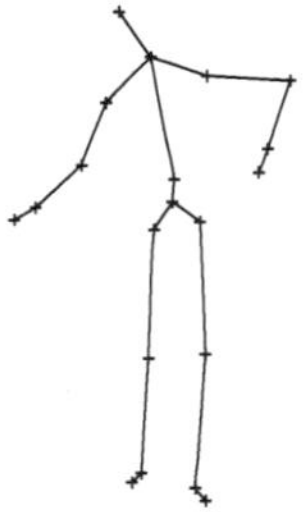
2b	Cauda equina syndrome			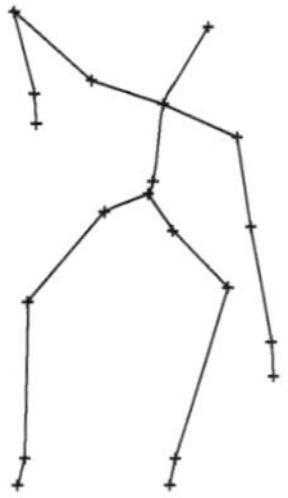	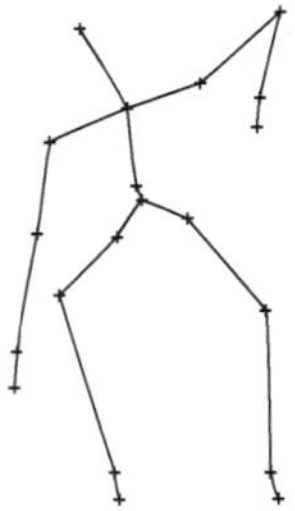	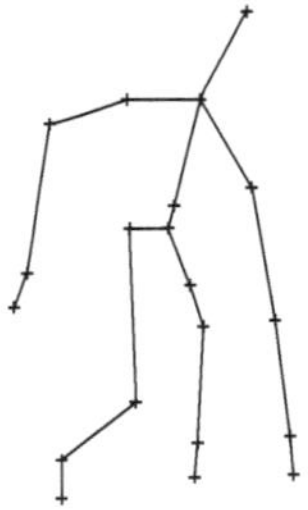	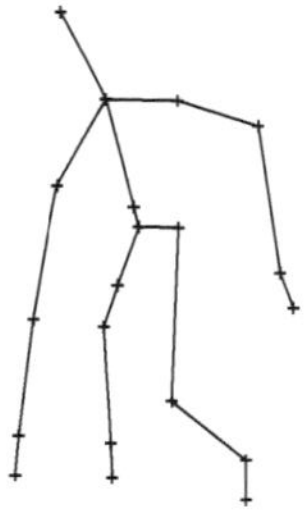

(*Continued*)

TABLE 25.2 (*Continued*)

RGB Images and the Skeletons of the Concerned Disorders

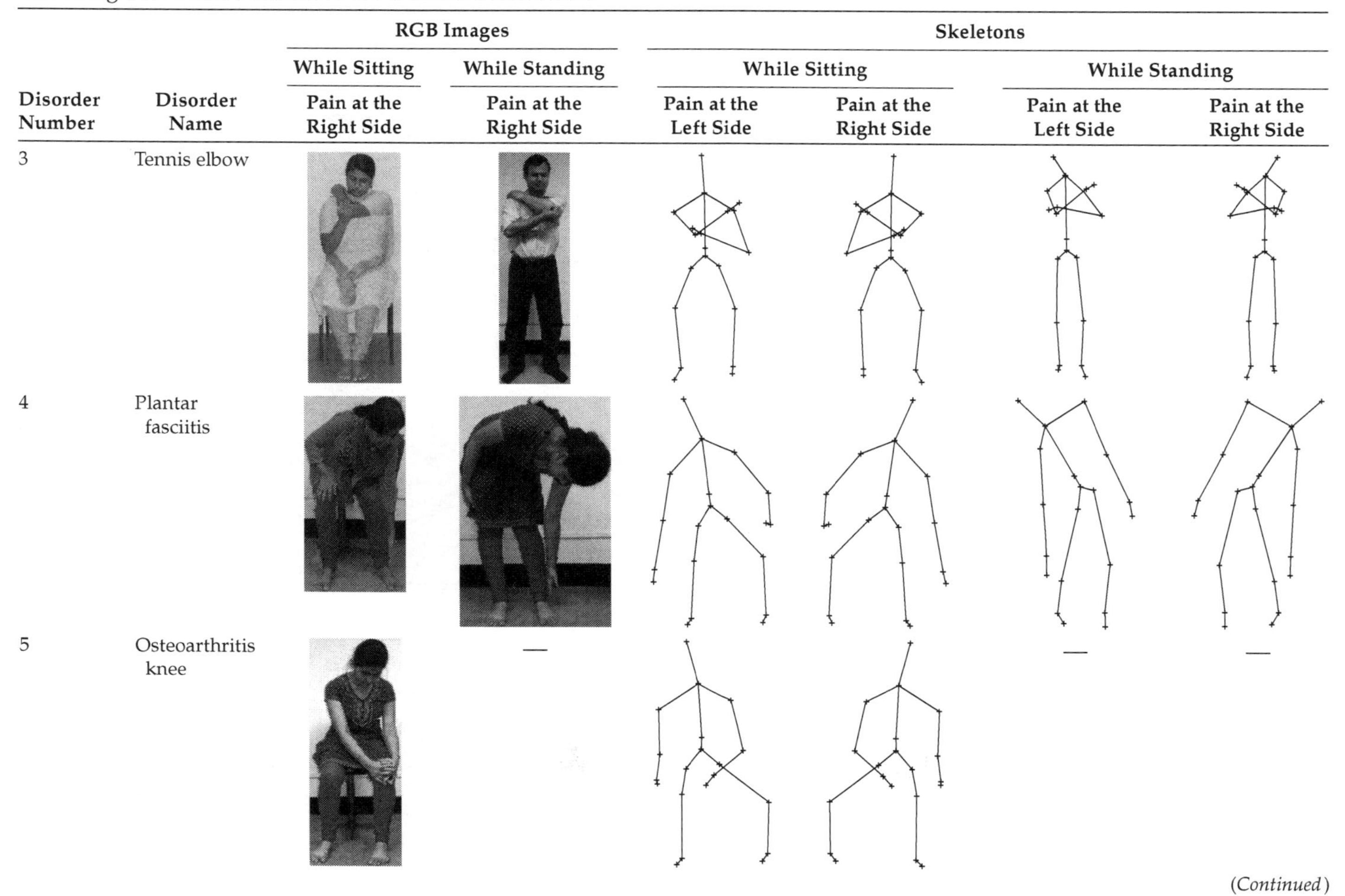

		RGB Images		Skeletons			
		While Sitting	While Standing	While Sitting		While Standing	
Disorder Number	Disorder Name	Pain at the Right Side	Pain at the Right Side	Pain at the Left Side	Pain at the Right Side	Pain at the Left Side	Pain at the Right Side
3	Tennis elbow						
4	Plantar fasciitis						
5	Osteoarthritis knee		—			—	—

(*Continued*)

TABLE 25.2 (*Continued*)

RGB Images and the Skeletons of the Concerned Disorders

		RGB Images		Skeletons			
		While Sitting	While Standing	While Sitting		While Standing	
Disorder Number	Disorder Name	Pain at the Right Side	Pain at the Right Side	Pain at the Left Side	Pain at the Right Side	Pain at the Left Side	Pain at the Right Side
6	Cervical spondylosis		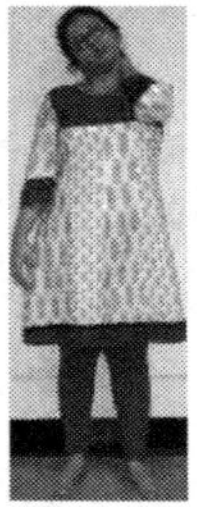	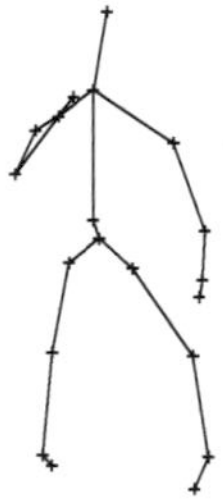	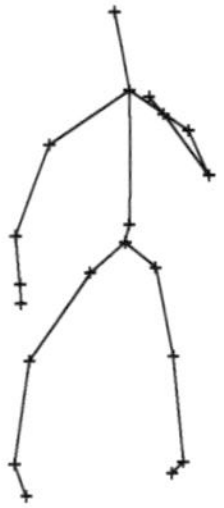	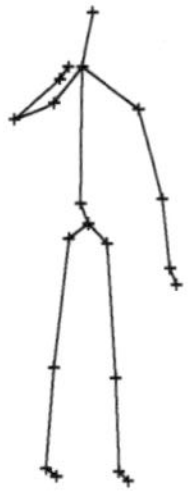	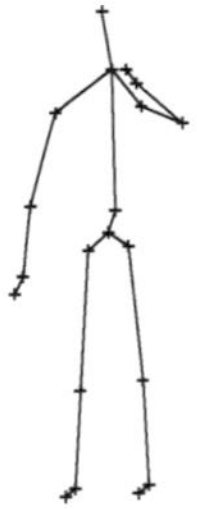
7	Osteoarthritis hand			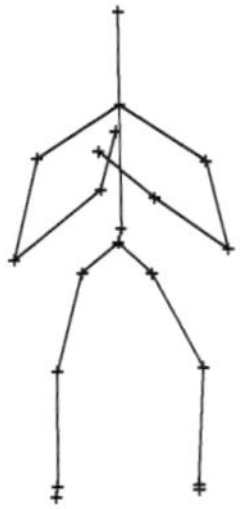	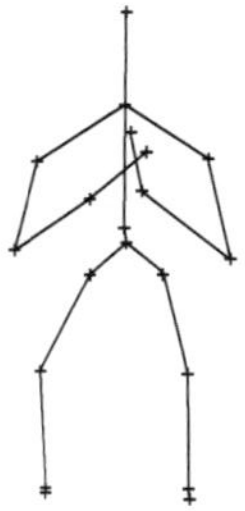	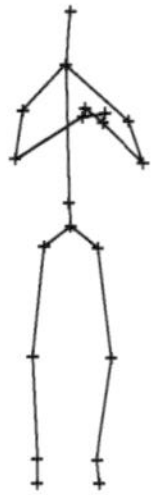	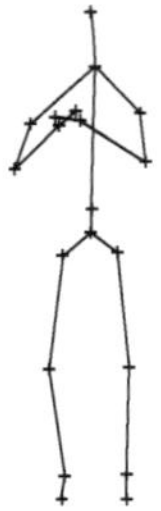

(*Continued*)

TABLE 25.2 (*Continued*)

RGB Images and the Skeletons of the Concerned Disorders

		RGB Images		Skeletons			
		While Sitting	While Standing	While Sitting		While Standing	
Disorder Number	**Disorder Name**	**Pain at the Right Side**	**Pain at the Right Side**	**Pain at the Left Side**	**Pain at the Right Side**	**Pain at the Left Side**	**Pain at the Right Side**
8	Frozen shoulder		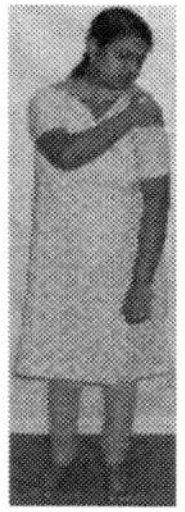			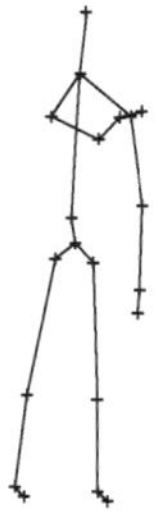	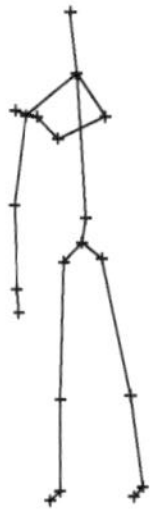
Disorder Number	Disorder Name	Pain at Both Sides		Pain at Both Sides		Pain at Both Sides	
1b	Lumbar spondylosis	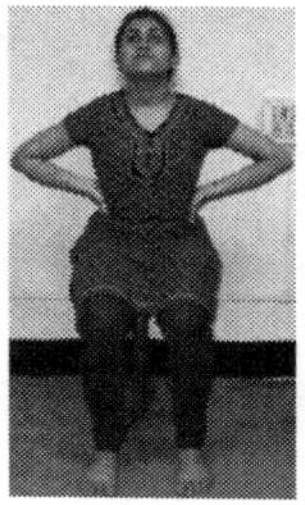		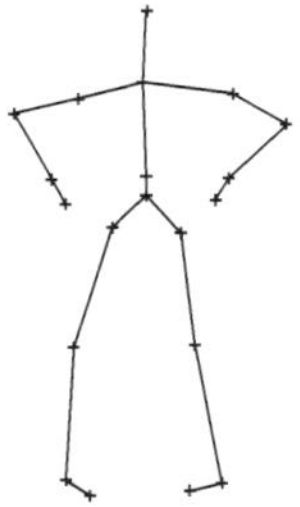		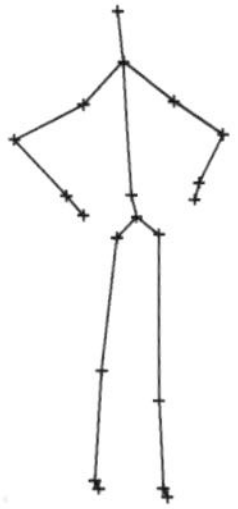	

for a long time. Even continuous reading can act as a stimulus for headache. Nowadays, most of the works are done on computers. Typing on the keyboard nonstop may give rise to pain in the palm. This is the next disorder category. Most of the buildings of the companies are multistoried. Although there are elevators, but climbing stairs and moving among different rooms frequently for people in profession like clerks and peons may generate pain in the knee, ankle, and calf muscle. These are the next few types of disorders considered. When knee pain persists, people find difficulty in transition from sitting to standing state and vice versa.

25.4 Methodology for Intelligent System Development

This section introduces the basic concepts about the ensemble classifier and the proposed algorithm.

25.4.1 Ensemble Classifier

Ensemble classifiers are constructed on the basis of the fact that the decision accuracy of a group of classifiers is far more reliable than a single classifier (Dietterich, 2000; Martínez-Muñoz and Suárez, 2007; Polikar, 2006). In our work, we have used "tree" classifier as the base classifier. The base classifier creates binary tree where each node operates on one of the features from the dataset. The predictions of the individual base classifier are combined (weighted voting) to decide the class of the test samples. Two important criteria must be satisfied in selecting the individual classifiers: they must be accurate (error rate better than random guess, also called weak learners) and diverse (different error on new dataset).

AdaBoost refers to adaptive boosting (Cao et al., 2012; Khreich et al., 2012; Meynet et al., 2007; Narasimha et al., 2009; Sun et al., 2007; Tie and Guan, 2009). If the process is iterated T times, each time AdaBoost creates a new weak classifier and the weights for all classified samples are updated as shown in Figure 25.4. The weights of the samples misclassified are increased, and the weights of the samples correctly classified are decreased. Again, the process is repeated with the new set of weighted samples. It is called adaptive because it is focuses on those samples that are misclassified in the previous iterations. AdaBoostM1 algorithm is used to implement the ensemble classifier in this work.

In the case of bagging (*b*ootstrap *agg*regat*ing*), classifiers are trained by different datasets that are obtained from bootstrapping the original dataset, that is, a subset of the dataset is created by randomly drawing (with replacement) n samples from the original dataset as provided in Figure 25.5. The diversity among the weak classifiers is explored by this resampling procedure, which is repeated T times. Finally, majority voting on the decision of the weak learners infers the class of an unknown sample. In this work, the value of T is considered to be 100, and the bootstrap size (n) is considered to be 30% of the total dataset.

The work of the tree classifier is stated as follows: Tree classifier (Lemon et al., 2003) forms a binary tree on the training set T with N samples and n classes where each node operates on a single feature yielding the smallest Gini's diversity index (Jost, 2006) as given by Equation 25.1 and splits the data into two sets $T1$ and $T2$ with $N1$ and $N2$ samples such that the condition in Equation 25.2 is satisfied. The term $p(i)$ stands for the relative

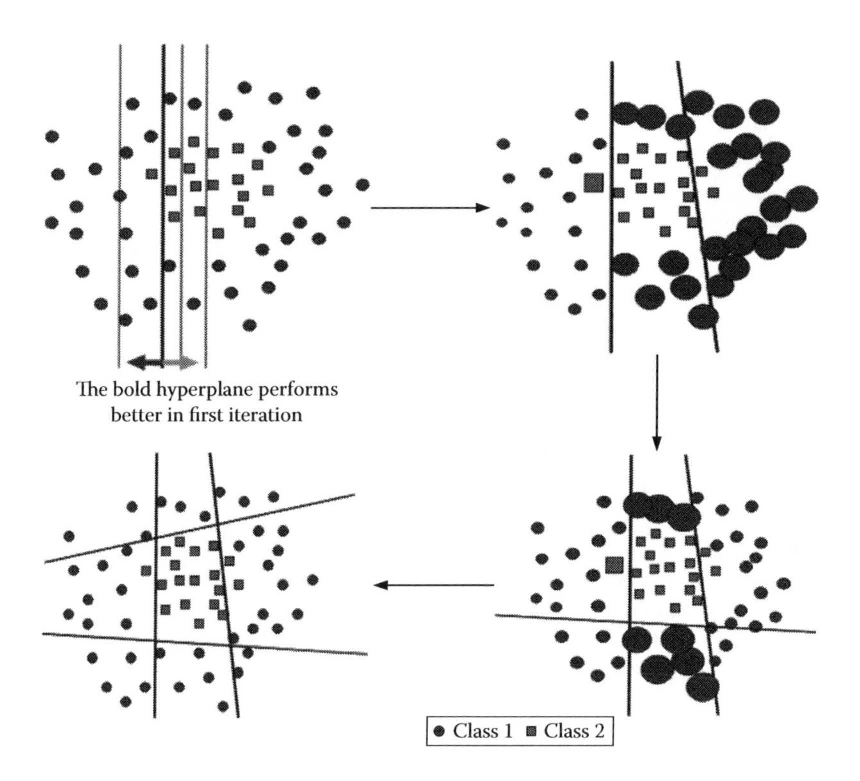

FIGURE 25.4
Pictorial view of boosting.

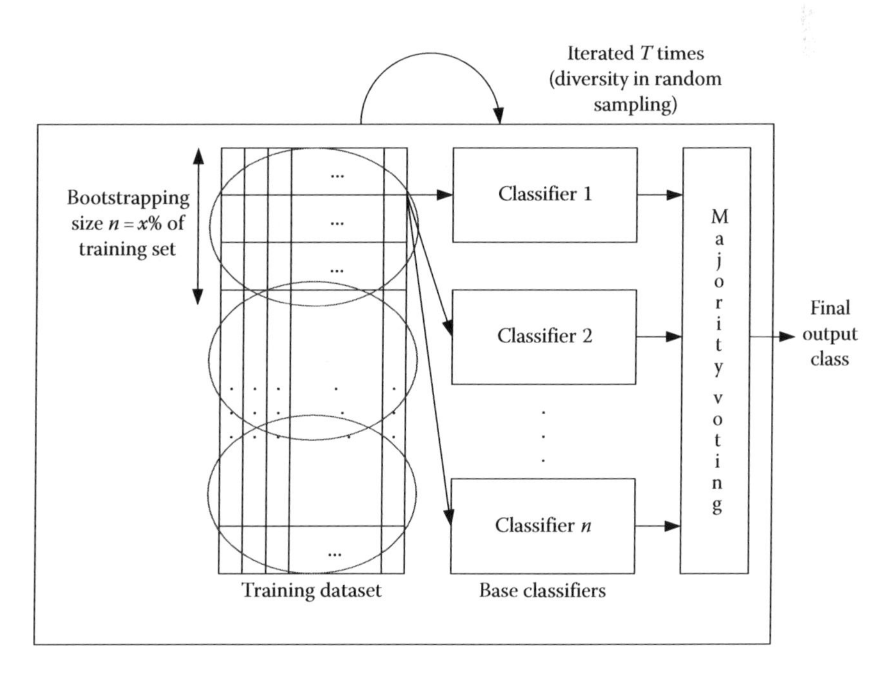

FIGURE 25.5
Pictorial view of bagging.

frequency of class i. For any unknown sample gesture, tracing a branch of the hierarchical tree for different values of the feature leads to a leaf node that decides the class of the unknown sample gesture:

$$gini(T) = 1 - \sum_{i=1}^{n} p(i)^2 \tag{25.1}$$

$$gini(T) = \frac{N1}{N} gini(T1) + \frac{N2}{N} gini(T2) \tag{25.2}$$

The performance of the classifier is judged by evaluating the classification accuracy from the confusion matrix. We have 32 classes. Every class is classified against every other class (one-vs.-one strategy), that is, class 1 is classified against class 2, again class 1 is classified against class 3, and so on. After classification, confusion matrix of order 2 × 2 is constructed. Row i of the confusion matrix represents samples actually belonging to class i, and column j represents samples classified as class j. So an element $c(i,j)$ of the confusion matrix depicts a sample classified as class j when it actually belongs to class i. Hence, the diagonal element $c(i,i)$ represents the number of correctly classified samples of the class i. The accuracy of a classification of samples from class i is calculated by the following formula:

$$Accuracy = \frac{\sum_{i} c(i,i)}{\sum_{i}\sum_{j} c(i,j)} \tag{25.3}$$

Thus, we have accuracies of every class classified against every other class. So we say the accuracy with which a class is classified is the average accuracy of the classification of that class against all other classes.

25.4.2 Algorithm for Gesture Recognition

The proposed work involves gesture recognition from elder persons. At first, a total of ($^{20}C_2$ – 19=) 171 features that are basically Euclidean distances between the 20 joints are extracted from each skeletal frame. Then the ensemble classifier is applied for the classification of disorder-revealing gestures. This total procedure is picturized in Figure 25.6. The feature extraction step along with the ensemble classifier forms the proposed cyber-physical system for health care.

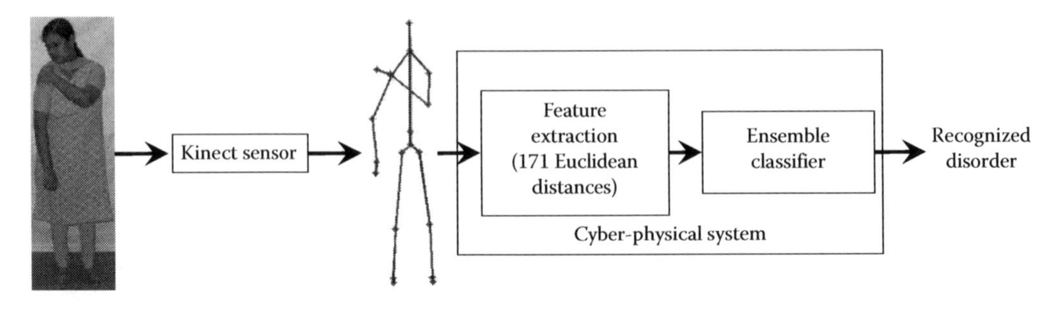

FIGURE 25.6
Pictorial view of proposed system.

25.5 Results and Discussions

In this section, we are going to explain the experimental results for the proposed work. First, the feature extraction procedure for an unknown gesture is provided, and then using statistical tests, the accuracy of the proposed method is validated. For this proposed work, we have prepared three datasets from Jadavpur University research group scholars of age group 25–30, 30–35, and 35–40 years. In each dataset, there are 20 subjects each imitating the gestures for 10 times.

As these scholars are mainly involved in research works in the domain of computer science engineering, thus they spend most of their time in the laboratory for around 7–8 h/day. As major part of their routine involves working while sitting constantly and also because of the adverse effect of age, their health is constantly decaying. We have noticed some typical gestures in those research scholars due to pain at different muscles and joints, and this gives us the idea to implement our proposed work in the artificial intelligence laboratory.

After regularly monitoring the scholars, we have acquired a total of 500 gestures from the scholars that belong to the testing dataset for our proposed work. For the sake of privacy, our system does not rely on the RGB image of the subjects, but for the sake of understanding, we are providing an unknown gesture using RGB and skeletal image in Figure 25.7.

25.5.1 Feature Extraction

For this proposed work, we are considering a total of 500 disordered gestures from the subjects as our testing dataset. From these, 500 gestures (each frame depicting a gesture using skeleton) are processed, and 171 Euclidean distances are calculated. The Euclidean

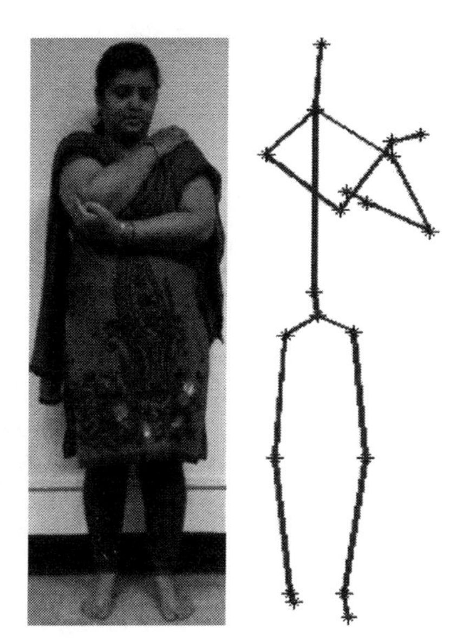

FIGURE 25.7
RGB image and the skeleton of unknown gesture.

TABLE 25.3

Sixteen Measured Euclidean Distances for Unknown Gesture

No.	Distance between	Distance Values	Maximum Distance Value	Normalized Distance Values
1	Hip center and shoulder center	0.5104	1.4802	0.3449
2	Hip center and head	0.6753		0.4562
3	Hip center and shoulder left	0.3785		0.2558
4	Hip center and elbow left	0.4269		0.2884
5	Hip center and wrist left	0.3465		0.2341
6	Hip center and hand left	0.3310		0.2236
7	Hip center and shoulder right	0.3974		0.2685
8	Hip center and elbow right	0.2551		0.1724
9	Hip center and wrist right	0.3494		0.2361
10	Hip center and hand right	0.3542		0.2393
11	Hip center and knee left	0.8149		0.5505
12	Hip center and ankle left	0.5888		0.3978
13	Hip center and foot left	0.3320		0.2243
14	Hip center and knee right	0.7244		0.4894
15	Hip center and ankle right	0.8113		0.5481
16	Hip center and foot right	0.5112		0.3454

distance *Dist* between two points with coordinates (x_1,y_1,z_1) and (x_2,y_2,z_2) is calculated using the following equation:

$$Dist = \sqrt{(x_1 - x_2)^2 + (y_1 - y_2)^2 + (z_1 - z_2)^2} \tag{25.4}$$

For the unknown skeleton, from these measured distances, 16 distances are given as example in Table 25.3. These 16 distances are the distance values between the hip center with all the other 16 joints excepting the spine, hip left, and hip right, which are neglected for their constant value for a subject that does not vary from gesture to gesture. After obtaining 171 distance values, the distances are normalized with respect to the maximum distance to avoid the effects of height, weight, and body type for different subjects. The maximum distance value obtained from all 171 features is 1.4802.

25.5.2 Classification Accuracy and Time Requirement

Figures 25.8 and 25.9 show the comparative study of two different techniques of ensemble classifiers, namely, AdaBoostM1 and Bagging, for three datasets prepared. From this comparison, it is evident that AdaBoostM1 is showing better result than Bagging.

25.5.3 Statistical Tests

The performance of the proposed method using the ensemble classifier for physical disorder recognition is examined here with respect to three databases considered in this chapter.

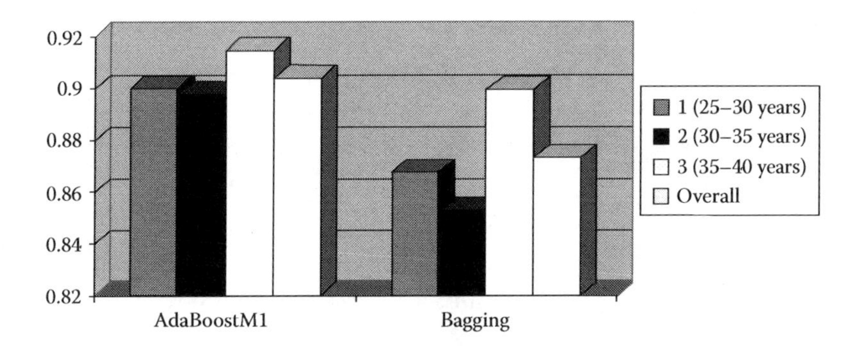

FIGURE 25.8
Comparison between two techniques of ensemble classifiers with respect to mean accuracy.

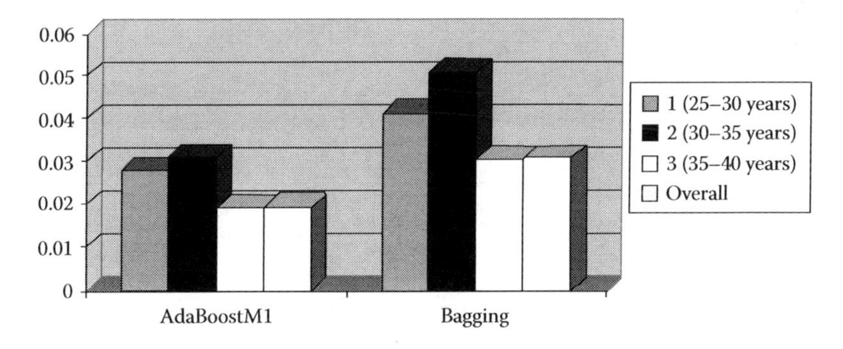

FIGURE 25.9
Comparison between two techniques of ensemble classifiers with respect to mean processing time.

25.5.3.1 Comparative Framework

This chapter compares the relative performance of the proposed algorithms with seven traditional gesture recognition algorithms/techniques. The comparative framework includes FCM (Cannon et al., 1986; Pal and Bezdek, 1995), SVM classifier (Cortes and Vapnik, 1995; Zhan and Shen, 2005), k-NN classification (Cunningham and Delany, 2007; Wu et al., 2002), Levenberg–Marquardt algorithm–induced neural network (LMA-NN) (Li et al., 2002; Yegnanarayana, 2004), (type 1) fuzzy relational approach (Chakraborty et al., 2009), and radial basis function network (RBFN) (Kim et al., 2005; Lisetti and Rumelhart, 1998). The parameters of all competing classifiers are tuned by noting the best performances after experimental trails. For FCM, the m value is empirically taken as 2. SVM has been used with a radial basis function kernel whose kernel parameter has a value of 1 and the classifier is tuned with a cost value of 100. The performance of k-NN has been reported for $k = 5$ using Euclidean distance as the similarity measure and majority voting to determine the class of the test samples. For LMA-NN, the number of neurons in the intermediate layer is taken as 10, the value of the blending factor between gradient descent and quadratic learning as 0.01, the increase and decrease factors of the blending factor as 10 and 0.1, respectively, and the training stopping condition as the attainment of minimum error gradient value of 1e−6.

25.5.3.2 Performance Metric

The physical disorder c of a subject (patient) in reality that is recognized by an algorithm may have four combinations as enlisted in Table 25.4.

Here, number of classes (C) is 32. Our comparative study uses the following performance metrics given in Table 25.5 to evaluate the performance of an algorithm.

25.5.3.3 Performance Analysis

The mean and the standard deviation (within parenthesis) of the performance metrics (Sokolova and Lapalme, 2009) for 30 independent runs are shown in Table 25.6. Here, "+" indicates that our algorithm provides better result than all the other algorithms, whereas "–" denotes that our algorithm is not better than other algorithms. The best result is marked in bold.

It is interesting to see that out of 5 performance metrics for all 3 databases, that is, out of 15 cases of performance measures, the ensemble classifier shows better results in 12 cases.

Three statistical tests are undertaken to analyze the relative performance of the proposed algorithm over the existing ones.

TABLE 25.4

Description of Classification Matrix

True positive of class c (TP_c): Number of physical disorders that are recognized as in class c and indeed belong to class c	*False negative of class c* (FN_c): Number of physical disorders that are recognized as in class $c' \in [1, C]$ provided $c' \neq c$ but indeed belong to class c
False positive of class c (FP_c): Number of physical disorders that are recognized as in class c but in fact belong to class $c' \in [1, C]$ provided $c' \neq c$	*True negative of class c* (TN_c): Number of physical disorders that are correctly recognized to not to belong to class c

TABLE 25.5

Explanation about Performance Metrics

Name	Description	Ideal Value Should Be	Equation
Recall	Effectiveness of a classifier to identify the class labels	1	$Recall = \frac{\sum_{c=1}^{C} TP_c}{\sum_{c=1}^{C} (TP_c + FN_c)}$
Precision	Degree of agreement of the data class labels with those of a classifier	1	$Precision = \frac{\sum_{c=1}^{C} TP_c}{\sum_{c=1}^{C} (TP_c + FP_c)}$
Accuracy	Overall correctness of the predictive model	1	$Accuracy = \sum_{c=1}^{C} \frac{TP_c + TN_c}{TP_c + TN_c + FP_c + FN_c}$
F1_score	Weighted average of the precision and recall	1	$F1\_score = \frac{2 \times Precision \times Recall}{Precision + Recall}$
Average error rate	Average per class classification error	0	$Average\ Error\ Rate = \sum_{c=1}^{C} \frac{FP_c + FN_c}{TP_c + TN_c + FP_c + FN_c}$

TABLE 25.6

Comparison of Different Physical Disorder Recognition Algorithms for 30 Runs

Dataset	Performance Metrics	Ensemble Classifier	FCM	SVM	*k*-NN	LMA-NN	Fuzzy Relation	RBFN	Significance
Research scholars of age 25–30 years	Recall	**0.9611 (0.0144)**	0.9625 (0.0786)	0.9036 (0.0309)	0.9856 (0.1584)	0.9104 (0.1428)	0.9663 (0.1383)	0.8527 (0.0113)	+
	Precision	**0.7825 (0.1173)**	0.7705 (0.0976)	0.7624 (0.0962)	0.7597 (0.0919)	0.7634 (0.0250)	0.7568 (0.0975)	0.7376 (0.1577)	+
	Accuracy	**0.8999 (0.0906)**	0.8395 (0.0114)	0.8023 (0.0664)	0.7907 (0.1652)	0.8066 (0.1300)	0.7783 (0.0793)	0.7028 (0.1786)	+
	F1_score	**0.9350 (0.0187)**	0.8933 (0.0117)	0.8665 (0.0781)	0.8581 (0.0577)	0.8697 (0.1807)	0.8488 (0.1557)	0.7910 (0.1557)	+
	Average error rate	0.1198 (0.1584)	**0.1104 (0.1979)**	0.1976 (0.1017)	0.2092 (0.0226)	0.1933 (0.1899)	0.2216 (0.1924)	0.2971 (0.1118)	–
Research scholars of age 30–35 years	Recall	0.9576 (0.1437)	**0.9983 (0.0483)**	0.9392 (0.1664)	0.9221 (0.1482)	0.9081 (0.1959)	0.9673 (0.1706)	0.9527 (0.1092)	–
	Precision	**0.7821 (0.0681)**	0.7751 (0.1427)	0.7674 (0.0245)	0.7650 (0.1869)	0.7630 (0.0970)	0.7570 (0.1542)	0.7547 (0.1884)	+
	Accuracy	**0.8978 (0.0848)**	0.8617 (0.0403)	0.8249 (0.0753)	0.8141 (0.1013)	0.8052 (0.0604)	0.7789 (0.0914)	0.7694 (0.1480)	+
	F1_score	**0.9335 (0.0345)**	0.9088 (0.0740)	0.8828 (0.1143)	0.8751 (0.1987)	0.8686 (0.0501)	0.8493 (0.1227)	0.8422 (0.0370)	+
	Average error rate	**0.1021 (0.0704)**	0.1382 (0.0100)	0.1750 (0.1220)	0.1858 (0.0818)	0.1947 (0.0338)	0.2210 (0.1703)	0.2305 (0.1709)	+
Research scholars of age 35–40 years	Recall	0.9818 (0.0265)	**0.9249 (0.1482)**	0.9028 (0.1677)	0.9085 (0.0116)	0.9473 (0.1155)	0.8989 (0.1414)	0.9935 (0.0617)	–
	Precision	**0.7849 (0.1986)**	0.7783 (0.1257)	0.7692 (0.1673)	0.7631 (0.1590)	0.7539 (0.0934)	0.7459 (0.1246)	0.7609 (0.0890)	+
	Accuracy	**0.9146 (0.1475)**	0.8779 (0.1088)	0.8334 (0.1328)	0.8054 (0.0218)	0.7659 (0.1293)	0.7339 (0.1336)	0.7958 (0.1916)	+
	F1_score	**0.9433 (0.1227)**	0.9200 (0.0423)	0.8889 (0.0690)	0.8688 (0.1420)	0.8396 (0.1808)	0.8153 (0.1883)	0.8618 (0.1654)	+
	Average error rate	**0.0876 (0.1186)**	0.1220 (0.0492)	0.1665 (0.1920)	0.1945 (0.1871)	0.2340 (0.1909)	0.2660 (0.0788)	0.2041 (0.0980)	+

25.5.3.4 McNemar's Statistical Test

Let f_A and f_B be two classifiers obtained by algorithms A and B, when both the algorithms have a common training set R. Let n_{01} be the number of examples misclassified by f_A but not by f_B, and n_{10} be the number of examples misclassified by f_B but not by f_A (Dietterich, 1998):

$$Z = \frac{\left(|n_{01} - n_{10}| - 1\right)^2}{n_{01} + n_{10}} \tag{25.5}$$

Let A be the proposed ensemble classifier and B is one of the other six algorithms. In Table 25.7, the null hypothesis has been rejected, if $Z_i > \chi^2_{1,\alpha=0.05} = 3.84$, where $\chi^2_{1,\alpha=0.05} = 3.84$ is the critical value of the chi-square distribution for 1 degree of freedom at an error probability of 0.05 (Bickel and Li, 1977). Here, the suffix i refers to the algorithm in row number i of Table 25.7. Only for this case, all the three datasets were merged while performing the McNemar's test.

25.5.3.5 Friedman Test

This test (García et al., 2009) is performed on the mean of *Accuracy* metric for 30 independent runs of each of the 7 algorithms as reported in Table 25.8. Let r_i^j be the ranking of the observed *Accuracy* obtained by the ith algorithm for the jth database. The best of all the k algorithms, that is, $i = [1, k]$, is assigned a rank of 1, and the worst is assigned the ranking k. Then the average ranking acquired by the ith algorithm over all $j = [1, N]$ algorithms is defined as follows:

$$R_i = \frac{1}{N}\sum_{j=1}^{N} r_i^j \tag{25.6}$$

Then χ_F^2 distribution with degree of freedom equals to $k - 1$ (García et al., 2009) is measured:

$$\chi_F^2 = \frac{12N}{k(k+1)}\left[\sum_{i=1}^{k} R_i^2 - \frac{k(k+1)^2}{4}\right] \tag{25.7}$$

In this chapter, N is the number of databases considered as 3 and k is the number of competitor algorithms considered as 7. In Table 25.8, it is shown that the null hypothesis has been rejected, as $\chi_F^2 = 16.0000$ is greater than $\chi^2_{6,\alpha=0.05} = 12.592$, where $\chi^2_{6,\alpha=0.05} = 12.592$ is the critical value of the χ_F^2 distribution for $k - 1 = 6$ degrees of freedom at an error probability of 0.05 (Zar, 1999). In Table 25.8, the second, third, and fourth columns represent the ranks of those datasets belonging to the three different age group (i.e., 25–30, 30–35, and 35–40) of the subjects for which the results are calculated.

25.5.3.6 Iman–Davenport Statistical Test

Another statistical test (Zar, 1999) using the results from Friedman's test (García et al., 2009) is obtained by the following equation:

$$F_F = \frac{(N-1)\times\chi_F^2}{N\times(k-1) - \chi_F^2} \tag{25.8}$$

TABLE 25.7

Performance Analysis Using McNemar's Test

Competitor Algorithm = B	Control Algorithm A = Ensemble Classifier											
	Jadavpur University Database of Research Scholars of Age 25–30				Jadavpur University Database of Research Scholars of Age 30–35				Jadavpur University Database of Research Scholars of Age 35–40			
	n_{01}	n_{10}	Z_i	Comment	n_{01}	n_{10}	Z_i	Comment	n_{01}	n_{10}	Z_i	Comment
FCM	39	52	1.5824	Accept	41	58	2.5859	Accept	27	66	15.5269	Reject
SVM	29	62	11.2527	Reject	33	60	7.2688	Reject	25	70	20.3789	Reject
k-NN	26	65	15.8681	Reject	30	64	11.5851	Reject	27	72	19.5556	Reject
LMA-NN	26	71	19.9588	Reject	31	68	13.0909	Reject	26	74	22.0900	Reject
Fuzzy relation	22	77	29.4545	Reject	21	70	25.3187	Reject	24	75	25.2525	Reject
RBFN	20	76	31.5104	Reject	14	76	41.3444	Reject	14	79	44.0430	Reject

TABLE 25.8

Performance Analysis Using Friedman and Iman–Davenport Tests

Algorithm	25–30 (r_i^1)	30–35 (r_i^2)	35–40 (r_i^3)	R_i	Friedman Test		Iman–Davenport Test	
					χ_F^2	Comment on Null Hypothesis	F_F	Comment on Null Hypothesis
Ensemble classifier	1	1	1	1.0000	16.0000	Reject	16.0000	Reject
FCM	2	2	2	2.0000				
SVM	4	3	3	3.3333				
k-NN	5	4	4	4.3333				
LMA-NN	3	5	6	4.6667				
Fuzzy relation	6	6	7	6.3333				
RBFN	7	7	5	6.3333				

It is distributed according to F distribution with $(k-1)$ and $(k-1) \times (N-1)$ degrees of freedom. For $k = 7$ and $N = 3$, it is shown in Table 25.8 that the null hypothesis has been rejected, as $F_F = 16.0000$ that is greater than $F_{6,12,\alpha=0.05} = 3.73$, where $F_{6,12,\alpha=0.05} = 3.73$ is the critical value of the F distribution for $k - 1 = 6$ and $(k-1) \times (N-1) = 12$ degrees of freedom at an error probability of 0.05 (Iman and Davenport, 1980; Picek et al., 2012; Zar, 1999).

25.6 Conclusion and Future Directions

The proposed hybrid system for gesture recognition is a simple and novel one. In this work, there is a closed loop of human gesture recognition by the use of Kinect interfaced computer and a subsequent feedback of corrective measure to the humans for assistance in maintaining a healthy lifestyle. A large dataset of different disorder-related gestures is created, and their diagnosis and treatment are implemented in daily life in home environment using the ensemble classifier. The decaying nature of age leads to some disorders arising out of muscle fatigue and joint decay.

Kinect detects only the skeleton of the subject thereby preserving the privacy of the subject. The scheme can be used for other applications as well like training in sports, different dancing forms, and teaching sign languages. Rehabilitation areas involve early detection of disorders in elders like osteoporosis, rheumatism, and arthritis.

The disadvantages of the proposed method include the limited range of the Kinect to a room as it uses the IR spectrum that detects only the line-of-sight range. Future works include the investigation of the possibility of any other data acquisition method that may overcome the limitation. The system should be cost-effective as well as be able to provide fast information in a suitable form to be processed, as speed is of vital importance in real-time applications. In future, the dataset will be augmented to include new gestures.

Acknowledgments

The work is supported by the University Grants Commission, India, and the University with Potential for Excellence Program (Phase II) in Cognitive Science, Jadavpur University.

Glossary

Adaptive boosting: The present work has utilized AdaBoost or adaptive boosting algorithm to implement a boosting ensemble classifier based on tree learning. Initially, the weights of all the samples in the dataset are equal. In each iteration, a new weak classifier is created using the whole dataset, and the weights for all samples are updated by increasing the weights of the samples misclassified and decreasing the weights of the samples correctly classified. Hence, it is of "adaptive" nature. A weighted voting mechanism determines the class of a new sample.

Cyber-physical system: After acquiring information about joint coordinates from the subject, the proposed work performs computations on those data and communicates the recognized disorder to the subject and thereby controlling the behavior of the subject to a certain extent. Just employing Kinect sensor, we can contribute toward the creation of a smart home for elderly persons in a simple manner. Hence, the proposed system qualifies to be a cyber-physical system.

Ensemble classifiers: An ensemble classifier or a multiple classifier system is made up of a number of "base" classifiers or "weak learners," each of which classifies the training dataset separately. The outcome of the ensemble classifier is a combination of the decisions of these base classifiers, and hence, its performance is usually better than the base classifiers, provided the individual classifier errors are uncorrelated. Two popular methods to implement the ensemble classifiers are bagging and boosting.

Feature extraction: Gestures corresponding to the different physical disorders are effectively coded from the 20 joint coordinates extracted from Kinect sensor in 3D space. Here, we have extracted 171 Euclidean distances and treated them as features. These features are essential to distinguish the 32 gestures corresponding to 8 disorders.

Gesture: A gesture can be generated by any state or motion of a body or body parts. Gesture recognition aims at processing of the information that is ingrained in the gesture and is not expressed by speech or text. This work identifies eight disorders involving pain in different parts of the body. After working in a fixed posture over a long duration, people have been observed to use certain typical gestures revealing pain at different muscles and body parts. These gestures are the subject of interest for our work and can be regarded as the patterns to be identified. Thus, the purpose of this work is to monitor the gesture of elder persons while in sitting/standing posture, thereby sensing the early-stage symptoms of certain disorders and making the subject aware about their health while suggesting an exercise as a precaution.

Health care: Aging cannot be stopped, but if the symptoms of the disorders are identified at an early stage, then the elderly health-related problems can be solved to a large extent. For recognition of disorders, the doctors mainly rely on a number of medical tests, for example, blood test, X-ray, and EEG. These procurements are costly and also time-consuming. The elders need to go to hospitals for routine checkup. Thus, the total procedure of recognizing disorders is nonflexible and for some elders nonaccessible due to cost and time constraints.

Kinect sensor: It automatically captures the 24 h activity of an elderly person. The early-stage symptoms of eight disorders are our concern in this work. As the persons project the gestures related to the disorders while sitting as well as standing, the total number of gestures obtained are 32 (some gestures are possible only while sitting or standing; some gestures occur due to pain in either side of the body). Whenever an elder person's gesture is matching with any of the 32 gestures, then an alarm is generated. After taking guidance from several doctors, we have made this proposed system. This system is not only flexible but also provides medical guidance at low cost. By the virtue of this system, we can treat the disorders at an early stage, and as a whole, the overall health of the elders becomes much better.

Tree classifier: A "tree" classifier has been used as the weak learner in this work. Each node of the hierarchical tree classifier decides on each of the features in the dataset to predict the class of a sample. The prediction can be made by tracing a branch to reach the leaf node (class) of the tree.

References

Bickel, P. J. and Li, B. (1977). Mathematical statistics. In *Test*. Citeseer.

Burba, N., Bolas, M., Krum, D. M., and Suma, E. A. (2012). Unobtrusive measurement of subtle nonverbal behaviors with the Microsoft Kinect. In *2012 IEEE Virtual Reality Workshops (VR)*, March 4–8, 2012, Costa Mesa, CA, pp. 1–4.

Cannon, R. L., Dave, J. V., and Bezdek, J. C. (1986). Efficient implementation of the fuzzy c-means clustering algorithms. *IEEE Transactions on Pattern Analysis and Machine Intelligence, 8*(2), 248–255.

Cao, J., Kwong, S., and Wang, R. (2012). A noise-detection based AdaBoost algorithm for mislabeled data. *Pattern Recognition, 45*(12), 4451–4465.

Chakraborty, A., Konar, A., Chakraborty, U. K., and Chatterjee, A. (2009). Emotion recognition from facial expressions and its control using fuzzy logic. *IEEE Transactions on Systems, Man and Cybernetics, Part A: Systems and Humans, 39*(4), 726–743.

Cortes, C. and Vapnik, V. (1995). Support vector machine. *Machine Learning, 20*(3), 273–297.

Cunningham, P. and Delany, S. J. (2007). k-Nearest neighbour classifiers. *Multiple Classifier Systems*, 1–17.

Dietterich, T. G. (1998). Approximate statistical tests for comparing supervised classification learning algorithms. *Neural Computation, 10*(7), 1895–1923.

Dietterich, T. G. (2000). An experimental comparison of three methods for constructing ensembles of decision trees: Bagging, boosting, and randomization. *Machine Learning, 40*(2), 139–157.

Diraco, G., Leone, A., and Siciliano, P. (2010). An active vision system for fall detection and posture recognition in elderly healthcare. In *Design, Automation & Test in Europe Conference & Exhibition (DATE)*, March 8–12, 2010, Dresden, Germany, pp. 1536–1541.

Dutta, T. (2012). Evaluation of the Kinect™ sensor for 3-D kinematic measurement in the workplace. *Applied Ergonomics, 43*(4), 645–649.

Freeman, W. T. (September 26, 1995). Dynamic and static hand gesture recognition through low-level image analysis. U.S. Patent 5,454,043.

Galna, B., Barry, G., Jackson, D., Mhiripiri, D., Olivier, P., and Rochester, L. (2014). Accuracy of the Microsoft Kinect sensor for measuring movement in people with Parkinson's disease. *Gait & Posture, 39*(4), 1062–1068.

García, S., Molina, D., Lozano, M., and Herrera, F. (2009). A study on the use of non-parametric tests for analyzing the evolutionary algorithms' behaviour: A case study on the CEC'2005 special session on real parameter optimization. *Journal of Heuristics, 15*(6), 617–644.

Henry, P., Krainin, M., Herbst, E., Ren, X., and Fox, D. (2010). RGB-D mapping: Using depth cameras for dense 3D modeling of indoor environments. In *The 12th International Symposium on Experimental Robotics (ISER)*, December 18–21, 2010, Delhi, India, Vol. 20, pp. 22–25.

Henry, P., Krainin, M., Herbst, E., Ren, X., and Fox, D. (2012). RGB-D mapping: Using Kinect-style depth cameras for dense 3D modeling of indoor environments. *The International Journal of Robotics Research, 31*(5), 647–663.

Iman, R. L. and Davenport, J. M. (1980). Approximations of the critical region of the fbietkan statistic. *Communications in Statistics-Theory and Methods, 9*(6), 571–595.

Itauma, I. I., Kivrak, H., and Kose, H. (2012). Gesture imitation using machine learning techniques. In *20th Signal Processing and Communications Applications Conference (SIU)*, April 18–20, 2012, Mugla, Turkey, pp. 1–4.

Jost, L. (2006). Entropy and diversity. *Oikos, 113*(2), 363–375.

Khoshelham, K. (2011). Accuracy analysis of kinect depth data. In *ISPRS Workshop Laser Scanning*, August 29–31, 2011, Calgary, Alberta, Canada, Vol. 38, p. 1.

Khoshelham, K. and Elberink, S. O. (2012). Accuracy and resolution of kinect depth data for indoor mapping applications. *Sensors, 12*(2), 1437–1454.

Khreich, W., Granger, E., Miri, A., and Sabourin, R. (2012). Adaptive ROC-based ensembles of HMMs applied to anomaly detection. *Pattern Recognition, 45*(1), 208–230.

Kim, D.-J., Lee, S.-W., and Bien, Z. (2005). Facial emotional expression recognition with soft computing techniques. In *The 14th IEEE International Conference on Fuzzy Systems, 2005, FUZZ'05*, May 22–25, 2005, Reno, NV, pp. 737–742.

Lai, K., Konrad, J., and Ishwar, P. (2012). A gesture-driven computer interface using Kinect. In *2012 IEEE Southwest Symposium on Image Analysis and Interpretation (SSIAI)*, April 22–24, 2012, Santa Fe, NM, pp. 185–188.

Le, T.-L., Nguyen, M.-Q., and Nguyen, T.-T.-M. (2013). Human posture recognition using human skeleton provided by Kinect. In *2013 International Conference on Computing, Management and Telecommunications* (*ComManTel*), January 21–24, 2013, Ho Chi Minh, Vietnam, pp. 340–345.

Lemon, S. C., Roy, J., Clark, M. A., Friedmann, P. D., and Rakowski, W. (2003). Classification and regression tree analysis in public health: Methodological review and comparison with logistic regression. *Annals of Behavioral Medicine*, *26*(3), 172–181.

Li, S., Kwok, J. T., and Wang, Y. (2002). Multifocus image fusion using artificial neural networks. *Pattern Recognition Letters*, *23*(8), 985–997.

Lisetti, C. L. and Rumelhart, D. E. (1998). Facial expression recognition using a neural network. In *FLAIRS Conference*, May 1998, Sanibal Island, FL, pp. 328–332.

Mak, C. M., Lee, Y., and Tay, Y. H. (2011). Causal Hidden Markov Model for view independent multiple silhouettes posture recognition. In *2011 11th International Conference on Hybrid Intelligent Systems* (*HIS*), December 5–8, 2011, Melacca, Malaysia, pp. 78–84.

Martin, C. C., Burkert, D. C., Choi, K. R., Wieczorek, N. B., McGregor, P. M., Herrmann, R. A., and Beling, P. A. (2012). A real-time ergonomic monitoring system using the Microsoft Kinect. In *2012 IEEE Systems and Information Design Symposium* (*SIEDS*), April 27, 2012, Charlottesville, VA, pp. 50–55.

Martínez-Muñoz, G. and Suárez, A. (2007). Using boosting to prune bagging ensembles. *Pattern Recognition Letters*, *28*(1), 156–165.

Metcalf, C., Robinson, R., Malpass, A., Bogle, T., Dell, T., Harris, C., and Demain, S. (2013). Markerless motion capture and measurement of hand kinematics: Validation and application to home-based upper limb rehabilitation. *IEEE Transactions on Biomedical Engineering*, *60*(8), 2184–2192.

Meynet, J., Popovici, V., and Thiran, J.-P. (2007). Face detection with boosted Gaussian features. *Pattern Recognition*, *40*(8), 2283–2291.

Mitra, S. and Acharya, T. (2007). Gesture recognition: A survey. *IEEE Transactions on Systems, Man, and Cybernetics, Part C: Applications and Reviews*, *37*(3), 311–324.

Murray, M. P., Kory, R. C., and Clarkson, B. H. (1969). Walking patterns in healthy old men. *Journal of Gerontology*, *24*(2), 169–178.

Narasimha, R., Ouyang, H., Gray, A., McLaughlin, S. W., and Subramaniam, S. (2009). Automatic joint classification and segmentation of whole cell 3D images. *Pattern Recognition*, *42*(6), 1067–1079.

Oszust, M. and Wysocki, M. (2013). Recognition of signed expressions observed by Kinect Sensor. In *2013 10th IEEE International Conference on Advanced Video and Signal Based Surveillance* (*AVSS*), August 27–30, 2013, Krakow, Poland, pp. 220–225.

Pal, M., Saha, S., and Konar, A. (2014). A fuzzy C means clustering approach for gesture recognition in healthcare. *International Journal of Enhanced Research in Science, Technology and Engineering*, *3*(4), 87–94.

Pal, N. R. and Bezdek, J. C. (1995). On cluster validity for the fuzzy c-means model. *IEEE Transactions on Fuzzy Systems*, *3*(3), 370–379.

Parajuli, M., Tran, D., Ma, W., and Sharma, D. (2012). Senior health monitoring using Kinect. In *2012 Fourth International Conference on Communications and Electronics* (*ICCE*), August 1–3, 2012, Hue, Vietnam, pp. 309–312.

Picek, S., Golub, M., and Jakobovic, D. (2012). Evaluation of crossover operator performance in genetic algorithms with binary representation. In Huang, D. S., Gan, Y., Premaratne, P., and Han, K., eds., *Bio-Inspired Computing and Applications* (pp. 223–230). Springer, New York.

Polikar, R. (2006). Ensemble based systems in decision making. *IEEE Circuits and Systems Magazine*, *6*(3), 21–45.

Rätsch, G., Onoda, T., and Müller, K.-R. (2001). Soft margins for AdaBoost. *Machine Learning*, *42*(3), 287–320.

Saha, S., Ghosh, S., Konar, A., and Nagar, A. K. (2013a). Gesture recognition from Indian classical dance using Kinect sensor. In *2013 Fifth International Conference on Computational Intelligence, Communication Systems and Networks* (*CICSyN*), June 5–7, 2013, Madrid, Spain, pp. 3–8.

Saha, S., Pal, M., Konar, A., and Janarthanan, R. (2013b). Neural network based gesture recognition for elderly health care using Kinect sensor. In Panigrahi, B. K., Das, S., Suganthan, P. N., and Dash, S.S., eds., *Swarm, Evolutionary, and Memetic Computing* (pp. 376–386). Springer, New York.

Sokolova, M. and Lapalme, G. (2009). A systematic analysis of performance measures for classification tasks. *Information Processing & Management*, *45*(4), 427–437.

Solaro, J. (2011). The kinect digital out-of-box experience. *Computer*, *44*(6), 97–99.

Stone, E. and Skubic, M. (2014). Fall detection in homes of older adults using the Microsoft Kinect. *IEEE Journal of Biomedical and Health Informatics*, *19*(1), 290–301.

Sun, Y., Kamel, M. S., Wong, A. K. C., and Wang, Y. (2007). Cost-sensitive boosting for classification of imbalanced data. *Pattern Recognition*, *40*(12), 3358–3378.

Tie, Y. and Guan, L. (2009). Automatic face detection in video sequences using local normalization and optimal adaptive correlation techniques. *Pattern Recognition*, *42*(9), 1859–1868.

Wu, Y., Ianakiev, K., and Govindaraju, V. (2002). Improved *k*-nearest neighbor classification. *Pattern Recognition*, *35*(10), 2311–2318.

Yegnanarayana, B. (2004). *Artificial Neural Networks*. PHI Learning Pvt. Ltd, Delhi, India.

Yu, X., Wu, L., Liu, Q., and Zhou, H. (2011). Children tantrum behaviour analysis based on Kinect sensor. In *2011 Third Chinese Conference on Intelligent Visual Surveillance* (*IVS*), December 1–2, 2011, Beijing, China, pp. 49–52.

Zar, J. H. (1999). *Biostatistical Analysis*. Pearson Education India, Karnataka, India.

Zhan, Y. and Shen, D. (2005). Design efficient support vector machine for fast classification. *Pattern Recognition*, *38*(1), 157–161.

Zhang, Z. (2012). Microsoft kinect sensor and its effect. *IEEE Multimedia*, *19*(2), 4–10.

Index

D

E

R

S

Z